BioMEMS and Biomedical Nanotechnology

Volume I

Biological and Biomedical Nanotechnology

BioMEMS and Biomedical Nanotechnology

Mauro Ferrari, Ph.D., Editor-in-Chief
Professor, Brown Institute of Molecular Medicine Chairman
Department of Biomedical Engineering
University of Texas Health Science Center, Houston, TX

Professor of Experimental Therapeutics
University of Texas M.D. Anderson Cancer Center, Houston, TX

Professor of Bioengineering
Rice University, Houston, TX

Professor of Biochemistry and Molecular Biology
University of Texas Medical Branch, Galveston, TX

President, the Texas Alliance for NanoHealth
Houston, TX

Volume I
Biological and Biomedical Nanotechnology

Edited by

Abraham P. Lee
Biomedical Engineering
University of California, Irvine

L. James Lee
Chemical and Biomolecular Engineering
The Ohio State University

Springer

Abraham P. Lee
University of California, Irvine
Irvine, California

James Lee
Ohio State University
Columbus, Ohio

Mauro Ferrari
Ohio State University
Columbus, OH

Volume I
ISBN-10: 1-4614-9815-X ISBN 10: 0-387-25842-6 (eBook) Printed on acid-free paper.
ISBN-13: 978-1-4614-9815-5 ISBN-13: 978-0387-25842-3 (eBook)
Set
ISBN-10: 0-387-25661-3 ISBN:10: 0-387-25749-7 (eBook)
ISBN-13: 978-0387-25561-3 ISBN:13: 978-0387-25749-5 (eBook)

9 8 7 6 5 4 3 2 1 SPIN 11406068

springer.com

Dedicated to Richard Smalley (1943–2005), in Memoriam

To Rick,

father founder of nanotechnology
prime inspiration for its applications to medicine
gracious mentor to its researchers
our light—forever in the trenches with us

(Rick Smalley received the 1996 Chemistry Nobel Prize for the co-discovery of carbon-60 buckeyballs)

Contents

List of Contributors

VOLUME I

Francesco Amato, Dept. of Experimental and Clinical Medicine, Università degli Studi Magna Graecia di Catanzaro, Catanzaro, Italy

Debjyoti Banerjee, Group Leader and Staff Mechanical Engineer, Applied Biosystems Inc. (formerly Microfluidics Engineer, NanoInk Inc.)

Phillip A. Barnes, Biomedical Engineering Center, The Ohio State University, Columbus, Ohio USA

L. Businaro, LILIT Group, National Nanotechnology Laboratory-TASC, Instituto Nazionale per la Fisica della Materia, Basovizza (Trieste) Italy

S. Cabrini, LILIT Group, National Nanotechnology Laboratory-TASC, Instituto Nazionale per la Fisica della Materia, Basovizza (Trieste) Italy

D. Cojoc, LILIT Group, National Nanotechnology Laboratory-TASC, Instituto Nazionale per la Fisica della Materia, Basovizza (Trieste) Italy

A.T. Conlisk, Dept. of Mechanical Engineering, The Ohio State University, Columbus, Ohio USA

Carlo Cosentino, Dept. of Experimental and Clinical Medicine, Università degli Studi Magna Graecia di Catanzaro, Catanzaro, Italy

Vittorio Cristini, Dept. of Biomedical Engineering/Mathetmatics, University of California, Irvine, Irvine, California USA

Liming Dai, Dept. of Chemical and Materials Engineering, University of Dayton, Dayton, Ohio USA

Paolo Decuzzi, CEMeC—Center of Excellence in Computational Mechanics, Dept. of Experimental Medicine, University Magna Gracia at Catanzaro, Italy

E. Di Fabrizio, LILT Group, National Nanotechnology Laboratory-TASC, Instituto Nazionale per la Fisica della Materia, Basovizza (Trieste) Italy

Mauro Ferrari, Ph.D., Professor, Brown Institute of Molecular Medicine Chairman, Department of Biomedical Engineering, University of Texas Health Science Center, Houston, TX; Professor of Experimental Therapeutics, University of Texas M.D. Anderson Cancer Center, Houston, TX; Professor of Bioengineering, Rice University, Houston, TX; Professor of Biochemistry and Molecular Biology, University of Texas Medical Branch, Galveston, TX; President, the Texas Alliance for NanoHealth, Houston, TX

Hermann B. Frieboes, Dept. of Biomedical Engineering, University of California, Irvine, Irvine, California USA

John P. Fruehauf, Medicine—Hematology/Oncology, University of California, Irvine, Irvine, California USA

Francesco Gentile, Dept. of Experimental Medicine, University Magna Gracia at Catanzaro, Italy

Pingang He, Dept. of Chemistry, East China Normal University, Shanghai, China

Lee Josephson, Center for Molecular Imaging Research, Massachusetts General Hospital/ Harvard Medical School, Charlestown, Massachusetts USA

Kyekyoon "Kevin" Kim, University of Illinois at Urbana-Champaign, Illinois USA

Rakesh Kumar, LILIT Group, National Nanotechnology Laboratory-TASC, Instituto Nazionale per la Fisica della Materia, Basovizza (Trieste) Italy

L. James Lee, Dept. of Chemical and Biomolecular Engineering, The Ohio State University, Columbus, Ohio USA

Stephen C. Lee, Dorothy M. Davis Heart and Lung Research Institute, Dept. of Cellular and Molecular Biology, The Ohio State University, Columbus, Ohio USA

Kam W. Leong, Dept. of Biomedical Engineering, Johns Hopkins School of Medicine, Baltimore, Maryland USA

Jun Li, NASA Ames Research Center, Center for Nanotechnology, Moffett Field, California USA

M. Meyyappan, NASA Ames Research Center, Center for Nanotechnology, Moffett Field, California USA

Orhan Nalcioglu, Radiological Sciences and Tu & Yuen Center for Functional Onco-Imaging, University of California, Irvine, Irvine, California USA

Dan V. Nicolau, Department of Electrical Engineering and Electronics, University of Liverpool, Liverpool, UK

Daniel W. Pack, University of Illinois at Urbana-Champaign, Illinois USA

F. Perennes, Sincrotrone Trieste ELETTRA, Basovizza (Trieste) Italy

R.Z. Proietti, LILIT Group, National Nanotechnology Laboratory-TASC, Instituto Nazionale per la Fisica della Materia, Basovizza (Trieste) Italy

Stanislav I. Rokhlin, Nondestructive Evaluation Program, The Ohio State University, Columbus, Ohio USA

F. Romanato, LILT Group, National Nanotechnology Laboratory-TASC, Instituto Nazionale per la Fisica della Materia, Basovizza (Trieste) Italy

Mark Ruegsegger, Dorothy M. Davis Heart and Lung Research Institute, Cardiology Divison, Biomedical Engineering Center, The Ohio State University, Columbus, Ohio USA

Jason Sakamoto, Biomedical Engineering, The Ohio State University, Columbus, Ohio USA

Mehmet Sarikaya, Molecular Biology & Genetics, Istanbul Technical University, Maslak, Istanbul, Turkey

John P. Sinek, Mathematics Dept., University of California, Irvine, Irvine, California USA

Piyush M. Sinha, Electrical and Computer Engineering, The Ohio State University, Columbus, Ohio USA

Bryan Ronain Smith, Biomedical Engineering Center, The Ohio State University, Columbus, Ohio USA

Candan Tamerler, Materials Science & Engineering, University of Washington, Seattle, Washington USA

M. Tormen, LILIT Group, National Nanotechnology Laboratory-TASC, Instituto Nazionale per la Fisica della Materia, Basovizza (Trieste) Italy

L. Vaccari, LILIT Group, National Nanotechnology Laboratory-TASC, Instituto Nazionale per la Fisica della Materia, Basovizza (Trieste) Italy

Lugen Wang, Nondestructive Evaluation Program, The Ohio State University, Columbus, Ohio USA

Bin Xie, Nondestructive Evaluation Program, The Ohio State University, Columbus, Ohio USA

Foreword

Less than twenty years ago photolithography and medicine were total strangers to one another. They had not yet met, and not even looking each other up in the classifieds. And then, nucleic acid chips, microfluidics and microarrays entered the scene, and rapidly these strangers became indispensable partners in biomedicine.

As recently as ten years ago the notion of applying nanotechnology to the fight against disease was dominantly the province of the fiction writers. Thoughts of nanoparticle-vehicled delivery of therapeuticals to diseased sites were an exercise in scientific solitude, and grounds for questioning one's ability to think "like an established scientist". And today we have nanoparticulate paclitaxel as the prime option against metastatic breast cancer, proteomic profiling diagnostic tools based on target surface nanotexturing, nanoparticle contrast agents for all radiological modalities, nanotechnologies embedded in high-distribution laboratory equipment, and no less than 152 novel nanomedical entities in the regulatory pipeline in the US alone.

This is a transforming impact, by any measure, with clear evidence of further acceleration, supported by very vigorous investments by the public and private sectors throughout the world. Even joining the dots in a most conservative, linear fashion, it is easy to envision scenarios of personalized medicine such as the following:

- patient-specific prevention supplanting gross, faceless intervention strategies;
- early detection protocols identifying signs of developing disease at the time when the disease is most easily subdued;
- personally tailored intervention strategies that are so routinely and inexpensively realized, that access to them can be secured by everyone;
- technologies allowing for long lives in the company of disease, as good neighbors, without impairment of the quality of life itself.

These visions will become reality. The contributions from the worlds of small-scale technologies are required to realize them. Invaluable progress towards them was recorded by the very scientists that have joined forces to accomplish the effort presented in this 4-volume collection. It has been a great privilege for me to be at their service, and at the service of the readership, in aiding with its assembly. May I take this opportunity to express my gratitude to all of the contributing Chapter Authors, for their inspired and thorough work. For many of them, writing about the history of their specialty fields of *BioMEMS and Biomedical Nanotechnology* has really been reporting about their personal, individual adventures through scientific discovery and innovation—a sort of family album, with equations, diagrams, bibliographies and charts replacing Holiday pictures....

It has been a particular privilege to work with our Volume Editors: Sangeeta Bhatia, Rashid Bashir, Tejal Desai, Michael Heller, Abraham Lee, Jim Lee, Mihri Ozkan, and Steve Werely. They have been nothing short of outstanding in their dedication, scientific vision, and generosity. My gratitude goes to our Publisher, and in particular to Greg Franklin for his constant support and leadership, and to Angela De Pina for her assistance.

Most importantly, I wish to express my public gratitude in these pages to Paola, for her leadership, professional assistance throughout this effort, her support and her patience. To her, and our children Giacomo, Chiara, Kim, Ilaria and Federica, I dedicate my contribution to BioMEMS and Biomedical Nanotechnology.

With my very best wishes

Mauro Ferrari, Ph.D.
Professor, Brown Institute of Molecular Medicine Chairman
Department of Biomedical Engineering
University of Texas Health Science Center, Houston, TX

Professor of Experimental Therapeutics
University of Texas M.D. Anderson Cancer Center, Houston, TX

Professor of Bioengineering
Rice University, Houston, TX

Professor of Biochemistry and Molecular Biology
University of Texas Medical Branch, Galveston, TX

President, the Texas Alliance for NanoHealth
Houston, TX

Preface

The growing demand for nanoscale structures and devices in the biomedical field presents significant career opportunities for future generations. Various novel materials and technologies have been developed in recent years. There, however, lacks a comprehensive book to systematically address this broad spectrum of new science and technologies. This volume is intended to provide an introduction to nanoscale devices for biological and biomedical applications. Sixteen chapters are included in this volume experts in the field of the nanobiotechnology have contributed to this work.

The volume is divided into three parts. The first part, *Synthetic Nanodevices for Biotechnology and Biomedicine*; covers the fabrication and characterization techniques of representative nanoscale structures such as carbon nanotubes, micro/nanospheres and particles, nanopores and nanochannels, and macro or microscale structures containing two-dimensional and three-dimensional nanoscale features made of polymers, silicon and other materials. The applications of these nanostructures and devices for biosensing, drug delivery and bioseparation are also introduced. The second part, *Hybrid Synthetic and Biomolecular Nanodevices*; focuses on the synthesis, interface structures, and medical applications of nanodevices made of biomolecule-polymer and biomolecule-inorganics hybrids. Finally, the third part, *Computation, Simulation, and Informatics for Bionanodevices*, provides nanoscale fluid and solid phase computation methodologies for selected biomedical applications.

We would like to thank all authors who devoted a great deal of time to make this volume possible. We hope the collected efforts from these distinguished professionals will present you a cohesive and balanced path into the intellectually exciting and fast evolving nanobiotechnology field.

Abraham P. Lee
Biomedical Engineering, University of California at Irvine

L. James Lee
Chemical and Biomolecular Engineering, The Ohio State University

Mauro Ferrari
Professor, Brown Institute of Molecular Medicine Chairman
Department of Biomedical Engineering
University of Texas Health Science Center, Houston, TX

Professor of Experimental Therapeutics
University of Texas M.D. Anderson Cancer Center, Houston, TX

Professor of Bioengineering, Rice University, Houston, TX

Professor of Biochemistry and Molecular Biology
University of Texas Medical Branch, Galveston, TX

President, the Texas Alliance for NanoHealth, Houston, TX

1

Biomolecular Sensing for Cancer Diagnostics Using Carbon Nanotubes

Jun Li and M. Meyyappan
NASA Ames Research Center, Center for Nanotechnology, Moffett Field, CA 94035

1.1. INTRODUCTION

The field of biomolecule sensing in the medical field is broad and rapidly evolving. The devices range in size from microns to centimeters across the sensing surface and rely on electronic, optical or other form of signals. If the sensing technology utilizes toxic reagents, then the use is limited to only *in vitro* application. In this chapter, biomolecule sensing using carbon nanotubes (CNTs) is discussed with specific application to cancer diagnostics.

Beyond the expected size advantages of the CNT-based sensors, there are other benefits as well. Conventional cytogenetic analysis and fluorescence *in situ* hybridization (FISH) take about three weeks for completion of the analysis. Molecular diagnostic arrays by PCR techniques take less time, still about a week. The sensitivity (i.e. ratio of the number of positive cells detected to all cells) of FISH is 5–10% and conventional cytogenetics is about 5%. Most cytogenetics, FISH and molecular diagnostic testing procedures involve bone marrow aspiration that causes pain. A biosensor that utilizes a nanoelectrode (such as CNT based electrode), in principle, can overcome many of these limitations. The CNT-based cancer diagnostics sensor discussed here can provide instantaneous results, facilitating rapid turn around time and chemotherapy dosing regimens. The detection ability can be 1 positive cell in 1000–10000 cells. Current testing is targetted at *in vitro* application and may be extended for *in vivo* diagnostics in the future, eliminating bone marrow aspiration.

The CNT based biosensor consists of a nanoelectrode array fabricated using conventional microfabrication techniques. In this array, each nanotube electrode is functionalized

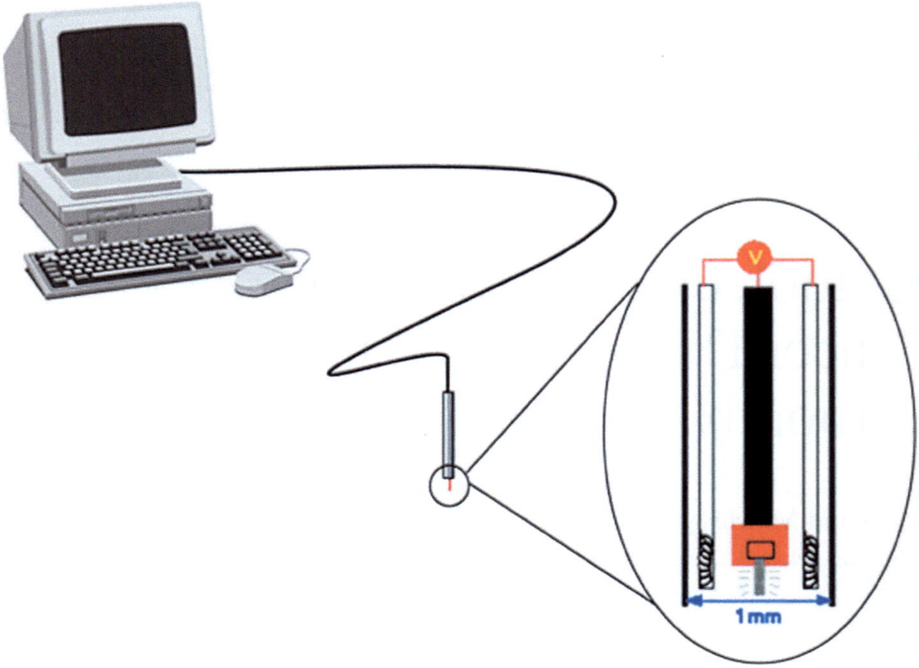

FIGURE 1.1. Schematic of a prototype catheter for cancer diagnostics.

with a probe molecule. The probe-target interaction is captured through the measurement of electrochemical signals amplified by the use of metal-ion mediators.

A proposed design for a biosensor catheter for cancer diagnostics is shown in Fig. 1.1 and the expected operating principle is as follows. The working end of the catheter consists of the carbon nanotube electrode functionalized with the probe molecules. The catheter is inserted into a soft tissue area suspicious for cancer and a pair of external electrodes (shown by the dark outline of the catheter), by applying a current to them, heats and lyzes the cells. The DNA from cancerous cell diffuses towards the stationary probe molecules and the hybridization is detected as an electrochemical signal.

In the sections below, discussion of carbon nanotubes and their interesting properties, nanoelectrode fabrication, testing and characterization are discussed, as a progress report in the fabrication of the biosensor Catheter.

1.2. CARBON NANOTUBES

For a detailed discussion on the properties, growth and applications of carbon nanotubes, the reader is referred to [1]; here, only a brief overview is provided. A carbon nanotube is an elongated fullerene molecule with diameter as small as 7 Å. Configurationally, a nanotube is equivalent to a sheet of graphite rolled into a tube. The resulting single-walled carbon nanotube (SWCNT) is denoted by its chiral vector (n, m) where n and m are indices in the graphene sheet lattice vector. When (n-m)/3 is an integer, the resulting tube is metallic; otherwise, it is a semiconductor. Besides this intriguing electrical property, the SWCNT is mechanically very strong. It exhibits an Young's modulus of over 1 TPa. The strength/weight ratio of SWCNT is about 600 times higher than that of steel. The maximum strain is about 10% which is higher than that of any other material. Thermal conductivity along the axis of the tube is very high, exhibiting a value upto 3000 W/mK. The conductivity and current carrying capacity are much higher than that of metals such as copper.

A MWCNT is a set of concentric cylinders with a central core where the wall separation is close to 0.34nm. Some of the properties discussed above drop off from the values of SWCNT, nevertheless high enough to create excitement in the research community for a variety of applications. Both SWCNTs and MWCNTs were first produced by an arc discharge synthesis [2]. But for most applications involving devices, electrodes, sensors etc, chemical vapor deposition (CVD) has emerged as a powerful alternative [3]. CVD allows *in situ* growth on a patterned substrate with possibility of subsequent processing steps in an assembly-like fashion. Variations of CVD with the use of a glow discharge have also emerged and this popular plasma enhanced CVD or PECVD has been used to grow individual, free-standing, vertically- aligned multiwalled tubes [3]. For the most part, the PECVD-grown nanostructures tend to have the hollow core periodically interrupted by a bamboo-like closure. The resulting structure is somewhat inferior when it comes to electron transport compared to an ideal MWCNT as electrons have to hop across these closures. In the literature, these structures are called multiwalled carbon nanofibers (MWCNFs) or simply carbon nanofibers (CNFs). In the remainder of this chapter, for generality, they are referred to as MWCNTs.

While the various forms of CNTs are chemically inert, their ends and sidewalls are amenable for attaching a variety of chemical groups. All these interesting structural, mechanical, electrical, thermal and other properties have led to an incredible array of application development. SWCNT based diodes and transistors have been constructed showing interesting electronic properties for memory and logic applications. Interconnects for wiring electronic circuits using nanotubes have been investigated. Potential for near term application with a mass market appeal exists with CNT based field emitters for flat panel TV displays. On the structural side there is active research in developing high strength, low weight polymer matrix, metal matrix and ceramic matrix composites with applications in automotive, aerospace, and construction industries. The ability to functionalize CNTs mentioned above opens up the possibilities to developing chemical sensors and biosensors. The biosensors can serve the needs in biomedical, homeland security, and astrobiology applications.

1.3. CARBON NANOTUBE ELECTRODES

1.3.1. Characteristics of a Good Electrode

Electrodes of all sizes using metals, various forms of carbon and other materials have been around for a variety of applications. Typical expectations of a good electrode and related issues are as follows:

- Appropriate level of conductivity for the chosen application
- The right size to meet the needs
- Ease of fabrication

- Reliability: lifetime, wear characteristics
- Compatibility with the environment
- Signal processing issues: integrity, signal-to-noise, cross-talk in an ensemble
- Approach: electrical vs. electrochemical
- Integration into a functional system

1.3.2. Why Use Nanoelectrode?

The sensitivity of an electrode is mainly determined by its ***signal to noise ratio***. The noise is the background current mainly due to the capacitive charging/discharging current at the electrode/electrolyte interface and thus proportional to the surface area (A) of the electrode as given by:

$$\mathrm{i_n} \propto \mathrm{C_d}^0\mathrm{A} \tag{1.1}$$

where $\mathrm{C_d}^0$ is the specific capacitance at the interface. In voltammetry measurements, the magnitude of the peak current of the redox signal is the sum of two terms: a linear diffusion as described in the Cottrell equation, and a nonlinear radial diffusion [4]:

$$\mathrm{i_{l,peak}} = nFAC_0 * \sqrt{\frac{D_0}{\pi t}} + nFAC_0 * \left(\frac{D_0}{r}\right) \tag{1.2}$$

where $\mathrm{i}_{l,peak}$ is the diffusion-limited electrical current, n is the number of electrons involved in the reaction with one electroactive species, F is the Faraday constant, C_o is the electroactive species concentration, D_o is the diffusion coefficient, t is time, and r is the radius of the electrode. Both terms are proportional to the concentration of the species present in the solution. The first term is proportional to the electrode surface area and decays to zero over time, whereas the second term is proportional to the inverse of the electrode radius and represents a steady state current due to a constant flux of material to the surface. The ratio of the second term to the first becomes larger as the radius is decreased. The second term dominates the measured peak current, $i_{l,peak}$ if the electrode size is less than 25 μm, which is commonly referred as ultramicroelectrode (UME) [5]. In this regime, the magnitude of the current decreases, but the signal-to-noise ratio is improved as the electrode size decreases, according to:

$$\mathrm{i_{l,peak}}/\mathrm{i_n} \propto nFC_oD_o/r \tag{1.3}$$

Clearly, ***the signal to noise will be improved by 1000 times*** if the electrode size is reduced from 20 μm to 20 nm.

The response time of an electrode is also a function of the electrode dimension. The cell time constant can be described as

$$\tau = \mathrm{R_uC_d} = \mathrm{rC_d}^0/4\kappa \tag{1.4}$$

where κ is the conductivity of the electrolyte. The electrode can ***respond 1000 times faster*** when the size is reduced from microns to nanometers so that fast electrochemical techniques can be applied. The electrochemical signal is defined by the total number of electrons that can be transferred between the electroactive species and the electrode. For high sensitivity analytical applications, this number is always limited. By employing fast electrochemical techniques, the same amount of electrons can be transferred to the measuring circuit in a much shorter time. As a result, the current, i.e. the real physical quantity being measured, will be much larger and can be differentiated much easier from the background noise.

From the above discussion, it is clear that the performance of an electrode with respect to temporal and spatial resolution scales inversely with the electrode radius. Therefore, the sensitivity can be dramatically improved by reducing the size of the electrodes to nanoscale. Indeed, a single redox molecule was detected using a Pt-Ir electrode with a diameter of 15 nm [6]. With the diameter approaching the size of the target molecules, nanoelectrodes can also interrogate biomolecules much more efficiently than conventional electrodes. There have been strong efforts in developing nanoelectrode based chemical and biosensors since 1980s. However, a reliable method to fabricate nanoelectrodes was lacking until the recent reports of CNT nanoelectrode fabrication using microfabrication approaches discussed below.

1.3.3. Why Use Carbon Nanotubes?

The outside diameter of a MWCNT varies from a few nanometers to about 200 nanometers and the length varies from a few microns to hundreds of microns. The physical dimension of MWCNTs is ideal for fabricating nanoelectrodes, with the dimension approaching the size of biomolecules. MWCNTs normally show highly conductive metallic properties. The open end is an ideal electrode similar to graphite edge plane, while the sidewall is very inert similar to graphite basal plane. The difference in electron transfer rate (ETR) between the open end and the sidewall differs by 5 to 6 orders of magnitude [7]. This makes MWCNT an ideal nanoelectrode which can pick up the signal at the tip and transfer it to the measuring unit connected at the other end with minimum interference by the surrounding environment.

From an electrochemical point of view, CNTs possess great properties similar to commonly used carbon electrodes (particularly for biosensors) such as fast electron transfer rate, wide potential window, flexible surface chemistry, and good biocompatibility. Only a few materials can provide such properties, necessary for maximizing the signal from detecting species while minimizing the noise from other species in the solution. For example, a metal electrode of Pt or Au will electrolyze water before reaching the electropotential needed for the detection of many biomolecules. This causes a large background current that masks out the real signal. Such problems can be avoided by using carbon as the electrode material. The dangling carbon bonds at the open end of a CNT can form various oxides similar to the edge of a graphite sheet, as shown in Fig. 1.2. With electrochemical etching or acid treatment, most dangling bonds can be further converted into -COOH for highly selective functionalization of biomolecules through the formation of amide bonds [8]. While it has been known for sometime that a CNT, particularly a MWCNT, is ideal for biosensing, *the major challenge is how to fabricate and integrate it as a nanoelectrode*. This is discussed in the next section.

1.3.4. Fabrication of CNT Nanoelectrodes

In principle, a single MWCNT can be grown on each individually addressed microelectrode indicated in Fig. 1.3 and used as a nanoelectrode. The microelectrode is more precisely referred to as a ***microcontact*** since it only provides an electrical contact with the MWCNTs. However, the use of individual CNTs is not reliable due to the large fluctuation

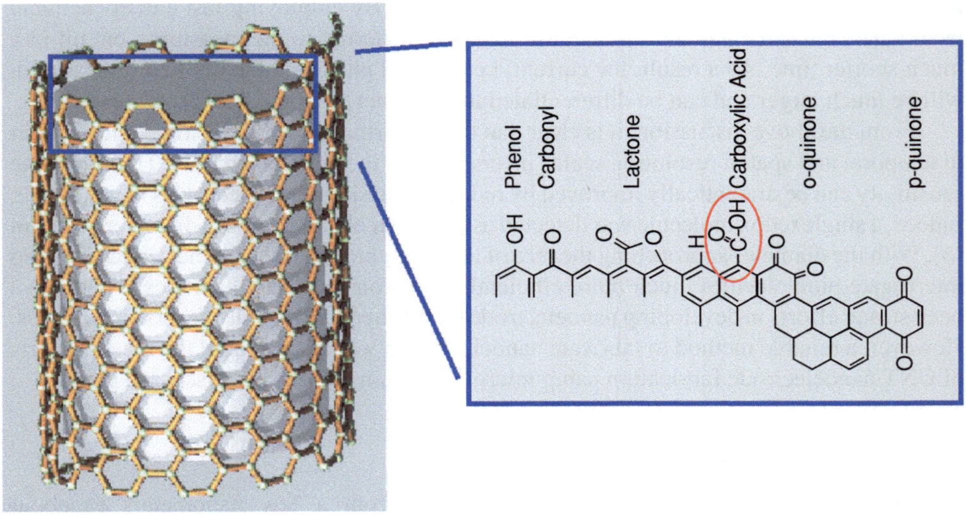

FIGURE 1.2. The functional groups at the open end of a carbon nanotube.

in the detected signal. The signal is also very weak that can be easily masked out by the electronic noises in the environment. These problems can be solved by using an array of nanoelectrodes on each microcontact, as shown in Fig. 1.3. The small circles within each microcontact in Fig. 1.3(a) represent a vertically aligned MWCNT. The side view indicates that such a MWCNT array is embedded in an insulating matrix such as SiO_2 exposing only the very end of MWCNTs to the solution. The other end of the MWCNT is attached to the micron sized metal substrate and wired out to the measuring circuit. For a microcontact with

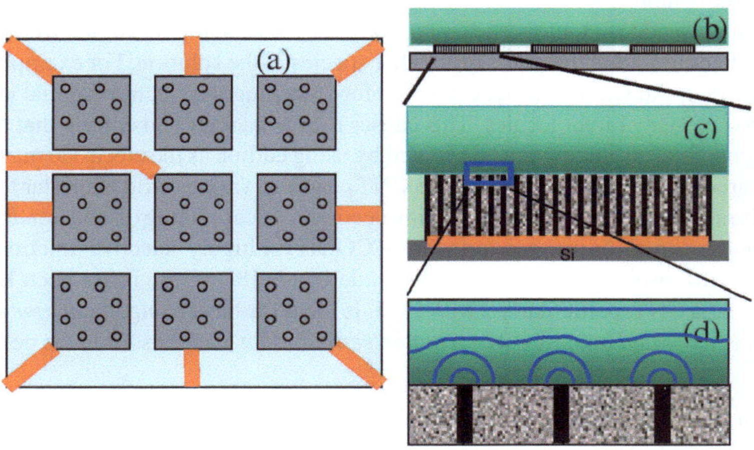

FIGURE 1.3. Schematic of a CNT nanoelectrode chip. (a) top view, (b) cross-sectional view, and (c) enlarged cross-sectional view of MWCNT nanoelectrode array on a single microcontact spot. (d) illustration of the diffusion layer around each MWCNT nanoelectrode.

a size of $20 \times 20\ \mu m^2$, hundreds to thousands of MWCNT can be grown, which provide enough statistics to improve the reliability. Such array-in-array can provide both extremely high sensitivity and reliability in a multiplexed system.

In a nanoelectrode array (see Fig. 1.3(c) and 3(d)), each electrode can generate a hemispherical layer with a concentration gradient (referred as the diffusion layer) if it is applied with an electropotential that can generate the electrochemical reaction of the species in the solution. The diffusion layer thickness is given by $6\ (Dt)^{0.5}$. The diffusion layers of the neighboring nanoelectrodes would overlap with each other in a high density array. High-density nanoelectrode arrays have been fabricated by filling metals in the channels of filtration membranes and even these high density nanoelectrodes have shown orders of magnitude lower detection limits compared to conventional macroelectrodes (1.6 nM vs 1.6 μM) [9]. Reducing the density can ensure that neighboring electrodes do not interfere with each other and each would truly behave as an individual nanoelectrode, which should give much higher sensitivity. However, this would require a spacing of over 1.5 μm for nanoelectrodes with the diameter of 100 nm, which was not possible with the fabrication method in previous studies. In contrast, a low density nanoelectrode array can be fabricated using MWCNTs with an average tube-tube spacing over 2.5 μm.

For CNT electrode sensors, the signal should ideally occur at the tip of the MWCNT and the transport is through the axial direction to underlying metal microcontacts. The underlying metal and the sidewall of MWCNTs need to be insulated with dielectric materials such as SiO_2. This provides the necessary electrical isolation of individual nanoelectrodes in addition to an overall mechanical robustness. A bottom-up approach to fabricate this electrode array is described below [10, 11]. The sequence of processing steps is shown in Fig. 1.4. A given electrode can have many vertically aligned MWCNTs at pre-specified

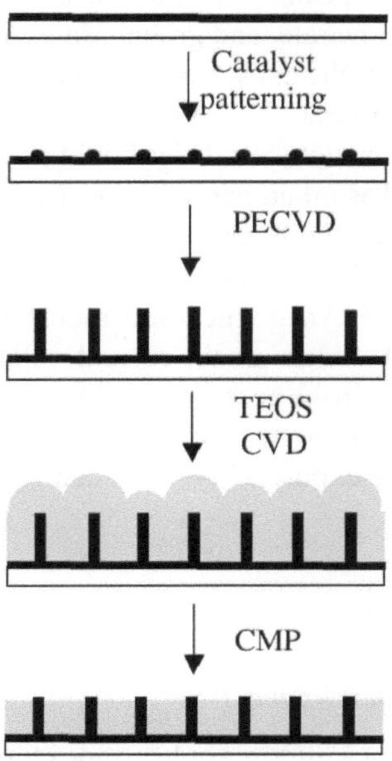

FIGURE 1.4. Schematic of the fabrication of CNT nanoelectrode array for biosensing.

locations to form an array electrode. Therefore, the first two steps would be the growth of the vertical MWCNTs and the insulation of individual MWCNTs with dielectric materials. A chemical mechanical polishing (CMP) step will be then applied to produce a smooth top surface and expose only the very ends of the MWCNTs. A dc plasma reactor with a hot filament (dc-HFCVD) has been used to obtain individual, freestanding, vertically aligned MWCNTs [12]. The feedstock for CNT growth is acetylene diluted in ammonia. Catalyst patterning is done using ion beam sputtering to deposit a thin layer of Ni on silicon substrates. Addition of an underlayer of Cr or similar metal first allows tuning the final conductivity as well as providing a barrier layer to prevent diffusion of Ni catalyst into the substrates.

The variables in the growth process include plasma power, choice of feedstock, total flow rate, ratio of active gas to diluents, growth temperature, additional bias on the substrate, and growth time. In addition, the choice for catalyst metal and its thickness, choice of underlayers and its thickness, and approach to deposit the catalyst also dictate the outcome [3]. The expected output parameters are CNT density, CNT diameter, CNT height, growth on pre-specified locations if desired, maintaining pre-specified distance between nanotubes through patterning if desired, material characteristics and conductivity, and good contact with the substrate.

Once the MWCNTs are grown on the substrate, a dielectric material can be used to isolate them from one another. The integrated circuit (IC) community has used SiO_2, some nitrides and spin-on-glass for similar purposes. Of these, SiO_2 meets the necessary isolation requirements and can be readily deposited using thermal CVD. Tetraethoxysilane (TEOS) is routinely used as a source gas in thermal CVD of SiO_2 at around 700–800° C and 100–500 mTorr. This process known as 'gap-filling' is common in IC manufacturing. The desirable outcome is to fill the gap completely without voids and provide the needed isolation characteristics. This depends on the inter-distance of the nanotubes in the array and the conformality of the dielectrics. Control on this can be exercised through growth temperature, reactor pressure, TEOS flow rate, and growth time. As seen in the process schematic (Fig. 1.4), the oxide dielectric fills the gaps, wraps around individual nanotubes (i.e. bury the nanotubes), and covers the substrate surface. A third step involving CMP is needed to remove the excess SiO_2 and expose the nanotubes. This should provide a flat surface with only the very end of MWCNTs exposed as inlaid nanodisk electrodes in the insulating SiO_2 matrix.

The fabricated electrodes can be characterized using a number of techniques. Scanning electron microscopy (SEM) provides top and perspective images of the embedded CNT array electrode as shown in Fig. 1.5. An atomic force microscope (AFM) with current sensing module (CS-AFM) can be used to measure the current-voltage (I-V) characteristics of individual nanotubes. As will be seen in the next sections, this characterication shows that MWCNTs are highly conductive metallic wires with good electrical contact with the underlying metal film. This ensures that the exposed MWCNT can be used as a good nanoelectrode.

1.4. PRELIMINARY RESULTS

1.4.1. Electronic Nano-Chip Development

As a first step for developing a multiplex electronic chip, a 3×3 array was successfully designed and fabricated using the procedure described above [8, 12]. Figures. 1.6(a) and (b)

FIGURE 1.5. Scanning electron microscopy (SEM) images of a planarized CNT nanoelectrode array embedded in SiO_2 matrix. (a) top view and (b) 45° perspective view.

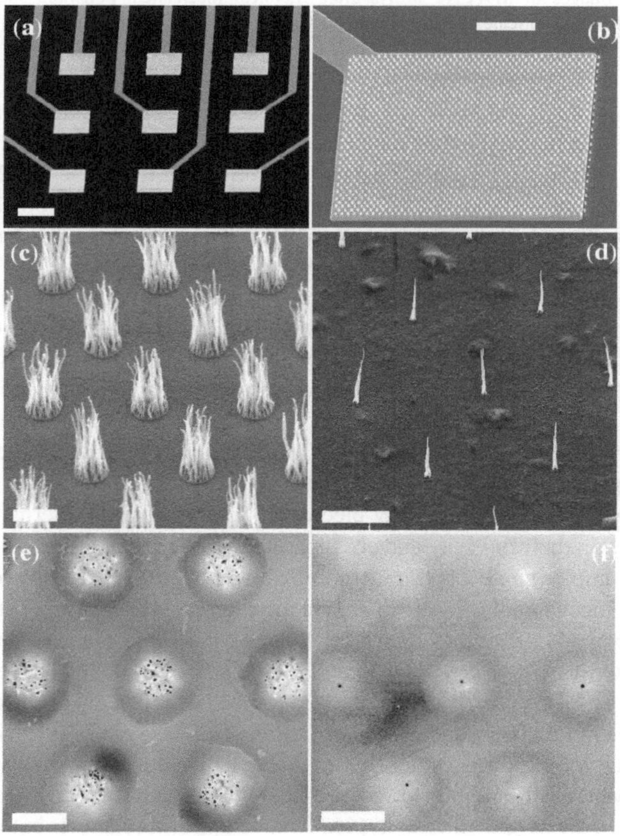

FIGURE 1.6. SEM images if (a) a 3 × 3 electrode array, (b) array of MWCNT bundles on one of the electrode pads, (c) and (d) array of MWCNTs on UV-lithography and e-beam patterned Ni spots respectively, (e) and (f) the surface of polished MWCNT array electrodes grown on 2 μm and 200 nm spots respectively. (a)–(d) are 45° perspective views and (e)–(f) are top views. The scale bars are 200, 50, 2, 5, 2, and 2 μm, respectively.

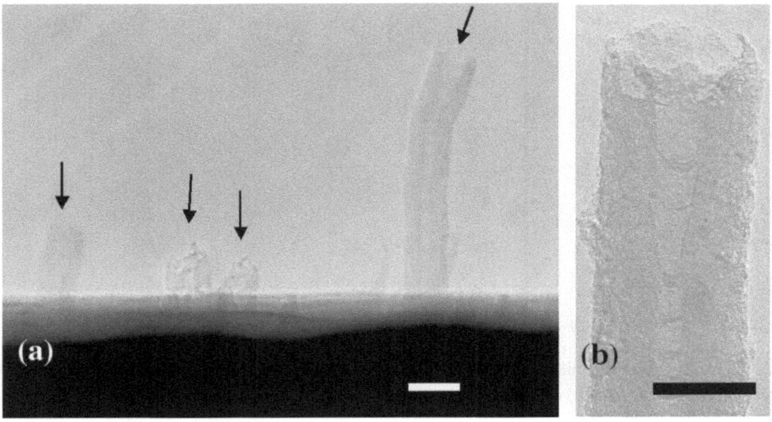

FIGURE 1.7. TEM images of MWCNT arrays after chemical mechanical polishing showing the graphite edge-plane like open ends. The scale bars are 50 and 20 nm.

show SEM images of a 3 × 3 array of individually addressed electrodes on a Si(100) wafer covered with 500 nm thermal oxide. The electrodes and contact lines are 200 nm thick Cr patterned with UV-lithography. Each electrode can be varied from 2 × 2 to 200 × 200 μm^2, consisting of a vertically aligned MWCNT array grown by PECVD from 10–20 nm thick Ni catalyst films. Figs. 1.6(c) and (d) show MWCNT arrays grown on 2μm and 200 nm diameter Ni spots defined by UV and e-beam lithography respectively. The spacing and spot size can be precisely controlled. The diameter of the MWCNTs is uniform over the whole chip and can be controlled between 30 to 100 nm by varying PECVD conditions. The number of CNTs at each spot can be varied as well by changing the thickness of the Ni film. Single nanotubes can be grown at each catalyst spots if their size is reduced below 100 nm. An SiO_2 gap-fill process described earlier has been used to encapsulate each nanotube and the substrate surface with a conformal SiO_2 film, resulting in a mechanically stable and well-insulated matrix, followed by CMP to expose the very end of the CNTs. Figs. 1.6(e) and (f) show the embedded CNT array electrodes with different patterns after polishing. Clearly, CNTs retain their integrity and are separated from each other. As shown in Fig. 1.7, the as-polished samples show CNTs protruding above the SiO_2 matrix by about 30–50 nm due to their high mechanical resilience. An electrochemical (EC) etching step is employed to shorten the exposed nanotubes and level them to the same plane as the SiO_2 matrix, exposing minimum CNT surface area. The PECVD grown MWCNTs have a bamboo-like structure with a series of closed shells along the tube, which seal off most of the hollow channel, leaving only the very end accessible by electrolytes (see Fig. 1.7b).

In general, it is too expensive to fabricate reliable nanoelectrode arrays with desired low-density using advanced lithography based on e-beam, EUV, and focused ion beam. However, since MWCNTs in the as-grown array typically have uniform diameter but varying heights, different densities can be easily obtained by stopping CMP at the proper stage. Figure. 1.8(a) shows a SEM image of a MWCNT array right after encapsulating in SiO_2 grains. During CMP, more and more MWCNTs are exposed as shown in Fig. 1.8(b)–(d), which can be easily monitored by measuring the electrical resistance (R) between two points at the surface. A

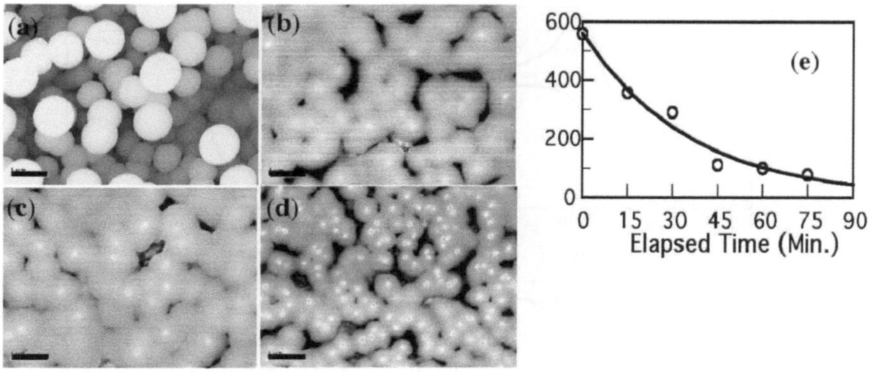

FIGURE 1.8. SEM images of (a) an as-encapsulated MWCNT array in SiO_2, and (b)–(d) the polished surface showing the number of exposed MWCNT increasing with the progress of CMP. All scale bars are 1 μm (e) The calibration curve used for controlling the density of exposed MWCNTs.

calibration curve in Fig. 1.8(e) shows the value of R vs. the elapsed time during the final stage of CMP. The resistance R drops exponentially with time and saturates at about 50 to 80 Ohm, where almost all MWCNTs are exposed. If R is controlled over 400 Ohm, a low-density MWCNT nanoelectrode array with an average tube-tube distance over 2.5 μm is readily fabricated. This method makes it possible to mass-produce nano-chips at minimum cost so that it can be used as a disposable cartridge for cancer diagnosis.

1.4.2. Electrochemical Properties of CNT Nanoelectrode Arrays

The MWCNT nanoelectrode array fabricated above has been carefully characterized with electrochemical (EC) measurements. The measurements are carried out in a three-electrode configuration by sealing the CNT array chip in a TEFLON cell using a 3 mm i.d. o-ring. A Pt coil and a saturated calomel electrode (SCE) are used as the counter and reference electrodes, respectively. Two MWCNT nanoelectrode arrays with average spacing of ~280 nm and 1.5 μm are compared side-by-side in measuring $K_4[Fe(CN)_6]$ in bulk solution and $Fc(CH_2)_2NH_2$ covalently attached to the exposed end of CNTs. As shown in Fig. 1.9, the high-density sample exhibits a typical cyclic voltammetry (CV) curve with a pair of redox peaks centered at the characteristic potential corresponding to $K_4[Fe(CN)_6]$ while the low-density sample gives a sigmoidal shaped CV curve. This confirms that the diffusion layer from an individual nanoelectrode in the high density sample completely overlaps with those of its neighbors, giving the characteristic CV similar to a solid macroelectrode. On the other hand, the majority of MWCNTs in the low-density sample remains as individual nanoelectrodes and are able to reach the characteristic steady state above certain potentials. As for surface attached ferrocene (Fc) molecules, the signal is much weaker because the total number of Fc decreases as the CNT density is lowered. A pair of redox waves separated by about 30 mV is observed with the high-density sample indicating a quasireversible behavior of surface binded redox species. The signal is so low for the low-density sample that it is masked out by the background noise. A special EC technique, AC voltammetry can be employed to maximize the signal contributed by each redox species as shown in

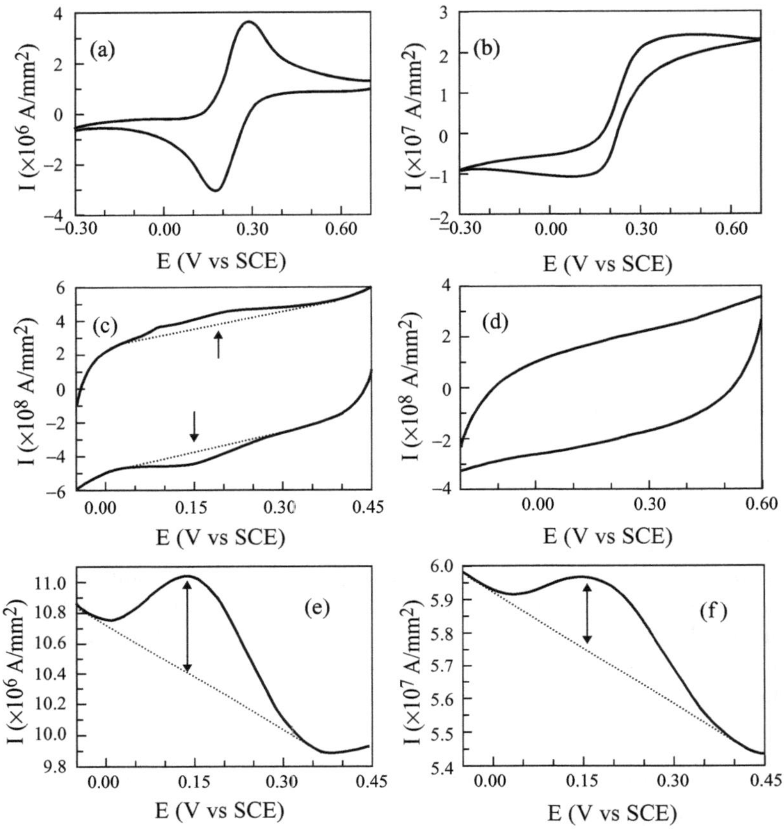

FIGURE 1.9. (a), (b) CV measurements in 1 mM $K_4Fe(CN)_6$ and 1.0 M KCI; (c), (d) CV measurements of Fc derivative functionalized MWCNT array electrodes in 1.0 M KCI solution; and (e), (f) ACV measurements at 50 Hz and an amplitude of 50 mV on the staircase DC ramp from −0.05 to 0.45 V (a), (c), (e) are measured with the high-density CNT array electrode and (b), (d), (f) are measured with the low density one, respectively. All CV measurements are taken with a scan rate of 20 mV/s. The arrows indicate the position of the redox waves of the Fc derivative functionalized on the high-density CNT array elecrode. The dotted lines indicate the baselines of the background current.

Figs. 1.9(e) and (f). Clearly, the signal is much bigger than that obtained using conventional DC voltammetry. Thus AC voltammetry will be the ideal technique for biosensors based on nanoelectrode array.

1.4.3. Functionalization of Oligonucleotide Probes

As mentioned earlier, the oxides at the CNT ends can be converted to carboxylic groups using EC etching. A ferrocene (Fc) derivative, $Fc(CH_2)_2NH_2$, can be selectively functionalized at the tube ends through amide bonds facilitated by the coupling reagents dicyclohexylcarbodiimide (DCC) and N-hydroxysuccinimide (NHS) [13] as illustrated in Fig. 1.10(a). Similar carbodiimide chemistry can be applied to functionalize primary amine

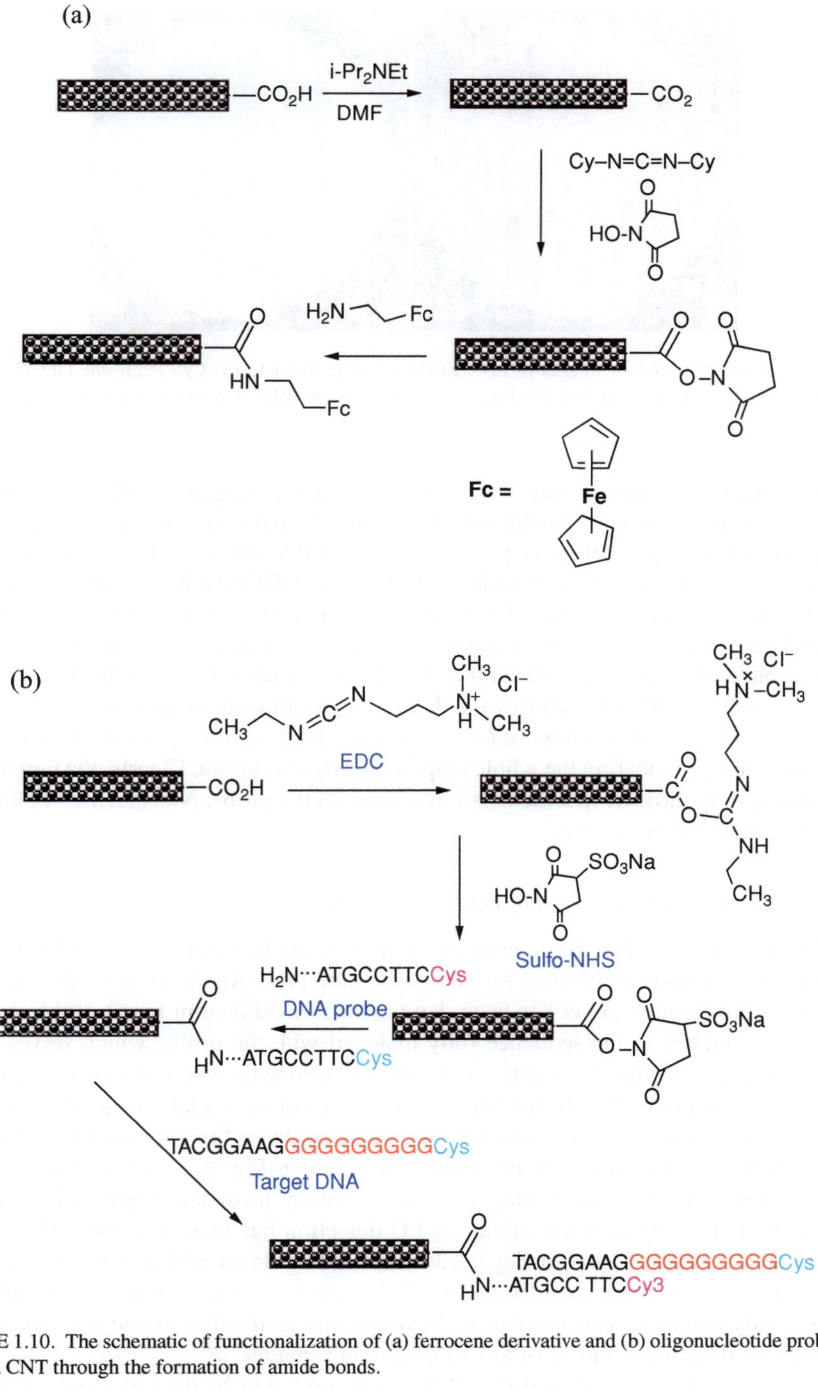

FIGURE 1.10. The schematic of functionalization of (a) ferrocene derivative and (b) oligonucleotide probe to the end of a CNT through the formation of amide bonds.

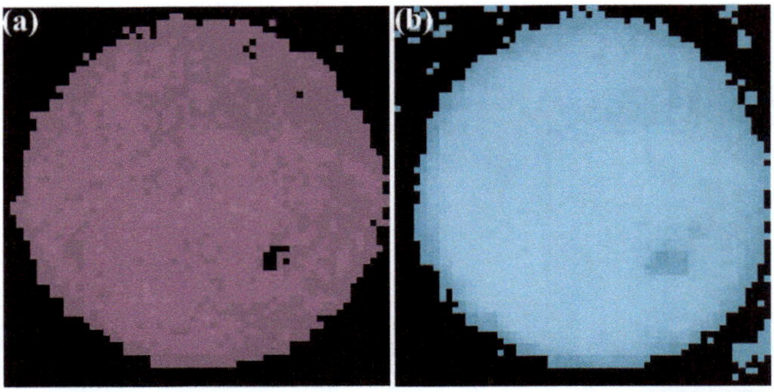

FIGURE 1.11. The fluorescence images of (a) Cy3 (from the probe) and (b) Cy5 (from the target) after the functionalization, hybridization, and washing steps. The probe functionalization was confined within a 3 mm i.d. o-ring.

terminated oligonucleotides using water-soluble coupling reagents 1-ethyl-3(3-dimethyl aminopropyl carbodiimide hydrochloride (EDC) and N-hydroxysulfo-succinimide (sulfo-NHS) as shown in Fig. 1.10(b). A probe [Cy3]5'-CTIIATTTCICAIITCCT-3'[AmC7-Q] and a target [Cy5]5'-AGGACCTGCGAAATCCAGGGGGGGGGG-3' are used, which are related to the wild type gene of BRCA1 Arg1443stop [14]. Fluorescence images of Cy3 and Cy5 labels are taken with a laser scanner after each step of functionalization, hybridization, and washing to confirm the attachment. Figure 1.11 shows the Cy3 and Cy5 images obtained on a MWCNT chip of $1 \times 1\ \text{cm}^2$ after hybridization and washing. The probe functionalization was confined within a 3 mm inner diameter o-ring but the hybridization was carried out by incubating the whole chip in the target solution. Clearly, the target binds to the surface only through specific hybridization with the probe. Nonspecific binding was removed by washing procedures.

1.4.4. Electrochemical Detection of DNA Hybridization

It has been reported that the guanine bases in DNA can be oxidized at ~1.05 V [15] and this oxidation reaction can be used for DNA detection [16]. An electrochemical detection using an oligonucleotide target has been demonstrated as shown in Fig. 1.10(b). A 10 bp polyG tag is attached to the sequence fully matched with the probe, which serves as the signal moiety. The guanine bases in the probe are replaced with inosine to ensure no guanine contribution from the probe. To further amplify the guanine oxidation signal, a mediator $Ru(bpy)_3^{2+}$ is introduced which can transfer electrons from the guanine bases dangling near the electrode surface as schematically illustrated in Fig. 1.12. Combining together the advantages of nanoelectrode array, guanine oxidation, mediator amplification, and AC voltammetry, extremely high sensitivity of EC detection has been achieved. The net AC voltammetry data is presented in Fig. 1.13(a). In scaling down with a MWCNT array on a $20 \times 20\ \mu\text{m}^2$ microcontact, *less than 10^6 of oligonucleotide targets* attached to this spot could be easily detected. This number is the maximum estimation, while the real number of target DNA detected could be orders of magnitude smaller.

It has also been demonstrated that PCR amplicon could be directly used for EC detection using CNT nanoelectrode arrays. The same BRCA1 gene probe is employed for

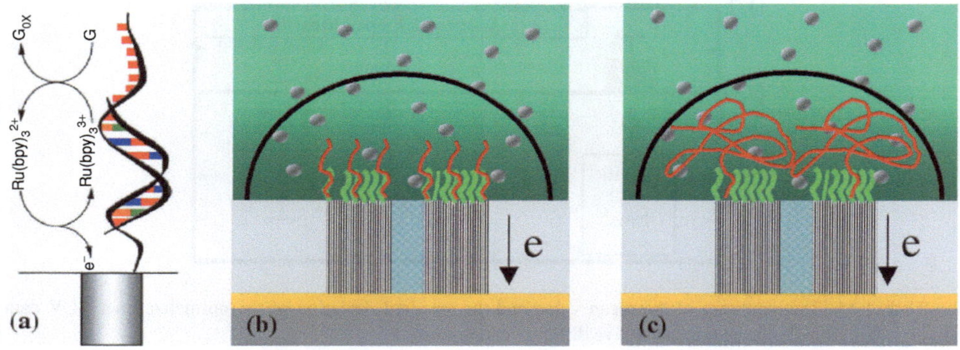

FIGURE 1.12. The schematic of (a) the mechanism of $Ru(bpy)_3^{2+}$ mediator amplified guanine oxidation, (b) the detection of oligonucleotide targets, and (c) PCR amplicon targets.

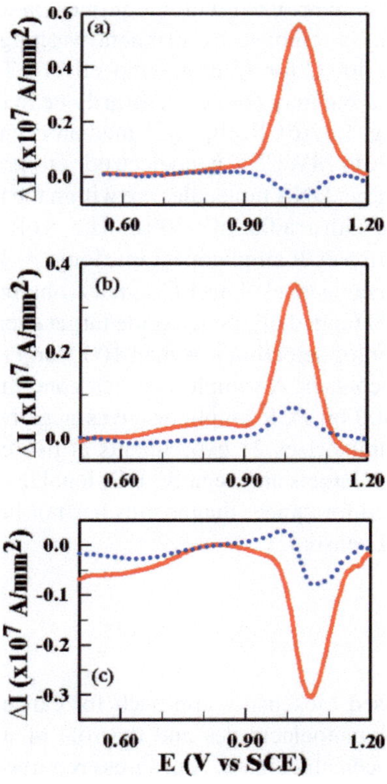

FIGURE 1.13. The net AC voltammetry data and control measurements in 5 mM Ru $(bpy)_3^{2+}$ and 0.20 M NaOAC (pH = 5.2) with an AC sinusoidal wave of 10 Hz and 25 mV amplitude on top of a staircase DC ramp after incubation in (a) specific polyG tagged oligonucleotide target, (b) specific PCR amplicon, and (c) non related PCR amplicon.

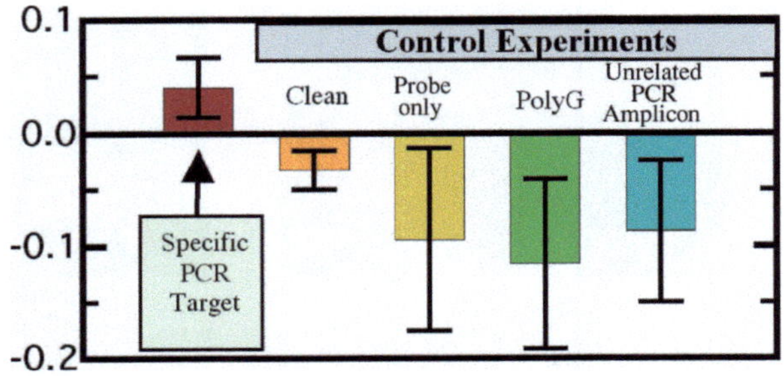

FIGURE 1.14. The summary of the mean value and the standard deviation of the normalized net ACV data $(I_{p1}-I_{p2})/I_{p1}$ from 21 measurements at five different conditions.

this. But the target is changed to ~300 bp normal allele within BRCA1 gene containing 5'-AGGACCTGCGAAATCCAG-3' which is complementary to the specific oligonucleotide probe. Genomic DNA from a healthy donor is used in PCR amplification. Considering the physical size (radius of gyration of ~6 nm for 300 bp ssDNA), it is expected that no more than ~70 targets can hybridized with the probes on each MWCNT of ~100 nm dia. The nonspecific binding is removed by stringent washing with 2 × SSC/0.1%SDS, 1 × SSC, and 0.1% SSC at 40° C for 15 min., respectively. The major part of the target molecules likely dangles near but may not be necessarily be in direct contact with the electrode surface as shown in Fig. 1.12(c). $Ru(bpy)_3^{2+}$ mediators can efficiently transport electrons from the guanine bases to MWCNT nanoelectrodes to provide an amplified guanine oxidation signal as long as target DNA molecules are within the three-dimensional diffusion layer (typically a hemisphere with a radius of ~300 nm, i.e. ~$6R_{ave}$, where R_{ave} is the average nanoelectrode radius). Since the PCR amplicon is quite long (~ 300 bp) and statistically contains ~75 guanine bases, the same level of net EC signal is obtained, as shown in Fig. 1.13(b) compared to the 10 bp polyG tagged oligonucleotide target even though the number of targets is much less. Sensitivity for detecting less than 1000 target molecules can be achieved using a 20 × 20 μm^2 microcontact. A sample that has gone through the same incubation condition in nonrelated ~400 bp PCR amplicon gives negative net EC signal as seen in Fig. 1.13(c). Figure 1.14 summarizes 21 experiments at five conditions. It clearly shows positive EC signal for specific targets and negative EC signal in all four control experiments. Such high reliability required for cancer diagnostics has not been reported by using other nanotechnology based DNA sensors.

1.5. SUMMARY

A carbon nanotube based biosensing approach for cancer diagnostics has been described. The motivation for nanoelectrodes and the role of nanotubes in developing an ultrasensitive platform has been elucidated. A progress report on the development of nanoelectrode, characterization, and electrochemical detection of DNA hybridization has been given. The preliminary results indicate the potential of this approach for cancer diagnostics with demonstrated characteristics of high sensitivity, reliability, and inexpensive microfabrication for cost effectiveness.

ACKNOWLEDGEMENTS

The authors are grateful to Alan Cassell, Hua Chen, and Jessica Koehne for their contributions to the biosensor development described here.

REFERENCES

[1] *Carbon Nanotube: Science and Applications*, M. Meyyappan (Ed.), CRC Press, Boca Raton, FL, 2004.
[2] A.P. Moravsky. E.M. Wexler, and R.O. Loutfy, Chapter 3 in ref. 1.
[3] M. Meyyappan. Chapter 4 in ref. 1.
[4] A.J. Bard and L.R. Faulkner. *Electrochemical Methods: Fundamental and Applications*, 2nd Ed. Wiley, New York, 216–218, 2001.
[5] R.M. Wightman. *Anal. Chem.*, 53: 1125A, 1981.
[6] F.-R.F. Fan and A.J. Bard. *Science*, 267: 817, 1995.
[7] L. McCreery. *In Electroanalytical Chemistry*, A.J. Bard (Ed.), Marcel Dekker, Inc., New York, 17: 221–374, 1991.
[8] J. Li, H.T. Ng, A. Cassell, W. Fan, H. Chen, Q. Ye, J. Kohene, J. Han, and M. Meyyappan. *Nano Lett.*, 3: 597, 2003.
[9] V. P. Menon and C.R. Martin. *Anal. Chem.*, 67: 1920, 1995.
[10] J. Li, R. Stevens, L. Delzeit, H.T. Ng, A. Cassell, J. Han, and M. Meyyappan. *Appl. Phys. Lett.*, 81 (5): 910, 2002.
[11] J. Li, Q.L. Ye, A.M. Cassell, H.T. Ng, R. Stevens, J. Han, and M. Meyyappan. *Appl. Phys. Lett.*, 82 (15): 2491, 2003.
[12] A.M. Cassell, Q. Ye, B.A. Cruden, J. Li, P.C. Sarrazin, H.T. Ng, J. Han, and M. Meyyappan. *Nanotechnology*, 15: 9, 2004.
[13] J.V. Staros. *Biochemistry*, 21: 3950, 1982.
[14] Y. Miki, J. Swensen, D. Shattuck-Eidens, P.A. Futreal, K. Harshman, S. Tavtigian, Q. Liu, C. Cochran, L.M. Bennett and W. Ding. *Science*, 266: 66, 1994.
[15] M.F. Sistare, R.C. Holmberg, and H.H. Thorp. *J. Phys. Chem. B*, 103: 10718, 1999.
[16] N.D. Popovich and H.H. Thorp. Interface, 11 (4): 30, 2002.

2

Microspheres for Drug Delivery

Kyekyoon "Kevin" Kim and Daniel W. Pack
University of Illinois at Urbana-Champaign

2.1. INTRODUCTION

With advances in biotechnology, genomics, and combinatorial chemistry, a wide variety of new, more potent and specific therapeutics are being created. Because of common problems such as low solubility, high potency, and/or poor stability of many of these new drugs, the means of drug delivery can impact efficacy and potential for commercialization as much as the nature of the drug itself. Thus, there is a corresponding need for safer and more effective methods and devices for drug delivery. Indeed, drug delivery systems—designed to provide a therapeutic agent in the needed amount, at the right time, to the proper location in the body, in a manner that optimizes efficacy, increases compliance and minimizes side effects—were responsible for $47 billion in sales in 2002, and the drug delivery market is expected to grow to $67 billion by 2006.

Controlled release drug delivery systems are being developed to address many of the difficulties associated with traditional methods of administration. Controlled release drug delivery employs devices—such as polymer-based disks, rods, pellets, or microparticles—that encapsulate drug and release it at controlled rates for relatively long periods of time. Such systems offer several potential advantages over traditional methods of administration. First, drug release rates can be tailored to the needs of a specific application; for example, providing a constant rate of delivery or pulsatile release. Second, controlled release systems provide protection of drugs, especially proteins, that are otherwise rapidly destroyed by the body. Finally, controlled release systems can increase patient comfort and compliance by replacing frequent (e.g., daily) doses with infrequent (once per month or less) injection.

While a variety of devices have been used for controlled release drug delivery, biodegradable polymer microspheres are one of the most common types and hold several advantages. Microspheres can encapsulate many types of drugs including small molecules, proteins, and nucleic acids and are easily administered through a syringe needle. They are generally biocompatible, can provide high bioavailability, and are capable of sustained release for long periods of time. Several commercial products are based on polymer microspheres including Lupron Depot® and Nutropin Depot®. Disadvantages of microspheres include difficulty of large-scale manufacturing, inactivation of drug during fabrication, and poor control of drug release rates. For example, Nutropin Depot, comprising Genentech's recombinant human growth hormone (rhGH) encapsulated within poly(D,L-lactide-co-glycolide) (PLG) microspheres using Alkermes' proprietary ProLease® encapsulation technology, was recently pulled from the market because manufacturing and production costs were too high.

In this chapter, the factors controlling drug release rates are reviewed, followed by several examples of controlled-release microsphere applications. More importantly, methods for microparticle fabrication will be described. In particular, the chapter will focus on a precision particle fabrication (PPF) method that has been reported recently. PPF provides unprecedented control of particle size, size distribution and morphology that may translate into enhanced control of drug release rates and improved understanding of the mechanisms controlling release.

2.2. BACKGROUND

2.2.1. Factors Affecting Release Rates

The microsphere fabrication method is a governing factor in the encapsulation and release of therapeutics. In addition, a complicated array of factors including the type of polymer, the polymer molecular weight, the copolymer composition, the nature of any excipients added to the microsphere formulation (e.g., for stabilization of the therapeutics), and the microsphere size can have a strong impact on the delivery rates.

First, the type of polymer used in microsphere fabrication and the way in which the polymer degrades obviously affect drug release rates. Depending on the rate of hydrolysis of their functional groups, polymers can be broadly categorized into two types: surface-eroding and bulk-eroding [29, 91, 169]. Bulk-eroding polymers, such as PLG, readily allow permeation of water into the polymer matrix and degrade throughout the microsphere matrix. In contrast, surface-eroding polymers, such as polyanhydrides, are composed of relatively hydrophobic monomers linked by labile bonds. In this way, they are able to resist the penetration of water into the polymer bulk, while degrading quickly into oligomers and monomers at the polymer/water interface via hydrolysis [148].

Bulk-eroding polymer microspheres are often characterized by a "burst" of drug—as much as 50% of the total drug load [131]—released during the first few hours of incubation, followed by a slow, diffusion-controlled release and sometimes a third phase in which the remaining drug is released quickly as a result of severe degradation of the polymer matrix. In microspheres composed of surface-eroding polymers, drug is released primarily at the surface as the polymer breaks down around it. Erosion of such polymers usually proceeds at a constant velocity [66]. If the drug of interest is homogeneously dispersed throughout a microsphere, the largest rate of release will occur at the beginning. As time proceeds, the surface area of the sphere and the release rate decrease asymptotically.

Polymer molecular weight can affect polymer degradation and drug release rates. As one might expect, an increase in molecular weight decreases diffusivity and therefore drug release rate [2, 95, 109, 118, 191]. In addition, a major mechanism for release of many drugs is diffusion through water-filled pores, formed as polymer degradation generates soluble monomers and oligomers that can diffuse out of the particle. These small products are formed more quickly upon degradation of lower molecular weight polymers. The decrease in release rates with increasing polymer molecular weight appears to hold for small molecules, peptides, and proteins [25, 123]. However, molecular weight typically has little effect on release rates from surface-eroding polyanhydride microspheres [69, 167].

The co-monomer ratios in many copolymers can also affect release rates. Most often, increasing the content of the more rapidly degrading monomer increases the release rate [112, 156, 161]. Similarly, when drug release is controlled by polymer erosion, release rate typically increases with higher concentration of the smaller and/or more soluble monomer [167]. However, the effect of the copolymer composition can be complicated by differences in the polymer phase behavior or the thermodynamics of the encapsulated drug [89].

A variety of excipients may be added to microsphere formulations to stabilize the drug during fabrication and/or release and may impact drug release through several different mechanisms. For example, to improve the encapsulation of bovine serum albumin (BSA) in microspheres of poly(ε-caprolactone) (PCL) and 65:35 PLG, Yang et al. included poly(vinyl alcohol) (PVA) in the BSA solution to stabilize the primary emulsion resulting in a more uniform BSA distribution in the microspheres [192]. Increasing concentrations of PVA decreased the initial burst of protein and the overall release rates. Jain et al. encapsulated myoglobin in PLG microspheres in the presence of a stabilizer, mannitol [77]. They report that mannitol increased the release rate and the final amount of drug released by increasing the initial porosity of the PLG matrix, leading to faster formation of the pore network within the sphere due to polymer erosion.

Clearly, microsphere size will strongly affect the rate of drug release. As size decreases, the surface area-to-volume ratio of the particle increases. Thus, for a given rate of drug diffusion through the microsphere, the rate of flux of drug out of the microsphere, per mass of formulation, will increase with decreasing particle size. In addition, water penetration into smaller particles may be quicker due to the shorter distance from the surface to the center of the particle. Also, while the decrease in surface area with particle size may lead to decreased rate of erosion of poorly water-permeable polymers like polyanhydrides, because surface area-to-volume ratio increases with decreasing particle size, drug release rates (per mass of polymer) will be faster for smaller polyanhydride microspheres.

2.2.2. Recent Applications of Controlled Release Microspheres

Controlled-release microspheres are in development for a number of interesting and important applications, especially for delivery of large, fragile drugs like proteins and nucleic acids. Several recent examples are described below.

2.2.2.1. Controlled-Release Vaccines Vaccination has been highly successful for controlling or even eradicating many important types of infectious diseases, and new or improved vaccines are being heavily investigated for AIDS [196], hepatitis B [170], anthrax [59], and SARS [197]. A frequent problem is the need for repeated administrations—usually injections—to ensure long-lasting immunity. For example, the current anthrax vaccine requires a series of boosters at 2 and 4 weeks, and at 6, 12, and 18 months following the first inoculation; and the Recombivax HB® vaccine for hepatitis B—required for most health-care workers in the U.S.—is administered in three injections at 0, 1, and 6 months. The need for multiple injections poses a serious problem for patients in developing countries with limited access to medical care, where awareness is lacking, or for transient populations. One promising alternative is a single-shot vaccine in which a drug delivery device provides the necessary boosters at specified times after administration [83]. Further, the ability to more precisely control the time course of vaccine delivery may lead to more effective vaccination with current antigens and may allow utilization of antigens that were previously ineffective [114].

Single-shot Vaccine delivery systems should provide the antigen(s) and adjuvant on a prescribed schedule and maintain the bioactivity of the antigen, both during fabrication of the delivery device and during the often prolonged residence time of the device in the body. In recent years, much work has focused on developing microsphere-based, single-administration, vaccine delivery vehicles [44, 45, 149, 158] using a variety of materials including hydroxypropyl cellulose/PLG [101], poly(ε-caprolactone) [160], PLA [141], chitosan [31], and collagen [74], though the majority have been fabricated with PLG [7, 44, 47, 158, 162]. Maintenance of antigen bioactivity has been problematic due to contact of the proteins with organic solvents and the hydrophobic polymer, and the use of strong physical forces to produce the microspheres [2, 36, 81]. To enhance vaccine stability, researchers have been focusing on several approaches, including the use of adjuvants to protect the protein antigens or by choosing different microsphere materials [125, 138, 139].

A major advantage of microspheres for vaccination is that they can be passively targeted to antigen-presenting cells (APCs) such as macrophages (MΦ) and dendritic cells. The ability of APCs to phagocytose particulates is dependent on the particle size. In particular, 1- to 10-μm diameter microspheres are optimally taken up by APCs in a number of tissues [9] and have been shown to enhance antigen-specific T-helper lymphocyte (Th) responses [142] (thus leading to an enhancement in antigen-specific antibody responses) and elicit a cytotoxic T lymphocyte (CTL) response (Nixon et al. 1996). T-cell activation in response to antigen-encapsulating microspheres has been shown to be 100-1000 fold better than antigen alone [179].

2.2.2.2. Stabilization of Encapsulated Protein Therapeutics A major problem with protein encapsulation in polymer particles is loss of protein bioactivity. Damage to proteins can occur during fabrication of the particles—via shear stresses or other physical forces, through contact with organic solvents, and by loss of water (e.g., upon lyophilization) [34, 76]—as well as during incubation and release in the warm, moist, in vivo environment. Two types of damage occur most often: (i) covalent or non-covalent intermolecular aggregation [34] and (ii) denaturation [51, 60]. Several studies have investigated the mechanisms of damage [153]. Protein stability can be enhanced by the addition of excipients to prevent aggregation and stabilize the folded protein structure or through judicious choice of polymer employed for fabrication of the devices.

Although a number of types of stabilizing excipients have been studied, poly(ethylene glycol) (PEG) and sugars have been the most common [46, 81, 88, 93, 163, 187]. For example, Perez-Rodriguez et al. very recently reported that co-dissolving PEG (M_w 5,000) and maltose with α-chymotrypsin in the primary emulsion greatly reduced protein aggregation and inactivation [163]. In a similar study, Castellanos et al. used PEG to enhance the stability of γ-chymotrypsin encapsulated in PLG microspheres formed by a solid-in-oil-in-water (s/o/w) emulsion process [35]. They reported that normal encapsulation procedures with protein lyophilized from buffer or a trehalose solution caused 30% protein aggregation and 50% loss in activity. Co-lyophilizing the protein with PEG prior to encapsulation reduced the activity loss to only 8%, but the protein still aggregated during encapsulation. By using PEG as an emulsifier in the secondary o/w emulsion, the resulting microspheres were free of protein aggregation and the enzyme completely maintained its activity upon extraction from the particles. Similar stabilization of horseradish peroxidase (HRP) [35] and BSA [1] was achieved with the same technique, suggesting that this may be a general approach for protein stabilization. Finally, Jiang and Schwendeman reported that upon blending of 20–30 wt% PEG in PLA microspheres encapsulating BSA, protein aggregation was avoided and in vitro protein release was sustained for 29 days [79].

Other types of excipients also are effective in the stabilization of proteins. Lysozyme inactivation, precipitation, and aggregation were largely prevented upon formation of the primary emulsion by the addition of 50 mM potassium phosphate, and the activity of the protein was largely unaffected [135]. Further, the acidic microclimate resulting from accumulation of polymer degradation products [63, 157] is a major factor causing protein instability in degrading microspheres [154]. Schwendeman and co-workers [84, 200] showed that addition of buffering salts can stabilize encapsulated proteins. Zhu et al. demonstrated that co-encapsulation of magnesium hydroxide ($Mg(OH)_2$), an antacid, increased the microclimate pH in PLG millicylinders and microspheres containing BSA [200]. The devices with $Mg(OH)_2$ showed minimal BSA aggregation and exhibited a smaller burst effect of non-aggregated BSA compared to cylinders without the antacid.

The type of polymer used for microsphere fabrication, its degradation rate, acidity of the degradation products, hydrophobicity, etc., can also impact stability of encapsulated proteins. Although PLG is the most common polymer used for polymer microspheres, PEG can be combined with polyesters to form poly(D,L-lactide)-co-poly(ethylene glycol) (PELA) diblock copolymers [113, 198], PLA-PEG-PLA triblock copolymers [24, 38, 111], and PLG-PEG-PLG triblocks [124]. It is thought that the more hydrophilic PEG may improve the affinity of protein for the matrix polymer and lead to better entrapment efficiency [103, 104, 199].

2.2.2.3. DNA Encapsulation Gene therapy holds tremendous potential for treating genetic diseases and acquired diseases including cancer, and as vaccines [8, 52, 53, 71, 176]. A major barrier to development of gene-based pharmaceuticals is safe and efficient DNA delivery. Much research has focused on development of gene delivery vectors including viruses [177], liposomes [55], and polymers [127]. However, parenteral administration of naked plasmid DNA (pDNA) leads to gene expression [28, 188], and controlled release of pDNA from polymeric matrices [26, 54, 116, 130, 155], microparticles [73, 116, 121, 171] and nanoparticles [48, 75, 119] has been reported recently. In particular, encapsulation and controlled release of pDNA from biodegradable microspheres may provide a number of advantages including protection from nuclease degradation, access to alternative routes of administration (e.g., nasal, pulmonary, oral, and mucosal), passive targeting to professional antigen-presenting cells [73, 115], and prolonged gene expression [116].

Problems currently facing the design of new DNA-loaded microparticle formulations include loss of bioactivity, poor encapsulation efficiency and low loading due to the difficulty of encapsulating such large hydrophilic macromolecules in the hydrophobic polymer matrix. Plasmid DNA is particularly susceptible to damage by physical forces, especially the shear forces employed in most microsphere fabrication processes [106]. In addition, loss of structural integrity can result from exposure to acidic environments [171, 180, 182]. Ando et al. devised a novel emulsion-solvent extraction/evaporation method for DNA-loaded microsphere encapsulation that prevents shear-induced damage of the plasmid [6]. In this method, the primary w/o emulsion, comprising the aqueous DNA solution and the polymer/solvent continuous phase, is formed by sonication and then quickly frozen in liquid nitrogen. The emulsion is allowed to warm slowly until reaching a temperature at which the organic phase melts, but the aqueous phase remains frozen. The primary emulsion—actually a suspension of ice particles—is then homogenized in water to form the microspheres. Because the DNA is entrapped in the frozen droplets, the plasmid does not experience the shear forces employed in homogenization. In microspheres prepared by the optimal procedure, 89% of the DNA retained its native conformation compared to only 39% using the conventional process.

2.3. FABRICATION OF POLYMER MICRO- AND NANOPARTICLES

Microsphere drug delivery systems have been fabricated by a variety of techniques including combinations of phase separation or precipitation (Young, 1999), emulsion/solvent evaporation [11, 43, 102, 117, 164, 186, 192], and/or spraying methods [58, 72, 126, 140, 183]. Variations of the fabrication parameters generally allow control of the particle size and size distribution. Drugs may be incorporated into the particles in several different ways depending on the properties of the drug. Hydrophobic therapeutics may be co-dissolved with the polymer in a solvent such as methylene chloride or ethyl acetate. Hydrophilic therapeutics, including proteins, may be suspended in the organic phase as a finely ground dry powder. Alternatively, an aqueous solution of a hydrophilic therapeutic may be mixed with the organic polymer solution to form a water-in-oil emulsion.

The emulsion-solvent extraction/evaporation methods are most commonly used, especially at the lab scale. In these processes, a solution containing the polymer (and possibly the drug to be encapsulated) is emulsified in a non-solvent phase (the continuous phase) containing a stabilizer. The emulsion can be prepared with any of a variety of physical methods including homogenization and sonication. The components are chosen such that the solvent is slightly soluble in the non-solvent. For example, to produce microspheres of PLG or polyanhydrides, common solvents are methylene chloride and ethyl acetate used in conjunction with an aqueous continuous phase containing poly(vinyl alcohol) (PVA) as a stabilizer [49]. After emulsification, the solvent is extracted into the continuous phase and allowed to evaporate. At the same time, the non-solvent may penetrate into the polymer-rich droplets. Due to loss of solvent, the dispersed phase is enriched in polymer until the droplets "harden" to become particles. The microspheres may then be filtered, washed, and lyophilized.

There are several disadvantages of the emulsion solvent-extraction/evaporation techniques that have limited their application. Because these methods are inherently batch operations, scale up of the processes is difficult and large-scale production can be costly. Another critical problem is that size distributions of particles are generally reproducible but non-uniform. Standard deviations of the distribution equal to 50% of the average size are not uncommon. Since the size of the spheres directly affects the drug release rate and syringability, it is important that size distributions be relatively narrow. In addition, as described above the presence of organic solvents and aqueous-organic interfaces may have adverse effects on encapsulated drugs [62, 147]—decreasing or even eliminating bioactivity. Organic solvents also may be very difficult to remove completely. Since many of the commonly used organic solvents (e.g., methylene chloride) are toxic, the concentration of residual solvent in the microsphere must be tightly regulated.

2.3.1. Techniques for Fabricating Uniform Microspheres

As pointed out in the preceding sections, an important limitation in the development of biodegradable polymer microspheres for controlled-release drug delivery applications has been the difficulty of specifically designing systems exhibiting precisely controlled release rates. Because microparticle size is a primary determinant of drug release [15, 17], it is worthwhile to develop a methodology for controlling release kinetics employing monodisperse microspheres.

There have been several reports of fabrication of uniform biodegradable polymer microspheres [4–5, 105, 152, 159]. Amsden and Goosen have used electrostatics to form EVAc spheres with diameters smaller than the nozzle from which they were sprayed [5]. The minimum size obtained, however, was only ~500 μm, and they concluded that "it is not possible to obtain small microbeads having a narrow size distribution." Amsden later reported production of uniform microspheres by extruding a polymer solution through a nozzle and into a stabilizing solution flowing perpendicular to the point of injection [3]. The PLG solution droplets exiting the nozzle are drawn into the PVA solution in a periodic fashion. The rate at which the droplets are removed from the nozzle, and therefore—under constant polymer flow rate—the size of the microspheres, are determined by properties of the stabilizing solution (density and viscosity), the diameter of the nozzle, and the interfacial tension between the polymer solution and the needle tip. Amsden reported PLG microspheres with diameters ranging from 67.7 to 295 μm. However, the size distributions varied widely, with standard deviations typically 10–30% of the average size (the minimum standard deviation was 5% of the average for 295-μm microspheres). Furthermore, formation of small particles may not be possible with this technique, as any further reduction of the particle size would apparently require smaller nozzles; as the nozzle size is decreased, shear forces at the orifice and the potential for clogging increase.

In summary, there have been several reports of the fabrication of biodegradable polymer microspheres with controlled, uniform size. However, none of these methods was successful in generating particles in the size range appropriate for drug delivery (~1–100 μm) while maintaining narrow size distributions. In addition, these conventional methods appear to be difficult to scale-up for commercial applications.

To remedy this situation, we have developed a novel microsphere fabrication technology which combines two techniques for generating monodisperse microspheres with precisely controlled sizes. This precision particle fabrication (PPF) technology also allows fabrication of predefined particle size distributions via continuous variation of the process parameters.

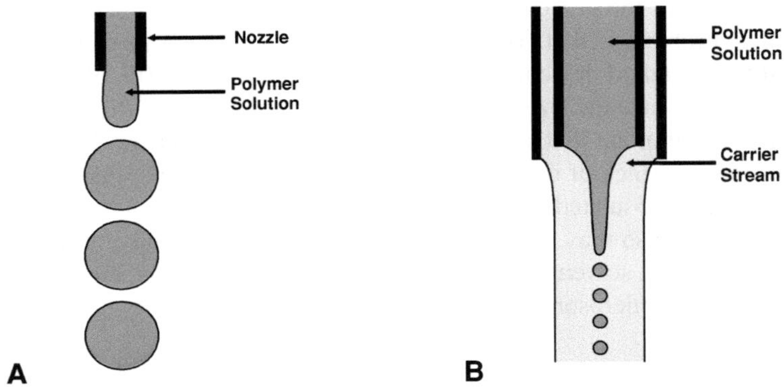

FIGURE 2.1. Schematic of the uniform-droplet generation scheme: (A) Single-nozzle approach and (B) Dual-nozzle approach with carrier stream.

The fundamental physical processes underlying the PPF technology are schematically described in Figure 2.1. The method basically comprises spraying a polymer-containing solution through a nozzle with (i) acoustic excitation to produce uniform droplets (Figure 2.1A), and (ii) an annular, non-solvent carrier stream allowing further control of the droplet size (Figure 2.1B). Using the PPF technology we have fabricated uniform solid microspheres of a variety of polymers including poly(D,L-lactide-co-glycolide) (PLG) [14, 17], polyanhydrides [18], ethylcellulose [39], chitosan [40], hetastarch [41], and gelatin hydrogel [42].

In general, the PPF apparatus (Figure 2.2) is designed to pass a solution containing the sphere material, and any drug to be encapsulated, through a small nozzle or other orifice (20-μm to a few millimeters in diameter) to form a smooth, cylindrical jet. The nozzle is

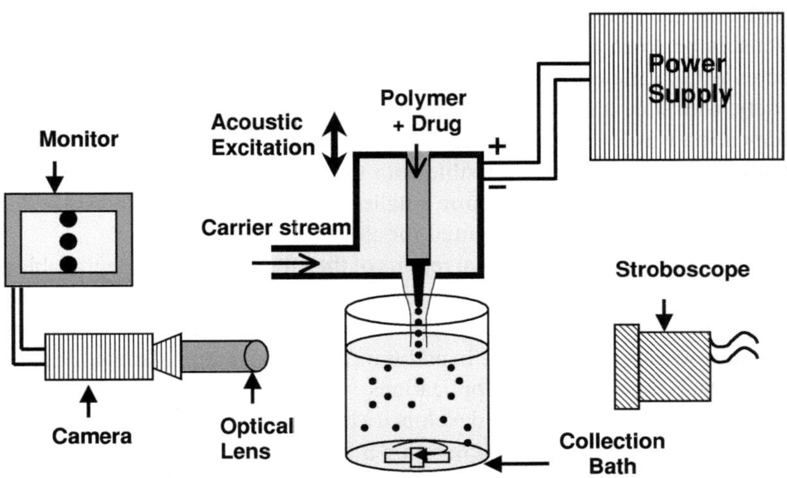

FIGURE 2.2. Schematic of the precision particle fabrication (PPF) system.

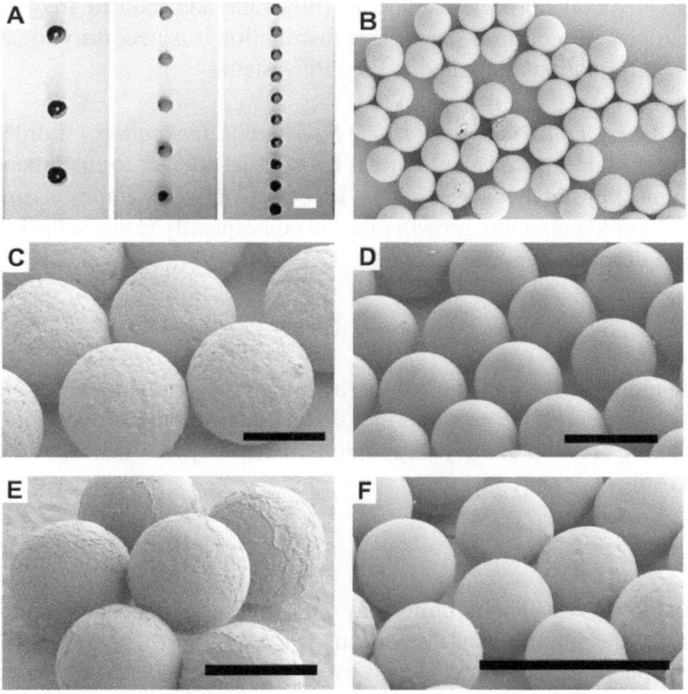

FIGURE 2.3. (A) Video micrographs showing PLG microdroplets (i.v. 0.19 dL/g; 50/50 lactide/glycolid) dissolved in methylene chloride (5% w/v) falling from the PPF nozzle at constant flow rate with increasing acoustic frequency. Also, scanning electron micrographs of (B) 40-μm PLG, (C) 60-μm ethyl cellulose (45 cp), (D) 20-μm chitosan, (E) 35-μm gelatin, and (F) 17-μm hetastarch PPF microscpheres.

vibrated by a piezoelectric transducer driven by a wave generator at a frequency tuned to match the flow rate and the desired drop size. The mechanical excitation launches an acoustic wave along the liquid jet generating periodic instabilities that, in turn, break the stream into a train of uniform droplets (Figure 2.3A). With this apparatus alone, the minimum particle size achievable is approximately twice the nozzle opening. This approach represents an improvement over conventional ultrasonic nozzles as the acoustic wave intensity is lower and we can tightly control the match between the frequency and solution flow rate.

We can further control sphere size, and in fact form droplets much smaller than the nozzle opening, by employing an annular flow of a non-solvent phase around the polymer jet (Figure 2.1B). The annular stream is pumped at a linear velocity greater than that of the polymer stream. Thus, frictional contact between the two streams generates an additional downward force that effectively "pulls" the polymer solution away from the tip of the nozzle. The polymer stream is accelerated by this force and, therefore, thinned to a degree depending on the difference in the linear velocities of the two streams. The carrier stream allows production of microspheres as much as 100-fold smaller than the orifice size. Thus, addition of the carrier stream accommodates high-viscosity materials and reduces the risk of clogging by allowing use of large nozzles.

In order to predict the process parameters (flow rates and acoustic frequency) required to generate particles of a desired size or size distribution it is necessary to understand the theory of droplet formation as applied to the PPF system.

2.3.1.1. Theory of the Precision Particle Fabrication Technology Lord Rayleigh first investigated the instabilities of a cylindrical inviscid jet subject to disturbance [145]. He found that the most unstable wavelength (λ_{max}) of a disturbance imposed on a jet surface, which will give rise to maximum growth rate and consequently result in the break-up of the jet into uniform droplets (Figure 2.1), is

$$\lambda_{max} = 9.016 r_j \tag{2.1}$$

where r_j is the radius of the undisturbed jet (approximately equal to, but typically slightly larger than, the diameter of the nozzle orifice). The practical range of acoustic wavelengths that give rise to the breakup of a liquid jet into uniform droplets was experimentally determined to be [146]

$$7r_j < \lambda < 36r_j. \tag{2.2}$$

Both upper and lower wavelength limits may vary somewhat depending on the noise level of the system and the amplitude of the acoustic wave.

The microsphere generator developed by us allows for control of the acoustic wave frequency and amplitude. The wavelength produced by a set frequency (f) is given by

$$f = v_j/\lambda \tag{2.3}$$

where v_j is the linear velocity of the liquid jet. Knowing that the volume of the resulting sphere should be equal to the volume of a cylindrical element of the jet, the length of which is defined by the acoustic wavelength, we find that the drop radius, r_d, is predicted to be

$$r_d = (3r_j^2 v_j/4f)^{1/3}. \tag{2.4}$$

At the optimum wavelength, $r_{d,max} = 1.891\, r_j$. Thus, by imposing a uniform, high-amplitude oscillation on the nozzle, which will dominate the random, natural instability, we can control the breakup of the stream into droplets and predict the orifice size ($\sim r_j$), solution flow rate (v_j), and acoustic frequency (f) needed to generate a desired sphere size or size distribution.

2.3.1.2. Uniform Polymer Microspheres Produced by PPF Technology Uniform microspheres of controlled sizes, both solid and hollow, were previously fabricated from a variety of non-polymeric materials using acoustic excitation [56, 64, 68, 78, 85, 87, 90]. More recently, we have demonstrated the PPF technology for fabricating monodisperse microspheres of various polymers such as PLG [14, 17], polyanhydrides [18], EC [39], chitosan [40], gelatin hydrogel [41], and hetastarch [42], with average diameters from ~4 μm to >500 μm (Figure 2.3B–F). These micrographs clearly demonstrate that the PPF technology: (a) can produce uniform polymeric microspheres with precisely controlled sizes,

(b) can be used to produce particles much smaller than the nozzle opening, and (c) is insensitive to the choice of materials.

We have also demonstrated encapsulation and in vitro release of model drug compounds such as rhodamine B, piroxicam, nifedipine, and felodipine [15, 17, 39–40]. The release kinetics of these compounds depended strongly on the microsphere size, as expected. Details of in vitro release profiles and mechanisms will be discussed below.

2.3.2. Techniques for Fabricating Uniform Core-Shell Microparticles

The ability to form microcapsules exhibiting a predefined diameter and shell thickness may offer several additional advantages in drug delivery. Higher drug loads may be realized by utilizing core materials offering increased drug solubility. In addition, fragile therapeutics such as proteins and DNA may be stabilized by generating a favorable environment in the core. Advanced drug release schedules such as delayed or pulsatile release, with the removal of drug "burst," may be possible by selectively varying the shell material or especially thickness [103–104]. For example, a surface-eroding polymer (e.g. polyanhydride) shell may be expected to protect a polyester core, and the encapsulated drug, for a prolonged time, the duration of which is governed by the shell thickness. However, efficient fabrication of such particles has not been previously reported to our knowledge. Finally, drugs could be released in tandem by selectively loading them into the core or shell phase thereby potentially enhancing drug efficacy [193–194].

Core-shell microparticles are significantly more difficult to manufacture than solid microspheres. A variety of techniques for fabricating microcapsules of varying sizes have been reported [32, 50, 65, 85, 87, 94, 120, 133, 134, 150, 172]. For example, variations of the common double-emulsion approach have also been used to prepare microcapsules by allowing the discontinuous phase of the primary emulsion to coalesce and form the core of the particle [50, 150, 172]. Such particles have shown an interesting pulsatile release profile that may be advantageous for vaccine delivery [150]. Important limitations of these approaches, however, are that the core and shell material must be immiscible and the microcapsule architecture is difficult to control; the core size and shell thickness depend strongly on the properties of the primary emulsion and the time over which the emulsion is allowed to coalesce.

It is also possible to generate microcapsules with solid cores by coating pre-formed microparticles with a second material. For example, Göpferich et al. [65] developed an ingenious method for coating microspheres with a second layer of polymer by resuspending preformed microspheres in a concentrated solution of a second (or the same) polymer followed by re-emulsifying the suspension in an aqueous continuous phase to form "microencapsulated microspheres." Presumably, the shell thickness can be controlled by varying the polymer concentrations and mass ratios, and the two materials can be miscible with one another. However, the thinnest shell reported was >50% of the overall particle diameter.

"Double-wall" particles comprising polymer cores and shells can be formed by controlling phase separation of the two sphere-forming materials [133–134]. One of the first studies consisted of adding two separate polymers dropwise to an aqueous phase and controlling precipitation rates such that one polymer has ample time to spread on the other [96, 97, 99, 122, 132]. Another technique utilized the phase separation of PLG and PLL at certain concentrations [103–104]. The radiosensitizer etanidazole was suspended as fine filaments (<20 μm) into the PLG phase and the two polymers were emulsified into each other forming a milky solution. Again, during the relatively prolonged extraction of solvent from these large, ~450-μm particles, the two polymers phase-separated, but some of each polymer still remained in both the core and shell phase. Etanidazole was entrapped primarily in the PLG core due to a higher affinity to the PLG phase. Most recently, a polyorthoester (POE) was encapsulated in a PLG shell [193–194]. In these studies, the overall polymer concentration remained constant at 5% (wt polymer/vol DCM) but the POE:PLG polymer ratio was adjusted to produce double-wall microspheres. When POE accounted for 50%, 60%, and 70% of the polymer weight, core-shell particles with PLG shells were formed while microspheres with intermingled polymer phases were formed at all other ratios.

These are simple and scalable processes, but several factors seem to influence the formation of complete microcapsules. First, the interfacial tension between the three phases (polymer 1, polymer 2, aqueous non-solvent) indicates which polymer will tend to spread on the other [133–134, 173]. The spreading coefficient in these cases, therefore, is expected to be a function of the polymer chemistry, solvent type, and polymer concentration [193–194]. Relying on these thermodynamically driven techniques to form microcapsules may necessitate extended solvent extraction times to allow the two polymer phases to come into intimate contact [96–100, 122, 132, 133] or phase separate [103–104] and then form a core/shell arrangement before the polymer solutions precipitate. An extended solvent extraction time may have been achieved in the reviewed work by creating large particles, >100 μm (and thus large nascent polymer droplets), lowering polymer concentration, lowering the temperature of the non-solvent bath, or adding solvent to the extraction phase. Thus, the achievable architectures, types of drugs that can be encapsulated, shell thicknesses obtainable, and release kinetics achievable may be limited.

2.3.2.1. Precision Core-Shell Microparticle Fabrication Following previous work reporting production of core/shell particles made from a variety of inorganic materials [68, 85, 87], the uniform solid microsphere fabrication methodology described in the preceding sections (Figures 2.1 and 2.2) has been further extended to produce uniform double-walled polymeric microspheres with controllable size and shell thickness. The method, as illustrated in Figure 2.4, employs three coaxial nozzles to produce a smooth coaxial jet of controllable size, comprising a carrier, annular shell and core streams, which is acoustically excited to break up into uniform core-shell droplets. As before, the non-solvent carrier stream surrounding the coaxial jet accelerates and makes it thinner before its breakup. The orientation of the jets, material flow rates, and rate of solvent extraction are controlled to vary the shell thickness.

Microcapsules have been fabricated with different arrangements of bulk-eroding poly(D,L-lactide-co-glycolide) (PLG) and surface-eroding poly[(1,6-bis-carboxyphenoxy) hexane] (PCPH) [20]. Variation of the fabrication parameters allowed complete encapsulation by the shell phase including the efficient formation of a PCPH shell encapsulating a PLG core. Utilizing this technology, microcapsule shell thickness can be varied from <2 μm to tens of microns while maintaining complete and well-centered core encapsulation for microcapsules near 50 μm in overall diameter. Scanning electron micrographs of microcapsules originally containing an oil or aqueous core are shown in Figure 2.5, and PCPH shell/PLG core and PLG core/PCPH shell microcapsules are shown in Figure 2.6. These micrographs are proof that the modified PPF technology is an effective single-step

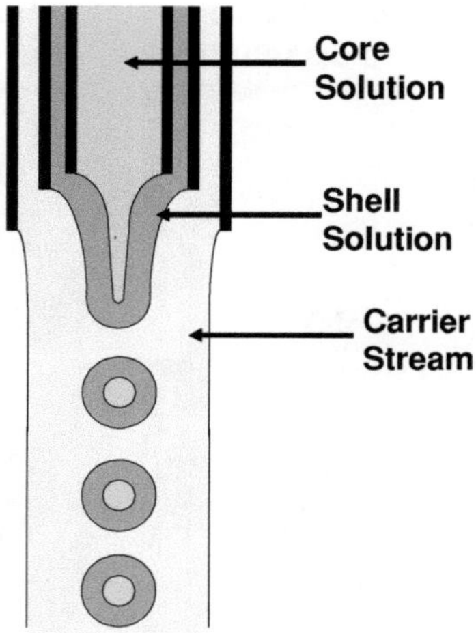

FIGURE 2.4. Schematic description of the triple-nozzle PPF methodology of producing uniform core-shell droplets using acoustic and carrier stream schemes.

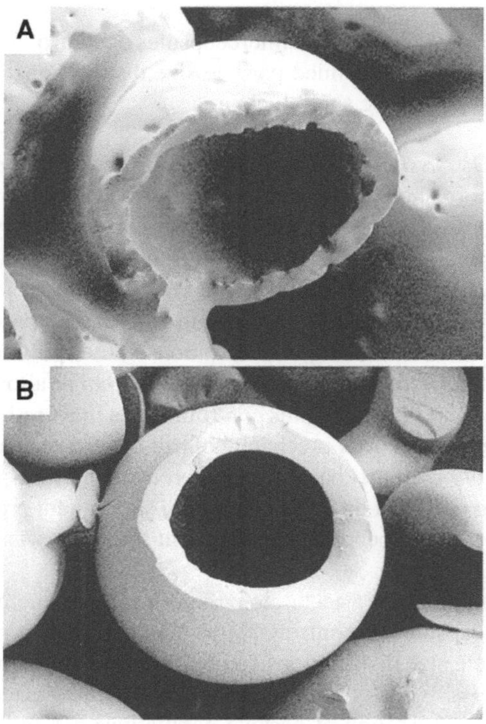

FIGURE 2.5. Scanning electron micrographs showing fractured (A) canola oil core/PLG shell microcapsule, and (b) microcapsule with an aqueous core containing 100 mg/mL dextran, 10 mg/mL BSA and a PLG shell.

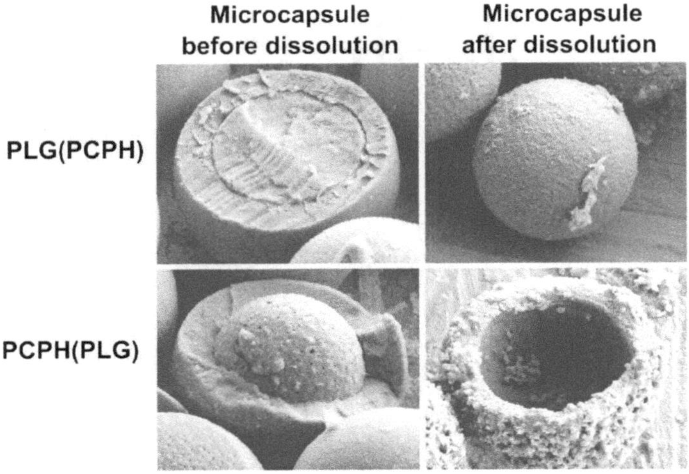

FIGURE 2.6. Scanning electron micrographs of double-wall microspheres with PLG shells encapsulating PCPH cores and PCPH shells encapsulating PLG cores, fabricated by the PPF method. Images show the effect of selectively dissolving PLG using tetrahydrofuran, which a poor solvent for PCPH. Scale bar = 25 μm.

method for producing uniform polymeric microcapsules of controllable size and shell thickness. Monodisperse or precisely defined particle size distributions can be achieved while maintaining the desired polymeric shell thickness.

2.3.2.2. Further Discussion of Microcapsule Fabrication Technique In comparison with the conventional core-shell particle methods described above, precision particle fabrication (PPF) technology sprays polymer solutions that are immediately brought into intimate contact as distinct and separate phases specifically arranged in the desired orientation. Utilizing this method for creating uniform microcapsules of controllable shell thickness provides a mechanical driving force to aid this thermodynamically driven process. Additionally, the present method is not limited to using concentrations where polymers are immiscible or specific conditions that slow the extraction of solvent but rather allows the exact control of polymer concentrations and flow rates thus controlling the mass ratios of the two polymers in each nascent droplet, resulting in precision control of microcapsule diameter and shell thickness. This flexibility can also be utilized to create conditions in the nascent droplet that facilitate spreading of PCPH on a PLG core until subsequent precipitation of the polymers kinetically traps the microcapsule arrangement that otherwise would be difficult to achieve.

In summary, the modified PPF technology is a single-step method for producing uniform polymeric microcapsules of controllable size and shell thickness. Monodisperse or precisely defined particle size distributions can be achieved while maintaining the desired polymeric shell thickness. Exact control of the volumetric flow-rates of the core and shell materials also allows the formation of particle populations exhibiting discretely or incrementally increasing shell thickness. In addition to advancing control of particle architecture, methods have been developed for modulating the locale of materials and compounds of interest. Specifically, polymer, oil, and aqueous cores have been encapsulated within uniform

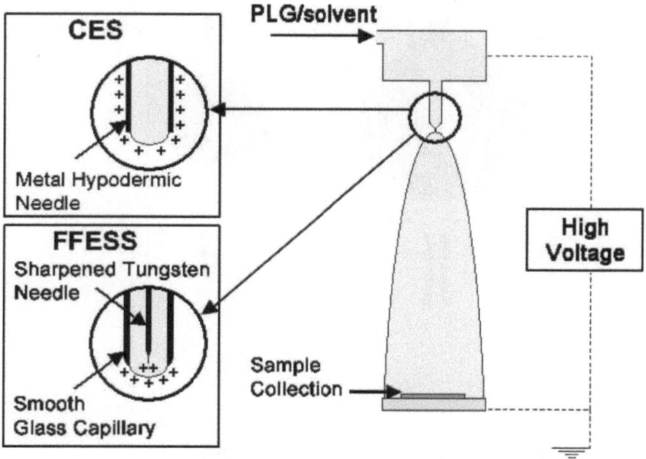

FIGURE 2.7. Schematic diagram describing the flow-limited field-injection electrostatic spraying (FFESS) technology. Shown for comparison is the conventional electrospraying (CES).

polymeric shells potentially enabling advancements in the areas of drug delivery, photonics, cell encapsulation, and catalysis to name a few. Monodisperse microcapsule advantages in reproducibility, control of transport phenomena, and consistency of physically derived properties may provide valuable assistance in a variety of research areas.

2.3.3. Use of Electrohydrodynamic Spraying for Fabrication of Uniform Micro and Nanospheres

A process known as electrohydrodynamic (EHD) or electrostatic (ESS) spraying can be used to fabricate both micro- and nanospheres of polymeric materials. In this method charge is injected at a controlled rate (i.e., at a controlled current) into a polymer solution contained in a nozzle such that the surface of the solution meniscus at the nozzle opening becomes highly charged. The charged meniscus surface is thus under the electrical tension forces that push the meniscus surface away from the nozzle, opposing the surface tension force that tries to minimize the meniscus surface area (Figure 2.7). The electrical forces increase with increasing charge injection, finally overcoming the surface tension force and ejecting the charged solution from the meniscus surface. The net result is that with an increase in the charge injection, there will be gradual reduction in the size of the drops that leave the nozzle. This so-called "drip mode" develops into a "single-jet mode" with further increase in the charge injection. The jet mode is invoked when the electrical tension forces at the charged meniscus surface are such that the charged solution is literally pulled away from the nozzle orifice as a thin jet, which in turn naturally breaks up into charged droplets due to the jet instability. As the charge injection is further increased, the "single-jet mode" develops into a "multijet mode" where more than one jet emanates from the charged meniscus surface at the nozzle opening. Since the electrical tension forces acting on the meniscus surface are a function of the surface charge density, the effect of increasing flow rate is similar to that of decreasing charge injection. Figure 2.8 shows this sequence of

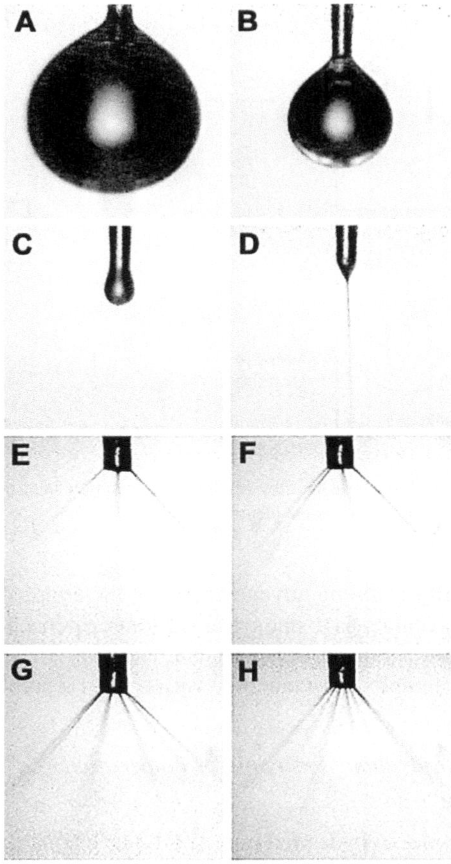

FIGURE 2.8. FFESS sequence with increasing voltage: (A–D) Jet formation using 5% w/v solution of PLG (0.20 dL/g) in methylene chloride flowing at 1 mL/h as voltage increases from 0 to 7 kV. (E–H) Multi-jet mode spryaing of 5% w/v PLG (0.20 dL/g) in acetonitrile flowing at 1 mL/h as voltage increases from 4 to 9 kV. Charge injection was achieved using a sharpened tungsten needle encased by a glass capillary having a 300-μm orifice positioned ~1–2 cm above a ground plate.

events that take place as the charging voltage (i.e., the charge injection) is increased, for a PLG solution. In the phenomena just described, generally known as EHD spraying, the typical flow rates used are necessarily very small. When fibers are produced as the final product, which is possible with polymeric solutions or material in molten state, the process is often referred to as electrospinning.

An alternative electrohydrodynamic method called flow-limited field-injection electrostatic spraying (FFESS) [86, 190] (Kim et al. 1994; Kim et al. 1995; Berkland et al. 2004) has been developed. FFESS can be used by itself alone or in tandem with the acoustic methods described in the preceding sections. In FFESS, charge injection is achieved by using a nano-sharpened tungsten needle connected to a high voltage in a process called field injection that is field emission or ionization depending on whether the polarity of the

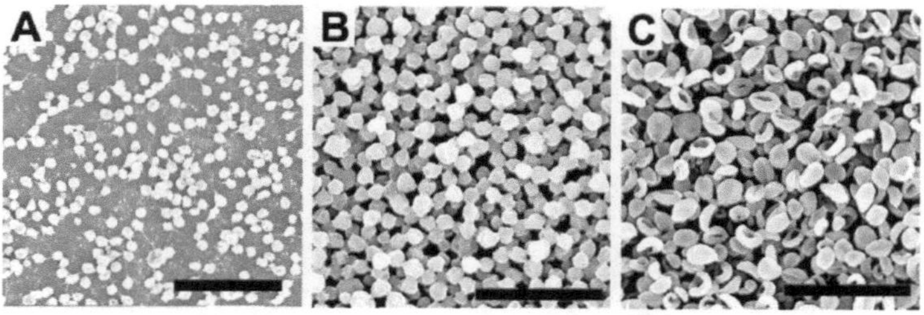

FIGURE 2.9. (A) Uniform 2.5-μm particles produced by spraying 5% w/v (0.24 dL/g) PLG in acetone flowing a 1 mL/h while applying 8 kV to a 26-guage needle positioned ~5 cm above a ground plate. Scale bar = 25 μm (B) Uniform ~300-nm particles produced by spraying 5% w/v (0.24 dL/g) PLG in methylene chloride flowing at 1 mL/h using FFESS at 15 kV appplied to a sharpened tungsten needle encased by a glass capillary having a 300-μm orifice positioned ~5 cm above a ground plate. Scale bar = 5 μm. (C) Uniform ~300-nm particles produced under same conditions as (B) except solvent was acetonitrile and applied voltage was 10 kV. Scale bar = 5 μm.

needle is negative or positive. Thus, the FFESS process allows one to produce sprays that are finer and more precisely controlled than those produced by conventional EHD techniques, which employ conducting hypodermic needles as the spray nozzle (Figure 2.7). The main reason is that the sharp charge-injection electrode used in FFESS renders more control and stability, and higher charge injection (Kim, 1994 & 1995). For applications involving advanced drug delivery, the effects of the key parameters controlling the FFESS process, such as applied voltage, polymer solution flow-rate, and solvent properties (surface tension, viscosity, vapor pressure) have been qualitatively evaluated [19]. For example, using the FFESS system a wide variety of uniform micro- and nanoparticles of PLG have been created. (Figure 2.9). By discretely varying production parameters, subtle changes in deposited polymer morphology are realized potentially resulting in enhanced performance of a medical or biological device. FFESS technology thus provides a simple and robust technique for fabricating devices with a precisely defined nano-structure from a broad range of biocompatible polymeric materials.

2.4. CONTROLLED RELEASE FROM PRECISION MICROSPHERES

Presented in this section are a brief review of the controlled-release literature and the results from several controlled release studies that have been carried out with PPF-produced microspheres. The data clearly show that release of model small-molecule drugs can be varied from typical diffusion-controlled profiles to slower, sigmoidal profiles as microsphere diameter is increased in the range of 10–100 μm. The data also show that drug release from mixtures of uniform microspheres corresponds to a weighted average of the release from individual uniform microspheres. As a result, it has been possible to choose appropriate mixtures to generate desired release rate profiles, in particular constant (i.e., zero-order) release. It is, therefore, concluded that microsphere mixtures with well-defined size distributions may provide a general methodology for controlling drug release rates [15, 17–18].

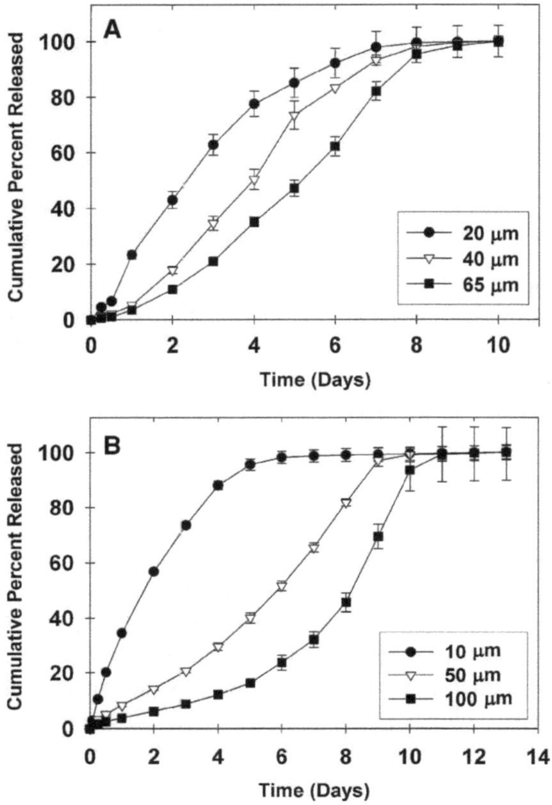

FIGURE 2.10. Effect of microsphere size on release of (A) rhodamine and (B) piroxicam from uniform PLG microspheres of various sizes.

2.4.1. In-vitro Release from Uniform Microspheres

Long-term zero-order release of small-molecule therapeutics from biodegradable microspheres has been difficult to achieve. Release of small compounds is often rapid and diffusion controlled [23, 82]. Often, release of small molecules encapsulated in polymeric particles typically is dominated by a large initial rate of release (or "burst") in the first 24 h, offering little advantage over conventional oral dosage forms [27, 67, 92, 137, 175, 178].

To examine the effect of microsphere size and size uniformity on drug release kinetics, we measured release profiles for two model drugs, rhodamine and piroxicam, from PLG spheres. Rhodamine release profiles are shown in Figure 2.10A. As expected, 20-μm microspheres exhibited a faster initial release than 65-μm microspheres, likely due to the increased surface-to-volume ratio of the smaller particles. Further, as drug loading increased, the initial rate of drug release increased. An interesting concave-upward (i.e., sigmoidal) profile was observed with the 65-μm particles and to a lesser extent with the 45-μm particles, wherein drug release was initially slow, then progressed to a more rapid release phase before leveling off [37, 165].

Piroxicam release profiles show similar trends (Figure 2.10B). Samples of 10-, 50- and 100-μm microspheres were studied. The microspheres span a broader size range than the rhodamine-loaded particles, resulting in a more pronounced difference in drug release profiles. The smallest microspheres (10-μm diameter) exhibited a rapid initial rate of release, with 40–60% of encapsulated piroxicam released within the first 24 hours. Initial release rates decreased with increasing microsphere diameter for all drug loadings examined. Further, the initial release rate decreased with increasing drug loading. Interestingly, the 50- and 100-μm particles exhibited sigmoidal release profiles similar to rhodamine release from 65-μm microspheres.

Modeling of release has elucidated the mechanisms controlling the varying shapes of the release profiles [16, 143]. The model assumes a simple diffusion-controlled release but incorporates initially non-uniform drug distribution in the microsphere and a time-dependent diffusivity. We showed that the effective drug diffusivity increased with decreasing polymer molecular weight caused by degradation. In this way, we are able to accurately predict the shapes and kinetics of the small-molecule release profiles. This model will facilitate design of microsphere systems to provide desired release rates.

2.4.2. In-vitro Release from Mixtures of Uniform Microspheres

Constant release is highly desirable for many drug delivery applications. Because there is a transition from the concave downward to sigmoidal release profiles as sphere size increases, it appears that nearly linear release may be achieved at a certain size. For example, between 10- and 50-μm, a microsphere size may exist that would provide zero-order piroxicam release over a 4- to 8-day duration (c.f. Figure 2.10). Others have reported linear or near-linear release profiles achieved with microspheres of similar size, ~30–50 μm in diameter [21, 110, 189]. For example, Woo et al. formulated a leuprolide delivery system using PLA microspheres with an average diameter of 51.7 μm achieving near-linear peptide release for 135 days following a 15-day period of "diffusion-controlled release" [189]. This early phase of release may result from the portion of the microspheres in this formulation under ~35-μm, which would be expected to release drug more rapidly. Further, Bezemer et al. used a poly(ethylene glycol)-poly(butylene terephthalate) (PEG-PBT) block copolymer to test the effects of microsphere size on drug release [21]. They also discovered that decreasing the average microsphere size from 108 μm to 29 μm causes the release kinetics to change gradually from zero-order release to release controlled by Fickian diffusion.

Other researchers have suggested that drug delivery rates may be controlled by mixing microspheres of varying sizes or characteristics. For example, Ravivarapu et al. mixed microspheres comprising 8.6-kD or 28.3-kD PLG encapsulating leuprolide acetate [144]. The low-molecular-weight polymer resulted in porous, quickly releasing microspheres while the high-molecular-weight formulation resulted in dense microspheres and produced a sigmoidal release profile. By mixing microspheres comprising the two polymers, release rates could be tailored, and the resulting profiles were linear combinations of those resulting from individual microspheres. Bezemer et al. produced linear lysozyme release over 25 days from PEG-PBT microspheres having a bimodal size distribution dominated by 50-μm and 110-μm particles (in essence a combination of two sizes) [21]. Finally, Narayani et al. combined gelatin microspheres of various size ranges producing zero-order release of methotrexate [128]. In general, however, zero-order release has been difficult to achieve

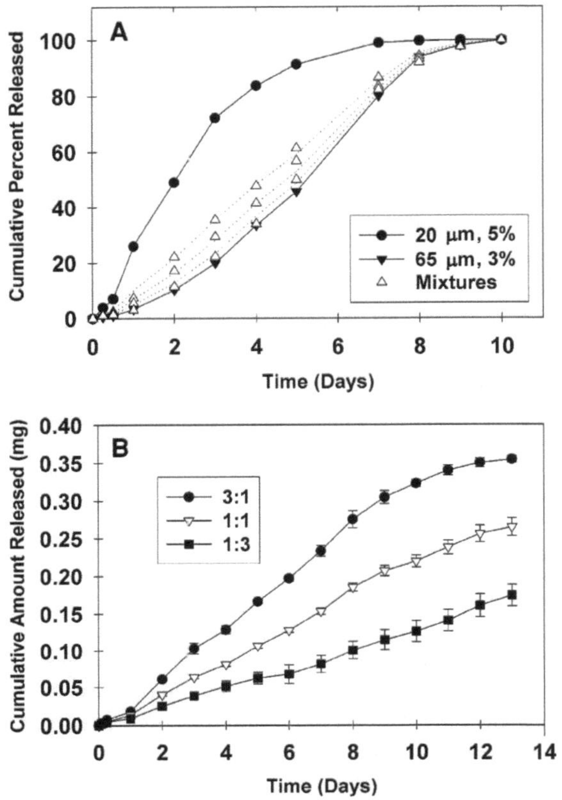

FIGURE 2.11. (A) In vitro release of rhodamine from mixtures of 20-μm and 65-μm diameter PLG microspheres in the ratios 1:4, 1:9 and 1:25 w:w. Filled symbos are experimental data points for individual microspheres and open symbols are experimental data points for the mixtures. The dotted lines are the weighted average of individual microsphere experimental release data. (B) In vitro release of piroxicam from mixtures of 10-μm and 50-μm diameter PLG microspheres.

with simple microsphere systems. Further, there is little generality, and unique systems are designed empirically for each drug.

Based on the different shapes of the uniform PPF microsphere release profiles, and given the reproducibility of the PPF methodology for uniform microsphere fabrication, it was reasoned that it may be possible to modulate release kinetics in a desired fashion by mixing appropriate proportions of two or more uniform microsphere preparations (Figure 2.11) [21–22, 128]. To investigate the generality of this hypothesis, we mixed uniform microspheres of different sizes to generate zero-order release of rhodamine and the clinically relevant NSAID, piroxicam. For example, multiple linear combinations of 10-, 50- and 100-μm piroxicam-containing microspheres at various drug loadings were examined computationally to identify a combination resulting in linear drug release. Two possible formulations were found. The first formulation combined 10-μm/15% and 50-μm/15% microspheres in ratios of 3:1, 1:1, and 1:3 wt:wt. This formulation resulted in slightly concave downward release profiles for the 3:1 and 1:1 ratios and a linear drug release profile for the 1:3 wt:wt mixture (Figure 2.11B).

In summary, because release kinetics from uniform spheres are very predictable and reproducible, our ability to fabricate uniform microspheres enhances this technique. We found that upon mixing uniform microsphere preparations, the resulting release profile is a mass-weighted average of the release profiles of the individual microspheres. This demonstrates that the microspheres release their payload independently; there is no interaction between the particles. In these experiments, the shapes of the rhodamine and piroxicam release profiles were such that it was possible to choose appropriate microsphere mixtures that provided zero-order release kinetics (Figure 2.11). However, it may not always be possible to generate a desired release profile from mixtures of only two microsphere sizes. Depending on the desired profile and the shape of the individual release curves, one may need to mix multiple microsphere samples or to fabricate complex microsphere size distributions. Because the PPF method provides a unique ability to generate predefined microsphere sizes [147], this technology may lead to enhanced control of release rates.

2.4.3. *In vitro Release with Double-Wall Microspheres*

Double-walled microspheres represent an increasingly important class of drug delivery devices that provide enhanced control of drug delivery schedules. Clearly, the overall particle size and shell thickness are important parameters in modulating the drug release rates. In one study, PPF was used to fabricate double-walled microspheres of predefined uniform diameters of 40–60 μm exhibiting a PLG core and poly(L-lactide) (PL) shell of controllable thickness from approximately 2 to 10 μm [13]. In vitro release of piroxicam from uniform microspheres of pure PLG and PL is compared to release from double-walled microspheres exhibiting different PL shell thicknesses in Figure 2.12. The benefit of utilizing a PL shell around the PLG core is threefold. First, the presence of the PL shell minimizes the initial drug "burst" so often associated with microparticle-based drug delivery systems. Secondly, by

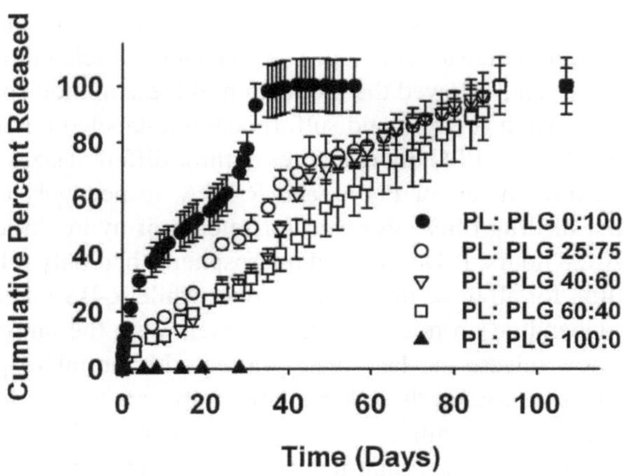

FIGURE 2.12. In vitro release of piroxicam from PPF particles with drug-free PLL shells and PLG cores containing piroxicam. The legend shows the weight ratios of PLL:PLG.

specifically controlling the mass ratio of PL:PLG and thus the resulting PL shell thickness, the release of piroxicam can be progressively modulated from biphasic (no shell) to zero-order (thickest shell in Figure 2.12). Lastly, the duration of piroxicam release was extended from one month to three months utilizing the same molecular weight PLG. The ability to specifically control double-walled microsphere shell thickness may provide a novel and precise approach to modulate drug release profiles.

2.4.4. Release of Macromolecules from Monodisperse Microspheres

Macromolecular therapeutics such as peptides, proteins and DNA are advancing rapidly toward the clinic. Because of typically low oral bioavailability, invasive delivery methods such as frequently repeated injections are required. Parenteral depots including biodegradable polymer microspheres offer the possibility of reduced dosing frequency but are limited by the inability to adequately control delivery rates.

The release of macromolecules typically exhibits an initial "burst" of drug, which can be as much as 10–50% of the drug load, followed by a "lag" phase exhibiting slow release and finally a period of steady release [10, 151, 184]. The initial burst of protein therapeutics has been attributed to their tendency to partition to the microsphere surface during the encapsulation process [61]. Multiple approaches have attempted to alleviate the burst by adding excipients to the polymer phase [80, 185], utilizing novel polymers [70, 166, 168, 181, 195], encapsulating particulate forms of the drugs [33], or exchanging the non-solvent aqueous phase used in the fabrication process with non-polar oils or alcohols to reduce the affinity of the encapsulated drugs for the bulk phase [57, 174]. The burst is usually followed by a lag period where diffusion is limited and little release of the macromolecule occurs. The duration and flatness of this phase is determined by the polymer degradation kinetics, particle size, and microsphere porosity [10]. In addition, the drug size, charge, and any potential interactions of the drug with the polymer can influence the lag phase [30]. Finally, the lag phase is followed by a period of steady release typically controlled by the polymer degradation rate.

To control release and investigate release mechanisms, model macromolecules were encapsulated in uniform poly(D,L-lactide-co-glycolide) (PLG) microspheres using PPF in combination with a double-emulsion method. The precision particle fabrication (PPF) technology described in Section 3 allowed the production of these monodisperse microspheres [12, 14]. Fluorescein-dextran (F-Dex) and sulforhodamine-labeled bovine serum albumin (R-BSA) were encapsulated in PLG microspheres of three different sizes; 31, 44 and 80 μm and 34, 47 and 85 μm diameter for F-Dex and R-BSA, respectively (Figure 2.13). Notably, the initial burst of drug often observed for release of hydrophilic macromolecules was not observed. At 28 hours, F-Dex-loaded microspheres had only released 1.2% of the total encapsulated drug for all three microsphere sizes while R-BSA-loaded microspheres released 1.9–4.8% depending on microsphere diameter. With the uniform microspheres, the duration of the slow release or "lag" phase varied with the microsphere diameter. In addition, the slow release phase for these formulations appears to be a diffusion-controlled process with rates that decrease with increasing sphere diameter. Plots of amount released versus the square root of time show that the diffusion-controlled phase lasts for 25–200 days, depending on the microsphere size.

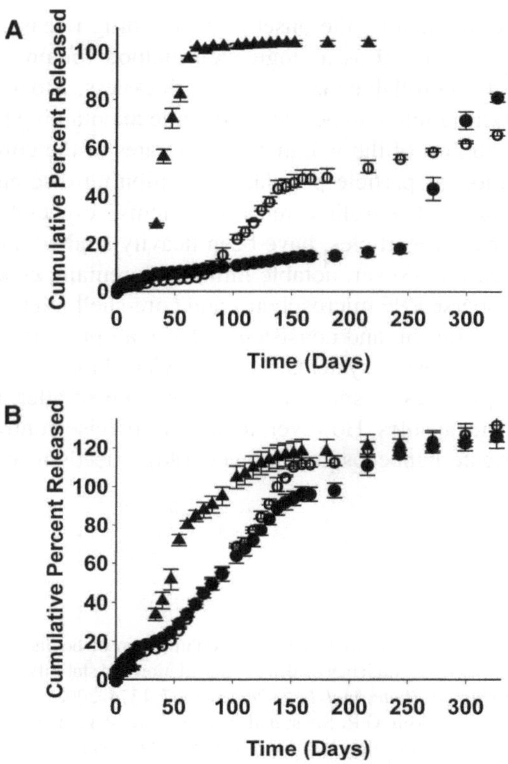

FIGURE 2.13. In vitro release of (A) fluorescein-labeled dextran and (B) rhodamine-labeled bovine serum albumin from PLG (0.17 dL/g) microspheres prepared by PPF with a double emulsion. In (A), spheres are 31 μm (filled circles), 44 μm (open circles) and 80 μm (triangles) in diameter. In (B), spheres are 34 μm (filled circles), 47 μm (open circles) and 85 μm (triangles) in diameter.

Release from uniform microspheres showed only two phases: a lag and more rapid release. F-Dex release profiles were very distinctive for the three different sizes. The diffusion-controlled lag period was followed by a sharp increase in release rate (Figure 2.13). The time at which the release rate increased was strongly dependent upon but inversely proportional to microsphere size. Again, after the lag periods, release rates of molecules increased dramatically. Thus, uniform microspheres represent a new delivery system for therapeutic proteins and DNA and provide unprecedented control of delivery rates using simple injectable depot formulations.

2.5. CONCLUSIONS

The development of PPF technology has allowed the production of uniform microspheres and double-wall microspheres capable of efficiently encapsulating model drugs. Of primary importance was the ability of monodisperse microsphere formulations to eliminate initial drug burst while modulating the onset of steady drug release. Modified PPF technology has also been established as a single-step method for producing uniform polymeric microcapsules of controllable size and shell thickness. Monodisperse or precisely defined particle size distributions can be achieved while maintaining the desired polymeric shell thickness. Exact control of the volumetric flow-rates of the core and shell materials also allows the formation of particle populations exhibiting discretely or incrementally increasing shell thickness. Controlled release systems, especially those comprising biodegradable polymer microparticles, have been heavily studied and have even reached the clinic in several cases. However, notable limitations remain, especially in controlling delivery rates. Monodisperse PPF microspheres and core-shell microparticles offer advantages in reproducibility, control, and consistency that may provide valuable assistance in designing advanced drug delivery systems. The FFESS technique is capable of producing nanometer-scale solid particles as small as 10 nm or even smaller, and may be applicable to fabrication of nanocapsules. However, to achieve precise control of the particle size and reproducibly fabricate nanocapsules the technology needs to be further refined and developed.

REFERENCES

[1] W.P. Al-Azzam, A. Emil, and Kai Griebenow. Co-lyophilization of bovine serum albumin (BSA) with poly(ethylene glycol) improves efficiency of BSA encapsulation and stability in polyester microspheres by a solid-in-oil-in-oil technique. *Biotechnol. Lett.*, 24(16):1367–1374, 2002.

[2] M.J. Alonso, R.K. Gupta, C. Min, G.R. Siber, and R. Langer. Biodegradable microspheres as controlled-release tetanus toxoid delivery systems. *Vaccine*, 12(4):299–306, 1994.

[3] B. Amsden. The Production of Uniformly Sized Polymer Microspheres. *Pharm. Res.*, 16(7):1140–1143, 1999a.

[4] B. Amsden. The production of uniformly sized polymer microspheres. *Pharm. Res.*, 16:1140–1143, 1999b.

[5] B.G. Amsden and M. Goosen. An examination of the factors affecting the size, distribution, and release characteristics of polymer microbeads made using electrostatics. *J. Control. Rel.*, 43:183–196, 1997.

[6] S. Ando, D. Putnam, D.W. Pack, and R. Langer. PLGA Microspheres Containing Plasmid DNA: Preservation of Supercoiled DNA via Cryopreparation and Carbohydrate Stabilization. *J. Pharmaceut. Sci.*, 88(1):126–130, 1998.

[7] B.B. Baras, M.A., and J. Gillard. Parameters influencing the antigen release from spray-dried poly(DL-lactide) microparticles. *Internat. J. Pharmaceut.*, 200(1):133–145, 2002.

[8] D.H. Barouch, S. Santra, J.E. Schmitz, M.J. Kuroda, T.-M. Fu, W. Wagner, M. Bilska, A. Craiu, X.X. Zheng, G.R. Krivulka and others. Control of viremia and prevention of clinical AIDS in rhesus monkeys by cytokine-augmented DNA vaccination. *Science*, 290:486–492, 2000.

[9] E.L.W. Barrow, G.A. Winchester, J.K. Staas, D.C. Quenelle, and W.W. Barrow. Use of microsphere technology for targeted delivery of rifampin to *Mycobacterium tuberculosis*-infected macrophages. *Antimicrob. Agents Chemothera.*, 42:2682–2689, 1998.

[10] R.P. Batycky, J. Hanes, R. Langer, and D.A. Edwards. A theoretical model of erosion and macromolecular drug release from biodegrading microspheres. *J. Pharm. Sci.*, 86:1464–1477, 1997.

[11] M.A. Beboit, B. Baras, and J. Gillard. Preparation and characterization of protein-loaded poly(ε-caprolactone) microparticles for oral vaccine delivery. *Int. J. Pharm.*, 184:73–84, 1999.

[12] C. Berkland. *Methods of controlling size distribution of polymeric drug delivery particles. [M.S.].* University of Illinois, Urbana, IL, 2001.

[13] C. Berkland, A. Cox, K.K. Kim, and D.W. Pack. Three-month, zero-order piroxicam release from monodispersed double-walled microspheres of controlled shell thickness. *J. Biomed. Mat. Res.*, 2004a (in press).

[14] C. Berkland, K. Kim, and D.W. Pack. Fabrication of PLG microspheres with precisely controlled and monodisperse size distributions. *J. Control. Rel.*, 73:59–74, 2001.

[15] C. Berkland, K. Kim, and D.W. Pack. PLG microsphere size controls drug release rate through several competing factors. *Pharm. Res.*, 20:1055–1062, 2003.

[16] C. Berkland, K. Kim, and D.W. Pack. Precision Polymer Microparticles for Controlled Release Drug Delivery. *ACS Symposium Series*, 879:197–213, 2004b.

[17] C. Berkland, M. King, A. Cox, K. Kim, and D.W. Pack. Precise control of PLG microsphere size provides enhanced control of drug release rate. *J. Control. Rel.*, 82:137–147, 2002.

[18] C. Berkland, M.J. Kipper, B. Narasimhan, K. Kim, and D.W. Pack. Microsphere size, precipitation kinetics, and drug distribution control drug release from biodegradable polyanhydride microspheres. *J. Control. Rel.*, 94:129–141, 2004c.

[19] C. Berkland, D.W. Pack, and K. Kim. Controlling surface nano-structure using flow-limited field-injection electrostatic spraying (FFESS) of poly(D,L-lactide-co-glycolide). *Biomaterials*, 25:5649–5658, 2004d.

[20] C. Berkland, E. Pollauf, D.W. Pack, and K. Kim. Uniform double-walled polymer microcapsules of controllable shell thickness. *J. Control. Rel.*, 96:101–111, 2004e.

[21] J.M. Bezemer, R. Radersma, D.W. Grijpma, P.J. Dijkstra, C.A. van Blitterswijk, and J. Feijen. Microspheres for protein delivery prepared from amphiphilic multiblock copolymers 2. Modulation of release rate. *J. Control. Rel.*, 67:249–260, 2000a.

[22] J.M. Bezemer, R. Radersma, D.W. Grijpma, P.J. Dijkstra, A.A. van Blitterswijk, and J. Feijen. Microspheres for protein delivery prepared from amphiphilic multiblock copolymers 2. Modulation of release rate. *J. Control. Rel.*, 67:249–260, 2000b.

[23] D.C. Bibby, N.M. Davies, and I.G. Tucker. Poly(acrylic acid) microspheres containing beta-cyclodextrin: loading and in vitro release of two dyes. *Int. J. Pharm.*, 187:243–250, 1999.

[24] B. Bittner, C. Witt, K. Mader, and T. Kissel. Degradation and protein release properties of microspheres prepared from biodegradable poly(lactide-co-glycolide) and ABA triblock copolymers: influence of buffer media on polymer erosion and bovine serum albumin release. *J. Control. Rel.*, 60(2–3):297–309, 1999.

[25] D. Blanco and M.J. Alonso. Protein encapsulation and release from poly(lactide-co-glycolide) microspheres. Effect of the protein and polymer properties and of the co-encapsulation of surfactants. *Euro. J. Pharm. Biopharm.*, 45(3):285–294, 1998.

[26] J. Bonadio, S.A. Goldstein, and R.J. Levy. Gene therapy for tissue repair and regeneration. *Adv. Drug Del. Rev.*, 33:53–69, 1998.

[27] S. Bozdag, S. Calis, H.S. Kas, M.T. Ercan, I. Peksoy, and A.A. Kincal. In vitro evaluation and intra-articular administration of biodegradable microspheres containing naproxen sodium. *J. Microencap.*, 18(4):443–456, 2001.

[28] V. Budker, G. Zhang, S. Knechtle, and J.A. Wolff. Naked DNA delivered intraportally expresses efficiently in hepatocytes. *Gene Ther.*, 3:593–598, 1996.

[29] F.V. Burkersroda, L. Schedl, and A. Gopferich. Why Degradable Polymers Undergo Surface Erosion or Bulk Erosion. *Biomaterials*, 23:4221–4231, 2002.

[30] S.M. Butler, M.A. Tracy, and R.D. Tilton. Adsorption of serum albumin to thin films of poly(lactide-co-glycolide). *J. Control. Rel.*, 58:335–347, 1999.

[31] P. Calvo, C. Remuñan-Lópex, J.L. Vila-Jato, and M.J. Alonso. Chitosan and chitosan/ethylene oxide-propylene oxide block copolymer nanoparticles as novel carriers for proteins and vaccines. *Pharm. Res.*, 14:1431–1436, 1997.

[32] F. Caruso. Nanoengineering of particle surfaces. *Adv. Mater.*, 13(1):11–22, 2001.

[33] I.J. Castellanos, K.G. Carrasquillo, J.D. Lopez, M. Alvarez, and K. Griebenow. Encapsulation of bovine serum albumin in poly(lactide-co-glycolide) microspheres by the solid-in-oil-in-water technique. *J. Pharm. Pharmacol.*, 53:167–178, 2001.

[34] I.J.C. Castellanos, and Kai. Gloydian; Crespo, Ruben; Griebenow. Encapsulation-induced aggregation and loss in activity of g-chymotrypsin and their prevention. *J. Control. Rel.*, 81(3):307–319, 2002.

[35] I.J.C. Castellanos and Kai. Ruben; Griebenow. Poly(ethylene glycol) as stabilizer and emulsifying agent: a novel stabilization approach preventing aggregation and inactivation of proteins upon encapsulation in bioerodible polyester microspheres. *J. Control. Rel.*, 88(1):135–145, 2003.

[36] A. Chang and R.K. Gupta. Stabilization of tetanus toxoid in poly(D,L-lactic-co-glycolic acid) microspheres for the controlled release of antigen. *J. Pharm. Sci.*, 85:129–132, 1996.

[37] Y. Cheng, L. Illum, and S.S. Davis. A poly(dl-lactide-co-glycolide) microsphere depot system for delivery of haloperidol. *J. Control. Rel.*, 55:203–212, 1998.

[38] Y. Choi, S.Y. Kim, S.H. Kim, K-S. Lee, C. Kim, and Y. Byun. Long-term delivery of all-trans-retinoic acid using biodegradable PLLA/PEG-PLLA blended microspheres. *Internat. J. Pharm.*, 215(1–2):67–81, 2001.

[39] Y.B. Choy, C. Berkland, H. Choi, D. W. Pack, and K. Kim. Fabrication and characterization of Uniform ethyl Cellulose Microspheres for use as Advanced Drug Delivery Vehicles, *30th Annual Meeting & Exposition of the Controlled Release Society*, 2003.

[40] Y.B. Choy, H. Choi, and K. Kim. A Novel Method of Fabricating Uniform Chitosan Microspheres of Precisely Controlled Size and Size Distribution, *31th Annual Meeting & Exposition of the Controlled Release Society*, 2004.

[41] Y.B. Choy, H. Choi, and K. Kim. Novel Fabrication Method for Uniform Gelatin Microspheres of Precisely Controlled Size and Size Distribution, *31th Annual Meeting & Exposition of the Controlled Release Society*, 2004.

[42] Y.B. Choy, H. Choi, and K. Kim, A novel method of fabricating uniform hydrogel microspheres of precise size and size distribution, manuscript in preparation (2004).

[43] L.Y. Chu, R. Xie, J.H. Zhu, W.M. Chen, T. Yamaguchi, and S.I. Nakao. Study of SPG membrane emulsification processes for the preparation of monodisperse core-shell microcapsules. *J. Colloid Interface Sci.*, 265: 187–196, 2003.

[44] J.L. Cleland. Solvent evaporation processes for the production of controlled release biodegradable microsphere formulations for therapeutics and vaccines. *Biotechnol. Progress*, 14(1):102–107, 1998.

[45] J.L. Cleland. Single-administration vaccines: controlled-release technology to mimic repeated immunizations. *Trends Biotechnol.*, 17(1):25–29, 1999.

[46] J.L. Cleland, E.T. Duenas, A. Park, A. Daugherty, J. Kahn, J. Kowalski, and A. Cuthbertson. Development of poly-(D,L-lactide-co-glycolide) microsphere formulations containing recombinant human vascular endothelial growth factor to promote local angiogenesis. *J. Control. Rel.*, 72(1–3):13–24, 2001.

[47] J.L. Cleland, A. Lim, A. Daugherty, L. Barron, N. Desjardin, E.T. Duenas, D.J. Eastman, J.C. Vennari, T. Wrin, P. Berman and others. Development of a single-shot subunit vaccine for HIV-1 with programmable in vivo autoboost and long-lasting neutralizing response. *J. Pharm. Sci.*, 87(12):1489–1495, 1998.

[48] H. Cohen, R.J. Levy, J. Gao, I. Fishbein, V. Kousaev, S. Sosnowski, S. Slomkowski, and G. Golomb. Sustained delivery and expression of DNA encapsulated in polymeric nanoparticles. *Gene Ther.*, 7:1896–1905, 2000.

[49] S. Cohen, T. Yoshioka, M. Lucarelli, L.H. Hwang, and R. Langer. Controlled delivery systems for proteins based on poly(Lactic/Glycolic Acid) microspheres. *Pharma. Res.*, 8(6):713–720, 1991.

[50] G. Crotts and Park TG. Preparation of porous and nonporous biodegradable polymeric hollow microspheres. *J. Controlled. Rel.*, 35:91–105, 1995.

[51] M. Diwan and T.G. Park. Pegylation enhances protein stability during encapsulation in PLGA microspheres. *J. Controlled Rel.*, 73(2–3):233–244, 2001.

[52] J.J. Donnelly, J.B. Ulmer, J.W. Shiver, and M.A. Liu. DNA vaccines. *Annu. Rev. Immunol.*, 15:617–648, 1997.

[53] V.J. Dwarki, P. Belloni, T. Nijjar, J. Smith, L. Couto, M. Rabier, S. Clift, A. Berns, and L.K. Cohen. Gene therapy for hemophilia A: production of therapeutic levels of human factor VIII *in vivo* in mice. *PNAS*, 92:1023–1027, 1995.

[54] J. Fang, Y.-Y. Zhu, E. Smiley, J. Bonadio, J.P. Rouleau, S.A. Goldstein, L.K. McCauley, B.L. Davidson, and B.J. Roessler. Stimulation of new bone formation by direct transfer of osteogenic plasmid genes. *Proc. Natl. Acad. Sci.*, 93:5753–5758, 1996.

[55] P.L. Felgner and G.M. Ringold. Cationic liposome-mediated transfection. *Nature*, 337:387, 1989.

[56] C.A. Foster, K. Kim, R.J. Turnbull, and C.D. Hendricks. Apparatus for producing uniform solid spheres of hydrogen. *Rev. Sci. Instrum.*, 48:625–631, 1977.

[57] M. Frangione-Beebe, R.T. Rose, P.T. Kaumaya, and S.P. Schwendeman. Microencapsulation of a synthetic peptide epitope for HTLV-1 in biodegradable poly(D,L-lactide-co-glycolide) microspheres using a novel encapsulation technique. *J. Microencapsul.*, 18:663–677, 2001.

[58] S. Freitas, H.P. Merkle, and B. Gander. Ultrasonic atomisation into reduced pressure atmosphere–envisaging aseptic spray-drying for microencapsulation. *J. Control. Rel.*, 95:185–195, 2004.

[59] A.M. Friedlander, S.L. Welkos, and B.E. Ivins. Anthrax Vaccines. *Curr. Topics Microbiol. Immunol.*, 271(Anthrax):33–60, 2002.

[60] K. Fu, K. Griebenow, L. Hsieh, A.M. Klibanov, and R. Langer. FTIR characterization of the secondary structure of proteins encapsulated within PLGA microspheres. *J. Control. Rel.*, 58(3):357–366, 1999.

[61] K. Fu, R. Harrell, K. Zinski, C. Um, A. Jaklenec, J. Frazier, N. Lotan, P. Burke, A.M. Klibanov, and R. Langer. A potential approach for decreasing the burst effect of protein from PLGA microspheres. *J. Pharm. Sci.*, 92:1582–1591, 2003.
[62] K. Fu, A.M. Klibanov, and R. Langer. Protein stability in controlled-release systems. *Nat. Biotechnol.*, 18:24–25, 2000a.
[63] K. Fu, D.W. Pack, A.M. Klibanov, and R. Langer. Visual evidence of acidic environment within degrading poly(lactic-co-glycolic acid) (PLGA) microspheres. *Pharm. Res.*, 17(1):100–106, 2000b.
[64] R.P. Gilliard, K. Kim, and R.J. RTurnball. Spherical hydrogen pellet generator for magnetic confinement fusion research. *Rev. Sci. Instrum.*, 52:183–190, 1981.
[65] A. Göpferich, M.J. Alonso, and R. Langer. Development and characterization of microencapsulated microspheres. *Pharm. Res.*, 11:1568–1574, 1994.
[66] A. Gopferich and R. Langer. Modeling of Polymer Erosion. *Macromolecules*, 26(16):4105–4112, 1993.
[67] B. Guiziou, D.J. Armstrong, P.N.C. Elliot, J.L. Ford, and C. Rostron. Investigation of in-vitro release characteristics of NSAID-loaded polylactic acid microspheres. *J. Microencapsul.*, 13(6):701–708, 1996.
[68] J.L. Guttman, C.D. Hendricks, K. Kim, and R.J. Turnbull. An investigation of the effects of system parameters on the production of hollow hydrogen droplets. *J. Appl. Phys.*, 50(6):4139–42, 1979.
[69] J. Hanes, M. Chiba, and R. Langer. Synthesis and characterization of degradable anhydride-co-imide terpolymers containing trimellitylimido-L-typrosine: novel polymers for drug delivery. *Macromolecules*, 29:5279–5287, 1996.
[70] J. Hanes, M. Chiba, and R. Langer. Degradation of porous poly(anhydride-co-imide) microspheres and implications for controlled macromolecule delivery. *Biomaterials*, 19:163–172, 1998.
[71] D. Hartigan-O'Connor, and J.S. Chamberlain. Progress toward gene therapy of Duchenne muscular dystrophy. *Seminars Neurol.*, 19(3):323–3332, 1999.
[72] P. He, S.S. Davis, and L. Illum. Chitosan microspheres prepared by spray drying. *Int. J. Pharma.*, 187:53–65, 1999.
[73] M.L. Hedley, J. Curley, and R. Urban. Microspheres containing plasmid-encoded antigens elicit cytotoxic T-cell responses. *Nat. Medicine*, 4:365–368, 1998.
[74] M. Higaki, Y. Azechi, T. Takase, R. Igarashi, S. Nagahara, A. Sano, K. Fujioka, N. Nakagawa, C. Aizawa, and Y. Mizushima. Collagen minipellet as a controlled release delivery system for tetanus and diphtheria toxoid. *Vaccine*, 19(23–24):3091–3096, 2001.
[75] S. Hirosue, B.G. Muller, R.C. Mulligan, and R. Langer. Plasmid DNA encapsulation and release from solvent diffusion nanospheres. *J. Control. Rel.*, 70(1–2):231–242, 2001.
[76] I. Jabbal-Gill, W. Lin, P. Jenkins, P. Watts, M. Jimenez, L. Illum, S.S. Davis, J.M. Wood, D. Major, P.D. Minor and others. Potential of polymeric lamellar substrate particles (PLSP) as adjuvants for vaccines. *Vaccine*, 18(3–4):238–250, 1999.
[77] R.A. Jain, C.T. Rhodes, A.M. Railkar, A.W. Malick, and N.H. Shah. Controlled release of drugs from injectable in situ formed biodegradable PLGA microspheres: effect of various formulation variables. *Europ. J. Pharm. Biopharm.*, 50(2):257–262, 2000.
[78] K.Y. Jang, K. Kim, and R.S. Upadhye. Study of sol-gel processing for fabrication of hollow silica-aerogel spheres. *J. Vac. Sci., Technol. A*, 8(3):1732–1735, 1990.
[79] W. Jiang and S.P. Schwendeman. Stabilization and controlled release of bovine serum albumin encapsulated in poly(D, L-lactide) and poly(ethylene glycol) microsphere blends. *Pharm. Res.*, 18(6):878–885, 2001a.
[80] W. Jiang and S.P. Schwendeman. Stabilization and controlled release of bovine serum albumin encapsulated in poly(D,L-lactide) and poly(ethylene glycol) microsphere blends. *Pharm. Res.*, 18:878–885, 2001b.
[81] W.S. Jiang and P. Steven. Stabilization of a model formalinized protein antigen encapsulated in poly(lactide-co-glycolide)-based microspheres. *J. Pharm. Sci.*, 90(10):1558–1569, 2001.
[82] B.G. Jones, P.A. Dickinson, M. Gumbleton, and I.W. Kellaway. The inhibition of phagocytosis of respirable microspheres by alveolar and peritoneal macrophages. *Int. J. Pharm.*, 236:65–79, 2002.
[83] K. Kane, J. Lloyd, M. Zaffran, L. Simonsen, and M. Kane. Transmission of Hepatitis B, Hepatitis C and Human Immunodeficiency Viruses through unsafe Injections in the Developing World: Model-Based Regional Estimates. Bulletin of the World Health Organization. Vol. 77, no. 10, pp. 801–807, 1999.
[84] J. Kang and S.P. Schwendeman. Comparison of the effects of Mg(OH)2 and sucrose on the stability of bovine serum albumin encapsulated in injectable poly(D,L-lactide-co-glycolide) implants. *Biomaterials*, 23(1):239–245, 2001.

[85] K. Kim, K.Y. Jang, and R.S. Upadhye. Hollow silica spheres of controlled size and porosity by sol-gel processing. *J. Am. Ceram. Soc.*, 74(8):1987–1992, 1991.

[86] K. Kim and R.J. Turnbull. Generation of charged drops of insulating liquids by electrostatic spraying. *J. Appl. Phys.*, 47(5):1964–1969, 1976.

[87] N.K. Kim, K. Kim, D.A. Payne, and R.S. Upadhye. Fabrication of hollow silica aerogel spheres by a droplet generation method and sol-gel processing. *J. Vac. Sci., Technol. A*, 7(3):1181–1184, 1989.

[88] T.W. King and C.W. Patrick, Jr. Development and in vitro characterization of vascular endothelial growth factor (VEGF)-loaded poly(DL-lactic-co-glycolic acid)/polyethylene glycol microspheres using a solid encapsulation/single emulsion/solvent extraction technique. *J. Biomed. Materials Res.*, 51(3):383–390, 2000.

[89] M.J. Kipper, E. Shen, A. Determan, and B. Narasimhan. Design of an Injectable System Based on Bioerodible Polyanhydride Microspheres for Sustained Drug Delivery. *Biomaterials*, 23:4405–4412, 2002.

[90] J.E. Kirwan, T.A. Lee, G.N. Schroering, H. Krier, J.E. Peters, J.P. Renie, and K. Kim. An experimental and theoretical study of a monodisperse spray. *AIAA J. Propulsion Power*, 4:299–307, 1988.

[91] N.L. Kumar, S. Robert, and J. Domb Abraham. Polyanhydrides: an overview. *Adv. Drug Delivery Rev.*, 54(7):889–910, 2002.

[92] J.K. Lalla and K. Sapna. Biodegradable microspheres of poly(dl-lactic acid) containing piroxicam as a model drug for controlled release via the parenteral route. *J. Microencapsul.*, 10(4):449–460, 1993.

[93] X.M.D. Lam, T. Eileen, and Jeffrey L. Cleland. Encapsulation and stabilization of nerve growth factor into poly(lactic-co-glycolic) acid microspheres. *J. Pharm. Sci.*, 90(9):1356–1365, 2001.

[94] G. Lambert, E. Fattal, A. Pinto-Alphandary, A. Gulik, and T. Couvreur. Polyisobutylcyanoacrylate nanocapsules containing an aqueous core as a novel colloidal carrier for delivery of oligonucleotides. *Pharm. Res.*, 17(6):707–714, 2000.

[95] P. Le Corre, P. Le Guevello, V. Gajan, F. Chevanne, and R. Le Verge. Preparation and characterization of bupivacaine-loaded polylactide and poly(lactide-glycolide) microspheres. *Internat. J. Pharm.*, 107(1):41–49, 1994.

[96] K. Leach, K. Noh, and E. Mathiowitz. Effect of manufacturing conditions on the formation of double-walled polymer microspheres. *J. Microencapsul*, 16(2):153–167, 1999.

[97] K.J. Leach and E. Mathiowitz. Degradation of double-walled polymer microspheres of PLLA and P(CPP:SA)20:80. I. In vitro degradation. *Biomaterials*, 19(21):1973–1980, 1998a.

[98] K.J. Leach and E. Mathiowitz. Degradation of double-walled polymer microspheres of PLLA and P(CPP:SA)20:80. I. In vitro degradation. *Biomaterials*, 19:1973–1980, 1998b.

[99] K.J. Leach, S. Takahashi, and E. Mathiowitz. Degradation of double-walled polymer microspheres of PLLA and P(CPP:SA)20:80. II. In vivo degradation. *Biomaterials*, 19(21):1981–1988, 1998a.

[100] K.J. Leach, S. Takahashi, and E. Mathiowitz. Degradation of double-walled polymer microspheres of PLLA and P(CPP:SA)20:80. II. In vivo degradation. *Biomaterials*, 19:1981–1988, 1998b.

[101] H.K. Lee, J.H. Park, and K.C. Kwon. Double-walled microparticles for single shot vaccine. *J. Control. Rel.*, 44(2–3):283–293, 1997.

[102] S.C. Lee, J.T. Oh, M.H. Jang, and S.I. Chung. Quantitative analysis of polyvinyl alcohol on the surface of poly(D,L-lactide-co-glycolide) microspheres prepared by solvent evaporation method: effect of particle size and PVA concentration. *J. Control. Rel.*, 59:123–132, 1999.

[103] T.H. Lee, J. Wang, and C. Wang. Double-walled microspheres for the sustained release of a highly water soluble drug: characterization and irradiation studies. *J. Control. Rel.*, 83:437–52, 2002a.

[104] T.H. Lee, J. Wang, and C.-H. Wang. Double-walled microspheres for the sustained release of a highly water soluble drug: characterization and irradiation studies. *J. Control. Rel.*, 83:437–452, 2002b.

[105] N. Leelarasamee, S.A. Howard, C.J. Malanga, and J.K.H. Ma. A method for the preparation of polylactic acid microcapsules of controlled particle size and drug loading. *J. Microencapsul.*, 5:147–157, 1988.

[106] C.S. Lengsfeld and T.J. Anchordoquy. Shear-induced degradation of plasmid DNA. *J. Pharm. Sci.*, 91:1581–1589, 2002.

[107] X. Li, Deng Xianmo, Yuan Minglong, Xiong Chengdong, Huang Zhitang, Zhang Yanhua, and W. Jia. In vitro degradation and release profiles of poly-DL-lactide-poly(ethylene glycol) microspheres with entrapped proteins. *J. App. Poly. Sci.*, 78(1):140–148, 2000.

[108] X.Z. Li, Y.R. Yan, W. Jia, M. Yuan, X. Deng, and Z. Huang. Influence of process parameters on the protein stability encapsulated in poly-DL-lactide-poly(ethylene glycol) microspheres. *J. Control. Rel.*, 68(1):41–52, 2000.

[109] R.T. Liggins and H.M. Burt. Paclitaxel loaded poly(L-lactic acid) microspheres: properties of microspheres made with low molecular weight polymers. *Internat. J. Pharm.*, 222(1):19–33, 2001.

[110] R.T. Liggins, S. D'Amours, J.S. Demetrick, L.S. Machan, and H.M. Burt. Paclitaxel loaded poly(l-lactic acid) microspheres for the prevention of intraperitoneal carcinomatosis after a surgical repair and tumor cell spill. *Biomaterials*, 21:1959–1969, 2000.

[111] D.W. Lim and T.W. Park. Stereocomplex formation between enantiomeric PLA-PEG-PLA triblock copolymers: characterization and use as protein-delivery microparticulate carriers. *J. App. Polymer Sci.*, 75(13):1615–1623, 2000.

[112] S.Y. Lin, K.S. Chen, H.H. Teng, M.J. Li. In vitro degradation and dissolution behaviours of microspheres prepared by three low molecular weight polyesters. *J. Microencapsul.*, 17(5):577–586, 2000.

[113] Y. Liu, X. Deng. Influences of preparation conditions on particle size and DNA-loading efficiency for poly(-lactic acid-polyethylene glycol) microspheres entrapping free DNA. *J. Control. Rel.*, 83(1):147–155, 2002.

[114] S. Lofthouse. Immunological aspects of controlled antigen delivery. *Ad. Drug Delivery Rev.*, 54(6):863–870, 2002.

[115] L. Lunsford, U. McKeever, V. Eckstein, M.L. Hedley. Tissue distribution and persistence in mice of plasmid DNA encapsulated in a PLGA-based microsphere delivery vehicle. *J. Drug Targeting*, 8:39–50, 2000.

[116] D.W.-M. Luo Kim, Nadya Belcheva, W. Mark Saltzman. Controlled DNA delivery systems. *Pharm. Res.*, 16(8):1300–1308, 1999.

[117] G.H. Ma, Z.G. Su, S. Omi, D. Sundberg, and J. Stubbs. Microencapsulation of oil with poly(styrene-*N*, *N*-dimethylaminoethyl methacrylate) by SPG emulsification technique: Effect of conversion and composition of oil phase. *J. Colloid Interface Sci.*, 266:282–294, 2003

[118] K. Mabuchi, A. Nakayama, and K. Iwamoto. Preparation and in vitro evaluation of poly(lactic acid) microspheres containing carmofur. *Yakuzaigaku*, 54(1):42–48, 1994.

[119] H.-Q. Mao, K. Roy, S.M. Walsh, J.T. August, and K.W. Leong. DNA-Chitosan Nanospheres for Gene Delivery. *Proceedeeings Intern. Symp. Control. Rel. Bioact. Mater.*, 23:401–402, 1996.

[120] S.M. Marinakos, J.P. Novak, L.C. Brousseau, A.B. House, E.M. Edeki, J.C. Feldhaus, and D.L. Feldheim. Gold particles as templates for the synthesis of hollow polymer capsules. Control of capsule dimensions and guest encapsulation. *J. Am. Chem. Soc.*, 121:8518–8522, 1999.

[121] E. Mathiowitz, J.S. Jacob, Y.S. Jong, G.P. Carino, D.E. Chickering, P. Chaturvedi, C.A. Santos, K. Vijayaraghavan, S. Montgomery, M. Bassett and others. Biologically erodable microspheres as potential oral drug delivery systems. *Nature*, 386:410–414, 1997.

[122] E. Mathiowitz and R. Langer. Massachusetts Institute of Technology, assignee. 1999. Multiwall polymeric microspheres. U.S.A. patent 5,912,017.

[123] R.C. Mehta, B.C. Thanoo, and P.P. DeLuca. Peptide containing microspheres from low molecular weight and hydrophilic poly(D,L-lactide-co-glycolide). *J. Control. Rel.*, 41(3):249–257, 1996.

[124] M. Morlock, T. Kissel, Y.X. Li, H. Koll, and G. Winter. Erythropoietin loaded microspheres prepared from biodegradable LPLG-PEO-LPLG triblock copolymers: protein stabilization and in-vitro release properties. *J. Control. Rel.*, 56(1–3):105–115, 1998.

[125] J.S. Moynihan, J. Blair, A. Coombes, F. D'Mello, and C.R. Howard. Enhanced immunogenicity of a hepatitis B virus peptide vaccine using oligosaccharide ester derivative microparticles. *Vaccine*, 20(13–14):1870–1876, 2002.

[126] L. Mu and S.S. Feng. Fabrication, characterization and in vitro release of paclitaxel(Taxol®) loaded poly (lactic-co-glycolic acid) microspheres prepared by spray drying technique with lipid/cholesterol emulsifiers. *J. Control. Rel.*, 76:239–254, 2001.

[127] R.J. Mumper and A.P. Rolland. Plasmid delivery to muscle: recent advances in polymer delivery systems. *Adv. Drug Del. Rev.*, 30:151–172, 1998.

[128] R. Narayani and K. Panduranga Rao. Gelatin microsphere cocktails of different sizes for the controlled release of anticancer drugs. *Int. J. Pharm.*, 143:255–258, 1996.

[129] D.F. Nixon, C. Hioe, and P.-D. Chen. Synthetic peptides entrapped in microparticles can elicit cytotoxic T cell activity. *Vaccine*, 14:1523–1530, 1996.

[130] M. Nof and L.D. Shea. Drug-release scaffolds fabricated form drug-loaded microspheres. *J. Biomed. Mat. Res.*, 59:349–356, 2002.

[131] P.B. O'Donnell and J.W. McGinity. Preparation of microspheres by the solvent evaporation technique. *Adv. Drug Del. Rev.*, 28:25–42, 1997.

[132] K.J. Pekarek, J.S. Jacob, and E. Mathiowitz. Double-walled polymer microspheres for controlled drug release. *Nature*, 367(6460):258–60, 1994a.

[133] K.J. Pekarek, J.S. Jacob, and E. Mathiowitz. Double-walled polymer microspheres for controlled drug release. *Nature*, 367:258–260, 1994b.

[134] K.J. Pekarek, J.S. Jacob, and E. Mathiowitz. One-step preparation of double-walled microspheres. *Adv. Mater.*, 6:684–687, 1994c.

[135] C. Perez and K. Griebenow. Effect of salts on lysozyme stability at the water-oil interface and upon encapsulation in poly(lactic-co-glycolic) acid microspheres. *Biotechnol. Bioeng.*, 82(7):825–832, 2003.

[136] C.M. Nashbly, Perez-Rodriguez, Karilys Gonzalez, Kai Griebenow. Stabilization of a-chymotrypsin at the CH2Cl2/water interface and upon water-in-oil-in-water encapsulation in PLGA microspheres. *J. Control. Rel.*, 89(1):71–85, 2003.

[137] D. Perumal, C.M. Dangor, R.S. Alcock, N. Hurbans, and K.R. Moopanar. Effect of formulation variable on in vitro drug release and micromeritic properties of modified release ibuprofen microspheres. *J. Microencapsul.*, 16(4):475–487, 1999.

[138] N. Puri, J.H. Kou, and P.J. Sinko. Adjuvancy enhancement of muramyl dipeptide by modulating its release from a physicochemically modified matrix of ovalbumin microspheres I. In vitro characterization. *J. Control. Rel.*, 69(1):53–67, 2000.

[139] N. Puri and P.J. Sinko. Adjuvancy enhancement of muramyl dipeptide by modulating its release from a physicochemically modified matrix of ovalbumin microspheres II. In vivo investigation. *J. Control. Rel.*, 69(1):69–80, 2000.

[140] F. Quaglia, G. De Rosa, E. Granata, F. Ungaro, E. Fattal, and M.I. La Rotonda. Feeding liquid, non-ionic surfactant and cyclodextrin affect the properties of insulin-loaded poly(lactide-co-glycolide) microspheres prepared by spray-drying. *J. Control. Rel.*, 86:267–278, 2003.

[141] R. Raghuvanshi, Y. Katare, K. Lalwani, M. Ali, O. Singh, and A. Panda. Improved immune response from biodegradable polymer particles entrapping tetanus toxoid by use of different immunization protocol and adjuvants. *Internat. J. Pharm.*, 245(1–2):109–121, 2002.

[142] L. Ramachandra, R. Song, and C.V. Harding. Phagosomes are fully competent antigen-processing organelles that mediate the formation of peptide class II MHC complexes. *J. Immunol.*, 162:3263–3272, 1999.

[143] C. Raman, C. Berkland, K.K. Kim, and D.W. Pack. Modeling small-molecule release from PLG microspheres: effects of polymer degradation and non-uniform drug distribution. 2004 (submitted).

[144] H.B. Ravivarapu, K. Burton, and P.P. DeLuca. Polymer and microsphere blending to alter the release of a peptide from PLGA microspheres. *Eur. J. Pharm.*, 50:263–270, 2000.

[145] L. Rayleigh. *Proc. London Math. Soc.*, 10:4, 1879.

[146] L. Rayleigh. *Phil. Mag. S.G.*, 14:184, 1882.

[147] H. Sah. Protein instability toward organic solvent/water emulsification: implications for protein microencapsulation into microspheres. *J. Pharm. Sci. Tech.*, 53(1):3–10, 1999.

[148] W.M. Saltzman. *Drug Delivery: Engineering Principles for Drug Therapy*. New York, Oxford University Press. 2001.

[149] A. Sanchez, R.K. Gupta, M.J. Alonso, G.R. Siber, and R. Langer. Pulsed controlled-release system for potential use in vaccine delivery. *J. Pharm. Sci.*, 85(6):547–552, 1996a.

[150] A. Sanchez, R.K. Gupta, M.J. Alonso, G.R. Siber, and R. Langer. Pulsed controlled-release system for potential use in vaccine delivery. *J. Pharm. Sci.*, 85:547–552, 1996b.

[151] M. Sandor, D. Enscore, P. Weston, and E. Mathiowitz. Effect of protein molecular weight on release from micron-sized PLGA microspheres. *J. Control. Rel.*, 76:297–311, 2001.

[152] P. Sansdrap and A.J. Moes. Influence of manufacturing parameters on the size characteristics and the release profiles of nifedipine from poly(DL-lactide-co-glycolide) microspheres. *Int. J. Pharm.*, 98:157–164, 1993.

[153] S.P. Schwendeman. Recent advances in the stabilization of proteins encapsulated in injectable PLGA delivery systems. *Crit. Rev. Therapeut. Drug Carrier Sys.*, 19(1):73–98, 2002.

[154] P.G. Shao and L.C. Bailey. Stabilization of pH-induced degradation of porcine insulin in biodegradable polyester microspheres. *Pharm. Devel. Tech.*, 4(4):633–642, 1999.

[155] L.D. Shea, E. Smiley, J. Bonadio, and D.J. Mooney. DNA delivery from polymer matrices for tissue engineering. *Nat. Biotechnol.*, 17:551–554, 1999.

[156] E. Shen, M.J. Kipper, B. Dziadul, M.-K. Lim, and B. Narasimhan. Mechanistic Relationships between Polymer Microstructure and Drug Release Kinetics in Bioerodible Polyanhydrides. *J. Control. Rel.*, 82:115–125, 2002.

[157] A. Shenderova, T.G. Burke, and S.P. Schwendeman. The acidic microclimate in poly(lactide-co-glycolide) microspheres stabilizes camptothecins. *Pharm. Res.*, 16(2):241–248, 1999.

[158] L. Shi, M.J. Caulfield, R.T. Chern, R.A. Wilson, G. Sanyal, and D.B. Volkin. Pharmaceutical and immunological evaluation of a single-shot hepatitis B vaccine formulated with PLGA microspheres. *J. Pharm. Sci.*, 91(4):1019–1035, 2002.

[159] K. Shiga, N. Muramatsu, and T. Kondo. Preparation of poly(D,L-lactide) and copoly(lactide-glycolide) microspheres of uniform size. *J. Pharm. Pharmacol.*, 48:891–895, 1996.

[160] L. Slobbe, N. Medlicott, E. Lockhart, N. Davies, I. Tucker, M. Razzak, and G. Buchan. A prolonged immune response to antigen delivered in poly (epsilon-caprolactone) microparticles. *Immun. Cell Biol.*, 81(3):185–191, 2003.

[161] G. Spenlehauer, M. Vert, J.P. Benoit, and A. Boddaert. In vitro and in vivo degradation of poly(DL-lactide/glycolide) type microspheres made by solvent evaporation method. *Biomaterials*, 10(8):557–563, 1989.

[162] C. Sturesson, P. Artursson, R. Ghaderi, K. Johansen, A. Mirazimi, I. Uhnoo, L. Svensson, A.-C. Albertsson, and J. Carlfors. Encapsulation of rotavirus into poly(lactide-co-glycolide) microspheres. *J. Control. Rel.*, 59(3):377–389, 1999.

[163] C.C. Sturesson J. Incorporation of protein in PLG microspheres with retention of bioactivity. *J. Control. Rel.*, 67(2–3):171–178, 2000.

[164] A. Supsakulchai, G.H. Ma, M. Nagai, and S. Omi. Preparation of uniform titanium dioxide (TiO_2) polystyrene-based composite particles using the glass membrance emulsification process with a subsequent suspension polymerization. *J. Microencapsul.*, 20:1–18, 2003.

[165] K. Suzuki and J.C. Price. Microencapsulation and dissolution properties of a neuroleptic in a biodegradable polymer poly(dl-lactide). *J. Pharm. Sci.*, 74:21–24, 1985.

[166] Y. Tabata, S. Gutta, and R. Langer. Controlled delivery systems for proteins using polyanhydride microspheres. *Pharm. Res.*, 10:487–495, 1993.

[167] Y. Tabata and R. Langer. Polyanhydride microspheres that display near-constant release of water-soluble model drug compounds. *Pharm. Res.*, 10(3):391–399, 1993a.

[168] Y. Tabata and R. Langer. Polyanhydride microspheres that display near-constant release of water-soluble model drug compounds. *Pharm. Res.*, 10:391–399, 1993b.

[169] J.A. Tamada and R. Langer. Erosion kinetics of hydrolytically degradable polymers. *Proc. Natl. Acad. Sci.*, 90(2):552–556, 1993.

[170] A. Thermet, C. Rollier, F. Zoulim, C. Trepo, and L. Cova. Progress in DNA vaccine for prophylaxis and therapy of hepatitis B. *Vaccine*, 21(7–8):659–662, 2003.

[171] A.M. Tinsley-Brown, R. Fretwell, A.B. Dowsett, S.L. Davis, and G.H. Farrar. Formulation of Poly(D,L-Lactic-Co-Glycolic Acid) Microparticles for Rapid Plasmid Plasmid DNA Delivery. *J. Control. Rel.*, 66:229–241, 2000.

[172] A.M. Tinsley-Brown, V.A. Mobsby, M.C. Outlaw, and G.H. Farrar. DNA Release from PLGA Microparticles for Vaccine Applications. *Proceed Int'l. Symp. Control. Rel. Bioact. Mater.*, vol. 26, pp. 344, 1999.

[173] S. Torza and G. Mason. Three-phase interactions in shear and electrical fields. *J. Colloid Int. Sci.*, 33:67–83, 1970.

[174] M.A. Tracy. Development and scale-up of a microsphere protein delivery system. *Biotechnol. Prog.*, 14:108–115, 1998.

[175] M. Tuncay, S. Calis, H.S. Kas, M.T. Ercan, I. Peksoy, and A.A. Hincal. Diclofenac sodium incorporated PLGA (50:50) microspheres: formulation considerations and in vitro/in vivo evaluation. *Int. J. Pharm.*, 195:179–188, 2000.

[176] R.G. Vile, S.J. Russell, and N.R. Lemoine. Cancer gene therapy: hard lessons and new courses. *Gene Therapy*, 7:2–8, 2000.

[177] R.G. Vile, A. Tuszynski, and S. Castleden. Retroviral vectors: from laboratory tools to molecular medicines. *Molec. Biotechnol.*, 5:139–158, 1996.

[178] B.W. Wagenaar and B.W. Muller. Piroxicam release from spray-dried biodegradable microspheres. *Biomaterials*, 15(1):49–54, 1994.

[179] M.C. Walsh, J.A. Banas, and S.P. Mudzinski. A two-component modular approach for enhancing T-cell activiation utilizing a unique anti-FcgRI-streptavidin construct and microspheres coated with biotinylated-antigen. *Biomolec. Eng.*, 20:21–33, 2003.

[180] E. Walter, K. Moelling, J. Pavlovic, and H.P. Merkle. Poly(D,L-lactide-co-glycolide)-Encapsulated DNA: Stability and Release Characteristics. *Proceedings Int'l. Symp. Control. Rel. Bioact. Mater.*, vol. 26, pp. 6407, 1999.

[181] J.-P. Wan, Y.-Y. Yang, T.-S. Chung, D. Tan, S. Ng, and J. Heller. POE-PEG-POE triblock copolymeric microspheres containing protein. II. Polymer erosion and protein release mechanism. *J. Control. Rel.*, 75:115–128, 2001.

[182] D.R. Wang, R. Deborah. Glen S. Kwon, and John Samuel. Encapsulation of plasmid DNA in biodegradable poly(D,L-lactic-co-glycolic acid) microspheres as a novel approach for immunogene delivery. *J. Control. Rel.*, 57(1):9–18, 1999.

[183] F.J. Wang and C.H. Wang. Sustained release of etanidazole from spray dried microspheres prepared by non-halogenated solvents. *J. Control. Rel.*, 81:263–280, 2002.

[184] J. Wang, B.M. Wang, and S.P. Schwendeman. Characterization of the initial burst release of a model peptide from poly(dl-lactide-co-glycolide) microspheres. *J. Control. Rel.*, 82(2–3):289–307, 2002.

[185] J. Wang, B.W. Wang, and S.P. Schwendeman. Mechanistic evaluation of the glucose-induced reduction in initial burst release of octreotide acetate from poly(D,L-lactide-co-glycolide) microspheres. *Biomaterials*, 25:1919–1927, 2004.

[186] G. Wei, G.J. Pettway, L.K. McCauley, and P.X. Ma. The release profiles and bioactivity of parathyroid hormone from poly(lactic-co-glycolic acid) microspheres. *Biomaterials*, 25:345–352, 2004.

[187] M. Wolf, M. Wirth, F. Pittner, and F. Gabor. Stabilisation and determination of the biological activity of -asparaginase in poly(,-lactide-co-glycolide) nanospheres. *Internat. J. Pharm.*, 256(1–2):141–152, 2003.

[188] J.A. Wolff, R.W. Malone, P. Williams, W. Chong, G. Acsadi, A. Jani, and P.L. Felgner. Direct gene transfer into mouse muscle in vivo. *Science*, 247:1465–1468, 1990.

[189] B.H. Woo, J.W. Kostanski, S. Gebrekidan, B.A. Dani, B.C. Tahanoo, and P.P. DeLuca. Preparation, characterization and in vivo evaluation of 120-day poly(dl-lactide) leuprolide microspheres. *J. Control. Rel.*, 75:307–315, 2001.

[190] J.P. Woosley, K. Kim, and R.J. Turnbull. Techniques for generating uniform charged particles of hydrogen isotopes. *J. Electrostat.*, 5:381–389, 1978.

[191] I. Yamakawa, Y. Tsushima, R. Machida, and S. Watanabe. In vitro and in vivo release of poly(DL-lactic acid) microspheres containing neurotensin analogue prepared by novel oil-in-water solvent evaporation method. *J. Pharm. Sci.*, 81(8):808–811, 1992.

[192] Y.Y. Yang, T.S. Chung, and N. Ng. Morphology, drug distribution, and in vitro release profiles of biodegradable polymeric microspheres containing protein fabricated by double-emulsion solvent extraction/evaporation method. *Biomaterials*, 22(3):231–241, 2001a.

[193] Y.-Y. Yang, M. Shi, S.-H. Goh, S. Moochhala, S. Ng, and J. Heller. POE/PLGA composite microspheres: formation and in vitro behavior of double walled microspheres. *J. Control. Rel.*, 88(2):201–213, 2003a.

[194] Y.-Y. Yang, M. Shi, S.-H. Goh, S.M. Moochhala, S. Ng, and J. Heller. POE/PLGA composite microspheres: formation and in vitro behavior of double walled microspheres. *J. Control. Rel.*, 88:201–213, 2003b.

[195] Y.-Y. Yang, J.-P. Wan, T.-S. Chung, P.K. Pallathadka, S. Ng, and J. Heller. POE-PEG-POE triblock copolymeric microspheres containing protein: I. preparation and characterization. *J. Control. Rel.*, 75:115–128, 2001b.

[196] K.R. Young and T.M. Ross. Particle-based vaccines for HIV-1 infection. *Current Drug Targets: Infectious Disorders*, 3(2):151–169, 2003.

[197] X.-J. Yu, C. Luo, J.-C. Lin, P. Hao, Y.-Y. He, Z.-M. Guo, L. Qin, J. Su, B.-S. Liu, Y. Huang, and others. Putative hAPN receptor binding sites in SARS_ CoV spike protein. *Acta Pharmacolog. Sinica*, 24(6):481–488, 2003.

[198] S. Zhou and X. Deng. In vitro degradation characteristics of poly–lactide-poly(ethylene glycol) microspheres containing human serum albumin. *Reac. Funct. Poly.*, 51(2–3):93–100, 2002.

[199] S.D. Zhou, Xianmo, Minglong Yuan, and Xiaohong Li. Investigation on preparation and protein release of biodegradable polymer microspheres as drug-delivery system. *J. App. Polymer Sci.*, 84(4):778–784, 2002.

[200] G. Zhu, S.R. Mallery, and S.P. Schwendeman. Stabilization of proteins encapsulated in injectable poly(lactide-co-glycolide). *Nature Biotechnol.*, 18(1):52–57, 2000.

3

Nanoscale Polymer Fabrication for Biomedical Applications

L. James Lee

Department of Chemical and Biomolecular Engineering, The Ohio State University 140 W.19th Ave., Columbus, Ohio 43210 USA

3.1. INTRODUCTION

Nanofabrication techniques for feature sizes less than 100 nm are available for silicon-based materials using high cost, cleanroom-based methods. The high cost may be acceptable for large-throughput manufacturing in the IC industry, but the broader needs and lower volumes in biomedicine require more cost-effective mass-production methods capable of replicating nanostructures in a wide range of materials. However, the properties of silicon (poor impact strength/toughness, poor biocompatibility) are inappropriate for many biomedical devices. For example, the conductivity of silicon is problematic in many micro/nanofluidic applications that require high voltage for electrokinetic flows. Non-conductive glass or quartz devices can be made using the same lithography/etching fabrication techniques. These materials, although less costly than silicon, are still much more expensive than most polymeric materials. In contrast, polymeric materials possess many attractive properties such as high toughness and recyclability. Some possess excellent biocompatibility, are biodegradable, and can provide various biofunctionalities. Proper combination of functional polymers and biomolecules can offer tailored properties for various biomedical applications, but the ability to process them at the nanoscale to form well-defined functional structures is largely underdeveloped. Nanofabrication techniques for feature sizes less than 100 nm are available for silicon-based materials using high cost, cleanroom-based methods. The high cost may be acceptable for large-throughput manufacturing in the IC industry.

Microscale fabrication methods of polymeric materials have been explored in recent years both in industry and in academia [1–3]. Although large-volume production is still rare, many manufacturing processes and commercial machines are available in the market. The major challenge now is to modify these processes and to optimize the processing conditions such that low-cost, high-speed, and high-quality mass production can be realized as in the macroscale production. On the other hand, the nanoscale polymer fabrication technology is still in its infancy. Many innovative concepts have been proposed and demonstrated, but industrial applications are rare. Since synthesis of monomeric organic species into polymeric materials can lead to the formation of various nanoscale structures (the so called "bottom-up" approach), a broad definition of nanoscale polymer fabrication should cover both "bottom-up" and "top-down" approaches. To narrow the scope, we focus on the "top-down" methods in this chapter. Traditional macroscale fabrication and IC-type microscale fabrication techniques are largely based on "top-down" manufacturing technologies. Because of its prior success in mass-production, this approach must play a key role in 3D manufacturing at the nanoscale and can be combined with "bottom-up" methods to develop new low-cost technologies.

In the following sections, we first briefly introduce potential biomedical applications of polymer nanostructures (Section 3.2). Various mold/master/stamper fabrication methods at the nanoscale (Section 3.3) will then be described. This is followed with a survey on nanoscale polymer replication methods (Section 3.4). Finally, connection of nanostructures with micro- and macro-world to form useful systems through parts assembly and bonding will be discussed (Section 3.5). The chapter ends with a conclusion suggesting future research and development directions (Section 3.6).

3.2. POTENTIAL BIOMEDICAL APPLICATIONS OF POLYMER NANOSTRUCTURES

3.2.1. Drug Delivery and Gene Therapy

One market for polymer biomedical devices that carries enormous potential economic and human benefits is that of advanced therapeutic delivery systems. The U.S. market for advanced drug delivery systems is approaching $20 billion and increasing rapidly [4]. Compared to an average of $250–300 million and 12–15 years to develop a new drug, an advanced delivery system can offer greatly enhanced performance at 20% of the cost and less than half the development time [5]. A primary driving force for drug delivery development is to meet the needs of many new protein-, gene-, and DNA-based drugs. They are potent but at the same time fragile molecules. Typically with a short half-life in vivo, these drugs require an effective delivery system to realize their therapeutic potential. Controlled release formulations, such as injectable polymeric microspheres and nanoparticles, often require fabrication conditions that damage the bioactivity of the proteins [6]. Delivering these potent drugs to the target tissues and cells in a controlled manner remains a significant challenge. One major issue is controlling the size and surface characteristics of these drug-loaded systems. As the systems go down in size, they can move easily across many biological barriers, such as gut epithelium or leaky endothelium in tumor tissues. When the surface characteristics of the drug-loaded systems can be well-controlled, the prospect of both passive and active targeting can be greatly improved. To reach the ultimate cellular targets, multiple biofunctionalities may have to be incorporated into the drug delivery system.

Another major issue is protecting the encapsulated drug, protein, or gene from degradation until it reaches the target tissue, cell, or nucleus. This requires an effective synthetic strategy to control the distribution of the encapsulant in the system.

Various techniques have been explored to tackle the drug-delivery challenges outlined above. Physical methods involve using mechanical and electrical energy to deliver drugs, mostly genes, by bioinjector, electroporation or gene gun. They are, however, invasive and limited only to the peripheral tissues. Chemical methods involve encapsulation by a polymeric or lipid carrier (i.e. non-viral vectors), while biological approaches often rely on viruses as a carrier (i.e. viral vectors). However, toxicity and poor transfection efficiency remain the significant drawbacks of these gene delivery systems. A chemical variation of the theme is to covalently conjugate the drug to a polymeric carrier. The first chemically synthesized polymer-protein conjugates like polyethylene glycol (PEG)-adenosine deaminase, and PEG-L-asparaginase and styrene maleic anhydride (SMANCS) earned market approval in the early 1990s. Many polymer-anticancer-drug conjugates have also provided promising results from clinical trials [7]. However, broader applications of polymer therapeutics require more sophisticated next-generation technologies that can produce biodegradable and biosoluble 3D biomimetic architectures with multi-level targeting ligands, sensing and imaging molecules, and bio/photo/sono/magneto responsive elements. Synthetic chemistry alone is insufficient to achieve all these functions. Drug delivery technology can be brought to the next level by the fabrication of smart materials into miniature devices that are responsive to the individual patient's therapeutic requirements [7, 8]. Nanoengineering techniques will be essential for the manufacture of such devices. Currently, there are no commercial products based on the miniaturized responsive drug-delivery approach, and only limited research.

3.2.2. Medical Diagnostics and Nanofluidics

Miniature medical diagnostic systems such as "lab-on-a-chip" (LOC) devices are a rapidly-growing market, with annual sales of over $1 billion and a growth rate of 65% [9]. Currently, the most prominent market niche is genetic, protein, and immunoassay analyses for pharmaceutical development. Applications on the horizon include advanced medical diagnostics, toxicology, and homeland security. Although microarray and microfluidic biosensors/chips have reached commercialization in recent years, mass production of devices with well-defined nanoscale features is still not available. The value of such devices in the field of biomedical engineering is enormous. For example, in LOC systems, single molecule detection and rapid molecular analysis require an environment that approaches molecular size. Small proteins such as albumin and glucose have a characteristic dimension on the order of 5–10 nm, while most gene and DNA molecules are in the range of 1–100 nm [5, 6], so nanoscale conduits are required to interrogate an individual molecule. Protection against biological warfare requires perfect exclusion of infectious agents and biotoxins. Microfiltration excludes only bacteria, while ultrafiltration excludes only infectious agents and possibly a few proteinaceous toxins. Nanofiltration is the most promising method for such applications, and it can also be used for cell immunoprotection and protein separation. Polymer membranes with a certain molecular weight (or pore size) cut-off [10] or track-etched nanoscale holes [11], and many ceramic membranes have non-uniform distribution of pore sizes [11–13]. These membranes often show an incomplete retention of viruses.

The flow of fluids through nanochannels may bring many functional components for biomedical diagnostic systems, as well as control of the transport of single biomolecules for drug delivery applications. This is due to the increasing ionic concentration within the nanochannels with decreasing channel height (resulting from electrochemical effects) and the increase in voltage drop across the membrane (resulting from far-field voltage and resistance effects) [14]. Understanding microscopic transport phenomena of biomolecules has been an active area of research with many microfluidic devices reaching commercialization in recent years [15, 16]. On the other hand, there have been relatively few nanofluidics studies and there are essentially no commercially available nanofluidic devices [17–22]. Future research in this area will deal with molecular transport within these nanochannels and the design of tailored electrolytes for specific nanofluidic systems. Actively controlled permeable/impermeable membranes/biocapsules for bioseparation, immunoprotection, and drug delivery can be achieved.

3.2.3. Tissue Engineering and Bioreactors

Tissue loss and organ failure from injury and disease account for about half of the total annual expenditures in health care in the U.S. [15] Many patients die because of the shortage of transplantable organs. For example, the American Heart Association reports that only 2,300 of the 40,000 Americans needing new hearts in 1997 received them [16]. Likewise, there are few available livers and kidneys for life-threatening situations, and even skin for burn victims is largely unavailable. Tissue engineering would allow the regeneration of tissues and organs by employing a polymer or inorganic biomaterial based scaffold. Even though the first clinical trial of skin regeneration was performed in the 1980's in the U.S., tissue engineering is still in its infancy. Many tissue regeneration attempts have been successful, but few have gained wide acceptance because the engineered tissues are too expensive and the production methods are not well-developed. Other major limitations are the lack of a sufficient supply of nutrients to cells under regeneration in thick layers and the inadequate resorption kinetics and poor mechanical strength of scaffolds [18]. Work to date has been largely devoted to clinical needs and focused on physiological aspects. Little directed effort has been given to identify or develop a single, multifunctional material platform that can be used as a controllable scaffold for the guided regeneration of a spectrum of different tissues and tissue systems.

Tissue engineering is the regeneration, replacement, or restoration of human tissue function by combining synthetic and bio molecules in appropriate configurations and environments [23]. The scaffold, the cells, and the cell-scaffold interaction are three major components in any tissue-engineered construct. Although many tissue scaffold materials, such as foams and non-woven fabrics, have been developed and used [24, 25], many challenges must be overcome in order for the promise of tissue engineering to become a practical reality. These include (1) low-cost fabrication of well-defined 3D scaffold configurations at both micro- and nano-scale; (2) incorporation of appropriate biocompatibility, bioactivity, and biodegradability in the scaffolding construct to manipulate cellular and subcellular functions; and (3) active control of transport phenomena and cell growth kinetics in order to mimic microvasculature functions. Micro-/nano-fabrication technology of polymers has tremendous potential in this field because it can achieve topographical, spatial, chemical, and immunological control over cells and thus create more functional tissue engineering constructs [26].

An ideal tissue scaffolding process should be able to produce well-controlled pore sizes and porosity, provide high reproducibility, and use no toxic solvents because these physical factors are associated with nutrient supply and vascularization of the cells in the implant as well as the development of a fibrous tissue layer that may impede nutrient access to cells. Current processing methods used for polymer scaffolds include solvent casting, plastic foaming, fiber bonding, and membrane lamination [25]. However, precise reproducible features in the micro- and nano-meter range are difficult to attain using these methods. By combining living cells and microfabricated two- and three-dimensional scaffolds with carefully controlled surface chemistry, investigators have begun to address fundamental issues such as cell migration, growth, differentiation and apoptosis, cell orientation and adhesion, as well as tissue integration and vascularization [27–29].

3.3. MOLD (MASTER) MAKING AND PROTOTYPING

3.3.1. Non-Cleanroom based Mold Making and Prototyping

The mold inserts (or masters) can be fabricated by a variety of techniques. For large features (>50 μm) with tolerances and repeatability in the range of about 10 μm, traditional computer numerically-controlled (CNC)-machining and wire electrodischarge machining (EDM) of materials like tool steel and stainless steel are often accurate enough. The advantage of this technique is that tool materials used are the same as those in conventional polymer molding, so their design, strength, and service life are well-established. Complicated 3D structures can also be machined easily. The main drawbacks are that it is difficult to make sharp corners or right angles, and the surface quality is usually poor (surface roughness around several μm) [30]. Diamond-based micro-milling/ micro-drilling [31], micro-EDM, and excimer or femtosecond laser-based [32] direct removal processes can reduce the surface roughness to 1 μm or less [33]. While diamond-based methods can also make features smaller than 10 μm, they are only applicable to "soft" metals such as nickel, aluminum, and copper. For prototyping, most of these methods can be used on polymeric materials directly to fabricate microfluidic devices.

Laser-based solid freeform fabrication (SFF) prototyping and tooling techniques can produce 3D structures by curing photosensitive polymers and sintering ceramic and metal particles [34–36]. However, diffraction, thermal diffusion and one-photon polymer interaction limit feature sizes to, at best, hundreds of nanometers. Using a femtosecond-pulsed-laser can limit the thermal diffusion lengths and allow multiple-photon polymer curing, both of which would allow creation of structures with feature sizes below the diffraction limit. Proximal probe-based optical processing may further reduce feature sizes to the order of tens of nm for the finest 3D tool features not achievable by IC type fabrication technology.

In laser processing technologies, the minimum achievable structure size is determined by the diffraction limit of the optical system and is on the order of the radiation wavelength. However, this is different for ultrafast laser pulses because there is a well-defined ablation threshold for a given solid material. By choosing the peak laser fluence slightly above the threshold value, one can overcome the diffraction limit [37]. In this case, only the central part of the laser beam can remove the material and it becomes possible to produce sub-wavelength structures. For direct ablative nanostructuring, it is advantageous to use ultrashort laser pulses like femtosecond lasers. A drawback of femtosecond laser ablation is that surface deformation appears around the edge of tightly focused laser beams because

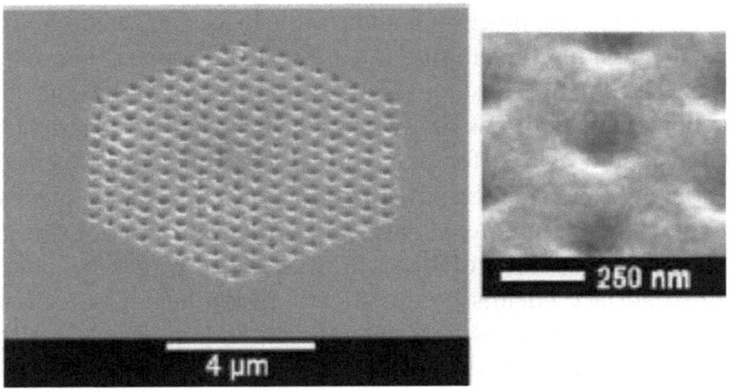

FIGURE 3.1. An example of a periodic nanostructure (with a defect cavity in the center) fabricated in a sapphire crystal with femtosecond laser pulses (left). On the right side, an enlarged fragment of a single hole is shown (with permission from [38]).

the applied laser fluence, although below the ablation threshold, exceeds the melting threshold. This effect can be minimized by using an imaging technique that better focuses the laser fluence [38]. For transparent materials like glass and some polymers, simultaneous absorption of several photons is needed to initiate ablation because of the nonlinear nature of the interaction between the femtosecond laser pulses and the transparent material. Multiphoton absorption produces initial free electrons that are accelerated by the femtosecond laser electric field to induce avalanche ionization and optical breakdown. This generates a microplasma, and its subsequent expansion results in the formation of a small structure on the solid surface. Figure 3.1 shows periodic nanowells fabricated in sapphire crystal with femtosecond laser pulses [38]. Currently, using femtosecond lasers for nanofabrication has a resolution on the order of 100–500 nm. Small structures may be achieved by a combination of femtosecond lasers with a scanning near-field optical microscope (SNOM) [39] or scanning probe lithography (SPL) [40].

Most femtosecond laser micro/nanofabrication is for metals and glass. It is a useful tool to produce die/molds for polymer replication. Direct femtosecond laser fabrication of polymer structures is still in the early stage. A two-photon photoreduction method on metal ion-doped polymer films has recently been developed [41]. The laser beam was split with a prism into two equal-power beams, one of which is optically delayed so that pulses from the two beams can coherently interact with the film, resulting in submicron grating lines. The laser-polymer interaction is initialized with an *in situ* photochemical reaction.

Two-photon polymerization of photosensitive resins initiated by femtosecond laser pulses has recently been demonstrated for the fabrication of 3D structures with a resolution better than 200 nm [37, 42, 43]. Typical UV photolithography resins containing photo-initiators sensitive to 390 nm radiation can be used. An example is SCR500 resin consisting of urethane acrylate monomer and oligomoers, and an initiator- triphenylamine-substituted phenylenevinylene derivative of EA4BPA-VB [1,4-bis(2-ethylhexyloxy)-2,5-bis(2-{4-[bis(4-bromophenyl)amino]phenyl}vinyl)benzene] (0.1 wt%) [43]. Since the resin is transparent to the incident laser light, the polymerization can be initiated inside the resin in a small volume tightly focused by laser pulses (e.g. 80 fs Ti:sapphire laser at 780 nm and a repetition rate of 80 MHz to initiate two-photon polymerization). When the laser focus is

FIGURE 3.2. SEM images of a three-dimensional periodic structure fabricated by two photon-polymerization in a hybrid polymer using femtosecond laser pulses (with permission from [38]).

moved through the resin, the polymerization occurs along the trace of the focus allowing the formation of any computer-generated 3D structures by direct laser writing. Figure 3.2 shows a 3D polymer structure produced by two-photon polymerization. In the future, two-photon polymerization will be used as a low-cost fabrication method to produce nanostructured 3D prototypes in medicine and biology applications.

3.3.2. Cleanroom based Mold Making

For smaller feature sizes (down to one micron or less), photolithographic methods, e-beam lithography, or scanning probe lithography (e.g. AFM dip pen lithography [44]) are often employed (i.e. surface machining). Here, a liquid photoresist is spin-coated or a self-assembly monolayer (SAM) is deposited on a galvanic starting layer. The micro/nanofeatures are formed after radiation exposure through a photomask and development or direct e-beam or scanning probe writing. Figure 3.3 shows a schematic of these surface machining methods. For prototyping, this photoresist structure can serve as a device itself or be used as a mold (called photoresist mold) in low temperature and low pressure molding processes. More generally, this structure is either used directly for electroplating or for wet/dry etching of silicon (which subsequently is electroplated) [17]. Both technologies yield a metal tool, usually nickel or nickel-cobalt. For features with a low aspect ratio (defined as the ratio of feature depth to width) or for rapid prototyping where the lifetime of mold inserts is not crucial, a silicon wafer etched by Reactive-Ion Etching (RIE) in SiO_2 can be utilized directly as a mold insert. For very small features (<1 μm) with high aspect ratios (up to 100 or higher), technologies like LIGA [17, 45] in thick resists (like EPON SU-8) or Deep RIE (DRIE) are needed to obtain the mold insert. In the following sections, we briefly introduce the three most widely used lithography methods.

3.3.2.1. LIGA LIGA is the German acronym for X-ray lithography (X-ray lithographie), electrodeposition (galvanoformung), and molding (abformtechnik). The process involves a thick layer of X-ray resist coated on a substrate (from microns to centimeters) followed by high-energy X-ray radiation exposure and development to arrive at a

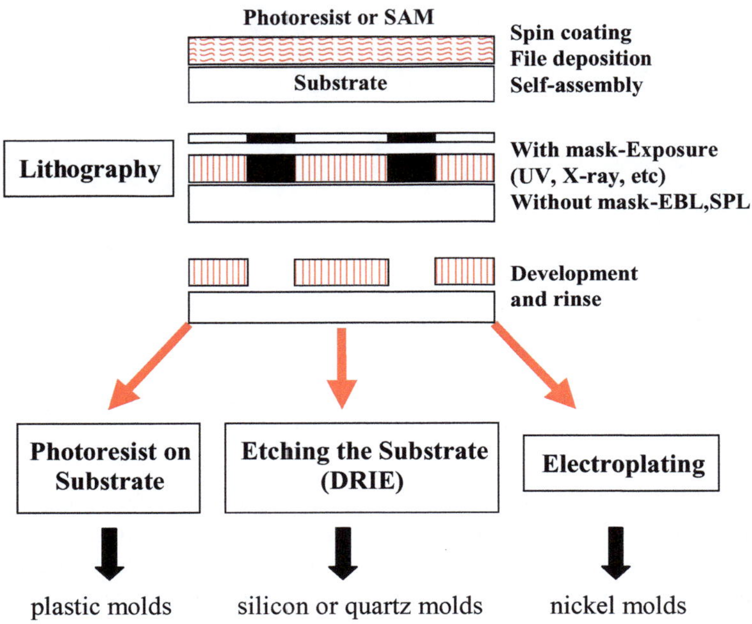

FIGURE 3.3. Schematic of surface machining methods of a master mold.

three-dimensional resist structure. Subsequent metal deposition fills the resist mold with a metal and, after resist removal, a freestanding metal structure results [45]. The metal shape may serve as a mold insert for precision injection molding at the microscale. The plastic mold retains the same shape, size, and form as the original resist structure, but is produced quickly and inexpensively as part of an infinite loop. Figure 3.4 shows a SEM micrograph of a high-density polyethylene (HDPE) microstructure [46].

The bandwidth of possible sizes in all three dimensions renders LIGA useful for manufacture of micro/nanostructures (micron and sub-micron dimensions), of packages for these microstructures (millimeter and centimeter dimensions), and even for the connectors from those packages to the "macro world".

X-ray mask production is one of the most difficult aspects of X-ray lithography. For highly transmissive materials for X-rays, the mask substrate by necessity must be a low Z (atomic number) thin membrane. X-ray masks should be rugged, alignable with respect to the sample, and capable of withstanding many exposures without distortion. The requirements for X-ray masks in LIGA differ substantially from those for the IC industry [47]. The main difference lies in the absorber thickness. In order to achieve a high contrast (>200), very thick absorbers (>10 μm vs. 1 μm) and highly transparent mask blanks (transparency >80%) must be used because of the low resist sensitivity and the great depth of the resist. Another difference involves the radiation stability of membrane and absorber. For conventional optical lithography, the supporting substrate is a relatively thick, optically flat piece of glass or quartz highly transparent to optical wavelengths. It provides a highly stable (>10^6 μm) basis for the thin (0.1 μm) chrome absorber pattern. In contrast, the X-ray mask consists of a very thin membrane (2 to 4 μm) of low-Z material carrying a high-Z thick

FIGURE 3.4. SEM micrograph of a HDPE molded microstructure (with permission from [46]).

absorber pattern [48]. A single exposure in LIGA results in an exposure dose a hundred times higher than in the IC case.

Due to excellent contrast and good process stability of e-beam lithography (a process often used to produce X-ray masks), PMMA is the preferred resist for deep-etch synchrotron radiation lithography. Two major concerns with PMMA as a LIGA resist are a rather low lithographic sensitivity and a susceptibility to stress cracking. To make throughput for deep-etch lithography more acceptable to industry, several avenues to more sensitive X-ray resists have been pursued. For example, copolymers of PMMA were investigated: methyl methacrylate combined with methacrylates with longer ester side chains show sensitivity increases of up to 32% (with tertiary butylmethacrylate). Unfortunately, deterioration in structure quality was observed [49].

LIGA depends critically on metal deposition to replicate the master photoresist mold. This additive process involves electroless metal deposition or electrodeposition. In order to continuously build thick deposits by chemical means without consuming the substrate in electroless metal deposition, it is essential that a sustainable oxidation reaction be employed instead of the dissolution of the substrate itself as in the case of immersion plating [50]. Metal alloys such as nickel-phosphorus, nickel-boron, cobalt-phosphorus, cobalt-boron, nickel-tungsten, copper-tin-boron, palladium-nickel, etc. can be produced by codeposition. Electroless plating may be preferred over electrodeposition due to the simplicity of the process since no special plating base is needed.

In the electroforming of micro-devices with LIGA, a conductive substrate, carrying the resist structures, serves as the cathode. The metal layer growing on the substrate fills the gaps in the resist configuration, thus forming a complementary metal structure. The use of a solvent-containing development agent ensures a substrate surface completely free of grease and ready for plating. The fabrication of metallic relief structures is a well-known art in the electroforming industry. The technology is used, for example, to make fabrication tools for compact discs where structural details in the submicron range are transferred. Because of the extreme aspect ratio, several orders of magnitude larger than in the crafting of CD masters, electroplating in LIGA poses new challenges.

Slight differences in metal layer thickness cannot be avoided in the electroforming process. Finish-grinding of the metal samples with diamond paste is used to even out micro roughness and slight variations in structural height. After finish grinding an electroplated LIGA workpiece, a primary metal shape results from removing the photoresist by ashing in an oxygen plasma or stripping in a solvent. In the case of cross-linked PMMA, the resist is exposed again to synchrotron radiation, guaranteeing sufficient solubility before being stripped. If the metal part needs to function as a mold insert, metal is plated several millimeters beyond the front faces of the resist structures to produce a monolithic micromold.

Once LIGA was established in the research community, interest in other micro- and nano-replication methods became more pronounced. Given the cost of the LIGA equipment, various LIGA-like processes took center stage. These pseudo-LIGA methods involve replication of masters created by alternate means such as deep reactive ion etching (DRIE) and novel ultraviolet (UV) thick photoresists. Among various pseudo-LIGA methods, UV-LIGA is low-cost and therefore widely used in microfluidic applications [30].

3.3.2.2. E-Beam Lithography Electron beam lithography (EBL) is probably the most widely used nanofabrication tool. Modern EBL technology can push resolution below 10 nm for isolated features and 30 nm for dense arrays of periodic structures [51]. Typically, a thin layer of PMMA resist layer (up to several hundreds nm) is spin-coated on the surface of thermally oxided silicon substrate. Exposures on PMMA is carried out by writing of single electron beam with a spot size as small as 1 nm. The writing speed is around 10 cm/h for nanoscale lines and 1,000 dots/s for nanoscale dots. The exposure dose range for which the desired pattern can be obtained increases with the feature size. For very dense arrays, the dose range is very narrow that a slight overdose may cause structures to collapse, while a slight underdose will not result in good resolution. Resist development after exposure and pattern transfer thereafter determine the resolution limitation of EBL. Usually, pure isopropylic alcohol (IPA) or a mixture of methylisobutylketone (MIBK) and IAB is used as the developer, and ultrasonic agitation during the development is used to remove the residual resist material at he bottom of the exposed area. The combination of ultrasonic agitation during the development and IPA as a developer for PMMA makes the openings in the resist as clean as possible for better pattern transfer. Pattern transfer is achieved by lift-off and RIE or DRIE. Figure 3.5 shows an EBL produced metal mold with nanoscale features. Direct writing of EBL to produce the final product is a slow process. But using EBL to generate a mold/master for polymer replication or a mask for lithography, e.g. nanoimprinting lithography (NIL) is a promising nanofabrication process.

3.3.2.3. Scanning Probe Lithography (SPL) In the last 10 years, SPL has attracted a great deal of attention in nanofabrication because of its simplicity and relatively low-cost. Most SPL techniques are negative printing methods that rely on scanning probe instruments to pattern substrate surface by selectively removing the pre-coated organic resist layer or self-assembled monolayers (SAMs). They include mechanical scratching [52, 53], anodization of Si surfaces [54, 55], electrochemical decomposition of SAM [56–59], electric

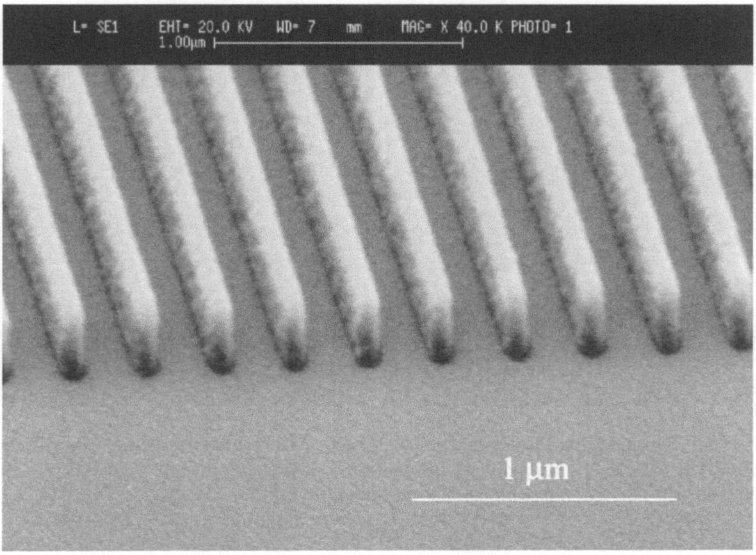

FIGURE 3.5. Nanoscale patterning by EBL: (a) AZPN114 resist profile of grating pattern showing 60 nm lines at 0.2 mm pitch exposed at 40 keV.

field-induced chemical reactions [60, 61], and electrochemical reactions in solution using electrochemical Scanning Tunnel Microscope (STM) tips [62–68]. Subsequent etching, adsorption or electroforming steps allow the transfer of nanoscale patterns to the substrate.

Dip-pen nanolithography (DPN) is a relatively new SPL technique. It uses an atomic force microscope (AFM) tip as a "pen", a surface coated solid as "paper", and self-assembly molecules with a chemical affinity for the solid surface as "ink" to directly write self-assembly molecules on the solid surface in a positive printing mode to form nanoscale patterns [44]. The process is schematically shown in Figure 3.6. A liquid meniscus forms

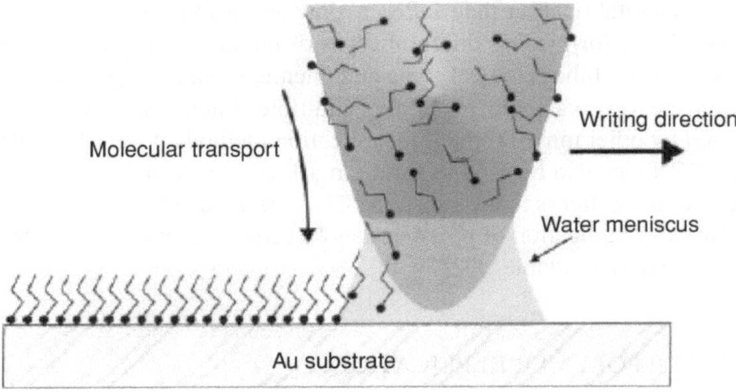

FIGURE 3.6. Schematic representation of DPL (with permission from [44]).

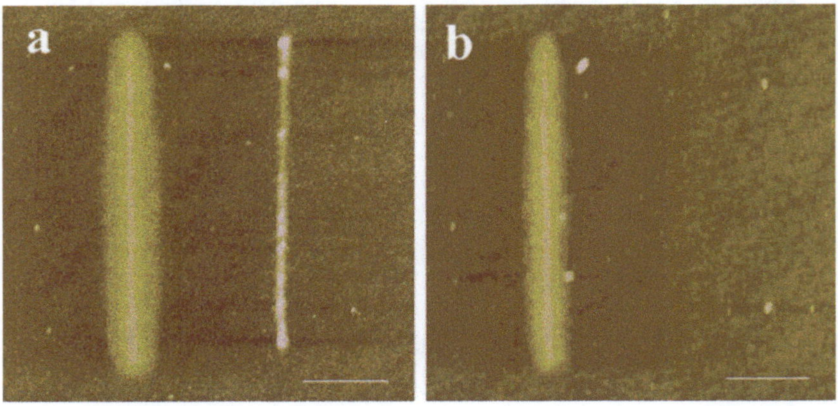

FIGURE 3.7. SiO_2 (left-hand line) and polymer (right-hand line) features (a) before and (b) after chemical oxidation with 2:1 H_2SO_4:H_2O_2. The polymer line was fabricated at 10 nm/s, 48% humidity, −12 V. Polymer line width: 30 nm; scale bar: 250 nm (with permission from [70]).

between the AFM tip, coated with self-assembly molecules and the solid surface. The meniscus size controls the transport rate of self-assembly molecules and DPN resolution. Typical self-assembly molecules arealkyl and aryl thiols like 1-octadecanethiol (ODT) and the most widely used solid surface is gold (Au). Other organic molecules and substrate materials such as hexamethyldisilazane (HMDS) on oxidized silicon (SiO2) and gallium arsenide (GaAs) surfaces [69], and poly(thiophene) on SiO_2 [70] have also been successfully applied in DPN. Using electrochemical deposition, DPN has also been extended to write metal and semiconductor nanostructures on the solid surface. In this E-DPN technique, metal salts dissolved in the water meniscus are reduced into metals electrochemically and deposited on the solid surface as shown in Figure 3.7. The advantage of DPN comparing to other SPL techniques is that it uses much less molecular substance.

The direct transport of molecules to solid surface in a positive printing mode can also be achieved by microcontact printing of soft lithography (μCP) [71]. μCP is a parallel methods capable of deposit molecules to the entire pattern simultaneously, which is more advantageous in nanofabrication than DPN, a slow serial technique, in nanofabrication. DPN, however, allows for placing different types of molecules at specific sites and does not need any expensive lithographical tools and chemicals such as photomask, resist and stamps. It could be very valuable for adding multiple functionalization onto nanoscale devices prepared by other mass-producible fabrication methods. Like EBL, NIL and other SPL methods, DPN can also be used to form a mask on the substrate surface and rely on etching to transfer the patterns to the substrate. The patterned substrate can then serve as a mold/stamp for mass-replication of polymer nanostructures. Commercial DPN equipment and technology are now available.

3.4. NANOSCALE POLYMER REPLICATION

Replication at the nanoscale includes repeated production of either macro and microscale parts with nano-sized structures, or independent nano-sized structures.

3.4.1. Soft Lithography

Micro-molding based on low viscosity liquid resins, instead of high viscosity polymer melts, has been attempted in recent years by several researchers [45, 72–74]. They combined the photolithography technique and molding of reactive resins for microfabrication. Whitesides et al [75] used a high-resolution transparency containing the design of the features as the mask in photolithography to produce a positive relief of photoresist on a silicon wafer. The PDMS resin was then cast onto the silicon wafer and cured at elevated temperatures. The polymer replica of the master, containing a negative relief of features, was easily peeled away from the silicon wafer and either used as the micro-device directly, or as a master for micro-contact printing, micro-molding in capillaries, or micro-transfer molding [76–79]. This method, called soft lithography, is an attractive way of rapid prototyping because of its simplicity [80, 81]. Soft lithography can also be extended to nanoscale replication when the mold (master) is prepared by EBL, SPL or LIGA. Figure 3.8 summarizes different soft lithography methods.

3.4.2. Nanoimprinting

For mass-producing polymeric nanostructures, the "open-mold" embossing (or imprinting) process is probably the most promising low-cost and high-throughput technology because the "closed-mold" injection molding of either thermoplastic or reactive resins is too difficult to perform at such a small size. In the case of embossing or imprinting, only a thin layer of polymer film coated onto a hard substrate is structured. The pattern on the mold surface can be easily transferred to the polymer film when the film is heated above its T_g or dissolved in a solvent in the case of thermoplastic polymers. By cooling or solvent evaporation, the pattern can be secured onto the polymer surface. In the case of liquid reactive resins, pattern transfer can also be achieved easily and the pattern is secured by either photo, thermal, or redox cure of the resin.

3.4.2.1. Hot Embossing Lithography The basic principle of hot embossing/imprinting is that a polymer substrate is first heated above its softening temperature (i.e. glass transition temperature, T_g for glassy polymers or melting temperature, T_m for semi-crystalline polymers). A mold (or master) is then pressed against the substrate, allowing the pattern to be fully transferred onto the substrate (embossing/imprinting). After a certain contact time between the mold and the substrate, the system is cooled down below the softening temperature, followed by separating the mold from the substrate (de-embossing/imprinting). The hot embossing process can be achieved in either a cyclic process or a continuous process.

In a cyclic process (Figure 3.9), a mold (master) is placed in a hydraulic or pneumatic press and a heated polymer sheet or film is stamped by applying the appropriate force thus replicating the structure from the mold to the polymer. This constitutes a low cost method for making prototypes. The CNC (computer numerical control)-machined tool steel molds have high strength and can last for thousands of embossing cycles without reconditioning. They can also be used to emboss metallic materials (e.g. Al and Cu) in addition to polymers [82]. For very simple features, like micron-sized channels on the surface of polymer sheets for measuring the electroosmotic mobility, metal wires used in the EDM process can serve as a mold in a wire-imprinting technique [83].

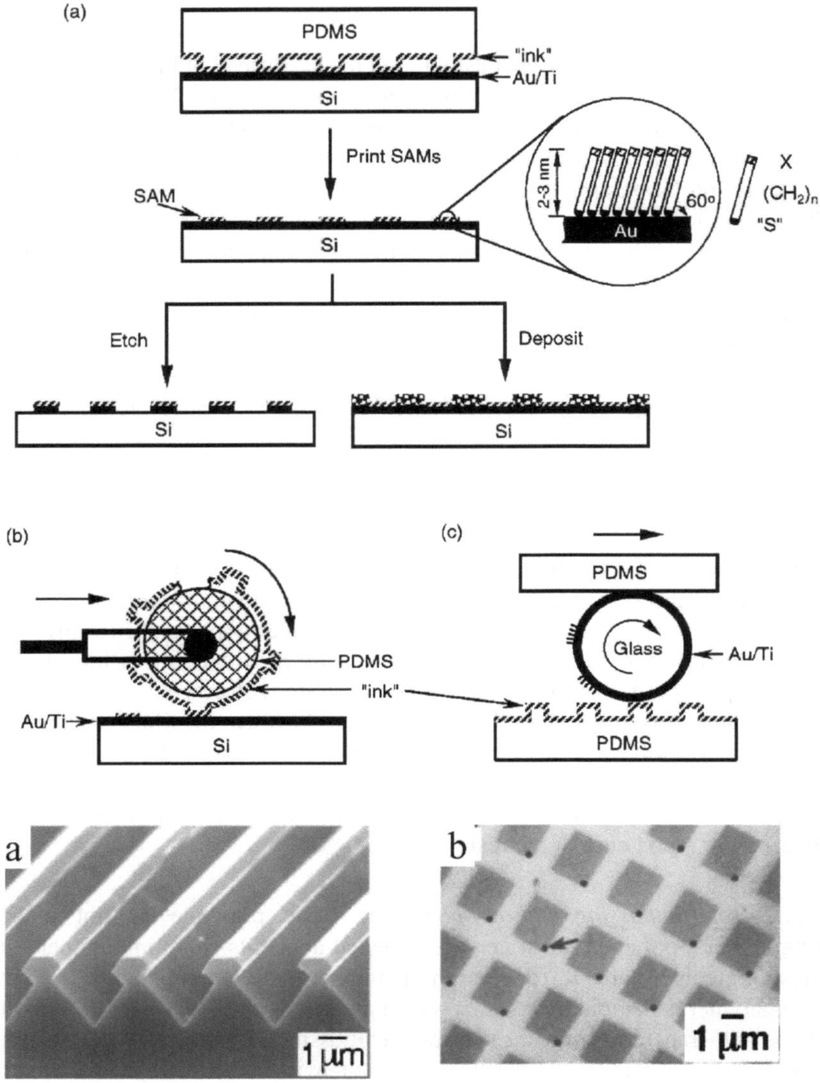

FIGURE 3.8. Examples of Soft lithography methods.

In many biomedical devices, the microscale and nanoscale features often are very small, have complicated patterns, or require a high-aspect ratio. Since performance of theses devices relies heavily on the quality of the micro/nanoscale features, most modern micro/nanoembossing techniques use LIGA or LIGA-like processes to prepare high resolution molds. Such "LIGA press" based hot embossing processes can produce high quality products. Electroplated nickel molds [84, 85] and RIE or DRIE made quartz molds [86, 87] have been used in many applications. Silicon [88–90] and polymeric [91, 92] molds have also been fabricated. Silicon molds provide advantages of decent material properties in

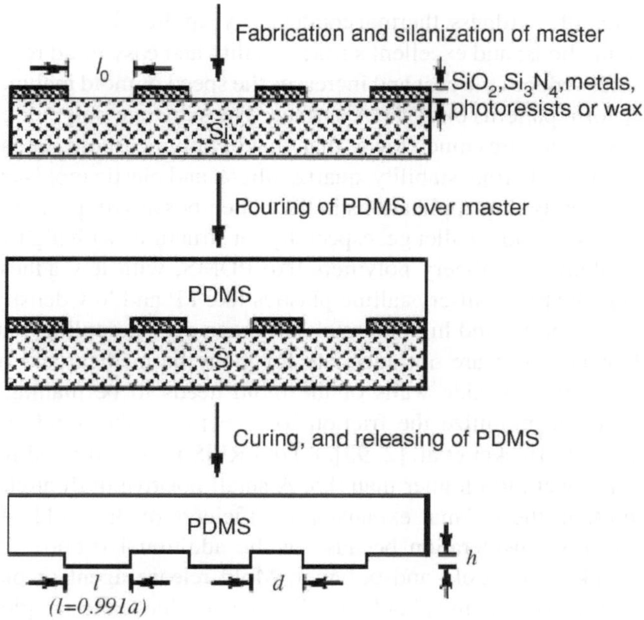

FIGURE 3.8. (*Cont.*)

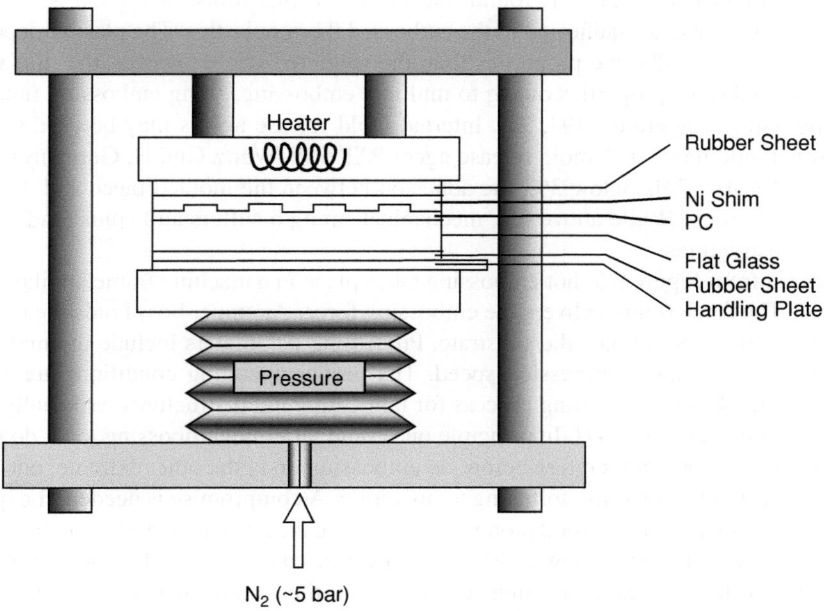

FIGURE 3.9. Schematic of a cyclic hot embossing lithography press.

terms of tensile strength, hardness, thermal conductivity; applicable to standard photolithography fabrication methods; and excellent surface quality and easy mold release. The use of thermoplastic molds reduces the cost and increases the speed of mold making. For example, EBL can directly form patterns on PMMA surface. The substrate polymers, however, need to have softening temperatures much lower than that of polymer mold. Because of low mechanical strength and/or thermal stability, quartz, silicon and plastic molds are only suitable for small quantity prototyping applications in the hot embossing/imprinting processes.

De-embossing is a major challenge, especially for structures with high-aspect ratio and made of glassy polymers. Rubbery polymers like PDMS, with low adhesive power and shape stability, and ductile semi-crystalline plastics like PP and low density PE, with low surface energy, low rigidity, and high thermal shrinkage during cooling, can be demolded much easier. They, however, are not adequate for many biomedical and optic-electronic applications. The surface of side walls of the mold needs to be maintained as smooth as possible in order to minimize the friction force between the mold and the polymer substrate. According to Becker et al. [2, 93], 80 nm RMS is an empirical limit for making structures with an aspect ratio higher than 0.5. A small positive draft angle may ease this constraint. In addition, the thermal expansion coefficients of the mold and the polymer need to be taken into consideration because of the additional friction force caused by differences in shrinkage of mold and polymer. Mold release agent or plasticizer is not desirable if fabricating a biochemical or biomedical device due to the sample contamination or an increased fluorescence background. Even without these concerns, it is not easy to apply external mold release agents to microstructures by the conventional liquid spraying methods, because they cannot be sprayed uniformly and may block microscale channels. Plasma deposition or sputtering methods have to be used. Applying a thin anti-adhesive film on the mold is helpful. It is found that the PTFE-like films have good adhesion to the metal surface and anti-adhesion to the embossed thermoplastics. The plasma-deposited film has better anti-adhesive properties than the sputtered one. However, the film would lose its anti-adhesive properties owing to multiple embossings, long embossing time, and high embossing temperature [94, 95]. Internal mold release agents may be used in some cases. It is found that 3–6% mold release agent PAT 665 (Wurtz GmbH, Germany) works well with PMMA [73]. Below 3%, the adhesion between the molded piece and the mold becomes too strong. While above 6%, mechanical strength suffers and voids tend to form in the microstructures.

As seen from Figure 3.9, hot embossing takes place in a machine frame similar to that of a press. The force frame delivers the embossing force. An upper boss holds the molding tool and the lower boss holds the substrate. Processing parameters include thermal cycle, compression force and compression speed. The proper operating conditions are critical when applying the hot embossing process for fabricating microstructures, especially those with high aspect ratio [96, 97]. In principle one could, after hot embossing, cool down the whole device to room temperature before de-embossing or, at the other extreme, one could de-emboss just below or at the softening temperature. A compromise is needed; the quality of the replication may not be good if one tries to remove the master when the polymer is still soft while cooling all the way down to room temperature takes too long. In order to minimize the process cycle time, thermally-induced stresses in the materials, and replication errors due to the thermal expansion coefficients of tool and substrate, the embossing temperature is set slightly above the softening temperature, while the de-embossing temperature is slightly below the softening temperature (i.e., the operating temperature range is near T_g or T_m). The temperature difference between embossing and de-embossing determines the thermal cycle time, typically from 25 to 40 °C. By actively heating and cooling the upper and lower bosses one could get cycle times of about 5 minutes for a typical hot embossing/imprinting process.

Because of the near T_g (T_m) processing, the polymer behaves more like a solid and cannot relax rapidly. This results in larger compression force required to emboss the material, which leads to higher flow-induced stresses. Furthermore, higher compression force may cause the destruction or wear of the mold insert more easily. Slow embossing speed is preferred for fabricating microstructures with freestanding columns or high aspect ratio because these types of structures are sensitive to lateral forces. Undercut structure cannot be constructed by hot embossing since the mold insert needs to be removed after processing. Automated de-embossing is crucial and required if the structures have vertical walls and a high-aspect ratio. A vacuum for hot embossing ensures longer lifetime for the mold (corrosion) and absorbs any water released during the embossing process from the polymeric material and prevents bubbles from entrapped gases [2, 93].

Embossing/Imprinting can also form nanoscale polymer structures. However, most nanoembossing/imprinting studies used polymers as a "mask" or sacrificial template to produce silicon nanostructures, following the same concept as the previously discussed relief imprinting process. Nanoimprinting lithography (NIL) developed by Chou and his coworkers at Princeton University [98–102] and Hot Embossing Lithography (HEL) studied by a number of researchers in Europe [103, 104] are typical examples. Figure 3.10 shows

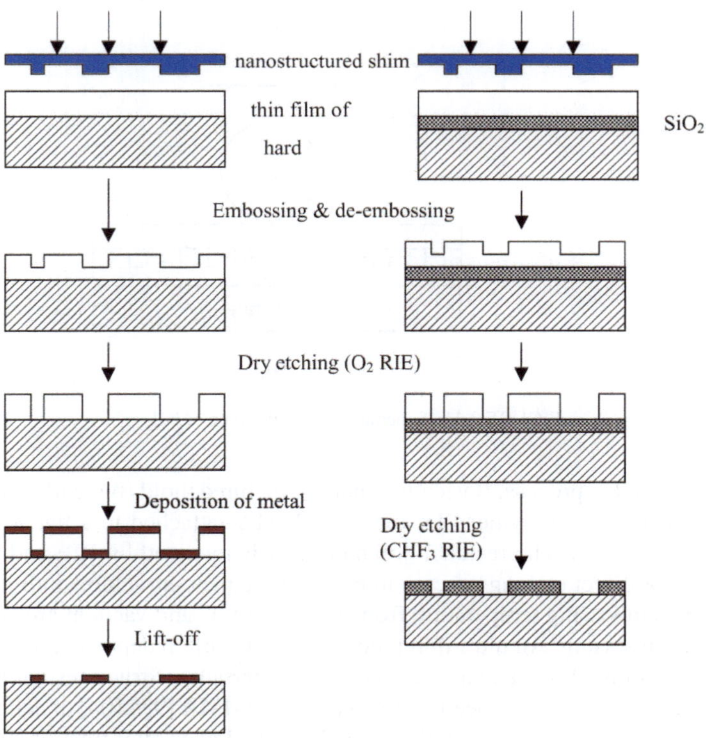

FIGURE 3.10. Schematic of the HEL process.

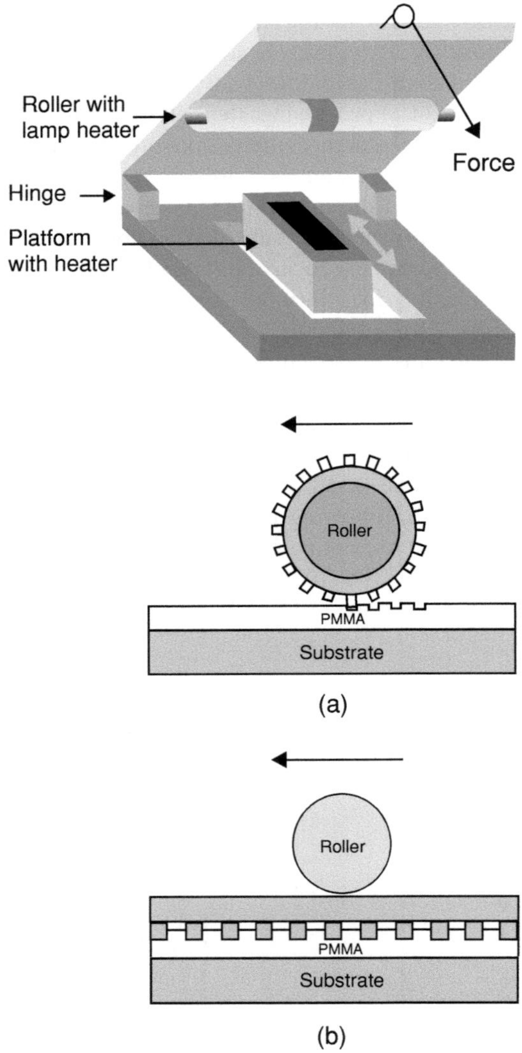

FIGURE 3.11. Schematic of a roller type of NIL.

schematics of the HEL process. Basically, a nano-structured mold (typically made by EBL) is used to imprint a thin polymer film (mostly PMMA) attached to a flat solid substrate (typically silicon wafer). The residual polymer layer is removed by RIE and followed by either a lift-off or direct etching process to produce the nano-structures into the substrate. For PMMA, an embossing temperature from 130 to 190 °C and vacuum are applied to ensure complete embossing. An ultra-thin teflon-like film is often deposited onto the mold to minimize the frictional force and avoid sticking dust particles. Structures with feature size ranging from 10 to 100 nm have been fabricated using PMMA [98–104] or cellulose acetate [105]. Figure 3.11 shows the schematics of both flat bed and roller type of NIL. Potential applications of these nano-structures include silicon quantum dots, wire and ring transistors, and a nano compact disk (CD) with 400 Gbits/in^2 storage density (near three orders of magnitude higher than the current CD). A major challenge of nanoscale embossing/imprinting is how to produce uniform features over a large surface area. A step-and-repeat process developed by Chou [106] is able to form nano-sized structures over large surface area. Another approach is to use a 'soft' mold and pneumatic embossing force. A diaphragm type press is used to provide good contact between the mold surface and the substrate surface [107].

Several "LIGA press" type cyclic embossing machines are available in the market. Figure 3.12 shows a hot embossing machine system HEX 03 manufactured by Jenoptik Mikrotechnik GmbH. Figure 3.13 shows an EVG520HE system from AV Groups in Austria. Both have similar capability. Recently, Obducat Company in Sweden developed a nanoimprinting system based on the diaphragm type press as shown in Figure 3.14. Molecular Imprints, Inc. in the U.S. has a micro/nanoembossing system based on UV cure of liquid resins. Their Imrpio 100 machine is shown in Figure 3.15.

3.4.2.2. Sacrificial Template Nanoimprinting In nanoscale replication, de-molding is the most challenging step because the stress resulting from the large contact surface between the tool and the polymer substrate tends to damage the molded polymer structures and/or the tool features. Using the simple soft lithography technique, low-cost sacrificial templates replicating the 3D tool features can be formed from water-soluble polymers reinforced with nanoparticles. These nanocomposite templates are strong enough to serve as molds during processing and can be easily removed and recycled in water.

In our laboratory, we are developing simple NIL techniques that can be used to mass-produce well-defined nanoparticles and nanofluidic devices with well-defined nano-pores for drug delivery and bioseparation. Figure 3.16 shows an example of this method. A nano-indenter or an AFM can be used to produce a master (mold), by first coating a photoresist on a metal-coated silicon wafer or quartz. Then, an array of holes is punched through the cured photoresist. Next, casting is carried out to produce a silicone rubber master with an array (e.g. 100×100 in 1 mm^2 area) of nano-sized probes. As an example, Figure 3.17 shows SEM photos of a silicone rubber master with a 5×5 array of pyramidal shape probes made by a Hysitron nano-indenter with a 90° diamond tip. The probe radius is less than 50 nm (the probe surface was covered by 20 nm-thick gold for SEM imaging), and can be made smaller by using different tips.

Figure 3.18 shows another example of using a group of differential etched optic fiber bundles (Sumitomo IGN-037/10) as the male master mold, followed with casting to make the poly(dimethyl siloxane) (PDMS) female daughter mold, re-casting in the polyvinyl alcohol (PVA)/water solution to make the male sacrificial template for nano-imprinting, and then the resulting nanoporous poly(lactic-co-glycite acid) (PLGA) layer [108]. Masters with an array of probes are fabricated on the distal faces of coherent fiber-optic bundles by differential wet etching [109]. This is a simple method without the need of a clean room facility [110–112]. By pulling the optic fibers, the fiber diameter can be reduced from 3 μm to less than 200 nm [113, 114]. This allows for the formation of high–density, small-sized nanotip arrays. The angle of the probe depends on the differential etching rates of the fiber and the surrounding cladding material. The master can be used to produce PVA sacrificial templates through a female PDMS daughter mold (resin curing), and then an array of well-defined nano-sized nozzles on a PLGA (solution casting) layer through a series of

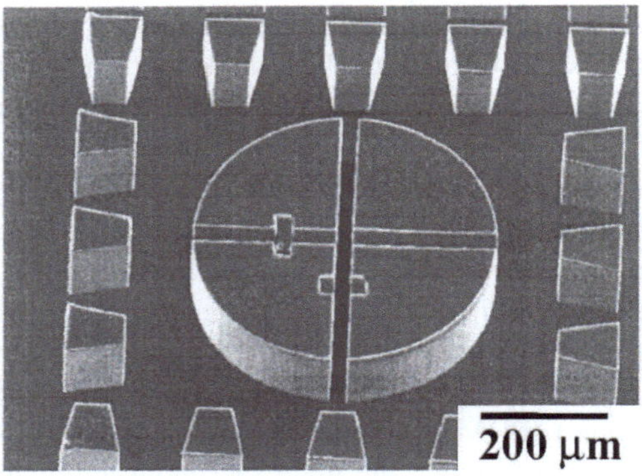

FIGURE 3.12. A hot embossing machine system HEX 03 (Jenoptik Mikrotechnik GmbH, Germany); Spec: press force: <200 kN; temperature inside chamber: <500 °C; max substrate size: f 150 mm with vacuum and cooling system.

nano-imprinting steps. The size of the small end of the nozzle depends on the tip shape and the imprinting depth. The size of the larger end of the nozzle depends on the angle of the tip and the layer thickness. Replication accuracy requires complete wetting between the mold and the casting fluid (i.e. between optic fiber and PDMS resin, or cured PDMS and PVA/water solution, or dry PVA and PLGA solution).

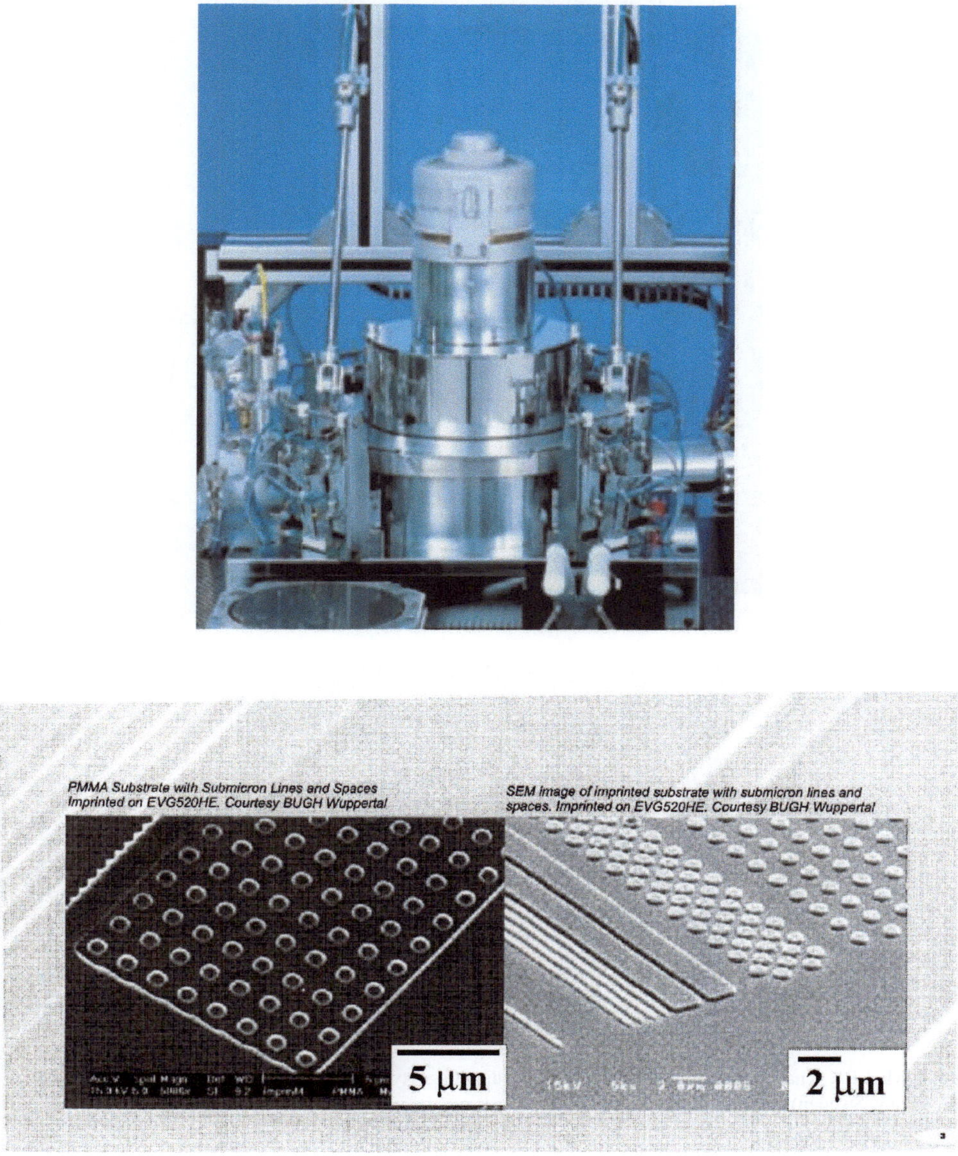

FIGURE 3.13. An EVG520HE system (AV Groups, Austria); Spec: press force: <40 kN; temperature inside chamber: <550 °C; max substrate size: f 205 mm with vacuum and cooling system.

3.4.3. Injection Molding at the Nanoscale

Nanostructures can also be replicated on large surfaces using injection molding of polymeric materials. A successful example is the production of high density CDs with structures down to 25 nm. A Si wafer was v-notched by EBL and subsequent dry etching,

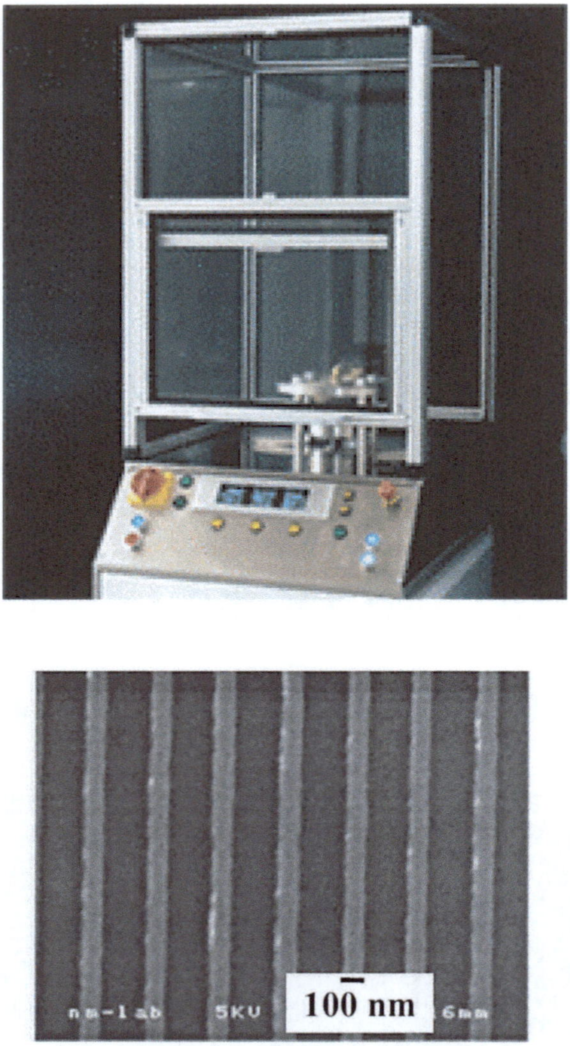

FIGURE 3.14. A nanoimprinting system based on the diaphragm type press: 2.5-inch imprint equipment (Obducat Company, Sweden).

and used as an injection molding stamp with two different PC materials [115–117]. Nano-injection molding was also used to convert collagen assembles into a very distinct fibrillar structure with a characteristic 65 nm periodic cross-striation [118]. With a single master, more than 600 structures have been replicated, thus proving that the procedure is suitable for the mass production of nanostructures. Such structures can serve as simple calibration scales for SPMs (Scanning Probe Microscopes) to calibrate the piezo scan mechanism before measurement. For this application a single master grating structure has to be produced with a precise period, from which a reference structure can be replicated, taking into account the

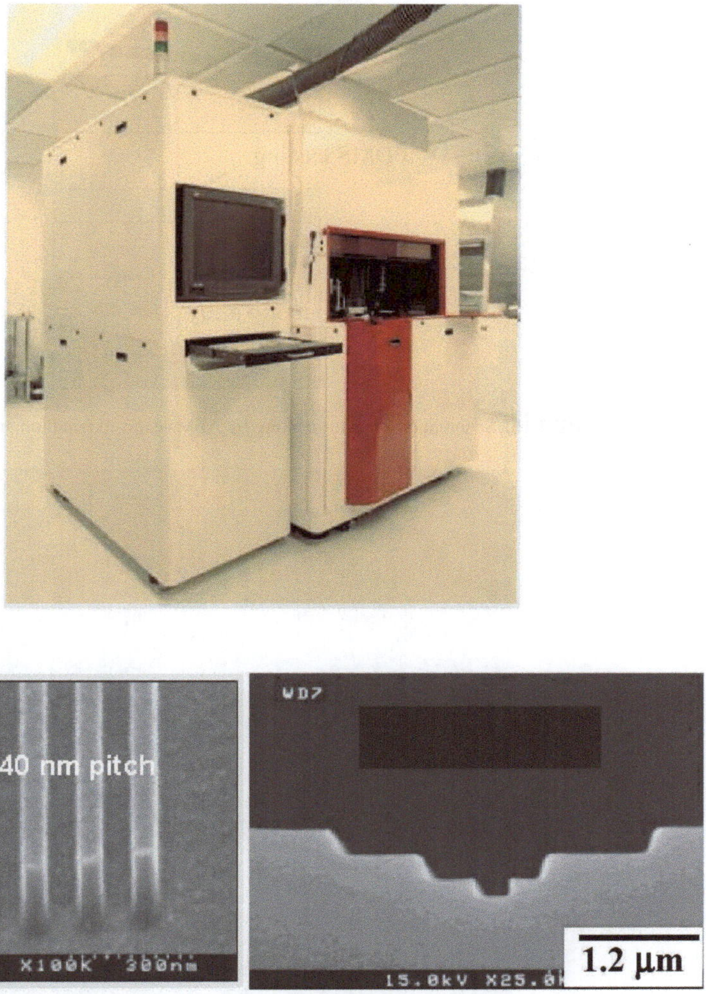

FIGURE 3.15. Imrpio 100 machine (Molecular Imprints, Inc., USA).

known shrinkage. The results also show that using the CD injection molding process, mass fabrication of storage media with even higher density is possible [119].

3.4.4. Other Technologies

The above mentioned fabrication techniques are capable of producing 2D, in some cases 3D, polymer nanostructures with complicated patterns. Most of them, however, require expensive tools that are substantially different from the macroscale polymer processing equipment. Consequently, their implementation to large-scale industrial production remains unproven. Although nanoscale injection molding is an extension of its macro- and

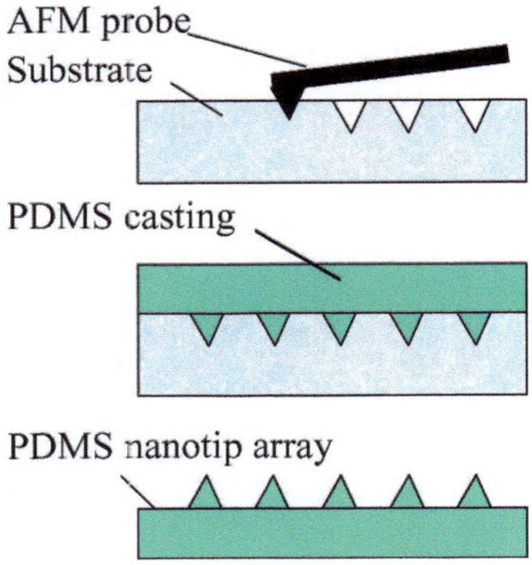

FIGURE 3.16. Schematic of master making by AFM (nano-indentation and soft lithography).

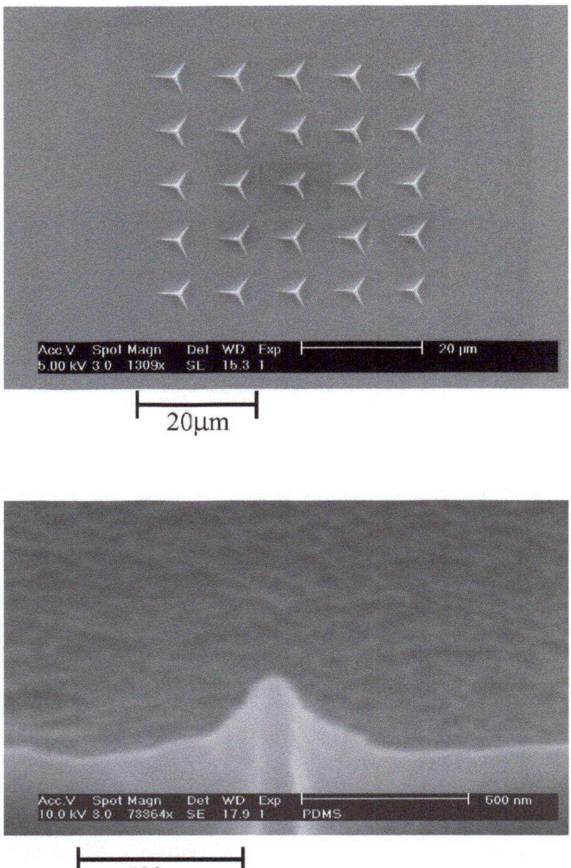

FIGURE 3.17. SEM photos of a silicone rubber master with a 5 × 5 array of pyramidal shape probes made by a Hysitron nano-indenter with a 90° diamond tip.

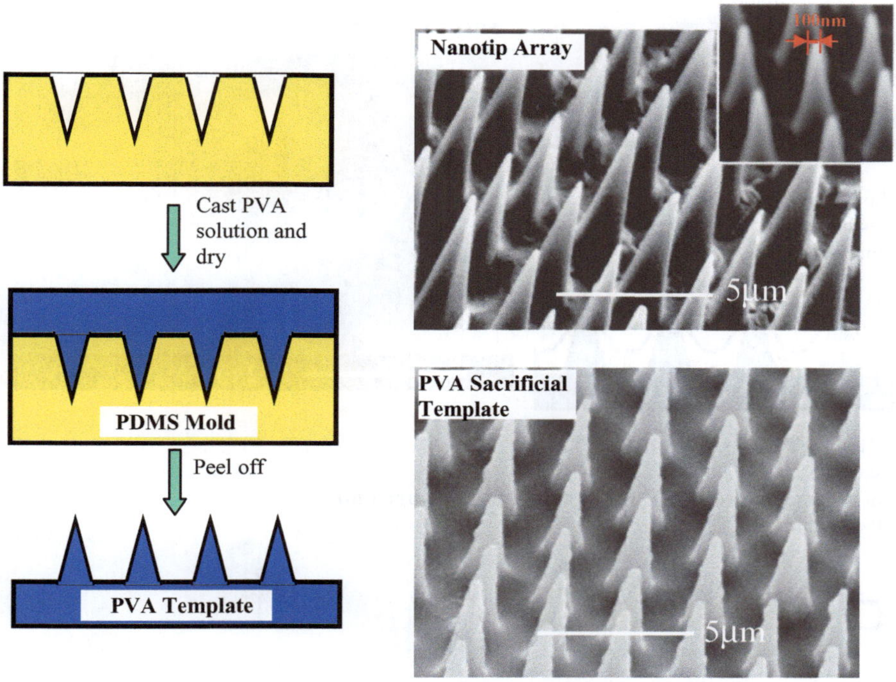

FIGURE 3.18. A sacrificial template lithography (STI) process to produce polymer nanonozzle array.

microscale counterparts, and nanoimprinting is, in principle, very similar to micro-embossing and macro-stamping, the conventional injection molding and embossing/stamping machines and molds cannot be used directly for producing nanoscale structures. There are, however, several well-known mass-producing polymer processing methods that can easily be modified to produce certain polymer nanostructures or products containing nanostructures. They are briefly introduced in the following sections.

3.4.4.1. Polymer Membranes with Nanopores Track etching of nanoporous polymer membranes is a well-established industrial process since the 1970s [120]. There are two methods of producing latent tracks in the polymer films to be etched into porous membranes-the irradiation with fragments from the nuclear reactors, or the use of ion beams from accelerators. Most membrane production is based on the latter method. Pore formation is achieved through chemical etching to remove the damaged polymer in the latent track. Several centers in the world manufacture track-etch membranes on a commercial scale [120]. Polycarbonate (PC) and poly(ethylene terephthalate) (PET) are the primary materials used. Polypropylene (PP), poly(vinyldiene fluoride) (PVDF), polyimides (PI) and several other high temperature polymers were also tried in the track-etching process. Difficulties in etching, high cost of the product and the existence of more competitive membranes limit their commercial uses. Pore diameter from 10 nm to tens of microns, pore density from 1 to 10^{10} cm^{-2}, and pore shape from cylindrical, conical to cigar-like can be produced for PC and PET. Figure 3.19

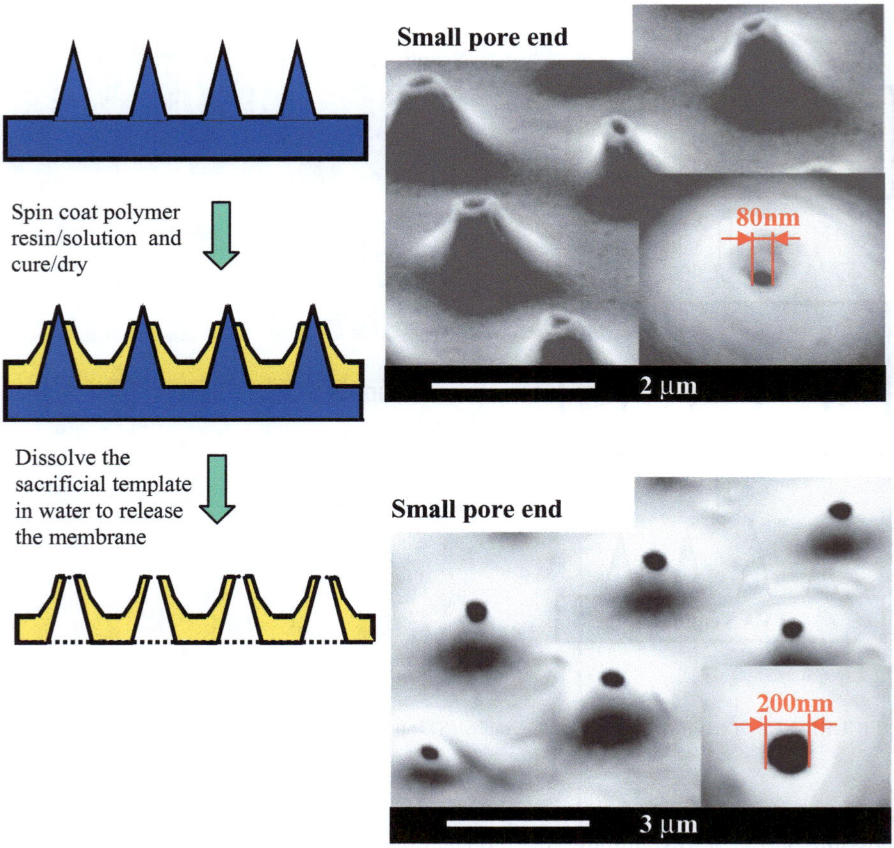

FIGURE 3.18. (*Cont.*)

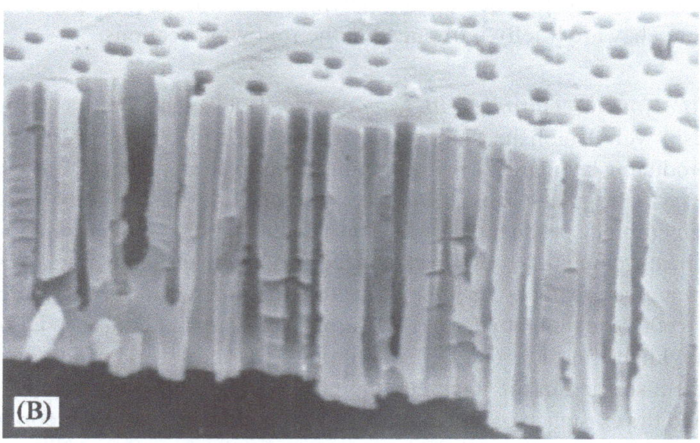

FIGURE 3.19. Pore structures of a track-etch membranes (with permission from [11]).

shows the pore structures of several track-etch membranes. Comparing to other polymer membranes, track-etch membranes provide two unique properties. First, the pore size is more uniform and secondly, the number of pores and the size of the pore can independently be varied in a wide range. Track-etched membranes are used for industrial process filtration, cell culture and laboratory filtration. The main applications are in biology because large-scale industrial processes are more cost-driven and less performance-sensitive. Another use of track-etched membranes is to serve as templates for making nanowires and nanotubes. The quality of nanoscale structures formed from track-etched membranes, however, is not as good as that from other template fabrication techniques such as "top down" LIGA and EBL approaches, and "bottom-up" block copolymers and self-assembled ceramic/metal approaches.

3.4.4.2. Nanofibers and Nanotubes Conventional fiber spinning techniques such as wet/dry spinning, melt spinning and gel spinning can produce large quantity of micron-sized polymer fibers. Electrostatic spinning or electrospinning is a process capable of producing polymer fibers with diameter less than 100 nm. It uses a high-voltage electric field to form solid fibers from a polymer solution [121] or melt [122] delivered through a millimeter-scale nozzle. This process was first studied by Zeleny in 1914 [123] and the first US patent was issued to Formhalls in 1934 [124]. Taylor [125] showed that the equilibrium shape of the suspended meniscus at the nozzle tip was a cone and when the applied voltage exceeded a critical voltage, a stable jet of liquid could be ejected. Electrospinning only became an interesting nanofabrication method for polymers in recent years. It is applicable to a wide range of polymers including those used in conventional fiber spinning (i.e. polyolefins, polyamides, polyesters, acrylic) [121, 126], as well as many specialty polymers like biodegradable polymers [127–134], electric conducting [135, 136] and photonic polymers [127, 137, 138], and biomolecules such as collagen [139, 140], proteins [141], DNA and polypeptides [142]. The fibers can be deposited onto a screen to form membrane/nest and even seamless 3D nanofiber structures. The nanofiber membrane/nest has been used for nanofiltration and [143, 144] as composite reinforcement [145], while 3D nanofiber structures have great potential to serve as various biomedical structural elements such as scaffolds in tissue engineering [127, 134, 139], wound dressing [130, 131], vascular grafts [142], drug delivery [132], and artificial organs (biocompatible films on prosthetic devices for implant applications) [141]. The scaffolds made of nanofibers show a morphologic similarity to the extracellular matrix of nature tissues, characterized by a broad range of pore size and distribution, high porosity and good mechanical strength.

Electrospinning of nanofibers can be easily achieved by applying a high voltage (several thousands volts/cm) to a capillary filled with the polymer fluid, which is spun onto a collector serving as an electrode (see Figure 3.20). When the charged jet of fluid is ejected from the tip of the taylor cone, it undergoes a whipping process whereby the solvent evaporates, leaving behind a charged polymer fiber that is highly stretched and reduced in diameter when traveling towards the collector surface. Splaying occurs in a region where the radical forces from the electric charges carried by the fiber become larger than the cohesive forces within the fiber. The single fiber divides into many smaller ones before landing on the collector. This splaying process together with large elongation due to the acceleration of the jet by electric forces is responsible for the nano-sized fibers produced by electrospinning. The solution must have a surface tension low enough, a charge density high enough, and a viscosity high enough to prevent the jet from collapsing into droplets before solvent

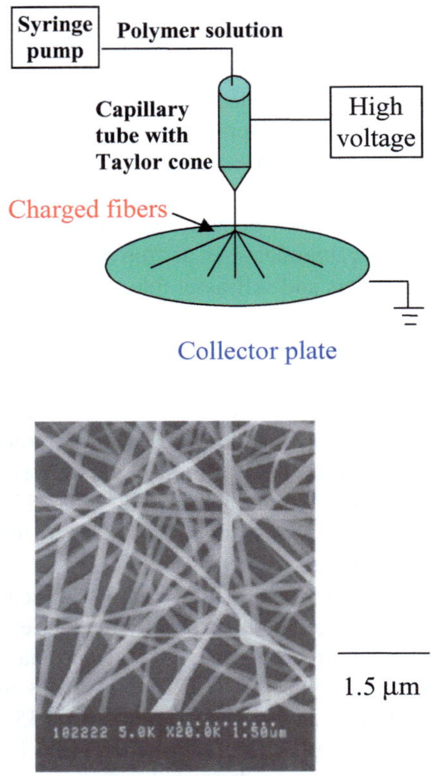

FIGURE 3.20. Polymer nanofibers by electro-spinning (with permission from [202]).

evaporation [126]. Bead formation is the main defect found on the spun fibers. In general, higher polymer concentration, charge density in the solution, and adding a small amount of salts tend to reduce bead-like defects, while increasing the electrospinning voltage causes bead formation. The diameter of the nanofibers increases along with the electrospinning voltage, polymer concentration and feeding rate of the solution, and decreases with the addition of salts [126, 133].

Nanotubes and nanochannels can also be formed using electrospun nanofibers as sacrificial templates. Bognitzki et al. [146] coated PLA nanofibers by chemical vapor deposition with poly(p-xylylene) (PPX). Subsequent annealing of these fibers above 250C under vacuum results in the degradation of the PLA core leading to PPX nanotubes. Czaplewski et al. [147] encapsulated heat depolymerizable polycarbonate (HDPC) by vapor deposition of SiO_2 or through the use of a spin-on glass. Although electrospinning is a very simple, low-cost, and versatile mass-production process to produce nanofibers, it is difficult to control their displacement, orientation, size and reproducibility. How to integrating electrospun nanofibers into more useful biomedical devices remains a major technical challenge.

Nanofibrillar surfaces on polymers can be formed through RIE using various gas compositions [148]. Figure 3.21 shows PET surface etched with an 50:50 O_2–N_2 gas mixture

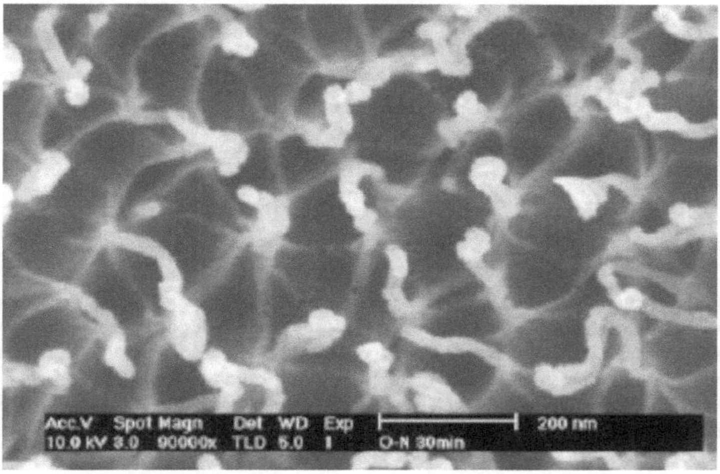

FIGURE 3.21. SEM image of the PET surface etched with a 50:50 O_2–N_2 gas mixture for 30 min (with permission from [148]).

for 30 min under a radio frequency power supply operating at 30 kHz. The dimensions of these fibrils are around 300 nm in length and 20 nm in diameter.

In addition to electrospinning and RIE, a variety of "bottom-up" synthetic methods have been reported to prepare nanofibers made of conductive polymers. They include template synthesis [149], interfacial polymerization [150, 151], bulk polymerization [152], polymerization guided by dialysis tubing [153], and using surfactants [154, 155] or coagulating media [149, 156]. Figure 3.22 shows polyaniline nanofibers deposited on Si wafer surface after guided polymerization by dialysis tubing.

3.4.4.3. Multiple Nanofilms/Layers Single or multi-layer thin polymer films can be formed by self-assembly of monolayers (SAM) or plasma deposition. These methods are mainly used for surface coating, not stand-alone multi-layer nanostructures. A relatively simple layer-multiplying coextrusion technique has recently been used to produce thousands

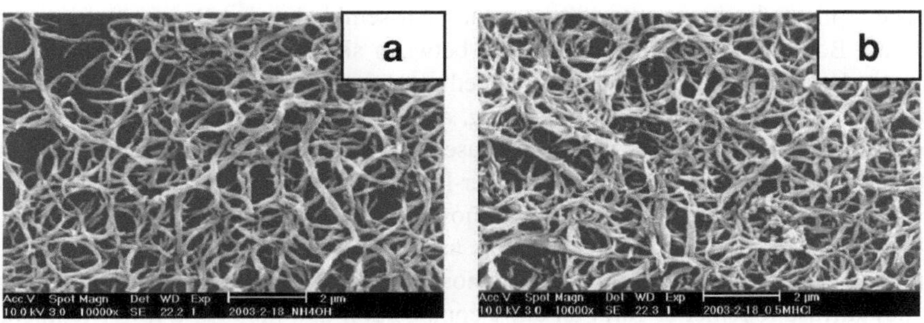

FIGURE 3.22. Scanning electron micrograph (SEM) of polyaniline nanofibers deposited on Si-wafer substrates (a) dedoped with 0.1M NH_4OH (b) redoped with 0.5M HCl (with permission from [153]).

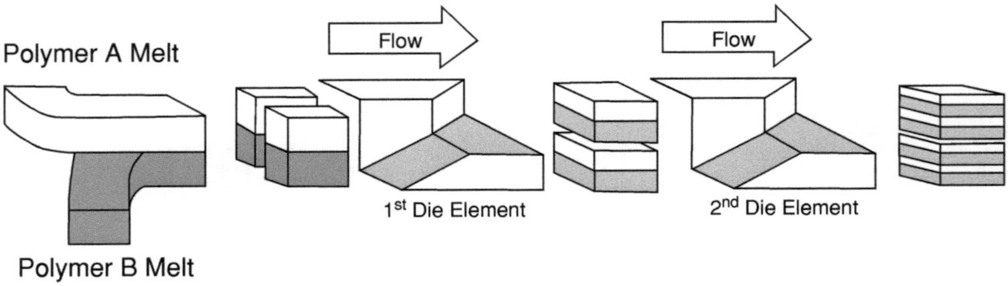

FIGURE 3.23. Layer-multiplying coextrusion for forced-assembly of polymer nanolayers. After the polymer melts are combined in the feedblock, the melt stream flows through a series of layer-multiplying die elements; each element splits the melt vertically, spreads it horizontally, and finally recombines it with twice the number of layers. The figure illustrates how two elements multiply the number of layers from 2 to 8. An assembly of n die elements produces 2(np1) layers. Eleven die elements produce 4096 layers. Finally, the melt is spread in a film die to further reduce the layer thickness (with permission from [157]).

of alternating layers of two immiscible polymers with each layer less than 100 nm [157–160]. The schematic of the process is shown in Figure 3.23. Two polymer melts are combined in the die entrance. The melt stream flows through a series of layer-multiplying die elements where each die element splits the melt vertically, spreads it horizontally, and then recombines it with twice the number of layers. An assembly of n die elements produces 2^{n+1} layers. Eleven die elements can produce 4096 layers. Experimental results of PC and PMMA [157] show that each layer maintains its original glass transition temperature when the layer thickness is larger than 100 nm. Shift of the glass transition temperature occurs when the layer thickness is below 100 nm. When the thickness is 10 nm or less, the layer shows an irregular periodicity and there is a single transition temperature indicating the entire structure is essentially the interphase of two immiscible polymers. Because of its low-cost and mass-production capability, this method has good potential to produce useful nanostructures for biomedical applications. For example, nanochannel arrays can easily be produced if one of the two polymers is removed after coextrusion.

3.5. ASSEMBLY AND BONDING

The replicated components often need to be assembled and sealed to perform certain functions. Bonding at the micro/nanoscale between silicon and silicon or other materials (e.g., glass, metal, etc.) is well developed and can be achieved by different methods such as adhesive bonding, eutectic bonding, anodic bonding, and fusion bonding [161]. Adhesive bonding has been in widespread use for over 30 years. Adhesives can be made electrically/thermally conducting (e.g., silver-loaded epoxy) or electrically isolating. Typical adhesive materials include epoxy thermoset resins, acrylic thermoplastic resins, and silicone resins. High pressure (>1 MPa) is usually necessary for adhesive bonding. An eutectic bond is formed by heating two (or more) materials (e.g., Au and Si) in a joint such that they diffuse together to form an alloy composition that melts at a lower temperature than the base materials (e.g., a Au-Si eutectic melts at 363 °C). For fusion bonding, temperatures above 1000 °C need to apply for silicon fusion bonding, while temperatures around

600 °C are necessary for glass fusion bonding. The bonding of silicon to glass (anodic bonding) is performed at a temperature ranging from 300 to 500 °C, with an applied voltage ranging from 500 to 1000 volts. These methods usually involve high temperature, high voltage, or high pressure. Several low temperature-bonding technologies (100 ∼ 200 °C) have been developed [162]. However, they usually need assistance from hydrofluoric acid (HF), plasma, or high pressure (up to 50 MPa). Therefore, among these bonding techniques for silicon/glass materials, only adhesive bonding may be applied to polymer-based micro/nano-sized devices.

Several techniques such as lamination (adhesive tape, thermal adhesive film), thermal (IR, hot-plate, laser) bonding, ultrasonic welding, and solvent bonding (i.e., partially dissolving the bonding surfaces, and evaporating the solvent) have been tried [2, 20, 33, 163–169] on polymer-based microdevices. McCormick et al. [170] successfully applied a thermal activated adhesive (Top flight MonoKote, Great Plane Model Distributors, Champaign, IL) to bond a Mylar sheet with an injection-molded acrylic microchip for DNA separations. Rossier et al. [166] used a 5-μm thick polyethylene adhesive layer to seal a UV laser photo-ablated PET microchannels with a PET film for electrophoretic separations. Dreuth and Heiden [167] applied a thin adhesive film on a substrate and the adhesive was then transferred to the elevated microstructures by stamp printing. Lamination by adhesive tapes or thermal adhesive films is probably the simplest and fastest method, but carries the risk of creating channels with different top and bottom surfaces. Moreover, when using the adhesive-bonding method, care needs to be taken in order to prevent the adhesive from flowing into the micro channels.

A solvent can be applied to partially dissolve the bonding surfaces, and evaporating the solvent will bring two halves together. This can solve the problem with using dissimilar materials as sidewalls. Lum and Greenstein [165] prepared microdepressions on one substrate and microprojections on the other, so that the substrates could be mated together to secure the relative position. A layer of monomer or pre-polymer was deposited on the microprojections before being mated and further polymerized to provide a bonding effect. Photoresist itself can also be used (e.g., SU-8 and Polyimide [171]) as a bonding agent to fabricate photoresist-based microfluidic devices. Glasgow et al. [169] introduced a solvent bonding technique in which a layer of polyimide precursor and solvent with dissolved precursor was placed in contact with patterned structures made of uncured polyimide precursor. The two halves were then cured with weights on top of the upper plate. They found that the bond quality was affected by the vent spacing for solvent evaporation, soft-bake duration, spin-coat speed during solvent application, and the concentration of the dissolved polyimide precursor in the solvent. However, as the solvent dissolves the bonding surface, it also partially dissolves the microchannel itself. Dimension control becomes quite difficult for very tiny channels. In addition, thermoset plastics or dissimilar materials cannot be effectively jointed by solvent bonding. Stress cracking of components is more likely to occur with this method than with other bonding methods. Furthermore, the use of solvents in a production line becomes more problematic, as environmental regulations and operator safety become prime issues for manufacturers throughout the world.

Ultrasonic welding is also a widely used welding method for joining plastic materials together due to its simplicity of use and rapid joining potential. In ultrasonic welding, high frequency ultrasonic energy is applied to produce mechanical vibration. Usually, horns and traps need to be fabricated around the designed joint area. The material at the joint surface

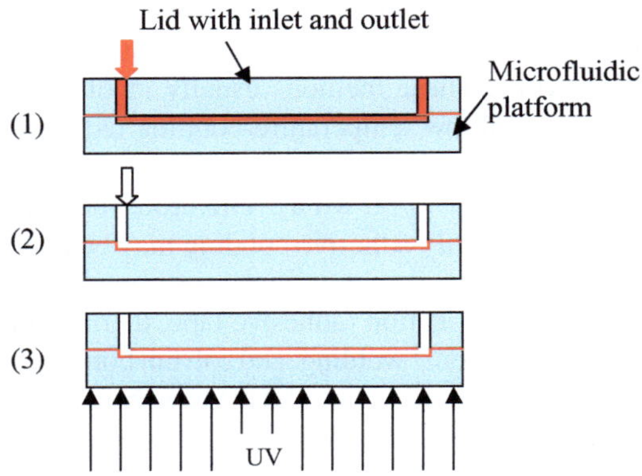

FIGURE 3.24. Schematic of a resin-gas injection bonding technique (1, resin injection; 2, gas injection; 3, resin curing).

(horn) melts from the heat generated due to the mechanical vibration. Traps are used to prevent molten plastic material from flowing into the designed joint area. This can pose a problem for bonding of microfluidic devices with very tiny channels (e.g., less than a hundred microns) because it would be difficult to fabricate even smaller horns and traps around the microchannels. This process can be done in a very fast manner (e.g., several seconds) depending on substrate materials, joint design, and the ultrasonic energy applied. However, ultrasonic welding is limited to thermoplastic materials with similar melting point. In addition, significant investment of capital equipment is necessary.

In general, these bonding techniques alter the surface of the microdevices by using external forces (e.g., solvent, adhesive, ultrasonic, laser) and applying pressure to bring two halves together. However, the same driving force that allows the bonding also tends to deform the features' shape. These methods have problems either with blocking the microchannels or with changing their dimensions and are mainly applicable for relatively large microchannels (several hundreds of microns to millimeters).

A microscale bonding and surface modification method has recently been developed in our laboratory [172]. The schematic of this resin-gas assisted bonding technique is shown in Figure 3.24. After a microfluidic platform and its lid are encapsulated in a holder, a photo-curable resin such as hydroxyethyl methacrylate (HEMA) mixed with a surface modification agent is injected into the platform, fillling the micron-sized channels and reservoirs, as well as the gap between the platform and the lid. Nitrogen gas is then pumped in to replace most of the resin inside the channels and reservoirs. A vacuum pump may also be used to replace the gas pressure and the resin inside the microchannel can be removed by applying a vacuum. The remaining resin is cured by UV light. HEMA is relatively hydrophilic, can bond acrylates well, and has been widely used in various medical applications. The resin-gas assisted bonding method can also achieve local surface modification, as shown in Figure 3.25. The experimental procedure is the same, except that prior to the curing of the remaining resin mixture, a mask (e.g., a photomask used in the lithography process) is placed between the lid plate and the UV source. The mask blocks out the UV light locally, so that the resin mixture in the locations under the mask will not cure. After the curing of

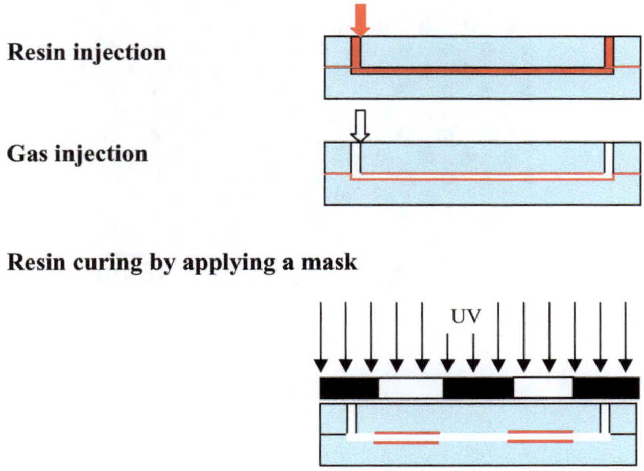

FIGURE 3.25. Schematic of the local surface modification method.

the un-masked parts is complete, the uncured resin mixture is removed by bleaching the surface with water or other solvents similar to the development step in the photolithography process. By adding surface modification agents, the interfacial free energy of the channel wall with the sample solution can be adjusted, and additional functionalities can be added to the channel wall.

At the nanoscale, gases at the supercritical or subcritical conditions can be used for polymer-polymer and polymer-inorganic surface bonding. Figure 3.26 shows polymer layers containing 5 μm sized microwells (Figures 3.26a and b) and 100–200 nm nanochannels (Figure 3.26c and d) before and after being bonded by CO_2 at low temperature and pressure. The micro- and nanostructures, including surface roughness, are well preserved and the bond strength is the same as the bulk material [173]. Although there are other methods to bond nanoscale polymer features [174–181], gas-assisted bonding is the most affordable and probably the only biologically benign and sterile method, in particular for simultaneous assembly of a large number of nanostructures.

3.6. CONCLUSIONS AND FUTURE DIRECTIONS

Figure 3.27 summarizes typical polymer micro/nanofabrication techniques including mold (master) making and molding (replication). Although many methods are available, none is as well-developed as in macrofabrication. Successful commercialization of many BioMEMS/NEMS products, such as disposable biochips/biosensors, drug delivery systems, and engineered particles, depends greatly on the availability of low-cost and high-throughput manufacturing methods.

Appreciable worldwide activity is underway to develop cost-effective methods for manufacturing polymeric microcomponents with well-controlled nanoscale (10^2–10^0 nm) features. Although several methods have recently been developed by combining various "top-down" lithography techniques with conventional manufacturing methods (e.g.,

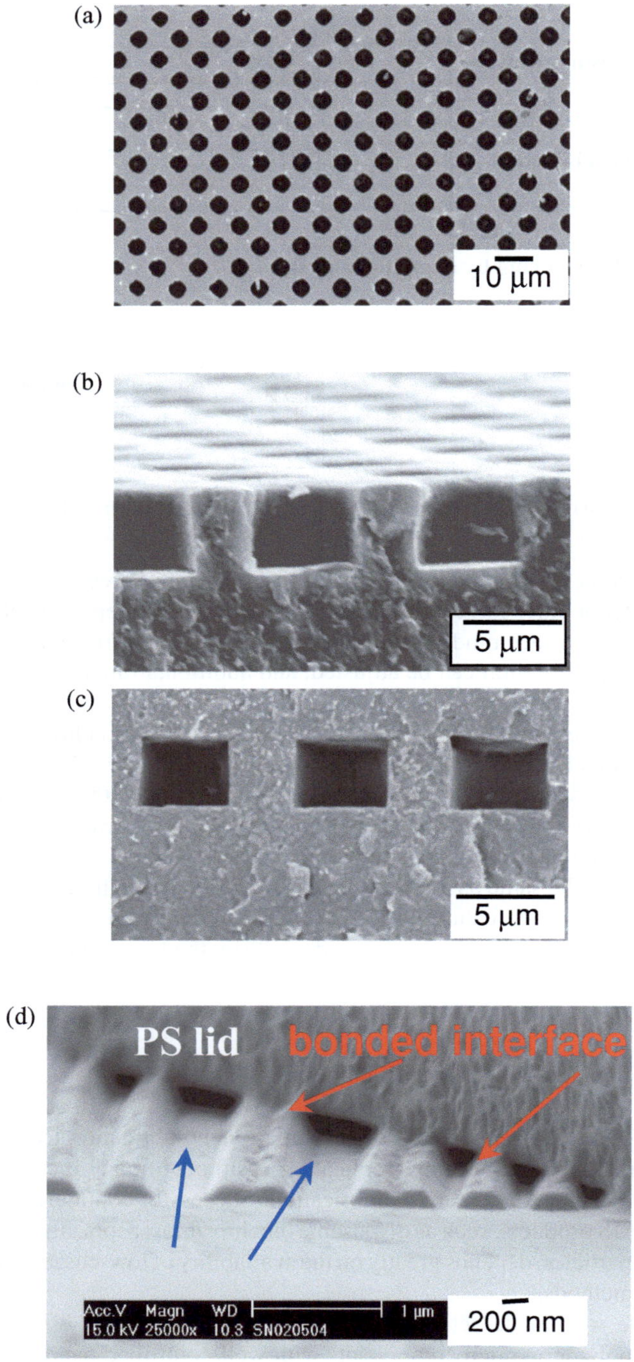

FIGURE 3.26. PLGA layers containing 5 µm sized microwells before CO_2 bonding (a) top view, (b) side view, (c) side view after CO_2 bonding at 35 °C, 100 psi, and (d) a 100–200 nm polystyrene nanochannels after being bonded by CO_2 at 70 °C, 200 psi.

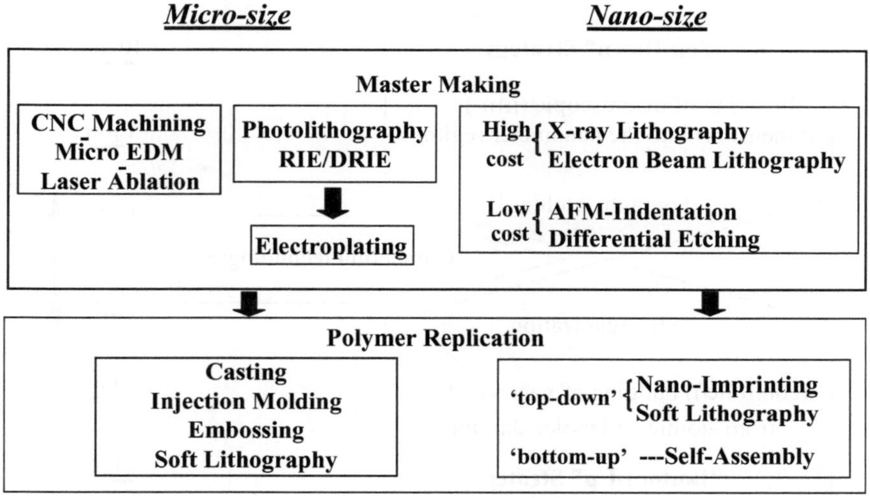

FIGURE 3.27. Polymer-based micro-/nanofabrication techniques.

embossing, silicone rubber casting), mass production of polymeric devices with well-defined nanoscale features and high aspect ratios remains a major challenge. "Bottom-up" self-assembly can be an inexpensive route to highly-ordered nanostructures with simple forms (thin films, particles, or fibers). However, self-assembly alone cannot yet yield ordered nanostructures with the precise, complicated 3D geometries needed for many biomedical devices. It is also difficult to scale-up self-assembled nanostructures without introducing significant defects.

These specific limitations to "top-down" and "bottom-up" processing can be addressed by developing hybrid ("top-down" + "bottom-up") manufacturing protocols for mass-producing precisely-designed, miniature biomedical devices comprised of biocompatible, mechanically-robust, polymer/ceramic composites. In these methods, precise and scalable micro-/nano-patterning of polymers is made by "top-down" processing, while self-assembly of nanostructured polymers or ceramics in a micro-/nano-patterned environment is carried out as the "bottom-up" processing. Such a combination covers a broad range of length scale as shown in Figure 3.28 and should be the best way to effectively mass-produce micro-sized BioMEMS/NEMS devices with nanoscale features and functions. Several approaches are briefly described here.

Non-Cleanroom Forming of 3D Polymer Nanostructures Current nanoscale mold making is based on various lithographical methods. They are limited to silicon, glass and soft metals, and are designed for producing final products or structures. In polymer nanofabrication, molds made of these materials may not be strong enough to survive repeated molding and de-molding cycles. Furthermore, these lithographical methods require expensive clean-room facilities and are often limited to planar 2D structures. On the other hand, tool steel and stainless steel have been proven mold materials in macroscopic polymer processing for many years. The traditional mold making methods can also produce 3D structures. These,

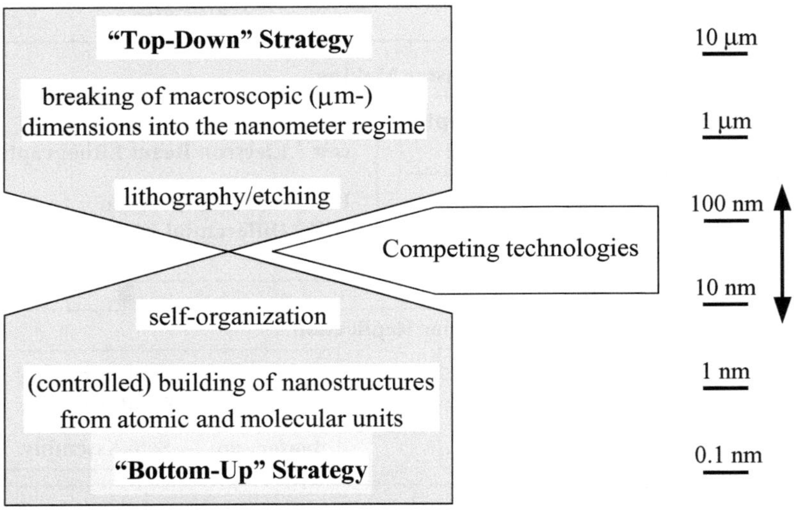

FIGURE 3.28. Fabrication of microdevices with nano-sized structures.

however, are not applicable to sub-micron or nanoscale structures. It would be highly desirable to develop low-cost, non-cleanroom methods that can produce steel molds with 3D nanostructures. This may be achieved by integrating several non-lithographical techniques. For example, microscale templates can be formed by micro-machining methods such as micro-milling, micro-EDM. Pulsed-laser-based ablation can then be used to fine-tune the microscale features into nanoscale structures. Diffraction, thermal diffusion limits feature sizes to, at best, hundreds of nanometers. One may use a femtosecond-pulsed-laser to limit the thermal diffusion lengths. Proximal probe-based optical processing may further reduce feature sizes to the order of tens of nm for the finest 3D tool features not achievable by IC type fabrication technology. This integrated "maskless" nanoscale processing protocol, however, has not yet been developed.

Dynamic Assembling of Polymer Nanostructures Numerous studies of electroosmotic flow (EOF) and electrophoresis (EP) have been carried out but they have seldom been used to control molecular self-assembly. On the other hand, two-dimensional, planar self-assembly has been demonstrated using externally-guided dynamic forces in the micro/nanoscale environment, such as in soft lithography with self-assembled monolayers [182], lithographically induced self-assembly [106], electric field assisted self-assembly of block copolymers [183], and holographic nanopatterning of biocatalytically formed silica [184]. The use of flow dynamics in such processes may lead to a breakthrough for 3D fabrication. Low-cost "top-down" fabrication methods can be used to form macro- and microscale polymer templates containing micro- and nanosized features (e.g., ~100 nm fluidic channels and circuits). Guided formation of smaller nanoscale structures and biofunctionality can be realized via external force-induced dynamic surface reactions. The surfaces of features on the templates can be modified by conductive polymers, surfactants, functional ceramics or biomolecules to make them electrically, magnetically, optically,

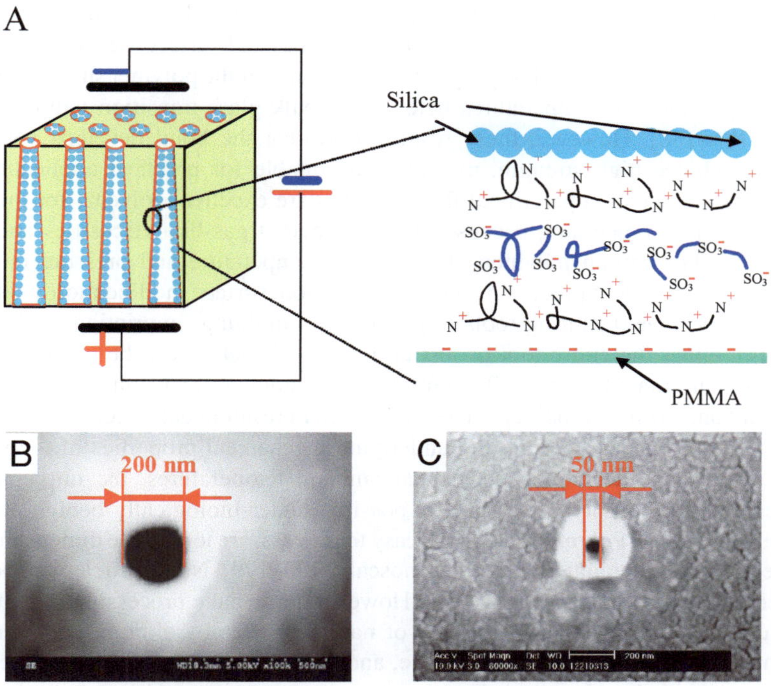

FIGURE 3.29. EOF based dynamic assembly of silica in STI nanonozzle array, (A) process schematic; nozzle size before (B) and after (C) dynamic assembly.

thermally, or biologically "conductive". These molecules or their precursors are driven into the small features by external forces such as EOF and EP. Their deposition can be controlled by the electric double layer, and their reactions by external energy forces such as UV triggered grafting or polymerization. Because of the small size, very high field strengths can easily be generated to control self-assembly and to guide the dispersion, orientation and interconnection of electrolyte molecules and nanoparticles. This is essentially a combined "top-down" and "bottom-up" manufacturing approach enhanced by nanofluidic dynamics. 3D nanostructures as small as several nanometers with well-defined biofunctionality on the surface can be achieved by this low-cost method. It has been demonstrated that a solid surface which is microscopically patterned with bioactive or catalytical molecules helps to "guide" the growth of nanostructured silica [184]. For nanofluidic devices, many nanoscale features we try to generate are in the form of long channels and complicated flow conduits. The self-assembly and reaction inside such channels and conduits will be guided by EOF and EP forces with conductive polymers coated on the channel surface. Our recent result (Figure 3.29) shows a silica nanopore with size around 50 nm formed at the entrance of a 200 nm diameter polymer channel by EOF-guided surface silicification [185].

Supercritical (Subcritical) Fluids-Enhanced Manufacturing of 3D Polymer Nanostructures When polymers are processed at the nanoscale, viscosity and surface tension can become a major limiting factor. Although organic solvents can be used to lower the viscosity, they are detrimental to many biomolecules and surface wetting may still lead to processing difficulties. Based on property depression near the polymer surface, interfacial bonding of polymers at temperatures below their bulk glass transition temperature was achieved [186–188]. However, these results show that the bond develops very slowly and cannot achieve high strength, making it unsuitable for practical applications. Our recent results show that supercritical fluids (SCFs) are excellent solvents for micro- and nanoscale polymer processing because of their inherent gas-like viscosities and surface tension [189]. For certain polymers, CO_2 at low temperature and pressure (subcritical conditions) is an excellent processing aid for fabrication, surface modification and bonding. For micro-/nano-manufacturing tools (e.g., injection molding, imprinting, and assembly or bonding), either the entire tool or just an individual section can be sealed to conduct gas-assisted nano-manufacturing. By controlling the amount of CO_2 through pressure, the degree of enhanced chain mobility can be controlled to result in entanglement or patterning of surface chains without altering the underlying mechanical properties of the substrate. Because this can be accomplished at near ambient temperatures, the impregnation of biomolecules into the polymer surface is possible under biologically benign and sterile conditions [190–192]. Polymers, although easy to process, are less stable dimensionally and mechanically than other materials at the nanoscale [193–199]. Nanoparticles can be used as reinforcements for polymeric composites. However, they make processing difficult at low temperatures. CO_2 may assist processing of nanostructures by altering polymer/particle interactions, lubricating the particle surface, and carrying additives into the structure. We have demonstrated the viscosity of nanocomposites containing CO_2 can be lower than pure polymer containing CO_2 [199], rendering better low temperature processability. The nanocomposites may provide better integrity to polymer nanostructures when CO_2 is removed after processing. Such CO_2-tunable polymer nanocomposites will play an important role in developing biologically benign materials and fabrication processes.

Micro-/nanofabrication of polymers for BioMEMS/NEMS applications opens a new paradigm for researchers in the chemical and polymer engineering field. Traditional principles of transport phenomena, polymer processing, and biology are still applicable in this area. The integration and interactions among them, however, bring up many challenges. For example, optically clear polymers have been processed and widely used in many applications for a very long time. Their mechanical and optic properties have also been investigated in detail. Yet, the processibility of these materials in micro- and nano-scale, and the resulting mechanical, optic, and biological properties remains largely unknown. Many biofunctional polymers, such as biodegradable polymers and functional hydrogels, have been used in many small-quantity biomedical applications; and dendrimers and conductive polymers may find BioMEMS/NEMS applications in the near future. The behavior of these materials (e.g., rheology, kinetics) in large quantity processing and BioMEMS/NEMS applications is still poorly understood. Micro-/nanofluidics has gained a great deal of interest in recent years. Most experimental work, however, is based on silicon and glass materials. Since polymers will be the primary material for future BioMEMS/NEMS applications, we need a better understanding of the micro- and nano-scale transport phenomena for microdevices made of polymers. Biocompatibility is an essential concern for materials used in BioMEMS/NEMS applications. Since the interaction between the biomolecules and micro/nano-devices occurs mainly on the solid surface, surface characterization and modification play critical roles.

They are also important in micro-/nanofluidics because surface forces and the manipulation of such forces are the most powerful way to control the flow. Although surface modification for biocompatibility has been addressed by many researchers [200, 201] and several surface characterization tools are available, their applicability to low-cost mass-production of BioMEMS/NEMS devices needs to be further studied.

REFERENCES

[1] M.R. Snyder. *Mod. Plast.*, 85, January 1999.

[2] H. Becker and C. Gartner. Polymer microfabrication methods for microfluidic analytical applications. *Electrophoresis*, 21(1):12–26, 2000.

[3] L.J. Lee. BioMEMS and micro-/nano-processing of polymers-an overview. *J. Chinese Inst. Chem. Eng.*, 34(1):25–46, 2003.

[4] R. Langer. Perspectives: Drug Delivery-Drug on Target. *Science*, (293):58–59, 2001.

[5] R. Panchagnula. Pharmaceutical Aspects of Paclitaxel. *Int. J. Pham.*, 172:1–15, 1998.

[6] R. Langer. Drug Delivery and Targeting. *Nature*, 392(6679 Suppl):5–10, 1998.

[7] R. Duncan. The dawning era of polymer therapeutics. *Nat. rev. Drug discov.*, 2(5):347–360, 2003.

[8] R. Langer and N.A. Peppas. Advances in biomaterials, drug delivery, and bionanotechnology. *AIChE J.*, 49(12):2990–3006, 2003.

[9] *Technical Insight*. San Antonio, TX, 2002.

[10] C.K. Colton. Implantable Biohybrid Artificial Organs. *Cell Transplant.*, 4(4):415, 1995.

[11] K.J. Kim and P.V. Stevens. Hydraulic and Surface Characteristics of Membranes with Parallel Cylindrical Pores. *J. Memb. Sci.*, 123(2):303–314, 1997.

[12] T.A. Desai, D.J. Hansford, L. Kulinsky, A.H. Nashat, G. Rasi, J. Tu, Y. Wang, M. Zhang, and M. Ferrari. Nanopore Technology for Biomedical Applications. *Biomed. Microdevuic.*, 2(2):11, 2000.

[13] T.A. Desai, D.J. Hansford, and M. Ferrari. Characterization of Micromachined Silicon Membranes for Immunoisolation and Bioseparation Applications. *J. Membrane Sci.*, 159:221, 1999.

[14] A.T. Conlisk, J. McFerran, Z. Zheng, and D.J. Hansford. Mass Transfer and Flow in Electrically Charged Micro- and Nano-channels. *Anal. Chemi.*, 74(9):2139–2150, 2002.

[15] R. Langer and J.P. Vacanti, Tissue Engineering. *Science*, 260(5110):920–926, 1993.

[16] D. Mooney and A.G. Mikos. *Growing New Organs*. Scientific American.com, 1999.

[17] M.J. Madou. *Fundamentals of Microfabrication: The Science of Miniaturization*. 2nd ed. New York: CRC Press LLC, 2002.

[18] Y. Ikada. *Tissue Engineering under Clinical Trials*. In *International Symposium on Fusion of Nano and Bio Technologies*. Tsukuba International Congress Center, Tsukuba, Japan, 2003.

[19] M.J. Madou, L.J. Lee, S. Daunert, K.W. Koelling, S. Lai, and C.-H. Shih. Design and Fabrication of CD-like Microfluidic Platforms for Diagnostics, Part I: Microfluidic Functions. *Biomedi. Microdevic.*, 3(3):245–254, 2001.

[20] L.J. Lee, M.J. Madou, K.W. Koelling, S. Daunert, S. Lai, C.G. Koh, Y.-J. Juang, Y. Lu, and L. Yu. Design and Fabrication of CD-like Microfluidic Platforms for Diagnostics, Part II: Polymer-Based Microfabrication. *Biomedi. Microdevic.*, 3(4):339–351, 2001.

[21] S. Lai, S. Wang, J. Luo, L.J. Lee, S.-T. Yang, and J. Madou Marc. Design of a compact disk-like microfluidic platform for enzyme-linked immunosorbent assay. *Anal. Chemi.*, 76(7):1832–1837, 2004.

[22] M. Nakamura, K. Decker, J. Chosy, K. Comella, K. Melnik, L. Moore, M. Zborowski, and J.J. Chalmers. Separation of Breast Cancer Cells by Quadrupole Magnetic Flow Sorter. *Biotech. Prog.*, 17:1145–1155, 2001.

[23] J.P. Vacanti and R. Langer. Tissue engineering: the design and fabrication of living replacement devices for surgical reconstruction and transplantation. *Lancet*, 354 Suppl 1:SI32–SI34, 1999.

[24] A.G. Mikos. Laminated Three-dimensional Biodegradable Foams for Use in Tissue Engineering. *Biomaterials*, 14(5):323, 1993.

[25] Y. Li and S.T. Yang. Effects of three-dimensional scaffolds on cell organization and tissue development. *Biotechnol. Bioprocess Eng.*, 6:311–325, 2001.

[26] T.A. Desai. Micro- and nanoscale structures for tissue engineering constructs. *Med. Eng. Phy.*, 22(9):595–606, 2000.
[27] L.G. Griffith and G. Noughton. Tissue engineering–current challenges and expanding opportunities. *Science*, 295(5557):1009, 2002.
[28] K.R. King, C.C. Wang, J.P. Vacanti, and J.T. Borenstein. Biodegradable Polymer Microfluidics for Tissue Engineering Microvasculature. *MRS Spring Symposium*, 2002.
[29] M. Gad-el-Hak. The fluid mechanics of microdevices—freeman scholar lecture. *J. Fluids Eng.*, 121:5, 1999.
[30] L. Yu, C.G. Koh, L.J. Lee, K.W. Koelling, and M.J. Madou. Experimental Investigation and Numerical Simulation of Injection Molding with Micro-Features. *Poly. Eng. Sci.*, 42(5):871–888, 2002.
[31] R.O. Warrington, in *Proceedings of Novel Microfabrication Options for BioMEMS Conference*. July 1999.
[32] C. Momma, S. Nolte, B.N. Chichkov, F. von Alvensleben, and A. Tuennermann. Precise laser ablation with ultrashort pulses. *Appl. Surf. Sci.*, 109/110(Laser Processing of Surfaces and Thin Films):15–19, 1997.
[33] M.A. Roberts, J.S. Rossier, P. Bercier, and H. Girault. UV Laser Machined Polymer Substrates for the Development of Microdiagnostic Systems. *Anal. Chemi.*, 69(11):2035–2042, 1997.
[34] M.J. Schifferer and A. Choi. Direct Materials Deposition: Designed Macro and Microstructure. *J. Mat. Res. Inno.*, 3(3):118, 1999.
[35] D. Hu, H. Mei, and R. Kovacevic. Improving Solid freeform Fabrication by Laser-based Additive Manufacturing. *P I Mech Eng B-J Eng*, 216(9):1253, 2002.
[36] H.B. Sun and S. Kawata. Two-photon Laser Precision Microfabrication and its Applicaitions to Micro-nano Devices and Systems. *J. Lightware Technol.*, 21(3):624, 2003.
[37] P.P. Pronko, S.K. Dutta, J. Squier, J.V. Rudd, D. Du, and G. Mourou. Machining of sub-micron holes using a femtosecond laser at 800 nm. *Optics Commun.*, 114(1,2):106–110, 1995.
[38] F. Korte, J. Serbin, J. Koch, A. Egbert, C. Fallnich, A. Ostendorf, and B.N. Chichkov. Towards nanostructuring with femtosecond laser pulses. *App. Phy. A: Mat. Sci. Process.*, 77(2):229–235, 2003.
[39] S.C. B. N. Nolte, H. Welling, Y. Shani, K. Lieberman, and H. Terkel. *Opt. Lett.*, 24:914, 1999.
[40] Farson, D.
[41] K. Kaneko, H.-B. Sun, X.-M. Duan, and S. Kawata. Two-photon photoreduction of metallic nanoparticle gratings in a polymer matrix. *Appl. Phy. Lett.*, 83(7):1426–1428, 2003.
[42] S. Kawata, H.-B. Sun, T. Tianaka, and K. Takada. Finer features for functional microdevices. *Nature*, 412(6848):697–698, 2001.
[43] H.-B. Sun, K. Takada, M.-S. Kim, K.-S. Lee, and S. Kawata. Scaling laws of voxels in two-photon photopolymerization nanofabrication. *Appl. Phy. Lett.*, 83(6):1104–1106, 2003.
[44] R.D. Piner, J. Zhu, F. Xu, S. Hong, and C.A. Mirkin, "Dip-pen" nanolithography. *Science*, 283(5402):661–663, 1999.
[45] W. Ehrfeld and H. Lehr. Deep X-ray lithography for the production of three-dimensional microstructures from metals, polymers and ceramics. *Rad. Phy. Chem.*, 45(3):349–365, 1995.
[46] M.S.K. Despa, W. Collier, Kevin, and R. John. *Microsys. Technol.*, 6:60, 1999.
[47] W. Ehrfeld, W. Glashauser, D. Muenchmeyer, and W. Schelb. Mask making for synchrotron radiation lithography. *Microelect. Eng.*, 5(1–4):463–470, 1986.
[48] R.A. Lawes. Submicron lithography techniques. *Appl. Surf. Sci.*, 36(1–4):485–499, 1989.
[49] J. Mohr, W. Ehrfeld, and D. Muenchmeyer, Requirements on resist layers in deep-etch synchrotron radiation lithography. *J. Vac. Sci. Technol., B: Microelect. Nanomet. Struct.*, 6(6):2264–2247, 1988.
[50] G.O. Mallory and J.B. Hadju. *American Electroplaters and Surface Finishers Society (AESF)*. Orlando, FL, 1990.
[51] C. Vieu, F. Carcenac, A. Pepin, Y. Chen, M. Mejias, A. Lebib, L. Manin-Ferlazzo, L. Couraud, and H. Launois. Electron beam lithography: resolution limits and applications. *Appl. Surf. Sci.*, 164:111–117, 2000.
[52] G.Y. Liu, S. Xu, and Y. Qian. Nanofabrication of self-assembled monolayers using scanning probe lithography. *Acc. Chem. Res.*, 33(7):457–466, 2000.
[53] U. Kunze and B. Klehn. Plowing on the sub-50 nm scale. Nanolithography using scanning force microscopy. *Adv. Mat.*, 11(17):1473–1475, 1999.
[54] H. Sugimura and N. Nakagiri. Scanning probe anodization: nanolithography using thin films of anodically oxidizable materials as resists. *J. Vacuum Sci. Tech., A: Vacuum, Surf., and Films*, 14(3, Pt. 1):1223–1227, 1996.

[55] B. Legrand and D. Stievenard. Nanooxidation of silicon with an atomic force microscope: A pulsed voltage technique. *Appl. Phys. Lett.*, 74(26):4049–4051, 1999.
[56] H. Sugimura, O. Takai, and N. Nakagiri. Multilayer resist films applicable to nanopatterning of insulating substrates based on current-injecting scanning probe lithography. *J. Vacuum Sci. Technol., B: Microelectron. Nanometer Struct.*, 17(4):1605–1608, 1999.
[57] H. Sugimura and N. Nakagiri. Electrochemical nanolithography using scanning probe microscopy: fabrication of patterned metal structures on silicon substrates. *Thin Solid Films*, 281–282(1–2):572–575, 1996.
[58] F.P. Zamborini and R.M. Crooks. Nanometer-scale patterning of metals by electrodeposition from an STM tip in air. *J. Amer. Chem. Soc.*, 120(37):9700–9701, 1998.
[59] C.B. Gorman, R.L. Carroll, Y. He, F. Tian, and R. Fuierer. Chemically Well-Defined Lithography Using Self-Assembled Monolayers and Scanning Tunneling Microscopy in Nonpolar Organothiol Solutions. *Langmuir*, 16(15):6312–6316, 2000.
[60] R. Maoz, E. Frydman, S.R. Cohen, and J. Sagiv. Constructive nanolithography: site-defined silver self-assembly on nanoelectrochemically patterned monolayer templates. *Adv. Mat.*, 12(6):424–429, 2000.
[61] R. Maoz, E. Frydman, S.R. Cohen, and J. Sagiv. "Constructive nanolithography": inert monolayers as patternable templates for in-situ nanofabrication of metal-semiconductor-organic surface structures-a generic approach. *Adv. Mat.*, 12(10):725–731, 2000.
[62] D.M. Kolb, R. Ullmann, and T. Will. Nanofabrication of small copper clusters on gold(111) electrodes by a scanning tunneling microscope. *Science*, 275(5303):1097–1099, 1997.
[63] F. Forouzan and A.J. Bard. Evidence for Faradaic Processes in Scanning Probe Microscopy on Mica in Humid Air. *J. Phy. Chem. B*, 101(50):10876–10879, 1997.
[64] D. Hofmann, W. Schindler, and J. Kirschner. Electrodeposition of nanoscale magnetic structures. *Appl. Phys. Lett.*, 73(22):3279–3281, 1998.
[65] R.J. Randler, D.M. Kolb, B.M. Ocko, and I.K. Robinson. Electrochemical copper deposition on Au(100): a combined in situ STM and in situ surface X-ray diffraction study. *Surf. Sci.*, 447(1–3):187–200, 2000.
[66] W. Schindler, D. Hofmann, and J. Kirschner. Nanoscale electrodeposition: A new route to magnetic nanostructures? *J. Appl. Phys.*, 87(9, Pt. 3):7007–7009, 2000.
[67] R. Yang, D.F. Evans, and W.A. Hendrickson. Writing and Reading at Nanoscale with a Scanning Tunneling Microscope. *Langmuir*, 11(1):211–213, 1995.
[68] C.Z. Li and N.J. Tao. Quantum transport in metallic nanowires fabricated by electrochemical deposition/dissolution. *Appl. Phys. Lett.*, 72(8):894–896, 1998.
[69] A. Ivanisevic and C.A. Mirkin. "Dip-Pen" Nanolithography on Semiconductor Surfaces. *J. Amer. Chem. Soc.*, 123(32):7887–7889, 2001.
[70] B.W. Maynor, S.F. Filocamo, M.W. Grinstaff, and J. Liu. Direct-Writing of Polymer Nanostructures: Poly(thiophene) Nanowires on Semiconducting and Insulating Surfaces. *J. Amer. Chem. Soc.*, 124(4):522–523, 2002.
[71] G.M. Whitesides, E. Ostuni, S. Takayama, X. Jiang, and D.E. Ingber. Soft lithography in biology and biochemistry. *Ann. Rev. Biomed. Eng.*, 3:335–373, 2001.
[72] R. Ruprecht, W. Bacher, J.H. Hausselt, and V. Piotter. Injection Molding of LIGA and LIGA-Similar Microstructures Using Filled and Unfilled Thermoplastics. *Proceedings of SPIE-The International Society for Optical Engineering*, (Micromachining and Microfabrication Process Technology): Vol. 2639, pp. 146–157, 1995.
[73] P. Hagmann and W. Ehrfeld. Fabrication of microstructures of extreme structural heights by reaction injection molding. *Internat. Poly. Process.*, 4(3):188–195, 1989.
[74] T. Hanemann, R. Ruprecht, and J.H. Hausselt. *Photomolding in microsystem technology*. Polymer Preprints (American Chemical Society, Division of Polymer Chemistry), 39(2):657–658, 1998.
[75] D. Qin, Y. Xia, J.A. Rogers, R.J. Jackman, X.-M. Zhao, and G.M. Whitesides. Microfabrication, microstructures and microsystems. *Topics Curr. Chem.*, 194(Microsystem Technology in Chemistry and Life Science):1–20, 1998.
[76] Y. Xia, E. Kim, X.-M. Zhao, J.A. Rogers, M. Prentiss, and G.M. Whitesides. Complex optical surfaces formed by replica molding against elastomeric masters. *Science*, 273(5273):347–349, 1996.
[77] Y. Xia, J.J. McClelland, R. Gupta, D. Qin, X.M. Zhao, L.L. Sohn, R.J. Celotta, and G.M. Whitesides . Replica molding using polymeric materials. A practical step toward nanomanufacturing. *Adv. Mat.*, 9(2):147–149, 1997.
[78] A. Kumar, N.L. Abbott, H.A. Biebuyck, E. Kim, and G.M. Whitesides. Patterned Self-Assembled Monolayers and Meso-Scale Phenomena. *Acco. Chem. Res.*, 28(5):219–226, 1995.

[79] E. Kim, Y. Xia, and G.M. Whitesides. Polymer microstructures formed by molding in capillaries. *Nature*, 376(6541):581–584,1995.

[80] C.S. Effenhauser, G.J.M. Bruin, A. Paulus, and M. Ehrat. Integrated Capillary Electrophoresis on Flexible Silicone Microdevices: Analysis of DNA Restriction Fragments and Detection of Single DNA Molecules on Microchips. *Anal. Chem.*, 69(17):3451–3457, 1997.

[81] D.C. Duffy, J.C. McDonald, O.J.A. Schueller, and G.M. Whitesides. Rapid Prototyping of Microfluidic Systems in Poly(dimethylsiloxane). *Anal. Chem.*, 70(23):4974–4984, 1998.

[82] M. Rode and B. Hillerich. Self Aligned Positioning of Microoptical components by Precision Prismatic Grooves Impressed into Metals. *IEEE J. Microelectromech. Sys.*, 8(1): March 1999.

[83] L.E. Locascio, C.E. Perso, and C.S. Lee. Measurement of electroosmotic flow in plastic imprinted microfluid devices and the effect of protein adsorption on flow rate. *J. Chromatograp.*, 857(1–2):275–284, 1999.

[84] L.W. Pan, L., W. L., and J. Ni. Cylindrical plastic lens array fabricated by a micro intrusion process. *IEEE*, 217, 1999.

[85] Y. Hirai, S. Harada, H. Kikuta, Y. Tanaka, M. Okano, S. Isaka, and M. Kobayasi. Imprint lithography for curved cross-sectional structure using replicated Ni mold. *J. Vacuum Sci. Tech., B: Microelectron. Nanomet. Struct.*, 20(6):2867–2871, 2002.

[86] G.B. Lee, S.H. Chen, G.R. Huang, Y.H. Lin, and W.C. Sun. Microfabricated plastic chips by hot embossing methods and their application for DNA separation detection. *SPIE*, 112:4117, 2000.

[87] S.Y. Chou, C. Keimel, and J. Gu. Ultrafast and direct imprint of nanostructures in silicon. *Nature*, 417(6891):835–837, 2002.

[88] L. Lin, C.-J. Chiu, W. Bacher, and M. Heckele. Microfabrication Using Silicon Mold Inserts and Hot Embossing. *Proceedings of the International Symposium on Micro Machine and Human Science, 7th, Nagoya*, pp. 67–71, Oct. 2–4, 1996

[89] H. Becker and U. Heim. Silicon as Tool Material for Polymer Hot Embossing. *IEEE International Conference on Micro Electro Mechanical Systems, Technical Digest*, 12th, Orlando, Fla., pp. 228–231, Jan. 17–21, 1999.

[90] G.A. Cirino, A.C. Arruda, R.D. Mansano, P. Verdonck, and L.G. Neto. Fabrication of PMMA Microlenses Using a Micromachined Silicon Mould. *Proceedings—Electrochemical Society*, (Microelectronics Technology and Devices): vol. 8 pp. 188–195, 2002.

[91] B.G. Casey, D.R.S. Cumming, I.I. Khandaker, A.S.G. Curtis, and C.D.W. Wilkinson. Nanoscale embossing of polymers using a thermoplastic die. *Microelectro. Eng.*, 46(1–4):125–128, 1999.

[92] R. Xing, Z. Wang, and Y. Han. Embossing of polymers using a thermosetting polymer mold made by soft lithography. *J. Vacuum Sci. Technol., B: Microelectron.Nanomet. Struct.—Process., Measure., Pheno.*, 21(4):1318–1322, 2003.

[93] H. Becker, U. Heim, and O. Roetting. Fabrication of Polymer High-Aspect-Ratio Structures with Hot Embossing for Microfluidic Applications. *Proceedings of SPIE-The International Society for Optical Engineering* (Microfluidic Devices and Systems II), vol. 3877, pp. 74–79, 1999.

[94] R.W. Jaszewski, H. Schift, P. Groening, and G. Margaritondo. Properties of thin anti-adhesive films used for the replication of microstructures in polymers. *Microelectr. Eng.*, 35(1–4, Micro- and Nano-Engineering 96):381–384, 1997.

[95] R.W. Jaszewski, H. Schift, and B. Schnyder, A. Schneuwly, and P. Groning. The deposition of anti-adhesive ultra-thin Teflon-like films and their interaction with polymers during hot embossing. *Appl. Surf. Sci.*, 143(1–4):301–308, 1999.

[96] M. Heckele, W. Bacher, and K.D. Muller. Hot embossing-the molding technique for plastic microstructures. *Microsys. Technol.*, 4:122, 1998.

[97] H. Becker and U. Heim. Hot embossing as a method for the fabrication of polymer high aspect ratio structures. *Sens. Actu., A: Phys.*, A83(1–3):130–135, 2000.

[98] S.Y. Chou, P.R. Krauss, and P.J. Renstrom. Imprint lithography with 25-nanometer resolution. *Science*, 272(5258):85–87, 1996.

[99] S.Y. Chou, P.R. Krauss, and P.J. Renstrom. Nanoimprint lithography. *J. Vacuum Sci. Technol., B: Microelect. Nanomet. Struct.*, 14(6):4129–4133, 1996.

[100] S.Y. Chou, P.R. Krauss, W. Zhang, L. Guo, and L. Zhuang. Sub-10 nm Imprint Lithography and Applications. *J. Vac. Sci. Technol. B*, 15(6):2897–2904, 1997.

[101] J. Wang, X. Sun, L. Chen, and S.Y. Chou. Direct nanoimprint of submicron organic light-emitting structures. *Appl. Phys. Lett.*, 75(18):2767–2769, 1999.

[102] H. Tan, A. Gilbertson, and S.Y. Chou, Roller nanoimprint lithography. *J. Vacuum Sci. Technol., B: Microelect. Nanomet. Struct.*, 16(6):3926–3928, 1998.

[103] R.W. Jaszewski, H. Schift, J. Gobrecht, and P. Smith. Hot embossing in polymers as a direct way to pattern resist. *Microelect. Eng.*, 41/42:575–578, 1998.

[104] H. Schift, R.W. Jaszewski, C. David, and J. Gobrecht. Nanostructuring of polymers and fabrication of interdigitated electrodes by hot embossing lithography. *Microelect. Eng.*, 46(1–4):121–124, 1999.

[105] B.G. Casey, W. Monaghan, and C.D.W. Wilkinson. Embossing of nanoscale features and environments. *Microelect. Eng.*, 35(1–4, Micro- and Nano-Engineering 96):393–396, 1997.

[106] S.Y. Chou. Nanoimprint lithography and lithographically induced self-assembly. *MRS Bulletin*, 26(7):512–517, 2001.

[107] B. Heidari, I. Maximov, and L. Montelius, Nanoimprint lithography at the 6 in. wafer scale. *J. Vacuum Sci. Technol., B: Microelec. Nanomet. Struc.*, 18(6):3557–3560, 2000.

[108] S. Wang, Changchun Zeng, Lai, Siyi, Yi-Je Juang, and Lames J. Lee, submitted to *Adv. Mat.*, 2004.

[109] T.H. Dam and P. Pantano, P. Nanotip array photoimprint lithography. *Rev. Sci. Inst.*, 70(10):3982–3986, 1999.

[110] T. Pangaribuan, K. Yamada, S. Jiang, H. Ohsawa, and M. Ohtsu. Reproducible fabrication technique of nonametric tip diameter fiber probe for photon scanning tunneling microscope. *Jap. J. Appl. Phys., Part 2: Lett.*, 31(9A):L1302–L1304, 1992.

[111] P. Pantano and D.R. Walt. Toward a near-field optical array. *Rev. Sci. Instru.*, 68(3):1357–1359, 1997.

[112] P. Pantano and D.R. Walt. Ordered Nanowell Arrays. *Chem. Mat.*, 8(12):2832–2835, 1996.

[113] W.P. Ambrose, P.M. Goodwin, J.C. Martin, and R.A. Keller. Alterations of single molecule fluorescence lifetimes in near-field optical microscopy. *Science*, 265(5170):364–367, 1994.

[114] B.G. Healey, S.E. Foran, and D.R. Walt. Photodeposition of micrometer-scale polymer patterns on optical imaging fibers. *Science*, 269(5227):1078—1080, 1995.

[115] S. Koeppel, H. Schift, M. Gabriel, and W. Kaiser. Nanostructures. Injection molding for small dimensions. *Kunststoffe-Syn.*, (2):11–14, 1999.

[116] H. Schift, C. David, M. Gabriel, J. Gobrecht, L.J. Heyderman, W. Kaiser, S. Koppel, and L. Scandella. Nanoreplication in polymers using hot embossing and injection molding. *Microelect. Eng.*, 53(1–4):171–174, 2000.

[117] A. D'Amore, D. Simoneta, W. Kaiser, H. Schift, and M. Gabriel. Injection molding in nano range. *Kunststoffe*, 90(6):52–55, 2000.

[118] N. Gadegaard, N.B. Larsen, and S. Mosler. Injection molding nanostructures. *Condensed Matter Phys. Chem. Dep. Riso natl. Lab.*, p. 66, 2000.

[119] *http://www.opticaldisc-systems.com/July-Aug99/47.htm.*

[120] R.L. Fleischer, P.B. Price, and R.M. Walker. Nuclear Tracks in Solids: Principles and Applications. p. 605, 1975.

[121] D.H. Reneker and I. Chun. Nanometer diameter fibers of polymer, produced by electrospinning. *Nanotechnology*, 7(3):216–223, 1996.

[122] L. Larrondo and R. St. John Manley. Electrostatic fiber spinning from polymer melts. I. Experimental observations on fiber formation and properties. *J. Poly. Sci., Poly. Phys. Ed.*, 19(6):909–920, 1981.

[123] J. Zeleny. *Phys. Rev.*, 3:69–91, 1914.

[124] A. Formhals. *Process and apparatus for preparing artificial threads*. 1934: US Patent No. 1,975,504.

[125] G.I. Taylor. *Proc. R. Soc. Lond. A*, 280:383–397, 1964.

[126] A. Frenot and I.S. Chronakis. Polymer nanofibers assembled by electrospinning. *Curr. Opin. Coll. Interface Sci.*, 8(1):64–75, 2003.

[127] G.L. Bowling, M.J., D.G. Simpson, E.R. Kenawy, and G.E. Wnek. Electrospinning Biomaterials. *J Textile Apparel, Technol. Manage. (Special issue: The Fiber Society Spring 2001 Conference, Raleigh NC)*, vol. 1, 2001.

[128] D., Z.X. Fang, W. Chen, S. Cruz, B. Hsiao, and B. Chu. Nano-Structured Electrospun poly-D,L-Lactide-co-glycolide Membranes for Anti-Adhesion Applications. *J Textile Apparel, Technol Manage, (Special issue: The Fiber Society Spring 2001 Conference, Raleigh NC)*, vol. 1, 2001.

[129] W., D.S. Kataphinan, D. Smith, and D.H. Reneker. Fabrication of electrospun and encapsulation into polymer nanofibers. *J Textile Apparel, Technol Manage, (Special issue: The Fiber Society Spring 2001 Conference, Raleigh NC)*, vol. 1, 2001.

[130] D. Smith, D. Reneker, W. Kataphinan, and S. Dabney. *Electrospun skin masks and uses thereof, in PCT Int. Appl.* University of Akron, USA. Wo. p. 14, 2001.

[131] D. Smith, D. Reneker, A. McManus, H. Schreuder-Gibson, C. Mello, M. Sennett, and P. Gibson. *Electrospun fibers and an apparatus therefor*, in *PCT Int. Appl.* University of Akron, USA. Wo. p. 44, 2001.

[132] E.-R. Kenawy, G.L. Bowlin, K. Mansfield, J. Layman, D.G. Simpson, E.H. Sanders, and G.E. Wnek. Release of tetracycline hydrochloride from electrospun poly(ethylene-co-vinylacetate), poly(lactic acid), and a blend. *J. Control. Rel.*, 81(1–2):57–64, 2002.

[133] X. Zong, K. Kim, D. Fang, S. Ran, B.S. Hsiao, and B. Chu. Structure and process relationship of electrospun bioabsorbable nanofiber membranes. *Polymer*, 43(16):4403–4412, 2002.

[134] W.-J. Li, C.T. Laurencin, E.J. Caterson, R.S. Tuan, and F.K. Ko. Electrospun nanofibrous structure: A novel scaffold for tissue engineering. *J. Biomed. Mat. Res.*, 60(4):613–621, 2002.

[135] I.D. Norris, M.M. Shaker, F.K. Ko, and A.G. MacDiarmid. Electrostatic fabrication of ultrafine conducting fibers: polyaniline/polyethylene oxide blends. *Syn. Met.*, 114(2):109–114, 2000.

[136] G.I. Taylor. *Proc. R. Soc. Lond. A*, 31:453–475, 1969.

[137] D., S.K. Ziegler, C. Dew, and L. Samuelson. Electrospun fibrous membranes of photovoltaic and conductive polymers. *J Textile Apparel, Technol Manage* (*Special issue: The Fiber Society Spring 2001 Conference, Raleigh NC*), vol. 1, 2001.

[138] Y, D.H. Zhang, I.D. Norris, A.G. MacDiarmid, and W.E. Jones. *High surface area chemosensor material by electrospinning of fluorescent conjugated polymer*. In *Abstracts of papers of the American Chemical Society, 222(2) August*. 2001.

[139] J.A. Matthews, G.E. Wnek, D.G. Simpson, and G.L. Bowlin. Electrospinning of Collagen Nanofibers. *Biomacromolecules*, 3(2):232–238, 2002.

[140] L. Huang, K. Nagapudi, R.P. Apkarian, and E.L. Chaikof. Engineered collagen-PEO nanofibers and fabrics. *J. Biomat. Sci.* Polymer edition, 12(9):979–993, 2001.

[141] C.J. Buchko, L.C. Chen, Y. Shen, and D.C. Martin, Processing and microstructural characterization of porous biocompatible protein polymer thin films. *Polymer*, 40(26):7397–7407, 1999.

[142] L. Huang, R.A. McMillan, R.P. Apkarian, B. Pourdeyhimi, V.P. Conticello, and E.L. Chaikof. Generation of Synthetic Elastin-Mimetic Small Diameter Fibers and Fiber Networks. *Macromolecules*, 33(8):2989–2997, 2000.

[143] P.P. Tsai, S.-G.H., Effect of Electrospinning Material and Conditions upon Residual Electrostatic Charge of Polymer Nanofibers. *J Textile Apparel, Technol Manage*, *Special issueSpecial issue: The Fiber Society Spring 2001 Conference, Raleigh NC*, vol. 1, 2001.

[144] J. Doshi. *Nanofiber-based nonwoven composites: properties and applications*. Nonwovens World, 64–68, 2001(augusti-september).

[145] J.-S. Kim and D.H. Reneker. Mechanical properties of composites using ultrafine electrospun fibers. *Poly. Comp.*, 20(1):124–131, 1999.

[146] M. Bognitzki, H. Hou, M. Ishaque, T. Frese, M. Hellwig, C. Schwarte, A. Schaper, J.H. Wendorff, and A. Greiner. Polymer, metal, and hybrid nano—and mesotubes by coating degradable polymer template fibers (TUFT process). *Adv. Mat.* (Weinheim, Germany), 12(9):637–640, 2000.

[147] D.A. Czaplewski, J. Kameoka, R. Mathers, G.W. Coates, and H.G. Craighead. Nanofluidic channels with elliptical cross sections formed using a nonlithographic process. *Appl. Phys. Lett.*, 83(23):4836–4838, 2003.

[148] H.M. Powell and J.J. Lannutti. Nanofibrillar Surfaces via Reactive Ion Etching. *Langmuir*, 19(21):9071–9078, 2003.

[149] C.R. Martin. Nanomaterials: a membrane-based synthetic approach. *Science*, 266(5193):1961–1966, 1994.

[150] C.R. Martin. Template Synthesis of Electronically Conductive Polymer Nanostructures. *Acco. Chem. Res.*, 28(2):61–68, 1995.

[151] L. Zhang and M. Wan. Synthesis and characterization of self-assembled polyaniline nanotubes doped with D-10-camphorsulfonic acid. *Nanotechnology*, 13(6):750–755, 2002.

[152] W.S. Huang, B.D. Humphrey, and A.G. MacDiarmid. Polyaniline, a novel conducting polymer. Morphology and chemistry of its oxidation and reduction in aqueous electrolytes. *J. Chem. Soc., Faraday Transact. 1: Phys. Chem. Cond. Phases*, 82(8):2385–2400, 1986.

[153] N.-R. Chiou. Controlled Synthesis of Polyaniline Nanofibers. In *ACS meeting*. 2004.

[154] J.C. Chiang and A.G. MacDiarmid. 'Polyaniline': protonic acid doping of the emeraldine form to the metallic regime. *Syn. Met.*, 13(1–3):193–205, 1986.

[155] A.G. MacDiarmid, J.C. Chiang, A.F. Richter, N.L.D. Somasiri, and A.J. Epstein. Polyaniline: synthesis and characterization of the emeraldine oxidation state by elemental analysis. *Conduct. Polym., Proc. Workshop*, 105–20, 1987.

[156] A.G. MacDiarmid, J.C. Chiang, W. Huang, B.D. Humphrey, and N.L.D. Somasiri. Polyaniline: protonic acid doping to the metallic regime. *Mole. Cry. Liq. Cry.*, 125(1–4):309–318, 1985.

[157] R.Y.F. Liu, Y. Jin, A. Hiltner, and E. Baer. Probing nanoscale polymer interactions by forced-assembly. *Macromole. Rapid Commun.*, 24(16):943–948, 2003.

[158] E. Baer, A. Hiltner, and H.D. Keith. Hierarchical structure in polymeric materials. *Science*, 235(4792):1015–1022, 1987.

[159] C.D. Mueller, S. Nazarenko, T. Ebeling, T.L. Schuman, A. Hiltner, and E. Baer. Novel structures by microlayer coextrusion—talc-filled PP, PC/SAN, and HDPE/LLDPE. *Poly. Eng. Sci.*, 37(2):355–362, 1997.

[160] C. Mueller, V. Topolkaraev, D. Soerens, A. Hiltner, and E. Baer. Breathable polymer films produced by the microlayer coextrusion process. *J. App. Poly. Sci.*, 78(4):816–828, 2000.

[161] M.A. Schmidt. Wafer-to-wafer bonding for microstructures formation. *Proceedings of the IEEE*, 86(8):1575–1585, 1998.

[162] A. Sayah, S. Solignac, T. Cueni, M.A.M. and Gijs. Development of novel low temperature bonding technologies for microchip chemical analysis applications. *Sen. Actu., A: Phys.*, A84(1–2):103–108, 2000.

[163] H. Becker and W. Dietz. Microfluidic Device for Micro-TAS Applications for Fabricated by Polymer Hot Embossing. in *Proc. SPIE*. 1998.

[164] K.D.R.S. Gandhi, and L.J. Bousse. Methods of fabricating polymeric structures incorporating microscale fluidic elements, US patent 6, 123, 798 (2000).

[165] P. Lum and M. Greenstein. *Microfluidic Structure Assembly with Mating Microfeatures*, in *U.S.* Hewlett-Packard Company, USA. Us. p. 17, 1999.

[166] J.S. Rossier, A. Schwarz, F. Reymond, R. Ferrigno, F. Bianchi, and H.H. Girault. Microchannel networks for electrophoretic separations. *Electrophoresis*, 20(4–5):727–731, 1999.

[167] H. Dreuth, and C. Heiden. A method for local application of thin organic adhesive films on micropatterned structures. *Mat. Sci. Eng., C: Biomimet. Mat., Sen. Sys.*, C5(3,4):227–231, 1998.

[168] M.J. Madou, L.J. Lee, K.W. Koelling, S. Lai, C.G. Koh, Y.-J. Juang, L. Yu, and Y. Lu. Design and Fabrication of Polymer Microfluidic Platforms for Biomedical Application. *Annu. Tech. Conf.—Soc. Plast. Eng. 59th*, Vol. 3: pp. 2534–2538, 2001.

[169] I.K. Glasgow, D.J. Beebe, and V.E. White. Design rules for polyimide solvent bonding. *Sens. Mat.*, 11(5):269–278, 1999.

[170] R.M. McCormick, R.J. Nelson, M.G. Alonso-Amigo, D.J. Benvegnu, and H.H. Hooper. Microchannel Electrophoretic Separations of DNA in Injection-Molded Plastic Substrates. *Anal. Chem.*, 69(14):2626–2630, 1997.

[171] S. Metz, R. Holzer, and P. Renaud. Polyimide-based microfluidic devices. *Lab on a Chip*, 1(1):29–34, 2001.

[172] S. Lai, X. Cao, and L.J. Lee. A packaging technique for polymer microfluidic platforms. *Anal. Chem.*, 76(4):1175–1183, 2004.

[173] Y. Yang, C. Zeng, and L.J. Lee. Three-dimensional assembly of polymer microstructures at low temperatures. *Adv. Mat.*, 16(6):560–564, 2004.

[174] L.J. Guo, X. Cheng, and C.-F. Chou. Fabrication of Size-Controllable Nanofluidic Channels by Nanoimprinting and Its Application for DNA Stretching. *Nano Lett.*, 4(1):69–73, 2004.

[175] M. Stjernstrom and J. Roeraade. Method for fabrication of microfluidic systems in glass. *J. Micromech. Microeng.*, 8(1):33–38, 1998.

[176] S.W. Turner, A.M. Perez, A. Lopez, and H.G. Craighead. Monolithic nanofluid sieving structures for DNA manipulation. *J. Vac. Sci. Technol., B: Microelect. Nanomet. Struct.*, 16(6):3835–3840, 1998.

[177] W. Li, J.O. Tegenfeldt, L. Chen, R.H. Austin, S.Y. Chou, P.A. Kohl, J. Krotine, and J.C. Sturm. Sacrificial polymers for nanofluidic channels in biological applications *Nanotechnology*, 14(6):578–583, 2003.

[178] C.K. Harnett, G.W. Coates, and H.G. Craighead. Heat-depolymerizable polycarbonates as electron beam patternable sacrificial layers for nanofluidics. *J. Vacuum Sci. Technol., B: Microelect. Nanomet. Struct.*, 19(6):2842–2845, 2001.

[179] H.-P. Chou, C. Spence, A. Scherer, and S. Quake. A Microfabricated Device for Sizing and Sorting DNA Molecules. *Proceedings of the National Academy of Sciences of the United States of America*, 96(1):11–13, 1999.

[180] H. Cao, Z. Yu, J. Wang, J.O. Tegenfeldt, R.H. Austin, E. Chen, W. Wu, and S.Y. Chou. Fabrication of 10 nm enclosed nanofluidic channels. *App. Phys. Lett.*, 81(1):174–176, 2002.

[181] H. Cao, J.O. Tegenfeldt, R.H. Austin, and S.Y. Chou. Gradient nanostructures for interfacing microfluidics and nanofluidics. *App. Phys. Lett.*, 81(16):3058–3060, 2002.

[182] Y. Xia, J.A. Rogers, K.E. Paul, and G.M. Whitesides. Unconventional Methods for Fabricating and Patterning Nanostructures. *Chem. Rev.*, 99:1823, 1999.

[183] T. Thurn-Albrecht, J. Schotter, G.A. Kastle, N. Emley, T. Shibauchi, L. Krusin-Elbaum, K. Guarini, C.T. Black, M.T. Tuominen, and T.P. Russell. Ultrahigh-density Nanowire Arrays Grown in Self-assembled Diblock Copolymer Templates. *Science*, 290:2126, 2000.

[184] L.L. Brott, R.R. Naik, D.J. Pikas, S.M. Kirkpatrick, D.W. Tomlin, P.W. Whitlock, S.J. Clarson, and M.O. Stone. Ultrafast Holographic Nanopatterning of Biocatalytically Formed Silica. *Nature*, 413:291, 2001.

[185] C. Zeng, S. Wang, and L.J. Lee. Dynamic Self-Assembly of Silica in Nanoporous Polymer Membrane. will be presented at *Polymer Processing Society-Americans Region Meeting*, 2004.

[186] Y.M. Boiko and R.E. Prud'homme. Bonding at Symmetric Polymer/Polymer Interfaces below the Glass Transition Temperature. *Macromolecules*, 30(12):3708–3710, 1997.

[187] Y.M. Boiko and R.E. Prud'homme. Strength development at the interface of amorphous polymers and their miscible blends, below the glass transition temperature. *Macromolecules*, 31(19):6620–6626, 1998.

[188] X. Zhang, S. Tasaka, and N. Inagaki. Surface Mechanical Properties of Low-Molecular-Weight Polystyrene Below Its Glass-Transition Temperatures. *J. Polym. Sci., Part B: Polym. Phys.*, 38:654, 2000.

[189] H. Li, D.L. Tomakso, and L.J. Lee. Interfacial Tension between PS and High Pressure CO_2-Experimental and Modeling. *Indus. Eng. Chem. Res.*, 42:6431, 2003.

[190] T.L. Sproule, J.A. Lee, H. Li, J.J. Lannutti, and D.L. Tomasko. Bioactive polymer surfaces via supercritical fluids. *J. Supercrit. Fluids*, 28(2–3):241–248, 2004.

[191] A.K. Dillow, F. Dehghani, J.S. Hrkach, N.R. Foster, and R. Langer. Bacterial Inactivation by Using Near- and Supercritical Carbon dioxide. *Proceedings of the National Academy of Sciences of the United States of America*, 96(18): pp. 10344–10348, 1999.

[192] J. Fages, B. Poirier, Y. Barbier, P. Frayssinet, M.L. Joffret, W. Majewski, G. Bonel, and D. Larzul. Viral inactivation of human bone tissue using supercritical fluid extraction. *ASAIO journal (American Society for Artificial Internal Organs : 1992)*, 44(4):289–293, Jul–Aug 1998.

[193] J.A. Forrest, K. Dalnoki-Veress, J.R. Stevens, and J.R. Dutcher. Effect of free surfaces on the glass transition temperature of thin polymer films. *Phys. Rev. Lett.*, 77(10):2002–2005, 1996.

[194] D.T. Hsu, F.G. Shi, B. Zhao, and M. Brongo. Theory for the Thickness Dependent Glass Transition Temperature of Amorphous Polymer Thin Films. *Electrochem. Soc. Proc.*, 99(7):53, 1999.

[195] J.L. Keddie, R.A.L. Jones, and R.A. Cory. Size-dependent depression of the glass transition temperature in polymer films. *Europhy. Lett.*, 27(1):59–64, 1994.

[196] V.M. Rudoy, O.V. Dement'eva, I.V. Yaminskii, V.M. Sukhov, M.E. Kartseva, and V.A. Ogarev. Metal nanoparticles on polymer surfaces: 1. a new method of determining glass transition temperature of the surface layer. *Coll. Jour.*, 64(6):746–754, 2002.

[197] J.H. Teichroeb and J.A. Forrest. Direct Imaging of Nanoparticle Embedding to Probe Viscoelasticity of Polymer Surfaces. *Phys. Rev. Lett.*, 91:016104, 2003.

[198] F. Xie, H.F. Zhang, F.K. Lee, B. Du, O.K.C. Tsui, Y. Yokoe, K. Tanaka, A. Takahara, T. Kajiyama, and T. He. Effect of Low Surface Energy Chain Ends on the Glass Transition Temperature of Polymer Thin Films. *Macromolecules*, 35(5):1491–1492, 2002.

[199] X. Han, C. Zeng, M.J. Wingert, L.J. Lee, K.W. Koelling, and D.L. Tomasko. Effect of clay surface modification on the polymer nanocomposite foam structure. *Annual Technical Conference—Society of Plastics Engineers*, vol. 2, no. 62, pp. 1723–1727, 2004.

[200] L.G. Griffith. Polymeric biomaterials. *Acta Materialia*, 48(1):263–277, 2000.

[201] B.D. Ratner. Surface modification of polymers: chemical, biological and surface analytical challenges. *Biosens. Bioelectro.*, 10(9–10):797–804, 1995.

[202] S. Zarkoob, R.K. Eby, D.H. Reneker, S.D. Hudson, D. Ertley, and W.W. Adams. Structure and morphology of electro-spun silk nanofibers. *Polymer*, 45(11):3973–3977, 2004.

4

3D Micro- and Nanofabrication and Their Medical Application

E. Di Fabrizio[(1)], F. Perennes[(2)], F. Romanato[(1)], S. Cabrini[(1)], D. Cojoc[(1)], M. Tormen[(1)], L. Businaro[(1)], L. Vaccari[(1)], R. Z. Proietti[(1),(3),(4)], and Rakesh Kumar[(1),(3)]

(1) LILIT group, National Nanotechnology Laboratory-TASC, Istituto Nazionale per la Fisica della Materia (INFM), S.S. 14, Km 163.5, I-34012 Basovizza (Trieste), Italy.
(2) Sincrotrone Trieste ELETTRA, S.S. 14, Km 163.500 I-34012 Bsovizza (Trieste), Italy.
(3) Department of Applied Physics, Osaka University, Osaka 565-0871 Japan.
(4) Japan Society for Promotion of Science.
(5) Physics, Department of Physics, Ch. Charan Singh University, Meerut-250 004, India.

4.1. INTRODUCTION

Nanotechnology is rapidly opening new possibilities in a wide variety of disciplines. Biomedical micro- and nano-devices, with components geometries as large as hundreds of microns to as small as a few nanometers, have potential uses that range from the analysis of biomolecules to disease diagnosis, prevention and treatment. Exciting technological advances have been made in nanofabrication and in micro- and nano-electromechanical systems [1] (MEMS, NEMS), microfluidic devices [2], micro- or nano-optics (diffractive optics [3], high efficiency multilevel zone plates [4], photonic crystals [5]), and innovations in the toolset of biology (microfluidic chips for DNA array [6]) and medical (microsurgical tools [7] and drug delivery), provide examples of advancement in nanotechnology. The application of carbon nanotube atomic force microscopy (AFM) imaging enabled multiplex detection of the positions of specific DNA sequences labelled with nanometer-size tags [8]. This method allowed the direct determination of haplotypes of patient samples, and provided a more accurate assessment of cancer risk from the UGT1A7 gene [8]. The phage display approach for isolating peptides with high binding affinities to specific crystal types discussed by Whaley *et al.* [9] might find use both in synthesis and self-assembly of hetero-structures made from nano-particles. The research in tissue engineering stands to benefit most from our growing ability to fabricate complex nano-structured materials. The enormous potential in nanotechnology is at least a few years from beginning to materialize the products and technologies involved in the medical application of nanotechnology. But the possibility of improved speed, greater sensitivity, reduced cost, and decreased invasiveness, has generated substantial interest in miniaturized devices.

In most cases, the advances have been aided by the highly engineered lithographic techniques. This technology is pushing towards devices miniaturization and complex patterning of a wide class of materials. New aspects are emerging from the capability and necessity of fabricating free shaping of 3D structures. They include an improvement of the patterning ability, an increasing of the functionality of a variety of material and an effective merging of nano-scale with micro scale objects. There are areas impacted by 3D micro and nano fabrication such as for example: MEMS, *microfluidics*, *photonics*, *tissue engineering*, *in situ drug delivery* and several others.

This contribution presents a review of brief descriptions of our various ongoing research activities at L*aboratory for* I*nterdisciplinary* LIT*hography* (LILIT) based on established lithography and in *emerging patterning techniques* for 3D micro- and nanofabrication that are designed to avoid limitations due to 2D fabrication techniques. The focus is on methods whose flexibility in patterning, and demonstrated technical strengths suggest potentially broad utility in biomedical studies.

4.2. 3D MICRO AND NANOFABRICATION

In the search for methods for 3D structuring at micron and nano resolution, several lithographic techniques and their combination are being investigated. Each technique has its own peculiarities and potentialities. Electron Beam Lithography (EBL) can generates "*gray-scale*" profiles by control of the exposure dose. High resolution X-Ray Lithography (XRL) can replicate multilevel mask amplifying the thickness profile [10] and its variant multiple exposure at tilted angles [11, 12] can generate complex 3D structures. Deep X-ray Lithography (DXRL) uses high energy photons (2–20 keV) to extrude 2D patterns in the out of plane direction with aspect ratio up to 100:1 (known as $2\,{}^1/_2$ D devices). As high energy photons allow a large proximity distance between mask and resist (up to few hundred of micrometers), DXRL can also be performed on pre-shaped resist layer to produce real 3D microstructures. Focused Ion Beam (FIB) Lithography has shown capability for direct milling and for growth of hard materials [13]. Nano-Imprint Lithography (NIL) can mould 3D profiles [14]. Hybrid lithography stands for any combination of the above-mentioned techniques. It has a good potential, exploiting the different abilities of each techniques, for the fabrication of even more complex 3D micro and nano structures. Some examples will be presented in the following section.

4.2.1. 3D Fabrication by X-ray and Deep X-ray Lithography for Biomedical Application

4.2.1.1. 3D Scaffolds for Tissue Engineering Tissue engineering offers the possibility to help in the regeneration of tissues damaged by disease or trauma and in some cases to create new tissues and replace failing or malfunctioning organs [15–21]. This is typically done through the use of *degradable biomaterials* to either induce surrounding tissue and cell in-growth or to serve as temporary *scaffolds* for transplanted cells to attach, grow, and maintain differentiated functions [17–19, 21–32]. In any case, the role of the biomaterial is temporary, but rather crucial to the success of the strategy. Therefore, the selection of a scaffold material is both a critical and difficult choice. There are many biocompatible materials available among metals, ceramics and polymers, but the criteria of biodegradability and non-brittle nature (ideally with tissue matching mechanical properties) excludes the use of all metals and most ceramics as *scaffolds* materials [17, 33] and gives preference to biodegradable polymers for most of the applications within the tissue engineering and regenerative medicine field.

Normally, these materials are produced in simple geometries like blocks, pins or splines. However, most types of tissue that are now in advanced stages of engineering are either amorphous or isotropic (such as bone and cartilage) or have a planar structure (the skin, blood vessels). However, some of the more complex organs and tissues, such as the liver, heart, and neural tissue, are proving more difficult to engineer because they have a very specific 3D cell distribution in which the 3D structure is inextricably linked to its function.

The definition of the most adequate *scaffold* design and the correspondent required properties is mainly determined by the tissue engineering approach selected for the regeneration of a specific tissue, as the scaffold must be able to induce the desired tissue response. 3D porous structures have been recognized as the most appropriate design to sustain cell adhesion and proliferation. For these reasons, it is considered essential to have a method of creating *biomaterial scaffolds* having a known and well-defined topology. In fact, as the demand for new and more sophisticated scaffolds develops, materials are being designed that have a more active role in guiding tissue development. Instead of merely holding cells in place, these matrices are designed to accomplish other functions through the combination of different format features and materials [30]. A good example of this is the use of drug delivery devices that can act simultaneously as scaffolds for cells growth. Other approaches include, for example, the combination (or incorporation) of *micro-spheres* or *nano-spheres* (with encapsulated cells, growth factors or other therapeutic agents) with a polymeric matrix. This type of multifunctional devices can also be designed to act as an injectable material, with the advantage of allowing minimal invasive surgery procedures for their implantation in the body. Another important field of current research in tissue engineering scaffolding is related to the development of external-stimuli-responsive matrices *i.e.*, matrices that have in their composition and structure certain elements that allow them to respond to a particular specific stimuli that can be produced by different mechanisms, such as magnetic, electric, ultrasound, irradiation or other effects. Other scaffolds are designed to respond to several physiological stimuli like pH, temperature or enzymatic concentrations changes, just to cite some examples. It is not our intention to review the subject on biodegradable polymers for their applications within the tissue engineering [33] and regenerative medicine field, but rather to present elementary information to make acquaint the basic reader about the subject.

Several methods for the deposition of biopolymers with controlled or *scaffolds* architecture have been described in the literature [35–38]. A hybrid technology combining the culture of autologous bone cells on 3D biodegradable scaffolds has been identified as a very promising solution to produce laboratory-made tissue-engineered living-bone replacement materials. For example, *Rapid Prototyping* (RP) combined with microstereolithography process [34] and *ceramic gelcasting technique* has successfully been used to generate porous ceramic matrices made directly from powder hydroxyapatite [35] offering mechanical stability and potential biodegradability. Microstereolithography is a microfabrication technique which is capable to produce small 3D structures with arbitrary shapes. The original CAD model of the structure architect is virtually decomposed into thin layers that are subsequently built up in the reaction chamber of the RP-machine by selectively polymerising a negative resin with UV-light curing. The binary image corresponding to each specific layer is generated by a micro-mirror array. After a layer is polymerised, another layer of resin monomers is applied and again selectively exposed. This procedure is repeated until the complete structure is built. Finally the unexposed resin is removed in an appropriate solution bath and then a final UV cure is applied to harden the complete structure. The resin scaffold is then used as a casting mould and filled by an aqueous suspension of hydroxyapatite particles with water soluble monomers and dispersing agents. A progressive thermal treatment up to 1300°C is used to successively dry and burn the resin and ensure the sintering of the ceramic particles resulting in an almost fully dense hydroxyapatite scaffold. However, as a draw back, the *Rapid Prototyping stereolithographic method* allows the fabrication of scaffold with strut size not smaller than 300 microns [37]. Although, this size is in the range of the trabeculae the cancellous bone consists of, this is dense hydroxyapatite, but not hydroxyapatite nano-particle reinforced collagen as in natural bone. Therefore scaling down the dimensions of the struts in our scaffolds, would enable the bone cells to remove the hydroxyapatite material and replace it by new bone material. The resorption lacuna (cavity) a single osteoblast generates, has an average depth of 50 μm, and for complete biodegradability the maximum thickness of the struts is about 100 μm.

4.2.1.2. 3D Scaffold Fabrication by Deep X-ray Lithography Our research interest in the area of tissue engineering is presently confined to create *biodegradable 3D scaffold structures* and devices to enable the in-vitro assembly of complex 3D cell-scaffold structures for tissue replacement and regeneration. These novel scaffolds structures should allow us to perfuse the growing tissue reducing nutrient limitations. We are exploring the use of DXRL techniques for the production of tissue engineering scaffolds. This approach should allow us to develop scaffold structures that will have a highly pseudo-vascularised network similar to that found in tissues, and also may allow the development of bespoke 3D scaffolds for individual patients based on data from medical scans.

DXRL is a method to produce casting moulds that offers on the one hand sufficient resolution and on the other hand the possibility to produce parts with reasonable outer dimensions within reasonable time. Polymer structures with aspect ratio up to 100:1 can be obtained with this fabrication technique. Polymethylmetacrylate (PMMA) is the standard polymer material used for deep X-ray lithography. When exposed through a mask to X-ray radiation with dose above 2 kJ/cm^3, it becomes soluble in a liquid solvent and can therefore be selectively etched away.

The requested *scaffold geometry* consists in a 5 mm side cube with interconnected circular pores of 50 μm diameters each separated by 180 μm. The mask fabrication requires two steps. A primary mask with gold dots (figure 4.1.a,b) is obtained by standard lithography process on SU-8 resist. This mask is then replicated into a negative tone (holes in gold foil, figure 4.1.c,d) by DXRL. The first DXRL exposition is performed in a 5 mm thick PMMA

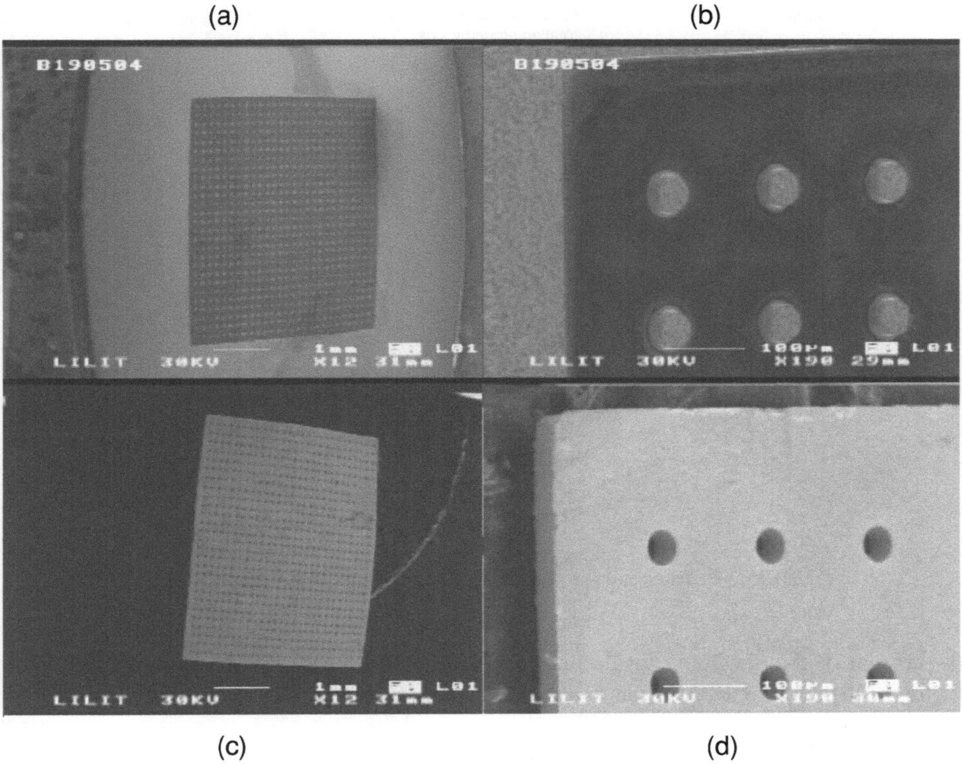

FIGURE 4.1. Primary X-ray mask (a,b) and final replica of X-ray mask (c,d) for the fabrication of 3D scaffold.

sheet to cut off the 5 mm side cube as well as partially etch the first array of pores. In order to deposit a sufficient dose at the depth of 5 mm, the exposure can only be performed when the Synchrotron light source is operated at 2.4GeV (instead of the standard 2GeV) with the right set of filters inserted into the beam path of the DXRL beamline [38]. The second and third exposures are then performed changing the orientation of the cube versus the beam direction as shown in figure 4.2. The 3D interconnection of the pores requires a perfect alignment of the mask with pre-etched pores for the second and third exposition.

Finally the cube is immersed in the developing solution for a total developing time of 60 hours. A SEM micrograph showing fabricated 3D scaffold in PMMA after the three DXRL exposures is shown in figure 4.3. A magnified view of the cube corner shows the three perpendicular axes in etched struts.

The main issue of concern was the ability to develop the 50 μm diameter aspect ratio pores over the entire length of 5 mm. The dynamic of the removed PMMA particles is slowed because they tend to be trapped inside the deep channels and block the further access of the developing solution to the remaining exposed PMMA. Visual inspection of cubes exposed only once showed that it was difficult to etch the pores over a depth higher than 1.5 mm. Therefore the depth of the cubes exposed in the 3 direction is likely to be in that range. It is expected that megasonic supported development that was shown to enhance

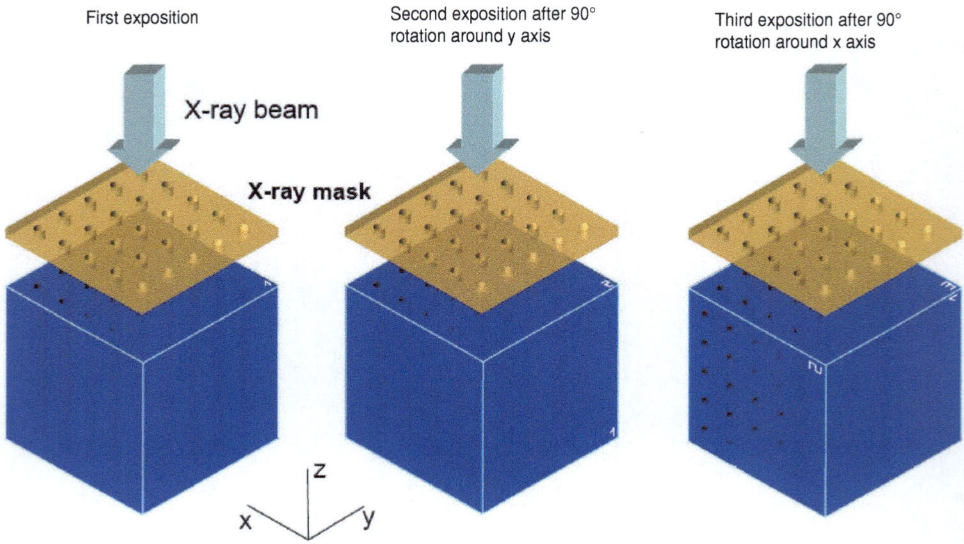

FIGURE 4.2. Fabrication process of the 3D scaffold in PMMA using DXRL.

the current of the developer inside the pores [39] and increase the developing rates by a factor 4 [40], will be the solution to achieve deeper struts.

4.2.2. 3D Microparts for Transdermal Drug Delivery System

4.2.2.1. Transdermal Drug Delivery System (TDS) Transdermal drug delivery is a viable administration route when drugs oral administration is not feasible due to poor drug absorption or enzymatic degradation in the gastrointestinal tract or liver. Injection using a hypodermic needle is the usual alternative but most often requires the intervention of a nurse or doctor and is a source of pain for the patient. A more appealing approach that

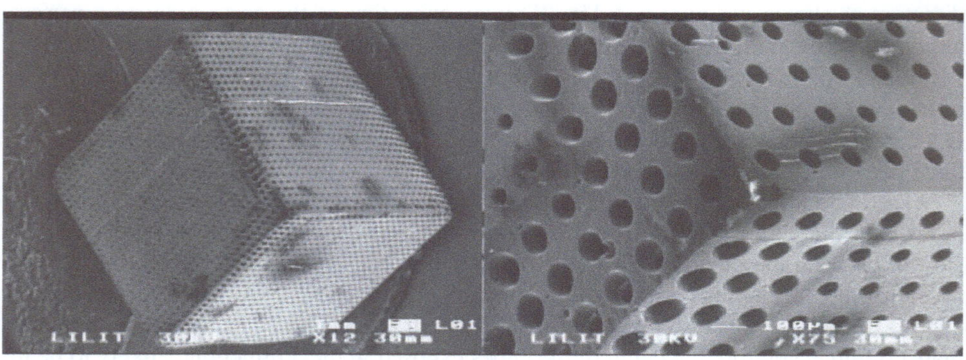

FIGURE 4.3. PMMA 3D scaffold.

offers the possibility of controlled release over time, is drug delivery across the skin using a patch [41, 42]. However, transdermal delivery is severely limited by the inability of the large majority of drugs to cross skin at therapeutic rates due to the great barrier imposed by skin's outer *stratum corneum layer*.

In a broad sense, the term transdermal delivery system includes all topically administered drug formulations intended to deliver the active ingredient into the general circulation. Transdermal therapeutic systems have been designed to provide controlled continuous delivery of drugs via the skin to the systemic circulation. The relative impermeability of skin is well known, and this is associated with its functions as a dual protective barrier against invasion by micro-organisms and the prevention of the loss of physiologically essential substances such as water. Elucidation of factors that contribute to this impermeability has made the use of skin as a route for controlled systemic drug delivery possible. Basically, four systems are available that allow for effective absorption of drugs across the skin. The microsealed system is a partition-controlled delivery system that contains a drug reservoir with a saturated suspension of drug in a water-miscible solvent homogeneously dispersed in a silicone elastomer matrix. A second system is the matrix-diffusion controlled system. The third and most widely used system for transdermal drug delivery is the membrane-permeation controlled system. A fourth system, recently made available, is the gradient-charged system. Additionally, advanced transdermal carriers include systems such as iontophoretic and sonophoretic systems, thermosetting gels, prodrugs, and liposomes. Many drugs have been formulated in transdermal systems, and others are being examined for the feasibility of their delivery in this manner (e.g., nicotine antihistamines, beta-blockers, calcium channel blockers, non-steroidal anti-inflammatory drugs, contraceptives, anti-arrhythmic drugs, insulin, antivirals, hormones, alpha-interferon, and cancer chemotherapeutic agents).

To increase skin permeability, a number of different approaches has been studied, ranging from *chemical/ lipid enhancers* [43, 44] to electric fields employing *iontophoresis* and *electroporation* [45] to pressure waves generated by ultrasound or *photoacoustic* effects [46, 47]. Although the mechanisms are all different, these methods share the common goal to disrupt *stratum corneum* structure in order to create "*holes*" big enough for molecules to pass through. The size of disruptions generated by each of these methods is believed to be of nanometer a dimension, which is large enough to permit transport of small drugs and, in some cases, macromolecules, but probably small enough to prevent causing damage of clinical significance.

An alternative approach involves creating larger transport pathways of microns dimensions using arrays of microscopic needles. These pathways are orders of magnitude bigger than molecular dimensions and, therefore, should readily permit transport of macromolecules, as well as possibly *supramolecular* complexes and *microparticles*. Despite their very large size relative to drug dimensions, on a clinical length scale they remain small. Although safety studies need to be performed, it is proposed that micron-scale holes in the skin are likely to be safe, given that they are smaller than holes made by hypodermic needles. Research also continues on various chemical penetration enhancers that may allow delivery of therapeutic substances. For example, penetration enhancers such as Azone may allow delivery of larger-sized molecules such as proteins and polypeptides.

Although the microneedles concept [48] was proposed in the 1970s, it was not demonstrated experimentally until the 1990s when the microelectronics industry provided the microfabrication tools needed to make such small structures. Since the first studies of transdermal drug delivery [49] in 1998, there has been rapidly increasing interest in the field, with most activity in the microfabrication community to develop novel needle fabrication technologies. The overarching motivation for microneedles is that they can provide a minimally invasive means to transport molecules into the skin. Guided by this goal, a number of specific strategies can be employed to use microneedles for transdermal delivery [50, 51].

4.2.2.2. Conceptual Design of a Transdermal Drug Delivery System Our approach for developing transdermal drug delivery system is based on the concept of by passing the *Stratum Corneum* using an array of sharp tip microneedles to inject the drug with the help of a micro pumping device. The system combines a microgear pump with an array of microneedles stacked on top of each other that can both be fabricated by LIGA technology. The conceptual layout of the system design is sketched in figure 4.4. The fluid is pumped from an external reservoir (not shown on the figure) and first fill-up the buffer reservoir before reaching the pump case. Activation of the gear allows the fluid to build up pressure to let it flow through each microneedle and reach the blood capillaries area under the skin.

The precision and small run-out of the deep X-ray lithography (DXRL) process provide the rotating sealing surface of the gear teeth on the cavity walls to create positive displacement, self priming pump [52] and minimize the back flow rate. A permanent magnet is embedded into one of the gear and coupled to a second magnet mounted on a micromotor. This magnetic coupling activation ensures a complete seal of the pump case thus eliminating the need for a complicated dynamic shaft seal assembly [53].

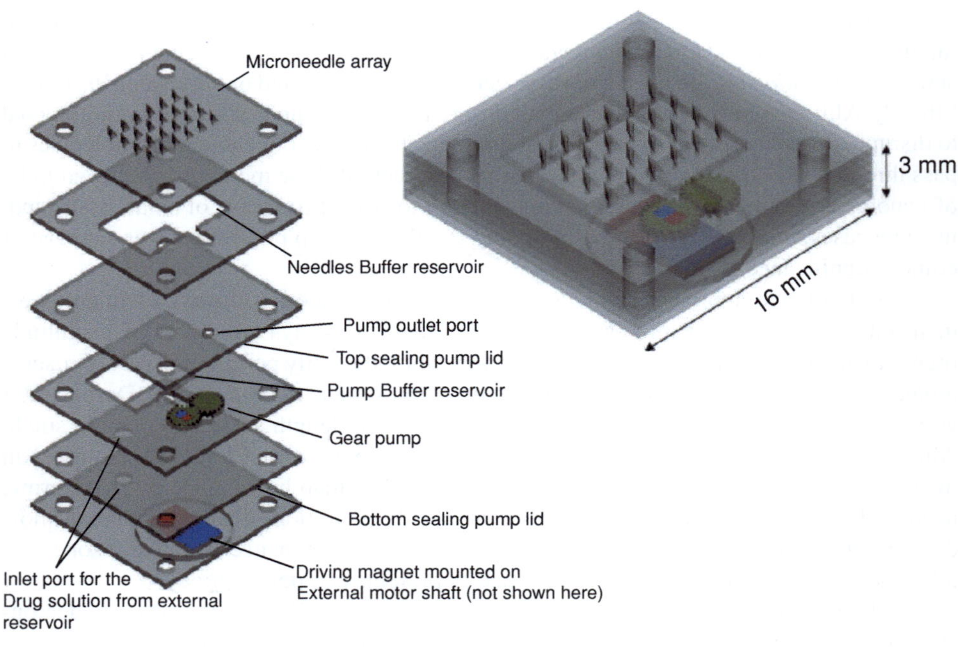

FIGURE 4.4. Schematic of a transdermal drug delivery system.

The majority of previous reported work on microneedles fabrication has been using silicon as material of construction [49–51]. However due to its brittle nature, metallic or hard polymer have gain considerable interest because of the range of material available and the possibility to fabricate by deep lithography, electroforming and moulding [54, 55]. Experiments had shown that a bevelled needle tip with a bevel angle between 30–75° greatly enhances the penetration into the skin [56]. Furthermore the density of the array must be limited to prevent a push-down effect of the skin. When pushed onto the skin the lateral force resulting in a bending moment on the needle shaft creates a maximum stress concentration at the base of the needle. Therefore a large base capable of handling this stress was also considered in our design approach.

4.2.2.3. Fabrication of the Drug Delivery System Parts The microgear pump is directly fabricated in PMMA by DXRL. The design of the gears and case is accomplished using a CAD tool minimising the clearance gap between both elements. Assuming the gears remain perfectly centered, the gap between the teeth tip and the case in the radial direction is fixed to 3 μm. This is done to minimise the back flow of the gearpump and made possible due to the extreme precision and resulting vertical sidewalls of the DXRL fabrication technique. The chromium mask is replicated by photolithography on 30 μm thick SU-8 resist and 20 μm of gold are electroplated to form the X-ray mask absorbing pattern. A self-standing 500 μm thick PMMA foil is exposed to X-ray through the mask and after development the pump case and both gears are released from the PMMA foil. Assembly is then performed by hand under a stereoscopic microscope. The figure 4.5 shows the fabrication process flow of the gear pump and a SEM photo of the assembled micropump is shown in figure 4.6. A circular magnet is to be mounted on the hollow gear to drive the pump.

The fabrication of beveled microneedles in hard polymer combines different processes like hot embossing, DXRL, and casting, which are described in details in figure 4.7. A PMMA sheet with a periodic 3D groove profile corresponding to the pitch size of the needle array is fabricated by hot embossing using a mould fabricated by conventional micromachining techniques. The embossed PMMA sheet is then glued on a substrate coated with a base plating layer. The X-ray mask including the cross-sectional design of the array and four circular alignment marks is used to expose the groove embossed PMMA sheet. Each needle cross-section is a 275 μm size equilateral triangle with a 40 μm diameter hole shifted toward the needle tip. The pitch of the array in both planar directions is fixed

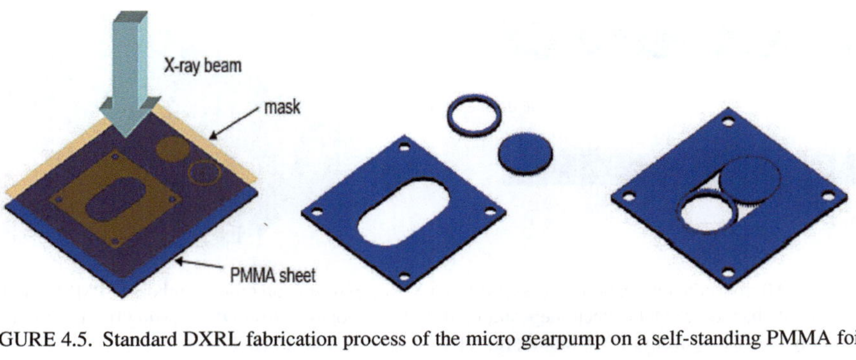

FIGURE 4.5. Standard DXRL fabrication process of the micro gearpump on a self-standing PMMA foil.

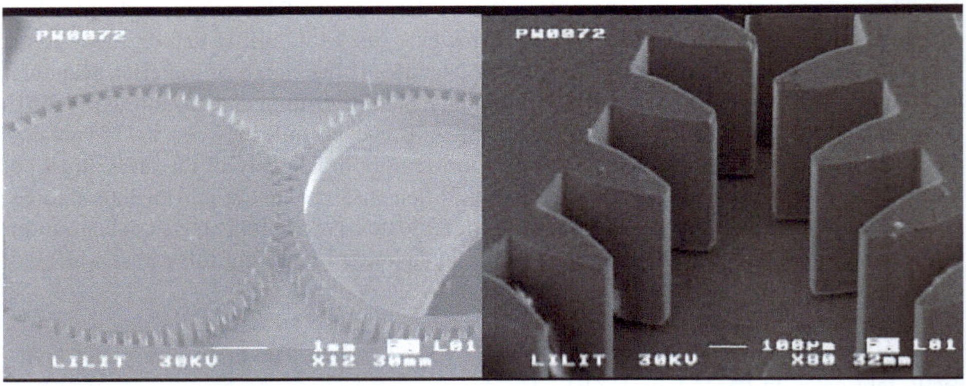

FIGURE 4.6. SEM images of the assembled gearpump.

to 800 μm. The gap that is necessary to leave between mask and sample because of the grooves does not affect the lithographic process due to the extreme low divergence of the X-ray beam and the negligible diffraction effect at these extremely short wavelengths. The X-ray exposure projects the triangular cross-section on the inclined wall of the PMMA grooves and after development the array of sharp-beveled microneedles is formed. The bevel

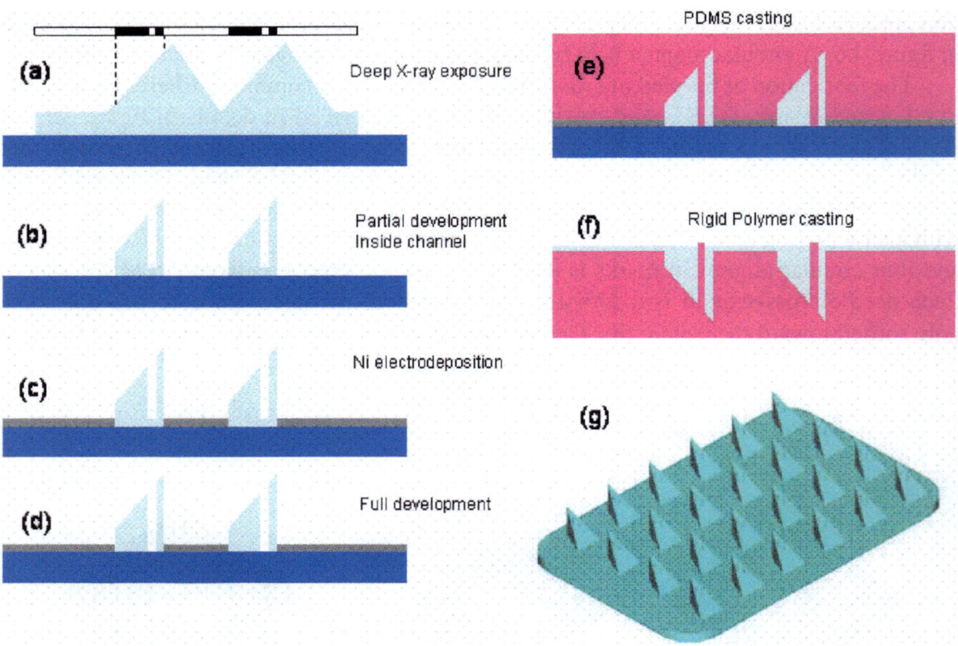

FIGURE 4.7. Microneedle fabrication process: (a) Deep X-Ray exposure on groove embossed PMMA (b) Partial development of channel (c) Metal electrodeposition, (d) Full development (e) PDMS casting from PMMA mould (f) Rigid Polymer casting from PDMS mould, (g) released microneedle array.

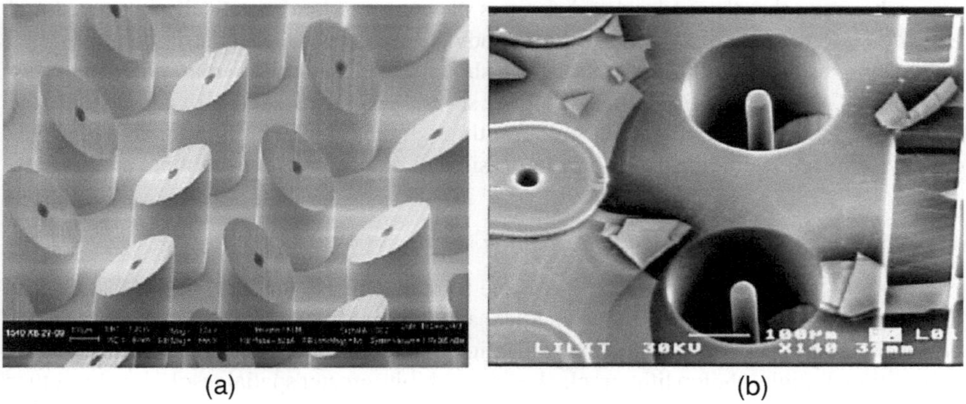

FIGURE 4.8. (a) Circular base PMMA microneedles obtained by deep X-ray lithography on embossed PMMA sheet, (b) PDMS inverse mould obtained by casting from the PMMA mould of (a).

angle is determined by the angle of the groove inclined wall versus the horizontal plane. Alignment marks are placed in order to obtain a needle height of 440 μm from tip to base. If the development is interrupted just after removal of the PMMA around the needles and alignment marks the needle channels remain partially etched due to the lower developing rate in the narrow holes. A Nickel layer is electroplated with a thickness corresponding to the height of the base of the future microneedle array. A second DXRL exposure is performed masking only the needle array to irradiate the PMMA alignment marks. During a second developing step the remaining PMMA inside the channels and the alignment marks are removed. An example of beveled microneedles with a circular cross-section obtained with the process described above is shown in figure 4.8.a.

PDMS (polydimethylsiloxane) is an elastomer that can fill microstructures whilst in its liquid phase and then cured at temperatures around 100°C. The cured PDMS has very weak adhesion on most materials and due to its elasticity can easily be peeled off the mould. Therefore the PMMA solid mould can be used to fabricate a PDMS negative imprinting of the needles with the internal pillar (corresponding to the needle channel) emerging from the top surface of a height equal to the thickness of the electroplated nickel. Figure 4.8.b shows an inverse PDMS mould obtained from the PMMA mould fabricated by deep X-ray lithography. The PDMS inverse mould is then used to cast the final microneedle array with a liquid polymer solution that solidifies in a few hours at room temperature (i.e. PMMA). This time the mould is peeled off and can be re-used making it suitable for an eventual mass-production of the needle array. Casted PMMA structures with aspect ratio up to 20:1 were obtained from a PDMS mould, allowing the use of PDMS as a soft mould to fabricate the hollow microneedle array in rigid polymer.

4.3. EMERGING METHODS FOR 3D MICRO AND NANOFABRICATION

Many of the basic principles of new emerging nano-patterning such as *nano-imprint lithography* (NIL) and *embossing* are, of course, not conceptually new. Nevertheless, recent research proves that these ideas, and variants of them, can be dramatically improved and extended into the nanometer range by using advanced materials, chemistries, and processing techniques. The resulting methods have shown to possess truly remarkable patterning capabilities. We present discussions from our several recent works here to demonstrate the power of these new lithographic techniques, through their application to functional prototype systems that would be difficult or impossible to realize with conventional approaches.

4.3.1. Two Photon assisted Microfabrication

Among the innumerable techniques able to realize micro and nano-devices, one that can play a very important role in bio-medical applications is the "*Two Photon assisted Microfabrication*". It not only allows the realization of structures that cannot be constructed by conventional single-photon lithography but also exhibits greater spatial resolution than other 3D micro-fabrication techniques as they are currently practiced. This technique, also called *Two Photon Photopolym erization* because of its characteristic to use polymers to realize the desired devices. The two-photon process is based on the idea of Maria Gppert-Mayer [57], who predicted that an atom or a molecule could interact with two photons simultaneously by absorbing them in the very same quantum event. As illustrated in figure 4.9, that the phenomena is energetically equivalent to the absorption of a single photon (γ_a) with energy equal to the sum of the energies carried by the two photons (γ_b and γ_c).

In term of wavelengths [58]:

$$\lambda_a = \left(\frac{1}{\lambda_b} + \frac{1}{\lambda_c}\right)^{-1} \tag{4.1}$$

Even if the Single-Photon Excitation (SPE) and the Two-Photon Excitation (TPE) are equivalent from the energetic point of view, there are some fundamental characteristics that make them very different. From the eq. 4.1, it is clear that with TPE it is possible to obtain fluorescence from the atoms that absorb UV just using visible light [59]; that is not possible by using SPE. Among the many advantages that we get is: *to have a radiation that can enter deeper in the sample* (the diffusion phenomena increases with the frequency of the radiation) and secondly the unmatched advantage that with visible-like emission it is possible to avoid damages to biological samples [60].

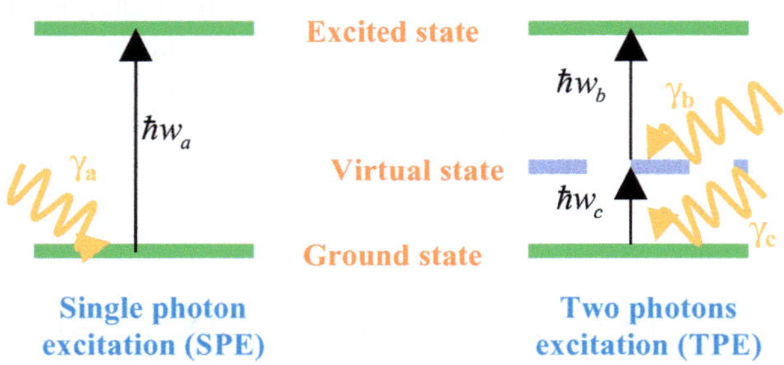

FIGURE 4.9. Scheme of single and two photon absorption process by an atom. The processes are equivalent as far as the sum of the energies $\hbar w_b$ and $\hbar w_b$ is equal to the energy of the photon γ_a ($\hbar w_a$).

Due to the quadratic dependence of two-photon absorption (TPA) rate on laser power (single photon absorption has a linear dependence on the laser power), photopolymerization (solidification) occurs only with high power sources. For this reason was possible to observe the two-photon excitation only in 1965 after the invention of laser sources [61, 62] and was utilized for micro-nanofabrication [63] in 1997.

Because of the square dependence on the light intensity, it is easy to guess that a TPE is much more difficult to obtain than a SPE process. Indeed, if we consider the Heisenberg principle giving the Two Photon Absorption Time, namely the interval of time when both the photons must be absorbed to realize a two-photon excitation,

$$t_{tp} = 10^{-16}\ \text{sec},$$

then the difficulties to see realized a TPE increase very much. To estimate the probability to have n photons excitation, a simple formula is needed:

$$P_{if}^{(n)}(t) = I^n \delta_n,$$

being I the intensity of light, δ_n the n-photons cross section and i and f the initial and final transition states]. For our purpose, just δ_1 and δ_2 are important. In general, they are quite different. Their magnitude orders are:

$$\delta_1 \approx 10^{-16} \div 10^{-17}\ cm^2$$

and

$$\delta_2 \approx 10^{-50}\ cm^4 \equiv 1GM,$$

where GM is for Goppert-Mayer [65].

As the small numbers just obtained say, to have a SPE we need big source intensity and/or a big single photon cross section, but for TPE they must be even bigger. To have a very high two-photon cross section, would allow the realization of TPE with low laser power hence low sample damage. Some materials got recently with these characteristics are E-stilbene with $\delta_2 \approx 12GM$ and pluto-E-stilbene that shows $\delta_2 \approx 1940GM$ [66–68].

A last ingredient must be introduced to understand completely why TPE can be much better of SPE in the field of microfabrication: it is about the *fabrication resolution*. By calculating the three-dimensional light distribution around a focal point, namely the Point Spread Function (PSF), it is possible to show that the intensity of the light $I(u, v)$, where u and v represent respectively the focal axis and the focal plane, has the form

$$I(u, 0) \propto \frac{1}{u^2} \tag{4.2}$$

$$I(0, v) \propto \frac{1}{v^2}. \tag{4.3}$$

The formulas (2) and (3) express a very important concept: the PSF depends on the square of the inverse of the distance from the focal point. Hence, if we consider the two-photon excitation phenomena, because of its quadratic dependence on the light intensity, we'll get that its PSF depends on the distance z from the focal point as:

$$I_{2\gamma} \propto \frac{1}{z^4}. \tag{4.4}$$

By comparing (2) with (3) and (4) it is immediately clear that the two-photon absorption process possesses intrinsic high spatial-localization characteristics, better than the SPA [69]. In fact, it is possible to show that the TPE has a spatial confinement of about *0.1 femto-liter* [59].

Two-photon photopolymerization [70, 71] has been hence recognized as a unique nanofabrication tool due to its intrinsic 3D fabrication capability [72, 73] and sub-diffraction limited spatial resolution [74, 75]. For its widespread use, a deep understanding on single voxels, the primitive building block for a 3D object [74–77], which can be approximated using following relation:

$$S = l_{diff}\alpha\sqrt{\frac{\ln(E_{re}/E_{th})}{4n\ln 2}}, \tag{4.5}$$

where S is the voxel size, l_{diff} is the diffraction limit of a laser focal spot and $\propto$ a constant reflecting characteristics of two-photon materials and exposure schemes and, therefore, describing the fact that diffraction limit is not the only factor to determine the voxel size; E_{re} and E_{th}, represent real (actually applied) and threshold laser power (P_{th}, P_{re}) at a given exposure time, or real and threshold exposure time (T_{th}, T_{re}) at a given exposure power; $n = 1, 2$ denotes single-photon and two-photon processes, respectively. It is interesting to note that low-numerical aperture (NA) focusing does not necessarily give rise to lateral spatial resolutions worse than high-NA focusing at a medium irradiation level [78]. This is because of threshold effect, and under a linearly polarized laser beam, voxels take shape of ellipsoid with three different axis lengths instead of a spinning ellipsoid. These considerations are important for properly designing and accurately depicting 3D structures with nanoscale features. Another significant merit of two-photon photopolymerization is the possibility of tuning voxel size according to requirements of device structures, conveniently either by adjusting the average irradiation laser power [71] (P scheme) or by changing the exposure time [70, 73] (T scheme). The two processes are equivalent to reach this end according to conventional exposure theory. However, at near-threshold condition, the voxel size scaling abides by different laws under these two approaches [79]. In two-photon absorption (TPA) assisted photopolymerization (solidification), a *solidified skeleton* is formed by the scanning locus. This *solidified skeleton* remains after the removal of unsolidified liquid resin [80]. With this method, various micromachines [71, 80] (figure 4.10 and figure 4.11) and photonic crystals [81, 82, 83] have been readily produced with near-diffraction-limit 3-D spatial resolutions. Photopolymerization is a photochemical reaction used for the creation of a polymer through a chain reaction initiated by light. Since most monomers and oligomers commonly employed do not possess initiating species with a sufficient quantum yield upon light exposure, it is necessary to introduce low-molecular-weight molecules called photoinitiators that start polymerization. A photosensitizer is also generally used, which has a large light absorbance and transfers the excitation to a photoinitiator. For many fabrications,

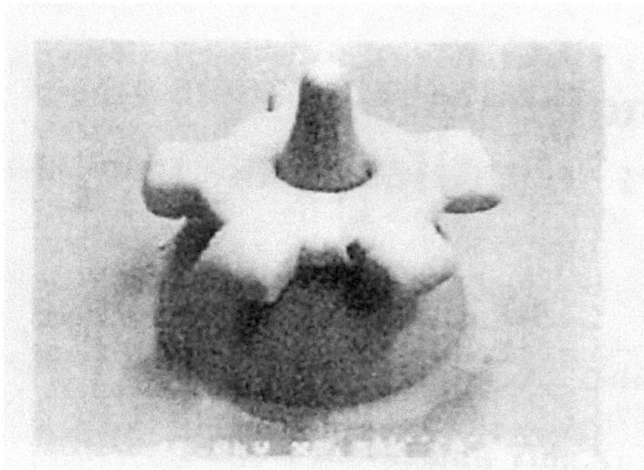

FIGURE 4.10. Micro-gear wheel.

resins consisting of urethane acrylate monomer/oligomer and radical initiators are used. Electronic transition energy for most initiator molecules corresponds to UV spectral range.

In particular, benzoyl chromophor is sensitive to near UV wavelength, and has a good photochemical reactivity, therefore utilized as common UV radical initiators. If the irradiation photon fluence is sufficiently high, e.g., by tightly focusing a fs laser, the probability of an electron simultaneously absorbing two photons is increased, and then TPA becomes practically useful. For producing sufficient photon flux density, it is essential to tightly confine laser pulses in both spatial and time domains. For the work presented hereafter a Titanium Sapphire laser that operated in mode-lock at 76 MHz and 780 nm with a 150-fs pulsewidth was utilized as the exposure source. A two-galvano-mirror set moves the laser beam in the two horizontal dimensions, and a piezo stage moved the laser focus vertically, both controlled by a computer. The laser is focused into the resin by a high numerical aperture (NA ~1.4, oil immersion) objective lens. The single-lens focusing geometry naturally

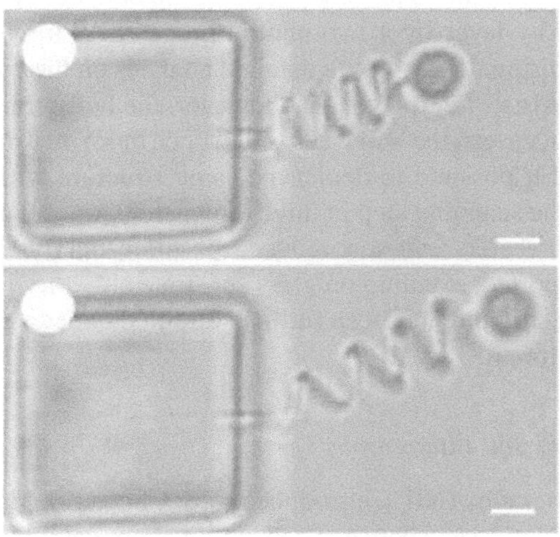

FIGURE 4.11. Micro-oscillator system.

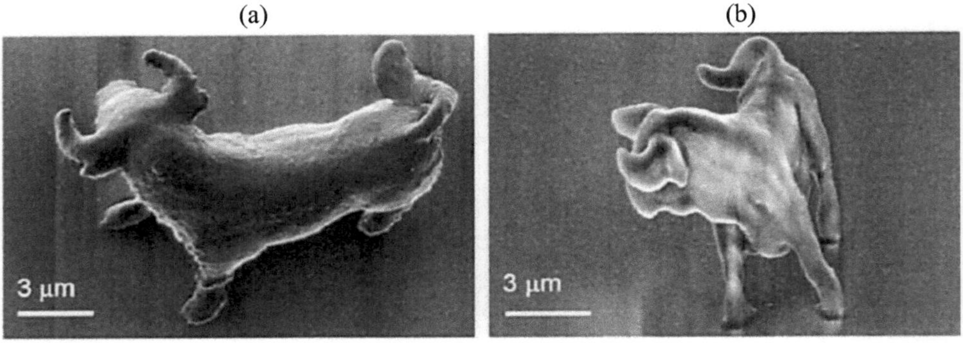

FIGURE 4.12. Example of a 3D structure obtained by means of two-photon lithography; (a) using the raster scan method, i.e. all the volume of the structure has been exposed, and (b) using the vector scanning method exposing only the microbull surface.

satisfies the requirement of pulse overlapping in both time and spatial domains. In this system, the utilization of short pulsewidth and tight focusing are critical for exciting sufficient amount of TPA and for achievement of high accuracy of fabrication. An average focal spot power of 1 mW under a 150 fs pulsewidth and 76 MHz repetition rate corresponds to a transient peak power of 20 GW/cm, or a photon flux density of $8 \cdot 10^{20}$ photon s^{-1} μm^{-2}.

Laser scanning is the step to convert pre-designed CAD pattern to a resist structure. Two basic modes for direct laser scanning has been used, i.e., raster-scan mode and vector-scan mode. In the raster mode, all voxels in a cubic volume that contains the microstructure are scanned by the actual/virtual focal spot, depending on the shutter ON/OFF (actual/ virtual). In the vector mode, the laser focus directly traces the profile to be defined, and requires a smaller number of voxels. Depending on structures, alternations and combination of the these two basic scanning modes could be used. Experimentally we fabricated the same object using the two modes. The microbull in figure 4.12.a was produced using a layer-by-layer raster-scanning scheme, i.e., all voxels consisting of the bull were exposed point-by-point, line-by-line, and layer-by-layer by a two-photon process. As a result, it took 3 hours to complete the manufacturing. If we make a detailed analysis on the bull structure, it is found that the entire bull consists of $2\ 10^6$ voxels. However, the bull profile can be well defined with only 5% of them. As a test, the bull was written once more by using the vector scanning. Astonishingly, we find it possible to depict the same structure within 13 minutes (figure 4.12.b). In both cases the scanning step in three dimensions was 50 nm, the latter, however, the fabrication time in vector scan was reduced by more than 90%. The TPA produced bull crust was self-supported, standing on glass substrate either in liquid or in air. To avoid possible distortion, the structure has been further solidified under a mercury lamp, which is a single-photon exposure process.

4.3.2. Nanoimprint and Soft Lithography

Nano-Imprint lithography (NIL), introduced by S. Chou et al. [84], is an attractive technique for filling the resolution gap between conventional optical lithography and techniques exploiting the self-assembling capabilities that matter often shows at the nanoscale. This

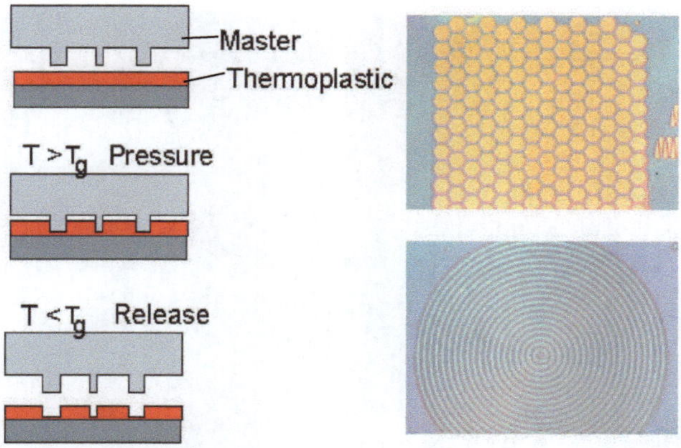

FIGURE 4.13. Left. Scheme of the nanoiprint process. A hard master is embossed into thin film of a thermoplastics, above the temperature transition T_g. The sample is cooled down below T_g before stamp release in order to avoid deformation of embossed structures. Right: Example of microstructures in a thin termoplasitics (PMMA) film.

low cost, high-resolution technique is based on a different principle from the main, already established, nanofabrication technologies. The process, depicted in figure 4.13, involves the hot embossing of a thin resist layer with a re-usable master etched in a hard material. In order to ensure low viscosity and easy flow of the resin during the embossing the temperature has to be raised to well above the glass transition temperature T_g of the polymer. The most common hot embossing process is carried out between the hot plates of a press, with a micro or nanostructured silicon substrate as a master and a silicon substrate spun with a thermoplastic resist layer like PolyMethylMethAcrylate (PMMA). Typical temperature and pressure parameters are in the range 150°C–240°C, 40–100 bars and imprinting times from 5 minutes to 1 hour. Nanoimprint is suitable for structuring large areas (8 inch substrates) in a single parallel process, and sub-10 nm resolution has already been demonstrated by several laboratories. Moreover, nanoimprint lithography is capable of replicating very accurately three-dimensional structures, which represents a strong competitive advantage over most of the existing lithographic techniques. Figure 4.14 shows an hexagonal array of micro lenses as an example of the 3D patterning potential of nano-imprint lithography. These arrays of micro lenses might find application to amplify fluorescent signals in the microfluidic channels when integrated with microfluidic devices for bio-analytical applications. A comprehensive review of the nanoimprint technology is given in [85].

Despite great merits of nanoimprint, few problems are limiting its spreading as an industrial technology for medium to mass volume manufacturing. Main limitations are related to rheology, accurate registration, and short stamp lifetime. Rheological problems have been identified in the flow of high viscosity polymer melt and in the patterning of structures with complex topologies, where the polymer transport occurs in narrow channels and over long distances.

Alignment steps have to be performed whenever the fabrication of a device is obtained in a multi step lithographic process. With the increasing miniaturization trends the registration becomes more and more demanding. Pressure and temperature represent additional difficulties for the accurate registration, and sets special design requirements on tool and

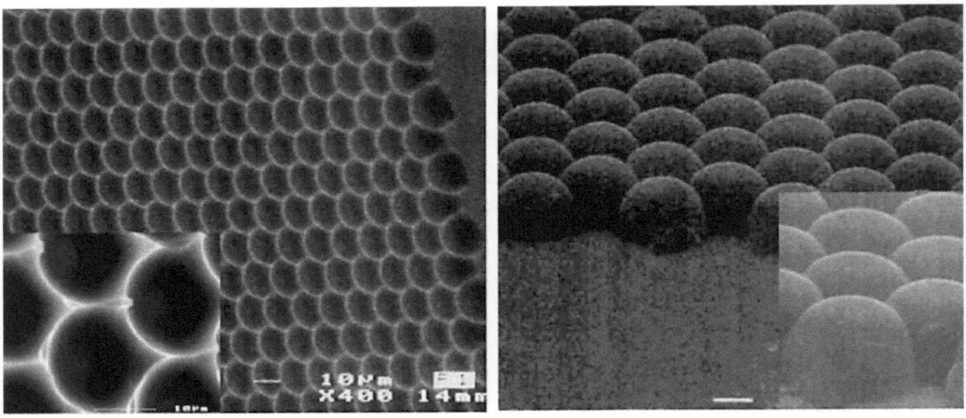

(a) (b)

FIGURE 4.14. Shown are (a) arrays of hemispherical cavities defined on the fused silica master by wet etching. The bar corresponds to 10 μm and, (b) Hexagonal array of microlenses imprinted in PMMA from the master.

stamp characteristics. Finally, the stamp is subject to contamination, and degradation due to the direct contact with a polymer coated substrate. This affects dramatically the number of possible imprinting cycles per stamp. A large international effort has been devoted to the resolution of all the above-mentioned problems.

The improvement of the polymer flow during embossing follows essentially very simple rules: the higher the temperature beyond the glass transition temperature (T_g) of the polymer, and the higher the imprinting pressure, the shorter is the imprinting time. These tendencies unfortunately are not always compatible with the industrial requirement in terms of throughput (due to the increase in the thermal cycling time) and overlay accuracy. Currently, the research has focused on the synthesis of new polymers, specially designed for the nanoimprint process, with the targets of reducing processing temperature (roughly 50–100 degrees above T_g), enhance the resistance to plasma as required for good pattern transfer and improving the stamp release. An alternative approach that overcomes the problems of thermal and pressure budget has gained attention in recent years. This approach consists of imprinting a monomer or a prepolymer fluid precursor of very low viscosity and then cure the film inside the mould cavities by thermal heating at a moderate temperature or using UV illumination.

This latter method is known as Step & Flash Lithography [86]. It requires the use a transparent mould as quartz for the UV curing of the polymer, which brings an additional benefit of greatly simplifies the alignment procedures. The polymer can be easily synthesized and engineered in order to minimize the viscosity of the precursor fluid and to obtain specific mechanical properties after cross-linking, in order to meet the optimal characteristics for subsequent pattern transfer by ion etching, lift-off or electroplating. A very well-known thermally or UV curable material, the polydimethylsiloxane (PDMS) has shown in particular convenient properties for application of nanoimprint and embossing techniques in the field of biomedicine. PDMS is a biocompatible, physiologically inert elastomer that is currently used to produce spare parts in human body (e.g. mammary prosthesis, contact lenses, intra-aortic balloon pumps). At the same time PDMS shows compatibility with classical microfabrication processes, and is a material of choice for micro and nano-fluidics, biochips, and optical components due to its transparency in the UV-visible spectral range. The process for structuring the PDMS, belonging to the class of techniques known as soft lithography [87], is applicable over a large range of dimensions, can be used to fabricate very high structures of resist (100 μm thick) and is also capable of nanometric resolution. A comprehensive review of the use of PDMS in soft lithography is given in chapters 8 and 10 of reference [85].

4.3.3. Focused Ion Beam Lithography for 3 Dimensional Structures

Focused Ion Beam Lithography FIB is a very powerful technique for writing direct patterns on many substrates, it is a mask-less and resist-less technique that allows a wide range of applications which provide good resolutions (down to 50 nm) [88–90]. FIBs can be used to pattern materials with nanometer dimensions by ion implantation, ion exposure of resist, ion milling, gas-assisted etching, and ion-induced deposition of material. This permits us to apply this technique to a wide range of applications where other techniques are not permitted or are too difficult and complex. If we couple the FIB with a SEM facility (figure 4.15) on the same chamber, it is possible to obtain prototype devices in a simple and fast way. In fact, the possibility to fabricate in the nanometer scale by the FIB and to align it, as well as to inspect in real time the final structure by the SEM,

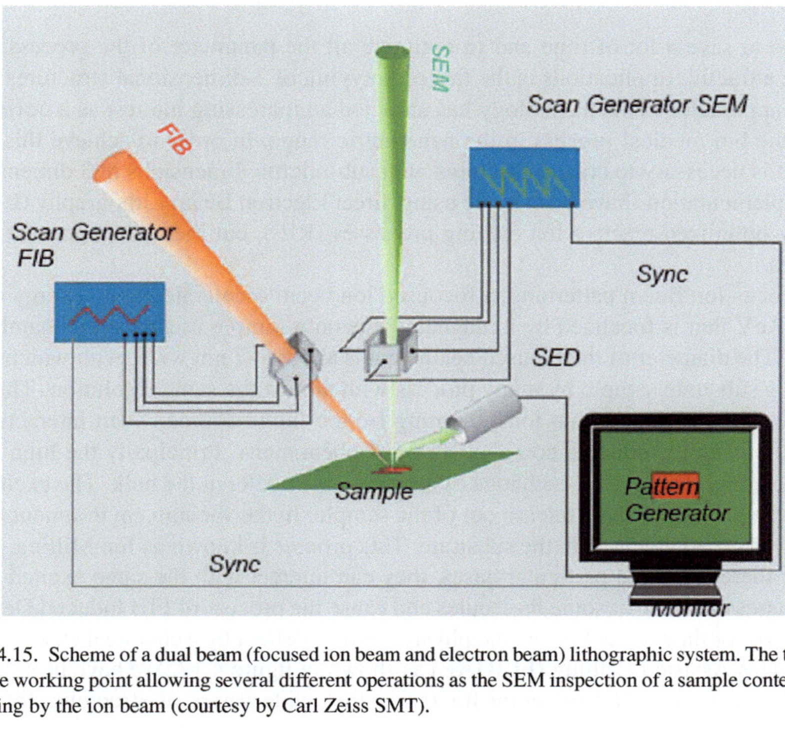

FIGURE 4.15. Scheme of a dual beam (focused ion beam and electron beam) lithographic system. The two beams cross at the working point allowing several different operations as the SEM inspection of a sample contemporarily to its milling by the ion beam (courtesy by Carl Zeiss SMT).

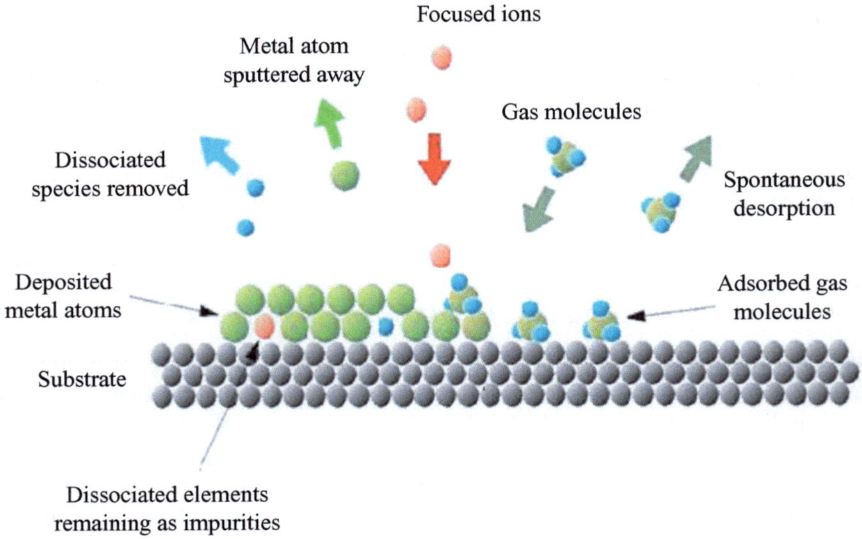

FIGURE 4.16. Scheme of focused ion beam induced etch. The milling action of the FIB is enhanced by the presence of ecthing gasses following those main steps: adsorption of the gas molecules on to the substrate surface, activation of a chemical reaction of the gas molecules with the substrate by the ion-beam generation of volatile reaction products (as GaCl3, SiCl4, SiF4, etc.), evaporation of volatile species and sputtering of non volatile species. (courtesy by Carl Zeiss SMT).

allows us to save a lot of time and to optimize all the parameter of the process. One of the most attractive applications is the fast prototyping of 3-dimensional structures for bio-medical application. This technology has attracted an increasing interest as a novel tool to control the bio-medical process in the nanometric range. In order to achieve this kind of device, it is necessary to create structures with sub-micron dimensions in 3 dimensionally. Most implementations have been made using direct Electron Beam Lithography (EBL) and carefully optimized reactive ion etching processes (RIE), but they don't provide real 3D structures.

In Focus Ion Beam patterning, a focoused ion beam accelerated to an energy of about 10–100 KeV that is focalized by a magnetic lens on a sample in a vacuum chamber (10–6 mbar). The diameter of the focused beam spot is about 5–7 nm wide, with which one can write on a substrate sample by many process with almost the same resolution. These high energetic ions (normally Ga+ ions), coming from a liquid source, when interacting with the sample matter produce a complex series of phenomena. Principally the high mass of the particle produces elastic exchange of energy to the nuclei in the bulk. The excitation of those nuclei can cause their release out of the sample. In the vacuum environment they can be removed creating a hole in the substrate. This process is known as Ion Milling. If in the chamber there are some particular gases, they can interact with the same excited surface. These gasses can deposit some molecules and cause the process of FIB Induced Deposition (figure 4.16) or they can enhanced the physical milling effect by a chemical etching process and FIB Gas Assisted Ecthing (FIBGAE) as shown in figure 4.17. We have been applying all such process for 3D fabrication for fabricating wide varies of structures for various

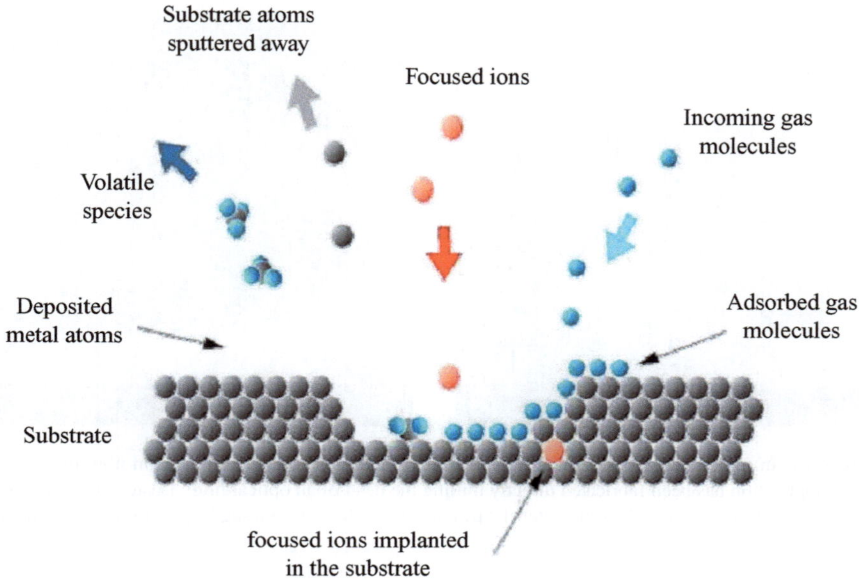

FIGURE 4.17. Scheme of focused ion beam induced deposition: adsorption of the precursor molecules on the substrate, ion beam / e-beam induced dissociation of the gas molecules, deposition of the material atoms and removal of the organic ligands. (courtesy by Carl Zeiss SMT).

applications including biomedical. Few of the fabricated devices and structure using FIB includes:

- *Micro-lens on top of fiber tip*
- *Nano-gaps*
- *Bio sensors devices*
- *Bio actuator devices*
- *Microfluidic micro channel*
- *3-Dimensional structures*

4.3.3.1. Micro-lens on Top of Fiber Tip Optical coupling using conventional optical elements placed at the Single Mode Fiber (SMF) end or between the fibre and the waveguide suffers with a poor coupling efficiency primarily due to optical axis alignment mismatch and optical mode mismatch. In order to improve the coupling efficiency between SMF and the optical elements, a large mode active optical device over the whole length of an active device is fabricated. Among such techniques, tapering the exit end of SMF or formation of a *micro-ball with higher refractive index melt on fiber end are commonly adopted so as to enhance the numerical aperture of the fiber*. However, the coupling losses are still large. In the last few years, various mode matching techniques using spot size transformers have been proposed and demonstrated [91–94]. The function of the mode transformer is to alter the shape and size of the beam from the active device to closely match that of the waveguide. *Diffractive Optical Elements* (DOEs) with continous relief fabricated as an *optical mode*

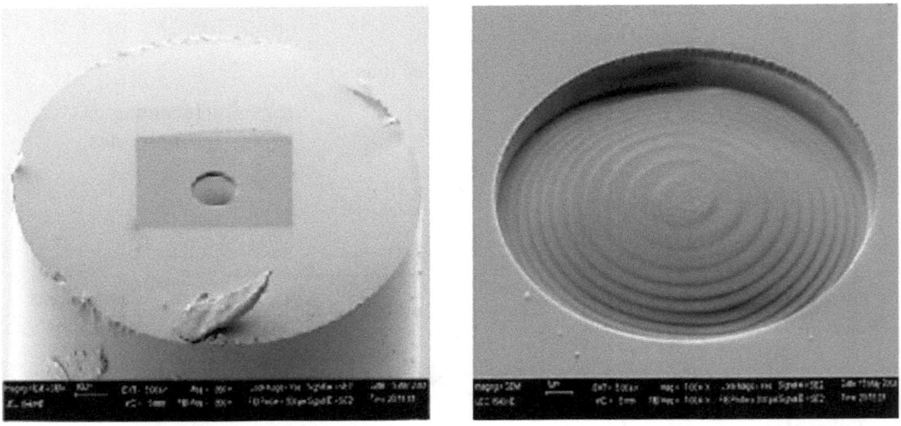

FIGURE 4.18. In some condition is difficult to work with "standard" lithographies. A on-fibre micro-lens for photonics application has been fabricated directly milling the head of an optical fibre. Designed microlens parameters: lens curvature: 28.5 μm, lens diameter: 10 μm; focal length: 58.6 μm, sag height, 1.0 μm; working under wavelength of 1550 nm.

converter to achieve efficient mode matched coupling had been fabricated on top of the fibre tip by e-beam lithography in a polymeric material and was reported by us recently [95]. The role of this element was to focus and shape the beam exiting the fiber into a desired intensity distribution. We adopted FIB technology to derive the required continuous relief in a *DOE-lens element for its use as an optical mode converter* fabricated on top-of-tip of the cleaved single mode fiber. The most obvious advantage of FIB is the simple procedure, with no need for pattern exchange from resist to substrate and direct milling of the required pattern on the substrate. Therefore, it is easier to control the relief form. In the FIB milling function, the high energy and small spot size of the ion beam (few nanometers) impinging the sample surface and facilitates easy material removal. The details of FIB fabrication of microlens on top of fibre tip by FIB and its characterisation was published recently [96]. FIB fabricated microlens profile on fibre tip is shown in figure 4.18. Designed microlens parameters were as follows: lens curvature: 28.5μm, lens diameter: 10μm; focal length :58.6μm, sag height, 1.0μm; working for wavelength of 1550 nm.

4.3.3.2. Nano-gaps for Molecular Conductivity Measurements The use of biological materials in the field of electronic or in the study of their physical property is a very new filed of application (electrical property like resistivity, thermal conductivity etc.). Sometime it is hard to apply a voltage to the extreme parts of a molecule or even cell. For example, to connect the two extremities of DNA chains in very narrow terminals would be required. In figure 4.19 we see the connection pad, made by electron beam lithography separated by a very small cut made by FIB milling. The dimension of the cut is around 10 nm and it was made by a very weak ion current (5 nA) and in a very short time. This is a very small distance not easy to be obtained by other technique; it is the desired dimension to be used in many experiments. Putting a molecule in between is not easy but possible and many bio-physical properties are now available to study [97].

FIGURE 4.19. Example of device fabrication using a Focused Ion Beam Milling: the electrical circuit and the connection pads, made by electron beam lithography, are separated by a very small cut made by FIB milling.

4.3.3.3. Bio Thermometer Devices Measuring the temperature is a very easy process. We can measure by using many principles applicable in different environment (mercury thermometer, thermocouple, liquid crystal thermometer, resistance thermometer RTD). But to enter inside a cell and check how the heating distribution changes during biological processes requires a special micro-structured device. A micropipette can enter cells without destroying too much; a nano-thermocouple or a nano-RTD detector could be fabricated on the cylindrical extremity of it (the final diameter could be around 1 micron). All those devices are already fabricated in small dimensions but only on a large and flat substrate. As stated before, the main characteristic of the FIB technique is the non required standard substrate. By using Ion milling as well as FIB Assisted Deposition it is possible to pattern a conductive circuit on the glass cylindrical lateral surface of the micro pipette as shown in figure 4.20. Few hundredsof a nanometer are enough to create a sufficient resistor sensitive at the temperature change. In this way, we can detect with precision (of less than 1 micron) where a cell produces heat and how much. In the same way is possible to fabricate a thermo detector and a localized heating system on an Atomic Force Microscope (AFM) tip. In

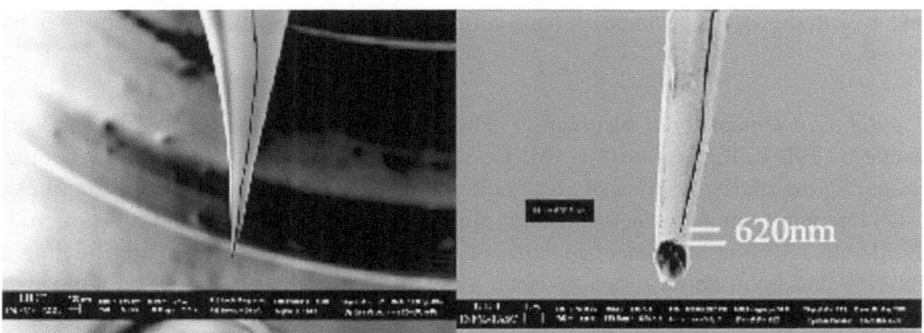

FIGURE 4.20. By using Ion milling as well as FIB Assisted Deposition it is possible to pattern a conductive circuit on the glass cylindrical lateral surface of the micro pipette. It was used as a intra-molecular thermometer.

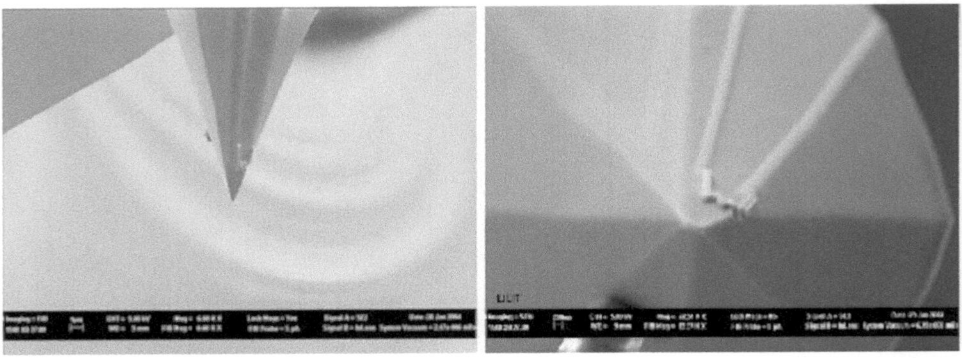

FIGURE 4.21. Example of device fabrication using a Focused Ion Beam Induced Deposition: a platinum induced deposition was used to add a thermometer and an heating devices on an AFM tip.

figure 4.21 shows a modified silicon tip and a resistance which can produce heating when a current passes, in addition a weak resistance at the side which can detect the real temperature; all of these can be used during the normal functioning of the topological imaging process of the AFM. It is a powerful instrument to study the shape of biological samples while obtaining information about heating distribution and deformation.

4.3.3.4. Bio Actuator Devices, Micro Tweezers Another particular application is the fabrication of mechanical tweezers on the top of a micro pipette. That's the case of a real actuator the can act directly at 1 micron range driven by electrical signals. This device is made by using a precursor gas diffused in the vacuum chamber contemporary at the Ion beam interaction with the substrate [98]. Three-dimensional (3D) nanostructures on a glass capillary have a number of useful applications such as manipulators, actuators, and sensors in the various microstructures. It was demonstrated a phenomenon that two diamondlike-carbon pillars on a tip of glass capillary fabricated by 30 keV Ga+ focused-ion-beam-chemical vapor deposition (FIB-CVD) with a precursor of phenanthrene vapor were able to work as a manipulator during FIB irradiation. It clearly works as 3D nanomanipulator and actuator, as it was demonstrated, by applying voltage onto an Au-coated glass capillary. In figure 4.22 is shown the principle of nano actuator and nanosensors and an example of nano tweezers.

4.3.3.5. 3-Dimensional Structures There are many different shapes that are made possible by FIB milling as well as FIB-CVD and FIB-GAE [99]. In figure 4.23 some of them and, particularly, a series of important diamond or amorphous carbon structures are shown, together with an array of metallic pins for capturing particles and SiO_2 micro box to hold cells. A complete list of 3D shapes is unlimited. It is only a matter of time and fantasy to find good applications for such a flexible technique. It should be remembered that it is a serial writing system, there is only one beam that writes sequentially all the patterns. Comparing the parallel lithographic system (like the optical lithography) all the serial systems are slower (not good for mass production) but incredibly more flexible (very good for the prototyping). The techniques FIB should represent one of the highest levels

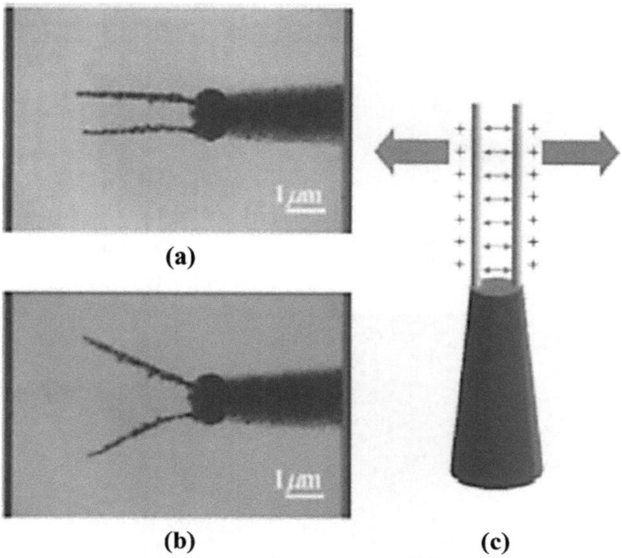

FIGURE 4.22. Example of device fabrication using a Focused Ion Beam Induced Deposition: SIM observation of two pillars moving during FIB irradiation. The surface of glass capillary was not coated by Au sputter. a) Two pillars did not move without FIB irradiation. b) Two pillars opened with FIB irradiation. c) Illustration of moving mechanism due to charge repulsion. (courtesy by Kometani et al. [98]).

of flexibility and associate it with a SEM facility, which constitutes the higher level for the fast prototyping of micro and nano structures.

4.4. HYBRID LITHOGRAPHY APPROACH

Hybrid lithography approaches are emerging as a class of fabrication methods that could potentially yield structures and devices with vastly complex 3D geometries, improved mechanical performance, and enhanced functionality. Hybrid approaches exploit the unique or most of the advantages and properties inherited by individual fabrication methodology. Particular promise is seen for fabrication methods that yield hybrid nanostructures akin to approaches that are commercially and clinically relevant.

4.4.1. X-ray and Nanoimprint Lithography for 3D Patterning

Most of the excellent characteristics that nanoimprint lithography (NIL) has shown as a two-dimensional patterning technique, such as high resolution, high throughput and low cost, do apply also to its three-dimensional structuring capabilities. Nevertheless, the potential of NIL for 3D remains largely unexplored and unexploited, hindered by the difficulties in the fabrication of high quality masters at the wafer scale level. This fact is due to the need of using advanced techniques such as Focused Ion Beam (FIB), or gray-tone Electron Beam (EB) lithography, that are intrinsically slow or require special know-how to be

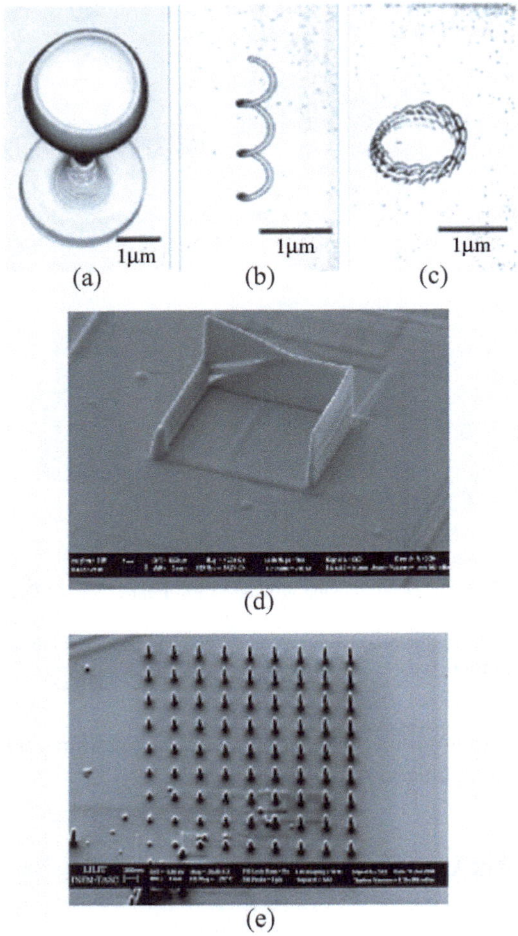

FIGURE 4.23. Example of device three-dimensional structures fabricated using a Focused Ion Beam Induced Deposition: (a) wineglass, (b) coil, (c) micro-crown (courtesy by J. Fujita et al. [99]), (d) SiO2 micro box to hold cells, (e) an array of metallic pins for capturing particles.

performed. In order to solve this difficulties we have developed a method for 3D patterning using nanoimprint, in combination with other techniques. Our aim is to demonstrate that there is a number of possibilities for fabricating complex 3D masters, using simple, robust and fast processes, offering a tight control on three-dimensional profiles in the scales from sub-100 nm to above 100 mm.

We present the results about an innovative lithographic approach (see figure 4.24) that combines nanoimprint (NIL) and X-ray lithography (XRL) for fabricating unconventional three-dimensional (3D) polymer structures [100]. The use of XRL for structuring a pre-patterned resist by NIL gives rise to high-resolution high-aspect-ratio structures whose overall profile is enveloped by the original 3D imprinted profile. The technological potential of this method has been demonstrated by patterning several different types of structures with

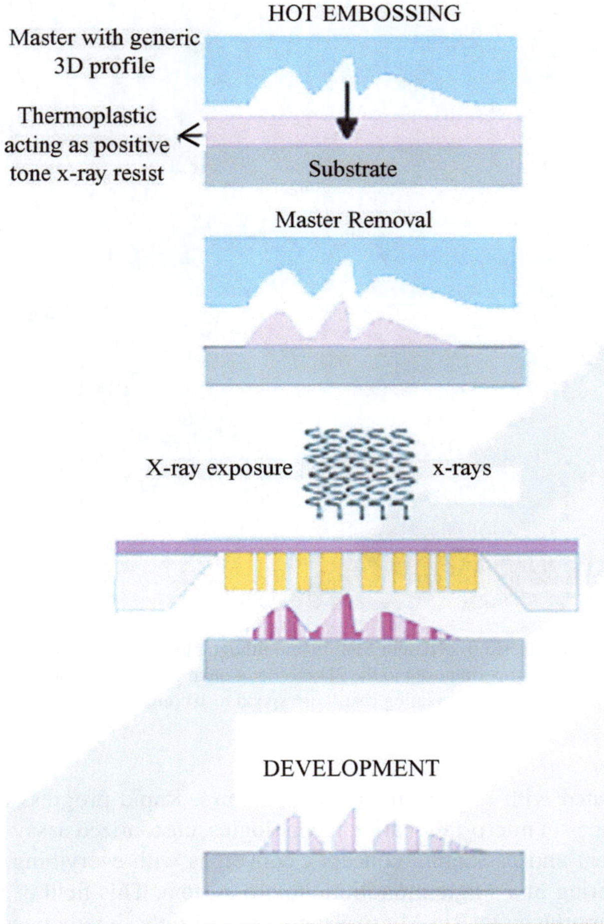

FIGURE 4.24. Scheme of the process combining nanoimprint and x-ray lithographies.

X-ray lithography on a hexagonal array of hemispheres previously obtained by nanoimprinting and is shown in figure 4.25.

4.4.2. Lithography at Interface-Binary Resist Process Combined with Multiple Tilted XRL and EBL Lithography

4.4.2.1. Multiple-tilted X-ray Lithography for 3-D Patterning The revolutionary miniaturization of analytical instrumentation and methodologies is nowhere more evident than in micro-fluidic systems. The development of miniaturized gas chromatographs and ink jet printer nozzles introduced the field of microfluidics [106, 107] applications have branched into many aspects of the biological and chemical sciences [108]. Many of these advantages stem from the possibility for significant device parallelization and the transport

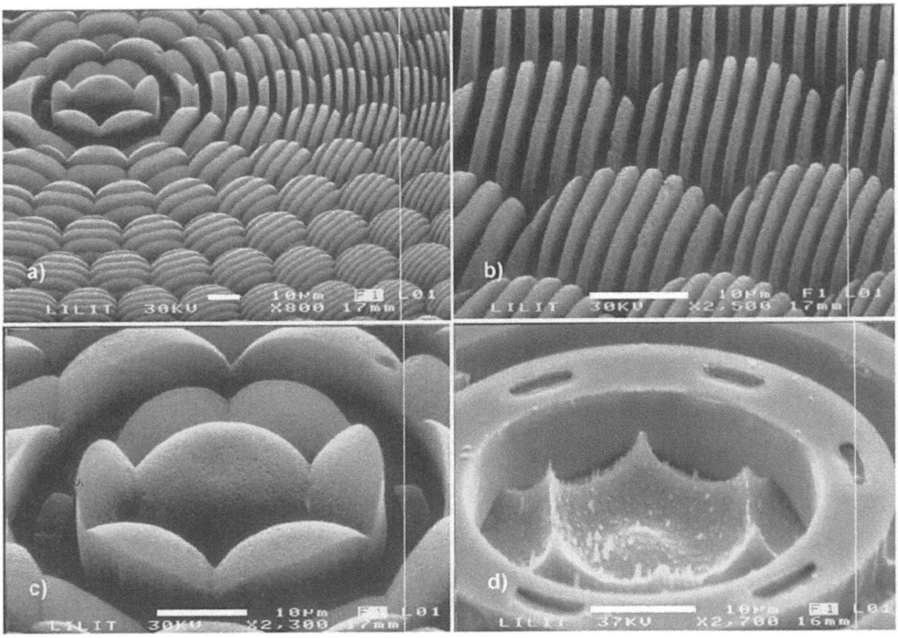

FIGURE 4.25. Structure fabricated by NIL and XRL hybrid lithography approach. Shown from top right (clockwise) are Zone Plate structure superimposed to the 3D profile of an array of hemispheres, (bottom extreme left) inverted 3D structure after nickel electroplating (bars correspond to 10 μm).

properties associated with laminar micro-fluidic flows. Rapid progress in micro-fluidics relies upon advances in micro-fabrication technologies, customized assay chemistries, materials development and packaging concepts, converges with everything from biology to electrical engineering in a single monolithic micro-system. This field of micro-fluidics is showing great importance to Bio-Microsystems. Based upon interconnected networks of micro-channels and reservoirs with tiny volumes, micro-fluidic devices are well matched with micro-electromechanical system (MEMS) and miniaturized optics and thus set a platform for the miniaturization of instrument for cellular manipulation and analysis, integrated micro-fluidics for bioassays and sample preparation, and micro- and nano-fluidic chip processors for the manipulation and self-assembly of bio-molecules, genetic.

A microfluidic platform fabrication is normally based on a 2-D multi-layer lamination method. Fluidic micro-channels are fabricated by sandwiching laser-machined adhesive-backed polyimide gaskets between layers of the device. As an example, the fabrication of micro-fluidic channel more often requires the covering of the patterned trenches by a ceiling-top realized by an aligned process of roof-layer sticking or soldiering. Individual fabricated components, such as micro-fabricated piezo-electrically actuated pumps and a micro-electrochemical cell are plugged into modules which in turn are implemented into the microfluidic platform.

In XRL, conventionally, the mask+wafer assembly is held perpendicularly to the beam. The mask shadows during one-to-one projection of the pattern on the resist providing a

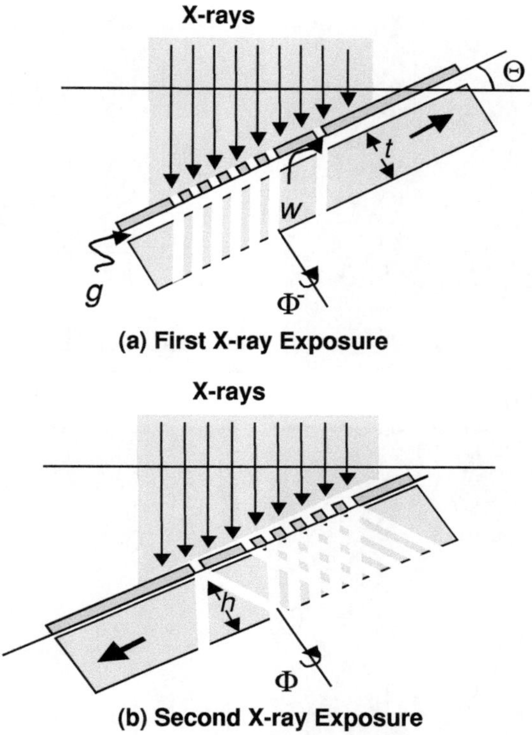

FIGURE 4.26. (a) Scheme of a tilted exposure. The X-ray mask is large as the beam width. The tilt angle q, the azimuth angle f, the pattern width w, the mask-sample gap g and the resist thickness t are shown. (b) Scheme of the exposure geometry after 180° azimuth rotation. The height of the channel is shown.

vertical, digital-like, lithographic profiles. However, the idea underlying for the realization of 3D pattern structuring by *multiple-tilted x-ray lithography* is based on unconventional exposure geometry. The schematics illustration of this geometry is shown in figure 4.26. In this scheme, the mask + wafer are mounted at a tilted angle with respect to the x-ray beam. Seen from the surface of the wafer, each opening of the x-ray mask behave as a collimated light source that exposes the resist along a tilted direction (figure 4.27.a). A 180° azimuth angular rotation around the axis perpendicular to the mask-sample system will generate a second exposure along another direction. The relative position of mask+wafer is kept fixed during the rotation and, due to the fact that no further alignment is required, the multiple-tilted-exposure at different angles can be regarded simply as an independent single step process. The resulted structure in this case is simply triangular shaped trench or a vertical network of intersecting planes useful for micro fluidic.

The concept of multiple-tilted x-ray lithography had successfully been implemented in fabrication of 3D structure [103, 104]. In figure 4.27.b, tilted pillar micro-fluidic channel (after gold electroplating and resist stripping) fabricated by a two multiple-tilt x-ray lithographic exposure is shown. Furthermore, it is possible to fabricated high aspect ratio microstructures with *multiple intersections*, such as grating structures comprising of

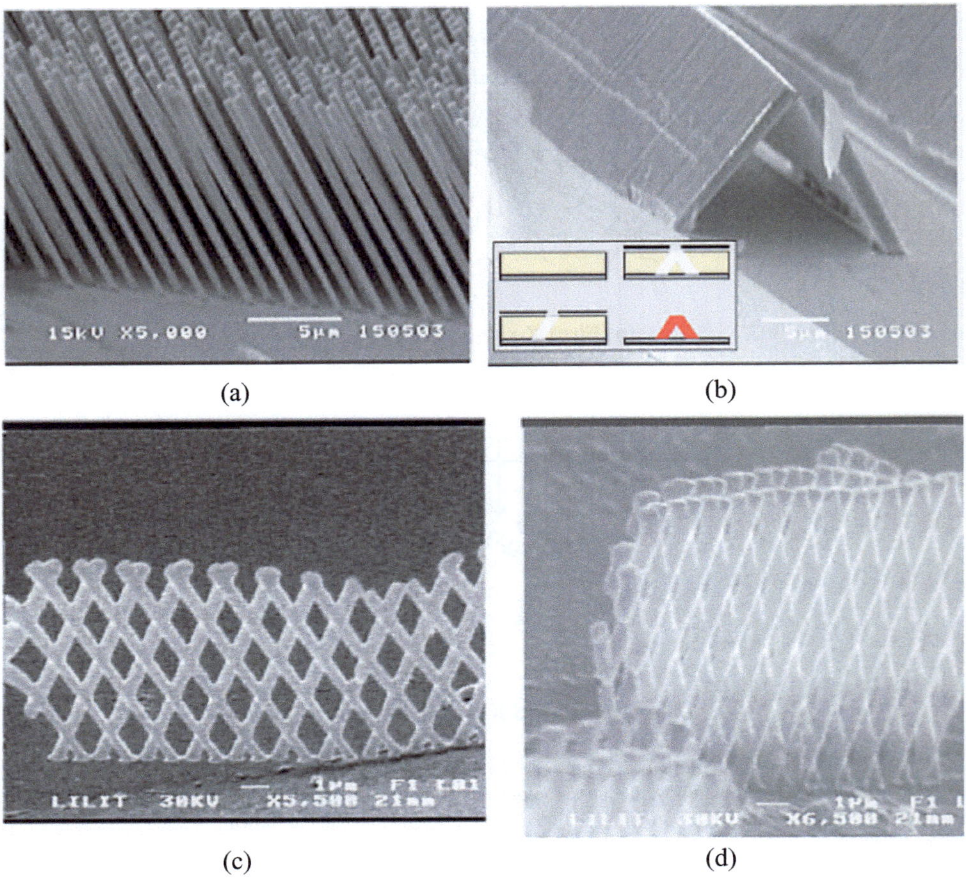

FIGURE 4.27. SEM micrograph showing a fabricated (a) high aspect ratio (45) arrays of tilted pillar obtained by single tilted x-ray lithography exposure. Shown are the arrays of 15 μm thick tilted pillars in gold obtained after electroplating in template generated in PMMA, (b) arrays of tilted pillar micro-fluidic channel (after gold electroplating and resist stripping) by a two multiple-tilt x-ray lithographic exposure (The scheme of the exposure procedure is also shown in the inset), (c) single vertical array (after gold electroplating) of intersecting pillars, which can work as filter in microfluidic channel, and (d) the intersecting pillars array can also be sequenced in several vertical layers of filters for their possible application as electrode in electrochemical micro sensors.

metallic filter sieves which can be used for filtration purposes. An example of such fabricated vertical grating is illustrated in figure 4.27.c. Also, shown in figure 4.27.d, a sequence of vertical gratings fabricated by the intersection of high aspect ratio inclined pillars. Metallic sieves could offer advantage in using them as cathodes in microfluidic devices for sorting different ionic species. Variety of metals can be used (nickel, chromium, copper, silver, etc.), but gold is preferable because of its high compatibility with biological fluids. Alternatively, in case a *sieve of insulator material* is required; the patterned resist can be infiltrated, for example, with silica or any other bio-compatible material by SOL-GEL technique.

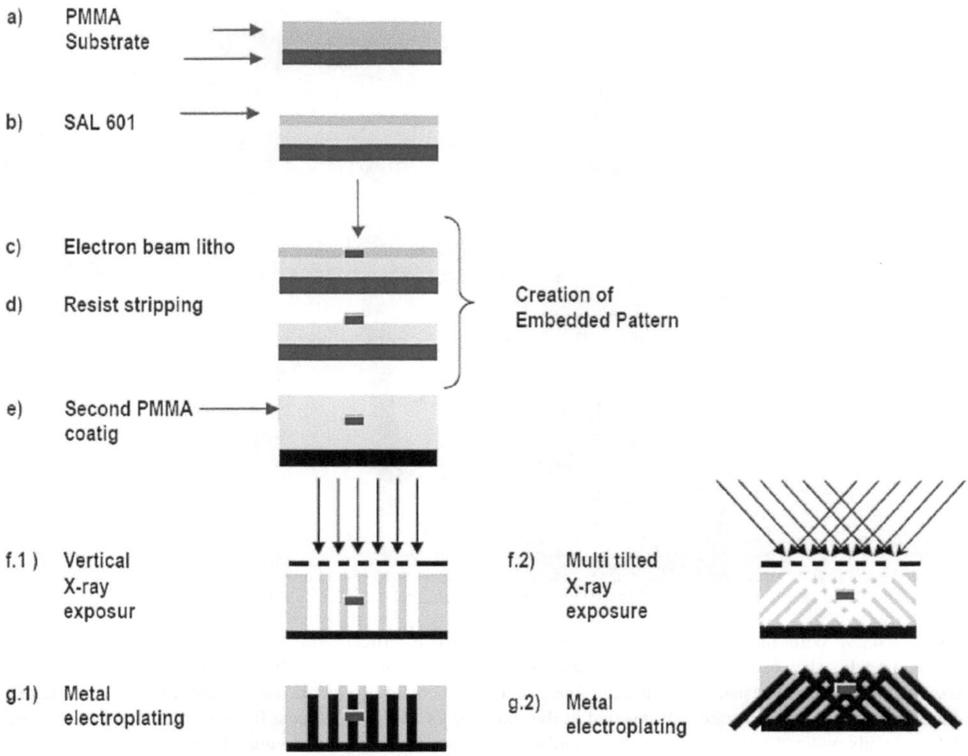

FIGURE 4.28. *Interface lithography* process flow scheme. Two resist are combined with two distinct lithography. After a first PMMA coating and baking (a), a second resist (SAL-607 ER7) is spun, baked (b), exposed by Electron beam lithography (c) and normally developed with MF312 (d). In the following a second PMMA coating is spun and normally baked (e). At this point x-ray lithography can be performed either in the usual vertical configuration (f.1) or in the multi tilted configuration (f.2). The SAL-607 ER7 structure at this point remains unchanged resulting completely embedded in the in the PMMA resist structure that can be used as template for a final electroplating metal growth (g).

4.4.2.2. Interface Lithography An hybrid lithography based on binary resist process [101] and combining two different lithographies, EBL and XRL, has been developed for the fabrication of suspended structures and for the realization of arbitrary shaped patterns embedded inside three dimensional structures. In figure 4.28, is shown the illustration of a process.

The developed process utilizes the combination of a low sensitivity positive tone resist, PMMA, and a high sensitivity negative resists, SAL-607 ER7 [102]. A layer of SAL-607 ER7 is spun and pre-baked on the top of a previously spun and baked PMMA layer (figure 4.28.a & figure 4.28.b). The low threshold dose of SAL-607 ER7 allows performing the electron beam lithography almost completely unaffecting the bottom layer PMMA resist owing to its low sensitivity. The process on the negative resist, completed with post baking and development, results in a chemically stabilized polymeric pattern of the designated pattern lying on the top of the first (bottom) PMMA layer are resulted. Subsequently, a

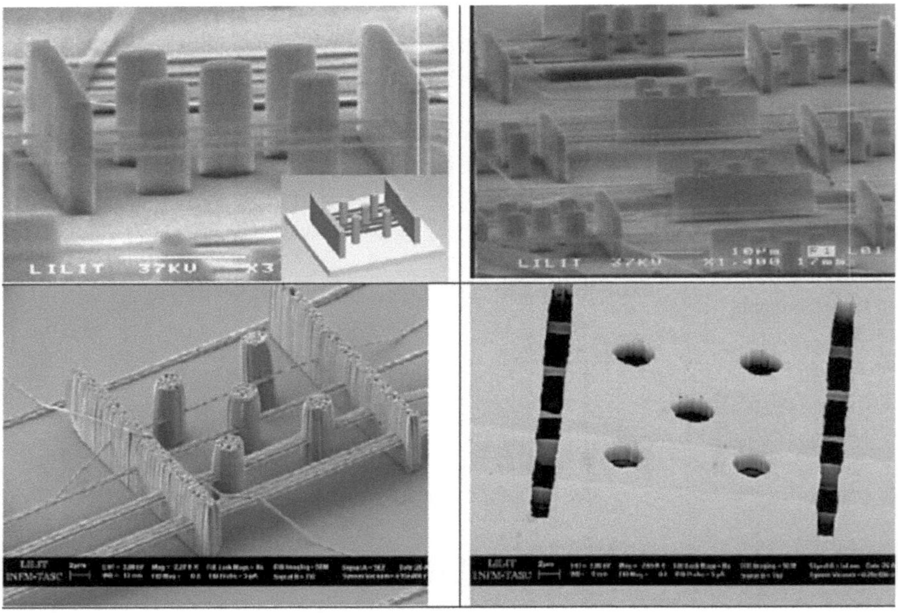

FIGURE 4.29. Wires made of SAL 601 resist suspended between vertical PMMA pillars representing micro tanks whose model is sketched in the inset. SEM micrographs showing in (a) & (b) the fabricated wire with different size made in SAL 601 resist, (c) illustrating linear defect (generated by electron beam lithography with different doses) parallel to the substrate and inserted at the interface of two 3-D lattices. Bottom inset (left), shows the defect details realized with the lowest dose, and the inset at top exhibit the alignment efficiency achieved, and (d) showing the detail of the waveguide entering the 3-D photonic crystal structure representing micro tanks whose model is sketched in the inset. The scale bar represent 10 μm.

second PMMA coating is spun and normally baked (figure 4.28.e). A slow baking process of the second PMMA layer assures a perfect homogeneity of the two PMMA layers embedding the predefined SAL pattern. At this juncture, XRL was performed either in the usual vertical configuration (figure 4.28.f1) or in the *multiple-tilted configuration* (figure 4.28.f2). The SAL-607 ER7 structure at this point is completely embedded in the PMMA resist structure which can be used as template for a final electroplating metal growth (figure 4.28.g.). The critical characteristic that allows performing the X-ray lithography on the whole resist structure is that SAL-607 is *transparent* to X-ray but remains *mechanically* and *chemically stable*. The successive developing of the structure in the PMMA does not invalidate the defect patterning at the interface.

The *interface lithography* has successfully been applied for the fabrication of several 3D test structure: a suspended resist wire (see figure 4.29). This structure represents a complex symbolic test pattern that, however, can be used as template to be converted into a *flyover channels* for micro-fluidic application. Several parallel lines of different thickness have been patterned by EBL on a 300nm thick SAL-607 ER7 film. The wire widths were decreased from 800nm to 200nm in step of 200nm. The exposure dose was 10 $\mu C/cm^2$. After developing SAL-607 ER7 and second PMMA resist layer coating, XRL was performed using a negative tone mask made of circles and rectangles few microns wide. The final structure resulted with two resist process: a suspended wires formed in SAL-607 ER7 (by EBL) bridging the PMMA rectangular pillar and cylinders fabricated by XRL (figure 4.29.a.b.c). We believe that the ultimate resolution in the case of wire could be better than 200nm we have achieved in this preliminary investigation and that, however, 200 nm is still much smaller than the required dimension for micro fluidic channels.

To obtain the flyover channel it is necessary to invert the tone of the template structure in order to transform the pillars into fluid reservoirs and wires into tubes connecting reservoirs. This can be easily obtained with metallic electrolytic growth exploiting as cathode the metallic film base plating on which the resist template has been realized. The gold electrolytic growth progressively overwhelms the resist wires that finally result embedded in a gold structure grown up to the top of the cylindrical PMMA micro reservoirs (figure 4.29.d). Removal of residual PMMA and SAL-601 resist was performed with hot acetone and Nanoremover[1] respectively. A top view of the fabricated structure is shown in figure 4.2.c where one can see the presence of left-over embedded SAL-601 wires that we have deliberately left for sake of clarity.

4.5. 3D TRAPPING AND MICRO MANIPULATION BY MEANS OF OPTICAL TWEEZERS

In 1986, Ashkin and colleagues reported the first single laser beam gradient trap, called optical tweezers [109]. A single-beam optical trap can be formed by highly focusing a laser beam using a microscope objective with a high numerical aperture. At the beginning, Ashkin and colleagues trapped micron-sized dielectric particles immersed in water. One year later, they reported also trapping of living biological samples in 3D trap using a single laser beam [110]. Laser trapping has been then used, for almost two decades, in applications as diverse as experiments on molecular motors in biology and the movement of Bose-Einstein condensates in physics [111]. During the last decade, the capabilities of optical tweezers have been greatly extended by the development of the so called 'tailored beams' (e.g. Laguerre-Gaussian and Bessel beams) [112, 113] and by schemes for generating large number of trapping sites and shapes simultaneously [114]. Even if trapping of dielectric micro spheres has been investigated mostly, trapping and manipulation of particles made from different materials, with different shapes and ranging in size from ten nanometers right up to tens of μm has been demonstrated as well [115–117].

4.5.1. Optical Tweezers Enabled 3D Trapping and Micromanipulation

Multiple trapping and independent manipulation of the trapped particles represents a challenging task required by many applications [118]. Multiple traps can be obtained either by rapidly *scanning* the laser beam or by *splitting* the *input beam* in several output beams. Scanned optical tweezers can trap multiple particles by dwelling briefly on each

[1] Nanoremover P. G. by Shipley

before moving onto the next [119]. However, its main drawbacks are the restriction to planar patterns of traps and the limitation on the extent and the complexity of the patterns. In the second technique above mentioned, the splitting of the beam is obtained by using a diffractive beamsplitter. Such a beamsplitter can be calculated as a computer generated diffractive optical element (DOE). Phase DOEs are known as optical elements that convert, by modulating only the phase, a specified illumination wave into a wave with a desired distribution of its amplitude, phase or polarization. Efficient iterative techniques, based on the thin-element approximation and the scalar diffraction theory, have been proposed to design phase DOEs with surface relief modulation for applications in visible light [120]. Successful manipulation in three dimensions was demonstrated using DOEs obtained by adding a lens term to the blazed gratings [121]. Similar techniques allow to obtain DOEs that, illuminated by a Gaussian beam, generate Laguerre-Gaussian or high-order Bessel beams [112, 113].

We have developed our own algorithms and code to design phase DOEs that can transform a single laser beam into an array of independent traps, each with individually specified characteristics, arranged in arbitrary 2D or 3D geometrical configurations. We have used two approaches: the first is based on phase retrieval using iterative algorithms (PRIA) [122], that can be followed by optimization using genetic algorithms, and the second is based on the propagation and superposition of spherical waves [123]. The fabrication is based on e-beam lithography, which allows to obtain phase DOEs characterized by high diffraction efficiency and very good signal to noise ratio (SNR) in the generated pattern. Phase modulation is accomplished by surface profile modulation in poly-methyl-methacrylate (PMMA) resist to generate array of traps. An example of such a DOE, obtained with PRIA is shown in figure 4.30.a. When illuminated with a Gaussian beam (wavelength 633 μm, beam diameter 3 mm) the DOE generates 5 × 5 spots in the second focal plane of a converging lens. The generated patterned is illustrated figure 4.30.b. Its size depends

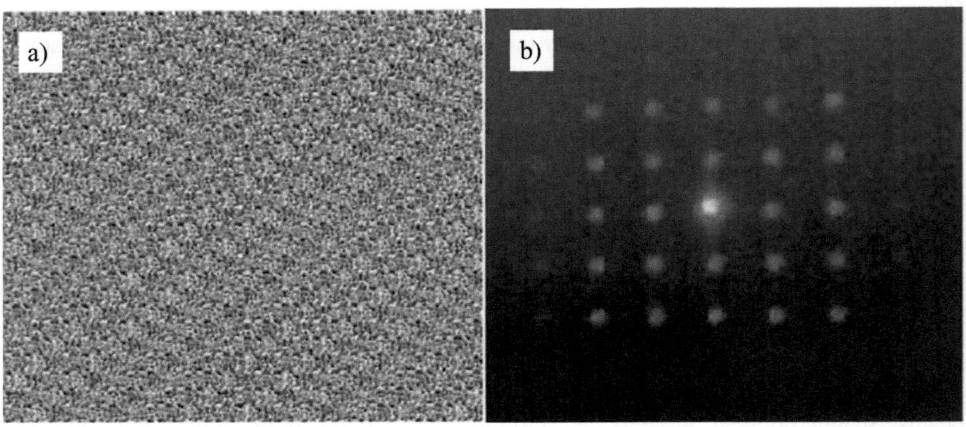

FIGURE 4.30. Showing (a) Diffractive optical element with four phase levels represented in grey levels; the DOE is 2 mm size defined by 512 × 512 pixels, (b) The intensity distribution of the diffracted pattern obtained from the DOE depicted in (a) in the focal plane of a converging lens (focal length 200 mm). The size of the pattern is about 2 mm [125].

on the focal length of the converging lens. One can note the presence of a bright zero order due to the mismatch of the phase levels in the fabricated DOE, from the calculated values (the contribution of the zero order can be reduced but can never be totally eliminated). Since the trap corresponding to zero order could be unwanted sometimes, the arrays of traps are displaced faraway from the zero order.

DOEs fabricated using e-beam lithography are static and therefore they can be used only for multiple trapping without any possibility to move the trapped particles together or independently. The only exception is the DOE that converts Gaussian to Laguerre-Gaussian beam, since this beam carries angular momentum that can be transferred to the trapped particles making them to spin or move on a circular orbit. The full utility of the DOEs in the field of laser trapping and micro manipulation is realized when a computer addressed spatial light modulator (SLM) is used to project sequences of trap-forming DOEs almost in real time.

An SLM imposes a desired amount of phase shift at each pixel in an array by varying the local optical path length. Typically, this is accomplished by controlling the local orientation of molecules in a layer of liquid crystal. Since at present SLMs have the minimum pixel size around 10 μm with a maximum fill factor of 90% and the number of pixels limited to 1 Mpixels, the quality of the diffracted pattern is lower than that obtained with a e-beam fabricated DOEs. This is the price to be paid for the dynamic control of the trapping array configuration. However, for a lot of applications there is a good compromise between dynamic control and the quality of the generated trapping pattern.

A typical optical setup used in experiments for multiple trapping and micro manipulation is illustrated in figure 4.31. The source is a single-mode CW laser fiber (YLM—10W from IPG Photonics) which generates a collimated linear polarized laser beam at $\lambda = 1064$ nm. The beam is expanded by a 3X telescope to match the active size of the SLM (Programmable Phase Modulator—PPM-X8267 Hamamatsu) on which we implement the DOEs. The PPM is an electrically-addressed phase SLM using an optical image transmitting element to couple an optically-addressed PAL-SLM (Parallel Aligned Nematic Liquid

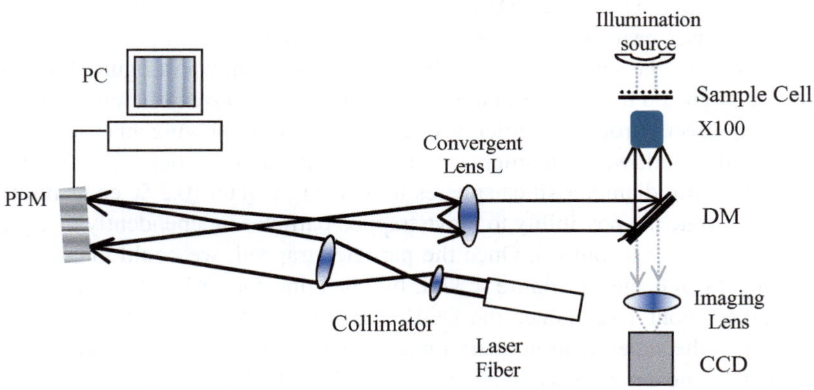

FIGURE 4.31. The optical setup used to implement multiple 3D dynamic trapping and manipulation: PC—Personal Computer, PPM—Programmable Phase Modulator, X100—microscope objective, DM—Dichroic Mirror, CCD—Charged Couple Device sensor [126].

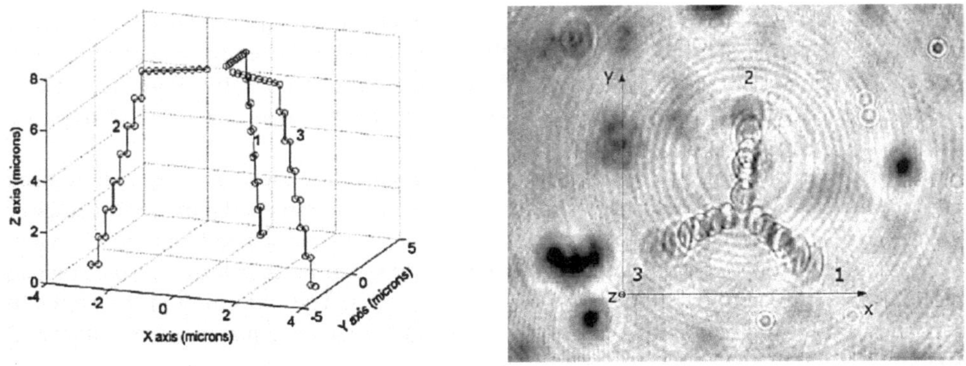

a) b)

FIGURE 4.32. Three trapped particles moved independently in x-y-z in a $8 \times 8 \times 8\ \mu m^3$ volume: a) the scheme of the trajectories followed by the particles; b) the superposition of 6 images taken during the movement from bottom to the top in 20s [125].

Crystal Spatial Light Modulator) with an electrically-addressed intensity modulator. Its main characteristics are: 1024×768 pixels on the LC display, 19 lp/μm maximum display spatial resolution on the PAL-SLM, $20 \times 20\ mm^2$ active area, 8 bit addressing, XVGA signal format and provides a phase shift higher than 2π at $\lambda = 1064$ nm. To generate 3D arrays of traps, we calculate DOEs which includes the lens effect, based on the spherical wave approach described in detail in references [124] and [125]. The desired pattern is thus obtained in a volume centered at a distance equal with the focal length of this lens. This pattern is then transferred by the convergent lens L and the microscope objective (X100, 1.4 NA) into the focal plane of the microscope objective, with a scale factor proportional to the ratio of their focal lengths. The sample cell is also positioned in the focus of the microscope objective. The dichroic mirror DM is used to properly fit the beam into the entrance pupil of the microscope objective. We have used an inverted microscope from Nikon (TE2000-E) and a Charged Couple Device (CCD) sensor to monitor the behavior of the particles interacting with the laser beam in the sample cell. One can consider, in practice, two slightly different versions of the setup described above. The first is shown in figure 4.31. The second, easier to align and handle, is obtained by removing the convergent lens L and sending the laser beam directly trough the microscope objective thus allowing an easier alignment. The sample cell is built with two microscope slides separated by sticky tape 120 μm thick and filled with 2 μm diameter silica spheres immersed in water (0.2 % concentration). In order to demonstrate the possibility to move trapped particles independently in x-y-z, three micro spheres were first trapped. Once the particles trapped, we could manipulate them independently as indicated in figure 4.32.a, by changing the DOE. The calculation of a new DOE takes about 1 sec (since the DOE is rather big: 1024×768 pixels). In figure 4.32.b, shown is the superposition of six images taken during the movement performed in 20 seconds. This movement was obtained with DOEs calculated using the spherical wave approach. The vertical displacement along the optical axis, z, is visible in the figure from different focusing of the micro spheres: spheres located on upper positions have smaller images than the spheres located in the focal plane of the microscope objective. The size of these images was correlated with the sphere vertical positions during a calibration procedure before the experiment. The particles are moved inside a volume of about $8 \times 8 \times 8\ \mu m^3$ with a 0.3 μm lateral increment for x-y displacements and 1 μm increment for vertical displacement along z axis. The number of trapped beads could be increased to hundreds for 2D and tens for 3D micromanipulation. The limitation comes from the available laser power (about 1 mW is required for each trapped particle). For 3D micromanipulation an additional limitation comes from the light scattered by the particles trapped in different planes.

4.5.2. 3D Micromanipulation of Cells by Means of Optical Tweezers

Following the technique described in the previous paragraph one can create an array of beads surrounding biological samples like cells. The beads can be used to trap and stress cells in order to understand their behavior under such conditions. In order to perform these investigations we surrounded the cell by beads and performed a preliminary experiment [126]. Nine latex microspheres were trapped and arranged as shown in figure 4.33. Eight of them were kept in a fixed position on a circle of 20 μm diameter. The one bead left has to be trapped and shifted axially for 6 μm. Since the manipulation conditions were known a priori, we calculated ten phases DOEs using the spherical wave approach. Each DOE corresponds to the same position of the beads trapped on the circle but to a different axial position for the axial trapped bead. A detail of such a DOE (512×512 pixels) is depicted in figure 4.34.a. The beads, trapped into the same plane are depicted in figure 4.34.b. After 10 sec, the central bead is shifted along the optical axis by 6 μm and the trapped beads in the new configuration are shown in figure 4.34.c and figure 4.34.d. One can notice that the central bead is not perfectly centered. This is due to the small misalignment between the optical axis of the MO and the optical axis defined by DOE implemented on the PPM.

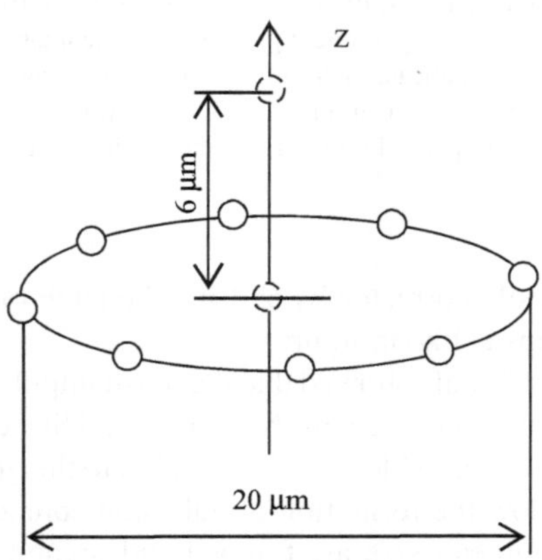

FIGURE 4.33. Microparticle manipulation on x-y.z. Eight beads are trapped on a circle situated in a plane orthogonal to the optical axis. Another particle is moved along the optical axis Three dimensional array consisting of eight beads disposed in two planes separated by 6 μm along the optical axis z [125].

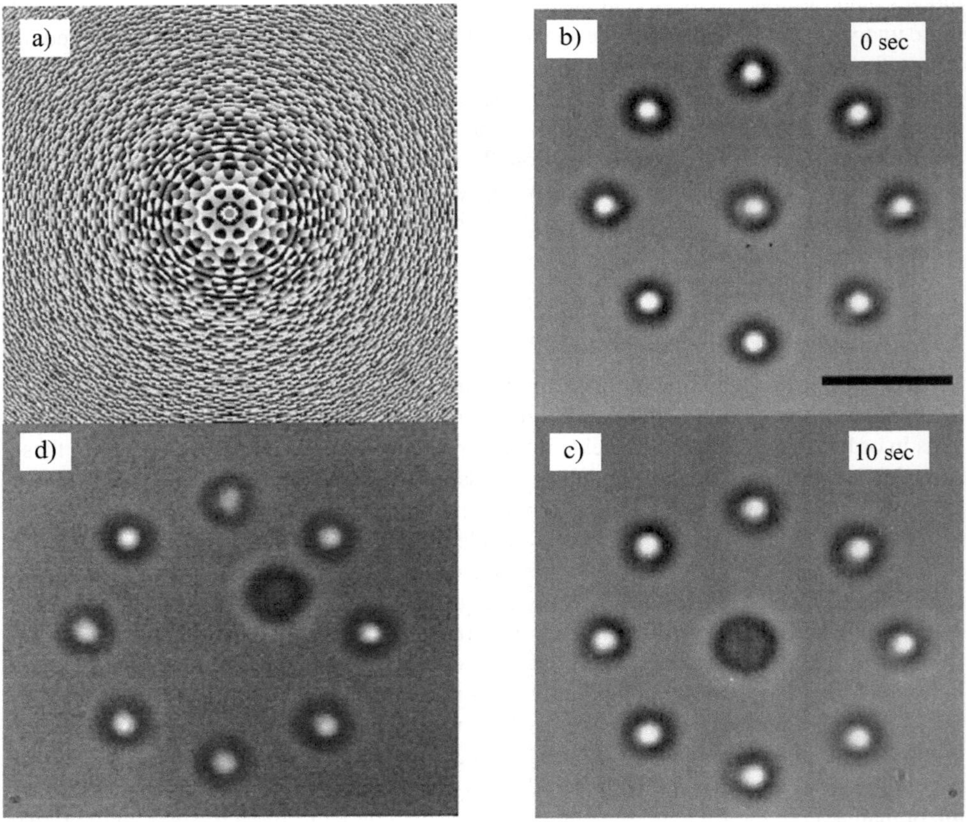

FIGURE 4.34. a) Detail (512 × 512 pixels) of the central part of one of the ten DOEs which generate an eight spots planar plus a an axial spot disposed in two different focal planes (f'1 = 400 mm and f'1 = 500 mm); there are implemented 256 phase levels, represented by different grey levels in the image; b) image of nine micro spheres trapped in the same plane; the scale bar indicates a distance of 10 μm; c) image of nine micro spheres trapped in two different planes: eight in the same plane as in Fig. 1b (z'2 = 10 μm) and one in a different plane at z'2 = 16 μm; d) misaligned central particle trapped almost in the centre of the circle [126].

The misalignment can be easily corrected by shifting the DOE on the PAL-SLM. After this correction the pattern of traps is like in figure 4.34.c.

The experiment with beads only showed that the 3D manipulation can be well controlled and hence we applied this technique to a cell (ND7 neuronal like cell) [127]. Cells are plated on non adherent substrate in order to block their adhesion to the glass. Beads are coated with fibronectin. This coating allow the formation of adhesion contact between cell and beads. Latex microbeads, 2 μm diameter size are trapped and organized first in a "three planes cage" as shown in figure 4.35. One cell is then moved together with the microscope stage at the cage location (figure 4.36.a.b.c). By dynamically varying the geometry of the trapping configuration we stressed the cell as shown in figure 4.36.d. The behavior of the cell under the mechanical action of the beads will be studied using fluorescence microscopy.

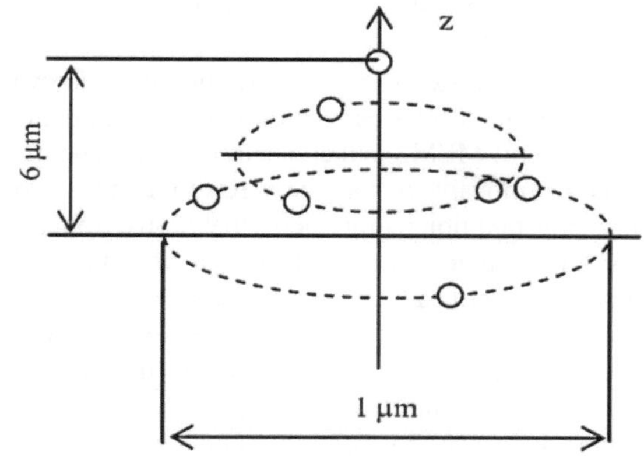

FIGURE 4.35. Seven microbeads organized in a "three planes cage": three beads are placed in the first plane, three beads in the second plane at about 3 microns from the first and one bead placed at 6 microns from the first plane [127].

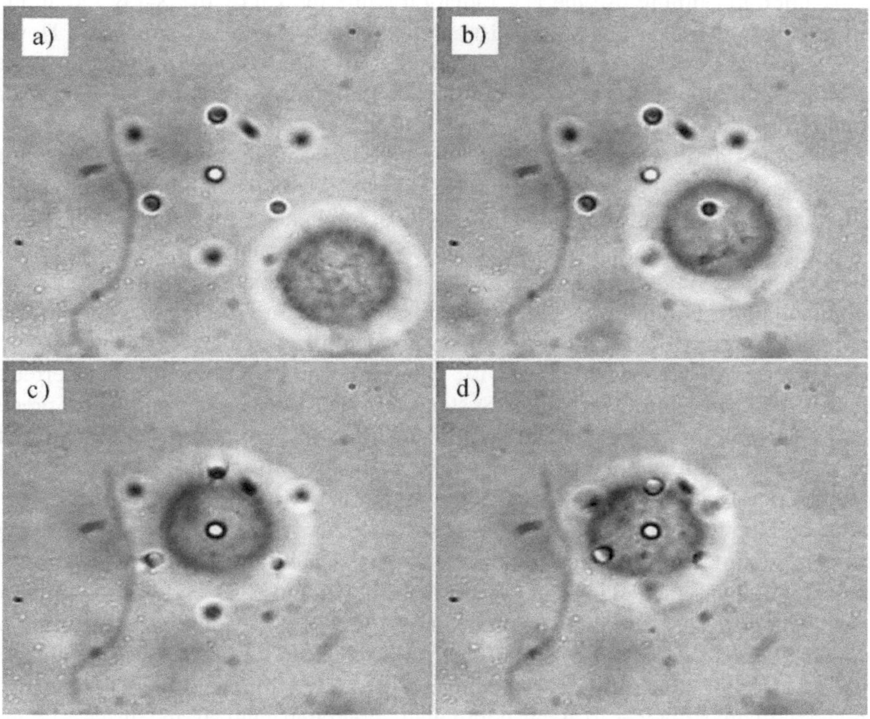

FIGURE 4.36. a–c) The microbeads cage built in the vicinity of a ND7 neuronal like cell: the cell is placed under the cage by moving the microscope stage, d) the cell is stressed by dynamically changing the geometry of the trapping configuration. The scale bar indicates a length of 15 microns [127].

4.6. MEMS DEVICES FOR BIOMEDICAL APPLICATIONS

Micro-Electro-Mechanical Systems technology, commonly known with the acronym MEMS, refers to the fabrication of devices with dimensions on the micrometer scale. The most important characteristic of MEMS is that electronic components are often integrated with micro-mechanical ones, movable or stationary, resulting in very complex miniaturized machine. The low power consumption requirements that derive from their small component size has made these system very precious, as evident from their numerous commercial applications, for instance ink jet printer heads, accelerometers for airbags, chemical sensors and many others. These structures are created via fabrication processes and equipment devoted for the Integrated Circuit (IC) industry. The fabrication of MEMS commonly involves bulk micromachining, surface micromachining or LIGA process.

Bulk micromachining defines microstructures by etching directly into the bulk material such as silicon crystal. In surface micromachining, microcomponent structures are defined via both additive (growth or deposition techniques) and subtractive process (etching process). Both techniques allow the use of integrated circuit technology. The LIGA process is very similar to bulk and surface micromachining; the bigger difference arises from the radiation source employed to transfer desired micro-features onto the substrate. Both surface and bulk micromachining use the UV radiation while the LIGA the X-ray radiation that permit to penetrate higher layers of resist and then to fabricate taller structures. However, specific techniques can be combined with the previous ones to create features in the nanoscale range, as it will be described later in the fabrication process of self-standing metallic nano-gap MEMS structures.

More importantly, the micromachinable materials and the range of processes now extend far beyond just those found in the IC industry. Medicine and biology are amongst the most promising and most challenging fields of application for MEMS. This does not come as a surprise considering that the technology has the capability to fabricate minimally invasive yet highly functional devices that match the size range of many structures found in the human body. Examples include pressure sensors that are small enough to fit through 1-μm catheters or pacemakers that have incorporated micro-scale accelerometers to pace the heart in proportion to the patient activity. These small and tiny devices, also referred to as Biomedical Micro-Electro-Mechanical Systems (Bio-MEMS), hold promises for precision surgery with micrometer control, for rapid screening of common diseases and genetic predispositions, and for implantable drug delivery systems for controlled release of drugs in dose and time.

The development of new, affordable, disposable analytic microchips is changing diagnostics. Examples of analytical functions that are benefiting from such developments include blood supply screening, analysis of biopsy samples and body fluids, minimally invasive and non-invasive diagnostic procedures, rapid identification of disease, and early screening. These systems will eventually perform diagnostic procedures in a multiplexed format that incorporates multiple complementary methods. Ultimately, these systems will be combined with other devices to create completely integrated analysis and disease-treatment systems. There are a number of mechanisms to provide timed release of drugs, such as micro-encapsulation and transdermal release (discussed earlier). Implantable Bio-MEMS are preferred for therapies that require several daily injections, such as for diabetes treatment. If the drug level is monitored in real time, it could also be adapted to metabolic variations. In treatments like chemotherapy, the device can be implanted where the drug is most needed. Among these techniques, implantable pumps have the advantage that the drug therapy can be delivered at the optimal time and concentration to a specific site.

4.6.1. Self-standing Metallic Nanogap MEMS Structures for Nano Trapping Application

The investigation of electronic transport through individual molecules or nano-scale objects has attracted considerable interest as an important step towards the fabrication of molecular devices whose functions are specifically defined by the electrical junctions in which single molecules or small molecular assemblies. An attractive approach to study single molecular properties is based on use of nano-gap electrodes, namely a pair of electrodes with a nano-meter gap [128–130]. By using this approach, single molecular samples are measured under conditions similar to those of real devices. In view of this, techniques for fabricating metal electrodes separated by nanometre-scale gaps are essential. Over the past few years, a number of methods for fabricating narrow-gap electrodes have been developed which includes mechanical break junctions [131–133] and electrochemical plating [130–134] for atomic-scale precision in the gap size. Alternatively, shadow evaporation [135, 136] and electro-migration-induced breaking of electron-beam lithography defined thin-metal nano-wires [137–139] permit only a crude control of the gap size. To use these nano-gaps as test-beds for single molecules requires the ability to tune the size of the gap.

In an alternative approach, we have focused our attention on NEMS like devices which are capable of adjusting the relative intra-electrode gap size of a pre-programmed gap by means of electro-thermal actuation which develop in adjacent conductors when currents flows into them. For this, we have fabricated nano-gap tip structures with intra-tip gap size falling into the sub-20 nm resolution. Electron beam lithography (EBL) and electroplating process have been used to fabricate two- and three-terminal nano-devices consisting of tip-shaped free standing electrodes with tip separation variable ranging between 10–100 nm. A silicon nitride (Si_3N_4) membrane, coated by 1.2 μm of PMMA, is exposed at 30 KeV with a beam current of 1 nA. The exposure dose was calculated including proximity correction effects. The patterned samples were then electroplated in a commercial Nickel bath at 55 °C. The thickness of the deposited layer was 0.8–0.9 μm. An ion milling process with Argon was carried out to remove the base plating film from the substrate. The final step was to create the self-standing structure by removing the Si_3N_4 membrane. This standard EBL process followed by electroplating had allowed us to obtain nano-tips with separations as smaller as 50 nm. In order to obtain separations in 10–20 nm range between nano-tips, before the EBL process, we used a defocused electron-beam to brush the PMMA layer for a time t ranging from 15 to 50s depending on the resist thickness. Thus the resist is partially exposed and, during the following EBL process, the Gaussian shape of the e-beam allows the resist near the edges of the tip to reach the right dose to be developed. We did not evaluate the effect of defocused e-beam brushing exposure time (***t***) on the reduction of separation between tips. However experimental results showed that for t ranges from 20–50 s, the inter-electrode distance reduced almost linearly from 40nm down to 10nm; for ***t*** higher than 60s, the tip separation could easily be reached 5 nm. A careful calibration of the brushing time between 40 and 60 s thus had allowed the reduction of *inter tip gap* to the 5–10 nm range. All the fabricated nano-devices were inspected by plane-view scanning electron microscopy (SEM) in order to establish the viability of the whole technological

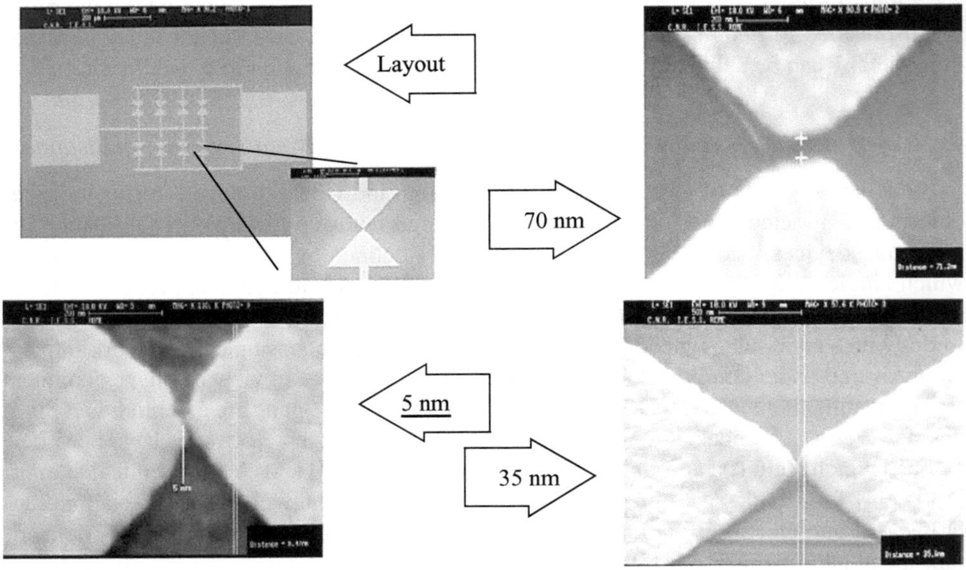

FIGURE 4.37. Fibrication of Nanodevices by means of EBL: self-standing electrodes separated by few nm gaps.

process and the separation between the nano-tips. In figure 4.37, SEM micrographs of several fabricated self-standing nano-gap electrodes tips with intra-tip gap in 5–70 nm range are shown.

Fabricated structures use resistive heating (Joule effect) to generate thermal expansion and movement. The power needed to adjust the relative distance between adjacent features is relatively small: for instance, by applying to the circuits a current of few tens of mA, the driving force overcomes the intrinsic elastic force of our nickel/gold microstructures, and a self standing electrodes change their distance by an amount of about 100 nm.

With our designed adjustable nano-electrode, we are exploring the possibility to make electrical measurements as a function of the length of stretched molecules. The electrodes, when covered or made by gold as shown in figure 4.38, can be useful to trap molecules functionalized with thiolic groups due to the highly stable bound between gold a sulphur.

CONCLUSIONS

Several emerging lithographic techniques for the fabrication of 3D micro and nano systems with application in the field of medicine have been reviewed. It was shown that they constitute the backbone technology for the realization of more and more complex micro and nano devices requested by the constant progress of the medical research.

So far microfabrication techniques based on layer-by-layer approach and developed specifically for microelectronics have been mostly used to produce 3D microsystems. However the high number of process steps required by the increasing grade of complexity renders those techniques time consuming and at the end very costly.

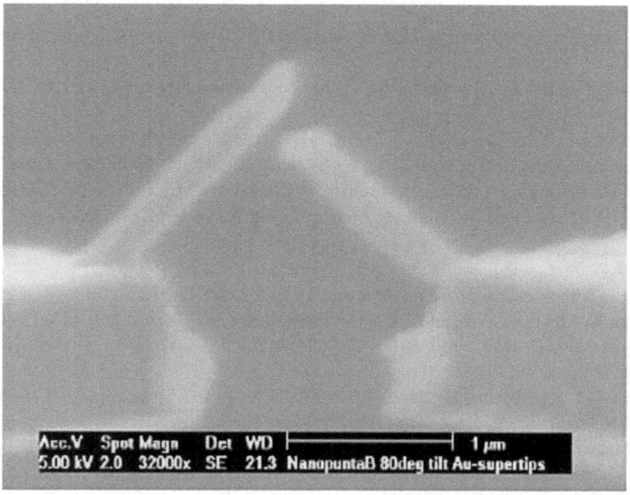

FIGURE 4.38. Nanoelectrodes for molecular measurements: in this case a reduction of the gap has been achieved by electron-beam induced deposition of Au.

Today a single technique has not yet emerged as the universal solution for the fabrication of 3D micro and nano devices. Therefore each problem can be solved identifying the most appropriate technique or combination of techniques to be used.

Clearly, trends in fabrication methodology look forward for more flexible lithographic techniques with intrinsic three-dimensional (3D) structuring capabilities. The two-photon lithography and the focused ion beam lithography seem to possess many of the required characteristics, but their study is still at its infancy.

Meanwhile the combination of several different lithographies is showing a great potential to improve the overall 3D patterning ability.

It's worth concluding with a remark on the opportunities raised by the interaction of such different disciplines like nanofabrication and medicine. Micro and nano technology has found an unexpected driving interest in the design and fabrication of devices for medicine promising great improvement in many medical fields. In turn the opportunity offered by these types of devices push other areas such as chemistry, physics science, biology, microfluidic, diagnostic techniques and instrumentation including microscopy and spectroscopy with unprecedented manner.

REFERENCES

[1] H.G. Craighead. *Science*, 290:1535, 2000.

[2] Hongkai Wu, Teri W. Odom, Daniel T. Chiu, and George M. Whitesides. *J. Am. Chem. Soc.*, 125:554, 2003.

[3] P. Ehbets, H.P. Herzig, D. Prongué, and M.T. Gale. *Opt. Lett.*, 17:908, 1992.

[4] E. Di Fabrizio, F. Romanato, M. Gentili, S Cabrini, B. Kaulich, J. Susini, and R. Barrett. *Nature*, 401:895, 1999.

[5] Yurii A. Vlasov, Xiang-Zheng Bo, James C. Sturm, and David J. Norris. *Nature*, 414:289, 2001.

[6] J. Khandurina and A. Guttman. *J. Chromatography A*, 943:159, 2002.

[7] Michael J. Vasile, Raja Nassar, and Jushan Xie. *J. Vac. Sci. Technol. B*, 16:2499, 1998.
[8] A.T. Woolley et al. Direct haplotyping of kilobase-size DNA using carbon nanotube probes. *Nat. Biotechnol.*, 18:760, 2000.
[9] S.R. Whaley et al. Selection of peptides with semiconductor binding specificity for directed nano-particle assembly. *Nature*, 405:665, 2000.
[10] Pantazis Mouroulis, Frank T. Hartley, Daniel W. Wilson, and Victor E. White. *Opt. Express*, 11:270, 2003.
[11] C. Cuisin, A. Chelnokov, J.-M. Lourtioz, D. Decanini, and Y. Chen. *Appl. Phys. Lett.*, 77:770, 2000.
[12] F. Romanato, L. Businaro, L. Vaccari, S. Cabrini, P. Candeloro, M. De Vittorio, A. Passaseo, M.T. Todaro, R. Cingolani, E. Cattaruzza, M. Galli, C.Andreani, and E. Di Fabrizio. *Microelect. Engin.* 67–68:679, 2003.
[13] M.J. Vasile, R. Nassar, and J. Xie. *J. Vac. Sci. Technol. B* 16:2499, 1998.
[14] Mingtao Li, Lei Chen, and Stephen Y. Chou. *Appl. Phys. Lett.*, 78:3322, 2001.
[15] C.M. Agrawal, K.A. Athanasiou, and J.D. Heckman. *Mater. Sci. Forum*, 250:115, 1997.
[16] J.H. Braybrook and L.D. Hall. *Prog. Polym. Sci.*, 15:715, 1990.
[17] V. Maquet, R. Jerome, and Porous Mater. *Tissue Eng.*, 15:250, 1997.
[18] D.J. Mooney, P.M. Kaufmann, K. Sano, K.M. McNamara, J.P. Vacanti, and R. Langer. *Transplant P.*, 26:3425, 1994.
[19] B. Saad, S. Matter, G. Ciardelli, G. K. Uhlschmid, M.Welti, P. Neuenschwander, U.W. Suter. *J. Biomed. Mater. Res.*, 32:355, 1996.
[20] M. Sittinger, J. Bujia, W.W. Minuth, C. Hammer, and G.R. Burmester. *Biomaterials*, 15:451, 1994.
[21] R.C. Thomson, M.J. Yaszemski, J.M. Powers, and A.G. Mikos. *J. Biomater. Sci. Polym. Ed.*, 7:23, 1995.
[22] M.G. Dunn, P.N. Avasarala, and J.P. Zawadsky. *J. Biomed. Mater. Res.*, 27:1545, 1993.
[23] W.W. Minuth, M. Sittinger, and S. Kloth. *Cell Tissue Res.*, 291:1, 1998.
[24] G. Mikos, G. Sarakinos, S.M. Leite, J.P. Vacanti, and R. Langer. *Biomaterials*, 14:23, 1993.
[25] L.E. Freed, G. Vunjak-Novakovic, R.J. Biron, D.B. Eagles, D.C. Lesnoy, S.K. Barlow, and R. Langer. *Biotechnology* (NY), 12:689, 1994.
[26] C.T. Laurencin, M.A. Attawia, H.E. Elgendy, and K.M. Herbert. *Bone*, 19: S93–S99, 1996.
[27] S.L. Ishaug, G.M. Crane, M.J. Miller, A.W. Yasko, M.J. Yaszemski, and A.G. Mikos. *J. Biomed. Mater. Res.*, 36:17, 1997.
[28] C.H. Rivard, C.J. Chaput, E.A. Desrosiers, L.H. Yahia, and A. Selmani. *J. Appl. Biomater.*, 6:65, 1995.
[29] Mikos, A. Thorsen, L. Czerwonka, Y. Bao, and R. Langer. *Polymer*, 35:1068, 1994.
[30] T. Tateishi, G. Chen, T. Ushida, T. Murata, and S. Mizuno. In: K.-U. Lewandrowski, D.Wise, D. Trantolo, J. Gresser, M. Yaszemski, D. Altobelli (eds.), *Tissue Engineering and Biodegradable Equivalents-Scientific and Clinical Applications*, Marcel Dekker Inc., New York 2002.
[31] J. Elisseeff, R. Langer, and Y. Yamada. In: K.-U. Lewandrowski, D. Wise, D. Trantolo, J. Gresser,M. Yaszemski, D. Altobelli (eds.), *Tissue Engineering and Biodegradable Equivalents-Scientific and Clinical Applications*, Marcel Dekker Inc., New York, p. 99, 2002.
[32] R. Thompson, M.C. Wake, M. Yaszemski, and A.G. Mikos. *Adv. Polym. Sci.*, 122:247, 1995.
[33] J. Pachence and J. Kohn. In: R. Lanza, R. Langer, W. Chick (eds.), *Principles of Tissue Engineering*, Academic Press, New York, p. 273, 1997.
[34] A. Bertsch, S. Zissi, J.Y. Jezequel, S. Corbel, and J.C. Andre. Microstereophotolithography using a liquid crystal display as dynamic mask-generator. *Microsys. Technol.*, 3:42, 1997.
[35] T.M.G. Chu, D.G. Orton, S.J. Hollister, S.E. Feinberg, and J.W. Halloran. Mechanical and in vivo performance of hydroxyapatite implants with controlled architectures. *Biomat.*, 23:1283, 2002.
[36] R. Landers, A. Pfister, U. Hübner, H. John, R. Schmelzeisen and R. Mülhaupt. Fabrication of soft tissue engineering scaffolds by means of rapid prototyping techniques. *J. Mat. Sci.*, 37:3107, 2002.
[37] A. Wöß, M. Rumpler, J. Stampfl, F. Varga, N. Fratzl-Zelman, P. Roschger, K. Klaushofer, and P. Fratzl. Towards Bone Replacement Materials from Calcium Phosphates via Rapid Prototyping and Ceramic Gelcasting, to appear in *Mat. Sc. Eng.*
[38] F. Pérennés, F. De Bona, and F.J. Pantenburg. Deep X-ray lithography beamline at ELETTRA. *Nucl. Inst. Meth. A*, 467–468:1274, 2001.
[39] J. Zanghellini, S. Achenbach, A. El-Kholi, J. Mohr, and F.J. Pantenburg. New development stategies for high aspect ratio structures. *Microsys. Technol.*, 4:94, 1998.
[40] P. Meyer, A. El-Kholi, and J. Schulz. Investigation of the development rate of irradiated PMMA microstructures in deep X-ray lithography. *Microelectro. Eng.*, 63:319, 2002.

[41] R.L. Bronaugh and H.I. Maibach. *Percutaneous Absorption: Drugs-Cosmetics-Mechanisms-Methodology*, Marcel Dekker, New York, 1999.
[42] E. Touitou. Drug delivery across the skin. *Expert Opin. Biol. Ther.*, 2:723, 2002.
[43] B. Barry and A. Williams. Penetration enhancers. *Adv. Drug Deliv. Rev.*, 56:603, 2003.
[44] G. Cevc. Lipid vesicles and other colloids as drug carriers on the skin. *Adv. Drug Deliv. Rev.*, 56:675, 2004.
[45] V. Preat and R. Vanbever. Skin electroporation for transdermal and topical delivery. *Adv. Drug Deliv. Rev.*, 56:659, 2004.
[46] Doukas. Transdermal delivery with a pressure wave. *Adv. Drug Deliv. Rev.*, 56:559, 2004.
[47] S. Mitragotri and J. Kost. Low-frequency sonophoresis: a review. *Adv. Drug Deliv. Rev.*, 56:589, 2004.
[48] M.S. Gerstel and V.A. Place. Drug delivery device, US Patent No.3,964,482 (1976).
[49] S. Henry, D. McAllister, M.G. Allen, and M.R. Prausnitz. Microfabricated microneedles: a novel method to increase transdermal drug delivery. *J. Pharm. Sci.*, 87:922, 1998.
[50] B.W. Barry. Novel mechanisms and devices to enable successful transdermal drug delivery. *Euro. J. Pharm. Sci.*, 14:101, 2001.
[51] M.R. Prausnitz. Microneedles for transdermal drug delivery. *Advn. Drug Deliv. Rev.*, 56:581, 2004.
[52] J. Döpper, M. Clemens, W. Ehrfeld, S. Jung, K-P Kämper, and H. Lehr. Micro gear pumps for dosing of viscous fluids. *J. Micromech. Microeng.*, 7:230, 1997.
[53] K. Deng, A.S. Dewa, D.C. Ritter, C. Bonham, and H. Guckel. Characterisation of gear pumps fabricated by LIGA. *Microsys. Technol.*, 4:163, 1998.
[54] S. Henry et al. Microfabricated Microneedles: A Novel Approach to Transdermal Drug Delivery. *J. of Pharma. Sci.*, 87:922, 1998.
[55] K. Kim, D.S. Park et al. A tapered hollow metallic microneedle array using backside exposure of SU-8. *J. Micromech. Microeng.*, 14:597, 2004.
[56] P.M. Wang, M.G. Cornwell, and M.R. Prausnitz, "Effect of Microneedle Tip Geometry on Injection and Extraction in the Skin". *Proc 2nd joint EMBS/BMES conf. Houston TX*, pp. 506–507, 2002.
[57] M. Goppert-Mayer. Uber Elementarakte mit zwei Quantensprungen. *Ann. Phys.*, 9:273, 1931.
[58] A. Diaspro and M. Robello. *J. Photochem. and Photobiol. B: Biol.*, 55:1–8, 2000.
[59] Stelzer et al. *Optics Commun.*, 104, 1994.
[60] W. Denk, J.H. Strickler, and W.W. Webb. *Science*, 248:73, 1990.
[61] Kaiser and Garret. *Phy. Rev. Lett.*, 7:229, 1961), 229
[62] Y.H. Pao and P.M. Rentzepis. *Appl. Phys. Lett.*, 6:93, 1965.
[63] S. Maruo, O. Nakamura, and S. Kawata. *Opt. Lett.* 22:132, 1997.
[64] O. Nakamura. *Micr. Res. Techn.*, 47:165–171, 1999.
[65] Chris Xu. *Confocal and Two-Photon Microscopy*, Wiley-Liss (Ed), New York, 2002.
[66] M. Albota et al. *Science*, 281:1653, 1998.
[67] B.H. Cumpston et al. *Nature*, 398:41, 1999.
[68] W. Zhou et al. *Science*, 296:116, 2002.
[69] S. Kawata, S. Shoji, and H.B. Sun. *IEICE Trans. Electron.*, E87-C:378, 2004.
[70] S. Kawata, H.B. Sun, T. Tanaka, and K. Takada. *Nature*, (London) 412:667, 2001.
[71] H.-B. Sun, T. Kawakami, Y. Xu, J.-Y. Ye, S. Matsuo, H. Misawa, M. Miwa, and R. Kaneko. *Opt. Lett.*, 25:1110, 2000.
[72] P. Galajda and P. Ormos. *Appl. Phys. Lett.*, 78:249, 2001.
[73] T. Tanaka, H.-B. Sun, and S. Kawata. *Appl. Phys. Lett.*, 80:312, 2002.
[74] S.M. Kuebler, M. Rumi, T. Watanabe, K. Braun, B.H. Cumpston, A.A. Heikal, L.L. Erskine, S. Thayumanavan, S. Barlow, S.R. Marder, and J.W. Perry. *J. Photopolym. Sci. Technol.*, 14:657, 2002.
[75] M. Miwa, S. Juodkazis, T. Kawakami, S. Matsuo, and H. Misawa. *Appl. Phys. A: Mater. Sci. Process.*, A73:561, 2001.
[76] H.-B. Sun, T. Tanaka, and S. Kawata. *Appl. Phys. Lett.*, 80:3673, 2002.
[77] R.J. DeVoe, H. Kalveit, C.A. Leatherdale, and T.R. Williams. *Proc. SPIE*, 4797:310, 2002.
[78] H.-B. Sun, M. Maeda, K. Takada, J.W.M. Chon, M. Gu, and S. Kawata. *Appl. Phys. Lett.*, 83:819, 2003.
[79] H.B. Sun, K. Takada, M.S. Kim, K.S. Lee, and S. Kawata. *Appl. Phys. Lett.*, 83:1104, 2003.
[80] S. Maruo and S. Kawata. *J. Microelectromech. Syst.*, 7:441, 1998.
[81] H.-B. Sun, S. Matsuo, and H. Misawa. *Appl. Phys. Lett.*, 74:786, 1999.

[82] B.H. Cumpston, S.P. Ananthavel, S. Barlow, D.L. Dyer, J.E. Ehrlich, L.L. Erskine, A.A. Heikal, S.M. Kuebuler, I.-Y. Sandy Lee, D.M. Maugon, J. Qin, H. Rokel, M. Rumi, X. Wu, S.R. Marder, and J.W. Perry. *Nature*, 398:51, 1999.

[83] H.-B. Sun, V. Mizeikis, Y. Xu, S. Juodkazis, J.-Y. Ye, S. Matsuo, and H. Misawa. *Appl. Phys. Lett.*, 79:1, 2001.

[84] S.Y. Chou, P.R. Krauss, and P.J. Renstrom. *Appl. Phys. Lett.*, 67: 3114, 1995 and *Science*, 272:85, 1996.

[85] C.M.S. Torres (ed.). *Alternative Lithography*, Kulwer Academic Plenum.

[86] P. Rucjjpeft, M. Colburn, B. Choi, Nounu, S. Johnson, and C.G. Wilson. *J. Vac. Sci. Thechnol. B*, 17:2965, 1999.

[87] Y. Xia and G.M. Whitesides, Angew. *Chem. Int. Ed.*, 37:550–575, 1998.

[88] K.A. Valiev. *The Physics of Sub-micron Lithography*, Plenum Press, New York, 1992.

[89] P.W.H. de Jager et al. *Microelectro. Eng.*, 30:353–356, 1996.

[90] K. Gamo. *Nucl. Instr. Meth Phys. Res.*, 121:464–469, 1997.

[91] N. Bouadma, A. Ougazzaden, M. Kamoun, C. Kazmierski, and L. Silverstre. *Electron. Lett.*, 32:1582, 1996.

[92] H. Sato, M. Aoki, M. Takahashi, M. Komori, K. Uomi, and S. Tsuji. *Electron. Lett.*, 31:1241, 1995.

[93] I.F. Lealman, L.J. Rivers, M.J. Harlow, and S.D. Perrin. *Electron. Lett.*, 30:1685, 1994.

[94] B. Mersali, H.J. Bruckner, M. Feulillade, S. Sainson, A. Ougazzaden, and A. Carenco. *J. Lightwave Tech.*, 13:1865, 1995.

[95] M. Prasciolu, P. Candeloro, R. Kumar, L. Businaro, E. Di Fabrizio, D. Cojoc, S. Cabrini, C. Liberle, and V. Degiorgio. *Jap. J. Appl. Phys.*, 42:4180, 2003.

[96] R. Schiappelli, M. Kumar, D. Prasciolu, S. Cojoc, M. Cabrini, G. De Vittorio, A. Visimberga, Degiorgio V. Gerardino, and E. Di Fabrizio. *Microelect. Eng.*, 73–74C:397, 2004.

[97] Klein et al. *Appl. Phys. Lett.*, 68(18):2575, 1996.

[98] R. Kometani et al. *Nucl. Instr. And Meth.* In *Phys. Res. B.*, 206:472–477, 2003.

[99] J. Fujita et al. *J. Vac. Sci. Technol. B.*, 22(1):257–263, 2004.

[100] M. Tormen, F. Romanato, M. Altissimo, L. Businaro, P. Candeloro, and E.M. Di Fabrizio. *J. Vac. Sci. Technol. B*, 22(2):766, 2004.

[101] F. Romanato, R. Kumar, and E. Di Fabrizio. Patent No. TO2003A000730 filed on 23.09.2003—Process for the fabrication of complex three dimensional structures on sub-micrometric scale by means two resist combined lithography.

[102] SAL-607 ER7 is trademark of Shipley Inc.

[103] W. Ehrfeld and A. Schmidt. Recent development in deep X-ray Lithography. *J. Vac. Sci. Technol. B*, 16.6: 3526–3534, 1998.

[104] F. Romanato, L. Businaro, L. Vaccari, S. Cabrini, P. candeloro, M. DeVittorio, A. Passaseo, M. T. Todoro, R. Cingolani, E. Cattaruzza, M. galli, C. Andreani, and E. DiFabrizio. *Microelectron. Eng.*, 67–68:479, 2003.

[105] F. Romanato, L. Businaro, L. Vaccari, S. Cabrini, P. candeloro, M. DeVittorio, A. Passaseo, M.T. Todoro, R. Cingolani, E. Cattaruzza, M. galli, C. Andreani, and E. DiFabrizio. *J. Vac. Sci. Tecjnol. B.*, 51:2912, 2003.

[106] L. Kuhn, E. Bassous, and R. Lane. *IEEE Trans. Electron Devices ED-25*, 1257, 1978.

[107] S.C. Terry, J.H. Jerman, and J.B. Angell. *IEEE Trans. Electron Devices ED-26*, 1880, 1979.

[108] C.M. Ho. In: *Proceedings of the 14th IEEE International Conference MEMS*, IEEE, New York, p. 375, 2001.

[109] A. Ashkin, J.M. Dziedzic, J.E. Bjorkholm, and S. Chu. Observation of a single-beam gradient force optical trap for dielectric particles. *Optics Lett.*, 11:288, 1986.

[110] A. Ashkin and J.M. Dziedzic. Optical Trapping and Manipulation of Viruses and Bacteria. *Science*, 235:1517, 1987.

[111] K. Dholakia, G. Spalding, and M. MacDonald. Optical Tweezers: The Next Generation. *Physics World*, 15:31, 2002.

[112] H. He, N.R. Heckenberg, and H. Rubinsztein-Dunlop. Optical Particle Trapping with Higher-order Doughnut Beams Produced Using High Efficiency Computer Generated Holograms. *J. Mod. Opt.*, 42:217, 1995.

[113] J. Arlt, V. Garces-Chavez, W. Sibbett, and K. Dholakia. Optical micromanipulation using a Bessel light beam. *Opt. Commun.*, 197:239, 2001.

[114] E.R. Dufresne, G.C. Spalding, M.T. Dearing, S.A. Sheets, and D.G. Grier. Computer-generated holographic optical tweezer arrays. *Rev. Sci. Instr.*, 72:1810, 2001.

[115] K. Svoboda and S.M. Block. Optical trapping of metallic Rayleigh particles. *Opt. Lett.*, 19:930, 1994.
[116] Y. Harada and T. Asakura. Radiation forces on a dielectric particle in the Rayleigh scattering regime. *Opt. Commun.*, 124:529, 1996.
[117] T. Tlusty, A. Meller, and R. Bar-Ziv. Optical Gradient Forces of Strongly Localized Fields. *Phys. Rev. Lett.*, 81:1738, 1998.
[118] D.G. Grier. A revolution in optical manipulation. *Nature*, 424:810, 2003.
[119] K. Sasaki, M. Kashioka, H. Misawa, N. Kitamura, and H. Masuhara. Pattern formation and flow control of fine particles by laser-scanning micromanipulation. *Opt. Lett.*, 16:1463, 1991.
[120] M. Bernhardt, F. Wyrowski, and O. Bryngdahl. *Appl. Opt.*, 30:4629, 1991.
[121] J. Liesener, M. Reicherter, T. Haist, and H.J. Tiziani. Multi-functional optical tweezers using computer-generated holograms. *Opt. Commun.*, 185:77, 2000.
[122] D. Cojoc, E. Di Fabrizio, L. Businaro, S.Cabrini, F. Romanato, L. Vaccari, and M. Altissimo. Design and fabrication of diffractive optical elements for optical tweezer arrays by means of e-beam lithography. *Microelectronic Eng.*, 61–62:963, 2002.
[123] E. Di Fabrizio, D. Cojoc, S. Cabrini, B. Kaulich, J. Susini, P. Facci, and T. Wilhein. Diffractive optical elements for differential interference contrast X-ray microscopy *Opt. Exp.*, 11:2278, 2003.
[124] D. Cojoc, S. Cabrini, E. Ferrari, R. Malureanu, M.B. Danailov, and E. Di Fabrizio. Dynamic multiple optical trapping by means of diffractive optical elements *Microelectro. Eng.*, 73–74:927, 2004.
[125] D. Cojoc, V. Emiliani, E. Ferrari, R. Malureanu, S. Cabrini, R. Z. Proietti, and E. Di Fabrizio. Multiple optical trapping by means of diffractive optical tweezers *Japanese J. Appl. Phys.*, 43:3910, 2004.
[126] D. Cojoc, V. Emiliani, E. Ferrari, V. Garbin, and E. Di Fabrizio. Dynamic multiple beads manipulation on x-y-z directions. *The International Symposium on Optical Science and Technology, Conference 5514: Optical Trapping and Optical Micromanipulation*, Denver, Colorado, USA, accepted for publication in SPIE Proc Vol. 5514, 2–6 August 2004.
[127] E. Ferrari, V. Emiliani, D. Cojoc, V. Garbin, C. Durieux, M. Coppey, and E. Di Fabrizio. Biological samples micro-manipulation by means of optical tweezers *J. Biomed. Opt.* (submitted 2004).
[128] A.F. Morpurgo, C.M. Marcus, and D.B. Robinson. *Appl. Phys. Lett.*, 74:2084, 1999.
[129] M.A. Reed, C. Zhou, C.J. Muller, T.P. Burgin, and J.M. Tour. *Science*, 278:252, 1997.
[130] C. Kergueris, J.P. Bourgoin, S. Palacin, D. Esteve, C. Urbina, M. Magoga, and C. Joachim. *Phys. Rev.* B, 59:12505, 1999.
[131] J.M. Ruitenbeek. *Rev. Sci. Instrum.*, 67:108, 1996.
[132] C.Z. Li, H.X. He, and N.J. Tao. *Appl. Phys. Lett.*, 77:3995, 2000.
[133] D.L. Klein, P.L. McEuen, J.E.B. Katari, R. Roth, and A.P. Alivisatos. *Appl. Phys. Lett.*, 68:2574, 1996.
[134] K. Liu, P. Avouris, J. Bucchignano, R. Martel, S. Sun, and J. Michl. *Appl. Phys. Lett.*, 80:865, 2002.
[135] C. Durkan, M.A. Schneider, and M.E. Welland. *J. Appl. Phys.*, 86:1280, 1999.
[136] H. Park, A.K.L. Lim, A.P. Alivisatos, J. Park, and P.L. McEuen. *Appl. Phys. Lett.*, 75:301, 1999.
[137] R. Sordan, M. Burghard, and K. Kern. *Appl. Phys. Lett.*, 79:2073, 2001.
[138] J. Park, A.N. Pasupathy, J.I. Goldsmith, C. Chang, Y. Yaish, J.R. Petta, M. Rinkoski, J.P. Sethna, H.D. Abruna, P.L. McEuen, and D.C. Ralph. *Nature*, 417:722, 2002.
[139] W. Liang, M.P. Shores, M. Bockrath, J.R. Long, and H. Park. *Nature*, 417:725, 2002.

5

Sacrificial Oxide Layer for Drug Delivery

Piyush M. Sinha[1] and Mauro Ferrari[2]

[1]*Electrical and Computer Engineering*

[2]*Professor, Brown Institute of Molecular Medicine Chairman, Department of Biomedical Engineering, University of Texas Health Science Center, Houston, TX; Professor of Experimental Therapeutics, University of Texas M.D. Anderson Cancer Center, Houston, TX; Professor of Bioengineering, Rice University, Houston, TX; Professor of Biochemistry and Molecular Biology, University of Texas Medical Branch, Galveston, TX; President, the Texas Alliance for NanoHealth, Houston, TX*

5.1. INTRODUCTION

Since the invention of silicon microfabrication technology in early 1960s, the integrated circuit (IC) has changed our world. During last 40 years, the semiconductor industry has come up as the fastest growing industry in our history. From a modest beginning, that allowed few transistors on a chip, we have reached integration level of tens of millions of components in a square cm of silicon. The minimum feature size on silicon is reducing and thus the number of devices per square cm is increasing. The observation made in 1965 by Gordon Moore [1], co-founder of Intel, that the number of transistors per square inch on integrated circuits had doubled every year since the integrated circuit was invented. Moore predicted that this trend would continue for the foreseeable future. In subsequent years, the pace slowed down a bit, but data density has doubled approximately every 18 months, and this is the current definition of Moore's Law.

This silicon fabrication technology was later extended to machining mechanical microdevices—that was later called microelectromechanical systems (MEMS). The pioneer work was done by Nathanson *et al.* in 1965 when they demonstrated the first micromachined structure to fabricate a free standing gold beam electrode used in a resonant gate transistor [2]. By late 1970s, there was an immense interest in silicon as a mechanical material [3, 4]. During 1980s and 1990s, many MEMS devices were fabricated, *e.g.* micromotors [5–7], deformable mirrors [8, 9], accelerometers [10–14], and comb-drive actuators [15].

In recent years, silicon fabrication technology has been extensively used for the development of microfluidic devices for biological and biochemical applications [16, 17]. Further, the integration of microfluidic devices and integrated circuits over last decade has revolutionized chemical and biological analysis systems, and has opened the possibility of fabricating devices with increased functionality and complexity for these applications [18, 19, 20]. Many of such devices that utilize electrically driven flow for liquid phase analysis such as capillary electrophoresis (CE) [21, 22], electrochromatography [23], capillary gel electrophoresis for DNA restriction and digestion and subsequent size-based separation [24, 25, 26], and DNA sequencing [27] have been successfully realized in a microchip. These integrated structures combine chemical and biological reactions such as sample preparation, separation and detection on a single microchip device to perform complex assays. This integration increases the speed and information content of analysis and reduces sample volume. Many names have been given to such devices that include bio-chip, lab-on a chip or μTAS (micro- or miniaturized total analysis system). Silicon microfabrication technology has also been used for the creation of drug delivery devices [28–32].

A very common method in silicon microfabrication is the use of a sacrificial layer. As the name suggests, a layer is created in surface or bulk micromachining in order to release different MEMS structures or in IC fabrication in order to improve the device performance. This layer is etched away later in the process. Silicon dioxide (SiO_2) is usually chosen in this process because of its availability in IC processes, its compatibility to the elevated temperature during polycrystalline silicon (poly-Si) deposition and anneal. SiO_2 adhere well to the silicon surface and has stable and reproducible bulk properties. The Si and SiO_2 interface can be achieved with very few mechanical and electrical defects and is stable over time. Further, SiO_2 is chemically resistant to most of the chemicals used in silicon processing and yet can be easily patterned and selectively etched with specific chemical or dry plasma etch.

Accounting for its multiple uses, different varieties of oxides have been fabricated for use as sacrificial layers. Examples are: thermally grown oxide, CVD (chemical vapor deposition) oxide, PSG (polysilicon on glass), SOI (silicon on insulator), and SOG (spin-on glass) [34]. This chapter first describes the different fabrication technologies for silicon dioxide. Oxides fabricated by these different techniques have a slightly different composition and thus different etch rates. These oxides are usually etched in order to release a structure, and therefore good etch selectivity of oxide against the structure material (usually silicon) are required. Further, high solubility of etch products are required in order to avoid etch residues that may result in stiction. The third section of this chapter reviews the different etching techniques and challenges encountered during etching. The last section reviews how sacrificial oxide has been used to fabricate different MEMS structures and ICs, followed by summary of the chapter.

5.2. SILICON DIOXIDE FABRICATION

When silicon is exposed to oxygen or ambient air at room temperature its surface immediately oxidizes forming a very thin layer of oxide of thickness 5–10 Å. The oxide growth then slows down and stops after a few hours with the oxide thickness of the order of 1–2 nm. This is native oxide layer [35]. Both the growth rate and the final oxide thickness depend on the chemical residue from the previous cleaning procedures. The sacrificial oxide layers are typically thicker than the native oxide, and are fabricated in a controlled way. Silicon dioxide fabrication can be divided into two categories—thermally grown oxide and deposited oxide. These fabrication technologies are described below.

5.2.1. Thermally Grown Oxide

Thermal oxidation of silicon can be achieved by heating the wafer to a high temperature, typically 850°C to 1200°C, in an atmosphere containing either pure oxygen (called dry oxidation) or water vapor (called wet oxidation). The chemical reaction occurring at the silicon surface for the case of dry oxidation is:

$$Si + O_2 \rightarrow SiO_2 \tag{5.1}$$

And for the case of wet oxidation it is:

$$Si + 2H_2O \rightarrow Si_2O + 2H_2 \tag{5.2}$$

Silicon is consumed when oxide grows. The final oxide layer is approximately 54% above the original surface and 46% below the original surface [35]. A simple model was developed by Deal and Grove in the early 1960s to explain and predict oxidation kinetics [36]. This model (also known as the linear parabolic model) assumes that an oxide of some thickness x_i is already present on the silicon surface and the structure is one dimensional so that this model will only apply to the oxide films grown on flat substrates. It also assumes that either equation (5.1) or equation (5.2) is taking place. Thus, this model cannot fully explain the oxidation of shaped surfaces, oxidation kinetics in mixed ambient or for very thin oxides.

In many cases, sacrificial oxides are thin films and are grown in dry conditions. This provides better control on oxide thickness. The initial oxidation is greatly enhanced by water vapor, and therefore generation of thin film oxides requires tight control of trace quantities of moisture. Quantities as low as 25 ppm of moisture in the supply gas can affect the oxide thickness [37]. The initial oxidation rate is also higher at higher temperatures, and therefore lower temperatures and reduced pressures are recommended [38].

Many models have been proposed to predict oxide growth rates in dry O_2 for thin oxides less than about 300 Å, each claiming an improvement over Deal-Grove Model. In 1987, Reisman *et al.* [39] suggested a simple power law of the form:

$$x_o = a(t + t_i)^b \text{ or } x_o = a\left(t + \left(\frac{x_i}{a}\right)^{\frac{1}{b}}\right)^b \tag{i}$$

Where a and b are constants for a given set of process conditions and t_i is the time corresponding to the oxide thickness x_i at the beginning of the growth process.

Han and Helms [40] came up with another approach and suggested that a wide body of oxidation data, including the dry O_2 thin regime, could be modeled by an expression of the form:

$$\frac{dx_o}{dt} = \frac{B_1}{2x_o + A_1} + \frac{B_2}{2x_o + A_2} \tag{ii}$$

Han and Helms found that by setting B_1/A_1 to infinity (i.e. $A_1 = 0$) agreed well with the experiments. B_2/A_2 values for different silicon crystal planes are given in reference [40]. All of these rate constants were found to fit Arrhenius expression.

Massoud [41–43] presented another model for thin film oxide growth rate on undoped silicon that is given by

$$\frac{dh}{dt} = \frac{B}{2h + A} + C_1 e^{-(h/L_1)} + C_2 e^{-(h/L_2)} \qquad \text{(iii)}$$

Where h is the oxide thickness, A and B are the linear and parabolic rate constants, and C_1, C_2, L_1 and L_2 are the characteristic constants and lengths respectively. This equation requires a limiting initial oxide thickness h_i necessary to fit experimental data, and therefore an oxide thickness below h_i cannot be computed using this model.

Many other models have been developed to explain the oxidation kinetics, all of them motivated by the shortcomings in the linear parabolic model. However, none of them have gained widespread acceptance. Most of the modeling effort today use Deal-Grove model as a starting model and then add or modify to it for the effect that is not considered or not modeled well by linear parabolic model.

5.2.2. Deposited Silicon Dioxide

The oxide film is deposited Most oxide films today are deposited at low temperature using LPCVD (low pressure chemical vapor deposition) or PECVD (plasma enhanced chemical vapor deposition) techniques [33]. Silane and oxygen are the most common source gases used for these CVD processes. The reaction that takes place is given below:

$$SiH_4 + O_2 \rightarrow SiO_2 + 2H_2 \qquad (5.3)$$

Deposition temperatures in LPCVD systems range from 300°C–900°C. However, low temperature deposition rate is usually quite low and the film quality may be poor. High quality film may be deposited in temperature range of 200°C–350°C using PECVD.

Sometimes other sources of O_2 are used, such as N_2O, NO, and CO_2. Silicon sources, other than silane have also been used. One such source is tetraethooxysilane or TEOS. It can form SiO_2 by decomposing via the reaction:

$$Si(OC_2H_5)_4 \rightarrow SiO_2 + \text{byproducts} \qquad (5.4)$$

The decomposition of TEOS requires higher deposition temperatures, but results in improved step coverage. When non-plasma, silane based deposition is done at temperature below 500°C, it is called low temperature oxide (LTO).

Depending upon the deposition method, large variation in density, chemical structure, stoichiometry, impurity contents and dopant contents are achieved. The requirements for microelectronics and sacrificial oxide are usually different. For example, while long-term stability and chemical purity are the required conditions for dielectric layers, high etch rate and etch selectivity are the main requirements for the sacrificial oxide layer. Chemical impurities in a deposited layer can significantly change the etch rate, *e.g.* water uptake decrease the etch rate of phosphorous doped oxides [44]. Conformal step coverage of sacrificial oxide results in a well-defined step or height of the cavity in the released structure, while

TABLE 5.1. Requirements for sacrificial oxide layer and general considerations [34].

Requirements for sacrificial oxide	General consideration
Etch selectivity against substrate and structural material	• Proper choice of etchants
High etch rate	• Proper choice of dopants, doing concentration and etchants
Large undercut	• High diffusion coefficient of etchant required
High solubility of etch products in order to avoid etch residues	• Appropriate dry-rinse methods
Little shrinkage	• Densification and anneal steps • Low stress oxide
Low thermal budget	• Low temperature deposition, reflow, and anneal for maximum compatibility to implants and metallization

planarizing sacrificial layer is desired for flat structures on top of steps in the topography. Conformal step coverage refers to uniform film coverage on both horizontal and vertical surfaces. Some requirements for sacrificial oxide and general considerations to achieve those are listed in table 5.1.

Deposited films are often doped with phosphorous or boron, or both in order to passivate or chemically stabilize the deposited oxide, lower the temperature of planarization reflow, and speed up the etching of the oxide. Depending upon the type of dopants these oxides are named as phosphosilicate glass (PSG), borosilicate glass (BSG) and borophosphosilicate glass (BPSG). LPCVD-BPSG has low reflow temperatures of 700–1000°C, and therefore can be used to overcome the problems of poor step coverage associated with LPCVD-LTO [34]. However, phosphorous content in excess of 6–8% results in a hygroscopic surface and potentially cause phosphoric acid at the oxide surface. Boron of concentration in excess of 5% makes the surface hygroscopic and unstable [46].

5.2.3. Thermally Grown Oxide vs Deposited Oxide

Deposited SiO_2 does not consume silicon, and therefore can be deposited on any surface (metal, nitride, or others). In fact, deposition is the only option to obtain an oxide layer on top of an underlying film where silicon is not available from the underlying film for SiO_2 growth. Further, when it is not possible to process the wafer at high temperature, oxide is deposited, since CVD processes can be used at lower temperature. However, deposited oxides are not used for thickness lower than 20 nm because control of deposition process is not as good as thermal oxidation process. When the control of oxide thickness is a major concern in thin film oxide, dry thermal oxidation is preferred over CVD oxide. Further, the interface between deposited oxide and silicon is not as good as in the case of thermally grown oxide either. Deposited oxide/silicon interface is annealed to improve the interface properties. Depending upon the requirement of sacrificial oxide, a suitable process is chosen.

5.2.4. Silicon-On-Insulator (SOI) as Sacrificial Layer

In some cases silicon-on-insulator is used as a starting material and buried oxide is used as a sacrificial layer. An example of such a wafer is shown in figure 5.1. Several approaches

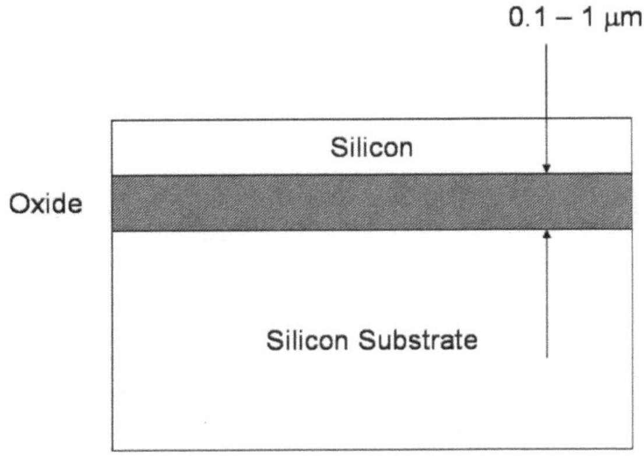

FIGURE 5.1. Silicon On Insulator (SOI) wafer technology.

are used to fabricate these wafers. One such process is known as SIMOX (separation by implanted oxygen) process. In this process, a large dose ($\approx 10^{18}$ cm^{-2}) of oxygen is implanted in a normal silicon wafer, followed by annealing. The silicon above the buried oxide is of reasonable crystal quality if the wafer is annealed properly, although the quality is generally not as good as the starting wafer. A second approach to achieve this kind of wafer is called BESOI (bonded and etch-back silicon-on-insulator). Two normal silicon wafers are oxidized first. After proper surface treatment, the two wafers are brought into contact with the two oxide surfaces facing each other. A good bonding as good as starting wafer is achieved after annealing. One of the two wafers is then etched or lapped away to leave the thin film of silicon on SiO_2. A combination of chemical mechanical polishing (CMP) and lapping is usually used for this purpose.

5.3. SACRIFICIAL OXIDE ETCHING

Sacrificial oxide is always etched away, sometimes during the process or mostly at the end of the process. A good etch selectivity is an important requirement for sacrificial oxide etch. The etch rate depends upon many factors, such as doping material, doping concentration, oxide fabrication method and geometry of sacrificial layer. Usually sacrificial oxides are buried layers and only small openings are available for the echants to reach the surface being etched as well as for the etched products to come of the cavity. In such case diffusion of etchants may become the rate limiting step. Because of these reasons, an understanding of etch kinetic is important.

5.3.1. Etch Mechanism

SiO_2 can be dissolved in bases such as alkaline hydroxides or carbonates, which are able to form alkali silicate that is soluble in water. However, these alkaline solutions dissolve silicon too and therefore are usually not used with silicon micromachining. Common oxide echants are hydrofluoric acid based compounds. Hydrofluoric species are able to break strong Si-O bond because of high electronegativity of the fluorine. The dissolution of SiO_2 in aqueous HF can be represented by following equation:

$$SiO_2 + 6HF \rightleftharpoons 2H^+ + SiF_6{}^{2-} + 2H_20 \tag{5.5}$$

When diluted in water, HF dissociates into H^+, F^- and various hydrofluoric species such as HF^{2-} and $(HF)_2$.

When Buffered HF (BHF) is used to etch oxide, ammonium fluoride get dissociated as

$$NH_4F \rightleftharpoons NH_4{}^+ + F^- \tag{5.6}$$

The overall chemical reaction can be given as

$$SiO_2 + 4HF + 2NH_4F \rightleftharpoons 2NH_4{}^+ + SiF_6{}^{2-} + 2H_20 \tag{5.7}$$

$(NH_4)_2SiF_6$ gets precipitated here due to its limited solubility.

Various models have been proposed to understand the role of HF, HF_2- and $(HF)_2$ in the etch reaction and to calculate the etch rate dependence on concentration and temperature. The model introduced by Blumberg in 1959 [46] is given below:

$$J_{react} = k_0(C_{HF})^n \tag{iv}$$

Where, k_0 and n are fitted to experimental data. C_{HF} is hydrofluoric acid concentration.

Many follow up models were developed to include HF^{2-} and $(HF)_2$, but none can predict the etch rate over a large range of concentration and materials. A good discussion on these models is given in reference [47]. A model incorporating second order effect of C_{HF} was proposed in reference [48]. According to this model

$$J_{react} = k_1(C_{HF}) + k_2(C_{HF})^2 \tag{v}$$

This uses vertical etch rate as well as underetch rate in sacrificial oxide channels to fit experimental parameters k_1 and k_2.

The etch rate dependent on temperature follows Arrehenius' law:

$$J = J_{react}\ .\ exp(E_0/RT) \tag{vi}$$

The activation energy E_0 was measured for various compositions of BHF for different oxides and is given in reference [49].

The sacrificial oxide etch rate is usually different than the regular oxide etch rate when the sacrificial oxide is not a top layer. In many cases the sacrificial oxide is a buried layer and there is a small opening that allows etchants to go to the oxide surface as well as allows the etch product to come out of the cavity as show in figure 5.2.

To model etch kinetics of a buried sacrificial oxide, it is important to know how the etch-front proceeds in the cavity formed by a large distant undercut. This is limited by the diffusion of reactive species to the etch front inside the cavity. The first order reaction

FIGURE 5.2. A small opening for sacrificial oxide etching that will release a mechanical structure.

kinetics is given by the Deal-Grove model [48, 50, 51]. The diffusion flux is proportional to the concentration difference of reacting species outside and inside of the cavity:

$$J_{diff} = D \, . \, (C_b - C_s)/\delta \qquad \text{(vii)}$$

Where, C_b and C_s are the reactant concentrations in the bulk and at the oxide surface, respectively. D is the diffusion coefficient and δ is the undercut distance. J_{diff} becomes equal to J_{react} at equilibrium when etch rate is limited by the reactant exchange rate required for etching. However, because of different sacrificial oxide structure in different MEMS devices, it is difficult to have a well defined model, and every different design needs to be considered differently.

5.3.2. *Etch Selectivity*

Selectivity is another crucial issue in sacrificial oxide etch. Usually a critical device structures exist surrounding the sacrificial oxide layer, and poor selectivity damages that structure. The damage could be homogeneous chemical etching, local dissolution or could be corrosion because of the formation of local electrochemical cells. As mentioned earlier, alkaline etchants react with silicon and therefore are not good for releasing silicon MEMS structure from sacrificial oxides. HNO_3/HF is not suitable either, because HNO_3 oxidizes the silicon surface and oxide is dissolved in HF.

5.3.3. *Stiction*

Choice of oxide etchant is also important in order to prevent stiction. This effect has been addressed theoretically as well as experimentally [52–58]. A conventional wet etching may result in two kinds of stiction failure. The first is temporary deformation due to capillary forces between micromachined superstructure and substrates during rinse-dry process after the sacrificial oxide etch. Another failure is the permanent stiction of the deformed superstructure to the substrate due to the residue products from wet etching. For example, in the case of microactuators, very bad stiction occurs during operation [59–62]. Small Si-H_2O contact angles at highly doped surfaces [60], native reoxidation of HF and BHF etched surface [61], and smooth surfaces [63] enhance the probability of the occurrence of sticking. Consequently, sacrificial oxide etchants resulting in rough and hydrophobic surfaces are preferable. Various techniques have been developed to alleviate the stiction problem. The solutions include micromechanical temporary support [64], sublimation of final liquid by free dry or solidification dry [65, 66], removing the final liquid by supercritical

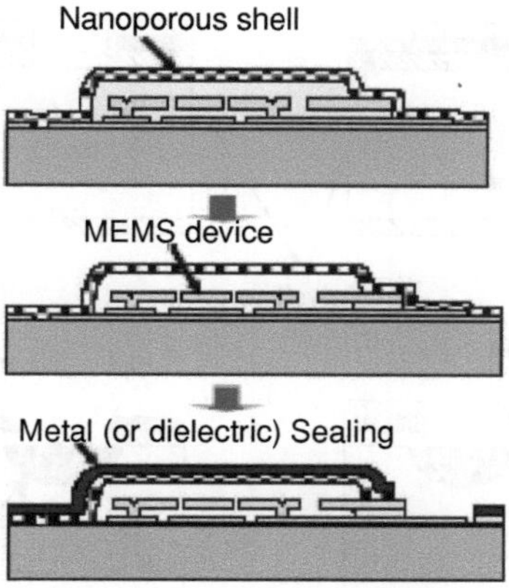

FIGURE 5.3. On-Chip encapsulation by porous polysilicon shell [79].

method [67], or using low surface tension liquid [63, 68, 69]. Researchers have also etched sacrificial oxides using anhydrous HF and CH_3OH in gas phase to achieve dry release [57, 70]. Methanol of low vapor pressure was used here as a catalyst instead of water vapor to minimize residue products and avoid capillary forces.

5.3.4. On-Chip Packaging

Another problem that is faced in sacrificial oxide etching is in the case of on-chip packaging of sealed cavities, robust membranes, and similar hollow structures. On-chip packaging is beneficial to many resonating devices, which frequency requires vacuum encapsulation of the device. In conventional approach, the sacrificial oxide is removed by allowing the etchant (*e.g.* HF) to reach the sacrificial oxide surface through photolithographically defined holes, followed by the sealing of carefully constructed etch access slits [71]. This causes long etching time, and extended exposure to the etchants raises concerns for material degradation [72]. Further, the sealing material on the surface of encapsulated device may affect the device properties. To overcome this problem a thin LPCVD deposited polysilicon was used as a shell material [73–75]. This polysilicon is deposited in a special condition [76–78] that is permeable to HF based etchants. This allows the underlying sacrificial oxide to be removed directly through the polysilicon layer without making photolithographically access holes. Process to achieve thick porous polysilicon shell layer that may be required in some application has also been reported [79]. A schematic of this approach is shown in figure 5.3.

5.4. APPLICATION OF SACRIFICIAL OXIDE IN DEVICES

Sacrificial oxide has been used in IC fabrication as well in MEMS fabrication. This section reviews the use of sacrificial layer in fabrication of different devices.

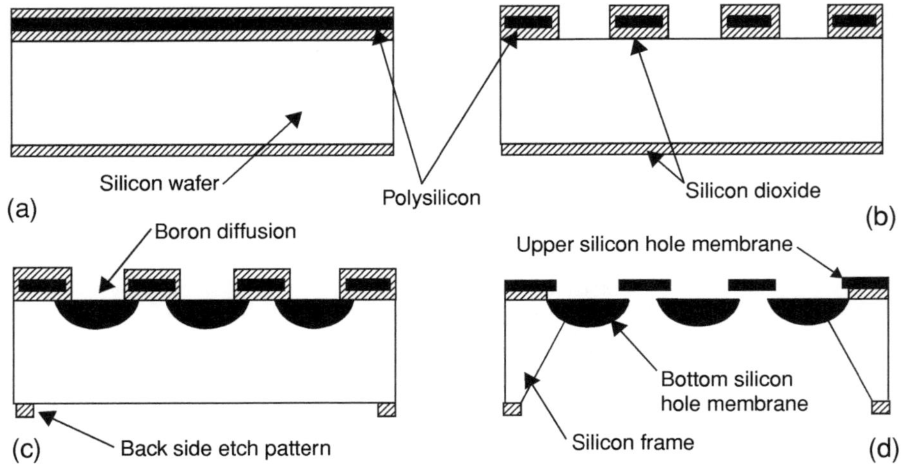

FIGURE 5.4. (a) Starting wafer, (b) the hole patterned before doping, (c) Boron diffusion through the patterned hole, and back side etching, (d) Complete fabricated single-crystal silicon membrane [83].

5.4.1. *Sacrificial Oxide for MEMS*

In 1965, Nathanson *et al.* [2, 80, 81] used sacrificial layer technique to fabricate resonant gate transistors consistent with silicon integrated technology. They electroplated gold beam electrode on top of sacrificial layer. The sacrificial layer thickness determined the nominal beam-to-substrate distance. The sacrificial layer was etched at the end to release the beam electrode.

Howe and Muller (in 1983) used sacrificial oxide to fabricate cantilever beams from polycrystalline silicon (polysilicon) [82]. They etched holes in sacrificial oxide layer fabricated on silicon surface. Next, a polysilicon layer was deposited and plasma etched, leaving the desired cross-section to form a cantilever. Then the underlying sacrificial oxide was etched off in HF. Since a gradual step was required in the polysilicon cantilever beam to minimize the stress concentration, a tapered oxide window edge was desired. In order to achieve this, the sacrificial oxide was constructed with both the wet thermal oxidation, and CVD oxide, followed by a densification process. The oxide layer consisted of 10% thermal SiO_2 and 90% phosphosilicate glass (about 8.75% phosphorous content). The thin rapidly etching surface layer needed for a tapered oxide window edge was created by low energy argon implant.

Kittilsland *et al.* [83] used sacrificial oxide to fabricate submicron particle filter in silicon. The fabrication steps are shown in figure 5.4. A sacrificial silicon dioxide was grown on (100) n-type silicon substrate. The thickness of this oxide determined the membrane separation distance. A polysilicon layer of 1.5 μm was deposited on top of this oxide. A silicon dioxide layer was grown on top of polysilicon layer to prevent damage during subsequent boron doping process [Fig 5.4(a)]. Holes in polysilicon and silicon dioxide were made using photolithography [Fig 5.4(b)]. Next, a high concentration of boron was diffused through the holes into silicon [Fig 5.4(c)]. The wafer was then etched from the back side to create opening from the bottom side. Boron doped layer acted as etch stop layer. The sacrificial oxide was then etched to form through hole in this structure [Fig 5.4(d)].

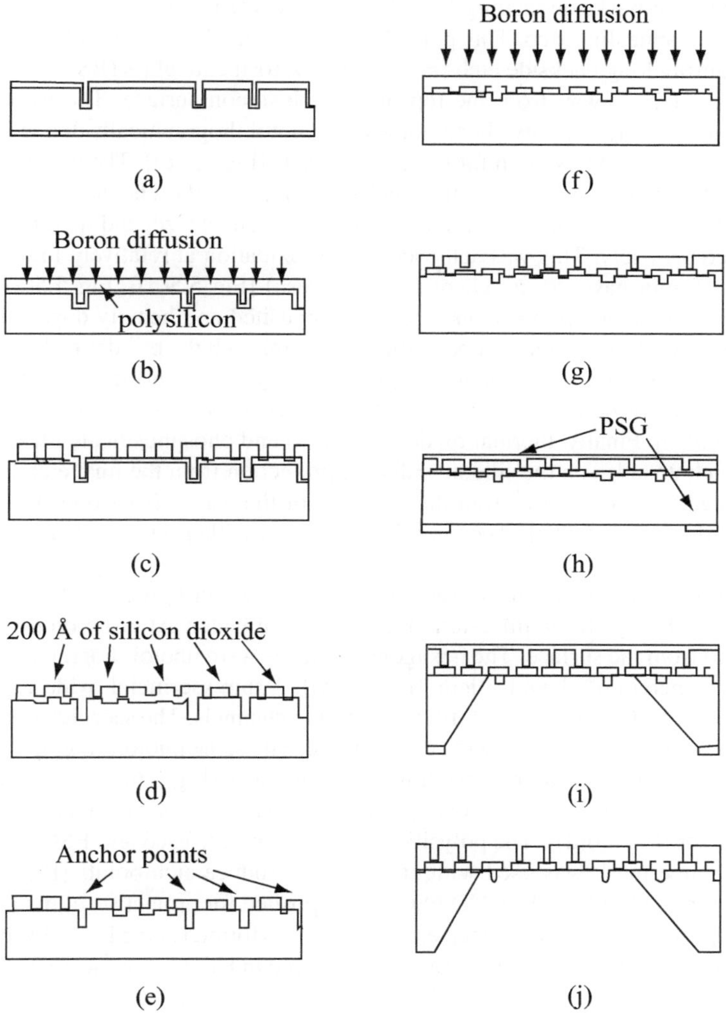

FIGURE 5.5. Fabrication protocol of filter Design I [43].

Chu *et al.* [43] used sacrificial layer to fabricate silicon membrane nano-filters. These filters can be used for immunoisolation. The filter channels were created by selective removal of sacrificial silicon dioxide sandwiched between silicon layers. They presented two sacrificial layer filter design that can achieve pore sizes as low as 10 nm in size. Sacrificial oxides were grown selectively in both the designs to consume silicon and then etched later. The consumed silicon defined the pore size of the filter, and therefore control of oxide growth was important in both the cases. This was achieved by dry oxidation.

Figure 5.5 shows the fabrication protocol of filter design 1. Hexagonal trenches were photolithographically defined and etched by plasma etching. These trenches were used to create support ridges. The wafers were then wet oxidized. This oxide served as an etch stop for final backside etching [Fig 5.5(a)]. LPCVD polysilicon was then deposited followed by an annealing step. This polysilicon was doped heavily with boron to improve selectivity during long backside etch [Fig 5.5(b)]. Borosilicate glass (BSG) formed during this step was etched away from the top of the polysilicon surface. The wafer was then reoxidized and rectangular polysilicon holes were photolithographically defined and etched into the deposited p^+ polysilicon for the exit channels [Fig 5.5(c)]. The wafer was cleaned and oxidized that was removed in the final step to create the channels. This sacrificial oxide growth consumed silicon and thus defined the channel size, and therefore needed to be controlled precisely. This was achieved by dry oxidation at relatively low temperature (850°C for 45 min gave 20 nm channel in this case) [Fig 5.5(d)]. Fig 5.5(e) shows the anchor points. A second polysilicon layer then deposited and heavily doped [Fig 5.5(f)]. This layer was anchored to the first polysilicon layer through the pre-defined anchor points. BSG was again removed from the top of this new polysilicon surface. Square holes were photolithograhically defined and etched into the second polysilicon layer to form entry pores [Fig 5.5(g)]. Finally, thermal oxide was grown and phosphosilicate glass (PSG) was deposited on both sides on the silicon wafer for protection from the final etch [Fig 5.5(h)]. A deep etched was performed from the backside of the wafer that stopped at the silicon dioxide interface [Fig 5.5(i)]. The sacrificial oxide was then etched in HF to open the pores.

The fabrication steps of the second design are shown in figure 5.6. Silicon etch stop was created by heavy Boron diffusion on (100) silicon wafer. After boron diffusion BSG was removed from the surface. The hexagonal trenches were photolithographically defined and plasma etched in p^+ doped silicon [Fig 5.6(a)]. A thin sacrificial oxide was grown on the top of the etched silicon [Fig 5.6(b)] to define the channels. The sacrificial oxide growth consumed silicon and thus defined the channel size. The same photolithographic mask was used to define anchor points by offsetting the alignment [Fig 5.6(c)]. The oxide in this region was then plasma etched. An updoped polysilicon was deposited on top of the etched thin oxide layer [Fig 5.6(d)]. The polysilicon was then p^+ diffused and BSG was removed. Finally, the same mask was used with an offset to produce etch points [Fig 5.6(e)]. The wafer was then oxidised and LPCVD PSG was deposited for protection of the filter against the final deep etch [Fig 5.6(f)]. A deep etch was then performed at the back side of the wafer [Fig (g)]. Sacrificial oxide and PSG was then dissolved in HF to generate pores for the filter [Fig 5.6(h)].

iMEDD Inc. [84] has fabricated nanopores in silicon using the sacrificial oxide technology. Through holes in silicon in the range of 10–100 nm were fabricated. Sacrificial oxide was grown to consume silicon and define the pores.

J. Tu and M. Ferrari developed microfabricated filter and immunoisolating capsule using sacrificial layer [85]. They used sacrificial oxide to create nanochannels in a sandwich structure. The oxides were grown in dry oxidation condition, and etched at the end of the process. The oxide growth consumed silicon and thus defined the nanochannels in the structure. Since the sacrificial oxide growth was the most important step in the filter fabrication, the oxide growth was controlled very accurately.

Sinha *et al.* [28, 86] fabricated nanochannels between the two silicon substrates by growing and subsequently etching sacrificial oxide. A cross-sectional view of the design and a top view of the bottom substrate are shown in figure 5.7. Figure 5.8 shows the process flow for the bottom substrate fabrication. In the first step entry flow chamber, exit flow

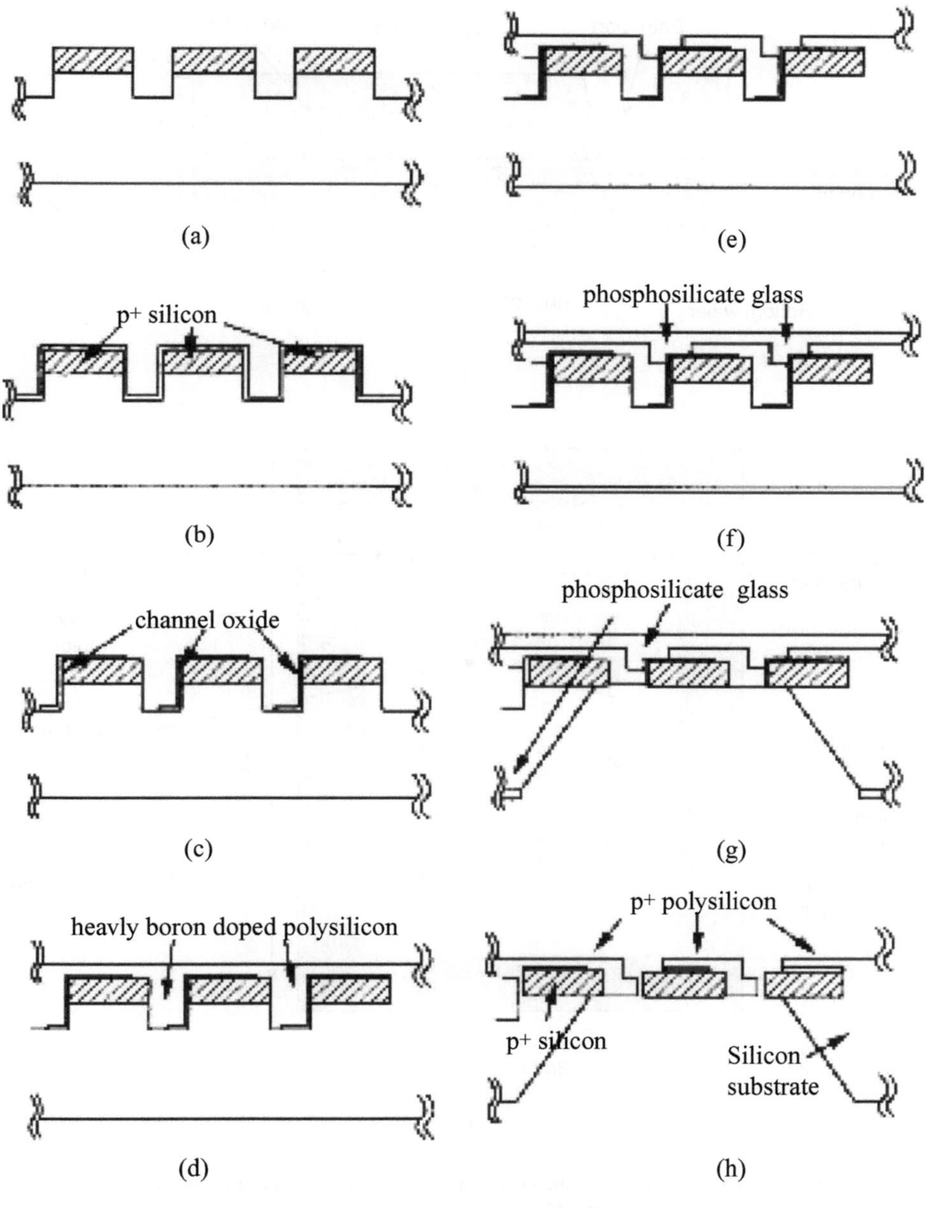

FIGURE 5.6. Fabrication protocol of filter Design II [43].

chamber and the fingers were photolithographically defined and etched into silicon. Low stress LPCVD nitride was used as a mask for the second photolithography step to define nanochannel regions. The defined regions were etched into the mask layer. The unetched regions protected by the nitride became the anchor points and the spacer regions. Next, a

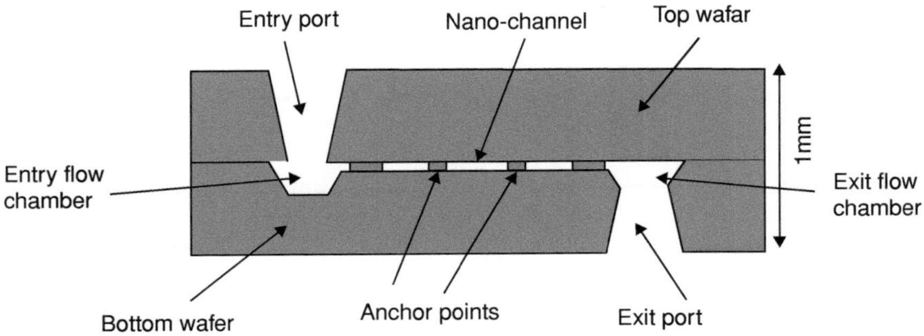

(a)

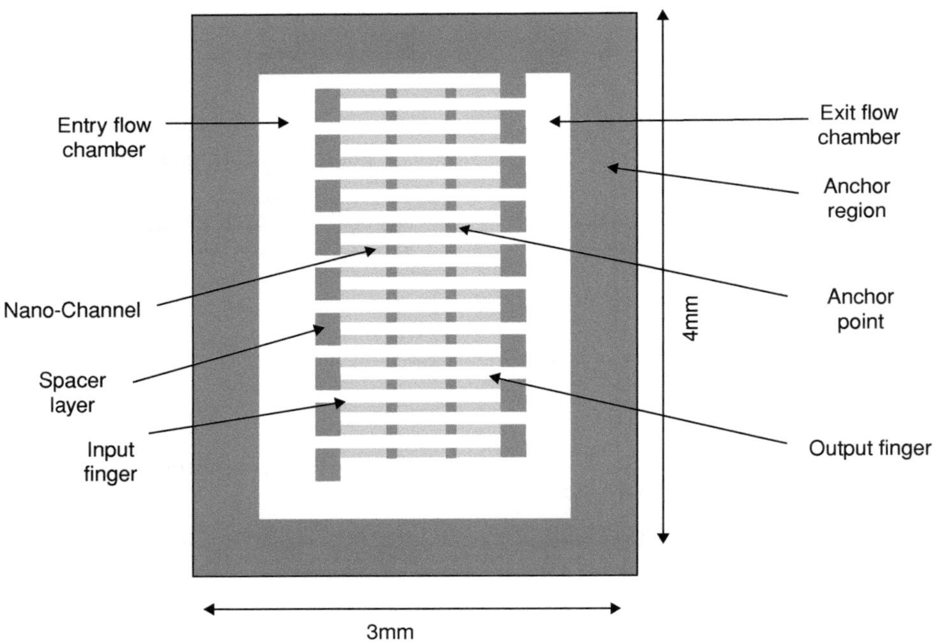

(b)

FIGURE 5.7. (a) A cross-sectional view of the sandwich structure nanoporous membrane. (b) A top view of the sandwich structure nanoporous membrane [86].

sacrificial oxide layer was grown using a thermal oxidation process that consumed substrate silicon from the channel region. This resulted into lowering the surface of the channel region with respect to the surface of the rest of the channel ridge, including the anchor points, thereby defining the height of the channel (H). A second layer of low stress LPCVD nitride was deposited to protect the sacrificial oxide, and then exit port was formed on the back of

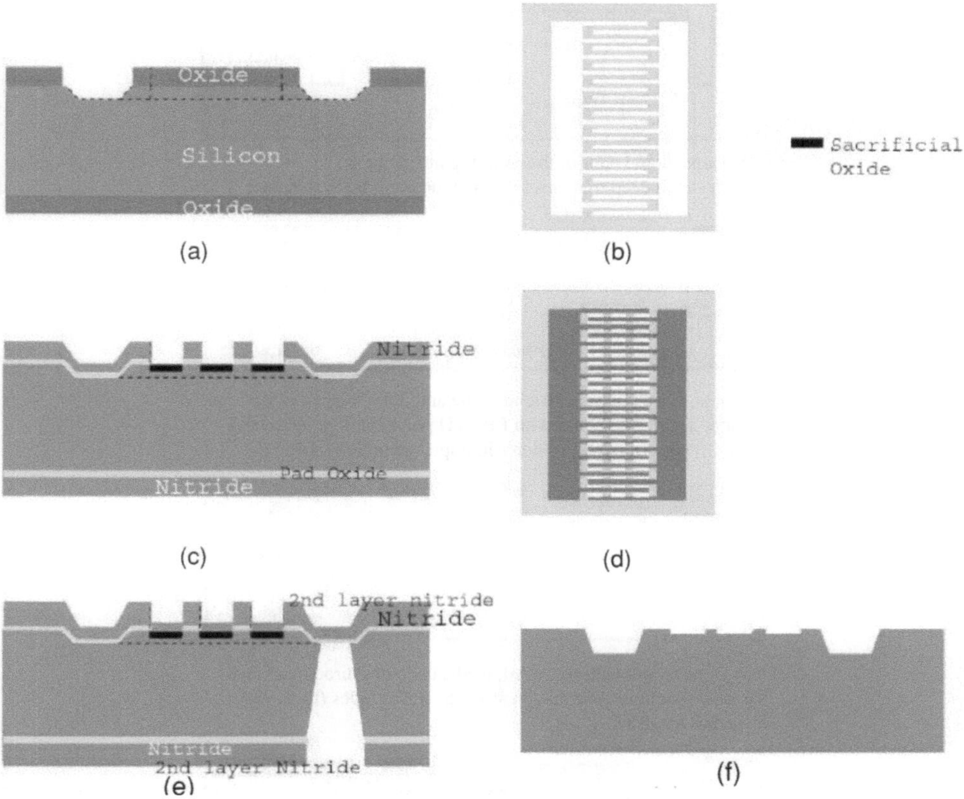

FIGURE 5.8. (a) Flow chambers and fingers, (b) mask 1 defining flow chambers and fingers, (c) nanochannels and anchor points, (d) mask 2 defining nanochannels and anchor points, (e) nanochannels protected by second layer of nitride and exit port (deep etch), and (f) complete bottom substrate fabrication [86].

the bottom substrate. The nitride mask and sacrificial oxide were removed at the end using concentrated HF solution. The entry port was formed in the top silicon substrate, and then bonded with the processed bottom substrate.

Ayazi and Najafi [87–89] developed sacrificial oxide based single wafer high aspect ratio combined poly and single-crystal silicon(HARPSS) MEMS technology, which is capable of producing tens of hundreds of micrometers thick electrically isolated poly and single-crystal silicon microstructures. High aspect ratio polysilicon structures were created by refilling deep trenches with polysilicon on top of sacrificial oxide layer. This technology provides features required for high-performance inertial sensors, as well as a range of actuators. The fabrication steps are described in figure 5.9.

Devoe and Pisano [90] used sacrificial oxide to fabricate surface micromachined piezoelectric accelerometers. The cross-section of this device is shown in figure 5.10. The fabrication protocols are summarized here. A passivation layer of silicon dioxide and low stress nitride was deposited on bare silicon, followed by 0.5 μm LPCVD phosphorous-doped polysilicon. The conductive polysilicon layer was patterned via reactive ion etching (RIE)

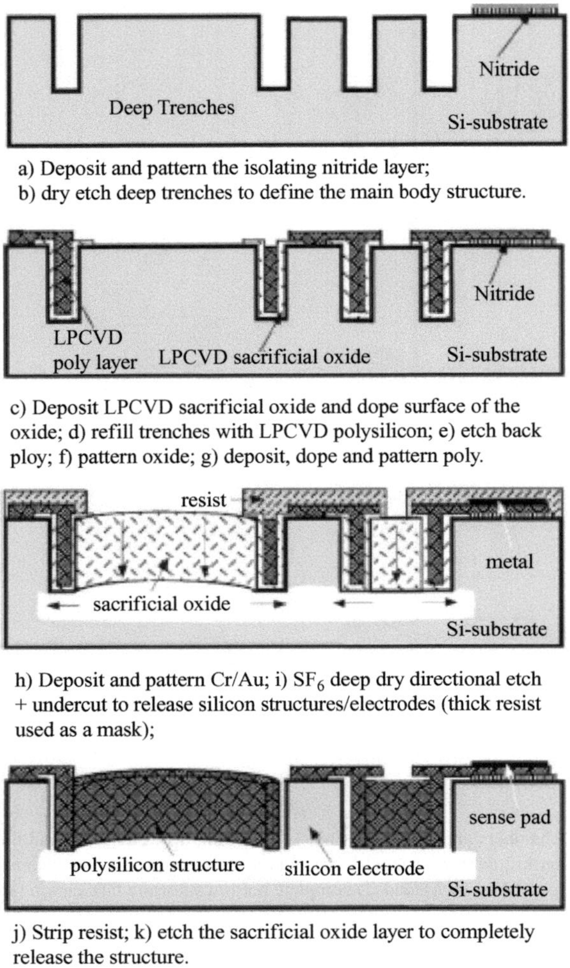

a) Deposit and pattern the isolating nitride layer;
b) dry etch deep trenches to define the main body structure.

c) Deposit LPCVD sacrificial oxide and dope surface of the oxide; d) refill trenches with LPCVD polysilicon; e) etch back ploy; f) pattern oxide; g) deposit, dope and pattern poly.

h) Deposit and pattern Cr/Au; i) SF_6 deep dry directional etch + undercut to release silicon structures/electrodes (thick resist used as a mask);

j) Strip resist; k) etch the sacrificial oxide layer to completely release the structure.

FIGURE 5.9. Fabrication process the HARPSS MEMS technology [87].

to define electrical contacts to the bottom electrode of the accelerometer. Next, a 2.0 μm layer of phosphosilicate glass (PSG) was deposited by LPCVD that served as a sacrificial layer. This was patterned to define regions where the accelerometer structure was later anchored to the substrate. A second layer of 2.0 μm thick phosphorous doped polysilicon was deposited by LPCVD on top of PSG and patterned by plasma etching to define mechanical accelerometer structure. This layer also acted as lower electrode for sensing film. A thin layer of stoichiometric silicon nitride was deposited followed by ZnO deposition. Finally, Pt was sputtered on top of ZnO. The wafer was annealed and then Pt, ZnO and Si_3N_4 layers were patterned in a single ion mill. The devices were released by passivating the ZnO film with photoresist, immersing in BHF to remove sacrificial PSG layer, and removing photoresist by O_2 plasma ashing.

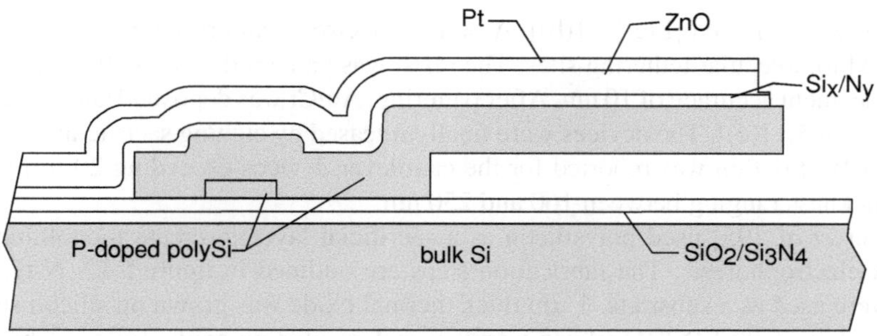

FIGURE 5.10. Cross-section of cantilever structure fabricated using sacrificial oxide [90].

Pisano and co-workers have extensively used sacrificial oxide to fabricate different MEMS structure, *e.g.* silicon based microneedles [91], angular microactuators for magnetic disk drives [92], polysilicon thin films containing through-pores measuring in 10–50 nm in diameter [71, 74, 75, 77, 78], and different microfluidic devices for sampling and analysis of biological fluids [93].

Craighead and co-workers used sacrificial oxide to fabricate polycrystalline silicon and silicon nitride resonators for sensitive detection of bound mass of selected chemical and biological species [94]. Figure 5.11 summarizes the fabrication steps of resonating device. First a 2 μm thick silicon dioxide sacrificial layer was thermally grown [Fig 5.11(a)]. The device layer consisted of either low stress silicon nitride or amorphous silicon deposited using LPCVD. Photolithography with a gold lift-off was performed to define 10 μm gold octagons used for alignment in the subsequent steps. E-bean lithography (EBL) was used to define the body of the oscillator. Hard etch mask chromium was deposited and lifted off. The device layer was etched down to the sacrificial oxide by plasma etch and the remaining

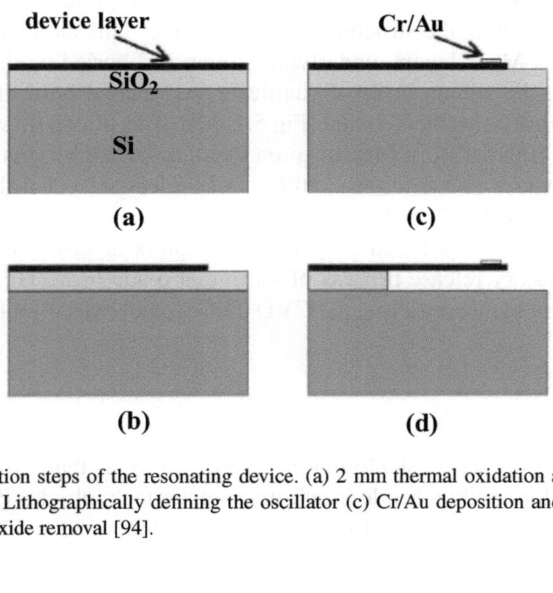

FIGURE 5.11. Fabrication steps of the resonating device. (a) 2 mm thermal oxidation and LPCVD deposition of the device layer. (b) Lithographically defining the oscillator (c) Cr/Au deposition and defined via EBL, and lift-off. (d) Sacrificial oxide removal [94].

chromium was removed [Fig 5.11(b)]. A bilayer electron beam resist process was used to define the biomolecular tethering sites. The resist was patterned using EBL and developed with a placement accuracy of 10 nm. After pattering, Au/Cr was deposited and subsequently lifted off [Fig 5.11(c)]. The devices were finally released by etching sacrificial oxide in HF [Fig 5.11(d)]. Stiction was reported for the cantilever devices exceeding 20 μm in length with a thickness ranging between 160 and 250 nm.

Turner *et al.* [95] used polysilicon as a sacrificial layer to create monolithic device for DNA electrophoresis. The fabrication steps are outlined in figure 5.12. N-type (100) wafers were used as a substrate. 1 μm thick thermal oxide was grown on silicon substrate. 190 nm of low stress LPCVD silicon nitride was deposited on oxide, followed by 500 nm of LPCVD polysilicon. 100 nm oxide hard mask was grown on this layer, followed by 40 nm of aluminum evaporation. EBL was used for patterning. The pattern was transferred to aluminum, and aluminum was used as a hard mask to pattern underlying SiO_2. Finally, oxide hard mask was used to pattern polysilicon sacrificial layer. The mask layers were then removed down to the polysilicon. Next, a low stress LPCVD nitride was deposited. This created a buried column of nitride between sacrificial polysilicon layers. Access holes were then photolithographically defined in order to remove sacrificial layer. The device was then released by etching the sacrificial layer. Turner *et al.* also used PECVD deposited sacrificial oxide to fabricate a prosthetic device for the central nervous system [96].

Sacrificial silicon dioxide has been used to fabricate small metal cantilevers with integrated silicon tips to achieve high resonant frequency associated with low spring constant for use in atomic force microscopy (AFM) [97]. Figure 5.13 describes the microfabrication process steps. The starting wafer was (100)—oriented silicon-on-insulator (SOI) wafer. The thickness of top silicon layer determined the height of the tip. The oxide layer served as an etch stop as well as sacrificial oxide. A pad oxide was grown followed by LPCVD nitride deposition to serve as a mask layer [Fig 5.13(a)]. The cantilever pattern was transferred into the mask at the bottom side of the wafer using photolithographic technique. Next, a deep etch was performed into silicon using nitride as a hard mask [Fig 5.13(b)]. A small circular pattern at the top side of the wafer was then defined photolithographically to form a tip. This was aligned with the structure at the back side. This circular pattern was at the end of the cantilever. Mask layers, underlying silicon and buried oxide were then etched [Fig 5.13(c)]. Next, the silicon layer of cantilever was etched until the oxide layer was reached leaving behind a sacrificial oxide [Fig 5.13(d)]. The silicon tip was then sharpened using KOH etching [Fig 5.13(e)]. Metallization was carried out by sputtering technique at the backside of the wafer, and then the cantilever was released by dissolving the sacrificial oxide in an RIE process [Fig 5.13(f)].

An example of micromachined polysilicon actuators is given in ref [70] that uses sacrificial oxide, and dry release process of sacrificial oxide using HF gas-phase etching to achieve no process induced stiction. LPCVD TEOS oxide was used here as a sacrificial layer.

5.4.2. Sacrificial Oxide in ICs

Sacrificial oxide is used in shallow trench isolation (STI) that has become the mainstream lateral isolation module for deep sub-micrometer technologies [98–102]. A typical STI process sequence includes the following process steps: pad oxide growth, LPCVD

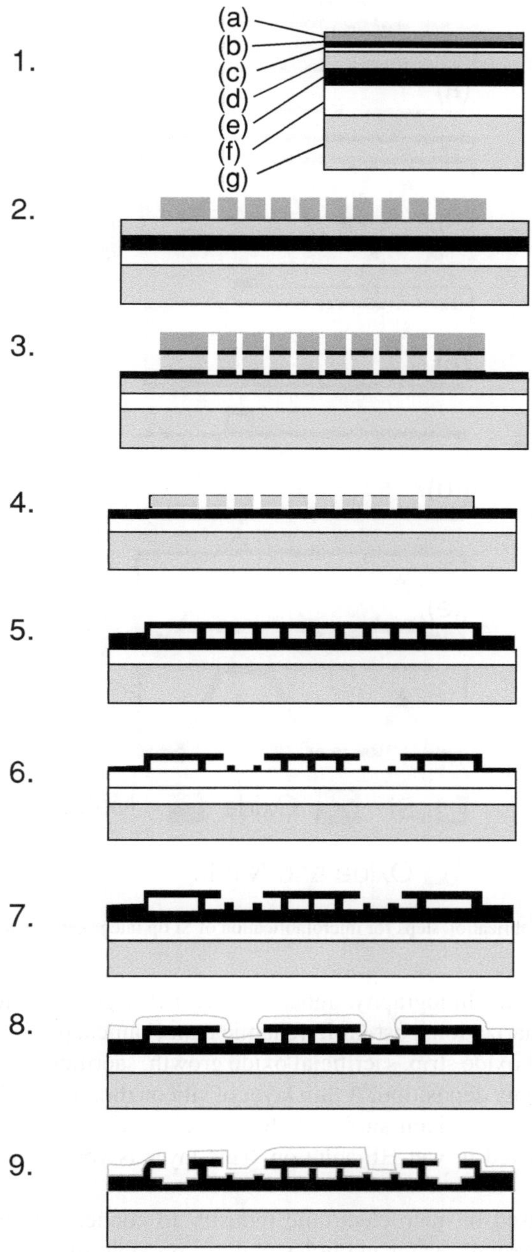

FIGURE 5.12. Fabrication steps of the device. (1) Starts with a thin film stack: (a) PMMA, (b) aluminum, (c) silicon dioxide, (d) polysilicon, (e) silicon nitride, (f) silicon dioxide, and (g) silicon. (2) Resist pattering using EBL. (3) Etching down to the polysilicon sacrificial layer. (4) Removal of the patterning layers down to the polysilicon. (5) Silicon nitride deposition. (6) Access holes etching in the nitride. (7) Removal of the sacrificial layer. (8) Resealing the device with LTO oxide. (9) Loading windows etch at both ends of the device [95].

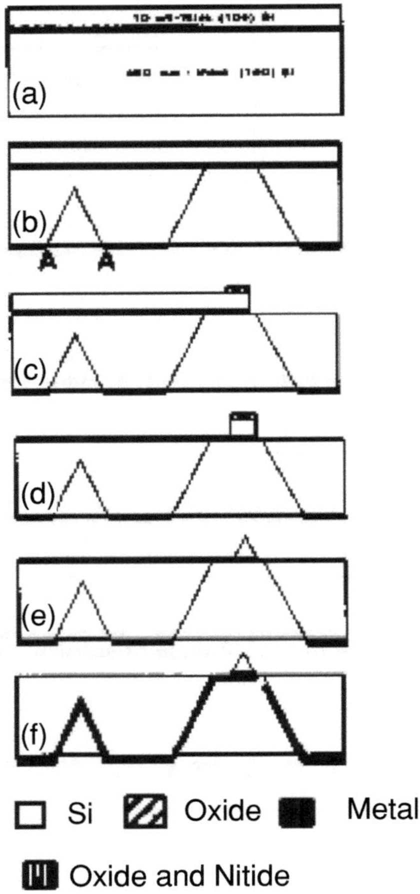

FIGURE 5.13. Fabrication steps for microfabrication of Si tip integrated metal cantilevers [97].

nitride deposition, trench lithography, trench etch, resist/strip/clean, linear oxidation, CVD oxide trench fill, planarization, post-CMP (chemical mechanical polishing) clean/light BHF dip, nitride strip, pad oxide strip, sacrificial oxide growth, sacrificial oxide strip, gate-oxide oxidation, and gate poly deposition. A thin layer of silicon dioxide sacrificial layer is grown to capture defects in the silicon surface before gate-oxide growth. This sacrificial layer is immediately removed in wet HF solution. This layer is very important for the device performance.

Via holes are used in microelectronic industry to connect different metal layers in integrated circuit. Through-wafer via holes make it possible to realize three dimensional structure by electrically connecting both sides of a wafer and manufacture microelectronic or micromechanical components within the bulk of the wafer, such as three-dimensional inductors [103]. In some application, it is necessary to cover the via hole with a thin electrically conductive membrane, *e.g.* a seed layer is electroplated through wafer via holes or an anode in microhollow cathode discharge [104–105]. In these cases, it is important

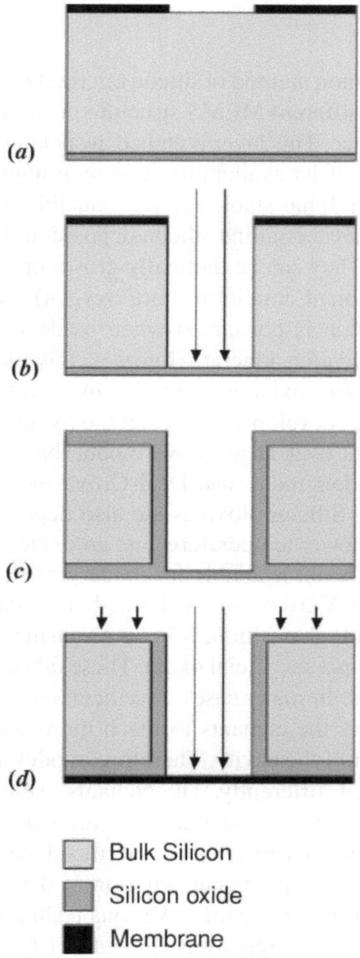

FIGURE 5.14. A schematic diagram of the fabrication steps. (a) The wafer is patterned (b) Etching down to the oxide layer (c) Thermal oxidation (d) Final membrane deposition onto the wafer and removal of the oxide layer above membrane by dry etching [106].

to achieve electrical isolation between the via holes and adjacent structure. It has been suggested that the sacrificial oxide can be used to achieve electrically isolated via structure covered by a thin conductive membrane [106]. The silicon dioxide sacrificial layer is used to cover the via structure during high-temperature steps and is removed at the end of the process after the membrane material has been deposited over the sacrificial oxide. This allows using any membrane material since any the membrane material is deposited at the end of the process. A schematic of the process steps in shown in figure 5.14.

Yee *et al.* [107] developed a fabrication process to integrate 2 μm thick polysilicon with CMOS analog/digital circuit. Polysilicon properties were reported for various deposition temperatures, doping and annealing conditions, and types of sacrificial oxides.

5.5. SUMMARY

Sacrificial oxide is a common method in silicon microfabrication technology. This layer is created in order to release different MEMS structure or in integrated circuits in order to improve the device performance. This layer is etched away later in the process. SiO_2 is used as a sacrificial layer because of its availability in silicon microfabrication processes and etch selectivity against silicon. It has stable and reproducible properties, and is compatible to high temperature during polycrystalline silicon deposition. Different varieties of oxides are used as sacrificial oxide. They can be thermally grown or CVD deposited. Thin oxides are usually grown in dry ambient condition (pure oxygen). Wet oxidation (water vapor) is used for thick oxide. Silicon is consumed when oxide is grown. Deal-Grove model explains and predicts the oxidation kinetic. However, this model does not fully explain the oxidation of shaped surface, oxidation kinetics in mixed ambient and for very thin oxide. Many models have been developed to predict the oxidation behavior for thin oxides, all motivated by improvement over Deal-Grove model, but none has gained widespread acceptance. Most of the models today use Deal-Grove model as a beginning equation and then add to or modify it. Silicon dioxides are also deposited using CVD techniques. Deposition usually occurs at lower temperature, and an oxide film can be deposited on top of any surface where silicon is not available for oxide growth. Silane and oxygen are the common gas sources used for CVD deposition. Tetraethooxysilane (TEOS) can also be used as a source of silicon for oxide deposition. Silicon-on-insulator is also used as a starting material and the buried oxide as a sacrificial oxide. These fabricated oxides are etched away sometimes during the process. In many cases, sacrificial oxide is a buried layer and there is a small opening that allows the etchants to reach the oxide surface as well as allows the etch products to come out of the cavity. Therefore, modeling the etch kinetics of every design needs to be considered differently. The etchants used in sacrificial oxide etching should have high selectivity against the device structure material that is typically silicon. Hydrofluoric acid based compounds are good for sacrificial oxide etching. Wet etching may result in stiction failure that is the permanent stiction of device structure to the substrate due to the residue products from wet etching. Various techniques have been developed to alleviate this problem. One is using anhydrous HF and CH_3OH in gas phase to achieve dry release of device structure. Application of sacrificial oxide includes MEMS as well as IC fabrication. Few examples are: micromotors, sensors, actuators and accelerometers as well as immunoisolating biocapsules, drug delivery devices, microneedles, DNA electrophoretic devices. Silicon dioxide sacrificial layer is used in IC fabrication in order to capture defects in the silicon surface before gate oxide growth.

REFERENCES

[1] G. Moore. VLSI, What Does the Future Hold, *Electron. Aust.*, 42(14): 1980.
[2] H.C. Nathanson and R.A. Wickstrom. A resonant-gate silicon surface transistor with high-Q bandpass properties. *Appl. Phys. Lett.*, 7:84, 1965.
[3] K.E. Petersen. Silicon as Mechanical Materials. *Proceedings of the IEEE*, vol. 70, no. 5, p. 420, 1982.
[4] R.T. Howe. Polycrystalline silicon micromechanical beams. *J. Electrochem. Soc.*, 130(6):1420, 1983.
[5] M. Mehregany, K.J. Gabriel, and W.S.N. Trimmer. Integrated fabrication of polysilicon mechanisms. *IEEE Trans. El. Dev.*, 35:719, 1988.

[6] S.F. Bart, T.A. Lober, R.T. Howe, J.H. Lang, and M.F. Schlecht. Design considerations for micromachined electric actuators Sensors. *Actuators*, 14:269, 1988.

[7] L.S. Fan, Y.C. Tai, and R.S. Muller. IC-processed electrostatic micromotors. *Sens. Act.*, 20:41, 1989.

[8] V.P. Jaecklin, C. Linder, N.F. de Rooij, J.M. Moret, and R. Vuilleumier. Line-addressable torsional micromirrors for light modulator arrays. *Sens. Act.*, A 41–42:324, 1994.

[9] M. Fischer, H. Graef, and W. von Münch, 'Electrostatically deflectable polysilicon torsional mirrors,' *Sens. Act.*, A, 44:83, 1994.

[10] T.A. Core, W.K. Tsang, and S.J. Sherman. Fabrication technology for an integrated surface-micromachined sensor. *Solid State Technol.*, 36:39, 1993.

[11] K.H.L. Chau, S.R. Lewis, Y. Zhao, R.T. Howe, S.F. Bart, and R.G. Marcheselli. An Integrated Force-Balanced Capacitive Accelerometer for Low-G Applications Digest Tech. Papers, *The 8th Int. Conf. Solid-State Sensors and Actuators (Transducers)* vol. 2, p. 593, 1995.

[12] C. Hierold, A. Hildebrandt, U. Näher, T. Scheiter, B. Mensching, M. Steger, and R. Tielert. A Pure CMOS Surface Micromachined Integrated Accelerometer, *Digest Tech. Papers, 9th Ann. Int. Workshop on MEMS*, San Diego, p. 174, 1996.

[13] M. Offenberg, F. Lärmer, B. Elsner, H. Münzel, and W. Riethmüller. Novel process for a monolithic integrated accelerometer. *Digest Tech. Papers, The 8th Int. Conf. Solid-State Sensors and Actuators, Transducers '95*, Stockholm vol. 2, p. 589, 1995.

[14] D.W. Burns, R.D. Horning, W.R. Herb, J.D. Zook. and H. Guckel. Resonant Microbeam Accelerometers, Digest Tech. Papers, *The 8th Int. Conf. Solid-State Sensors and Actuators, Transducers '95*, Stockholm, vol. 2, p. 659, 1995.

[15] T. Hirano, T. Furuhata, K. J. Gabriel, and H. Fujita. Design, fabrication, and operation of submicron gap comb-drive microactuators. *J. Microelectromech. Syst.*, 1:52, 1992.

[16] H. Suzuki. Advances in the microfabrication of electrochemical sensors and systems, *Electroanalysis*, 12(9):703, 2000.

[17] J. Wang. Electrochemical detection for microscale analytical systems: a review. *Talanta*, 56:223, 2002.

[18] P.C. Simpson, A.T. Woolley, and A. Mathies. Microfabrication technology for the production of capillary array electrophoresis chip. *Biomed. Microdev.*, 1(1):7, 1998.

[19] S. Liu and A. Guttman. Electrophoresis microchips for DNA analysis. *Trends Anal. Chem.*, 23(6):422, 2004.

[20] A.T. Woolley, K. Lao, A.N. Glazer, and R.A. Mathies. Capillary Electrophoresis chips with integrated electrochemical detection. *Anal. Chem.*, 70:684, 1998.

[21] A. Manz, D.J Harrison, E.M.J Verpoorte, J.C Fettinger, A Paulus, H. Lödi, and H.M. Widmer. *J. Chromatogr.*, 593:253, 1992.

[22] D.J. Harrison, K. Fluri, K. Seiler, Z.H. Fan, C.S. Effenhauser, and A. Manz. *Science*, 261:895, 1993.

[23] S.C. Jacobson, R. Hergenröder, L.B. Koutny, R.J. Warmack, and J.M. Ramsey. *Anal. Chem.*, 66:1107, 1994.

[24] C.S. Effenhauser, A. Paulus, A. Manz, and H.M. Widmer. *Anal. Chem.*, 66:2949, 1994.

[25] M. DeSilva, Y. Zhang, P. Hesketh, G. Maclay, S. Gendel, and J. Stetter. *Biosen. Bioelectron.*, 10:675, 1995.

[26] V. Mirsky, M. Riepl, and O. Wolfbeis. *Biosens. Bioelectron.*, 12(9–10):977, 1997.

[27] A.T. Woolley and R.A. Mathies. *Proc. Natl. Acad. Sci. U.S.A.*, 91:11348, 1994.

[28] P.M. Sinha, M. Nocentini, M. Calderale, and M. Ferrari. Design, fabrication and mechanical strength analysis of sandwich structure nanoporous silicon filter for application in medicine. *J. Micromech. Microeng.*

[29] J.R. Lewis and M. Ferrari. BioMEMS for drug delivery application Lab-on-a-chip: Chemistry in Miniaturized Synthesis and Analysis System. R.E. Oosterbroek, and A. van den Berg (Ed.), New York, Elsevier Science 373, 2003.

[30] D.D. Breimer. Future challenges for drug delivery. *J. Controll. Rel.*, 62:3, 1999.

[31] R. Guy, M. Powell, J. Fix, and K. Park. Controlled release technologies: current status and future prospects. *Pharmacol. Res.*, 13:1759, 1996.

[32] B. Narasimhan and R. Langer. Zero-order release of micro-and macromolecules from polymeric devices: the role of the burst effect. *J. Contro. Rel.*, 47:13, 1997.

[33] J.D. Plummer, M. Deal, and P.B. Griffin. *Silicon VLSI Technology: Fundamental, Practices, and Modeling*. Prentice Hall, Upper Saddle River, NJ, 2000.

[34] J. Buhler, F.-P. Steiner, and H. Baltes. Silicon dioxide sacrificial layer etching in surface micromachining. *J. Micromech. Microeng.*, 7:R1, 1997.

[35] R.C. Jaeger. *Introduction to Microelectronic Fabrication: Modular Series on Solid State Devices*, G.W. Neudeck and R.F. Pierret (Ed.), Addison-Wesley Publishing Co. vol. 5, 1993.

[36] B.E. Deal and A.S. Grove. General relationship for the thermal oxidation of silicon. *J. of Appl. Phys.*, 36:3770, 1965.

[37] E.A. Irene. The effects of trace amount of water of the thermal oxidation of Si in oxygen. *J. Electrochem. Soc.*, 121(9):1613, 1974.

[38] W.-H. Chu, R. Chin, T. Huen, and M. Ferrari. Silicon Membrane Nanofilters from Sacrificial Oxide Removal. *J. Microelectromech. Syst.*, 8(1):34, 1999.

[39] A. Reisman, E.H. Nicollian, K.C. Williams, and C.J. Merz. The modeling of silicon oxidation from 1×10^{-5} to 20 atmosphere. *J. Elec. Mat.*, 16:45, 1987.

[40] C.J. Han and C.R. Helms. Parellel oxidation mechanism for Si oxidation in dry O_2. *J. Electrochem. Soc.*, 134:1297, 1987.

[41] H.Z. Massoud, J.D. Plummer, and E.A. Irene. Thermal oxidation of silicon in dry oxygen growth-rate enhancement in the thin regime—I: Experimental results. *J. Electrochem. Soc.*, 132(11):2685, 1985.

[42] ——— Thermal oxidation of silicon in dry oxygen. *J. Electrochem. Soc.*, 132(7):1745, 1985.

[43] ——— Thermal oxidation of silicon in dry oxygen growth-rate enhancement in the thin regime—II: Physical mechanisms. *J. Electrochem. Soc.*, 132(11):2693, 1985.

[44] Y. Shioya and M. Maeda. Comparison of phosphosilicate glass films deposited by three different chemical vapor deposition methods. *J. Electrochem. Soc.*, 133:1943, 1986.

[45] S. Wolf and R.N. Tauber. *Silicon processing for the VLSI era Process Technology*. (Sunset Beach, CA, Lattice Press vol.1, 1986.

[46] A.A. Blumberg. Differential thermal analysis and heterogeneous kinetics: the reaction of vitreous silica with hydrofluoric acid. *J. Phys. Chem.*, 63:1129, 1959.

[47] D.J. Monk, D.S. Soane, and R.T. Howe. A review of the chemical reaction mechanism and kinetics for hydrofluoric acid etching of silicon dioxide for surface micromachining applications. *Thin Solid Films*, 232:1, 1993.

[48] J. Liu, Y.C. Tai, J. Lee, K.C. Pong, Y. Zohar, and C.M. Ho. In Situ Monitoring and Universal Modelling of Sacrificial PSG Etching using Hydrofluoric Acid. *Digest Tech. Papers, 6th Ann. Int. Workshop on MEMS '93*, (Fort Lauderdale vol. 71,) 1993.

[49] H. Proksche and G. Nagorsen. The influence of NH_4F on the etch rates of undoped SiO_2 in buffered oxide etch. *J. Electrochem. Soc.*, 139:521, 1992.

[50] D.J. Monk, D.S. Soane, and R.T. Howe. Hydrofluoric acid etching of silicon dioxide sacrificial layers, I. Experimental observations. *J. Electrochem. Soc.*, 141:264, 1994.

[51] D.J. Monk, D.S. Soane, and R.T. Howe. Hydrofluoric acid etching of silicon dioxide sacrificial layers, II. Modeling. *J. Electrochem. Soc.*, 141:270, 1994.

[52] C.M. Mastrangelo and C.H. Hsu. Mechanical stability and adhesion of microstructures under capillary forces—part I: basic theory. *J. Microelectromech. Syst.*, 2:33, 1993.

[53] C.M. Mastrangelo and C.H. Hsu. Mechanical stability and adhesion of microstructures under capillary forces—part II: experiments. *J. Microelectromech. Syst.*, 2:44, 1993.

[54] R. Legtenberg, J. Elders, and M. Elwenspoek. Stiction of surface micromachined structures after rinsing and drying: model and investigation of adhesion mechanisms. *Sens. Act.*, A, 43:230, 1994.

[55] T. Abe, W.C. Messner, and M.L. Reed. Effects of elevated temperature treatments in microstructure release procedures. *J. Microelectromech. Syst.*, 4:66, 1995.

[56] G.T. Mulhern, D.S. Soane, and R.T. Howe. Supercritical Carbon Dioxide Drying of Microstructures Digest Tech. Papers. *7th Int. Conf. Solid-State Sensors and Actuators Transducers '93*, Yokohama, 296 (1993).

[57] J.H. Lee, Y.I. Lee, W.I. Jang, C.S. Lee, and H.J. Yoo. Gas-Phase Etching of Sacrificial Oxides Using Anhydrous HF and CH_3OH. In *Proceedings of IEEE Tenth Annual International Workshop on Micro Electro Mechanical Systems*. p. 448, January 1997.

[58] R. Legtenberg, J. Elders, and M. Elwenspoek. Stiction of Surface Micromachined Structures after Rinsing and Drying : Model and Investigation of Adhesion Mechanism. *Int. Conf. on Solid-State Sensors and Actuators transducers 93*, Yokohama, Japan, p. 198, June 1993.

[59] A. Torii, M. Sasaki, K. Hane, and S. Okuma. Adhesion of microstructures investigated by atomic force microscopy. *Sens. Actu.*, A, 40:71, 1994.

[60] Y. Sato and M. Maeda. Study of HF-treated heavily-doped Si surface using contact angle measurement. *Japan. J. Appl. Phys.*, 33:6508, 1994.

[61] M.R. Houston and R. Maboudian. Stability of ammonium fluoride-treated Si(100). *J. Appl. Phys.*, 78:3801, 1995.

[62] F.S.A. Sandejas, R.B. Apte, W.C. Banyai, and D.M. Bloom. Surface Microfabrication of Deformable Grating Light Valves for High Resolution Displays. *Proceedings of the 7th Int. Conf. Solid-State Sensors and Actuators (Transducers '93)*, Yokohama, Japan, p. 6, 1993.

[63] R.L. Alley, G.J. Cuan, R.T. Howe, and K. Komvopoulos. The Effect of Release-etch Processing on Surface Microstructure Stiction. *Digest Tech. Papers, Solid-State Sensor and Actuator Workshop*, Hilton Head, p. 202, 2002.

[64] C.-J. Kim, A.P. Pissano, R.S. Muller, and M.G. Lim. *Polysilicon Microgripper*, IEEE Solid-State Sensor and Actuators Workshop, Hilton Head Island, South California, USA, p. 48, June 1990.

[65] H. Guckel, J.J. Sniegowski, and T.R. Christenson. Fabrication of Micromechanical Devices from Polysilicon Films with Smooth Surfaces. *Sens. Act.*, 20(1&2):117, 1989.

[66] D. Kobayashi, T. Hirano, T. Furuhata, and H. Fujita. An Integrated Lateral Tunneling Unit. In *Proc. of IEEE Micro Electro Mechanical Systems*, Germany, p. 214, 1992.

[67] G.T. Mulhern, D.S. Soane, and R.T. Howe. Supercritical Carbon Dioxide Drying of Microstructures. *Int. Conf. on Solid-State Sensors and Actuators Transducers '93*, Yokohama, Japan, p. 296, 1993.

[68] T. Abe, W.C. Messner, and M.L. Reed. Effects of elevated temperature treatments in microstructure release procedures. *J. Microelectromech. Syst.*, 4:66, 1995.

[69] M. Ohtsu, H. Shirohata, K. Minami, and M. Esashi. Drying of Self-Supported Thin Film Structures after Rinsing in Small Surface Tension Liquid. *Tech. Digest 13th Sensor Symp.*, p. 33, 1995.

[70] J.H. Lee, H.H. Chung, S.Y. Kang, J.T. Baek, and H.J. Yoo. Fabrication of surface micromachined polysilicon actuators using dry release process of HF gas-phase etching. IEDM 96, San Francisco, USA, p. 761, 1996.

[71] L. Lin, K.M. McNair, R.T. Howe, and A.P. Pisano. Vacuum-encapsulated lateral microresonators. *Proceedings of the 7th International Conference on Solid-State Sensors and Actuators (Transducers'93)*, Yokohama, Japan, p. 270, 1993.

[72] J.A. Walker, K.J. Gabriel, and M. Mehregany. Mechanical Integrity of Polysilicon Films Exposed to Hydrofluoric Acid Solutions. *Proceedings of the IEEE Micro Electro Mechanical Systems*, Napa Valley, California, USA, p. 56, 1990.

[73] G.M. Dougherty, T.D. Sands, and A.P. Pissano. Microfabrication using one-step LPCVD porous polysilicon films. *J. Micromech. Systems*, 12(4):418, 2003.

[74] K.S. Lebouitz, R.T. Howe, and A.P. Pisano. Permeable Polysilicon Etch Access Windows for Microshell Fabrication. *Proceedings of the 8th International Conference on Solid-State Sensors and Actuator*, Stockholm, Sweden, p. 224, 1995.

[75] K.S. Lebouitz, A. Mazaheri, R.T. Howe, and A.P. Pisano. Vacuum Encapsulation of Resonant Devices Using Permeable Polysilicon. *Proceedings of the IEEE MEMS Conference*, Orlando, FL, p. 470, 1999.

[76] M.W. Judy and R.T. Howe. Polysilicon Hollow Beam Lateral Resonators. *Proceedings of the IEEE Micro Electro Mech. Syst. (MEMS)*, Ft. Lauderdale, FL, p. 265, February 1993.

[77] K.S. Lebouitz, R.T. Howe, and A.P. Pisano. Permeable Polysilicon Etch-Access Windows for Microshell Fabrication. *Proceedings of the 8th Intl. Conf. Solid-State Sensors and Actuators (Transducers '95)*, Stockholm, Sweden, p. 224, June 1995.

[78] K.S. Lebouitz, A. Mazaheri, R.T. Howe, and A.P. Pisano. Vacuum Encapsulation of Resonant Devices using Permeable Polysilicon. *Proceedings of the IEEE Micro Electro Mech. Syst. (MEMS)*, Orlando, FL, p. 470, 1999.

[79] R. He, Li Fan, M.C. Wu and C.-J. Kim. Porous Polysilicon Shell formed by Electrochemical Etching for On-chip Vacuum Encapsulation. *Solid-State Sensors, Actuators and Microsystems Workshop*, South Carolina, p. 332, June 2004.

[80] H.C. Nathanson, W.E. Newell and R.A. Wickstrom. Tunning forks sound a hopeful note. *Electronics*, 38:84, 1965.

[81] H.C. Nathanson, W.E. Newell, R.A. Wickstrom, and J.R. Davis Jr. The resonant gate transistor. *IEEE Trans. Electron Devices*, 14(3):117, 1967.

[82] R.T. Howe and R.S. Muller. Polycrystalline silicon micromechanical beams. *J. Electrochem. Soc.*, 130(6):1420, 1983.

[83] G. Kittilsland, G. Stemme, and B. Norden. A sub-micron particle filter in silicon. *Sens. Actu.*, A21–23:904, 1990.

[84] F.J. Martin and C. Groove. Microfabricated drug delivery systems:Concept to improve clinical benefits. *Biomed. Microdev.*, 3(2):97, 2001.

[85] J. Tu and M. Ferrari. Microfabricated Filter and Capsule Using A Substrate Sandwich, U.S. Patent No. 5,938,923, August 17, 1999.

[86] P. Sinha, G. Valco, S. Sharma, X. Liu, and M. Ferrari. Nanoengineered device for drug delivery application. *Nanotechnology*, 15:S585, 2004.

[87] F. Ayazi and K. Najafi. High aspect-ratio combined poly and single-crystal silicon (HARPSS) MEMS technology. *J. Microelectromech. Syst.*, 9(3):288, 2000.

[88] F. Ayazi and K. Najafi. High aspect-ratio polysilicon micromachining technology. *Sen. Actu.*, A87:46, 2000.

[89] F. Ayazi. The HARPSS process for fabrication of precision MEMS intertial sensors. *Mechatronics*, 12:1185, 2002.

[90] D.L. DeVoe and A.P. Pisano. Surface micromachined piezoelectric accelerometers (PiXLs). *J. Microelectromech. Syst.*, 10(2):180, 2001.

[91] L. Lin and A.P. Pisano. Silicon-processes microneedles. *J. Microelectromech. Syst.*, 8(1):78, 1999.

[92] D.A. Horsley, M.B. Cohn, A. Singh, R. Horowitz, and A.P. Pisano. Design and fabrication of an angular microactuator for magnetic desk drives. *J. Microelectromech. Syst.*, 7(2):141, 1998.

[93] J.D. Zahn, A.A. Deshmukh, A.P. Papavasiliou, A.P. Pisano, and D. Liepmann. An Integrated Microfluidic Device for the Continuous Sampling and Analysis of Biological Fluids. *Proceedings of the ASME International Mechanical Engineering*, NewYork, NY, 2001.

[94] B. Llic, H. . Craighead, S. Krylov, W. Senaratne, C. Ober, and P. Neuzil. Attogram detection using nanoelectromechanical oscillators. *J. Appl. Phys.*, 95(7):3694, 2004.

[95] S.W. Turner, A.M. Perez, A. Lopez, and H.G. Craighead. Monolithic nanofluid sieving structures for DNA manipulation. *J. Vac. Sci. Technol.*, B 16(6):3835, 1998.

[96] J.N. Turner, W. Shain, A. Spence, S. Retterer, H. Craighead, and M. Isaacson. A prosthetic device for the central nervous system. *NNUN Abs. Biol. Chem.*, 22, 2002.

[97] A. Chand, M.B. Viani, T.E. Schaffer, and P.K. Hansma. Microfabricated small metal cantilevers with silicon tip for atomic force microscopy. *J. Microelectromech. Syst.*, 9(1):112, 2000.

[98] J. Tony Pan, Dennis Ouma, P. Li, D. Boning, F. Redeker, J. Chung, and J. Whitby. *Planaqization and integration of shallow trench isolation*. VMIC Salta Clara, CA June p. 1, 1998.

[99] H.-S. Lee. The effect of a sacrificial oxide process on metal-oxide-semiconductor field effect transistors fabricated with ion-beam-nitridation technology. *Solid State Electron.*, 29(1):25, 1986.

[100] D. Shain and R. Badilo. A Trench Isolated SOI Bipolar Process. *IEEE SOS/SOI Technology Conference*, p. 83, October 1990.

[101] K. Rho, J. Burkhardt, J. Schweigert, and D. Fertig. Effect of Trench Spacer Etch on PMOS Threshold Voltage. *5th International Symposium on Plasma Process-Induced Damage*, p. 57, May 2000.

[102] S. Nakashima, T. Ohno, S. Nakamura, T. Ueki, Y. Kado, T. Tsuchiya, T. Takeda, and T. Sakai. Sacrificial Oxidation Techniques of Top Si Layer to Reduce Source-to-Drain Leakage Current in 0.25-μm MOSFETs/SIMOX. *Proceedings of the of IEEE International SOI Conference*, p. 124, 30 Sept.–3 Oct.

[103] H.T. Soh, C.P. Yue, A. McCarthy, C. Ryu, T.H. Lee, S.S. Wong, and C.F. Quate. Ultra-low resistance, through-wafer via (TWV) technology and its applications in three-dimensional structures on silicon. *Japan. J. Appl. Phys.*, 38:2393, 1999.

[104] K.H. Schoenbach, R. Verhappen, T. Tessnow, F.E. Peterkin, and W.W. Byszewski. Microhollow cathode discharges. *Appl. Phys. Lett.*, 68:13, 1996.

[105] K.H. Schoenbach, A. El-Habachi, W. Shi, and M. Ciocca. High-pressure hollow cathode discharges plasma sources. *Sci. Technol.*, 6:468, 1997.

[106] D. Rosen, J. Olsson, and C. Hedlund. Membrane covered electrically isolated through-wafer via holes. *J. Micromech. Microeng.*, 11:334, 2001.

[107] Y. Yee, M. Park, W. Lee, S. Kim, and K. Chun. An integration process of micro electro mechanical polysilicon with CMOS analog/digital circuits. *Sens. Actu.*, 78:120, 1999.

6

Carbon Nanotube Biosensors

Pingang He[1] and Liming Dai[2]

[1]*Department of Chemistry, East China Normal University, 3663 N Zhongshan Rd, Shanghai 200062, China*

[2]*Department of Chemical and Materials Engineering College of Engineering, University of Dayton, 300 College Park, Dayton, Ohio 45469, USA*

6.1. INTRODUCTION

Owing to their nano-dimensions, rich electronic states, large surface area, high mechanical strength, and excellent chemical and thermal stability, carbon nanotubes have attracted a great deal of interest [1]. Among the many potential applications [1, 2], carbon nanotubes have recently become promising functional materials for the development of advanced biosensors with novel features. It has been demonstrated that carbon nanotubes could promote electron-transfer with various redox active proteins, ranging from glucose oxidase [3, 4] with a deeply embedded redox center to cytochrome *c* [5, 6] and horseradish peroxidase [7, 8] with surface redox centers. For the use of carbon nanotubes in biosensing applications, however, the ability to immobilize biomolecules on the carbon nanotube structure without diminishing their bioactivity is indispensable. Therefore, a number of intriguing physicochemical approaches have recently been devised for functionalization of carbon nanotubes with biomolecules [9–15]. Consequently, many biological species, such as DNA, proteins and enzymes, have been immobilized onto carbon nanotubes either on their sidewalls or at the end-caps [16–20]. While the carbon nanotube bioconjugates are functional materials essential for the development of advanced nanotube biosensors, the device design and fabrication also play an important role in regulating their biosensing performance. For many electrochemical biosensing applications, randomly entangled carbon nanotubes have been physically coated onto conventional electrodes [17, 21–24]. The use of vertically aligned carbon nanotubes [25], coupled with well-defined chemical functionalization, should offer additional advantages for facilitating the development of advanced biosensors with a high sensitivity and good selectivity [20, 26–28].

In this chapter, we present an overview of the recent progress in carbon nanotube functionalization and electrode fabrication for biosensing applications by discussing some selected examples from the research and development carried out by many research groups, including our own work.

6.2. THE STRUCTURE AND CHEMICAL REACTIVITY OF CARBON NANOTUBES

As can be seen in Figures 6.1A&B, carbon nanotubes may be viewed as a graphite sheet that is rolled up into a nanoscale tube form (single-walled carbon nanotubes, SWCNTs)

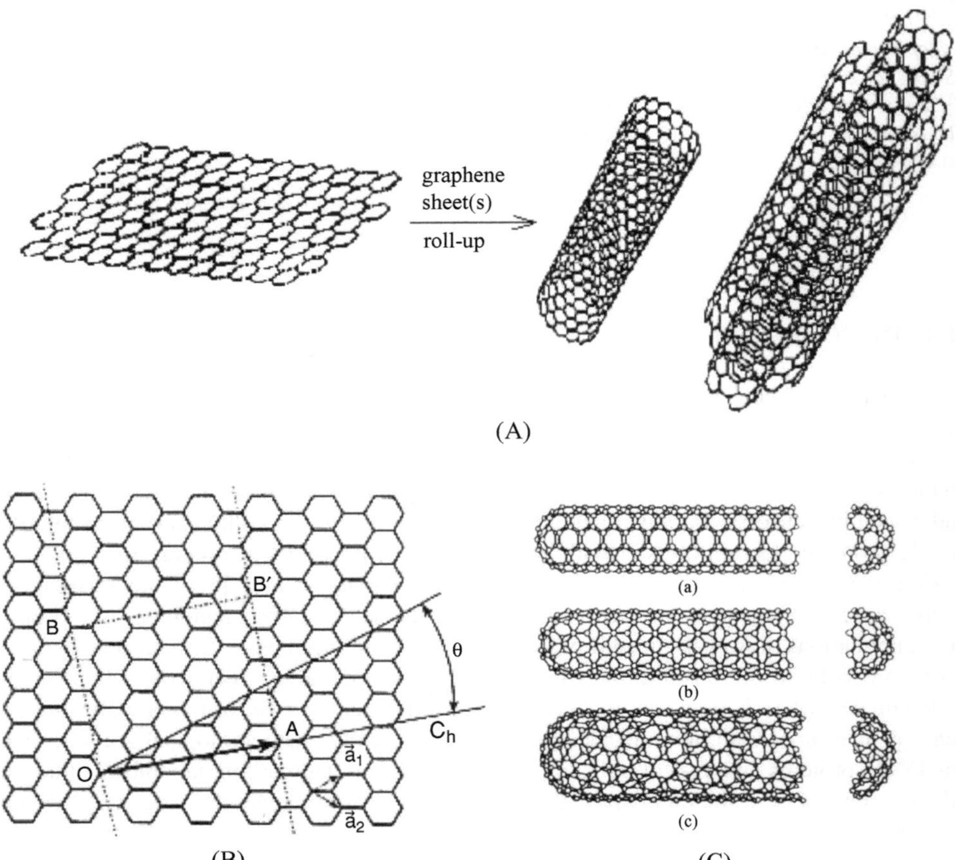

FIGURE 6.1. Schematic representation of (A) single-/multi-walled carbon nanotube formation by rolling up graphene sheet(s). (B) carbon nanotube formation based on a 2D graphene sheet of lattice vectors a_1 and a_2, the roll-up chiral vector $C_h = na_1 + ma_2$, and the chiral angle θ between C_h and a_1. When the graphene sheet is rolled up to form the cylindrical part of the nanotube, the chiral vector forms the circumference of nanotube's circular cross-section with its ends meeting each other. The chiral vector (n, m) defines the tube helicity. (C) Schematic representation of SWCNTs. (a) (5,5) armchair nanotube; (b) (9,0) Zigzag nanotube; (c) (10,5) chiral nanotube.

or with additional graphene tubes around the core of a SWCNT (multi-walled carbon nanotubes, MWCNTs) [1, 29, 30]. These elongated nanotubes consist of carbon hexagons arranged in a concentric manner, with both ends of the tubes normally capped by fullerene-like structures containing pentagons (Figure 6.1C). They usually have a diameter ranging from a few Ångstroms to tens of nanometers and a length of up to centimeters. Because the graphene sheet can be rolled up with varying degrees of twist along its length, carbon nanotubes have a variety of chiral structures [1, 30]. Depending on their diameter and helicity of the arrangement of graphitic rings in the walls, carbon nanotubes can exhibit interesting electronic properties attractive for potential applications in areas as diverse as sensors, actuators, molecular transistors, electron emitters, and conductive fillers for polymers [1, 2].

In order to meet specific requirements demanded for particular applications (*e.g.* biocompatibility for nanotube biosensors and interfacial strength for blending with polymers), chemical modification of carbon nanotubes is essential. The pentagon-containing tips of carbon nanotubes are more reactive than the sidewalls so that a variety of chemical reagents have been attached at the nanotube tips. Recently, some interesting reactions have also been devised for chemical modification of both the inner and outer nanotube walls, though the seamless arrangement of hexagon rings has rendered the sidewalls relatively unreactive. Judicious application of these *site-selective* reactions to non-aligned and aligned carbon nanotubes has opened up a rich field of carbon nanotube chemistry, which has been reviewed in several excellent review articles [9–15]. Below, we summarize the recent progress in functionalization of carbon nanotubes with biomolecules—a research area which has been so crucial to the development of carbon nanotube biosensors.

6.3. FUNCTIONALIZATION OF CARBON NANOTUBES

6.3.1. Non-covalent Functionalization

Many biological species, such as carbohydrates [31–34], nucleic acids [35, 36], peptides [37, 38], and proteins [39, 40], can be non-covalently adsorbed on the carbon nanotube surfaces through hydrophobic, π-π stacking, and electrostatic interactions or even be trapped inside of the nanotube hollow cavity through supramolecular inclusion [16, 40]. For instance, SWCNTs have been dissolved in an aqueous solution of starch and iodine by forming starch-wrapped nanotube supramolecular complex [31]. Similarly, SWCNTs were also found to be soluble in an amylase-encapsulated form in an aqueous solution of amylase [32]. More interestingly, Dodziuk *et al.* [34] have recently reported the solubilization of SWCNTs in an aqueous solution of η-cyclodextrins (η-CD), which has a 12-membered ring structure with an inner cavity of ~1.8 nm in diameter (Figure 6.2a), by encapsulating individual nanotubes within the CD ring structure, as schematically shown in Figure 6.2b. However, it is unlikely to solubilize SWCNTs in an aqueous solution of γ-CD due to its smaller ring [33].

In addition to the aforementioned non-covalent surface modification of carbon nanotubes by macromolecular wrapping or ring inclusion, physical adsorption of small molecular moieties has also been exploited as an alternative approach for noncovalent functionalization of the nanotube sidewalls. In this context, Chen *et al.* [40] reported a simple approach to noncovalent functionalization of the inherently hydrophobic surface of SWCNTs by irreversibly adsorbing a bifunctional molecule, 1-pyrenebutanoic acid succinimidyl ester, from dimethylformamide (DMF) or methanol solvent. The π-π stacking interaction between the

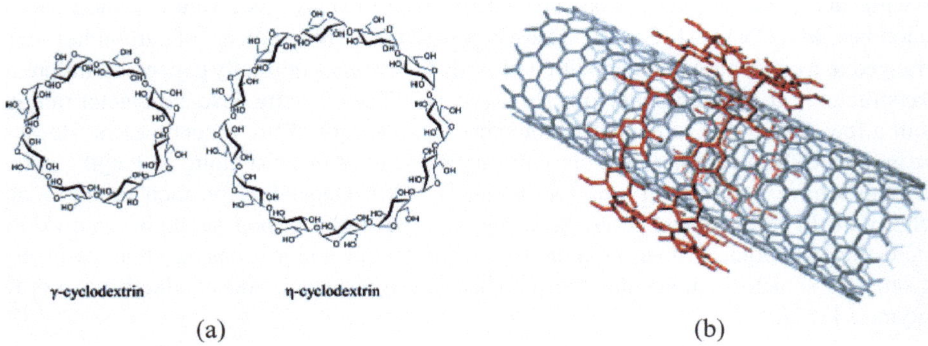

FIGURE 6.2. (a) Structures of γ, η-cyclodextrins. (b) two η-CDs in the head-to-head arrangement threaded on SWCNT (from Reference 34).

pyrenyl group of 1-pyrenebutanoic acid and the graphitic nanotube sidewall serves as the driving force for the irreversible adsorption to occur. The succinimidyl ester groups thus anchored onto SWCNTs are highly reactive for nucleophilic substitution by primary and secondary amines that exist in abundance on the surface of many proteins (Figure 6.3).

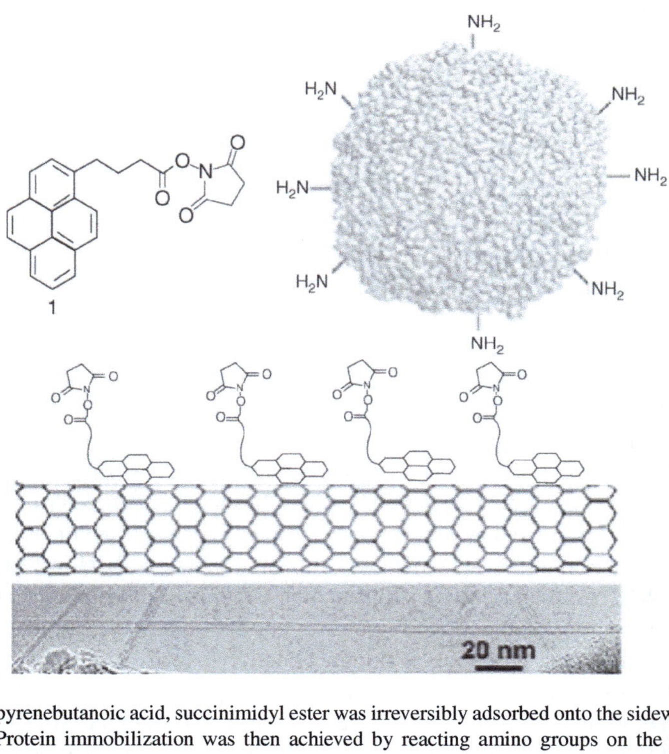

FIGURE 6.3. 1-pyrenebutanoic acid, succinimidyl ester was irreversibly adsorbed onto the sidewall of a SWCNT via π-stacking. Protein immobilization was then achieved by reacting amino groups on the protein with the succinimidyl ester to form amide bonds. Lower panel shows a TEM image of an as-grown SWCNT on a gold TEM grid (from Reference 40).

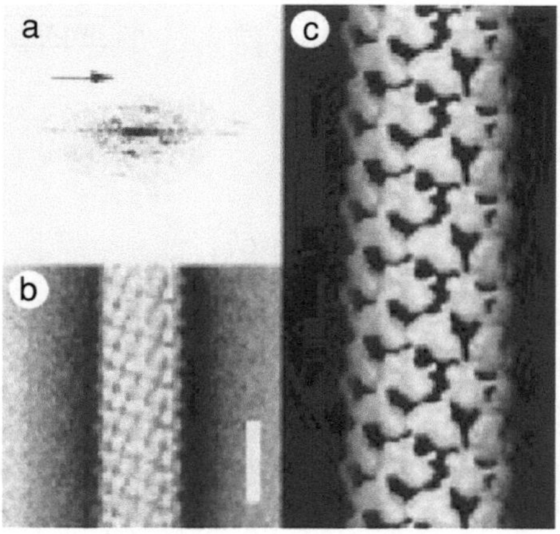

FIGURE 6.4. Analysis of the streptavid helices formed on MWCNTs. a) computed power spectrum of the Fourier transform of a helical array of strepavidin molecules. The helical repeat is 12.8 nm and the arrow indicates 1/2.5 nm^{-1}. b) Noise-free view of the helical repeat obtained by correlation. c) Three-dimensional model of streptavidin assemblies calculated by retroprojecting the noise-free helical repeat show in (b) along the Euler angles deduced from the analysis of the diffraction pattern. The bar represents 25 nm in (b) and 12.5 nm in (c) (from Reference 42).

This study, therefore, opened up an avenue for immobilizing a wide range of biomolecules, including ferritin, streptavidin and biotinyl-3,6-dioxaoctanediamine (biotin-PEO-amine), onto carbon nanotube structures [40]. Indeed, Besteman *et al.* [41] have successfully adopted this concept for immobilizing glucose oxidase (GOx) on *as-grown* SWCNTs supported by a silicon wafer. The GOx-decorated SWCNTs were then used for sensing enzyme activities.

Similarly, a large number of proteins have also been demonstrated to strongly bind to the exterior surface of MWCNTs via non-specific adsorption. For instance, Balavoine *et al.* [42] observed that proteins, such as streptavidin and HupR, could be spontaneously adsorbed onto MWCNTs to form a close-packed helical structure (Figure 6.4). Based on the same principle, Dieckmann *et al.* [37] have created a peptide structure that folds into an amphiphilic R-helix to coat carbon nanotubes. Given the fact that the hydrophobic face of the helix can interact noncovalently with the aromatic surface of carbon nanotubes whilst the hydrophilic face can promote self-assembly through the charged peptide-peptide interactions, the amphiphilic peptides could be used not only to solubilize SWCNTs but also to assemble the peptide-coated nanotubes into supramolecular structures with a controllable size and geometry. It is also possible to assemble the peptide-wrapped nanotubes onto substrate surfaces pre-patterned with antibodies for creating architectures useful in electrical circuits for molecular sensing applications.

6.3.2. *Chemically Covalent Modification*

Compared with the non-covalent surface modification discussed above, covalent chemical modification should provide not only a stronger linkage between the functional groups and nanotubes but also additional advantages for controlling the attachment site and

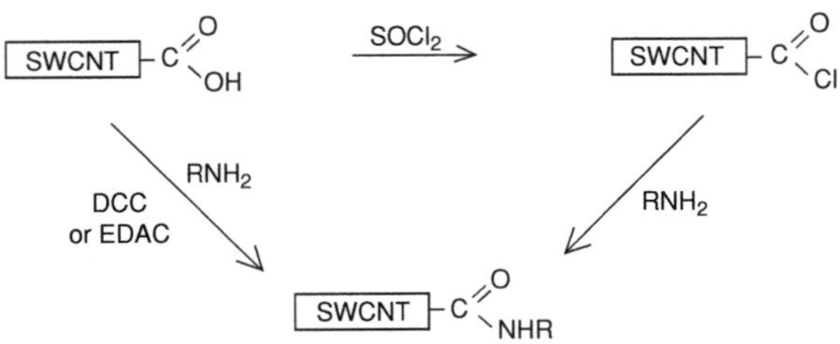

FIGURE 6.5. A typical functionalization protocol for COOH-containing carbon nanotubes.

accessibility to the attached biomolecules, leading to an improved selectivity and stability for sensing applications. Covalent functionalization of carbon nanotubes with biomolecules has been achieved either through chemical derivatization of the carboxylic acid groups on oxidized nanotubes or by direct chemical modification of the pristine carbon nanotubes via addition reactions.

6.3.2.1. Chemical Functionalization via Carboxylic Acid Groups It is well-known that a variety of oxygenated groups, including carboxylic acid moieties, can be introduced at the nanotube ends and/or the structural defect sites along the tube walls through strong acid oxidization (*e.g.* sonication of SWCNTs in mixtures of sulfuric/nitric acids or sulfuric acid/hydrogen peroxide [43–46]) and ozone treatment [47–49] of SWCNTs, respectively. The oxidized carbon nanotubes can be further functionalized with thionyl chloride [50, 51] or via diimide-activated amidation with EDAC or DCC under ambient conditions (Figure 6.5).

Of particular interest, Sun and co-workers [52–55] have chemically attached poly (propionylethylenimine-*co*-ethylenimine), an aminopolymer that is widely used in peptide synthesis, to the acid-oxidation-induced carboxylic acid groups on carbon nanotubes via diimide-activated amidation with EDAC under ambient conditions. On the other hand, various biomolecules, including DNA [56–58], PNA [59], and proteins [60, 61], have also been successfully coupled to the acid-oxidized carbon nanotubes through the carboxylic acid functional linkers.

Figure 6.6 shows a typical procedure for attaching amine-containing proteins onto carbon nanotubes. As can be seen, carbon nanotubes were first acid-oxidized to introduce surface carboxylic acid groups. The carboxylic acid groups were then activated by EDAC to form a highly reactive O-acylisourea active intermediate, which can be converted into a more stable active ester (succinimidyl intermediate) in the presence of N-hydroxysuccinimide (NHS). Finally, the active ester underwent nucleophilic substitution reaction with the amine groups of biomolecules (*e.g.* proteins). With this approach, Jiang *et al.* [61] have successfully attached proteins of ferritin or bovine serum albumin (BSA) onto the acid-oxidized carbon nanotubes.

Using the carboxylic acid group as a functional linker, Baker *et al.* [62] have also developed a multistep route towards the covalent grafting of DNA oligonucleotides onto SWCNTs. As shown in Figure 6.7, the modification procedure involved the activation of the carboxylic acid groups of oxidized SWCNTs at the nanotube ends and sidewalls by thionyl chloride, followed by the reaction with ethylenediamine to produce amine-terminated sites. The amines were then reacted with the heterobifunctional cross-linker succinimidyl 4-(*N*-maleimidomethyl)-cyclohexane-1-carboxylate (SMCC), leaving the

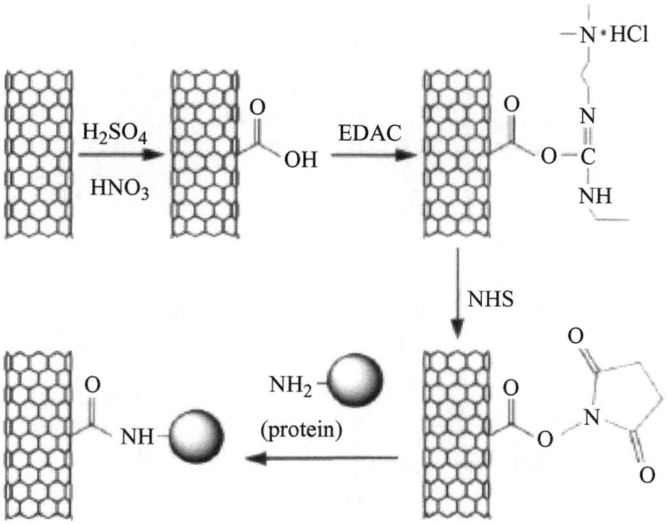

FIGURE 6.6. Schematic view of the attachment of proteins to carbon nanotubes via a two-step process of diimide-activated amidation (from Reference 61).

surface terminated with maleimide groups. Finally, thiol-terminated DNA chains were reacted with the maleimide groups, resulting in the formation of DNA-grafted SWCNTs.

6.3.2.2. Chemical Functionalization via Direct Addition Reactions In view of the presence of unsaturated π-electron systems in carbon nanotubes, Smalley *et al.* [63–65] and Holzinger *et al.* [66] have functionalized the sidewall of SWCNTs directly by fluorination and azide-thermolysis, respectively. Generally speaking, direct addition of functionalities into the carbon nanotube structure was feasible by nitrene cycloaddition, nucleophilic carbine addition, and radical addition.

As shown in Figure 6.8a, the addition of nitrenes could be carried out by heating SWCNTs up to 160°C in the presence of a 200-fold excess of alkyl azidoformate as the nitrene precursor. After thermally-induced N_2 extrusion from **1**, the nitrene addition led to the formation of alkoxycarbonylaziridino-SWCNTs. On the other hand, nucleophilic carbene addition onto the sidewall of SWCNTs has been achieved by reacting the nanotube electrophilic π system with dipyridyl imidazolidene (Figure 6.8b). To demonstrate radical addition, Holzinger *et al.* [66] have exploited the photoinduced addition of perfluorinated alkyl radicals onto SWCNTs by illuminating the nanotube suspension with a medium-pressure mercury lamp (150 W) for 4 h in the presence of a 200-fold excess of heptadecafluorooctyl iodide (Figure 6.8c). Subsequent derivatization of the fluorine-containing side groups allowed the introduction of various sidewall functionalities into the nanotube structure [12]. Moghaddam *et al.* [67] have also used the azide-photochemistry to functionalize carbon nanotubes by attaching single-stranded DNA chains onto the nanotube sidewalls and tips.

Besides, Prato and co-workers [68] have reported that carbon nanotubes underwent 1.3-diplar cycloaddition when heated in DMF in the presence of α-amino acid and aldehyde. This reaction provides a versatile and powerful methodology for functionalization of carbon nanotubes with different functionalities, which can be further coupled with amino acids and

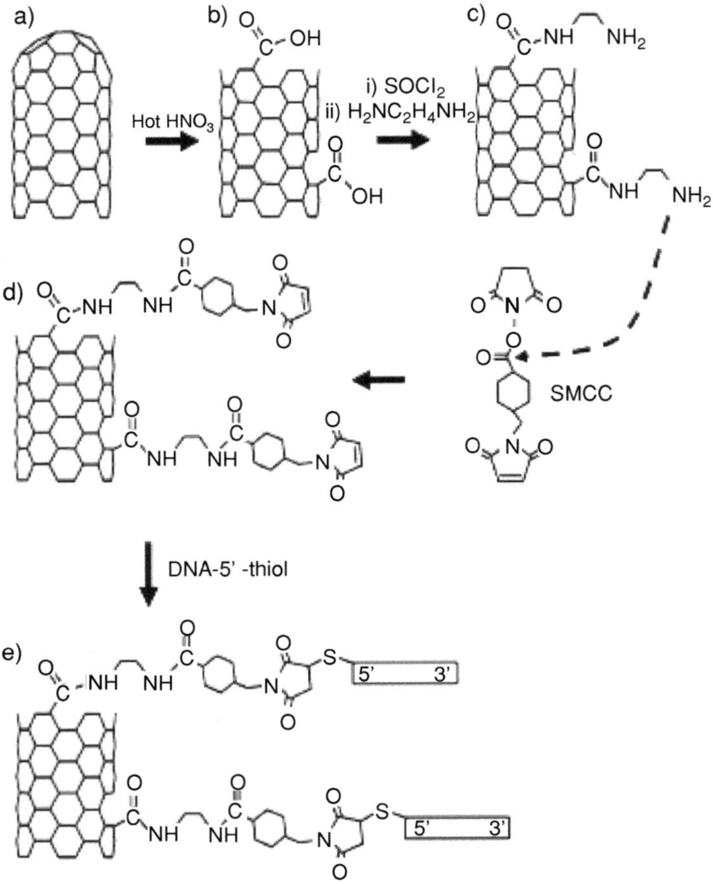

FIGURE 6.7. Overall scheme for the preparation of covalently linked DNA-nanotubes (from Reference 62).

bioactive peptides. It has been demonstrated that some of the peptide functionalized carbon nanotubes can enhance the immunization to virus-specific neutralizing antibody responses [69–73].

6.4. FABRICATION OF CARBON NANOTUBE ELECTRODES

6.4.1. Non-aligned Carbon Nanotube Electrodes

As electrode materials, carbon nanotubes possess many advantages, including their excellent electrical property, high chemical stability, strong mechanical strength, and large surface area. As a result, carbon nanotube based electrodes have recently been used in electrochemical- and bio-sensors [5, 21, 23]. In particular, Britto *et al.* [21] have used unmodified carbon nanotubes as electrodes to study the oxidation of dopamine and observed

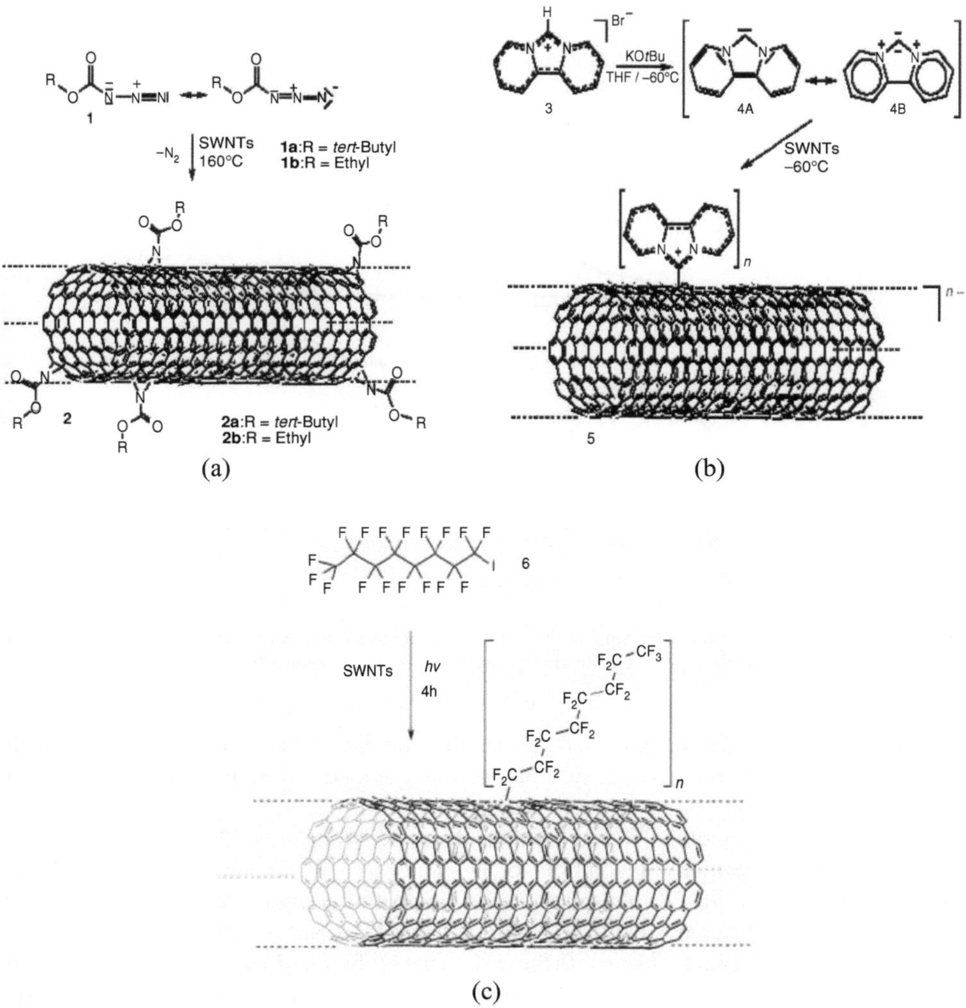

FIGURE 6.8. Schematic presentations of (a) nitrene cycloaddition; (b) nucleophilic carbine addition; and (c) radical addition (from Reference 66).

a high oxidation reversibility with the pristine carbon nanotube electrodes. Alternatively, Luo *et al.* [22] prepared a SWCNT-modified glassy carbon electrode by spin-casting a thin layer of the nitric acid oxidized SWCNTs (see Section 3.2.1) on a glassy carbon electrode. The glassy carbon electrode modified with the oxidized SWCNTs was found to show favorable electrocatalytic behavior toward the oxidation of biomolecules, such as dopamine (DA), epinephrine and ascorbic acid (AA), with a very stable electrochemical reactivity.

Furthermore, Davis *et al.* [17] reported a significantly enhanced faradic response from redox-active sulfonated anthraquinone using an SWCNT-modified glassy carbon

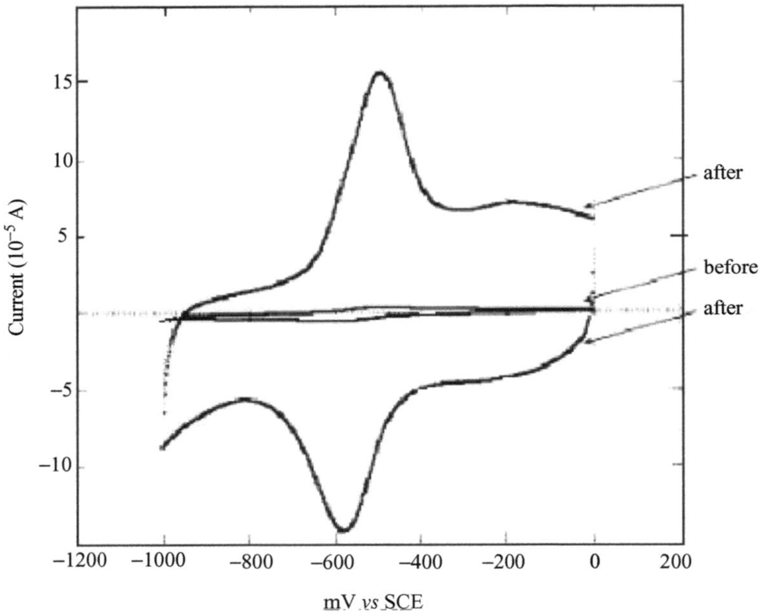

FIGURE 6.9. Comparative faradic responses (100 mVs^{-1}) of a glassy carbon macroelectode to 2 mM 1,5-AQDS before and after modification with the acid-oxidized SWCNTs (from Reference 17).

electrode. As can be seen in Figure 6.9, the nanotube-modified electrode clearly shows much stronger voltammetric responses to sulfonated anthraquinone with respect to a glassy carbon electrode.

Using a carbon nanotube-coated graphite electrode, Wang *et al.* [24] have also demonstrated a good voltammetric resolution for DA and AA with an anodic potential difference of 270 mV in a pH 5.0 phosphate buffer, indicating a high selectivity for DA over AA. The observed difference in electrocatalysis toward DA and AA was attributed to a 3D porous interfacial layer associated with the carbon nanotube-modified electrode. The carbon nanotube-modified electrodes were also found to exhibit a strong and stable electrocatalytic response to NADH with a low redox potential [74], and to be useful as electrochemical detectors on a capillary electrophoresis microchip [75] for in-situ determination of oxidizable amino acids in ion chromatography [76].

Along with the development of carbon nanotube electrodes through modification of conventional electrodes by carbon nanotubes with and without chemical modification, carbon nanotube composite electrodes have also been investigated for biosensing applications. For instance, Hrapovic *et al.* [77] reported that electrochemical sensors made from SWCNTs mixed with platinum nanoparticles of 2–3 nm in diameter displayed a remarkably improved sensitivity toward hydrogen peroxide. In this study, Nafion, a perfluorosulfonated polymer, was used to solubilize SWCNTs and interact with Pt nanoparticles to form a network that connected Pt nanoparticles to the nanotube electrode surface (Figure 6.10).

Using carbon nanotube and Teflon composite materials without metal nanoparticles, Wang *et al.* [78] developed a simple method for preparing effective carbon nanotube

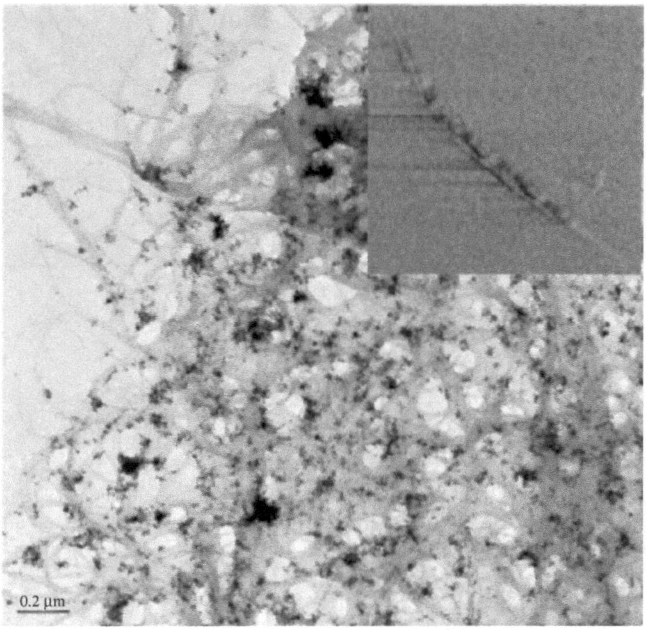

FIGURE 6.10. TEM micrograph of SWCNT in the presence of Pt nanoparticles. Inset: AFM tapping mode phase image (size, 1 µm × 1 µm; data scale, 20 nm) of one SWCNT in the presence of Pt nanoparticles (from Reference 77).

electrochemical biosensors. These authors first produced the carbon nanotube and Teflon composite material by hand-mixing desired amounts of carbon nanotubes with granular Teflon in the dry state. They then added a desired amount of the enzyme (GOx or ADH) and NAD^+ cofactor into the carbon nanotube/Teflon composite (30/70 wt %). Like its graphite-based counterpart, the bulk of the resulting carbon nanotube/Teflon electrode serves as a "reservoir" for the enzyme. Unlike its graphite-based counterpart, the carbon nanotube/Teflon composite electrode combines the major advantages of carbon nanotubes with those of bulk composite electrode; displaying a remarkably high electrocatalytic activity for hydrogen peroxide and NADH.

Nonfluorine-containing polymers have also been used for preparing composite electrodes with carbon nanotubes. In this regard, Ajayan *et al.* [79] prepared an enzyme-containing polymer and SWCNT composite biosensor from a solution of suspending SWCNT, α-chymotrypsin, and poly(methyl methacrylate) in toluene. While Gavalas *et al.* [80, 81] used carbon nanotube sol-gel composite as electrochemical sensing materials and enzyme-friendly platforms for the development of stable biosensors, Wohlstadter *et al.* [82] compounded carbon nanotubes with poly(ethylene vinylacetate), EVA, to produce a carbon nanotube and EVA composite sheets. These carbon nanotube composite electrodes showed the properties of nanoscopic materials with the advantages of macroscopic systems.

More interestingly, the SWCNT-modified electrode was found to possess capabilities to promote direct electron transfer with certain redox-active enzymes [83]. To demonstrate the direct electron transfer capability, Crooks *et al.* [84] designed a new microelectrode

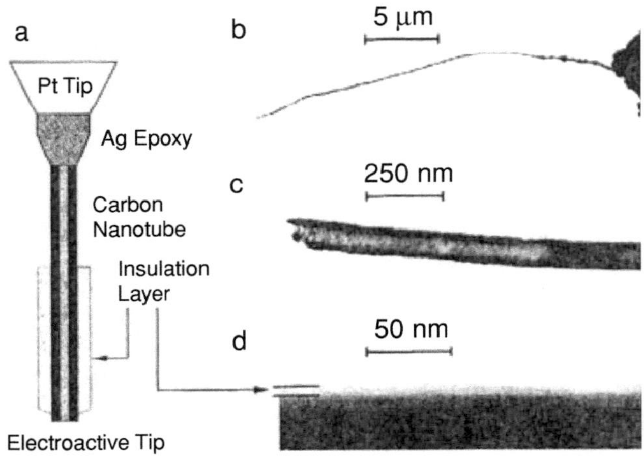

FIGURE 6.11. a) Schematic representation of a partially insulated carbon nanotube electrode. TEM images of mounted nanotubular electrodes showing b) a 30 μm long electrode; c) the tip of a ~100 nm diameter uninsulated nanoelectrode; d) a ~10 nm thick insulation layer of polyphenol on a ~220 nm diameter nanotube (from Reference 84).

consisting of a carbon nanotube attached to a sharpened Pt wire (Figure 6.11). While the high-resolution TEM image given in Figure 6.11c clearly shows an opened nanotube end, Figure 6.11d shows the sidewall of the nanotube was coated by a thin layer of electrically insulating polyphenol. The nanotube electrode thus prepared shows a sigmoidal voltammetric response to $Ru(NH_3)_6^{3+}$ solution, characteristic of direct electron transfer with a long cylindrical ultra-microelectrode.

6.4.2. Aligned Carbon Nanotube Electrodes

The use of ordered (aligned/patterned) carbon nanotubes as biosensing electrodes should provide additional advantages than the aforementioned electrodes based on the non-aligned, randomly entangled carbon nanotubes. Many research groups have used porous membranes (*e.g.* mesoporous silica, alumina nanoholes) as the template for preparing well-aligned carbon nanotubes with uniform diameters and lengths [25], as exemplified by Figure 6.12 [85].

Without using a template, Dai and co-workers [86, 87] have also prepared large-scale multiwalled carbon nanotubes aligned perpendicular to quartz substrates by pyrolysis of iron (II) phthalocyanine, FePc, under Ar/H_2 at 800 ± 1100 °C. As can be seen in Figure 6.13a, the constituent carbon nanotubes have a fairly uniform length and diameter. The same group has also developed microfabrication methods for patterning the aligned carbon nanotubes with a sub-micrometer resolution (Figure 6.13b) [25, 86, 88].

To construct aligned carbon nanotube electrodes, Dai *et al.* [86, 89] have developed several techniques for transferring the aligned carbon nanotube arrays, in either patterned or nonpatterned fashion, to various other substrates of particular interest (*e.g.* polymer films for organic optoelectronic devices or metal substrates for electrochemistry). Just like their

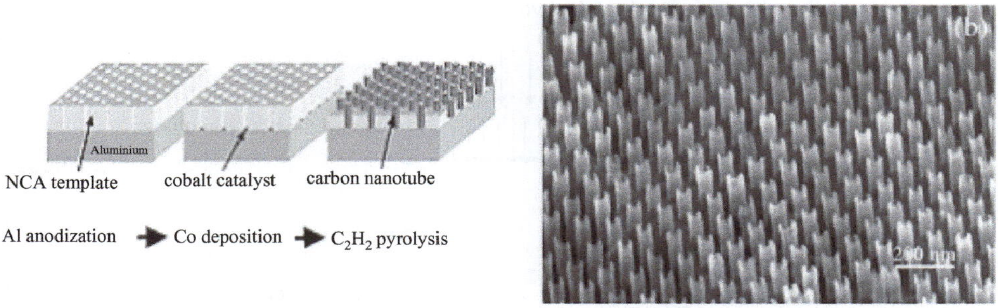

(a) (b)

FIGURE 6.12. (a) Schematic of fabrication process. (b) SEM image of the resulting hexagonally ordered array of carbon nanotubes (from Reference 85).

nonaligned counterpart, the use of aligned carbon nanotubes for biosensing will inevitably require modification of their surface characteristics to meet the specific requirements for this particular application (*e.g.* biocompatibility). Given that the alignment is an additional advantage for the use of carbon nanotubes in many devices, including biosensors, a particularly attractive option is the surface modification of carbon nanotubes whilst largely retaining their aligned structure.

In order to prevent the aligned nanotubes from collapsing during subsequent chemically modification, Meyyappan and co-workers [57, 90] have filled the gaps between the aligned nanotubes with a spin-on glass (SOG) and then modified the aligned nanotube tips through, for example, the acid oxidative reaction for further grafting appropriate biomolecules to the carboxylic acid groups on the acid-oxidized aligned carbon nanotubes for biosensing applications [91].

In addition to the activation of aligned carbon nanotubes by acid oxidation, Dai and co-workers [92] have developed a simple, but effective, approach for activating the aligned

(a) (b)

FIGURE 6.13. (a) A typical SEM image of the aligned carbon nanotube film prepared by pyrolysis of FePc. The misalignment seen for some of the nanotubes at the edge was caused by the peeling action used in the SEM sample preparation (from Reference 86). (b) Aligned carbon nanotube micropattern (from Reference 88).

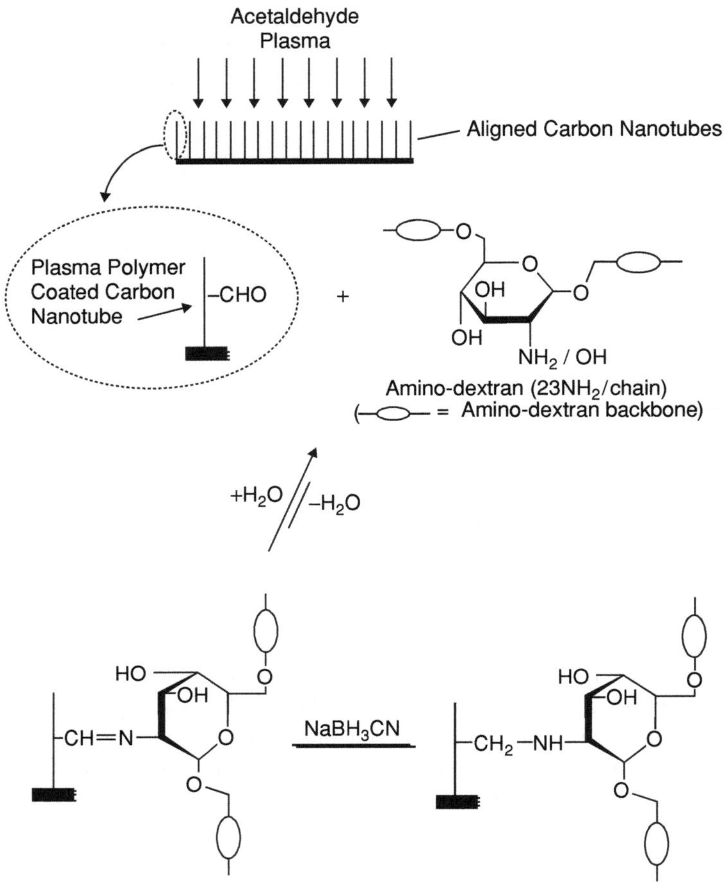

Scheme 1. Covalent immobilization of amino-dextran chains onto acetaldehyde-plasma activated carbon nanotubes. For reasons of clarity, only one of the many plasma-induced aldehyde surface groups is shown for an individual nanotube (from reference 92).

carbon nanotubes with appropriate plasma treatment, followed by further chemical modification under benign conditions through reactions characteristic of the plasma-generated functional groups. By so doing, these authors have successfully immobilized aminodextran chains onto acetaldehyde-plasma-treated aligned carbon nanotubes through the formation of an Schiff-base linkage, which was further stabilized by reduction with sodium cyanoborohydride (Scheme 1). The polysaccharide-grafted nanotubes are very hydrophilic; potentially useful for various biological-related applications (*e.g.* biosensing).

Subsequently, Dai *et al.* [93] have also demonstrated that H_2O-plasma etching can be used to open the aligned carbon nanotubes, and hence allow the chemical modification of the inner, outer, or both surfaces of the aligned nanotubes.

Apart from the plasma surface modification, Dai and co-workers [27, 89] have also modified aligned carbon nanotube electrodes by electrochemically depositing a thin layer

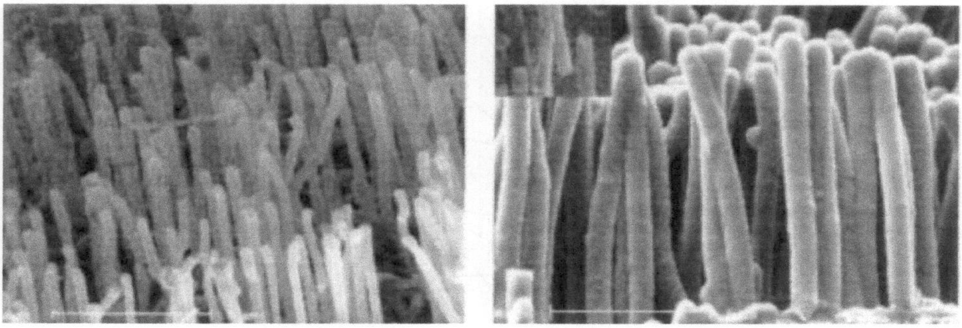

(a) (b)

FIGURE 6.14. SEM images of (a) pure aligned carbon nanotube array before PPy deposition and (b) aligned PPy-NT coaxial nanowires, inset shows clear image of single tube coated with PPy. Potentiostatic oxidation (1 V) of 0.10 M pyrrole was carried out in an electrolyte solution containing 0.1 M $NaClO_4$ in pH 7.45 buffer for 1 min (from Reference 27).

of appropriate conducting polymer (*e.g.* polyaniline, polypyrrole) onto individual nanotubes to form the *so-called* conducting polymer-carbon nanotube, CP-NT, aligned coaxial nanowires. The SEM image for the CP-NT coaxial nanowires given in Figure 6.14b shows the same features as the pristine aligned nanotube array of Figure 6.14a, but with a larger diameter due to the presence of the newly-electropolymerized polypyrrole coating in this particular case [27]. The uniform conducting polymer coating was also evidenced by TEM imaging [89].

6.5. CARBON NANOTUBE BIOSENSORS

6.5.1. Protein and Enzyme Biosensors

To further support the observation that carbon nanotubes can promote electron-transfer with certain proteins and enzymes [6, 8, 83]. Li and co-workers[6] have investigated the electrochemical behavior of cytochrome *c* on a SWCNT film electrode (Figure 6.15).

As can be seen in Figure 6.15, cytochrome *c* gave no obvious electrochemical response at the bare GC electrode (Figure 6.15a) whilst showed irreversible behavior with a difference between the anodic peak potential and the cathodic peak potential ($\Delta E = 265$ mV) at the unactivated SWCNT-modified electrode (Figure 6.15b). However, a pair of well-defined redox peaks with $\Delta E = 73.7$ mV was observed for cytochrome *c* at the same electrode after modification with an activated SWCNT film (Figure 6.15c). These results clearly indicate a significantly improved electron-transfer process for the carbon nanotube-modified electrode.

On the other hand, Wang *et al.* [94] have demonstrated that Nafion-coated carbon nanotube electrodes could enhance the redox activity of hydrogen peroxide so dramatically that a remarkable decrease in the over oxidation potential for hydrogen peroxide was observed. The accelerated electron transfer from hydrogen peroxide allowed for glucose measurements at very low potentials (−0.05 V), where interfering reactions of acetaminophen,

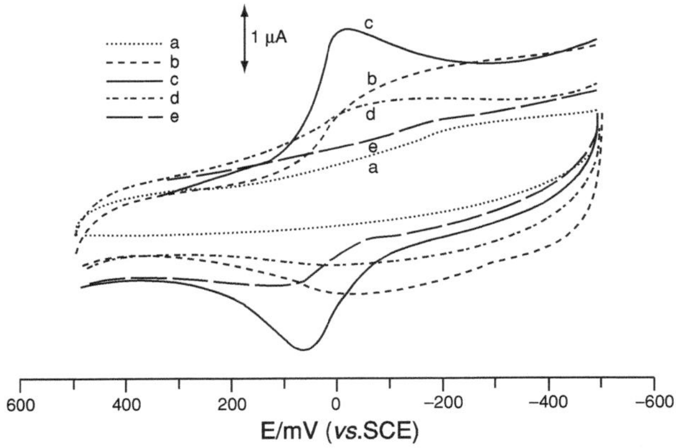

FIGURE 6.15. Cyclic voltammograms at a bare GC electrode (a), an unactivated SWCNT-modified GC electrode (b, e), and an activated SWCNT-modified GC electrode (c, d) in the absence of cytochrome c (d, e) and in the presence of 5.0×10^{-4} M cytochrome c (a–c) in 0.1M phosphate buffer solution (pH 6.24). The potential scan rate is $0.02V\ s^{-1}$. The activation was achieved by scanning the SWCNT-modified electrode in a fresh 0.1M phosphate buffer solution (pH 6.24) over +1.5 to −1.0 V at a scan rate of $1\ Vs^{-1}$ for 1.5 min (from Reference 6).

uric acid, and ascorbic acid were minimized. Similarly, Luong *et al.* [95] have used 3-aminopropyltriethoxysilane (APTES) as both a solubilizing agent for carbon nanotubes and an immobilization matrix for GOx to construct a mediatorless, highly efficient glucose sensor. In this study, a well-defined redox response from glucose was observed at −0.45 V (*vs.* Ag/AgCl) whereas no response was seen for three common interfering species, namely uric acid, ascorbic acid, and acetaminophen, at a concentration corresponding to the physiological level of 0.1 mM.

Unlike the above glucose biosensors based on nonaligned carbon nanotubes, Ren and co-workers [96–98] developed a glucose biosensor by covalently immobilizing glucose oxidase onto an aligned carbon nanotube nanoelectrode ensemble (CNT-NEE) through the amide formation between the GOx and carboxylic acid groups (Section 3.2.1) on the aligned carbon nanotube tips (Figure 6.16).

Due to the high surface packing density of the individual nanoelectrodes within the CNT-NEE assembly, the aligned carbon nanotube glucose biosensor thus prepared was found to be ultrasensitive with a high selectivity for electrochemical analysis of glucose over common interferents (*e.g.* acetaminophen, uric and ascorbic acids) (Figure 6.17).

Rusling and co-workers [99] have also covalently linked myoglobin and horseradish peroxidase to an aligned SWCNT electrode through the amide linkage. Quasi-reversible Fe^{3+}/Fe^{2+} voltammetry was observed for the iron heme enzymes myoglobin and horseradish peroxidase. The results demonstrated that the "trees" in the nanotube forest behaved electrically similar to a metal; conducting electrons from the external circuit to the redox sites of the enzymes.

Starting with nonaligned carbon nanotubes, Gooding *et al.* [18] have demonstrated that a self-assembly of the oxidation-shortened SWCNTs aligned normal to an electrode surface could act as aligned molecular wires to allow electrical communication between the underlying electrode and redox proteins covalently attached onto the top ends of the

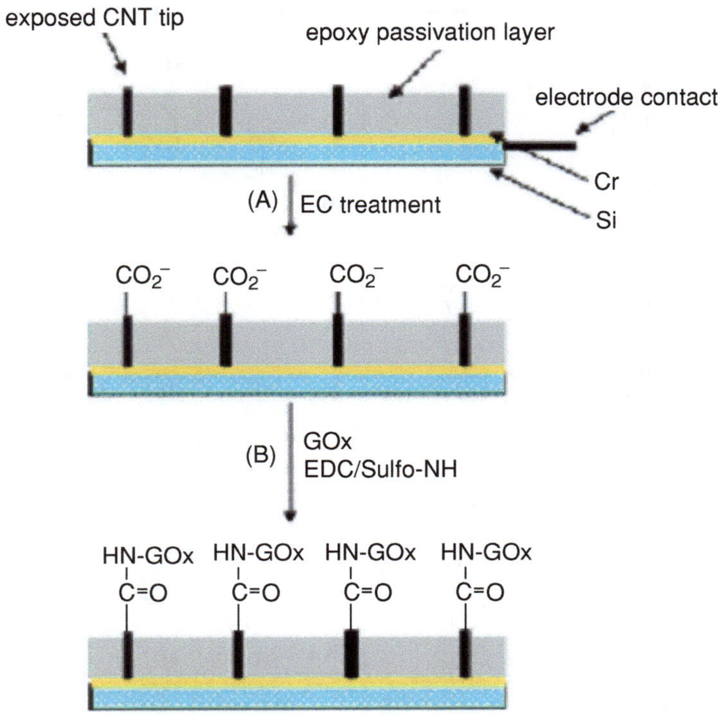

FIGURE 6.16. Fabrication of a glucose biosensor based on CNT nanoelectrode ensembles: (A) Electrochemical treatment of the CNT-NEE for functionalization (B) Coupling of the enzyme (GOx) to the functionalized CNT-NEE (from Reference 96).

SWCNTs. As schematically shown in Figure 6.18, a small redox protein, Microperoxidase MP-1, obtained by proteolytic digestion of horse heart cytochrome *c*, was attached to the free ends of the aligned nanotubes.

In this particular case, electron transfer could occur easily as the redox active center (*i.e.* an iron protoporphyrin IX) in MP-11 was not shielded by a polypeptide, and it was demonstrated that the rate of electron transfer was not affected by the tube length. The rate constant for electron transfer for the nanotube-modified electrodes was found to be similar to that for MP-11 attached directly to a cystamine-modified gold electrode, though the redox peak intensity is much higher for the former due to its large surface area (Figure 6.19).

Willner and co-workers[4] have also investigated the long-range electron transfer from redox enzymes chemically bound onto the aligned SWCNT structure.

Figure 6.20 shows the procedure used by these authors to assemble the aligned SWCNTs onto a gold electrode. As can be seen, the oxidized SWCNTs were first covalently coupled with 2-thioethanol/cystamine mixed monolayer (3:1 ratio) assembled on an Au electrode. The incorporation of 2-thioethanol in the mixed monolayer was anticipated to prevent nonspecific adsorption of the surfactant-protected SWCNTs onto the electrode surface. The amino derivative of flavin adenine dinucleotide (FAD) cofactor was then coupled to the carboxylic acid groups at the free top-ends of the aligned SWCNTs. Finally, glucose oxidase

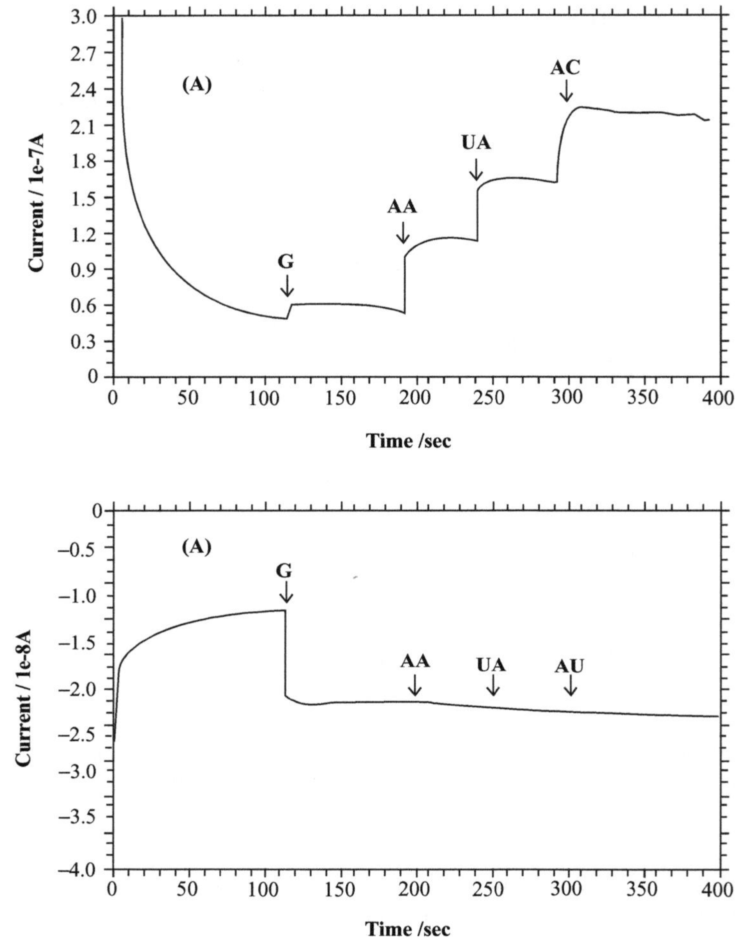

FIGURE 6.17. Amperometric responses of the CNT-NEE glucose biosensor to glucose (G), ascorbic acid (AA), uric acid (UA), and acetaminophen (AC) at potentials of +0.4 V (A) and −0.2 V (B) (from Reference 96).

(GOx) was reconstituted on the FAD units linked to the ends of the aligned SWCNTs for glucose sensing.

Figures 6.21A&B show that the bioelectrocatalytic oxidation of glucose on the Au electrode modified with GOx—attached carbon nanotubes (average nanotube length of 25 nm) occurred at $E > 0.18$ V, and the electrocatalytic anodic current became higher as the concentration of glucose increased. While Figure 6.21C reproduces the calibration curves corresponding to the anodic currents generated by the GOx-reconstituted SWCNT electrodes of different SWCNT lengths in the presence of variable glucose concentrations. Figure 6.21D shows a strong linear dependence between the turnover rate of electron transfer and L^{-1} (L is the nanotube length). The observed length-controlled electron transfer was further supported by examining the interfacial electron transfer to the FAD sites at the ends

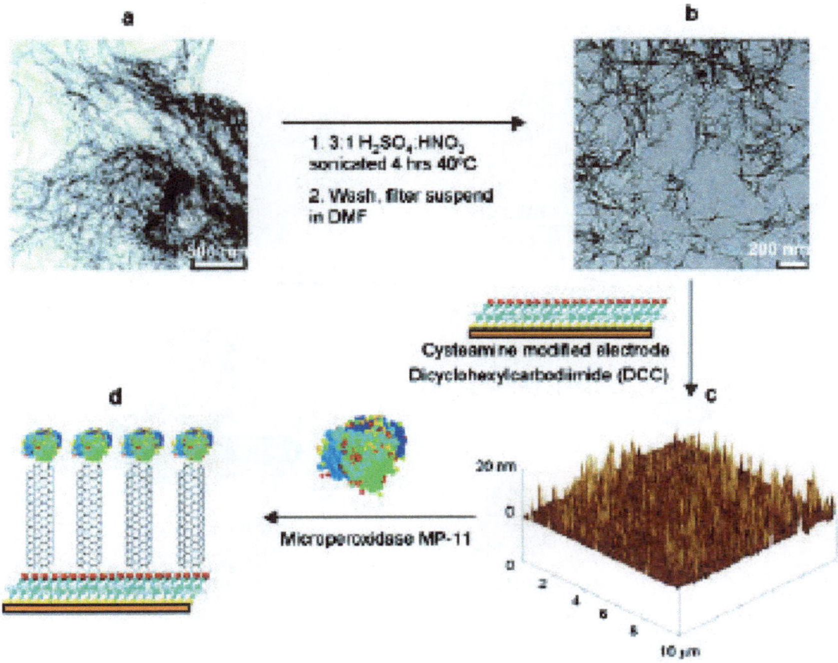

FIGURE 6.18. A schematic representation showing the steps involved in the fabrication of aligned SWCNT arrays for direct electron transfer with enzymes such as microperoxidase MP-11 (from Reference 18).

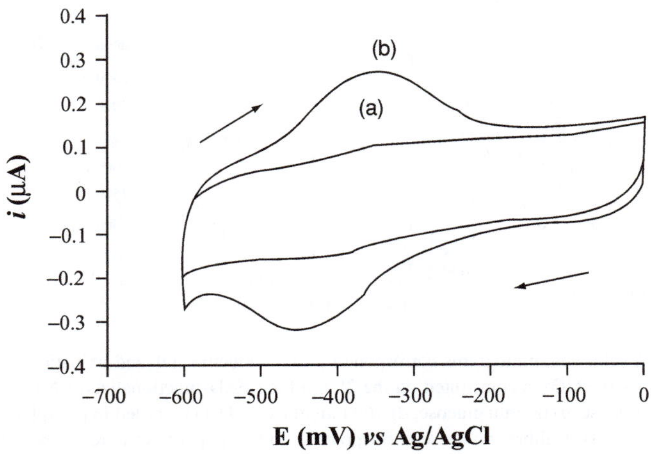

FIGURE 6.19. Cyclic voltammograms of (a) Au/Cysteamine after being immersed in DMF and MP-11 solution and (b) Au/cysteamine/SWCNTs/MP-11 in 0.05 M phosphate buffer solution pH 7.0 containing 0.05 M KCl under argon gas at a scan rate of 100 mV s^{-1} versus Ag/AgCl (from Reference 18).

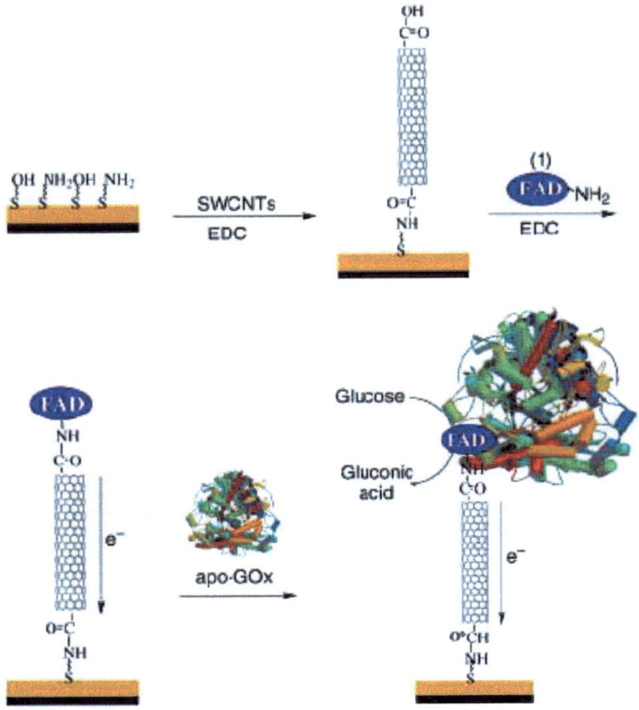

FIGURE 6.20. Assembly of the SWCNT electrode with chemically bound glucose oxidase (from Reference 4).

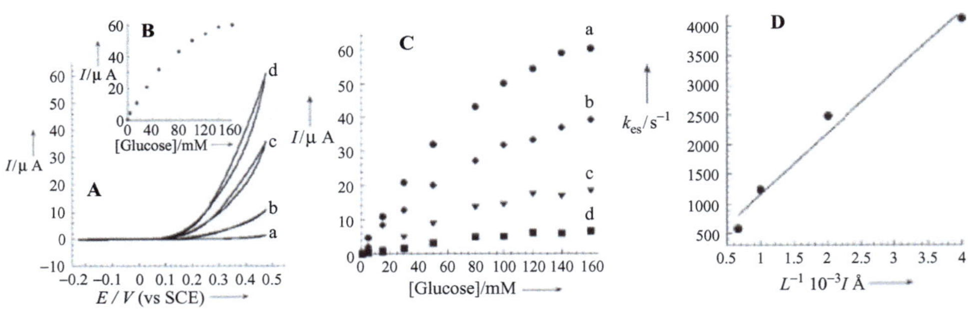

FIGURE 6.21. A) Cyclic voltammograms corresponding to the electrocatalyzed oxidation of different concentrations of glucose by the GOx reconstituted on the 25 nm long FAD-functionalized CNTs assembly: a) 0 mm glucose. b) 20 mm glucose, c) 60 mm glucose, d) 160 mm glucose. Data recorded in phosphate buffer, 0.1 m, pH 7.4, scan rate 5 mVs^{-1}. B) Calibration curve corresponding to the amperometric responses of the reconstituted GOx/CNTs (25 nm) electrode (at E = 0.45 V) in the presence of different concentrations of glucose. C) Calibration curves corresponding to the amperometric responses (at E = 0.45 V) of reconstituted GOx–CNTs electrodes in the presence of variable concentrations of glucose and different CNT lengths as electrical connector units: a) about 25 nm SWCNTs. b) about 50 nm SWCNTs. c) about 100 nm SWCNTs. d) about 150 nm SWCNTs. D) Dependence of the electron-transfer turnover rate between the GOx redox center and the electrode on the lengths of the SWCNTs comprising the enzyme electrodes (from Reference 4).

of the aligned SWCNTs, though the mechanism for the tube length-dependent electron transfer has not been fully understood. Given that no tube length dependence was observed in Gooding's study [18] (*vide supra*), the length-dependent charge transfer observed in Figure 6.21 is most likely resulted from defect sites introduced along the nanotube wall during the oxidative shortening process. However, the detailed mechanism of the tube-length-dependent charge transfer deserves further investigation.

To avoid the tedious processes for preparing aligned SWCNTs, Dai *et al.* [27, 100] have used the FePc-generated aligned multiwalled carbon nanotubes (Section 4.2) as a novel electrode platform for biosensing. In a typical experiment, these authors coated the aligned MWCNTs with conducting polymers by electrochemical depositing a thin layer of polypyrrole or polyaniline onto individual carbon nanotubes along their tubular length (Figure 6.14) in the presence of glucose oxidase. It was found that the detection of H_2O_2 at these aligned MWCNT electrodes coated with GOx-containing conducting polymers can also be achieved at low anodic potentials, leading to glucose sensors with a very high sensitivity and selectivity [101, 102].

6.5.2. DNA Sensors

A major feature of the Watson-Crick model of DNA is that it provides a vision of how a base sequence of one strand of the double helix can precisely determine the base sequence of the partner strand for passing the genetic information in all living species. The principle learned from this breakthrough has now been applied to the development of biosensors for DNA analysis and diagnosis through the very specific DNA pairing interaction [103, 104]. Owing to their high sensitivity, facial data processing, great simplicity, and good compatibility with electronic detection technologies, DNA electrochemical biosensors are of great interest. The unique electronic properties and large surface area have made carbon nanotubes ideal electrodes for constructing advanced biosensors. This, coupled with various chemical reactions reported for the attachment of DNA chains onto carbon nanotubes (Sections 3.2.1 and 3.2.2) has facilitated significantly the development of carbon nanotube DNA sensors.

Using a glass carbon electrode (GCE) modified with MWCNTs, Fang *et al.* [105] observed an enhanced sensitivity for electrochemical DNA biosensor based on carbon nanotubes. Figure 6.22 shows schematically the steps for constructing the nanotube-DNA biosensor. To start with, carboxylic acid functionalized multiwalled carbon nanotubes (COOH–MWCNTs) were dropped on a GCE electrode, single-strand DNA oligonucleotides (ss-DNAs) were then covalently bonded onto the COOH–MWCNTs via the amide formation. The hybridization reaction on the electrode was monitored by differential pulse voltammetry (DPV) analyses using an electroactive daunomycin intercalator as the indicator.

Compared with conventional DNA sensors with oligonucleotides being directly incorporated onto carbon electrodes, the carbon nanotube based DNA sensor showed a dramatically increased DNA attachment density and complementary DNA detection sensitivity due to the large surface area and good charge-transport properties characteristic of carbon nanotubes. These promising results have further prompted these authors to develop novel biosensors for direct electrochemical detection of DNA hybridization by AC impedance measurements [106, 107]. They electrodeposited polypyrrole and oligonucleotide probe onto a MWCNT-coated GCE electrode. The large surface area of carbon nanotubes allowed a large volume of polypyrrole/oligonucleotide to be deposited onto the electrode whilst the

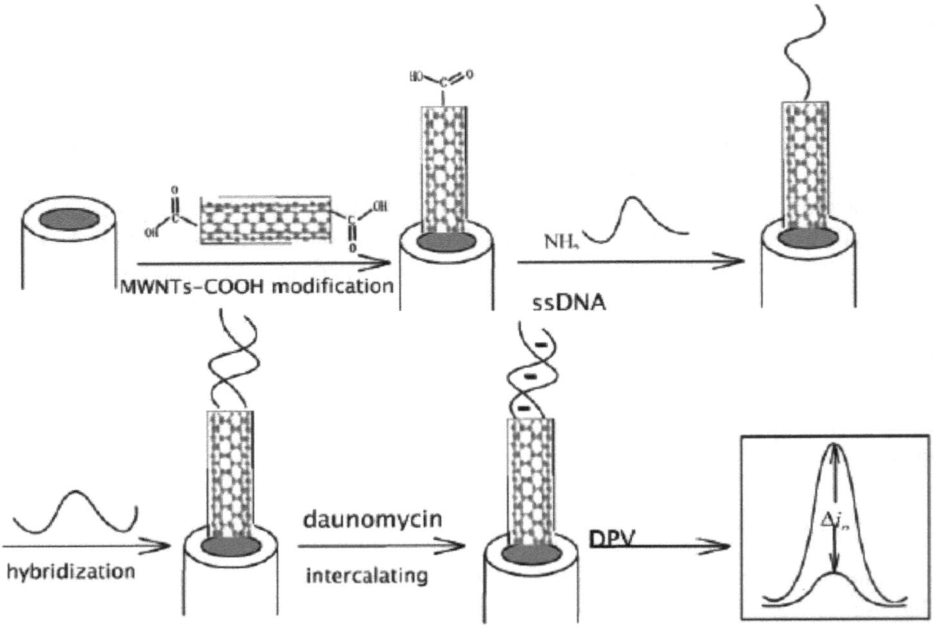

FIGURE 6.22. Schematic representation of the enhanced electrochemical detection of DNA hybridization by DNA biosensor based on the COOH-MWCNTs (from Reference 105).

polypyrrole coating on individual carbon nanotubes remained sufficiently thin. Owing to their high electrical conductivity, carbon nanotubes can clearly reflect any resistance change in the polypyrrole/oligonucleotide thin film caused by hybridization reaction (Figure 6.23).

In a somewhat related but separate study, Guo *et al.* [108] used electrochemical impedance spectroscopy to investigate the process of the electrostatic assembly of calf thymus DNA on MWCNTs-modified gold electrodes in the presence of a cationic polyelectrolyte PDDA. The gold electrode modified with carboxylic acid functionalized MWCNTs was dipped alternately in aqueous solutions of PDDA and calf thymus DNA to form a multilayer film. Electrochemical impedance spectroscopy was then used to characterize the interfacial properties for the modified electrode.

A typical impedance spectrum presented in the form of the Nyquist plot includes a semicircle portion at higher frequencies corresponding to the electron-transfer-limited process and a linear part at a lower frequency range associated with the diffusion-limited process. The semicircle diameter in the impedance spectrum equals the electron-transfer resistance, *R*et, which is related to the electron-transfer kinetics of the redox probe at the electrode surface. As can be seen in Figure 6.24A, *R*et decreased from 8.01 to 6.75 kΩ upon surface modification of the electrode with MWCNTs, attributable to the good electrical conductivity and large surface area of MWCNTs. However, the semicircle diameter increased significantly to 26.72 kΩ after the adsorption of PDDA/DNA layer due to its non-conducting nature.

The impedance spectra for the MWCNT-modified microelectrode after adsorption of different layer numbers of PDDA/DNA are given in Figure 6.24B, which shows an increase in the semicircle diameter with increasing layer number for the adsorbed multiplayer film.

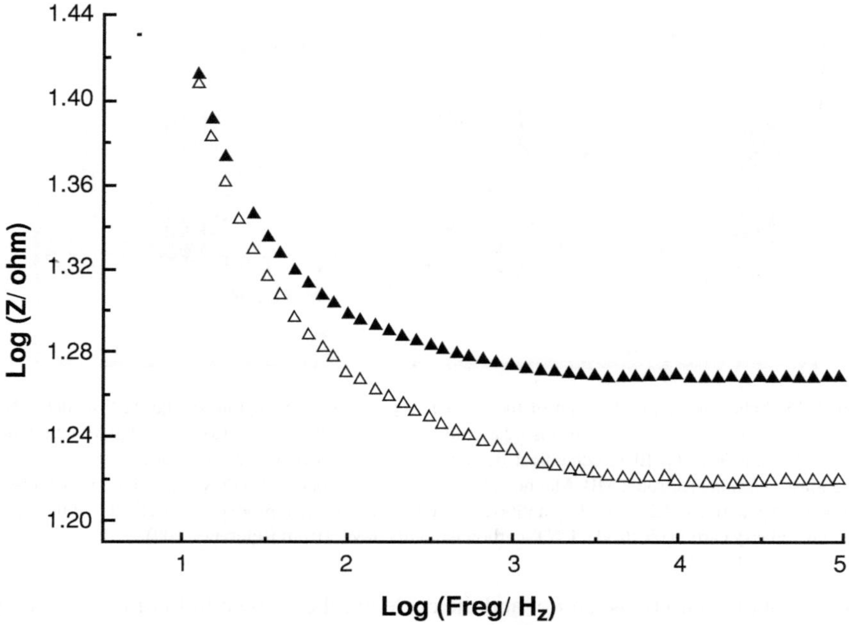

FIGURE 6.23. Impedance curves before (solid triangle) and after (hollow triangle) hybridization in 0.3 M PBS solution on an open circuit voltage. Frequency range: 10–10^5 Hz; amplititude: 5 mV vs. SCE (from Reference 106).

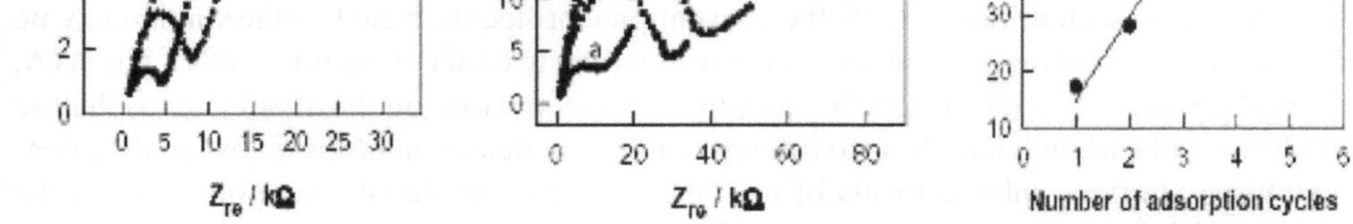

FIGURE 6.24. (A) Nyquist plots (Zim vs. Zre) of: (a) bare gold electrode, (b) MWCNTs-modified gold electrode and (c) MWCNTs-modified gold electrode after the assembly of PDDA and DNA in 10mM $K_4Fe(CN)_6$ + 10mM $K_3Fe(CN)_6$ + 0.1M KCl solutions. (B) Nyquist plots (Zim vs. Zre) in 10mM $K_4Fe(CN)_6$ + 10mM $K_3Fe(CN)_6$ + 0.1M KCl solutions at MWCNT-modified gold electrode after different numbers of PDDA/DNA adsorption cycles: (a) 1; (b) 2; (c) 3; (d) 4; (e) 5. (C) Relationship between the electron-transfer resistance (*R*et) and the number of PDDA/DNA adsorption cycles (from Reference 108).

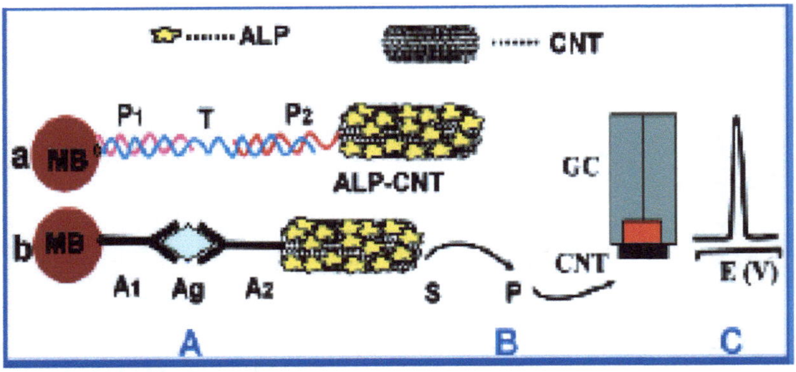

FIGURE 6.25. Schematic representation of the analytical protocol: (A) Capture of the ALP-loaded CNT tags to the streptavidin-modified magnetic beads by a sandwich DNA hybridization (a) or A1-Ag-A2 interaction (b). (B) Enzymatic reaction. (C) Electrochemical detection of the product of the enzymatic reaction at the CNT-modified glassy carbon electrode; MB, Magnetic beads; P1, DNA probe 1; T, DNA target; P2, DNA probe 2; A1, first antibody; Ag, antigen; A2, secondary antibody; S and P, substrate and product, respectively, of the enzymatic reaction; GC, glassy carbon electrode; CNT, carbon nanotube layer (from Reference 109).

The linear relationship between *R*et and the cycle number shown in Figure 6.24C suggests the formation of a uniform PDDA/DNA multiplayer film on the MWCNT-modified gold electrode.

In order to amplify electrical detection of DNA hybridization, Wang and co-workers [109–111] developed a carbon nanotube-based dual amplification route, in which carbon nanotubes play a dual amplification role (namely carrying numerous enzyme tags and accumulating the product of the enzymatic reaction), for dramatically amplifying enzyme tag numbers in sensing of DNAs and proteins (Figure 6.25). The process shown in Figure 6.25 involved the sandwich DNA hybridization (a) or antigen-antibody (b) binding along with magnetic separation of the analyte-linked magnetic-bead/CNT assembly (A), followed by enzymatic amplification (B), and chronopotentiometric stripping detection of the product at the CNT-modified electrode (C).

Figure 6.26 shows the dramatic signal enhancement associated with the carbon nanotube-based dual amplification route for DNA hybridization (A) and Ag-Ab (B) bioassays. As can be seen in Figure 6.26, the conventional protocols, based on the single-enzyme tag and a glassy-carbon electrode, were not responding to either 10 pg mL^{-1} DNA target (A, a) or 80 pg mL^{-1} IgG (B,a). The first amplification step based on the alkaline phosphatase (ALP) loaded carbon nanotubes (b) offered convenient detection of these low analyte concentrations. Further enhancements of the DNA and protein signals were observed in the second amplification path, employing the carbon nanotube-modified electrode (c). The latter reflects the strong adsorptive accumulation of the liberated α-naphthol on the carbon nanotube layer. Two series of six repetitive measurements of 1 pg mL^{-1} DNA target or 0.8 ng mL^{-1} IgG yielded reproducible signals with relative standard deviations of 5.6 and 8.9%, respectively.

Using aligned carbon nanotube electrodes, He and Dai [20] have recently developed novel nanotube–DNA sensors of a high sensitivity and selectivity by grafting single-strand DNA chains onto aligned carbon nanotubes generated from FePc [27]. In this study, aligned

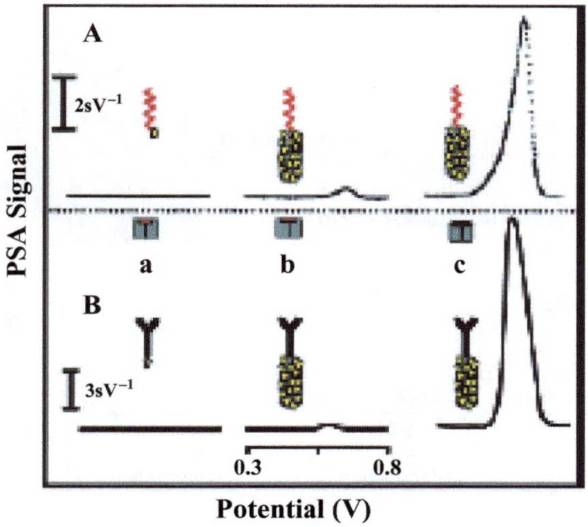

FIGURE 6.26. Chronopotentiometric signals for 10 pg mL^{-1} target oligonucleotide (A) and 80 pg mL^{-1} IgG (B) using the glassy carbon (GC) transducer and (a) a single ALP tag and (b) CNT-loaded with multiple ALP tags; (c) same as (b) but using the CNT-modified GC electrode. Amount of magnetic beads, 50 μg; sandwich assay with 20 and 30 min for each hybridization event and Ag/Ab association, respectively; sample volume, 50 μL. Detection, addition of 50 μL R-naphthyl phosphate (50 mM) solution with a 20-min enzymatic reaction. Measurements of the R-naphthol product were performed at the bare or modified GC electrodes, using a 2-min accumulation at +0.2 V in a stirred phosphate buffer solution (0.05 M, pH 7.4; 1 mL), followed by a 10-s rest period (without stirring) and application of an anodic current of +5.0 μA. See Supporting Information for the concentrations of the oligonucleotide probes and antibody, and sequence of oligonucleotide probes, levels and preparation of the ALP-DNA-CNT and ALP streptavidin-CNT conjugates (from Reference 109).

carbon nanotubes supported by gold substrate were first treated with acetic acid-plasma to introduce the surface carboxylic acid groups for grafting single-strand DNA (ssDNA) chains with an amino group at the 5′-phosphate end (*i.e.* [AmC6]TTGACACCAGACCAACTGGT-3′, **I**). Complementary DNA (cDNA) chains labeled with ferrocenecarboxaldehyde, FCA, (*i.e* [FCA-C6]ACCAGTTGGTCTGGTGTCAA-3′, **II**) were then used for hybridizing with the surface-immobilized oligonucleotides to form the double-strand DNA (dsDNA) helices on the aligned carbon nanotube electrodes.

The performance of the aligned carbon nanotube–DNA sensors for sequence-specific DNA diagnoses was demonstrated in Figure 6.28. The strong oxidation peak seen at 0.29 V in curve a of Figure 6.28 is attributable to ferrocene and indicates the occurrence of hybridization of FCA-labeled cDNA (**II**) chains with the nanotube-supported ssDNA (**I**) chains, leading to a long-range electron transfer from the FCA probe to the nanotube electrode through the DNA duplex. In contrast, the addition of FCA labeled non-complementary DNA chains (*i.e.* [FCA-C6]CTCCAGGAGTCGTCGCCACC- 3′, **III**) under the same conditions did not show any redox response of FCA (curve b of Figure 6.27). Subsequent addition of target DNA chains (*i.e.* 5′-GAGGTCCTCAGCAGCGGTGGACCAGTTGGTCTGGTGTCAA-3′, **IV**) into the above solution, however, led to a strong redox response from the FCA-labeled DNA (**III**) chains (curve c of Figure 6.28) because the target DNA (**IV**) contains complementary sequences for both DNA (**I**) and DNA (**III**) chains.

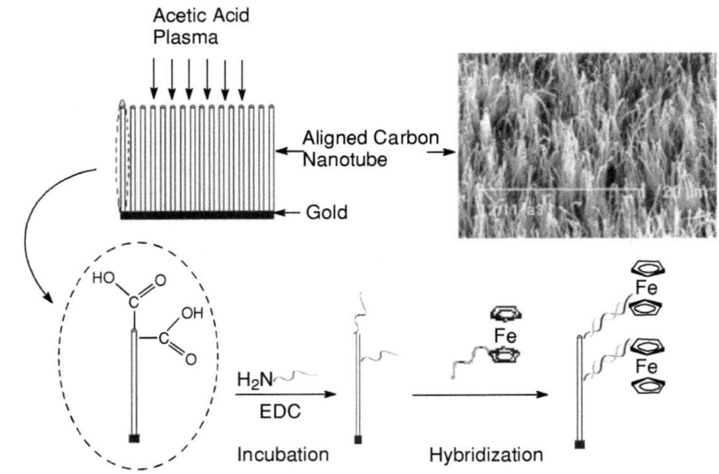

FIGURE 6.27. A schematic illustration of the aligned nanotube–DNA electrochemical sensor. The upper right SEM image shows the aligned carbon nanotubes after having been transferred onto a gold foil. For reasons of clarity, only one of the many carboxyl groups is shown at the nanotube tip and wall, respectively (from Reference 20).

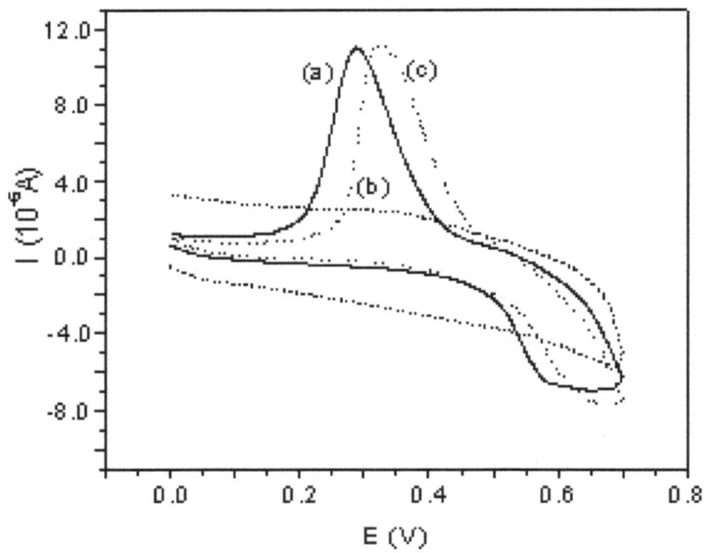

FIGURE 6.28. Cyclic voltammograms of the ssDNA (**I**)-immobilized aligned carbon nanotube electrode after hybridization with FCA-labeled complementary DNA (**II**) chains (a), in the presence of FCA-labeled noncomplementary DNA (**III**) chains (b), and after hybridization with target DNA (**IV**) chains in the presence of the FCA-labeled noncomplementary DNA (**III**) chains (c). All the cyclic voltammograms were recorded in 0.1 M H_2SO_4 solution with a scan rate of 0.1 V s^{-1}. The concentration of the FCA-labeled DNA probes is 0.05 mg ml^{-1} (from Reference 20).

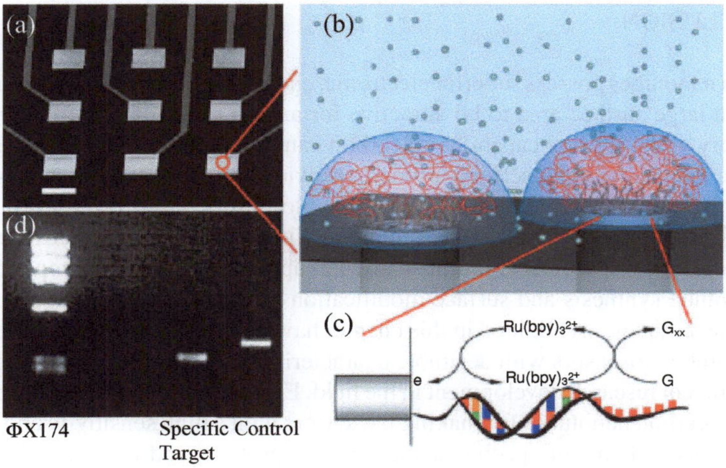

FIGURE 6.29. (a) An SEM image of an individually addressable 3 × 3 microcontact array with a MWCNT nanoelectrode array on each site. The scale bar is 200 μm. (b) Schematic of the mechanism to detect DNA hybridization using a MWCNT nanoelectrode array. The long single-stranded DNA PCR amplicons are hybridized to the short oligonucleotide probes which are functionalized at the very end of the MWCNTs. Ru(bpy)$_3^{2+}$ mediators are used to transfer electrons from the guanine groups to the MWCNT nanoelectrode for all target molecules within the hemispherical diffusion layer of the nanoelectrodes. (c) The schematic mechanism for the guanine oxidation amplified with Ru(bpy)$_3^{2+}$ mediators. (d) The gel electrophoresis. The lanes from left to right are DNA molecular weight standard (*Φ*X174RFDNA-HaeIII digest), a specific PCR amplicon target with ~300 bases, and a control sample with an unrelated PCR amplicon with ~400 bases, respectively (from Reference 28).

Meyyappan and co-workers [28, 57, 91] have developed a more advanced micropatterned ultrasensitive DNA biosensor based on aligned carbon nanotubes. As shown in Figure 6.29a, aligned MWCNTs were directly grown on individual metal microcontacts, followed by encapsulating the MWCNT arrays and the substrate surface with a spin-on glass (SOG) layer to only expose the nanotube ends at the surface. Each of the SOG-encapsulated individual MWCNTs is alike a nanoelectrode, and there are about 100 MWCNT nanoelectrodes or more on each microcontact. Abundant carboxylic acid groups were produced at the end of MWCNTs by electrochemical etching, and then functionalized with a specific oligonucleotide probe through the amide formation. Figures 6.29b&c show schematically the mechanism of the MWCNT nanoelectrode array for the DNA detection. A specific probe [Cy3]5′-CTIIATTTCICAIITCCT-3′-[AmC7-Q] was used in that study, which contains the sequence of the normal allele of the BRCA1 gene associated with the occurrence of several cancers. Electroactive guanine groups in the probe are replaced with nonelectroactive inosines. Genomic DNA from a healthy donor was used in PCR to amplify.

Figure 6.29d shows the electrophoresis results of a DNA molecular weight standard (*Φ*X174RFDNA-HaeIII digest) and the two PCR amplicons, respectively. The nonspecific binding is removed through stringent washing in three steps using 2 × SSC/0.1%SDS, 1 × SSC, and 0.1% SSC respectively, by shaking the sample in each of the solutions at 40°C for 15 min. Ru(bpy)$_3^{2+}$ mediators were used to efficiently transport electrons from the guanine bases to the MWCNT nanoelectrode and to provide an amplified guanine oxidation signal as long as target DNA molecules are within the three-dimensional diffusion layer.

6.6. CONCLUSION

Carbon nanotubes possess superior electronic, chemical, thermal, and mechanical properties with a large surface area to be attractive for a wide range of potential applications. The discovery of the sensing capability of carbon nanotubes is significantly intriguing. The high surface area and good electronic property provided by carbon nanotubes is an attractive feature in the advancement of chemical and biosensors. However, the use of carbon nanotubes for biosensing will inevitably require modification of their surface characteristics to meet the specific requirements for this particular application (*e.g.* biocompatibility). The carbon nanotube synthesis and surface modification methods, together with the electrode fabrication techniques, highlighted in this chapter have allowed the development of various carbon nanotube biosensors with desirable characteristics. We have given a brief summary of the most recent research development in the field. Even this brief account has revealed the versatility of carbon nanotubes for making biosensors with a high sensitivity and selectivity for probing DNA chains of specific sequences and proteins and enzymes with different redox properties. With continued research and development in this field, the possibility of producing carbon nanotube biosensors for practical applications will soon be in sight.

ACKNOWLEDGEMENTS

We thank our colleagues whose work is reviewed in this article. We are also grateful for financial support from the American Chemical Society (ACS-PRF 39060-AC5M), National Science Foundation (NSF-CCF-0403130), and the Materials and Manufacturing Directorate of the Air Force Research Laboratory, The Dayton Development Coaltition, and Wright Brothers Institute for supporting the WBI Endowed Chair Professorship in Nanomaterials to LD.

REFERENCES

[1] P.J.F. Harris. *Carbon Nanotubes and Related Structures—New Materials for the Twenty-First Century*, Cambridge University Press, Cambridge, 2001.
[2] L. Dai. *Intelligent Macromolecules for Smart Devices: From Materials Synthesis to Device Applications*, Springer-Verlag, London, 2004.
[3] K. Yamamoto, G. Shi, T.S. Zhou, F. Xu, J.M. Xu, T. Kato, J.Y. Jin, and L. Jin. *Analyst*, 128:249, 2003.
[4] F. Patolsky, Y. Weizmann, and I. Willner. *Angew. Chem. Int. Ed.*, 43:2113, 2004.
[5] J.J. Davis, R.J. Coles, and H.A.O. Hill. *J. Electroanal. Chem.*, 440:279, 1997.
[6] J.X. Wang, M.X. Li, Z.J. Shi, N.Q. Li, and Z.N. Gu. *Anal. Chem.*, 74:1993, 2002.
[7] G. Wang, J.J. Xu, and H.Y. Chen. *Electrochem. Commun.*, 4:506, 2002.
[8] Y.D. Zhao, W.D. Zhang, H. Chen, Q.M. Luo, and S.F.Y. Li. *Sens. Actuators, B* 87:168, 2002.
[9] A. Hirsch. *Angew. Chem., Int. Ed.*, 41:1853, 2002.
[10] J.L. Bahr and J.M.Tour. *J. Mater. Chem.*, 12:1952, 2002.
[11] L. Dai, T. Lin. T. Ji, and V. Bajpai. *Aust. J. Chem.*, 56:635, 2003.
[12] V.N. Khabashesku, W.E. Billups, and J.L. Margrave. *Acc. Chem. Res.*, 35:1087, 2002.
[13] Y.-P. Sun, K. Fu, Y. Lin, and W. Huang. *Acc. Chem. Res.*, 35:1096, 2002.
[14] S. Niyogi, M.A. Hamon, H. Hu, B. Zhao, P. Bhowmik, R. Sen, M. Itkis, and R.C. Haddon. *Acc. Chem. Res.*, 35:1105, 2002.
[15] D. Tasis, N. Tagmatarchis, V. Georgakilas, and M. Prato. *Chem. Eur. J.*, 9:4000, 2003.

[16] S.C. Tsang, J.J. Davis, M.L.H. Green, H.A.O. Hill, Y.C. Leung, and P.J. Sadler. *Chem. Commun.*, 1803, 1995.
[17] J.J. Davis, K.S. Coleman, B.R. Azamian, C.B. Bagshaw, and M.L.H. Green. *Chem. Eur. J.*, 9:3732, 2003.
[18] J.J. Gooding, R. Wibowo, J. Liu, W. Yang, D. Losic, S. Orbons, F.J. Mearns, J.G. Shapter, and D.B. Hibbert. *J. Am. Chem. Soc.*, 125:9006, 2003.
[19] A. Callegari, S. Cosnier, M. Marcaccio, D. Paolucci, F. Paolucci, V. Georgakilas, N. Tagmatarchis, E. Vázquez, and M. Prato, *J. Mater. Chem.*, 14:807, 2004.
[20] P. He and L. Dai. *Chem. Commun.*, 348, 2004.
[21] P.J. Britto, K.S.V. Santhanam, and P.M. Ajayan. *Bioelectrochem. Bioenerg.*, 41:121, 1996.
[22] H. Luo, Z. Shi, N. Li, Z. Gu, and Q. Zhuang. *Anal. Chem.*, 73:915, 2001.
[23] Q. Zhao, Z. Gan, and Q. Zhuang. *Electroanalysis*, 14:1609, 2002.
[24] Z. Wang, J. Liu, Q. Liang, Y. Wang, and G. Luo. *Analyst*, 127:653, 2002.
[25] L. Dai, A. Patil, X. Gong, Z. Guo, L. Liu, Y. Liu, D. Zhu, *Chem. Phys. Chem.*, 4:1150, 2003. and references cited therein.
[26] P. Diao, Z. Liu, B. Wu, X. Nan, J. Zhang, and Z. Wei. *Chem. Phys. Chem.*, 10:898, 2002.
[27] M. Gao, L. Dai, and G.G. Wallace. *Electroanalysis*, 15:1089, 2003.
[28] J. Koehne, H. Chen, J. Li, A. MCassell, Q. Ye, H.T. Ng, J. Han, and M. Meyyappan. *Nanotechnology*, 14:1239, 2003.
[29] P.M. Ajayan. *Chem. Rev.*, 99:1787, 1999.
[30] R. Saito, G. Deesselhaus, and M.S. Dresselhaus. *Physical Properties of Carbon Nanotubes*, Imperial College Press, London, 1998.
[31] A. Star, D.W. Steuerman, J.R. Heath, and J.F. Stoddart. *Angew. Chem., Int. Ed.*, 41:2508, 2002.
[32] O.-K. Kim, J. Je, J.W. Baldwin, S. Kooi, P.E. Pehrsson, and L.J. Buckley. *J. Am. Chem. Soc.*, 125:4426, 2003.
[33] J. Chen, M.J. Dyer, and M.-F. Yu. *J. Am. Chem. Soc.*, 123:6201, 2001.
[34] H. Dodziuk, A. Ejchart, W. Anczewski, H. Ueda, E. Krinichnaya, G. Dolgonos, and W. Kutner. *Chem. Commun.*, 986, 2003.
[35] M. Zheng, A. Jagota, E.D. Semke, B.A. Diner, R.S. McLean, S.R. Lustig, R.E. Richardson, and N.G. Tassi. *Nat. Mater.*, 2:338, 2003.
[36] N. Nakashima, S. Okuzono, H. Murakami, T. Nakai, and K. Yoshikawa. *Chem. Lett.*, 32:456, 2003.
[37] G.R. Dieckmann, A.B. Dalton, P.A. Johnson, J. Razal, J. Chen, G.M. Giordano, E. Munoz, I.H. Musselman, R.H. Baughman, and R.K. Draper. *J. Am. Chem. Soc.*, 125:1770, 2003.
[38] S. Wang, E.S. Humphreys, S.-Y. Chung, D.F. Delduco, S.R. Lustig, H. Wang, K.N. Parker, N. W.Rizzo, S. Subramoney, Y.-M. Chiang, and A. Jagota. *Nat. Mater.*, 2:196, 2003.
[39] B.F. Erlanger, B. Chen, M. Zhu, and L.E. Brus. *Nano Lett.*, 1:465, 2001.
[40] R.J. Chen, Y. Zhang, D. Wang, and H. Dai. *J. Am. Chem. Soc.*, 123:3838, 2001.
[41] K. Besteman, J.-O. Lee, F.G.M. Wiertz, H.A. Heering, and C. Dekker. *Nano Lett.*, 3:727, 2003.
[42] F. Balvlavoine, P. Schultz, C. Richard, V. Mallouh, T.W. Ebbeson, and C. Mioskowski. *Angew. Chem., Int. Ed. Eng.*, 38:1912, 1999.
[43] J. Chen, M.A. Hamon, H. Hu, Y. Chen, A.M. Rao, P.C. Eklund, and R.C. Haddon. *Science*, 282:95, 1998.
[44] J. Liu, A.G. Rinzler, H. Dai, J.H. Hafner, R.K. Bradley, P.J. Boul, A. Lu, T. Iverson, K. Shelimov, C.B. Huffman, F. Macias-Rodriguez, Y.-S. Shon, T.R. Lee, D.T. Colbert, and R.E. Smalley. *Science*, 280:1253, 1998.
[45] Y. Lin, A.M. Rao, B. Sadanadan, E.A. Kenik, and Y.-P. Sun. *J. Phys. Chem. B*, 106:1294, 2002.
[46] A.G. Rinzler, J. Liu, H. Dai, P. Nikolaev, C.B. Huffman, F.J. Rodriguez-Macias, P.J. Boul, A.H. Lu, D. Heymann, D.T. Colbert, R.S. Lee, J.E. Fischer, A.M. Rao, P.C. Eklund, and R.E. Smalley. *Appl. Phys. A*, 67:29, 1998.
[47] D.B. Mawhinney, V. Naumenko, A. Kuznetsova, and J.T. Yates, Jr. *J. Am. Chem. Soc.*, 122:2383, 2000.
[48] D.B. Mawhinney, V. Naumenko, A. Kuznetsova, J.T. Yates, Jr.; J. Liu, and R.E. Smalley. *Chem. Phys. Lett.*, 324:213, 2000.
[49] S. Banerjee and S.S. Wong. *J. Phys. Chem. B*, 106:12144, 2002.
[50] M.A. Hamon, J. Chen, H. Hu, Y. Chen, M.E. Itkis, A.M. Rao, P.C. Eklund, and R.C. Haddon. *Adv. Mater.*, 11:834, 1999.
[51] J. Chen, A.M. Rao, S. Lyuksyutov, M.E. Itkis, M.A. Hamon, H. Hu, R.W. Cohn, P.C. Eklund, D.T. Colbert, R.E. Smalley, and R.C. Haddon. *J. Phys. Chem. B*, 105:2525, 2001.

[52] W. Huang, Y. Lin, S. Taylor, J. Gaillard, A.M. Rao, and Y.-P. Sun. *Nano Lett.*, 2:231, 2002.
[53] Y. Lin, D.E. Hill, J. Bentley, L.F. Allard, and Y.-P. Sun. *J. Phys. Chem. B*, 107:10453, 2003.
[54] J.E. Riggs, Z. Guo, D.L. Carroll, and Y.-P. Sun. *J. Am. Chem. Soc.*, 122:5879, 2000.
[55] R. Czerw, Z. Guo, P.M. Ajayan, Y.-P. Sun, and D.L. Carroll. *Nano Lette.*, 1:423, 2001.
[56] K.A. Williams, P.T.M. Veenhuizen, B.G. de la Torre, R. Eritja, and C. Dekker. *Nature*, 420:761, 2002.
[57] C.V. Nguyen, L. Delzeit, A.M. Cassell, J. Li, J. Han, and M. Meyyappan. *Nano Lett.*, 2:1079, 2002.
[58] C. Dwyer, M. Guthold, M. Falvo, S. Washburn, R. Superfine, and D. Erie. *Nanotechnology*, 13:601, 2002.
[59] M. Hazani, R. Naaman, F. Hennrich, and M.M. Kappes. *Nano Lett.*, 3:153, 2003.
[60] W. Huang, S. Taylor, K. Fu, Y. Lin, D. Zhang, T.W. Hanks, A.M. Rao, and Y.-P. Sun. *Nano Lett.*, 2:311, 2002.
[61] K. Jiang, L.S. Schadler, R.W. Siegel, X. Zhang, H. Zhangc, and M. Terronesd. *J. Mater. Chem.*, 14:37, 2004.
[62] S.E. Baker, W. Cai, T.L. Lasseter, K.P. Weidkamp, and R.J. Hamers. *Nano Lett.*, 2:1413, 2002.
[63] E.T. Michelson, C.B. Huffman, A.G. Rinzler, R.E. Smalley, R.H. Hauge, and J.L. Margrave. *Chem. Phys. Lett.*, 296:188, 1998.
[64] E.T. Michelson, I.W. Chiang, J.L. Zimmerman, P.J. Boul, J. Lozano, J. Liu, R.E. Smalley, R.H. Hauge, and J.L. Margrave. *J. Phys. Chem. B*, 103:4318, 1999.
[65] P.J. Boul, J. Liu, E.T. Mickelson, C.B. Huffman, L.M. Ericson, I.W. Chiang, K.A. Smith, D.T. Colbert, R.H. Hauge, J.L. Margrave, and R.E. Smalley. *Chem. Phys. Lett.*, 310:367, 1999.
[66] M. Holzinger, O. Vostrowsky, A. Hirsch, F. Hennrich, M. Kappes, R. Weiss, and F. Jellen. *Angew. Chem. Int. Ed.*, 40:4002, 2001.
[67] M.J. Moghaddam, S.Taylor, M. Gao, S. Huang, L. Dai, and M.J. McCall. *Nano Lett.*, 4:89, 2004.
[68] V. Georgakilas, K. Kordatos, M. Prato, D.M. Guldi, M. Holzinger, and A. Hirsch. *J. Am. Chem. Soc.*, 124:760, 2002.
[69] D. Pantarotto, J. Hoebeke, R. Graff, C.D. Partidos, J-P. Briand, M. Prato, and A. Bianco. *J. Am. Chem. Soc.*, 125:6160, 2003.
[70] D. Pantarotto, C.D. Partidos, J. Hoebeke, F. Brown, E. Kramer, J-P. Briand, S. Muller, M. Prato, and A. Bianco. *Chem. Biol.*, 10:961, 2003.
[71] A. Bianco and M. Prato. *Adv. Mater.*, 15:1765, 2003.
[72] D.M. Guldi, M. Marcaccio, D. Paolucci, F. Paolucci, N. Tagmatarchis, D. Tasis, E. Vázquez, and M. Prato. *Angew. Chem. Int. Ed.*, 42:4206, 2003.
[73] D. Pantarotto, J.-P. Briand, M. Prato, and A. Bianco. *Chem. Commun.*, 16, 2004.
[74] M. Musameh, J. Wang, A. Merkoci, and Y. Lin. *Electrochem. Commun.*, 4:743, 2002.
[75] J. Wang, G. Chen, M.P. Chatrathi, and M. Musameh. *Anal. Chem.*, 76:298, 2004.
[76] J. Xu, Y. Wang, Y. Xian, L. Jin, and K. Tanaka. *Talanta*, 60:1123, 2003.
[77] S. Hrapovic, Y. Liu, K.B. Male, and J.H.T. Luong. *Anal. Chem.*, 76:1083, 2004.
[78] J. Wang and M. Musameh. *Anal. Chem.*, 75:2075, 2003.
[79] K. Rege, N.R. Raravikar, D.-Y. Kim, L.S. Schadler, P.M. Ajayan, and J.S. Dordick. *Nano Lett.*, 3:829, 2003.
[80] V.G. Gavalas, R. Andrews, D. Bhattacharyya, and L.G. Bachas. *Nano Lett.*, 1:719, 2001.
[81] V.G. Gavalas, S.A. Law, J.C. Ball, R. Andrews, and L.G. Bachasa. *Anal. Biochem.*, 329:247, 2004.
[82] J.N. Wohlstadter, J.L. Wilbur, G.B. Sigal, H.A. Biebuyck, M.A. Billadeau, L. Dong, A.B. Fischer, S.R. Gudibande, S.H. Jameison, J.H. Kenten, J. Leginus, J.K. Leland, R.J. Massey, and S.J. Wohlstadter. *Adv. Mater.*, 15:1184, 2003.
[83] A. Guiseppi-Elie, C. Lei, and R.H. Baughman. *Nanotechnology*, 13:559, 2002.
[84] J.K. Campbell, Li. Sun, and R.M. Crooks. *J. Am. Chem. Soc.*, 121:3779, 1999.
[85] J. Li, C. Papadopoulos, and J.M. Xu. *Appl. Phys. Lett.*, 75:367, 1999.
[86] S. Huang, L. Dai, and A.W.H. Mau. *J. Phys. Chem. B*, 103:4223, 1999.
[87] D.C. Li, L. Dai, S. Huang, A.W.H. Mau, and Z.L. Wang. *Chem. Phys. Lett.*, 316:349, 2000.
[88] Y. Yang, S. Huang, H. He, A.W.H. Mau, and L. Dai. *J. Am. Chem. Soc.*, 121:10832, 1999.
[89] M. Gao, S. Huang, L. Dai, G. Wallace, R. Gao, and Z. Wang. *Angew. Chem. Int. Ed.*, 39:3664, 2000.
[90] J. Li, A. Cassell, L. Delzeit, J. Han, and M. Meyyappan. *J. Phys. Chem. B*, 106:9299, 2002.
[91] J. Li, H.T. Ng, A. Cassell, W. Fan, H. Chen, Q. Ye, J. Koehne, J. Han, and M. Meyyappan. *Nano Lett.*, 3:597, 2003.
[92] Q. Chen, L. Dai, M. Gao, S. Huang, and A.W.H. Mau. *J. Phys. Chem. B*, 105:618, 2001.
[93] S. Huang, L. Dai, and A.W.H. Mau. *J. Phys. Chem. B*, 3543:106, 2002.

[94] J. Wang, M. Musameh, and Y. Lin. *J. Am. Chem. Soc.*, 125:2408, 2003.
[95] J.H.T. Luong, S. Hrapovic, D. Wang, F. Bensebaa, and B. Simard. *Electroanalysis*, 16:132, 2004.
[96] Y. Lin, F. Lu, Y. Tu, and Z. Ren. *Nano Lett.*, 4:191, 2004.
[97] Y. Tu, Y. Lin, and Z.F. Ren. *Nano Lett.*, 3:107, 2003.
[98] Y. Tu, Z.P. Huang, D.Z. Wang, J.G. Wen, and Z.F. Ren. *Appl. Phys. Lett.*, 80:4018, 2002.
[99] X. Yu, D. Chattopadhyay, I. Galeska, F. Papadimitrakopoulos, and J.F. Rusling. *Electrochem. Commun.*, 5:408, 2003.
[100] M. Gao, L. Dai, and G.G. Wallace. *Synth. Met.*, 137:1393, 2003.
[101] S.G. Wang, Q. Zhang, R. Wang, and S.F. Yoona. *Biochem. & Biophys. Res. Commun.*, 311:572, 2003.
[102] S. Sotiropoulou and N.A. Chaniotakis. *Anal. Bioanal. Chem.*, 375:103, 2003.
[103] W.C.I. Homs. *Anal. Lett.*, 35:1875, 2002.
[104] J.J. Gooding. *Electroanalysis*, 14:1149, 2002.
[105] H. Cai, X. Cao, Y. Jiang, P. He, and Y. Fang. *Anal. Bioanal. Chem.*, 375:287, 2003.
[106] H. Cai, Y. Xu, P. He, and Y. Fang. *Electroanalysis*, 15:1864, 2003.
[107] Y. Xu, Y. Jiang, H. Cai, P. He, and Y. Fang. *Anal. Chim. Acta.*, 516:19, 2004.
[108] M. Guo, J. Chen, L. Nie, and S. Yao. *Electrochim. Acta.*, 49:2637, 2004.
[109] J. Wang, G. Liu, and M.R. Jan. *J. Am. Chem. Soc.*, 126:3010, 2004.
[110] J. Wang, G. Liu, M.R. Jan, and Q. Zhu. *Electrochem. Commun.*, 5:1000, 2003.
[111] J. Wang, A.-N. Kawde, and M. Musameh. *Analyst*, 128:912, 2003.

7

Characterization Methods for Quality Control of Nanopore and Nanochannel Membranes

Carlo Cosentino[1,*], Francesco Amato[2], and Mauro Ferrari[3]

[1]*Department of Experimental and Clinical Medicine, Università degli Studi Magna Græcia di Catanzaro, via T. Campanella 115, 88100 Catanzaro, Italy*

[2]*Dept. of Experimental and Clinical Medicine, Università degli Studi Magna Graecia di Catanzaro, Catanzaro, Italy*

[3]*Professor, Brown Institute of Molecular Medicine Chairman, Department of Biomedical Engineering, University of Texas Health Science Center, Houston, TX; Professor of Experimental Therapeutics, University of Texas M.D. Anderson Cancer Center, Houston, TX; Professor of Bioengineering, Rice University, Houston, TX; Professor of Biochemistry and Molecular Biology, University of Texas Medical Branch, Galveston, TX; President, the Texas Alliance for NanoHealth, Houston, TX*

7.1. INTRODUCTION

Nanotechnology is considered a fascinating subject not only by scientists, but also by people not involved in research. Likely, the appeal derives from common people thinking of nanotech devices as "invisible, mysterious objects", capable of accomplishing complex tasks. Such a mysterious feeling can be explained by the fact that nanodevices features cannot be entrapped, because of their dimensions, by the common experience of human sensing, like other systems, exhibiting much more complex structures or functions, but macroscopic dimensions (e.g. airplanes, robots, skyscrapers).

The issue of sensing, measuring, testing, controlling nanometric features is actually not only a concern of common people, but a challenging scientific matter, which is referred to as the problem of characterization. One typically thinks of technology like the set of methods and processes which allow to fabricate a certain device; however, also the availability of technologies for testing the structure and properties of a certain device is of paramount importance for a successful application, both in the research and industrial fields.

* E-mail: carcosen@unina.it

Obviously, life is rather easier for people involved in the characterization of devices belonging to the macroscopic world, for at least two reasons: 1) there is a huge availability of technologies, because macroscopic systems typically have been studied for longer time; 2) structural features and input/output signals of the systems can be often measured in a direct way, i.e. there is no need to use other measurements and/or complex processes (physical or computational) to obtain the desired data.

Although important in every scientific application, characterization of devices becomes a fundamental issue in the nanotechnology field, as important as fabrication itself, because it often represents not only a way of testing the integrity and quality of a system, but the only way to fully and correctly understand its behavior and potentialities.

In this chapter we will try to give an overview on the current state of the art of characterization methods for nanopore and nanochannels synthetic membranes. In our work we will focus especially on ultrafiltration (UF) membranes, which tipically have pore sizes ranging from few nanometers to several tens of nanometers.

The main structural parameters of a membrane are the shape, area and length of the pores, their distribution and density, the porosity, the cross-sectional structure and tortuosity. As one could expect, there is not a single method which can yield all of this data, but each technique is most suitable for the measurement of certain parameters. Imaging techniques, for example, can provide accurate data on surface porosity and pore shape, but give no indication on pore length and tortuosity, which can be better derived by diffusion tests.

The most used testing methods in the industrial community, especially in the biomedical field, are those based on rejection performance (also known as challenge tests), mainly because the first applications of membranes were in the filtration and culturing of microorganisms. Thus, a typical characterization is given in terms of the ability to retain specific organisms or synthetic particles, and this characteristic is expressed by the "Beta ratio" (number of organisms challenging the membrane/number of organisms passing through). This method, though very popular, cannot be considered as a good candidate for quality control, because it presents two serious disadvantages: 1) it is a destructive test, since the membrane cannot be reused after being in contact with; 2) the dimensions of biological challenge particles are not "well-defined", e.g. two different molecules of the same molecular weight can have different configurations such that their molecular diameter is different; moreover the interaction with the membrane itself can cause folding or unfolding, and thus erroneous interpretation of the data. On the other hand, in order to use synthetic particles in the challenging test, one should be certain that the particles have the desired characteristics and are uniform, which could be as difficult as testing the membrane itself with other methods.

An ideal characterization method should be non-destructive, accurate, repeatable, fast and should give the maximum possible number of data. Toward this end, many methods have been devised, since the beginning of last century, which can be classified according to the physical mechanisms they exploit:

- Imaging
  - ✓ Scanning Electron Microscopy (SEM)
  - ✓ Transmission Electron Microscopy (TEM)
  - ✓ Atomic Force Microscopy (AFM)
- Fluid Transport
  - ✓ Bubble point
  - ✓ Gas Transport

- Fluid Adsorption
  - ✓ Nitrogen Adsorption/Desorption (BEH)
  - ✓ Mercury porosimetry
  - ✓ Liquid-vapor equilibrium (BJH)
  - ✓ Gas-liquid equilibrium (Permoporometry)
  - ✓ Liquid-solid equilibrium (Thermoporometry
- Electrical Conductance
- Ultrasonic spectroscopy
- Molecular Transport

By now, some of these methods, like mercury porosimetry, BEH, BJH, can be conceivably considered classical, in the sense that they have been used by the scientific community for many years and are well documented in literature and university textbooks (e.g. [1, 2]); therefore they will not be discussed in the following.

7.2. MICROSCOPY OBSERVATION

At present, microscopy observation and image processing of micrographs can be considered a reliable source of information for many membrane morphological characteristics, such as pore shape and size, their distribution and density, surface porosity, cross-sectional structure. Nevertheless, these tools are not suitable for obtaining information about pore length and tortuosity.

Microscopic techniques, like scanning electron microscope (SEM) or transmission electron microscope (TEM), present significant difficulties, mainly concerning the preparation of membrane sample. In order to observe, via SEM, the cross sections of the membrane without causing a collapse of the structure, the sample must be dried and then fractured at liquid nitrogen temperature.

In TEM observation, instead, the dried sample is first embedded, if necessary, and then cut by microtome. Care has to be given in the choice of embedding material, to avoid interaction with the membrane, and then to cut very thin sections (in the order of tens of nm).

Although the first electron micrographs of a membrane documented in literature can be attributed to Riley et al. [4], only in the 1980 Merin and Cheryan [5] succeeded to visualize membrane surface pores by using a replica technique and TEM.

Both SEM and TEM expose the membrane to an high electron beam energy, which causes damages, thus hindering the correct inspection of the surface. A significant improvement has been brought in the 1980's by the introduction of field emission scanning electron microscopy (FESEM), which can achieve very high resolution (<1 nm) even with low beam energy/accelerating voltage.

Kohtake et al. [6] have first used FESEM imaging technique to observe pores on UF membrane surfaces. Kim et al. [7] exploited FESEM micrographs to calculate the pore size, density and surface porosity of various types of UF membranes, using an higher resolution than Kohtake et al. and analyzing the differences between membrane surfaces before and after protein filtration [8].

Although it is possible to measure and calculate the mean pore size and pore size distribution from SEM micrographs, it also results an extremely boring and time consuming

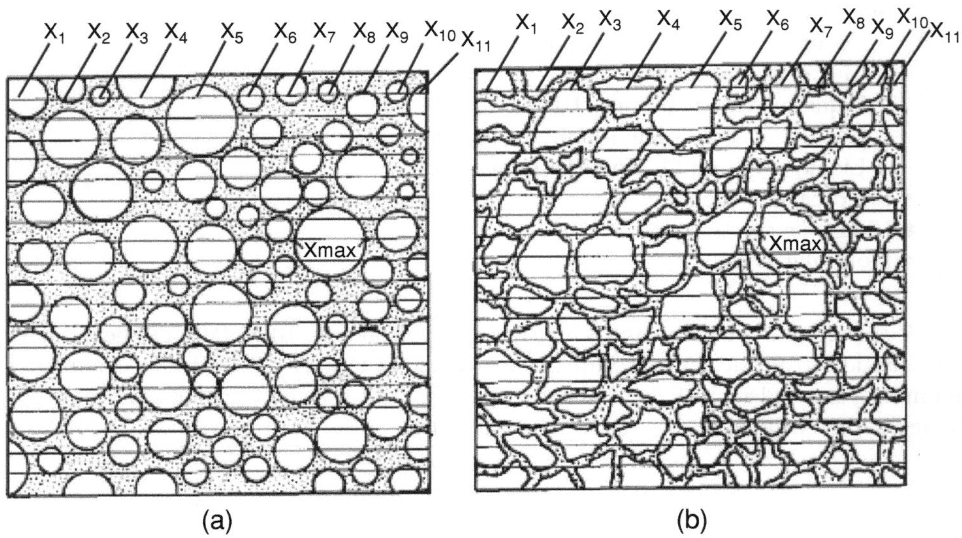

FIGURE 7.1. Schematic representation of electron micrographs of the plane parallel to the membrane surface. The vacant regions surrounded by dotted area indicate pores: (a) a membrane which has ellipsoidal pores; (b) an actual membrane which has complicated pores; $x_1, x_2, x_3, \ldots, x_{11}$ and x_{max} indicate cut-off length (Reprinted from [3] with permission from Elsevier).

work. Manabe et al. [9] have developed a method for characterizing the pores of a polymeric membrane having a mean pore diameter larger than 10 nm via electron micrographs. They devised ellipsoidal, spherical and straight-through cylindrical pore models and formulated theoretical equations relating the pore radius distribution function $N(r)$ to the distribution function $F(x)$, where x is the length of test segments cut off by pores in an electron micrograph as illustrated in Figure 7.1.

The theoretical relations are illustrated in Table 7.1; results obtained with a regenerated cellulose (RC) membrane are reported in [9] along with a comparison with other methods, such as mercury intrusion (MI) and bubble pressure (BP).

Image analysis techniques allow fast evaluation of the pore size and distribution from micrographs. Vivier et al. [10] have exploited quantitative image analysis for characterizing a microporous PTFE hollow fiber membrane. More recently, several software tools have been developed for this purpose, e.g. the NIH Image (developed by the National Institute of

TABLE 7.1. Pore radius distribution function for various pore typologies. Φ_e is a shape factor.

Ellipsoidal pores	$N_e(r) = -\frac{4}{\pi \Phi_e} \frac{d(F(x)/x)}{dx}$
Spherical pores	$N_s(r) = -\frac{4}{\pi} \frac{d(F(x)/x)}{dx}$
Straight-through cylindrical pores	$F(x) = x \int_{x/2}^{\infty} \frac{N(r)}{(4r^2 - x^2)^{1/2}} dr$

Health, USA, Division of Computer Research and Technology). This program allows the determination of the porosity (A_K), the pore density (N), the mean pore radius (r_p) and the pore size distribution of a membrane. This tool has been exploited by Masselin et al. [11] to measure the porosity, the pore density, the mean pore radius and the pore size distribution of five asymmetric ultrafiltration membranes by means of FESEM images. The analysis is based on the digitalization of the original grey-scale image, followed by measurement of the area of each pore. In [11], calculations were performed with a 0.5 nm interval. This value appeared to be a good compromise between the calculation precision and the number of intervals. The following parameters have been evaluated:

- The *maximum frequency* gives an information on the gathering of data in the same radius interval. This means that it gives an idea on the confidence that can be attributed to the modal pore interval and radius.
- The *modal interval* is the pore radius interval for which the pore radii are the most frequent. A 10% variation around the maximum frequency value is accepted in order to take into account wide distribution. The center of this interval is called the *modal pore radius* (its frequency is assumed to be maximum).
- The pore radius *distribution extent* is the maximum difference between the superior boundary of the pore radius interval for which the frequency is minimum (except zero), and the inferior boundary of the modal interval.

Finally, the membrane thickness has been also measured on images. Results allowed the comparison between the $A_K/\Delta x$ values obtained from image analysis and the $A_K/\Delta x$ values obtained by diffusion experiments.

In the last years, atomic force microscopy (AFM) has become a popular method for investigating the surface microstructure of polymer and inorganic membranes [12]. AFM characterization of membranes has focused on measurement of nanometer sized surface pores [13–16] and comparison with pore size analysis from scanning electron microscopy (SEM) [17–18], correlation of surface structure with membrane properties (such as molecular weigh cut-off, MWCO) [19], the effects of surface roughness on fouling [20–21], and adhesion force between model particles and the membrane surface [22].

It is worth mentioning the studies made by Smorgonskaya et al. [23] on the structural organization of bulk nanoporous carbon (NPC) materials produced from carbides. The studies were carried out by the small angle X-ray scattering (SAXS), X-ray diffraction (XRD), and high resolution transmission electron microscopy (HRTEM) techniques, which will not be described further in this survey.

7.3. BUBBLE POINT

Determining pore size distributions in membranes by monitoring liquid permeation has been used for many years [24]. Several testing methods, such as Coulter porosimetry, bubble point test, permoporometry, biliquid permoporometry and thermoporometry, rely on the same basic physical principle, the displacement of a wetting liquid. First a wetting liquid is introduced in the pores and is retained by capillary forces (fluid A), then a less wetting fluid (fluid B), liquid or gas, is put in contact with one side of the membrane and an increasing pressure is applied until it eventually displaces the former. The pressure

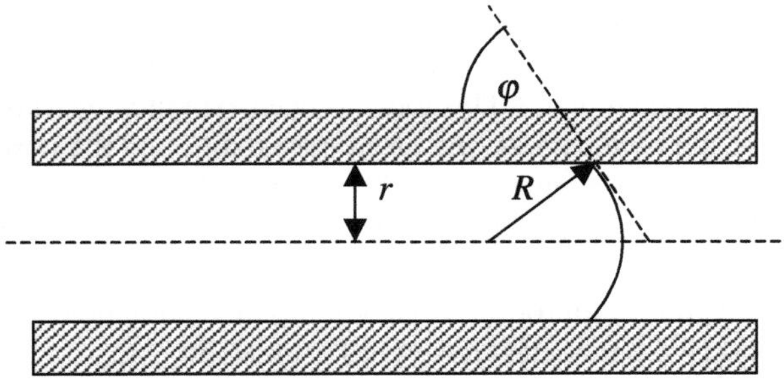

FIGURE 7.2. Theoretical model of interfacial meniscus in a cylindrical pore.

difference Δp needed to expel the wetting liquid from a pore with radius r is given by Laplace's equation

$$\Delta p = 2\sigma H \tag{7.1}$$

where σ is the interfacial tension of the fluid-fluid system, and H is the mean curvature of the meniscus. The phenomenon is generally modeled as a spherical meniscus in a cylindrical pore, the principal curvatures $c_1 = c_2 = 1/R$, and the mean curvature can be expressed in this case as $H = c_1 + c_2/2 = 1/R = \cos\varphi/r$. The parameter φ, the contact angle, relates the pore radius, r, to the meniscus radius, R, as illustrated in Figure 7.2. The pressure difference can be expressed as

$$\Delta p = \frac{2\sigma\cos\varphi}{r} \tag{7.2}$$

Therefore, a larger pressure difference is needed to expel fluid A from smaller pores letting fluid B permeate into them. The ideal flow versus pressure drop curve obtained by this method exhibits an S shape as shown in Figure 7.3 and will hereafter be referred to as the flow-pressure curve.

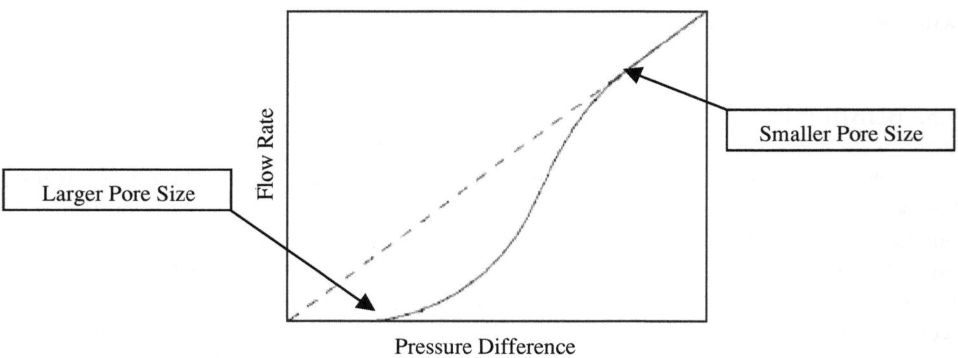

FIGURE 7.3. Ideal flow-pressure curve obtained from the bubble point test.

The Hagen-Poiseuille equation

$$\dot{V} = \frac{\pi r^4 \Delta p}{8 \mu L} \tag{7.3}$$

express the volume flow rate of a liquid in capillaries of circular cross-section, as a function of the viscosity, μ, of the fluid B and of the pore length, L. A number pore size distribution can be obtained by analyzing the flow-pressure curve. Initially this has been done graphically, more recently, a common statistical distribution is assumed a priori and the characteristic parameters are obtained by fitting the flow-pressure curve exploiting by using common mathematical software tools.

As mentioned earlier, different combination of fluids can be used in the bubble point technique. Jakobs et al. [24] have analyzed two kinds of commercial ceramic membranes, Anodisc® by Anotec, which are made by anodic oxidation resulting in straight, non-intersecting almost cylindrical pores with a very sharp pore size distribution (nominal pore size 200 Å), and Membralox® byUS Filter, which are fabricated by the sol-gel slipcasting method (nominal pore size 40–50 Å). Anodisc® membranes are not suitable for industrial applications, because of their structural fragility and small surface areas; nevertheless, due to the very well defined and uniform pore morphology, these membranes are an ideal model system to benchmark characterization methods. Membralox® membranes, on the contrary, exhibit a much more complicate structure and morphology, which does not agree well with the simplifying models used in most characterization methods. However, they possess a much higher mechanical strength and ratio of surface area to module volume, which make them preferable for practical applications.

The analysis has been conducted by using three different wetting/non-wetting liquid combination: 1) ethanol-helium, 2) water-butanol and 3) water-butanol-methanol. Jakobs et al. have evidenced a common drawback of this kind of characterization technique: nanopore membranes are often fragile systems, which cannot be exposed to high pressures required for the displacement of the wetting fluid. This is especially true when using a liquid-gas combination, for this reason they were unable to perform the ethanol-helium test on Anodisc® membranes, whereas testing by water-butanol and water-butanol-methanol was possible for both kind of membranes, because of lower pressures required for liquid-liquid tests.

An interesting phenomenon, evidenced in [24], occurs when testing membranes having high porosity and highly interconnected network structure of the pores, such as Membralox® ones: the resulting experimental flow-pressure curve does not follow the theoretical S shape, due to the phase separation of the mixture inside the pores, therefore it results almost impossible to determine the minimum and maximum pore size. This phenomenon could be explained by considering the device as a ternary system, where the two solvents are in equilibrium until they get in contact with the third phase, i.e. the membrane material itself. The addition of this third phase creates a new interface and, from a thermodynamic point of view, the equilibrium is perturbed. Another possible explanation is based on the increasing importance of wall effects in systems with a very high surface to volume ratio, i.e. pores with very small dimensions. Due to the molecular scale pore dimensions, long range intermolecular forces from the pore wall are acting on fluid molecules at any position within the pore. One liquid may be attracted stronger by these forces and forms a continuous film on the pore walls, while the other liquid collects in the middle of the pore channel. The resulting two-phase cocurrent flow is described by Dullien [25].

Moreover, a crucial point in the bubble point method is the estimation of the contact angle for the system at hand, i.e. fluid-fluid-solid. If the effect of this parameter is neglected the pore diameters might easily be overestimated. The contact angle values can be measured employing a Wilhelmy balance as described by Good and Stromberg [26].

7.4. GAS PERMEABILITY

The measurement of the flow rate of a gas through a porous membrane can provide a means of determining a mean pore radius of the porous membrane, rather than a pore size distribution [27]. Darcy's law gives the gas flux through the membrane

$$J = K\frac{\Delta P}{l} \tag{7.4}$$

where K is the permeability coefficient and $\Delta P = P_1 - P_2$ is the pressure difference across the membrane. When the mean free path of the permeating gas is comparable in size with the mean pore diameter, K contains both viscous and non-viscous terms and is expressed by

$$K = K_0 + \frac{B_0}{\eta}\overline{P} \tag{7.5}$$

where $\overline{P} = (P_1 + P_2)/2$ is the mean pressure and K_0 and B_0 are characteristics of the membrane and the gas.

The average molecular velocity of a gas is given by Carmanand equation

$$\overline{v} = \left(\frac{8RT}{\pi M}\right)^{1/2} \tag{7.6}$$

where M is molecular weight. The equivalent pore radius of the membrane can be expressed as

$$r_p = \frac{16}{3}\left(\frac{B_0}{K_0}\right)\left(\frac{2RT}{\pi M}\right)^{1/2} \tag{7.7}$$

Using this theoretical formalism, several authors have calculated experimentally the mean pore radius r_p [27–31]. However, it seems that the technique does not yield reliable results, especially for membrane having very small pore sizes (few nanometers). In fact, fairly high pore radii have been found for polysulfone [27, 29] and cellulose acetate RO [30] membranes. Altena et al. [31] have shown that the resulting mean pore size is also dependent on the kind of gas one uses for the test, which is understandably not a good point.

More recently, Meixner and Dyer [32] stressed the importance of accuracy in the characterization of gas transport through membranes, particularly in the measurement of pure gas permeance as a function of mean pressure. They assert that this method, due to demanding limits on experimental error and misinterpretation of data, has been often applied incorrectly and propose a novel theoretical formalism for describing gas transport through both monolithic and multilayer porous systems.

7.5. PERMOPOROMETRY

Permoporometry is a well established technique for the characterization of porous membranes. There are two variants of the method [39]:

- Liquid displacement permoporometry (LDP), introduced by Capanelli et al. [33];
- Diffusion permoporometry (DP), which was first used by Katz and Baruch [35].

The physical mechanism underlying this method is the capillary condensation of liquids in nanopores. Indeed, the vapor pressure of a liquid is dependent on the radius of curvature of its surface. This dependence is expressed by Kelvin equation for a liquid in a capillary tube

$$RT \ln \frac{p}{p_0} = \frac{2\gamma \overline{V} \cos\theta}{r_K} \tag{7.8}$$

where p and p_0 are the vapor pressures in the capillary and under standard conditions, respectively, γ is the surface tension between the capillary liquid and air, $\overline{V}$ is the molar volume of the liquid, θ the contact angle, r_K the Kelvin radius, R is the gas constant, and T the absolute temperature. From Eq. (8) it is clear that the vapor pressure of the liquid in a capillary varies with the radius of the capillary.

The first step consists in blocking all the pores by capillary condensation; this can be achieved by applying an appropriate pressure. Then the pressure is reduced, so that pores having a size corresponding to the vapor pressure applied are emptied and become available for gas transport. The distribution of the size of the active pores can be derived from the analysis of the gas flow through the membrane during the pressure decrease.

Katz and coworkes [35, 36] applied this technique to measure the pore size distribution of Millipore VS and VM type membranes, which have a nominal pore diameter of 25 and 50 nm, respectively. Thay could not observe significant differences between the two kind of membranes, moreover the pore size distribution had a peak around 10 nm.

Cuperus et al. measured the pore size distribution of γ-alumina membranes, Nuclepore 15 nm membranes, DDS GR61PP (MWCO 29 kDa) membranes and lab-made PPO and polysulfone (PSf) membranes [34].

They obtained a very sharp pore size distribution both for the alumina and Nuclepore membrane and the results were in good agreement with nominal data provided by the manufacturer.

Cao et al. [38] have measured the pore size distribution of porous alumina membranes by both permoporometry and the nitrogen adsorption/desorption technique. They found that the results obtained by the two methods were in a good agreement, with an experimental pore size around 10 nm. Furthermore, they evidenced that permoporometry is more accurate, since pore size distribution is more narrow than with nitrogen adsorption/desorption. A possible explanation could rely on the fact that permoporometry measures only the active pores, whereas nitrogen adsorption/desorption cannot distinguish between active and dead-end pores.

Hazri and Vayda [39] have described a modified technique for the analysis of diffusion permoporometric data obtained with asymmetric membranes. The modifications yield better estimates of pore sizes and pore size distributions. They show how erroneous conclusions could be drawn when treating asymmetric pore membrane under the general assumption of right cylindricity. In this case, while there is little error in the estimation of pore radii, the pore numbers are grossly overpredicted by the cylindrical pore model. The extent of this overprediction is directly proportional to the asymmetry of the membrane and the thickness of its active layer.

7.6. THERMOPOROMETRY

Thermoporometry has been introduced by Brun et al. [40] in 1973. It is particularly suitable for the structural analysis of membranes, because the measures are taken in a wet environment, thus reproducing the same operating conditions in which the membrane is normally used.

The technique is based on the lowering of the triple point temperature of a liquid filling a porous material. The phase transitions (crystallization or melting) for a liquid confined within a pore are observed to shift to lower temperatures that are determined by the pore size. This difference in transition temperature, ΔT, between confined and bulk solvent can be detected calorimetrically by differential scanning calorimetry (DSC).

The Gibbs-Thomson equation express the relationship between ΔT and the pore radius

$$\frac{\Delta T}{T_0} = \frac{2\gamma_{ls}}{\Delta H_0} \frac{v_1}{R_P} \tag{7.9}$$

where γ_{ls} is the liquid-solid interfacial tension, ν_1 the liquid phase molar volume, ΔH_0 the molar heat of fusion, R_P the pore radius, and $\Delta T = T - T_0$ the triple point depression. During the state change, there is a layer of thickness t which does not participate to the freezing or melting, but remains adsorbed on the surface of the pore. Taking this into account, the pore radius can be calculated as

$$R_P = -\frac{A}{\Delta T} + t \tag{7.10}$$

where A is a constant depending on the solvent used.

Smolders and Vugteveen [41] and Zeman et al. [42] have applied this method for characterizing poly(2,6-dimethyl-1,4-phenylene) oxide (PPO) UF membranes which were asymmetric and the mean pore radius obtained was in between 2 and 3 nm. They have also measured the distribution by means of a nitrogen adsorption/desorption technique and the obtained mean radius well agreed with that calculated from thermodiagram.

In [42, 43] Zeman et al. exploited thermoporometry and BJH technique for the characterization of 3 types of polysulfone UF and found similar results from the methods. However, when they measured the mean pore radius by SEM imaging, they obtained much smaller values. A further characterization conducted by challenge test, using dextran having a solute radius of 5 nm confirmed the validity of SEM-determined radius. Consequently,

Zeman et al. [43] suggested that the erroneous results by both thermoporometry and nitrogen adsorption/desorption were due to the measurement of pores in the sublayer of the asymmetric, which have much larger pore sizes than those in the active surface.

A further inconvenient in the use of thermoporometry is the fact that polymeric UF membranes exhibit less stability against chemicals, anisotropic structure and small pole volume in the skin . Thus liquids which can be used in this technique are quite limited—water is usually used—and highly sensitive calorimetric measurement is required [3].

Wulff [44] used well characterized polystyrene/DVB samples with varying pore sizes (10–40 nm) in order to analyze the application of acetonitrile, C_2H_3N as an alternative solvent for thermoporometric measurements.

7.7. ELECTRICAL CONDUCTANCE

The need for non-destructive characterization and testing methods of nanoporous membrane has led to the investigation of novel non-invasive techniques. The common aim of this technique is to correlate the filtration properties of the membrane to some other easily measurable properties of the device, in order to avoid long and complicate tests, like those based on diffusion or TEM, which can take many days, expensive materials, complex procedures.

Among these new techniques, the analysis of electrical conductance properties seems to be the most likely to become a standardizable control quality method in large scale industrial production of nanofiltration membranes, because of its easiness, rapidity, repeatability and reliability.

A valuable example of how diffusion properties can be correlated to electrical conductance has been given by Carbonaro et al. [45]. They have assumed the following relationship between membrane electrical resistance, R_M, and the structural parameters

$$R_M = \frac{l\tau}{\chi_{pore} A_M \varepsilon} \tag{7.11}$$

where l is the membrane thickness, τ the tortuosity coefficient, κ_{pore} pore the conductivity inside pores, A_M the membrane area, ε the membrane porosity. The electrical conductivity inside pores, κ_{pore}, is assumed to be different from nominal conductivity κ_0, because, when the channel dimensions are in the nanometer range, surface conduction effects into the channel cannot be neglected [46]. Since the pore walls are in contact with an aqueous solution, they acquire an electric surface charge, which leads to a reorganization of the ions of the solution in order to maintain the neutrality of the system. Consequently, a diffuse space charge, electrical double layer (EDL), appears in the solution adjacent to the pore walls. The EDL creates an excess conduction which is the reason of the difference between the conductivity inside the pores and that of the bulk solution. The effect is expected to become more and more significant when the pore dimensions shrink.

Carbonaro et al., following the methods described in [47] by Fievet et al., and in [48] by Szymczyk et al., measured the electrical conductivity and the diffusion coefficients of several membranes, produced by iMEDD, Inc. (Columbus, OH), which exhibit straight rectangular channel having 45 μm width, 4 μm length and height in the range 5–50 nm.

The electrical resistance was measured by means of an EVOM epithelial voltohmmeter, which is a device specifically designed to perform non-destructive routine trans epithelial electric resistance (TEER) measurements. Two pairs of electrodes (Ag/AgCl) were used to supply an alternating current at 12.5 Hz and to measure the voltage gradient respectively. The resistance can be derived from potential difference measurement by simply applying Ohm's law. Note that the potential difference depends on both the composition of the solution and the membrane structure.

The measurement have been eventually correlated with diffusion data and a linear relationship has been drawn between the diffusion rate and the measured conductance for several pore sizes.

This general approach is likely to be followed in the characterization of other kinds of nanoporous membranes, also in the industrial field, because it allows to obtain an experimental reference curve, which is then applicable for inexpensive and rapid quality control.

7.8. ULTRASONIC SPECTROSCOPY

Broadband ultrasonic spectroscopy (BUS) is a well-known technique used to measure velocity and attenuation of ultrasonic waves in materials. The technique is based on frequency-domain analysis, by using the Fourier transform, of broadband pulses transmitted through a sample, or reflected from it. Density, velocity (or thickness) and attenuation of sound waves and the frequency dependence of these variables can be obtained [49]. The application of BUS to membrane can be traced back to the work of Haines et al. [51], who applied the technique to thin layered media.

In the last years, ultrasonic time-domain reflectometry (TDR) has emerged as a new technique for characterization of filtration membranes [52–56], in particular, for real-time non-invasive measurement of membrane compaction and fouling. It is based on the reflection of an ultrasonic pulse reflected on the membrane surface and time-delayed measurements to determine changes in the membrane due to compaction or fouling. A disadvantage of this technique, however, consists in the pulses overlap in the time-domain, so direct time-delay measurements are not possible.

BUS does not require any contact with the sample, thus avoiding contamination and allowing very rapid inspection and on-line and real-time measurements. An advantage of BUS over TDR is represented by the fact that the latter provides information only about one surface of the membrane, whereas the former yields also data on the whole volume and the back surface and can be used for precise time-delay measurements even when pulses overlap in the time-domain and when strong dispersion takes place.

The influence of membrane structural parameters like pore size, porosity, tortuosity, permeability, and flux resistivity on the velocity and attenuation of sound waves is well documented in literature [57–62]. Moreover, attenuation of sound is influenced by the presence of any external agent in the pore space as moisture [63], and has been used to monitor wetting and drying processes [64].

Gómez Álvarez-Arenas [49] has applied an air-coupled ultrasonic broadband spectroscopy technique to the characterization of a wide set of commercial filtration membranes. Ultrasonic wideband pulses are transmitted to and received from a gas by

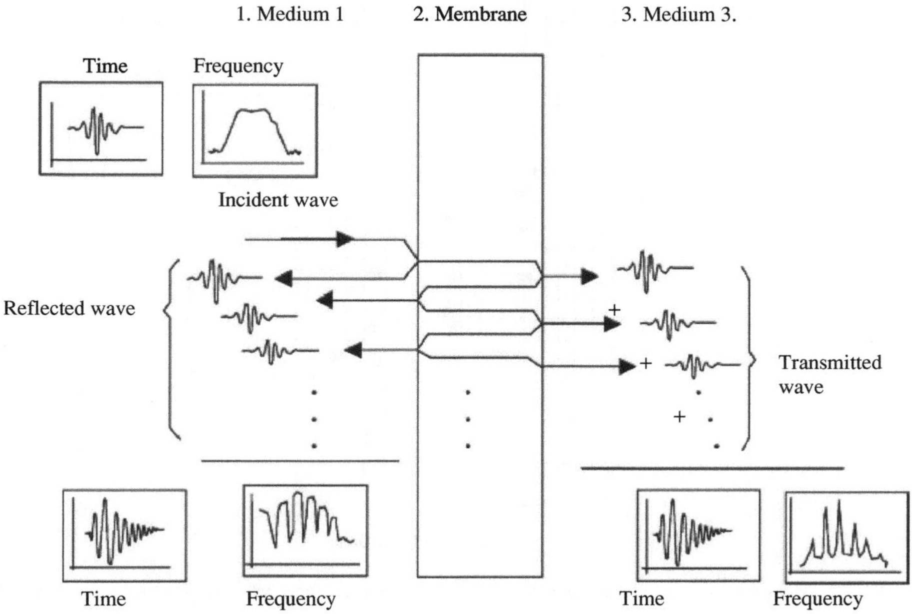

FIGURE 7.4. Schematic illustration of sound transmission through a plate and multiple reverberations within it (Reprinted from [49] with permission from Elsevier).

gas-coupled transducers and the pulse propagates through a gas gap between sample and transducers.

The basic working principle of the technique is illustrated in Figure 7.4: an ultrasonic signal travels through a gas gap and impinges normally on the membrane surface; at this first interface a certain amount of the energy is reflected back, while the rest of the energy is transmitted into the membrane. Transmitted wave propagates through the membrane until it reaches the back surface where, again a part of the energy is reflected back, and the rest is transmitted through the interface and received at receiver transducer after travelling through a gas gap. This process is repeated for each of the multiple internal reflections in the membrane. When time-delay between consecutive internal reflections at a sample surface equals $1/f$ (being f the frequency of the wave) a constructive interference between multiple reflections within the sample is built up, when this constructive interference takes place at the rear surface of the sample, maximum value of the transmitted energy is observed: this is a thickness resonance of the sample.

The reader is referred to [49] for the details on the theoretical modelling of the problem of transmission of ultrasonic waves through a membrane, and for explanation of the experimental set-up. The ultrasonic spectroscopy method has been applied to 23 different kind of membrane having nominal pore sizes ranging from 25 nm to 5 μm. Results show that there is a clear correlation between wave velocity and membranes characteristic properties, like pore size (see Figure 7.5), water flow and bubble point, whereas attenuation is independent of them. On the other hand, attenuation is very sensitive to membrane fouling, wetting or damage, therefore it can be suitable for integrity test.

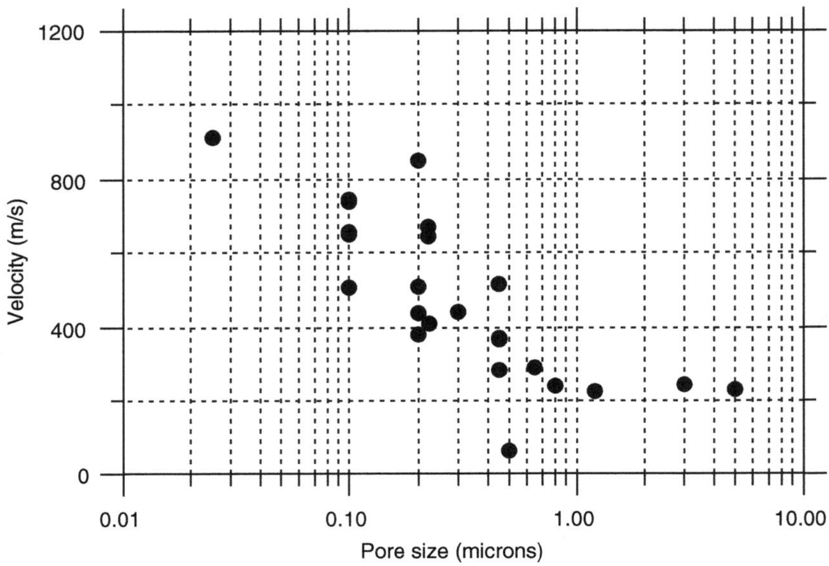

FIGURE 7.5. Measured velocity of ultrasonic waves (m/s) in the membranes vs. pore size (Reprinted from [49] with permission from Elsevier).

7.9. MOLECULAR TRANSPORT

The characterization methods based on molecular transport relies on the consideration that measurable flux and rejection quantities are closely related to membrane structural characteristics, such as thickness, pore size, length, density, distribution, tortuosity. Therefore, assuming to have an accurate and reliable transport model based on these characteristic parameters, it should be possible to derive their values by fitting the experimental data on permeate flux and solute rejection.

A first classification of molecular transport models distinguishes two class of methods: black-box and hydrodynamic models. For the first category we can mention the irreversible thermodynamics approach [65, 66] and the Stefan-Maxwell multicomponent diffusion equations [68]. This methods derive the equations phenomenologically, the focus is not on the transport mechanisms internal to the membrane, but rather on the relationships between inputs and outputs.

On the contrary, hydrodynamic models, whose origin can be traced back to 1936 [69], are based on the fundamental hydrodynamic equation for the transcapillary transport of rigid spheres. Since they are the most commonly applied method for computational membrane characterization we will focus on this second class of models.

7.9.1. Classical Transport Models

Ferry's hydrodynamic model describes the diffusion of rigid spheres, of radius r_s, in cylindrical pores of uniform radius r_p and length $\Delta x \gg r_p$, assuming a steady-state flow of solvent and neglecting possible molecular interaction in the pores.

He represented the velocity of the permeating solution as

$$v(r) = v_{\max}\left\{1 - \left(\frac{r}{r_p}\right)^2\right\} \tag{7.12}$$

where r_p is the pore radius. The volume of the solution entering the pore in time dt is given by

$$dV = dt\int_0^{r_p} 2\pi r v(r)dr \tag{7.13}$$

Ferry also introduced the steric hindrance factor: it accounts for the fact that a molecule has its own volume, thus, in order to enter the pore, should not bump on the edges of the pore entrance. Considering the molecule as a rigid sphere, neglecting Brownian motion, and assuming the velocity $v(r)$ as the velocity of its geometrical center, the number of molecules entering the pore in time dt is given by

$$dn = C_m dt\int_0^{r_p - r_s} 2\pi r v(r)dr = C_m \pi v_{\max}\left\{(r_p - r_s)^2 - \frac{(r_p - r_s)^2}{2r_p^2}\right\}dt \tag{7.14}$$

The concentration C_m is the concentration at the membrane surface; indeed, it is well-known that this concentration is usually higher than the average solution concentration in the source volume.

Thus the concentration of the permeate is

$$C_p = \frac{dn}{dV} = C_m\left\{2\left(1 - \frac{r_s}{r_p}\right)^2 - \left(1 - \frac{r_s}{r_p}\right)^4\right\} \tag{7.15}$$

and the actual rejection is

$$R = 1 - \frac{C_p}{C_m} = 1 - \left\{2(1-q)^2 - (1-q)^4\right\} \tag{7.16}$$

where

$$q = \frac{r_s}{r_p} \tag{7.17}$$

Then Pappenheimer et al. [71, 72] introduced the wall drag factor, *f(q)*, which express the frictional resistance to diffusion in free in the pore, and was derived by Ladenburg [73].

At a macroscopic level, the solute flux is described by

$$J_s^d = -D_0 f(q) S_D A_K \frac{dC}{dx} = -D_{\mathit{eff}} A_k \frac{dC}{dx} \tag{7.18}$$

where D_0 and D_{eff} are the diffusion coefficients in an unconstrained solution and in a pore, respectively, and A_k is the ratio of available cross-sectional pore area to total membrane area, which corresponds to the porosity of the membrane surface. Note that eq. (7.18) is essentially Fick's first law of diffusion, but with a reduced diffusion coefficient. The steric hindrance factor, S_D, and the wall correction factor, $f(q)$, are

$$S_D = (1 - q)^2 \tag{7.19}$$

$$f(q) = \frac{1}{1 + 2.4q} \tag{7.20}$$

Accordingly, the ratio of D_{eff} to D_0 can be represented as follows

$$\frac{D_{eff}}{D_0} = f(q)S_D = \frac{(1 - q)^2}{1 + 2.4_q} \tag{7.21}$$

By analogous reasoning, they also derived the relation for rectangular slits

$$\frac{D_{eff}}{D_0} = \frac{\left(1 - \frac{r_s}{w/2}\right)}{1 + 3.4\left(\frac{r_s}{w/2}\right)^2} \tag{7.22}$$

where w is the width of the slit.

Note that to this point only the diffusion has been considered. An accurate model of transport phenomena should take into account the effect of convection flow. The following flux relation

$$J_s = -D_0 f(q) S_D A_k (C_m - C_p) + f(q) S_F J_v C_m \tag{7.23}$$

was first introduced by Renkin [70] (S_F is the convection hindrance factor). A second term has been added, which describes the flux due to convection.

Note that, in Renkin's model, the same $f(q)$ is used for diffusive and convective flow. This not a suitable assumption, as proved in other works [76, 77], when the flow is not stationary. Moreover, when the concentration difference at the two sides of the membranes is not small, the concentration term C_m in eq. (23) needs to be corrected and the transport equation must be integrated across the membrane.

7.9.2. Diffusion Through Nanochannels

As discussed above, the classical theory of diffusion is based on Fick's law, and the related diffusion coefficient, D_0. Historically, the effect of pores on the molecular transport phenomena has been always described in mathematical terms via an effective (reduced) diffusion coefficient D_{eff}, which depends on the nominal coefficient, D_0, and on the ratio between molecular size and channel dimension, i.e. r_s/r_p.

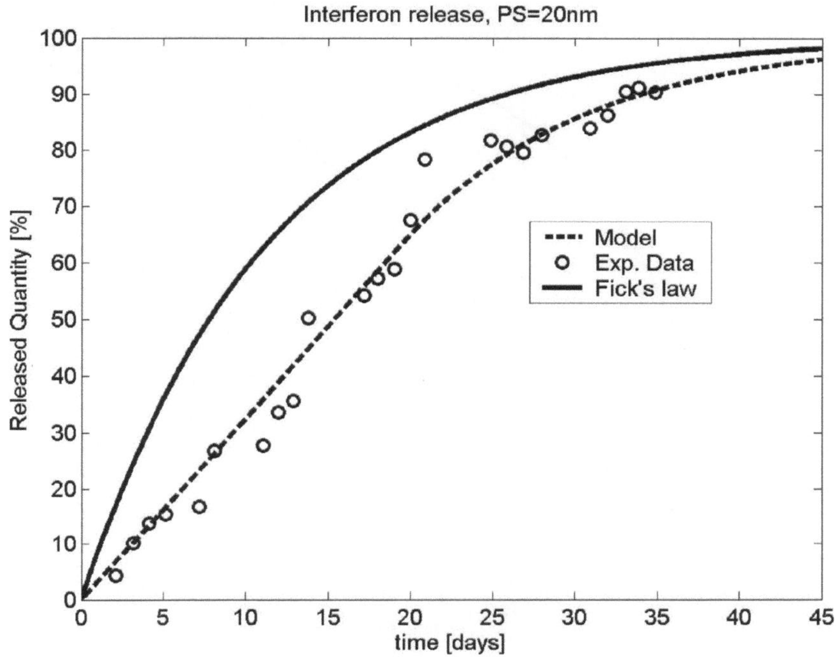

FIGURE 7.6. In vitro interferon diffusion through nanopore membrane (20 nm pore size) under sink conditions: experimental data (o), Fick's law prediction (-), model based simulation (- -).

Cosentino et al. [80] laid the basis of a novel theoretical formalism for the modeling of biomolecules diffusion through two-dimensional nanochannels. The starting point is the work of Martin et al. [79], who conducted several experiments of diffusion of biomolecules through microfabricated silica membranes, produced by iMEDD, Inc. (Columbus, OH).

The membranes used, which are an optimized variant of the membranes described by Chu et al. [81] and Desai et al. [82–83], are produced by photolithograpy, thin film deposition and selective etching of silicon wafers. They are made up of highly uniform rectangular channels, having a length of about 45 μm, a width of 45 μm, and a selectable height in the range 5–50 nm.

Measurements of diffusion kinetics of solutes across these membranes under sink conditions revealed non-Fickian behavior as the nanopore height approaches the hydrodynamic diameter of the solute. In particular, Martin et al. observed a zero-order kinetics in the release profiles of human recombinant interferon α2b, lysozyme, and bovine serum albumine (BSA), using 13 nm and 20 nm channels. After a certain percentage of the total amount of solute has been released, the linear release profiles turns back to an exponential shape (Figure 7.6–Figure 7.8).

Experimental evidences of non-Fickian diffusion in various kinds of materials are well documented in literature, especially concerning the single file diffusion (SFD) and wall drag effect (WDE) phenomena [84–92]. However, the case analyzed in [79] and [80] differs from those in previous literature, because 1) the membrane is made up of silicon and

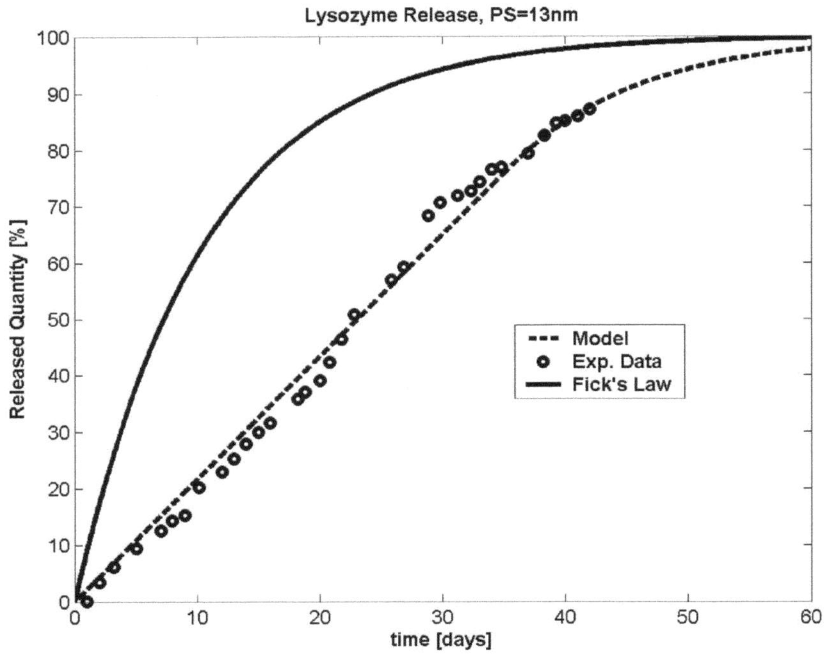

FIGURE 7.7. In vitro Lysozyme diffusion through nanopore membrane (13 nm pore size) under sink conditions: experimental data (o), Fick's law prediction (-), model based simulation (- -).

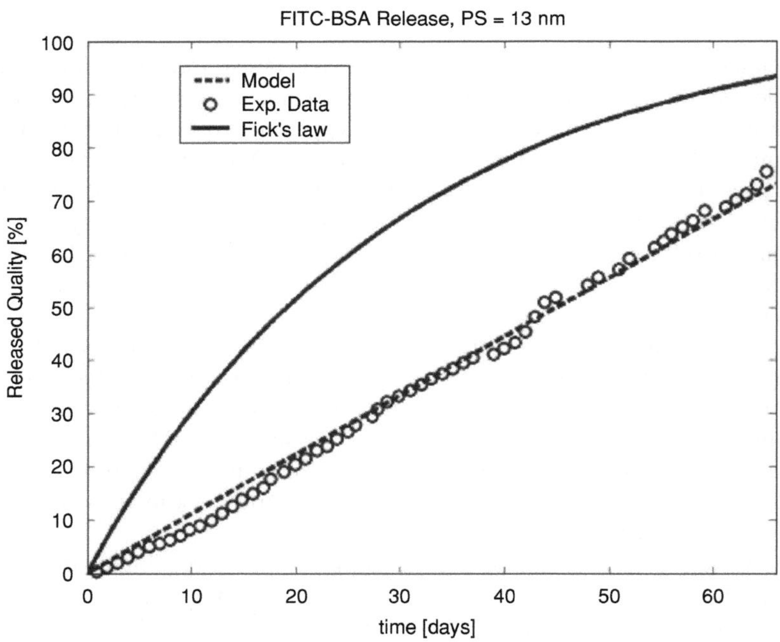

FIGURE 7.8. In vitro diffusion kinetics of fluorescein isothiocyanate (FITC) labeled-BSA through 13 nm pore size membrane under sink conditions: experimental data (o), Fick's law prediction (-), model based simulation (- -).

fabricated by photolithographic techniques and 2) the pores are rectangular and nanometric only in one dimension (two-dimensional channels).

On the basis of these experimental results, Cosentino et al. [80] proposed a novel formalism for the interpretation of the zero-order release. The classical modeling approach, which makes use of the reduced diffusion coefficient D_{eff}, is not suitable to explain the zero-order kinetics, because, combining mass conservation equation with Fick's law, it results

$$J_A(t) = \left(c_{A1}^0 - c_{A2}^0\right) \frac{D_{AB}}{L} e^{-\lambda_A t}, \quad \lambda_A = \frac{D_{AB} S}{V_1 L}\left(1 + \frac{V_1}{V_2}\right) \tag{7.24}$$

where c_{Ai}^0 is the concentration in the i-th compartment, D_{AB} is the diffusion coefficient of solute A in solvent B, S is the membrane active surface, L the length and V_i the volume of the i-th compartment.

Clearly, in the case expressed by (24) the solute release from reservoir (volume 1) to sink (volume 2) follows an exponential law, independently of the value of D_{AB}.

The interpretation proposed by Cosentino et al. is based on the assumption of a flux saturation effect, which can be expressed in mathematical terms as

$$J_A(t) = sat\left[\frac{D_{AB}}{L}(c_{A1}(t) - c_{A2}(t))\right]_{-\hat{J}_A}^{\hat{J}_A} \tag{7.25}$$

which is a modified formulation of Fick's first law.

The concentration gradient is approximated by the difference of the concentrations, divided by the channel length, L. The mass flux saturates at the threshold level, $\hat{J}_A$, which is intuitively dependent on the molecular and channel dimensions. Fitting of the experimental data, obtained via this modeling approach, are shown in Figure 7.6–Figure 7.8.

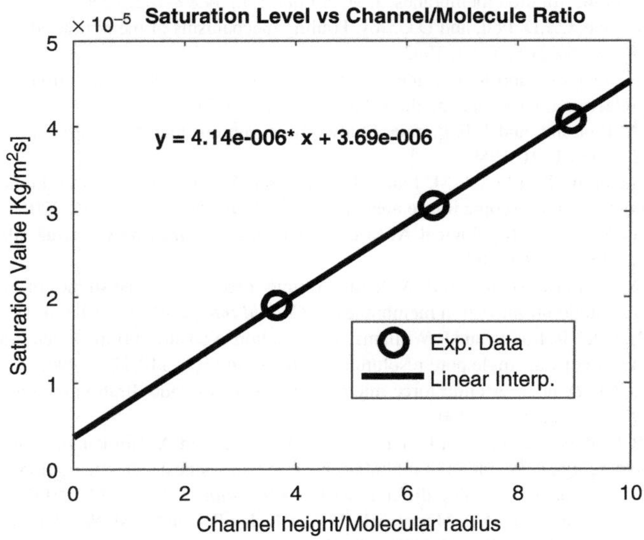

FIGURE 7.9. Saturation threshold value vs ratio of channel height and molecular hydrodynamic radius.

The foremost point in this novel formalism is the relation between the flux saturation threshold and the system parameters (pore size and structure, molecular dimension). A first experimental investigation data yielded the linear relation reported in Figure 7.9, which expresses the flux threshold as a function of channel dimension to molecular size ratio. Since the volume of a sphere increases with the cube of the radius one would expect such a function to have a third order polynomial profile; however the linearity could be explained by the fact that the channels are nanometric only in height, and thus constrain the diffusion in only one dimension. Of course, in order to assess the modeling hypotheses and to construct a reliable diffusion prediction tool, which could be a quality control tool as well, more experimental data need to be analyzed, with various proteins of different molecular weight. Moreover, a physical interpretation should be given, starting from first principles, of the flux saturation effect, so that it could be directly be expressed as a function of membrane structural characteristics.

REFERENCES

[1] M. Mulder. *Basic Principles of Membrane Technology*, Kluwer Academic Publishers, Dordrecht, 1991.
[2] R.W. Baker. *Membrane Technology and Applications*, John Wiley & Sons, 2004.
[3] S. Nakao. Determination of pore size and pore size distribution 3. Filtration membranes, *J. Membrane Sci.*, 96:131–165, 1994.
[4] R.L. Riley, J.O. Gardner, and U. Merten. *Science*, 143:801, 1964.
[5] U. Merin and M. Cheryan. Ultrastructure of the surface of a polysulfone ultrafiltration membrane. *J. Appl. Polym. Sci.*, 25:2139–2142, 1980.
[6] M. Koutake, Y. Uchida, T. Kimura, Y. Sagara, A. Watanabe, and S. Nakao. Observation of UF membranes pores through a scanning electron microscope and their pure water fluxes. *Maku*, 10:310–312, 1985.
[7] K.J. Kim, A.G. Fane, C.J.D. Fell, T. Suzuki, and M.R. Dickson. Quantitative microscopic study of surface characteristics of ultrafiltration membranes. *J. Membrane Sci.*, 54:89–102, 1990.
[8] K.J. Kim, A.G. Fane, C.J.D. Fell, and D.C. Joy. Fouling mechanisms of membranes during protein ultrafiltration. *J. Membrane Sci.*, 68:79–91, 1992.
[9] S. Manabe, Y. Shigemoto, and K. Kamide. Determination of pore radius distribution of porous polymeric membranes by electron microscopic method. *Polym. J.*, 17:775–785, 1985.
[10] H. Vivier, M.-N. Ponsand, and J.-F. Portala. Study of microporous membrane structure by image analysis. *J. Membrane Sci.*, 46:81–91, 1989.
[11] I. Masselin, L. Durand-Bourlier, J.-M. Laine, P.-Y. Sizaret, X. Chasseray, and D. Lemordant. Membrane characterization using microscopic image analysis. *J. Membrane Sci.*, 186:85–96, 2001.
[12] W. Yoshida and Y. Cohen. Topological AFM characterization of graft polymerized silica membranes. *J. Membrane Sci.*, 215:249–264, 2003.
[13] W.R. Bowen, N. Hilal, R. Lovitt, and P. Williams. Atomic force microscope studies of membranes: surface pore structures of diaflo ultrafiltration membranes. *J. Colloid Interface Sci.*, 180:350, 1996.
[14] W.R. Bowen, N. Hilal, R. Lovitt, and P. Williams. Visualisation of an ultrafiltration membrane by non-contact atomic force microscopy at single pore resolution. *J. Membrane Sci.*, 110:229, 1996.
[15] W.R. Bowen and T. Doneva. Atomic force microscopy studies of nanofiltration membranes: surface morphology. *Desalination*, 129:163, 2000.
[16] N.A. Ochoa, P. Pradános, L. Palacio, C. Pagliero, J. Marchese, and A. Hernández. Pore size distributions based on AFM imaging and retention of multidisperse polymer solutes: characterization of polyethersulfone UF membranes with dopes containing different PVP. *J. Membrane Sci.*, 187:227, 2001.
[17] A.K. Fritzsche, A.R. Arevalo, M.D. Moore, V.B. Elings, K. Kjoller, and C.M. Wu. The surface structure and morphology of polyvinylidene fluoride microfiltration membrane by atomic force microscopy. *J. Membrane Sci.*, 68:65, 1992.

[18] J.Y. Kim, H.K. Lee, and S.C. Kim. Surface structure and phase separation mechanism of polysulfone membranes by atomic force microscope. *J. Membrane Sci.*, 164:159, 1999.

[19] J.M. Tan and T. Matsuura. Effect of nonsolvent additive on the surface morphology and the gas separation performance of poly(2,6-dimethyl-1,4-phenylene)oxide membranes. *J. Membrane Sci.*, 160:7, 1999.

[20] E.M. Vrijenhoek, S. Hong, and M. Elimelech. Influence of membrane surface properties on initial rate of colloidal fouling of reverse osmosis and nanofiltration membranes. *J. Membrane Sci.*, 188 (1):115, 2001.

[21] W.R. Bowen and T. Doneva. Atomic force microscopy studies of membranes: effect of surface roughness on double-layer interactions and particle adhesion. *J. Colloid Interface Sci.*, 229:544, 2000.

[22] W.R. Bowen, N. Hilal, R. Lovitt, and C.J. Wright. A new technique for membrane characterisation: direct measurement of the force of adhesion of a single particle using an atomic force microscope. *J. Membrane Sci.*, 139:269, 1998.

[23] E. Smorgonskaya, R. Kyutt, A. Danishevskii, C. Jardin, R. Meaudre, O. Marty, S. Gordeev, and A. Grechinskaya. X-ray and HRTEM structural studies of bulk nanoporous carbon materials produced from carbides. *J. Non-Cryst. Solids*, 299–302:810–814, 2002.

[24] E. Jakobs and W.J. Koros. Ceramic membrane characterization via the bubble point technique. *J. Membrane Sci.*, 124:149–159, 1997.

[25] F.A.L. Dullien. *Porous Media, Fluid Transport and Structure*. Academic Press, New York, 1992.

[26] R.J. Good and R.R. Stromberg. *Surf. Colloid Sci.*, Vol. 11, p. 1, 1979.

[27] H. Yasuda and J.T. Tsai. Pore size of microporous polymer membranes. *J. Appl. Polym. Sci.*, 18:805–819, 1974.

[28] Y. Shimizu, H. Akabane, A. Tanioka, K. Miyasaka, and K. Ishikawa. Effects of estension on the gas permeability of hard elastic polypropylene films. *J. Polym. Sci., Polym. Phys. Ed.*, 17:1495–1506, 1979.

[29] I. Cabasso, K.Q. Robert, E. Klein, and J.K. Smith. Porosity and pore size determination in polysulfone hollow fibers. *J. Appl. Polym. Sci.*, 21:1883–1900, 1977.

[30] A. Kakuta, M. Kuramoto, M. Ohno, H. Kushida, A. Tanioka, and K. Ishikawa. Freeze-dried cellulose acetate membrane fine-structure observation. *J. Polym. Sci., Polym. Chem. Ed.*, 18:3229–3243, 1980.

[31] F.W. Altena, H.A.M. Knoef, H. Heskamp, D. Bargeman, and C.A. Smolders. Some comments on the applicability of gas permeation methods to characterize porous membranes based on improved experimental accuracy and data handling. *J. Membrane Sci.*, 12:313–322, 1983.

[32] D.L. Meixner and P.N. Dyer. Characterization of the transport properties of microporous inorganic membranes. *J. Membrane Sci.*, 140:81–95, 1998.

[33] G. Capanelli, F. Vigo, and S. Munari. Ultrafiltration membranes—Characterization methods. *J. Membrane Sci.*, 15:289, 1983.

[34] F.P. Cuperus, D. Bargeman, and C.A. Smolders. Permoporometry. The determination of the size distribution of active pores in UF membranes. *J. Membrane Sci.*, 71:57–67, 1992.

[35] M.G. Katz and G. Baruch. New insights into the structure of microporous membranes obtained using a new pore size evaluation method. *Desalination*, 58:199–211, 1986.

[36] A. Mey-Marom and M.G. Katz. Measurement of active pore size distribution of microporous membranes—a new approach. *J. Membrane Sci.*, 27:119–130, 1986.

[37] F.P. Cuperus, D. Bargeman, and C.A. Smolders. Thermoporometry and Permoporometry Applied to UF-Membrane Characterization. *Proceedings of the Internation Workshop on Characterization of Ultrafiltration Membranes*, pp. 115–124, 1987.

[38] G.Z. Cao, J. Meijerink, H.W. Brinkman, and A.J. Burgraaf. Permoporometry study on the size distribution of active pores in porous ceramic membranes. *J. Membrane Sci.*, 83:221–235, 1993.

[39] A. Hazri and A.M. Vayda. A new computational technique for the analysis of diffusion permporometric data for asymmetric membranes. *J. Membrane Sci.*, 101:61–66, 1995.

[40] M. Brun, A. Lallemand, J.-F. Quinson, and C. Eyraud. A new method for the simultaneous determination of the size and the shape of pores: the thermoporometry. *Thermochim. Acta*, 21:59–88, 1977.

[41] C.A. Smolders and E. Vugteveen. New characterization methods for asymmetric ultrafiltration membranes. *Mater. Sci. Synth. Membr.*, 327–338, 1985.

[42] L. Zeman, G. Tkacik, and P. Le Parlouer. Pore volume distribution in UF membranes. *Mater. Sci. Synth. Membr.*, 339, 1985.

[43] L. Zeman, G. Tkacik, and P. Le Parlouer. Characterization of porous sublayers in UF membranes by thermoporometry. *J. Membrane Sci.*, 32:329–337, 1987.

[44] M. Wulff. Pore size determination by permoporometry using acetonitrile, *Thermochim. Acta*, 419:291–294, 2004.

[45] A. Carbonaro, R. Walczak, P.M. Calderale, and M. Ferrari. Nano-pore silicon membrane characterization by diffusion and electrical resistance. *J. Membrane Sci.*, 241:249–255, 2004.

[46] A. Szymczyk, B. Aoubiza, P. Fievet, and J. Pagetti. Electrokinetic phenomena in homogenous cylindrical pores. *J. Colloid Interf. Sci.*, 216:285–296, 1999.

[47] P. Fievet, A. Szymczyk, B. Aoubiza, and J. Pagetti. Evaluation of three methods for the characterization of the membrane-solution interface: streaming potential, membrane potential and electrolyte conductivity inside pores. *J. Membrane Sci.*, 168:87–100, 2000.

[48] A. Szymczyk, P. Fievet, B. Aoubiza, C. Simon, and J. Pagetti. An application of the space charge model to the electrolyte conductivity inside a charged microporous membrane. *J. Membrane Sci.*, 161:275–285, 1999.

[49] T.E. Gómez Álvarez-Arenas. Air coupled ultrasonic spectroscopy for the study of membrane filters. *J. Membrane Sci.*, 213:195–207, 2003.

[50] W. Sachse and Y. Pao. On the determination of phase and group velocity of dispersive waves in solids. *J. Appl. Phys.*, 49(8):4320–4327, 1978.

[51] N.F. Haines, J.C. Bell, and P.J. McIntyre. The application of broadband ultrasonic spectroscopy to the study of layered media. *J. Acoust. Soc. Am.*, 64(6):1645–1651, 1978.

[52] R.A. Peterson, A.R. Greenberg, L.J. Bond, and W.B. Krantz. Use of ultrasonic TDR for real-time non-invasive measurement of compressive strain during membrane compaction. *Desalination*, 116:115–122, 1998.

[53] A.P. Mairal, A.R. Greenberg, W.B. Krantz, and L.J. Bond. Real-time measurement of inorganic fouling of RO desalination membranes using ultrasonic time-domain reflectometry. *J. Membrane Sci.*, 159:185–196, 1999.

[54] A.P. Mairal, A.R. Greenberg, and W.B. Krantz. Investigation of membrane fouling and cleaning using ultrasonic time-domain reflectometry. *Desalination*, 130:45–60, 2000.

[55] V.E. Reinsch, A.R. Greenberg, S.S. Kelley, R. Peterson, and L.J. Bond. A new technique fort he simultaneous, real-time measurement of membrane compaction and performance during exposure to high-pressure gas. *J. Membrane Sci.*, 171:217–228, 2000.

[56] J. Li, R.D. Sanderson, and E.P. Jacobs. Non-invasive visualization of the fouling of microfiltration membranes by ultrasonic time-domain reflectometry. *J. Membrane Sci.*, 201:17–29, 2002.

[57] M.E. Delany and E.N. Bazley. acoustical properties of fibrous absorbent materials. *Appl. Acoust.*, 3:105–116, 1970.

[58] W. Qunli. Empirical relations between acoustical properties and flow resistivity of porous plastic open-cell foam. *Appl. Acoust.*, 25:141–148, 1988.

[59] R.N. Chandler and D. Linton-Johnson. The equivalence of quasi-static flow in fluid-saturated porous media and the Biot's slow wave in the limit of zero frequency. *J. Appl. Phys.*, 52:3391–3395, 1981.

[60] D. Linton-Johnson, T. Plona, C. Scala, F. Pasieb, and H. Kojima. Tortuosity and acoustic slow waves. *Phys Rev. Lett.*, 49:1840–1844, 1982.

[61] D. Linton-Johnson, J. Koplik, and L-M- Schwartz. New pore-size parameter characterizing transport in porous media. *Phys Rev. Lett.*, 57:2564–2567, 1986.

[62] P. Nagy and L. Adler. Slow wave propagation in air-filled porous materials and in natural rocks. *Appl. Phys. Lett.*, 56:2504–2506, 1990.

[63] T. Schlief, J. Gross, and J. Fricke. Ultrasonic attenuation in silica aerogels. *J. Non-Cryst. Solids*, 145:223–226, 1992.

[64] J. Stor-Pellinen, E. Haeggström, and M. Luukkala. Measurement of paper-wetting processes by ultrasound transmission. *Meas. Sci. Technol.*, 1:406–411, 2000.

[65] O. Kedem and A. Katchalsky. Thermodynamic analysis of the permeability of biological membranes to nonelectrolytes. *Biochim. Biophys. Acta*, 27:229, 1958.

[66] K.S. Spiegler and O. Kedem. Thermodynamics of hyperfiltration (reverse osmosis): criteria for efficient membranes. *Desalination*, 1:311, 1966.

[67] N.A. Peppas and D.L. Meadows, Macromolecular structure and solute diffusion in membranes: an overview of recent theories. *J. Membrane Sci.*, 16:361–377, 1983.

[68] B.C. Robertson and A.L. Zydney. A Stefan-Maxwell analysis of protein transport in porous membranes. *Sep. Sci. Technol.*, 23:1799–1811, 1988.

[69] J.D. Ferry. Statistical evaluation of sieve constants in ultrafiltration, *J. Gen. Physiol.*, 20:95, 1936.

[70] E.M. Renkin. Filtration, diffusion and molecular sieving through porous cellulose membranes, *J. Gen. Physiol.*, 38:225, 1954.
[71] J.R. Pappenheimer, E.M. Renkin, and L.M. Borrero. Filtration, diffusion and molecular sieving through peripheral capillary membranes. *Am. J. Physiol.*, 167:12, 1951.
[72] J.R. Pappenheimer. Passage of molecules through capillary walls. *Am. J. Physiol.*, 22:387, 1953.
[73] R. Ladenburg. Über den Einfluβ von Wänden auf die Bewegung einer Kugel in einer reibenden Flüssigkeit. *Ann. Phys.*, (Leipzig), 23:447, 1907.
[74] H. Faxen. Die Bewegung einer starren Kugel längs der Achse eines mit zäher Flüssigkeit gefüllten Rohres. *Ark. Mat. Astron. Fys.*, 17:1, 1923.
[75] A.K. Solomon. Characterization of biological membranes by equivalent pores. *J. Gen. Physiol.*, 15:355, 1968.
[76] W.L. Haberman and R.M. Sayre. Motion of rigid and fluid spheres in stationary and moving liquids inside cylindrical tubes, David Taylor Model basin Report No. 1143, Department of the Navy, U.S., 1958.
[77] T. Bohlin. On the drag on a rigid sphere moving in a viscous liquid inside a cylindrical tube. *Trans. R. Insti. Technol.* Stockholm, 155, 1960.
[78] A. Verniory, R. Du Bois, P. Decoodt, J.P. Gassee, and P.P. Lambert. Measurement of the permeability of biological membranes. *J. Gen. Physiol.*, 62:489, 1973.
[79] F. Martin, R. Walczak, A. Boiarski, M. Cohen, T. West, C. Cosentino, and M. Ferrari. Tailoring width of microfabricated nano-channels to solute size can be used to control diffusion kinetics, accepted for publication on *J. Control. Rel.*
[80] C. Cosentino, F. Amato, A. Boiarski, and M. Ferrari. A dynamic model of biomolecules diffusion through two-dimensional nanochannels, submitted to *J. Phys. Chem. A.*
[81] W.H. Chu, R. Chin, T. Huen, and M. Fermi, Silicon membrane nanofilters from sacrificial oxide removal. *J. MicroelectroMech. Syst.*, 8(1): 1999.
[82] T.A. Desai, D.J. Hansford, L. Kulinsky, A.H. Nashat, G. Rasi, J. Tu, Y. Wang, M. Zhang, and M. Ferrari. Nanopore technology for biomedical applications. *J. Biomed. Microdev.*, 2(1):11–40, 1999.
[83] T.A. Desai, D. Hansford, and M. Ferrari. Characterization of micromachined membranes for immunoisolation and bioseparation applications. *J. Membrane Sci.*, 159:221–231, 1999.
[84] T. Meersmann, J.W. Logan, R. Simonutti, S. Caldarelli, A. Comotti, P. Sozzani, L.G. Kaiser, and A. Pines. Exploring Single-File Diffusion in One-Dimensional Nanochannels by Laser-Polarized ^{129}Xe NMR Spectroscopy. *J. Phys. Chem. A*, 104:11665–11670, 2000.
[85] Q. Wei, C. Bechinger, and P. Leiderer. Single-file diffusion of colloids in one-dimensional channels. *Science*, 287:625–627, 2000.
[86] V. Kukla, J. Kornatowski, D. Demuth, I. Girnus, H. Pfeifer, L.V.C. Rees, S. Schunk, K.K. Unger, and J. Karger. NMR studies of single-file diffusion in unidimensional channel zeolites. *Science*, 272:702–704, 1996.
[87] V. Gupta, S.S. Nivarthi, D. Keffer, A.V. McCormick, and H.T. Davis. Evidence of single-file diffusion in zeolites. *Science*, 274:164, 1996.
[88] K. Hahn, J. Karger, and V.V. Kukla. Single file-diffusion observation. *Phys. Rev. Lett.*, 76:2762–2765, 1996.
[89] P. Nelson and S. Auerbach. Self-diffusion in single file zeolite membranes is Fickian at long times. *J. Chem. Phys.*, 110:9235–9244, 1999.
[90] J.M.D. MacElroy and S.H. Suh. Self-diffusion in single-file pores of finite length. *J. Chem. Phys.*, 106:8595–8597, 1997.
[91] D.G. Levitt. Dynamics of a single-file pore: non-fickian behaviour. *Phys. Rev. A*, 8:3050–3054, 1973.
[92] B. Lin, J. Yu, and S.A. Rice. Direct measurements of constrained Brownian motion of an isolated sphere between two walls. *Phys. Rev. E*, 62:3909–3919, 2000.

8

Magnetic Nanoparticles for MR Imaging

Lee Josephson

Center for Molecular Imaging Research, Massachusetts General Hospital/Harvard Medical School, Building 149, 13th Street, Charlestown, MA 02129

8.1. INTRODUCTION

The combination of a nonmaterial, magnetic nanoparticles, with magnetic resonance imaging, is yielding major advances in diverse areas of biology and medicine. This review will present a short history of iron oxide based nanoparticles, and review important new developments in the fields of magnetic nanoparticles and MRI. Magnetic nanoparticles are currently used in approved MRI contrast agents for imaging hepatic metastases and show considerable potential in clinical testing for imaging nodal metastases. New applications of magnetic nanoparticles include (i) ex-vivo labeling of cells with nanoparticles, followed by MR imaging in vivo, (ii) magnetic nanoparticles as biosensors termed magnetic relaxation switches, to measure a wide range of analytes in vitro, (iii) magneto/optical nanoparticles providing a fluorescent signal in addition to their magnetic character and (iv) biomolecule targeted magnetic nanoparticles for the imaging of specific molecular targets by MRI. This review will cover each of these diverse developments.

8.2. A BRIEF HISTORY OF POLYMER COATED IRON OXIDE NANOPARTICLES AS PHARMACEUTICALS

Unlike many of the probes and chemistries spawned by the field now termed molecular imaging, magnetic nanoparticles are chemically similar to a large number of materials with a long history of clinical use. Magnetic nanoparticles consist of a core of superparamagnetic (highly magnetic) iron oxide and polymer coating, and to this basic design biomolecules

TABLE 8.1. Polymer coated iron oxides as pharmaceuticals.

Drug	Composition	Major Indication	Manufacturer/Status
InFeD® iron dextran	Paramagnetic iron/dextran	Anemia treatment	Watson Pharmaceuticals/ Approved US
Dexferrum® iron dexran	Paramagnetic iron/dextran	Anemia treatment	Amercan Reagent/ Approved US
Feridex IV® ferumoxides®	Superparamagnetic iron/ dextran	Suspected liver metastases	AMI, Berlex/ Approved EC, USA, Japan
Resovist® ferucarbotran®	Superparamagnetic iron/ carboxydextran	Suspected liver metastasis	Schering AG Approved EC & Japan
Gastromark® ferumoxsil®	Superparamagnetic iron/ silane	Bowel marker	AMI, Mallinckrodt/ Approved USA & EC
Combidex® Ferumoxtran	Superparamagnetic iron/ dextran	Suspected lymph node metastases	AMI, Cytogen/ Post Phase III trials
Supravist® SHU555C	Superparamagnetic iron/ carboxydextran	Suspected liver metastases	Schering AG/ In trials
Ferumoxtrol® AMI-228	Superparmagnetic iron/ carboxymethyldextran	Anemia treatment. MR angiography, lymph node mets	AMI/ Phase III (anemia)
Clariscan® feruglose® NC 10050	Superparamagnetic iron/ PEG Coating	MR angiography Blood pool imaging	Nycomed,Amersham/ Phase I/II trials

are attached. Polymer-coated iron oxides have been used in the treatment of iron anemias since the 1960's [1, 2]. Dextran has often been used as polymeric coating of the core iron oxide because of its prior use as a plasma expander and because of its affinity for iron oxides. The iron oxides used for anemia treatment are paramagnetic (weakly magnetic) and are injected in widely varying amounts depending on the severity of the anemia. The affinity between dextran and iron oxide was then used to develop a nanoparticle termed "dextran magnetite," which exhibited a far stronger form of magnetism than the paramagnetic iron used in the treatment of anemia, that is superparamagnetism. Since it was first recognized that superparamagnetic iron oxide nanoparticles could shorten water relaxation times, a large literature on the use of magnetic nanoparticles with MRI has developed. This recognition was far from obvious, since image distortions and spectral degradation were known to occur when many different types of magnetic materials occurred accidentally in the applied fields of MR imagers or MR spectrometers. One of the first to demonstrate the ability of magnetic nanoparticles to shorten the T2 relaxation times in a controlled and useful fashion was Oghushi, who in 1978 demonstrated this effect with a dextran-magnetite [3]. The field has advanced slowly but continuously, with magnetic nanoparticles forming the active ingredient of several approved drugs and others in clinical trials. Table 8.1 provides examples of polymer-coated iron oxides used clinically, both paramagnetic and superparamagnetic.

Magnetic nanoparticles can be considered to be of two basic types: polymer coated and molecularly targeted. Polymer coated magnetic nanoparticles are those that are not designed to be recognized by a specific biomolecule (receptor, antigen). "Molecularly targeted" magnetic nanoparticles are those that are designed to bind a specific target, typically by the attachment of a biomolecule (antibody, peptide, polysaccharide). It should be noted that polymer coated nanoparticles, though not designed to be recognized by a biomolecule, may nevertheless be recognized, either through weak interactions with slow internalization, or by reacting with a pre-existing material such as plasma opsonins, followed by recognition by a biological receptor [4]. Nevertheless this definition of materials according to the intent of its maker is useful in understanding the design of magnetic nanoparticles and their fate in vivo.

The first polymer-coated magnetic nanoparticles used as MR contrast agents were polydisperse (size heterogeneous) superparamagnetic iron oxides that were rapidly cleared from the blood by macrophages of the reticuloendothelial system (RES), principally in the liver and spleen [5–7]. With T2 weighted images, these materials darken normal liver tissue preferentially over cancerous lesions, and therefore serve as contrast enhancing agents to detect primary or metastatic hepatic tumors. The approved drugs Feridex IV and Resovist work in this manner [8, 9]. It was soon recognized that by decreasing nanoparticle size, blood half-life could be extended, and associated with the longer blood half-life was an increased uptake by the macrophages of lymph nodes, which had a very low uptake when large nanoparticles were injected [9–11]. The decrease in size was achieved by using nanoparticles that were monodisperse (one crystal per nanoparticle in solution); earlier preparations of superparamagnetic iron oxides were polydisperse (consisted of nanoparticles bearing 1, 2, or more crystals per particle in solution) [10, 12]. This change in biodistribution, which reflects a change in the physical properties of the nanoparticle, other than the attachment of an antibody or peptide to achieve binding to a molecular target, might be termed "nonspecific, physical property-mediated, pharmacokinetic targeting." The realization that monodisperse nanoparticles had a far longer blood half-life and more uptake by lymph nodes has led to extensive clinical trials indicating magnetic nanoparticles darkened normal, but not metastatic lymph nodes in a highly useful manner [13, 14]. Long circulating, monodisperse, polymer-coated superparamagnetic iron oxides include MION (clinical form Combidex), AMI-228 (ferumoxtol), NC 10050 and SHU555C. The development of magnetic nanoparticles is represented in Figure 8.1, which includes only the major materials used as MR contrast agents.

A wide variety of superparamagnetic particles and nanoparticles have been described and tested in animals, including lipid [15], and PEG coated [16, 17] materials. All rely on superparamagnetic iron oxide crystals (3–10 nm in diameter) for their magnetic properties and T2 shortening effects. The most widely used polymer-coated magnetic nanoparticles (Feridex IV, Resovist, Combidex, AMI-228/ferumoxytol) feature a core superparamagnetic iron oxide and a polymeric coating of a dextran or a dextran derivative.

As shown in Figure 8.1 a nanoparticle termed amino-CLIO (amine functionalized, CrosslLinked Iron Oxide) was developed by crosslinking and aminating the dextran coating of MION (ref), and has been used in diverse applications discussed below.

An important area of current clinical research with monodisperse, polymer-coated magnetic nanoparticles is their use to detect mm-sized occult nodal metastases in non-enlarged lymph nodes [13]. The method may have a particularly great impact in prostatic cancer, where the nodes at risk of metastasis are poorly accessible, and where nodal spread can lead an avoidance of prostatectomy, since there is no evidence that prostatectomy

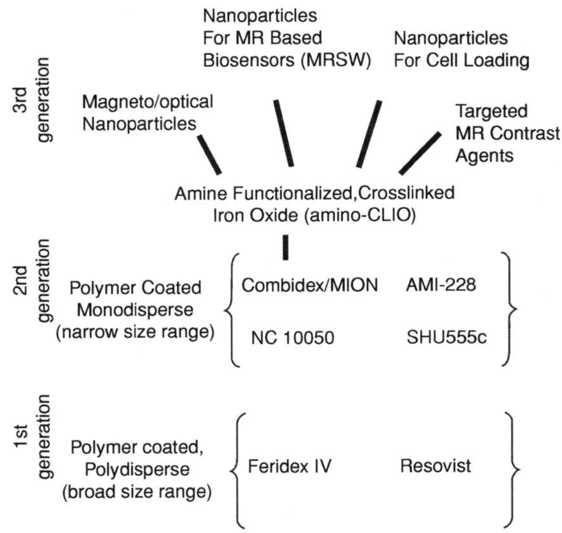

FIGURE 8.1. **History of magnetic nanoparticle development.** Early polymer coated nanoparticles were recognized to be polydisperse, which led to narrower sizes range monodisperse materials now in the clinic. Combidex/MION can be crosslinked and aminated to provide amino-CLIO used in a variety of applications.

with nodal involvement is of therapeutic benefit. The method relies on imaging nodal macrophages in lymph nodes and detecting alterations when metastatic spread occurs. In one trial, the noninvasive method demonstrated sensitivities and specificities of over 95% for different types of primary cancers. Additional advances of the enhanced MR technique include simultaneous measurements of tumor burden, angiogenesis and local host response, all being important predictors of clinical outcomes. Sophisticated image analysis is now being applied to improve visualization of nodal metastases including automatic detection of lymph nodes using autosegmentation, automated statistical analysis of tissue parameters, and 3D rendering in context of anatomical landmarks. These tools are particularly useful in planning surgical cancer resection and planning of intensity modulated radiation therapy.

8.3. MAGNETO/OPTICAL NANOPARTICLES AS OPTICAL PROBES

A simple but important modification of polymer-coated magnetic nanoparticles is afforded by the attachment of a near infrared fluorochrome to the polymeric coating [18–20]. When the fluorochrome Cy5.5 is attached to amino-CLIO, the resulting nanoparticle, denoted CLIO-Cy5.5, is similar to materials in clinical trials in its metabolism [21], but the near infrared fluorochrome can be used to monitor nanoparticle disposition by FACS, fluorescent microscopy or small animal optical imaging. An interesting potential clinical application for magneto/optical nanoparticles is in the area of brain tumors, where it has been shown that the Combidex/MION nanoparticles are internalized when the blood brain barrier is disrupted [22]. CLIO-Cy5.5 can serve as a pre-contrast MR agent or as an intraoperative optical agent as shown in Figure 8.2. The ability to "see" the same nanoparticle

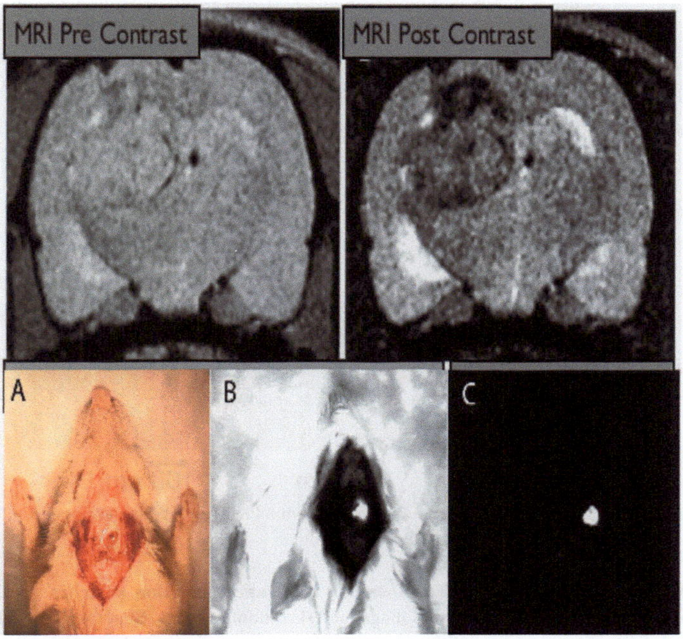

FIGURE 8.2. **Magneto/optical Nanoparticles as an MRI contrast agent and optical probe.** Top: Pre and post MR images of a rat bearing a 9L gliosarcoma tumor stably transfected to express GFP. The tumor darkens due to uptake of the magneto/nanoparticle (CLIO-Cy5.5) by microglia in tumor. Bottom; (A) White light image of tumor, cranium removed. (B) GFP fluorescence showing tumor margins. (C) NIRF image showing tumor margins. Figure from [22].

both in a pre-contrast MR image and intra-operatively may help the surgeon, since the tumor may shift position with surgery and the relationship between the pre-contrast MR and presentation of the tumor during surgery become unclear. In part because of these issues prototype intra-operative MR imagers have been designed.

8.4. MAGNETIC NANOPARTICLES AS BIOSENSORS

When multivalent magnetic nanoparticles bind to multivalent targets in solution they form stable nanoassemblies, and associated with nanoassembly formation is an increase in R2 relaxivity and decrease in the spin-spin relaxation time (T2) of surrounding water molecules [23], see Figure 8.3. Nanoassemblies can be disassembled and returned to their original dispersed state by a number of methods (heat, enzyme cleavage, disulfide bond reduction). Thus nanoparticles switch between a dispersed and nanoassembled states and are termed "magnetic relaxation switches (MRSW). MRSW are unique as a biosensor system because they do not employ a biomolecule immobilized on a solid phase, and because they use radiofrequency radiation at the Larmour precession frequency of water protons rather than light. Therefore they are homogeneous type assays (no separation of bound and free), and can be performed in media that absorb, scatter or fluoresce when they interact with

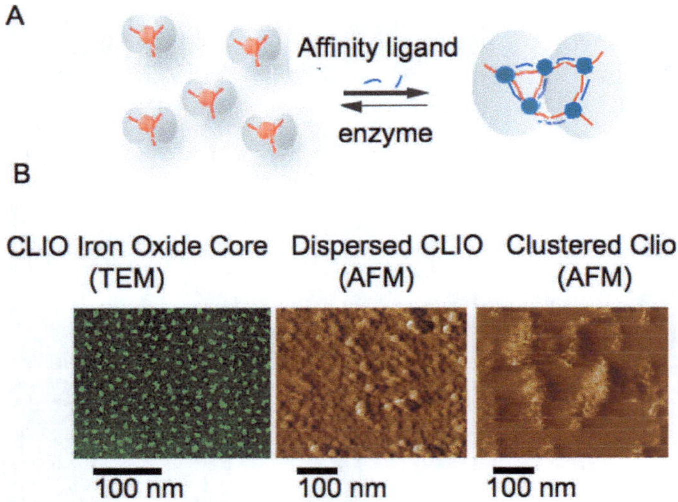

*Reprinted with permission from ChemBioChem 5:261—264, Wiley, Weinheim, Germany (2004)

FIGURE 8.3. **An Nanoparticle MRI based biosensor, magnetic relaxation switches.** (A) Diagram of the magnetic nanoparticles (nanosensors) acting as magnetic relaxation switches. Superparamagnetic nanoparticles self-assemble in the presence of a target with a corresponding decrease in the solution T2 relaxation time. Self-assembled nanoparticles can be dispersed by the action of an enzyme, temperature or pH change depending on the nature of the bond holding the nanoassembly together. (B) Transmission electron microscopy (TEM) and atomic force microscopy (AFM) images of a monolayer of nanosensors. TEM shows the iron oxide crystal cores with an average diameter of 8 nm. AFM of a similar monolayer shows the cross-linked dextran shell of nanoparticles, with an average particle size of 50 nm. Upon target recognition, CLIO forms nanoassemblies of 200–300 nm by reaction with targets in solution. Figure from [26].

incident light. MRSW have now been used as the basis sensitive, homogenous assays to detect a variety of different molecular interactions in biological samples with minimal or no sample preparation. Detection thresholds for DNA and proteins at the low femtomole level (0.5 fmol) [24], but with large high multivalent targets like viruses as few as five viral particles per 10 uL of herpes simplex virus-1 (HSV-1) and adenovirus-5 (ADV-5) have been detected in serum solutions [25]. The assay format is highly flexible and has been used to hydrolytic enzyme activities like proteases, endonucleases, DNA methylases and peroxidases. For a recent review see [26].

8.5. MAGNETIC NANOPARTICLES FOR CELL LOADING AND TRACKING BY MRI

The ability to track therapeutic entities in vivo, whether the entity be as small as an organic molecule (molecular weight less than 1000 daltons) or as large as a cell (diameter greater than 1 micron), is essential for development of all new therapies. The ability to label cells ex-vivo and track them may advance four broad areas of cell based-therapy.

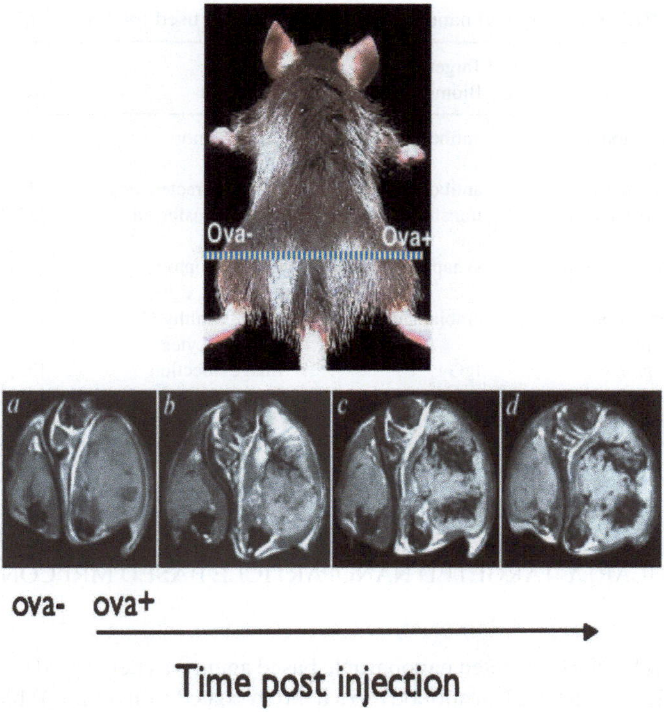

FIGURE 8.4. **Time course of Tat-CLIO loaded T cell homing to B16-OVA tumor.** Serial MR imaging was performed after adoptive transfer into a mouse carrying both B16F0 (*left*) and B16-OVA (*right*) melanomas. *A–D*, axial slices through the mouse thighs at *A*, before adoptive transfer; *B*, 12 h; *C*, 16 h; *D*, 36 h after adoptive transfer of cells. Figure from [29].

There are (i) stem cell therapy (bone marrow reconstitution after IV administration), (ii) cell implantation therapy (differentiation followed by recovery of function in damaged areas), (iii) immunopotentiation therapies for cancer treatment, (iv) transfusion related therapies (platelets, red blood cells, etc). Loading cells ex-vivo with magnetic nanoparticles and monitoring their disposition by MRI is attractive because cells loaded with nanoparticles will introduce small amounts of iron compared to the amounts injected with MR contrast agents, and because of the clinical availability and high spatial resolution of MRI. Only cells that phagocytize nanoparticles, internalize nanoparticles appreciably in vitro. A nanoparticle for loading cells that are not phagocytic was synthesized by attaching the membrane translocating sequence of the tat protein of the HIV virus to amino-CLIO nanoparticles [27]. The nanoparticle, tat-CLIO has been used to label stem cells [28] and, more recently cytotoxic T- lymphocytes [29]. Figure 8.4 shows the attack of Tat-CLIO loaded cytotoxic T cells (sensitized to ovalbumin), on a tumor (expressing ovalbumin). When the dose of T cells was split, and administered at different time points, cells attacked completely different segments of the tumor, suggesting dose division may enhance the potency of T cell based immunopotentiation therapies clinically. The study is an example of how cell labeling and tracking may lead to improved therapies.

TABLE 8.2. Targeted nanoparticle contrast agents used for MR imaging.

Target	Targeting Biomolecule	Comment	Reference
Neuroblastoma antigen	antibody	Image tumor	[30]
Cardiac myosin	antibody	Image infracted area	[31]
Transferrin receptor	transferrin	Image transfected gene	[37, 38]
Phosphatidyl serine	synaptotagmin	Image apoptotic cells	[42]
Asialoglycoprotein Receptor	arabinogalactan	Image healthy hepatocytes	[43]
Fc receptor	IgG	Image infection	[32]
Platelet RGD Receptor	RGD	Thrombus	[44]

8.6. MOLECULARLY TARGETED NANOPARTICLE BASED MRI CONTRAST AGENTS

The first molecularly targeted nanoparticle based agent was developed by Renshaw and coworkers [30], who attached an antibody to a dextran coated iron oxide. Table 8.2 provides selective examples of the large literature on targeted nanoparticle based MR contrast agents where the molecular target was imaged in vivo. References where a molecular target was imaged solely using cell based phantoms are excluded.

A wide variety of conjugation chemistries have been used for the attachment of biomolecules to superparamagnetic iron oxide nanoparticles. One of the first and most commonly used is the periodate oxidation of dextran, followed by formation of Schiff base between the amine of the biomolecule and aldehyde of the dextran, followed by borohydride reduction to stabilize the Schiff base [31, 32]. This approach to attaching biomolecules to nanoparticles was described earlier by Molday, who synthesized antibody-conjugated nanoparticles as probes for transmission electron microscopy [33]. A second approach for the attachment of biomolecules crosslinks and aminates the dextran of MION, a dextran coated iron oxide, a procedure that stabilizes the coating around the iron oxide and provides an amino group. (Dextran contains only hydroxyl groups.) The resulting material is known as amino-CLIO (amino-CrossLinked Iron Oxide). Thiol bearing biomolecules can then be attached to the amino groups of amino-CLIO through a wide variety of commercially available bifunctional agents in a site specific manner [23, 24, 27].

A targeted imaging modality must be able to detect molecular targets present in low concentrations in tissues. Targeted magnetic nanoparticles are highly detectable by MRI because they contain thousands of iron atoms per nanoparticle and because the relaxivities per mole of metal are substantially higher than gadolinium chelates [12, 34]. Magnetic nanoparticles can have relaxivities higher than 150 $(\text{mM sec})^{-1}$ [35] and are composed of two thousand iron atoms per nanoparticle [36]. Thus the per particle R2 relaxivity of magnetic nanoparticles can be 300,000 $(\text{mM sec})^{-1}$. Thus nanoparticles alter the signal intensity of nanoparticle sensitive T2 or T2* weighted pulse sequences at nanomolar and subnanomolar concentrations of nanoparticle.

The transferrin receptor provides an excellent example of a molecule to which magnetic nanoparticles have been targeted. Transferrin, an 80 kDa plasma protein to which ferric ions are bound, is internalized by receptor-mediated endocytosis, and provides a mechanism for supplying the cell with iron. Since the rodent and the human transferrin/transferrin receptor systems exhibit a high degree of species specificity, the expression of the human transferrin receptor can be imaged in rats or mice using human transferrin based magnetic nanoparticles [37–39]. The gene for human transferrin receptor can be incorporated into gene therapy vectors and its expression used as marker of successful gene therapy. Here the uptake of transferrin-magnetic nanoparticles imaged by MRI provides a non-invasive method of monitoring the efficacy of gene therapy, see Figure 8.3.

8.7. THE FUTURE

Polymer coated magnetic nanoparticles and targeted magnetic nanoparticles have several key advantages as platforms for the design of the next generation of nanoparticle based diagnostic pharmaceuticals. As noted (Table 8.1) many pharmaceuticals employ a similar design to molecularly targeted nanoparticles, which use a core of iron oxide and a polymeric coating. While every new drug must undergo a toxicological evaluation before clinical trials can commence, a large body of clinical and pre-clinical knowledge will aid the design, manufacture and formulation of the next generation of magnetic nanoparticle based pharmaceuticals. In this regard it is of interest to compare magneto/optical magnetic nanoparticles with a different type of fluorescent nanoparticle, the quantum dot, which has been the subject of recent studies [40, 41]. Magneto/optical polymer coated and targeted nanoparticles are multimodal (optical detection and MR detection), while the quantum dot is only detectable optically. Quantum dots usually contain cadmium, a metal that is an established carcinogen in humans due to its ability to cause cancer in workers involved with battery manufacture. Magneto/optical nanoparticles, whether polymer coated or targeted, are composed of materials, (iron oxides, polymers, peptides, fluorochromes) which have long histories of use in safe, injectable pharmaceuticals Magnetic nanoparticles technology is well positioned to play an important role in the development of the next generation of targeted imaging agents.

REFERENCES

[1] D.L. Burns, E.A. Mascioli, and B.R. Bistrian. Parenteral iron dextran therapy: a review. *Nutrition*, 11(2):163–168, 1995.

[2] S.T. Callender. Treatment of iron deficiency. *Clin. Haematol.*, 11(2):327–338, 1982.

[3] M. Ohgushi, K. Nagayama, and A. Wada. Dextran-magnetite: a new relaxation reagent and its application to T2 measurements in gel systems. *J. Mag. Res.*, (1969–1992), 29(3):599–601, 1978.

[4] A. Moore, R. Weissleder, and A. Bogdanov. Jr., Uptake of dextran-coated monocrystalline iron oxides in tumor cells and macrophages. *J. Magn. Reson. Imaging*, 7(6):1140–1145, 1997.

[5] S. Saini et al. Ferrite particles: a superparamagnetic MR contrast agent for the reticuloendothelial system. *Radiology*, 162(1 Pt 1):211–216, 1987.

[6] A. Hemmingsson et al. Relaxation enhancement of the dog liver and spleen by biodegradable superparamagnetic particles in proton magnetic resonance imaging. *Acta. Radiol.*, 28(6):703–705, 1987.
[7] J.T. Ferrucci and D.D. Stark. Iron oxide-enhanced MR imaging of the liver and spleen: review of the first 5 years. *AJR Am. J. Roentgenol.*, 155(5):943–950, 1990.
[8] A. Chachuat and B. Bonnemain. European clinical experience with Endorem. A new contrast agent for liver MRI in 1000 patients. *Radiologe.*, 35(11 Suppl 2):S274–S276, 1995.
[9] P. Reimer and B. Tombach. Hepatic MRI with SPIO: detection and characterization of focal liver lesions. *Eur. Radiol.*, 8(7):1198–1204, 1998.
[10] R. Weissleder et al. Ultrasmall superparamagnetic iron oxide: an intravenous contrast agent for assessing lymph nodes with MR imaging. *Radiology*, 175(2):494–498, 1990.
[11] R. Weissleder et al. Ultrasmall superparamagnetic iron oxide: characterization of a new class of contrast agents for MR imaging. *Radiology*, 175(2):489–493, 1990.
[12] C.W. Jung and P. Jacobs. Physical and chemical properties of superparamagnetic iron oxide MR contrast agents: ferumoxides, ferumoxtran, ferumoxsil. *Magn. Reso. Imaging*, 13(5):661–674, 1995.
[13] M.G. Harisinghani et al. Noninvasive detection of clinically occult lymph-node metastases in prostate cancer. *N. Engl. J. Med.*, 348(25):2491–2499, 2003.
[14] R. Sigal et al. Lymph node metastases from head and neck squamous cell carcinoma: MR imaging with ultrasmall superparamagnetic iron oxide particles (Sinerem MR)—results of a phase-III multicenter clinical trial. *Eur. Radiol.*, 12(5):1104–1113, 2002.
[15] J.W. Bulte and M. De Cuyper. Magnetoliposomes as contrast agents. *Methods Enzymol*, 373:175–198, 2003.
[16] L. Illum et al. Development of systems for targeting the regional lymph nodes for diagnostic imaging: in vivo behaviour of colloidal PEG-coated magnetite nanospheres in the rat following interstitial administration. *Pharm. Res.*, 18(5):640–645, 2001.
[17] *NC 100150. (Clariscan, PEG-Ferron). Drugs R D*, 3(5):303–304, 2002.
[18] L. Josephson et al. Near-Infrared fluorescent nanoparticles as combined MR/optical imaging probes. *Bioconjug. Chem.*, 13(3):554–560, 2002.
[19] M.F. Kircher, L. Josephson, and R. Weissleder. Ratio imaging of enzyme activity using dual wavelength optical reporters. *Mol. Imaging*, 1(2):89–95, 2002.
[20] M.F. Kircher, R. Weissleder, and L. Josephson. A dual fluorochrome probe for imaging proteases. *Bioconjug. Chem.*, 15(2):242–248, 2004.
[21] P. Wunderbaldinger, L. Josephson, and R. Weissleder. Tat Peptide Directs Enhanced Clearance and Hepatic Permeability of Magnetic Nanoparticles. *Bioconjug. Chem.*, 13(2):264–268, 2002.
[22] M.F. Kircher et al. A multimodal nanoparticle for preoperative magnetic resonance imaging and intraoperative optical brain tumor delineation. *Cancer. Res.*, 63(23):8122–8125, 2003.
[23] L. Josephson, J.M. Perez, and R. Weissleder. Magnetic nanosensors for the detection of oligonucleotide sequences. *Angewandte Chemie, Internat. Ed.*, 40(17):3204–3206, 2001.
[24] J.M. Perez et al. Magnetic relaxation switches capable of sensing molecular interactions. *Nat. Biotechnol.*, 20(8):816–820, 2002.
[25] J.M. Perez et al. Viral-induced self-assembly of magnetic nanoparticles allows the detection of viral particles in biological media. *J. Am. Chem. Soc.*, 125(34):10192–10193, 2003.
[26] J.M. Perez, L. Josephson, and R. Weissleder. Use of magnetic nanoparticles as nanosensors to probe for molecular interactions. *Chembiochem*, 5(3):261–264, 2004.
[27] L. Josephson et al. High-efficiency intracellular magnetic labeling with novel superparamagnetic-Tat peptide conjugates. *Bioconjug. Chem.*, 10(2):186–191, 1999.
[28] M. Lewin et al. Tat peptide-derivatized magnetic nanoparticles allow in vivo tracking and recovery of progenitor cells. *Nat. Biotechnol.*, 18(4):410–414, 2000.
[29] M.F. Kircher et al. In vivo high resolution three-dimensional imaging of antigen-specific cytotoxic T-lymphocyte trafficking to tumors. *Cancer. Res.*, 63(20):6838–6886, 2003.
[30] P.F. Renshaw et al. Immunospecific NMR contrast agents. *Magn Reson Imaging*, 4(4):351–357, 1986.
[31] R. Weissleder et al. Antimyosin-labeled monocrystalline iron oxide allows detection of myocardial infarct: MR antibody imaging. *Radiology*, 182(2):381–385, 1992.
[32] R. Weissleder et al. Polyclonal human immunoglobulin G labeled with polymeric iron oxide: antibody MR imaging. *Radiology*, 181(1):245–249, 1991.
[33] R.S. Molday and D. MacKenzie. Immunospecific ferromagnetic iron-dextran reagents for the labeling and magnetic separation of cells. *J. Immunol. Methods*, 52(3):353–367, 1982.

[34] L. Josephson et al. The effects of iron oxides on proton relaxivity. *Magn. Reso. Imaging*, 6(6):647–653, 1988.
[35] D. Hogemann et al. High throughput magnetic resonance imaging for evaluating targeted nanoparticle probes. *Bioconjug. Chem.*, 13(1):116–121, 2002.
[36] T. Shen et al. Monocrystalline iron oxide nanocompounds (MION): physicochemical properties. *Magn. Reso. Med.*, 29(5):599–604, 1993.
[37] M. Kresse et al. Targeting of ultrasmall superparamagnetic iron oxide (USPIO) particles to tumor cells in vivo by using transferrin receptor pathways. *Magn. Reso. Med.*, 40(2):236–242, 1998.
[38] R. Weissleder et al. In vivo magnetic resonance imaging of transgene expression. *Nat. Med.*, 6(3):351–355, 2000.
[39] J.W. Bulte et al. Neurotransplantation of magnetically labeled oligodendrocyte progenitors: magnetic resonance tracking of cell migration and myelination. *Proc. Natl. Acad. Sci. U.S.A.*, 96(26):15256–15261, 1999.
[40] X. Gao et al. In vivo cancer targeting and imaging with semiconductor quantum dots. *Nat. Biotechnol.*, 22(8):969–976, 2004.
[41] W.C. Chan et al. Luminescent quantum dots for multiplexed biological detection and imaging. *Curr. Opin. Biotechnol.*, 13(1):40–46, 2002.
[42] M. Zhao et al. Non-invasive detection of apoptosis using magnetic resonance imaging and a targeted contrast agent. *Nat. Med.*, 7(11):1241–1244, 2001.
[43] L. Josephson et al. A functionalized superparamagnetic iron oxide colloid as a receptor directed MR contrast agent. *Magn. Reso. Imaging*, 8(5):637–646, 1990.
[44] L.O. Johansson et al. A targeted contrast agent for magnetic resonance imaging of thrombus: implications of spatial resolution. *J. Magn. Reso. Imaging*, 13(4):615–618, 2001.

9

Polymer Design for Nonviral Gene Delivery

Kam W. Leong

Department of Biomedical Engineering, Johns Hopkins School of Medicine, Baltimore, MD 21205

9.1. INTRODUCTION

Gene therapy continues to hold promise in treating a variety of inherited and acquired diseases. The great majority of gene therapy trials rely on viral vectors for gene transduction because of their high efficiency. Viruses remain the vectors of choice in achieving high efficiency of gene transfer in vivo. Viral vectors, however, pose safety concerns unlikely to abate in the near future [1–3]. Issues of immunogenicity and toxicity remain a challenge. Limitations of cell mitosis for retrovirus, contamination of adenovirus, and packaging constraints of adeno-associated virus (AAV) also lessen their appeal. Non-viral vectors, although achieving only transient and lower gene expression level, may be able to compete on potential advantages of ease of synthesis, low immune response, and unrestricted plasmid size [4–9]. They have the potential to be administered repeatedly with minimal host immune response. They can also satisfy many of the pharmaceutical issues better than the viral vectors, such as scale-up, storage stability, and quality control. However, non-viral gene delivery is still too inefficient to be therapeutic for many applications. Development of safe and effective non-viral gene carriers is still critical to the ultimate success of gene therapy.

The most promising non-viral gene delivery systems, other than the "gene gun" in DNA vaccine applications, are comprised of ionic complexes formed between DNA and polycationic liposomes [6, 10, 11]. Polycation polymers have been increasingly proposed as potential vectors because of their versatility. Rigidity, hydrophobicity/ hydrophilicity, charge density, biodegradability, and molecular weight of the polymer chain are all parameters that in principle can be adjusted to achieve an optimal complexation with DNA. It is likely that for different cells or tissues, or different routes of administration in vivo, the desirable characteristics of the DNA nanoparticles, or polyplexes, would differ. Polymeric carriers with their versatility are well positioned to meet the challenges.

Design of new polymeric gene carriers has therefore been an intensively pursued research area in recent years. It would be beyond the scope of this review to cover all the gene carrier designs. Rapid advance of the field also necessitates coverage of only recent findings. Given the disparate conditions between in vitro and in vivo, cell culture studies accompanied by in vivo experiments generally produce more valuable insights. Conclusions derived solely from cell culture studies about the potency of a certain gene carrier may have little relevance in vivo. Nevertheless, mechanistic studies are mostly limited to the reproducible in vitro conditions. This review will therefore concentrate on the progress of innovation in polymeric gene carriers in the past few years.

9.1.1. Barriers for Nonviral Gene Transfer

Before the discussion of the different polymer designs, it would be informative to first highlight the potential barriers of nonviral gene transfer. We will also discuss the general principles of incorporating features into a polymer to overcome these barriers. With reference to Figure 9.1, many obstacles have to be overcome for an exogenous gene to reach the nucleus of the cell from the extracellular space.

1. Condensation of the plasmid DNA. The DNA must be condensed to a size permitting internalization by the cells. The charge density and the ionicity of the polymer is therefore an important parameter. The optimal size of the polymer-DNA complex (polyplex) remains debatable. Conventional wisdom favors smaller sizes, below 150 nm. This is also the general threshold below which receptor-mediated endocytosis is believed to be operative. Larger sizes, beyond 1 μm, may facilitate phagocytosis by professional immune cells for DNA vaccination applications. Contact of the polyplexes with biological fluid would easily lead to aggregation, which may severely limit the ability of the polyplexes to be endocytosed and reaching the final site, the nucleus. Minimizing aggregation of the polyplexes is therefore an important design consideration. Incorporation of hydrophilic segments

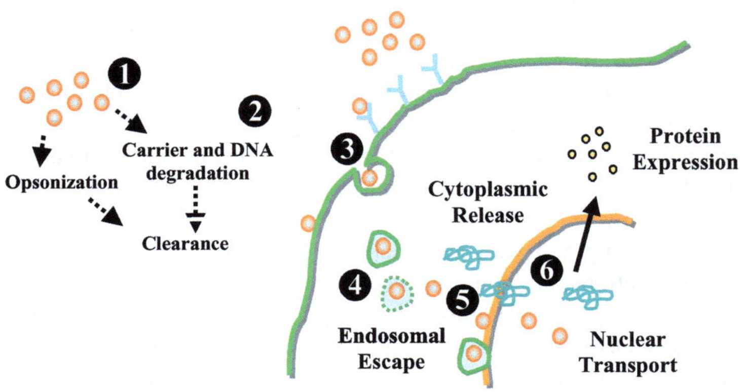

FIGURE 9.1. Barriers for nonviral gene transfer.

into the polyplexes, such as conjugation of PEG to the polymer, which shows benefit of reducing nonspecific interactions with plasma proteins and lessening complement system activation, is a popular and effective strategy.

2. *Transport of polyplex to cell.* The task of the polyplex is to reach the cell surface without being cleared from the system and to protect the DNA from degradation by nucleases. Most polyplexes are formed by electrostatic interaction. They must remain stable in the extracellular fluid where a high concentration of polyelectrolytes might disrupt this interaction. Polymers that can bind the DNA tightly are therefore preferred. Excessive net positive charge on the surface however tends to lead to aggregation and coating by negatively charged proteins such as albumin. Strong cationicity of the polymer is nevertheless needed to produce a compact polyplex structure that can impede enzymatic degradation of the embedded DNA.

3. *Internalization by cell.* Once the nanoparticles reach the target cells, they bind with the negatively charged cell membrane and are subsequently endocytosed, by mechanisms specific or non-specific. Achieving gene transfer to the target cells is highly desirable. Cell-specific targeting ligands, such as monoclonal antibodies, peptides and sugars can be conjugated to the gene carriers to promote receptor-mediated endocytosis. Polymers that contain functional groups amenable to ligand conjugation would be desirable. The conjugation step can be executed pre- or post-complex formation. The former allows for fine control of the degree of substitution, but the presence of the ligand, such as a polypeptide, may interfere with the complexation. The ligand may also be buried in the polyplex and not presented on the surface properly. Post-complexation conjugation would unquestionably decorate the polyplex surface. The stability of the polyplex must survive the reaction, and often the need to isolate the reacted polyplex from unreacted ligands would significantly diminish the yield of the polyplex.

4. *Escape from the endosome.* After the non-specific or receptor-mediated endocytosis, most polyplexes are sequestered in the endosomal compartment, which at some point fuses with the lysosomes. Potent enzymes and low pH in these compartments either degrade or swell the cationic polymer, freeing the DNA, and also breaking down the DNA. Some anionic lipids in the endosome may also compete with DNA to bind with the cationic polymer, dissociating the complex and somehow release the DNA into the cytoplasm [12]. Escape of the polyplex or the DNA into the cytoplasm is a significant step in the gene transfer process. Efficient destabilization of the endosomes/lysosomes by endosomolytic reagents such as chloroquine [13], lipids [14], and peptides [15] would enhance the gene transfer efficiency. Polymer that can possess this endosomal disruptive property has figured prominently in the design consideration [16, 17]. This can take the form of lipophilic sidechain, or attachment of peptides to the sidechains to interact specifically with the vesicle membrane. Alternatively, polymers with excessive amino groups have been hypothesized to reduce the acidification of the endosome, eventually causing endosomal swelling and collapse [18]. This "proton sponge" effect has been first postulated for the potency of polyethylenimine in gene delivery. According to this hypothesis, PEI and PEI-DNA complexes sequestered into the endosome absorb protons that are pumped into this organelle. Protonation of the amine groups on PEI leads to swelling of the polymer, concomitant with an influx of chloride ions to neutralize the build-up of a charge gradient. The net effect is an increase in osmotic pressure to destabilize and rupture the endosome, leading to release of the endosomal content into the cytoplasm. Many polymers have been synthesized to mimic the pH buffering

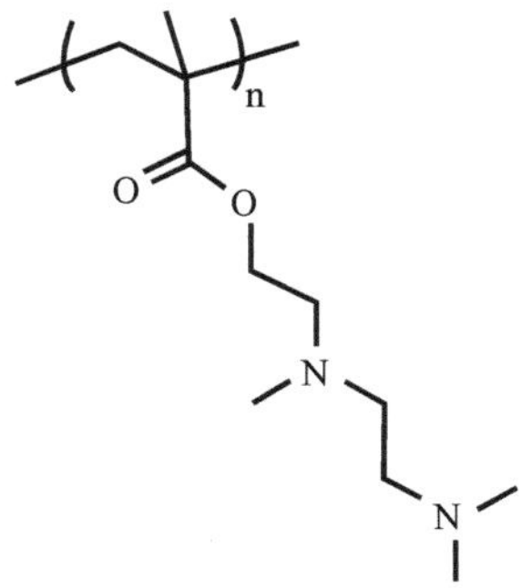

FIGURE 9.2. Chemical structure of poly(2-methyl-acrylic acid 2-[(2-(dimethylamino)-ethyl)-methyl-amino]-ethyl ester).

capacity of PEI for this reason. However, the beneficial effects of buffering capacity must be weighed against the other colloidal properties of the polyplex. A number of studies have shown that polymers mimicking the buffering behavior of PEI do not always produce similar enhancement effects. For instance, a polymer with two tertiary amine groups in each monomeric unit poly(2-methyl-acrylic acid 2-[(2-(dimethylamino)-ethyl)-methyl-amino]-ethyl ester) (Figure 9.2) [19], does not improve in vitro transfection despite showing potent DNA-condensing ability and buffering capacity within the pH range of 5–7. Lack of endosomal escape is identified as one of the reasons for the poor transfection efficiency of these polyplexes.

5. Transport to the peri-nuclear space. Polyplex escaped from the vesicles into the cytoplasm intact can diffuse to the nuclear membrane. There is also evidence the polyplex-containing vesicles can be transported to the nuclear membrane via the molecular motors. The property of the polymer that would accelerate this active and passive diffusion process is currently unknown. The polyplex can also be dissociated to release the DNA, which is degraded in both the endosome and the cytoplasm. Microinjected plasmid DNA is rapidly degraded in the cytoplasm with an apparent half-life of 50–90 min [20]. The translocation of unpacked DNA from cytoplasm to nucleus is probably through diffusion, a relatively slow process compared to cytoplasmic degradation, which leads to an inefficient nuclear transport. It is understandable that DNA would be more stable in the cytoplasm if it were still complexed with the cationic polymer. However, only certain synthetic polymers, such as PEI, show protection of DNA in the cytoplasm and promote nuclear transport. The pathway of the transport of polyplex from cytoplasm to nucleus is not well understood, with some hypothesis leaning towards the involvement of anionic phospholipids.

6. Translocation into nucleus. There is evidence that intact polyplex can be present in the nucleus of even non-dividing cells. Again this phenomenon remains poorly understood. If the property of the polyplex that promotes this translocation can be identified, it would be a breakthrough for nonviral gene transfer. It is generally believed that it is the unpacked DNA that translocates into the nucleus for ultimate gene expression. The inclusion of viral nuclear localization signals (NLS) has been demonstrated to be an efficient strategy to facilitate nuclear transport [21, 22]. NLS are generally peptide or protein sequences that are recognized by the cellular machinery, which can complex with DNA to promote nuclear accumulation. It is not clear if attachment of NLS to the polyplex can efficiently transport the polyplex into the nucleus. The same characteristics of polymer that allows attachment of ligands can be used for the conjugation of NLS.

It becomes clear even with this over-simplified picture delineating the barriers for gene transfer that the correlations between physicochemical properties of the polyplex and their transfection potency would be multifactorial. It would be challenging to design polymers incorporating all the desirable features into a single structure, particularly in the absence of biological guidance. For instance, stable complexes would prevent DNA release while unstable ones would be vulnerable to rapid DNA degradation. Both are undesirable. Polyplexes with an intermediate stability may exhibit the best transfection efficiency. But how to strike the balance remains a trial-and-error process. While this review will focus on polymer design, the structural considerations must also be considered in the context of formulation, which can substantially alter the efficacy of the resulting polyplex. Constraint by the scope of this review, the formulation aspect is only briefly covered when appropriate, but the importance of formulation must be kept in mind. Important parameters include the molar ratio of polycation nitrogen to DNA phosphate (N/P ratio), the order of addition of the polyelectrolytes, the mode of mixing, and the addition of excipients.

9.2. SYNTHETIC POLYMERIC GENE CARRIERS

9.2.1. Polyethyleneimine

Polyethylenimine (PEI) is one of the most potent synthetic gene carriers and the most widely studied [23–27]. It can mediate gene transfer in a variety of cells in vitro and tissues in vivo. PEI is available in linear and branched forms. The most commonly studied structure is the 25KD branched PEI, with a combination of primary, secondary, and tertiary amines (Figure 9.3). Every third atom on PEI is a primary amino nitrogen, which can be protonated at weakly acidic pH. As a result of this high density of amines, the high transfection efficiency of PEI has been postulated to relate to its buffering capacity and the "proton sponge" effect described above. The molecular weight of PEI is an important factor in gene delivery efficiency. PEI with molecular weight below 2000 Da is essentially ineffective [28, 29]. There are varying reports of the effects of increasing PEI size, but polymers >10 kDa

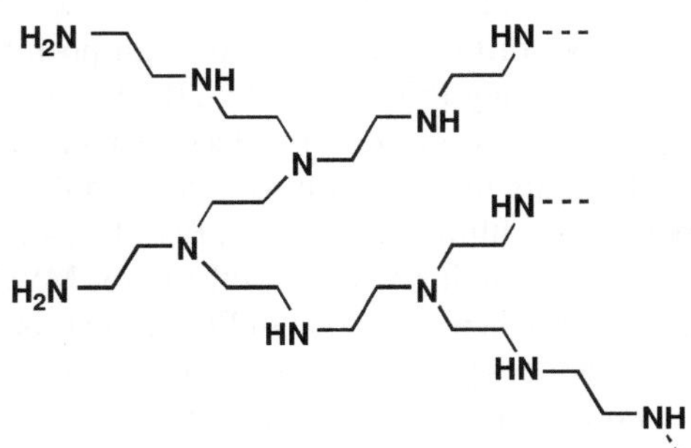

FIGURE 9.3. Chemical structure of branched polyethylenimine (PEI).

appear to be effective. Although early studies suggest the greater potency of the branched structure [30], recent studies have compared linear vs. branched PEI and concluded that the linear polymer can be as effective as the branched PEI for transfection *in vitro* and *in vivo* [31, 32]. Toxicity is one of the main concerns for PEI, particularly for in vivo delivery. The cytotoxicity is postulated to derive from its ability to permeabilize cell membranes. Moreover, PEI is not degradable and high molecular weight PEI may accumulate in the body. Many approaches have therefore been explored to reduce the cytotoxicity, such as PEGylation [33] and conjugation of low molecular weight PEI with cleavable cross-links such as disulfide bonds [34]. In one approach, reaction of low molecular weight PEI with difunctional PEG succinimidyl succinate (SS-PEG) produces a copolymer with biodegradable ester linkages in the main backbone. By carefully controlling the reaction conditions it is possible to generate water soluble PEI-PEG network before the gelation point [35]. These polymers show lower toxicity than the 25-kDa branched PEI and three times more efficient than the low molecular weight PEI starting material. However, the transfection efficiency is inferior to the 25-kDa PEI. In another approach, degradable PEI derivatives are synthesized by reacting 800-Da PEI with 1,3-butanediol diacrylate to generate the ester-cross-linked polymers, resulting in a highly branched structure [28]. The resulting polymers are degradable and similar in size and DNA-binding properties to the standard 25-kDa PEI, but less toxic and can mediate gene expression two- to 16-fold more efficiently.

A different approach to generate biodegradable PEI is to link low molecular weight PEI (MW 1,200) with an oligomer of lactic acid and succinic acid [36]. Instead of relying on cleavage of the ester or disulfide bonds in the other designs, the biodegradability of this copolymer derives from the hydrolytic cleavage of the amide bonds, which is significantly slower than the ester bonds. With the hydrolysis of this polymer even lower at acidic pH, where nucleases exhibit maximal activity, the hypothesized advantage of this gene carrier is that the polyplex would be more stable and protective against nuclease degradation. While the cytotoxicity profile is favorable, the low transfection efficiency of this polymer compared to the standard 25-KD branched PEI highlights the fact that dissociation of the polyplex is an important rate-limiting step and suggests that this polymer might have been too stable.

Accessibility of the primary amino groups of PEI to convenient covalent attachment has been extensively exploited for optimizing the characteristics of the complexes, from lowering the charge density, improving cytotoxicity, reducing aggregation to enhancing cellular uptake via receptor-mediated endocytosis. In a study systematically examining the effect of PEG conjugation on PEI, grafting many short PEG blocks (550D) onto PEI (25 KD) forms larger and more diffuse complexes of high positive surface charge [37]. Copolymers with only a few but long PEG blocks (5–20 KD) in contrast would condense DNA into small and compact complexes of lower surface charge. Increasing the substitution of long PEG blocks on PEI leads to irregularly-shaped complexes of near-neutral zeta-potential. The graft copolymers with more than six PEG blocks exhibit low cytotoxicity. The transfection efficiency correspondingly varies, with the low MW PEG-grafted PEI being the most potent. In another approach, methylation of PEI (25 kDa branched) converts all the charged groups to quaternary amines [24]. Acylation using pendant palmitic acid chains and attachment of PEG chains produce amphiphilic PEI derivatives that can efficiently condense DNA. The modification improves biocompatibility markedly over PEI but also reduces the transfection efficiency in vitro and in vivo. In contrast, acetylation of PEI, which reduces the buffering capacity of PEI, results in enhanced transfection up to 21-fold compared to unmodified PEI, both in the presence and absence of serum [29]. This demonstrates the imperfect understanding of the mechanism of the effectiveness of PEI. The improvement has been speculated to derive from more effective polyplex unpackaging, altered endocytic trafficking, and/or increased lipophilicity of the acetylated PEI-DNA complexes.

Pluronic (P123) has been grafted to PEI (2K) to improve the potency of the polyplex, demonstrated in higher transgene expression in liver when delivered intravenously and compared to 25 KD branched PEI [38]. The P123-g-PEI(2K) system requires unmodified Pluronic P123 as an excipient to stabilize the dispersion, the function of which appears to be optimizing the size of the polyplex. Conjugation of cholesterol to branched PEI of different molecular weights via the secondary amines generates water-soluble lipopolymers that show higher in vitro transfection efficiency than PEI, attributed to altered buffering capacity of PEI and increased cellular uptake due to the presence of the cholesterol [39, 40].

Evidence of the benefit of modifying PEI with PEG has been demonstrated in vivo. In one study, a conjugate with one PEG chain per PEI molecule (2 KDa) is able to mediate transgene expression in the spinal cord up to 11-fold higher than PEI homopolymer after intrathecal administration of the polyplex into the lumbar spinal cord subarachnoid space [41]. The benefit of this PEG grafting is hypothesized to come from smaller complexes, lower degree of aggregation, and reduced interaction with serum components. In a follow-up study, the feasibility of achieving prolonged transgene expression in the rat spinal cord through repeated intrathecal administration of this polyplex into the lumbar subarachnoid space is demonstrated [42]. With a single injection, PEI polyplexes produce transgene expression in the spinal cord 40-fold higher than that of naked DNA. The transgene expression at the initial level persists for about 5 days, with a low-level expression still detectable at week 8. With repeated dosing at a 2-week interval, the gene expression is attenuated by 70%. Using the PEG-grafted PEI polyplexes, no attenuation of gene expression is observed after repeated intrathecal injections administered 2 weeks apart, even in those rats receiving three doses. Given that lumbar puncture is a routine and relatively nontraumatic clinical procedure, this PEG-grafted PEI polyplex has been proposed as a viable delivery system for the gene therapy of spinal cord disorders.

Sugars such as galactose and mannose have been attached to PEI for increasing targeting efficiency to hepatocytes and macrophages, respectively. Epidermal growth factor (EGF) attached to PEI via PEG has been shown to significantly improve transfection efficiency in relevant tumor cell lines with increases of 10 to 100 folds compared to PEGylated complexes without EGF [23]. Similar approach of using PEG as a spacer to conjugate the antigen binding fragment (Fab') of the OV-TL16 antibody to target human ovarian carcinoma cell lines has also been successful [43]. Similar to other studies, presence of the PEG and antibody does not appear to interfere with the condensing ability of PEI. Transfection of the target cell lines is highly specific and shows an improvement of 80-fold in efficiency compared to PEG-PEI/DNA complexes. In another study, folate-PEG-folate-grafted-PEI (FPF-g-PEI) is superior to PEI alone in terms of cytotoxicity and transfection efficiency of IFN-γ gene against cancer cells [44, 45].

However, conversion of the amino groups of PEI would affect the properties of the resulting polyplex. With transfection efficiency multi-factorial as it is, improving cellular uptake may not necessarily improve transfection. A systematic study on examining the degree of substitution of galactose in PEI shows that only at low N/P ratios might there

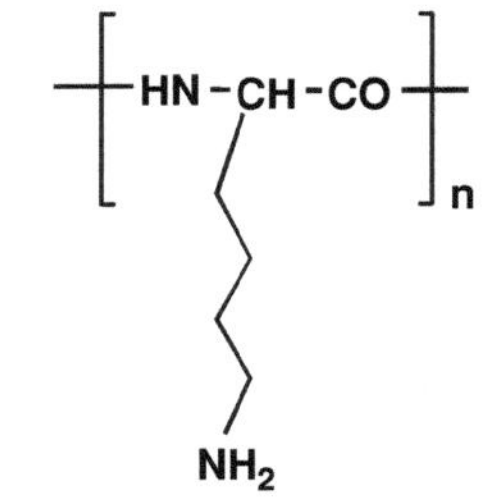

FIGURE 9.4. Chemical structure of poly(L-lysine) (PLL).

be a slight positive effect on transfecting asialoglycoprotein-expressing HepG2 cells. The efficiency is actually reduced at N/P ratios greater than 2–3 [26]. Larger size and less compaction of the nanoparticles have been hypothesized as the cause of this decline.

9.2.2. Polylysine

Poly-L-lysine (PLL) is one of the earliest investigated gene carriers because of its excellent DNA condensation ability and efficient protection of DNA from nuclease digestion (Figure 9.4). However, it is quite cytotoxic and transfects poorly both *in vitro* and *in vivo*. Similar to the approaches of modifying PEI, efforts have focused on reducing cytotoxicity and introducing biodegradability into PLL by copolymerization with PEG [46]. The block copolymer shows substantially lower toxicity, and biodegradation is observed in a decrease of 80% in molecular weight within 72 h. Resistance to inhibition by the serum components in transfection is also achieved. However, the transfection efficiency is still relatively low. Crosslinking PLL by disulfide bonds and surface modification with PEG improves the solubility of the polyplexes and shows a 10-fold increase in plasma circulation time following intravenous administration to Balb/c mice [47]. In modification through the sidechain instead of the backbone, covalent attachment of palmitoyl and methoxypoly(ethylene glycol) to PLL produces amphiphilic polymers that are one to 2 orders of magnitude less cytotoxic and can improve gene transfer to human tumor cell lines in comparison to the parent PLL [48]. Conjugation of imidazole as an endosomal escape moiety to the sidechains of PLL has been shown to increase transfection efficiency [49]. The polymer with a high imidazole content (86.5 mol %) can mediate a transgene expression comparable to that of PEI and with less cytotoxicity. A recent study shows that conjugation of receptor-associated protein (RAP) to poly-D-lysine (PDL) can significantly enhance transfection in HepG2 cells compared to nonconjugated PDL, RAP-PLL, and LipofectAMINE/DNA complexes in the presence of 10% fetal bovine serum [50]. The enhancement appears to be LDL receptor-mediated. A polymeric micelle, PLGA-g-PLL has also been proposed as an effective carrier for gene delivery [51].

9.2.3. Poly(α -(4-aminobutyl)-L-glycolic acid)

Poly(α-(4-aminobutyl)-L-glycolic acid) (PAGA) is an analogue of PLL with a degradable ester linkage in the main chain [52] (Figure 9.5). PAGA efficiently condenses and protects DNA, and shows enhanced transfection efficiency in culture compared to other

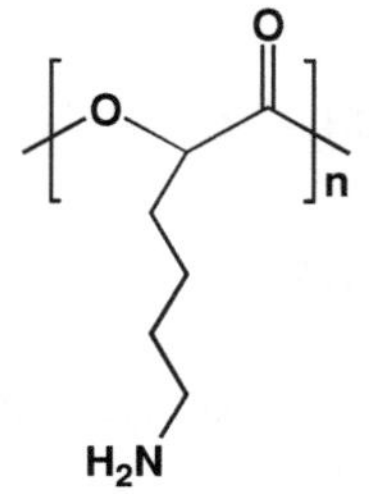

FIGURE 9.5. Chemical structure of poly(α-(4-aminobutyl)-L-glycolic acid) (PAGA).

PLL-based systems. It is also less cytotoxic than Lipofectamine and SuperFect (a dendrimer). PAGA generates positive results in producing cytokines as an immunoadjuvant in a murine cancer model [53]. Recent studies using these polyplexes to deliver cytokine genes also show promising results in therapy of cancer and autoimmune insulitis [54, 55].

9.2.4. Polyamidoamine Dendrimer

Starburst polyamidoamine dendrimers with either ammonia or ethylenediamine as core molecules have been used for gene delivery [56, 57] (Figure 9.6). These molecules are highly defined in terms of molecular weight and geometry. They are spherical molecules, with primary amine groups on the surface, which can be used to condense DNA. Dendrimers can transfect a wide variety of cells in culture, with efficiency matching that of PEI under optimal conditions. The higher generation dendrimers (G6 and G7) are effective at disrupting anionic

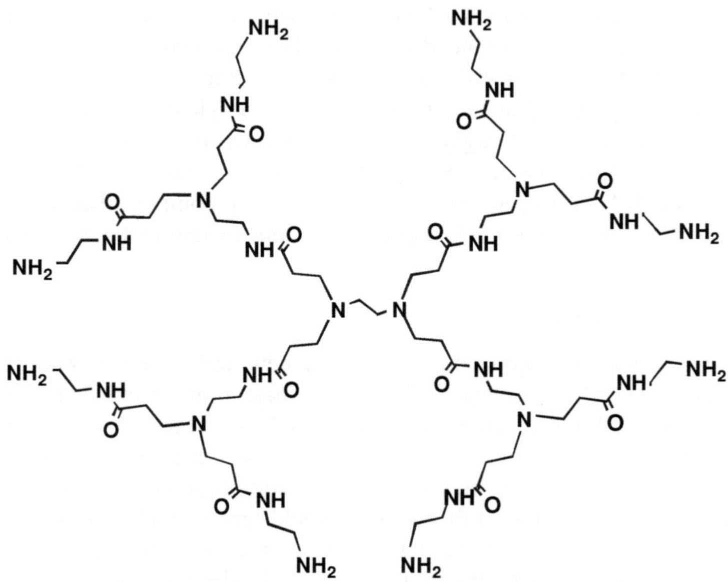

FIGURE 9.6. Chemical structure of polyamidoamine dendrimers.

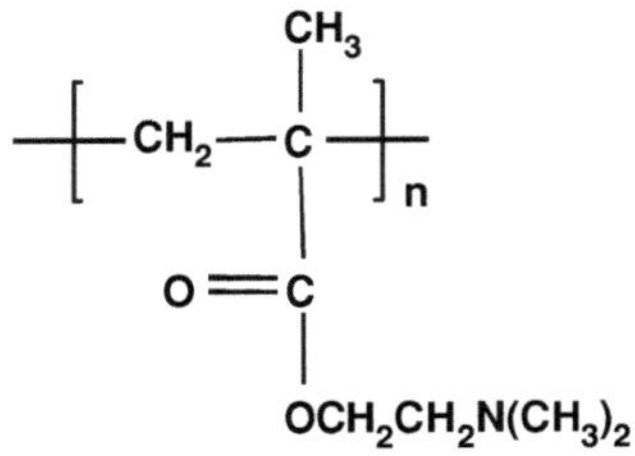

FIGURE 9.7. Chemical structure of poly((2-dimethylamino)ethyl methacrylate).

vesicle membranes, which may explain the high transfection efficiency [58]. However, the advantage of these dendrimers over PEI remains to be established. A recent study proposes a simple synthetic approach of producing dendritic polyamines with different molecular weights and adjustable degrees of branching by using hyperbranched PEI as the core [56]. Polyamidoamine (PAMAM) functionalization of PEI with MW of 5 and 21 KD produces polymers with little or no cytotoxicity in the cell lines investigated. In another effort to reduce toxicity, the interior tertiary amine groups of PAMAM-OH dendrimers are methylated to impart the cationicity [59]. This polymer can condense DNA efficiently and is less cytotoxic compared with starburst PAMAM and 25 KD branched PEI. Shielding of the interior positive charges by the surface hydroxyls is postulated to be the mechanism for the reduced toxicity.

9.2.5. Poly((2-dimethylamino)ethyl methacrylate)

Poly((2-dimethylamino)ethyl methacrylate) is a water-soluble cationic polymer capable of binding plasmid DNA and transfecting different cell types [60] (Figure 9.7). The high transfection efficiency of PDMAEMA is probably due to its high buffering capacity to destabilize endosomes, combined with its ability to release the DNA in the cytosol and/or the nucleus. Its strength and shortcomings are similar to those of PEI. A recent study links PEG with poly(2-(dimethylamino)ethyl methacrylate (DMAEMA)-co-N-vinyl-2-pyrrolidone (NVP)) and attaches galactose to the PEG terminal [61]. The polyplex shows a slightly negative surface charge, allowing the coating of a cationic, pH sensitive, endosomolytic peptide, KALA, to generate a positively charged complex. The ternary complex can produce a transgene expression level in HepG2 cells matching that of Lipofectamine.

9.2.6. Poly(β-amino ester)

One of the promising approaches to discover new and effective gene carriers is to use a combinatorial synthetic strategy [62, 63]. Poly(beta-amino esters) can be synthesized by the conjugation addition reaction of diamines and diacrylates (Figure 9.8). The wide variety of commercially available diamines and diacrylates facilitates parallel synthesis of a large number of polymers and study of their structure-property relationship. In general, small particle size and positive surface charge would lead to higher internalization rates of these polyplexes. The high-throughput, semi-automated methodology has been used to synthesize a library of 2350 structurally unique, degradable, cationic polymers [64].

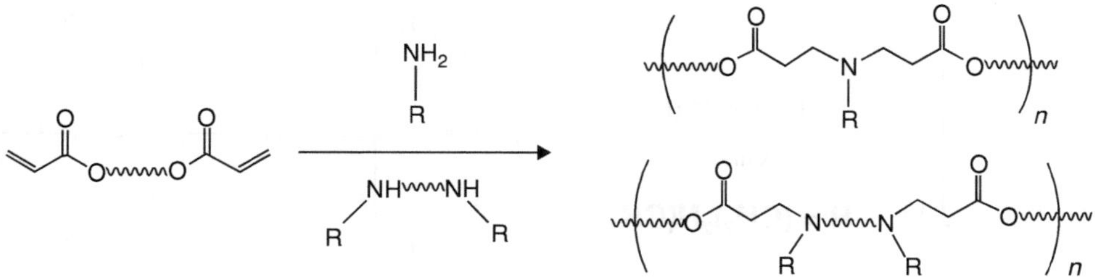

FIGURE 9.8. Synthetic scheme for poly(β-amino ester)s (PAE).

High-throughput, cell-based screening has identified 46 new polymers that can transfect with a higher efficiency than PEI. Interestingly, a study related to the PAE shows that in addition to factors such as chemical structure and polymer weight, the end-groups of the PAE, either amine- or acrylate-terminated, also play a significant role in determining the transfection efficiency [63]. In two of the PAE structures investigated, synthesized by reacting 1,4-butanediol diacrylate or 1,6-hexanediol diacrylate with 1-aminobutanol, only the amine- but not the acrylate-terminated end groups show significant transfection efficiency.

A similar class of polymers potentially amenable to the same combinatorial synthetic strategy has been proposed by the same group [65]. Based on the one-step reaction of bis(secondary amines) with diglycidyl ester of dicarboxylic acid, a number of poly(amino alcohol esters) can be synthesized from the commercially available library of diacids and diamines. The preliminary study on structures based on the reaction of N,N'-dimethyl-1,3-propanediamine and N,N'-dimethyl-1,6-hexanediamine with diglycidyl adipate has shown the desirable properties of DNA-binding ability, lower toxicity than PEI, and rapid degradation of the polymer ($t_{1/2} \sim 5$ days). However, the transfection efficiency has yet to be determined. A related polycation developed by a different group, named by the authors as network poly(amino ester) (n-PAE), has also shown comparable transfection efficiency to that of PEI, but with lower cytotoxicity [66]. The potential advantage of this polymer is postulated to be the enhanced stability over linear cationic polyesters. However, the biodegradability of this cationic network has not been reported.

9.2.7. *Polyphosphazene*

Polyphosphazenes can be biodegradable and can exhibit a wide range of physicochemical properties by varying the structures of its sidechains [67]. Polyphosphazenes

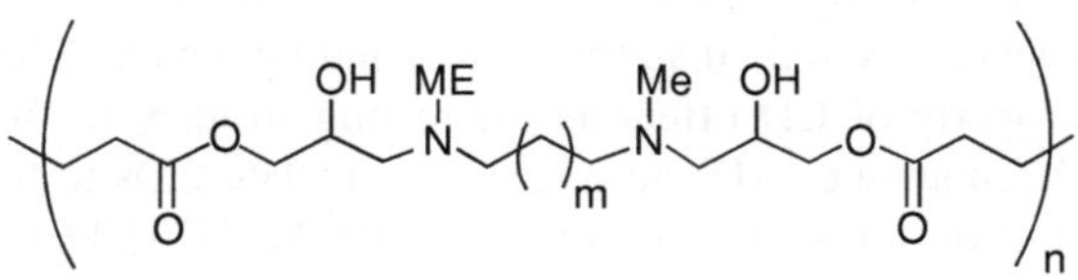

FIGURE 9.9. Chemical structure of poly(amino alcohol ester)s.

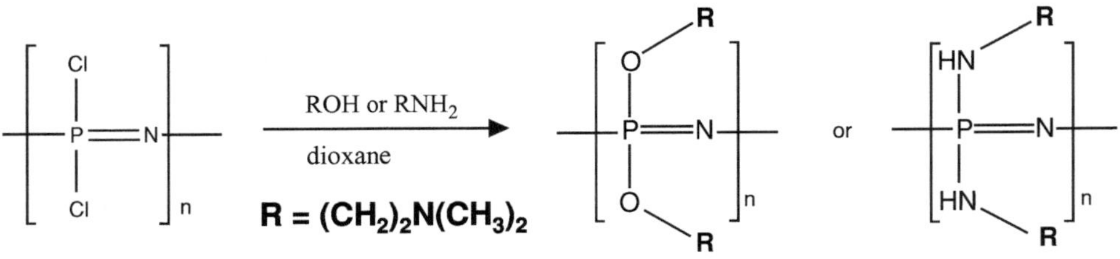

FIGURE 9.10. Synthetic scheme for polyphosphazenes.

with a 2-dimethylaminoethanol (DMAE) or 2-dimethylaminoethylamine (DMAEA) side group can bind DNA efficiently to form nanocomplexes (Figure 9.10). The half-life for the biodegradation of the two polymers at pH 7.5 and 37°C is 7 and 24 days, respectively. Both polyphosphazenes show comparable transfection efficiencies but lower toxicity compared to pDMAEMA. Only a tertiary amine structure can be used in the sidechain because the use of primary and secondary amines would lead to a crosslinked polyphosphazene in the substitution reaction of poly(dichloro)phosphazene. This would limit the flexibility of the polymer design as other studies have shown that primary amines, when the other structural parameters are kept constant, generally produce the highest transfection efficiency.

9.2.8. Cyclodextrin-containing Polycation

A class of polycations containing cyclodextrins (CDs) in the backbone has shown interesting gene properties [7, 68–73]. A recent excellent review covers the structure-property relationship of a large number of these polymers [74]. The linear water-soluble polymers are generally synthesized by the polycondensation of a difunctionalized CD monomer (either hydroxyl- or amine- derivatized) and a difunctionalized comonomer. A systematic study on a series of amidine-based polycations containing the carbohydrates d-trehalose and beta-cyclodextrin (CD) within the polycation backbone shows that the cytotoxicity is related to the carbohydrate size (trehalose vs. CD) and its distance from the charge centers. Further studies examine the charge type in the cationic segment. The quaternary ammonium analogues exhibit transfection efficiency and similar cytotoxicity compared to their amidine analogues [72]. Other structural parameters influencing the gene delivery efficiency are the type of CD and nature of the spacer between the CD-ring and the primary amines. The transfection efficiency as well as toxicity against BHK-21 cells increases with the length of the alkyl chain between the CD and the charge centers in the polycation backbone, while the toxicity decrease with the hydrophilicity of the spacer (alkoxy versus alkyl). Gamma-CD-based polycations are less toxic than the corresponding beta-CD-based polycations.

The inclusion property of CD offers an interesting strategy to functionalize or decorate the polyplex. Adamantane (AD) derivatives intercalate CDs to form inclusion complexes with association constants on the order of 10^4–10^5 M^{-1}[74]. PEGylated-adamantane can therefore bind to the polyplex and reduce aggregation, and ligands conjugated at the distal end of PEG can be used for targeting. This leads to a suitable formulation for systemic administration in vivo and galactosylated polyplexes can transfect HepG2 cells with 10-fold higher efficiency than glucosylated complexes (control). To take advantage of this surface modification scheme, β-CD is grafted to linear- and branched- PEI and combined with PEGylated adamantane [71]. These polyplexes show reduced cytotoxicity, improved tissue biocompatibility, and increased stability against aggregation in biological fluid. Tail-vein administration in mice shows low but positive transgene expression in the liver.

9.2.9. Polyphosphoester

The motivation for developing polyphosphoesters as gene carriers comes from the desire to overcome the barrier of releasing the DNA from the polyplex and to use this series of polymers for systematic mechanistic studies [75–81]. In our earlier work we have observed through intracellular trafficking studies that chitosan polyplexes may be too stable for releasing the DNA. A biodegradable polycation would ensure the "unpacking" of the polyplex. We have synthesized and studied two series of biodegradable polyphosphoesters, polyphosphates (PPEs) and polyphosphoramidates (PPAs) for gene delivery (Figure 9.11) [76–80]. Similar in rationale to the design of other polycationic gene carriers, the biodegradability of PPEs offers several advantages. It can provide a function of extracellular sustained release, and conceivably even intracellular sustained release. Such a gene delivery system

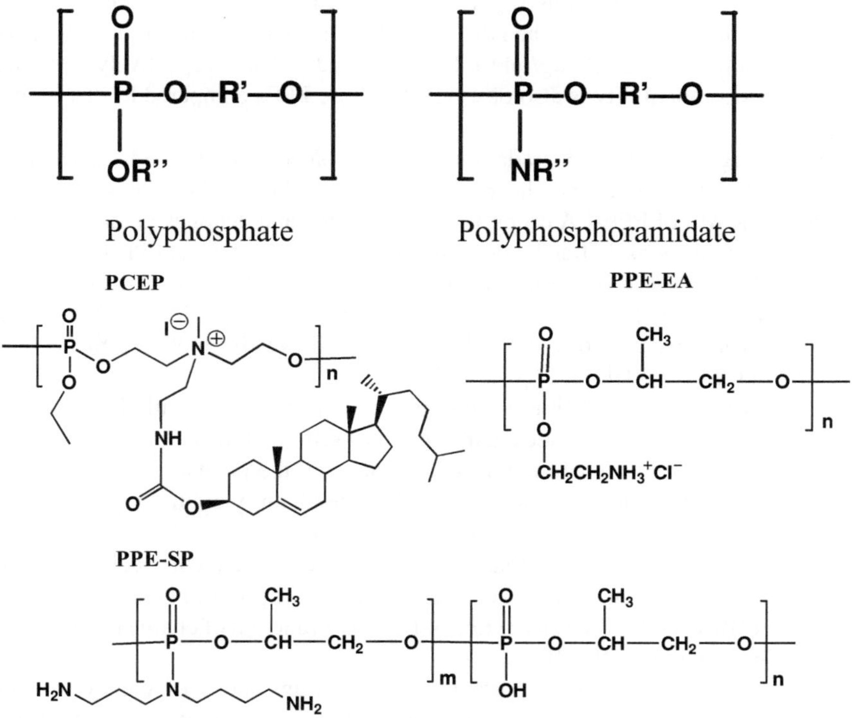

FIGURE 9.11. Chemical structure of polyphosphoesters.

can therefore substantially improve the bioavailability of DNA both inside and outside of cells. This controlled release property can be adjusted by varying the polycationic polymer and DNA ratio (N/P) ratio as well as the composition of the backbone and the sidechain structure, which in turn can also influence the transfection efficiency *in vivo*.

In contrast to the tetravalency of carbon atoms in polyesters, the pentavalency of a phosphorus atom in the backbone of polyphosphoesters makes it possible to conjugate functional groups, including charged groups through a phosphate (P-O) or a phosphoramide (P-N) bond as side chains. Starting from the parent polyphosphite, the P-H bonds can be readily converted for conjugation of different chemical structures. We have synthesized a series of cationic PPEs and PPAs from the precursor polymer poly(1,2-propylene H-phosphonate), which is obtained by ring opening polymerization of the cyclic monomer, 4-methyl-2-oxo-2-hydro-1,3,2-dioxaphospholane. This structure is chosen with consideration of biocompatibility. Comprised of propylene oxide in the backbone, the potential breakdown products should be relatively innocuous. Conjugation of charged groups to this parent polymer produces water-soluble PPEs or PPAs with a range of physicochemical properties. In keeping the chemical structure of the backbone and the molecular weight constant, we can then systematically study the effect of the charge groups on toxicity and transfection efficiency.

The cytotoxicity of the PPEs and PPAs are all lower than that of PLL and PEI. For instance, a PPE with an ethanolamine sidechain (PPE-EA) (Figure 9.11) would not affect the viability and morphology of HEK293 and COS-7 cells at a dose of up to 100 μg/ml. In contrast, the two most widely used polymeric gene carriers, poly-L-lysine (PLL, Mw 27 KDa) and polyethylenimine (branched PEI, Mw 25 KDa) exhibit a high toxicity with a LD_{50} below 40 μg/ml for PLL and 20 μg/ml for PEI to HEK 293 cells, respectively. Significantly better tissue biocompatibility is also observed for these PPEs compared to PEI in muscles of mice.

PPEs undergo hydrolytic degradation when incubated in PBS at 37°C, which is due to the hydrolytic cleavage of the phosphoester bonds in the backbone as well as the side chains [82]. The M_w of PPE-EA drops 12% in 24 hours, and declines gradually to 10,450 from 30,300 (65% decrease) after 7 days of incubation. After ten days, PPE-EA degrades to oligomers with a decreased net positive charge, which fails to condense plasmid DNA. PPE-HA (a hexylamine sidechain) degrades much slower in PBS. The Mw of PPE-HA drops only 50% in 20 days from 3.76×10^4 to 1.87×10^4, followed by another 20% decrease (to 1.48×10^4) in the next 30 days.

The relatively fast degradation of these polycations compared with that of polyphosphate with ethoxy groups in the side chain suggests a self-catalytic degradation mechanism involving nucleophilic attack of the phosphate bonds by the pendant amino groups. This mechanism also likely leads to a cleavage of the side chain, yielding negatively charged phosphate ions instead. Such a negatively charged backbone would shield further attack of OH^- on $P = O$, and result in a decreased degradation in subsequent incubation. Interestingly, PPE-EA polyplex can provide a controlled release for plasmid DNA over a period of two weeks in PBS at 37°C. The release rate of DNA is a function of charge ratio, decreasing with increasing N/P ratio [79, 82].

Polymer structure plays an important role in the transfection efficiency. Among the polyphophates examined, PPE-EA shows good cell transfection ability, but PPE-MEA (a methylethylamine sidechain) performs poorly [78]. This is probably due to the fact that

PPE-MEA is quickly dissociated from DNA in the media as the sidechain is cleaved rapidly. The transfection ability of PPA-DMA (a dimethylamine sidechain), which contains tertiary amino groups, is quite poor for all the cell types.

To further understand the influence of charge groups in transfection mediated by these polyplexes, we hypothesize that a combination of PPA-SP (a spermidine sidechain) (Figure 9.11) and PPA-DMA, with primary and tertiary amino groups, would impart stronger pH buffering capacity, and in turn enhance the transfection efficiency. Transfection of COS-7 cells using PPA-SP/PPA-DMA mixture shows that a mixture at certain ratios mediates significantly higher levels of gene expression than either polymer alone. Under optimal conditions (at a PPA-SP/PPA-DMA molar ratio of 4), the transfection efficiency achieves by PPA-SP/PPA-DMA mixture is 20 and 160 times higher than PPA-SP and PPA-DMA mediated transfection, respectively. The mechanism of enhancement in transfection efficiency by the mixture carriers appears to be unrelated to the particle size or zeta potential or DNA uptake, or lysosomal escape. This approach represents a simple strategy to enhance the transfection efficiency of polymeric gene carriers.

The *in vivo* transfection ability of PPAs has been evaluated using a therapeutic gene bcl-2 using a protocol for gene transfer to CNS neurons by peripheral intramuscular injection. This non-invasive approach achieves gene expression in the central nervous system after intramuscular injections of the PPA-SP and PPA-DMA polyplexes [83]. The controlled release characteristics of the PPEs is attractive for intramuscular gene transfer. The polyplexes may act as a depot for sustained and local DNA release at the injection site and potentially enhance and prolong the gene expression in the muscle cells. In our study, intramuscular administration of PPE-EA/p43-LacZ complexes results in significantly higher and delayed β-gal expression in muscle at low DNA doses [79]. At a dose of 2μg of DNA per muscle, complexes with an N/P ratio of 0.5 produces a 17-fold higher β-gal expression on day 7, and the level persists until at least day 14. In contrast, naked DNA injection shows a peak expression of 280 pg of β-gal per muscle at day 3, and the expression levels off at days 7 and 14. At high DNA doses, the enhancement effect becomes less significant. A similar improvement is also seen with the delivery of the interferon-α2b gene by the PPE-EA polyplexes. The efficacy of PPE-EA polyplexes has also been investigated in the central nervous system by intracisternal injection into the mouse cerebrospinal fluid [84]. Transgene expression mediated by naked DNA is mainly detected in the brain stem, a region close to the injection site. With either PPE-EA or PEI as a carrier, higher levels of gene expression could be detected in the cerebral cortex, basal ganglia, and diencephalons. Transgene expression in the brain mediated by PPE-EA polyplexes at an N/P ratio of 2 persists for at least 4 weeks, with a significant higher level than that produced by either naked plasmid DNA or PEI at the 4-week time point. Furthermore, PPE-EA displays much lower toxicity in cultured neural cells as compared to PEI and does not cause detectable pathological changes in the CNS.

To further understand the effect of the charge group residing in the backbone of the polyphosphoester instead of the sidechain, a poly[[(cholesteryl oxocarbonylamido ethyl) methyl bis(ethylene) ammonium iodide] ethyl phosphate] (PCEP) is synthesized [76] (Figure 9.11). Carrying a positive charge in its backbone and a lipophilic cholesterol structure in the side chain, PCEP self-assembles into micelles in aqueous buffer at room temperature with an average size of 60–100 nm. It is analogous to the polymeric version of a cationic lipid. Intramuscular administration of the polyplexes shows a luciferase expression in muscle increasing with time during 3 months, although the expression level is lower than that by direct injection of naked DNA. In addition to biodegradability and lower toxicity than PEI, the PCEP micelle carrier offers structural versatility. The backbone charge density and the side chain lipophilicity are two parameters that can be varied through copolymerization and monomer variation to optimize the transfection efficiency.

9.3. NATURAL POLYMERIC GENE CARRIERS

9.3.1. *Chitosan*

Chitosan is a biodegradable polysaccharide comprised of D-glucosamine repeating units and is extracted from crustacean shells (Figure 9.12). It has been shown to be non-toxic and with good soft tissue compatibility in humans. It has been widely used in pharmaceutical research as a carrier for drug delivery and as biomedical materials for artificial skin and wound healing applications. Chitosan in the soluble form has been shown to increase the trans and para cellular transport of drugs across mucosal epithelium. It has been investigated as a gene carrier in recent years [85–91]. Chitosan, with a degree of deacetylation above 60%, has the interesting property that it is insoluble at physiological pH, partly due to its high crystallinity. In acidic pH, the amino groups are protonated, rendering the polysaccharide soluble. Complexation with DNA can easily take place at pH 5.5 to form nanoparticles. Afterwards, upon resuspension in physiological pH, the chitosan-DNA nanoparticles are stable without any need for crosslinking. Ligands can be conjugated to the nanoparticle to stimulate receptor-mediated endocytosis and potentially to target specific cell/tissue, in a convenient post-complexation step after the nanoparticles are formed.

The properties of the chitosan polyplexes are sensitive to the molecular weight and degree of deacetylation (DA) of the polysaccharides. With a molecular weight of 390 KD and DA around 87% for the chitosan, the transfection efficiency of the polyplexes is cell-type dependent. Typically, it is three to four orders of magnitude, in relative light units, higher than background level in HEK293 cells, and two to ten times lower than that achieved by LipofectAMINE-DNA complexes [89]. The presence of 10% fetal bovine serum does not interfere with their transfection ability. Chloroquine can be co-encapsulated in the nanoparticles at 5.2%, but with negligible enhancement effect despite the fact that chitosan only shows limited buffering capacity compared with PEI. The feasibility and benefit of attaching transferrin, KNOB, and PEG on the surface of the polyplexes to improve the cellular uptake and storage stability have been demonstrated [89]. Another targeting system is the attachment of lactobionic acid (LA) bearing the galactose group to chitosan polyplexes. The

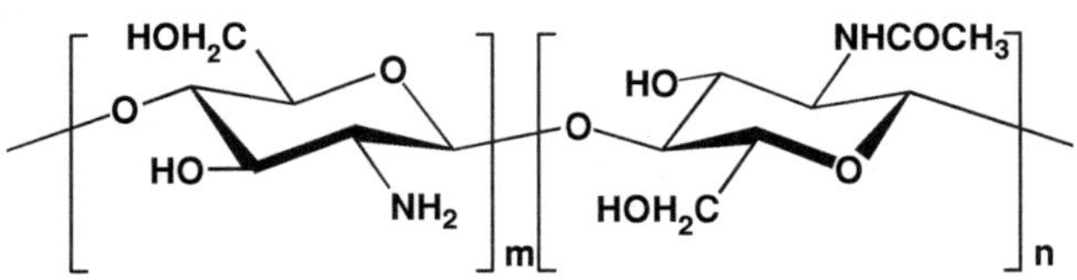

FIGURE 9.12. Chemical structure of chitosan.

resulting polyplexes show efficient targeting to hepatocytes and the transfection efficiency increases with the degree of galactosylation [92].

Intracellular studies indicate that these polyplexes are quite stable even after residing in HEK293 cells for over 24 hours. Release of the DNA is believed to be a significant rate-limiting step for these polyplexes. This is supported by findings that intramuscular administration of these polyplexes would produce poor transfection compared with naked DNA. A subsequent study evaluates the potential benefit of decreasing the stability of chitosan to enhance the unpacking of the polyplex [93]. Decreasing the degree of deacetylation of chitosan from 90 to 62% leads to a decrease in transfection efficiency against HEK293, HeLa, and SW756 cells due to particle destabilization in the presence of serum proteins. However, in vivo study of intramuscular injection in mice shows that the transgene expression increases with decreasing deacetylation over the time points tested.

A recent study examines the effect of Mw and DA of chitosan on cytotoxicity and cellular uptake of the polymer and the polyplexes [94]. Transformation of chitosan into nanoparticles does not modify the cytotoxicity profile of the polymer but facilitate their endocytosis by A549 cells. Uptake of the chitosan polyplexes is a saturable event for chitosan, and decreases with Mw ranging from 213 to 10 kDa and with DA ranging from 88% to 46%. This may also explain in part the lower in vitro transfection efficiency observed with chitosan polyplexes of lower DA in other studies.

To address the issue of poor unpacking characteristics of chitosan polyplexes, a recent approach relies on well-defined chitosan oligomers [95, 96]. Chitosan with MW ranging from 10 to 50-mers and with polydispersities of 1.01 to 1.09 produce polyplexes that can more readily dissociate to free the DNA than a high-MW ultrapure chitosan (UPC, approximately a 1000-mer). The more easily dissociated polyplexes mediate a faster onset of action and produce a higher gene expression both in HEK293 cells in vitro and after lung administration in vivo as compared to the more stable UPC polyplexes. A 120- to 260-fold higher luciferase gene expression can be observed compared to UPC in the lung tissue of mice after intratracheal administration. Another study modifies the oligomeric chitosan (<20 monomer units) by quarternization [97]. Trimethylated chitosan (TMO) is superior to oligomeric chitosan in transfection efficiency and with favorable cytotoxicity profile, but does not appear to be as potent compared to other chitosan polyplexes.

The in vivo efficacy of chitosan polyplexes has been demonstrated in other animal models. We have demonstrated that oral immunization using chitosan-DNA nanoparticles carrying the gene for a major peanut allergen, Arah2, can modulate Ag-induced hypersensitivity in a murine model of peanut allergy [90]. Four weeks after the first immunization, significant differences in the levels of secreted and serum antibodies can be observed between polyplex-immunized and control (PBS fed or naive) or naked DNA (pCMVArah2) immunized mice. Mice immunized with pCMVArah2-polyplexes show increased levels of secretory IgA in fecal extracts, indicating induction of mucosal immune response. In addition, a significant induction of serum anti-Arah2 IgG2a is observed in polyplex-treated animals, indicating a Th1 type T cell response. At all time points, mice receiving pCMVArah2 alone (naked DNA) without chitosan shows no detectable levels of either fecal IgA or serum IgG2a response. Mice receiving sensitization alone without prior immunization or treated with naked DNA (pCMVArah2) shows physical signs of anaphylaxis within 10–20 min of challenge. In contrast, significantly less severe and delayed anaphylactic responses are seen in chitosan-pCMVArah2-immunized animals following challenge.

In another study we have explored the use of chitosan polyplexes for oral immunisation to induce immune responses specific to both the left and right domains of Der p1 [98]. Der p 1 is the most important *Dermatophagoides pteronyssimus* allergen and is a major triggering factor for mite allergy worldwide. Intramuscular immunization with full-length *Der p 1* gene induces significant humoral response to the left domain but not to the right domain of Der p 1 allergen. Oral delivery of chitosan polyplexes followed by intramuscular boosting 13 weeks later manages to prime Th1-skewed immune responses against both domains of Der p1. Chitosan polyplex has also been investigated as a prophylactic intranasal gene transfer strategy to deliver a cocktail of plasmid DNAs encoding RSV antigens [99]. A single administration of these polyplexes induces expression of the mRNA and proteins of all RSV antigens in the lung and results in a significant reduction of viral titers and viral antigen load after acute RSV infection of these mice. The polyplex-treated mice show no significant change in airway reactivity to methacholine and no apparent pulmonary inflammation. There is also significant induction of RSV-specific IgG antibodies, nasal IgA antibodies, cytotoxic T lymphocytes, and interferon-gamma production in the lung and splenocytes compared with controls. In a related study, chitosan polyplexes delivering the IFN-gamma gene has been tested for their potential to modulate ovalbumin (OVA)-induced inflammation and airway inflammation and hyperresponsiveness (AHR) [100]. Mice treated with the chitosan polyplexes exhibit significantly lower AHR to methacholine challenge and less lung histopathology. Production of IFN-gamma is increased after polyplex administration while the Th2-cytokines, IL-4 and IL-5, and OVA-specific serum IgE are reduced compared to control mice. The study suggests that mucosal delivery of these chitosan IFN-gamma-polyplexes may reduce established allergen-induced airway inflammation and AHR.

Other cationic gene carriers based on natural biopolymers are gelatin and collagen [91, 101–103]. Gelatin as the denatured form of collagen carries positive charge at acidic pH, where complexation with DNA can take place. Unlike chitosan, however, the resulting polyplex is not stable at physiological pH, requiring crosslinking of the gelatin and compromising the quality of the polyplex in terms of size, degradability, and aggregation. Collagen in its native form also is not cationic enough to bind DNA efficiently. Methylation of the carboxyl groups of collagen raises the pKa of the polypeptide and enables the modified collagen to form complexes with DNA in PBS. However, both transfection of cell culture and the muscle produce poor results [103]. Failure to protect the complexed DNA from enzymatic degradation is postulated as one of the reasons for this poor outcome.

9.4. BIOMATERIALS APPROACH TO GENE DELIVERY

Polymers can also play the role of a gene carrier by serving as a matrix for sustained release instead of a DNA-condensing polycation, or as an excipient in the naked DNA or polyplex formulation. Collagen gel represents the early example of a depot delivering naked DNA [104]. Polymers without charge can encapsulate instead of condense the DNA in nano- and microparticles. Examples are DNA-containing poly(lactide-co-glycolide) (PLGA) nanoparticles fabricated by double-emulsion and solvent evaporation techniques [9]. The DNA loading level is typically low and the transfection efficiency poor. Promising results have nevertheless been obtained by judicious application of this type of nanoparticles, for example, in DNA vaccination where transgene expression may not need to be high.

Poly(ortho esters) (POE), synthesized by condensation copolymerization of a diketene acetal and diols, undergo hydrolysis to eventually generate monomeric diols and a small amount of acid. The hydrolysis of this class of polymer is pH dependent, being relatively stable at physiological pH, but hydrolyze rapidly at around pH 5. This property is exploited for a DNA vaccine delivery system [105]. Encapsulating the antigen gene in the POE microspheres 1–10 μm in size, the relatively stable polymer limits the release and protects the DNA from enzymatic degradation in the extracellular space. Once internalized and sequestered into the phagosomal compartment, rapid degradation of the polymer releases the DNA. The efficacy of this DNA vaccine delivery system is demonstrated in suppressing the growth of tumor cells bearing a model antigen in mice.

To improve encapsulation efficiency of the DNA, an interesting polymer has been synthesized by carbonyldiimidazole (CDI)-mediated coupling of diamines to poly(vinyl alcohol) (PVA), with subsequent grafting of D,L-lactides and glycolides [106]. The resulting amine-modified branched polyesters show rapid biodegradation. Their amphiphilic character also facilitates the encapsulation of DNA. Transfection efficiency can be higher than PEI and increases exponentially with the amount of amine substitution in the polymer.

Formulation approaches have also yielded significant improvements. Ternary complex can be formulated by coating the polyplex with molecules addressing any of the rate-limiting steps discussed in Figure 9.1. For instance, PEI-PEG complexes have been mixed with lipiodol to produce stable water-in-oil emulsion [107]. When injected into an occluded femoral artery via a catheter, the emulsion can localize the complexes at the injection site and produce transgene expression mainly in the muscle rather than other organs. The expression level is 3-fold higher than that of PEI at one day post-injection. Adding human serum albumin to PEI-DNA during the complex formation leads to an intravenous formulation that can produce a positive, although very transient, transgene expression in the lung tissue of mice [108]. The administration can be repeated without inhibition of gene expression. It is speculated that as a major protein component of serum, HSA helps to minimize rapid opsonization of the PEI-DNA complexes by serum proteins. A formulation of adding folic acid to the PEI-DNA complexes produces enhancement that cannot be reproduced with other anionic compounds such as cholic acid, citric acid, EDTA, or glutamic acid [109]. It is speculated that the anionic and amphiphilic folic acid leads to a larger and more hydrophobic complex that improves its interaction with the cell membrane for a more effective uptake.

An interesting approach of formulating polyplexes is to synthesize nanogels by crosslinking ionic and nonionic hydrophilic polymers. One example is the nanosized cationic network of cross-linked PEO and PEI (PEO-cl-PEI) [110]. Complexation of this cationic network with polyanions leads to the formation of a hydrophobic core from the neutralized polyions and a hydrophilic shell from PEO. Successful delivery of oligonucleotides has been demonstrated with these nanogels in vitro and in vivo. Delivery of DNA requires a target ligand to induce cellular internalization. Replacing PEO with the membrane-active Pluronics improves the transfection efficiency [111]. Such nanogels can produce positive transgene expression in liver following systemic administration in mice. The appealing feature of these nanogels is their resistance to aggregation and stability in storage. Along the same line, a recent study demonstrates the potential of using inverse microemulsion polymerization to produce relatively monodisperse and nontoxic nanogels [112]. A terpolymer built on 2-acryloxyethyltrimethylammonium chloride (AETMAC), 2-hydroxyethylacrylate (HEA), and poly(ethylene glycol)diacrylate (PEGdiA) can trap oligonucleotides in the nanogels for efficient cellular delivery. This technique provides control over the nanogel composition, size, and swelling behavior by varying the crosslinker and charged monomer concentrations in the polymerization. The nanogels resist aggregation on storage in aqueous solution at 4°C for up to six months. The inverse microemulsion polymerization is applicable to a wide range of vinyl polymers. With monomers containing a higher charge density and basicity, this technique should be able to produce similarly attractive nanogels that can entrap plasmid DNA.

The polymer may not need to have direct interaction with the DNA in order to have an enhancement effect. Pluronic block copolymers at low concentration of 0.01% (w/v) can increase the level and duration of transgene expression 5- to 20-fold compared to naked DNA in skeletal muscle of mice [113, 114]. The effect is not driven by condensation of the DNA, but rather hypothesized to derive from an adjuvant effect where the Pluronic copolymer acts as a biological response modifier to influence the transcriptional control of the transgene expression in an unspecified manner. Another example is pH-sensitive polymers. Poly(alkylacrylic acid)s are membrane-disruptive polymers that can enhance the release of contents from the acidic endosomal compartment to the cytoplasm [16, 17]. Poly(propylacrylic acid) (PPAA), for example, significantly enhances in vitro transfection of lipoplex formulations in cell culture. Injection of PPAA-lipoplex formulation into healing wounds in mouse also shows enhanced transgene expression over lipoplex alone. Inclusion of polyacrylic acid into PEI polyplexes has also been reported to improve gene delivery in intravenous administration, with the mechanism of improvement postulated to involve flocculation of the polyplexes and electrostatic shielding from opsonizing serum proteins.

9.5. SUMMARY

Undoubtedly there remain significant challenges to improve nonviral transgene expression to the therapeutic levels. This review highlights the barriers known to limit the nonviral gene transfer process. Considerable progress has been made in the design of polymers to address these barriers. Some innovations highlighted in this review point to the promising front of dealing with the problems of toxicity and aggregation. The wide variety of polymeric gene carriers available bodes well for tackling different gene therapy applications. Cell types with different metabolic characteristics would probably require polyplexes of different properties for optimal transfection. Route of administration would likewise impose its unique constraints. For instance, polyplexes optimized for intravenous injection may not be suitable for intramuscular administration, where ready dissociation of DNA from the polyplexes appears to be crucial for high transgene expression. The issue of optimal unpacking of the polyplex remains unresolved, which is again cell- and tissue- dependent. The different biodegradable gene carrier designs however should be able to cater to the needs. Of critical importance is gaining the mechanistic insights from in vitro intracellular trafficking and in vivo biodistribution studies to guide the rationale design of these biodegradable gene carriers. Advances in nanotechnology such as quantum dot synthesis and imaging will bolster this effort. Developments in micro- and nano- fabrication may also provide the support for breakthrough. One can envisage the use of micro- and nanofluidics to synthesize a multi-component polyplex with precision and control in the future.

With the tremendous promise of genetic medicine, nonviral gene therapy is a worthy goal of biomaterials, nanotechnology, and cell and molecular biology research.

REFERENCES

[1] J. Kaiser. Gene therapy. Side effects sideline hemophilia trial. *Science*, 304(5676):1423–1425, 2004.
[2] D.A. Williams and C. Baum. Medicine. Gene therapy–new challenges ahead. *Science*, 302(5644):400–401, 2003.
[3] M. Cavazzana-Calvo, A. Thrasher, and F. Mavilio. The future of gene therapy. *Nature*, 427(6977):779–781, 2004.
[4] J.C. Fratantoni, S. Dzekunov, V. Singh, and L.N. Liu. A non-viral gene delivery system designed for clinical use. *Cytotherapy*, 5(3):208–210, 2003.
[5] H. Ma and S.L. Diamond. Nonviral gene therapy and its delivery systems. *Curr. Pharm. Biotechnol.*, 2(1):1–17, 2001.
[6] M. Nishikawa and L. Huang. Nonviral vectors in the new millennium: delivery barriers in gene transfer. *Hum. Gene. Ther.*, 12(8):861–870, 2001.
[7] S.H. Pun and M.E. Davis. Development of a nonviral gene delivery vehicle for systemic application. *Bioconjug. Chem.*, 13(3):630–639, 2002.
[8] M. Thomas and A.M. Klibanov. Non-viral gene therapy: polycation-mediated DNA delivery. *Appl. Microbiol. Biotechnol.*, 62(1):27–34, 2003.
[9] J. Panyam and V. Labhasetwar. Biodegradable nanoparticles for drug and gene delivery to cells and tissue. *Adv. Drug. Deliv. Rev.*, 55(3):329–347, 2003.
[10] K. Ewert et al. Cationic lipid-DNA complexes for gene therapy: understanding the relationship between complex structure and gene delivery pathways at the molecular level. *Curr. Med. Chem.*, 11(2):133–149, 2004.
[11] M.C. Pedroso de Lima, S. Neves, A. Filipe, N. Duzgunes, and S. Simoes. Cationic liposomes for gene delivery: from biophysics to biological applications. *Curr. Med. Chem.*, 10(14):1221–1231, 2003.
[12] Y. Xu and F.C. Szoka, Jr. Mechanism of DNA release from cationic liposome/DNA complexes used in cell transfection. *Biochemistry*, 35(18):5616–5623, 1996.
[13] P. Midoux et al. Specific gene transfer mediated by lactosylated poly-L-lysine into hepatoma cells. *Nucleic. Acids Res.*, 21(4):871–878, 1993.
[14] A. El Ouahabi, M. Thiry, V. Pector, R. Fuks, J.M. Ruysschaert, and M. Vandenbranden. The role of endosome destabilizing activity in the gene transfer process mediated by cationic lipids. *FEBS Lett.*, 414(2):187–192, 1997.
[15] B.A. Sosnowski, A.M. Gonzalez, L.A. Chandler, Y.J. Buechler, G.F. Pierce, and A. Baird. Targeting DNA to cells with basic fibroblast growth factor (FGF2). *J. Biol. Chem.*, 271(52):33647–33653, 1996.
[16] R.A. Jones et al. Poly(2-alkylacrylic acid) polymers deliver molecules to the cytosol by pH-sensitive disruption of endosomal vesicles. *Biochem. J.*, 372(Pt 1):65–75, 2003.
[17] T.R. Kyriakides, C.Y. Cheung, N. Murthy, P. Bornstein, P.S. Stayton, and A.S. Hoffman. pH-sensitive polymers that enhance intracellular drug delivery in vivo. *J. Control Rel.*, 78(1–3):295–303, 2002.
[18] O. Boussif et al. A versatile vector for gene and oligonucleotide transfer into cells in culture and in vivo: polyethylenimine. *Proc. Natl. Acad. Sci. U.S.A.*, 92(16):7297–7301, 1995.
[19] A.M. Funhoff, C.F. van Nostrum, G.A. Koning, N.M. Schuurmans-Nieuwenbroek, D.J. Crommelin, and W.E. Hennink. Endosomal escape of polymeric gene delivery complexes is not always enhanced by polymers buffering at low pH. *Biomacromolecules*, 5(1):32–39, 2004.
[20] D. Lechardeur et al., Metabolic instability of plasmid DNA in the cytosol: a potential barrier to gene transfer. *Gene Ther.*, 6(4):482–497, 1999.
[21] L.J. Branden, A.J. Mohamed, and C.I. Smith. A peptide nucleic acid-nuclear localization signal fusion that mediates nuclear transport of DNA. *Nat. Biotechnol.*, 17(8):784–787, 1999.
[22] A. Ziemienowicz, D. Gorlich, E. Lanka, B. Hohn, and L. Rossi. Import of DNA into mammalian nuclei by proteins originating from a plant pathogenic bacterium. *Proc. Natl. Acad. Sci. U.S.A.*, 96(7):3729–3733, 1999.

[23] T. Blessing, M. Kursa, R. Holzhauser, R. Kircheis, and E. Wagner. Different strategies for formation of pegylated EGF-conjugated PEI/DNA complexes for targeted gene delivery. *Bioconjug. Chem.*, 12(4):529–537, 2001.

[24] A. Brownlie,, I.F. Uchegbu, and A.G. Schatzlein. PEI-based vesicle-polymer hybrid gene delivery system with improved biocompatibility. *Int. J. Pharm.*, 274(1–2):41–52, 2004.

[25] A. Gautam, J.C. Waldrep, F.M. Orson, B.M. Kinsey, B. Xu, and C.L. Densmore. Topical gene therapy for pulmonary diseases with PEI-DNA aerosol complexes. *Methods Mol. Med.*, 75:561–572, 2003.

[26] K. Kunath, A. von Harpe, D. Fischer, and T. Kissel. Galactose-PEI-DNA complexes for targeted gene delivery: degree of substitution affects complex size and transfection efficiency. *J. Control Rel.*, 88(1):159–172, 2003.

[27] P. Lampela, P. Soininen, A. Urtti, P.T. Mannisto, and A. Raasmaja. Synergism in gene delivery by small PEIs and three different nonviral vectors. *Int. J. Pharm.*, 270(1–2):175–184, 2004.

[28] M.L. Forrest, J.T. Koerber, and D.W. Pack, A degradable polyethylenimine derivative with low toxicity for highly efficient gene delivery. *Bioconjug. Chem.*, 14(5):934–940, 2003.

[29] M.L. Forrest, G.E. Meister, J.T. Koerber, and D.W. Pack. Partial acetylation of polyethylenimine enhances in vitro gene delivery. *Pharm. Res.*, 21(2):365–371, 2004.

[30] L. Wightman et al. Different behavior of branched and linear polyethylenimine for gene delivery in vitro and in vivo. *J. Gene. Med.*, 3(4):362–372, 2001.

[31] S.M. Zou, P. Erbacher, J.S. Remy, and J.P. Behr. Systemic linear polyethylenimine (L-PEI)-mediated gene delivery in the mouse. *J. Gene Med.*, 2(2):128–134, 2000.

[32] B. Brissault, A. Kichler, C. Guis, C. Leborgne, O. Danos, and H. Cheradame. Synthesis of linear polyethylenimine derivatives for DNA transfection. *Bioconjug. Chem.*, 14(3):581–587, 2003.

[33] H.K. Nguyen et al. Evaluation of polyether-polyethyleneimine graft copolymers as gene transfer agents. *Gene Ther.*, 7(2):126–138, 2000.

[34] M.A. Gosselin, W. Guo, and R.J. Lee. Efficient gene transfer using reversibly cross-linked low molecular weight polyethylenimine. *Bioconjug. Chem.*, 12(6):989–994, 2001.

[35] C.H. Ahn, S.Y. Chae, Y.H. Bae, and S.W. Kim. Biodegradable poly(ethylenimine) for plasmid DNA delivery. *J. Control Rel.*, 80(1–3):273–282, 2002.

[36] H. Petersen, T. Merdan, K. Kunath, D. Fischer, and T. Kissel. Poly(ethylenimine-co-L-lactamide-co-succinamide): a biodegradable polyethylenimine derivative with an advantageous pH-dependent hydrolytic degradation for gene delivery. *Bioconjug Chem.*, 13(4):812–821, 2002.

[37] H. Petersen et al. Polyethylenimine-graft-poly(ethylene glycol) copolymers: influence of copolymer block structure on DNA complexation and biological activities as gene delivery system. *Bioconjug. Chem.*, 13(4):845–854, 2002.

[38] C.L. Gebhart, S. Sriadibhatla, S. Vinogradov, P. Lemieux, V. Alakhov, and A.V. Kabanov. Design and formulation of polyplexes based on pluronic-polyethyleneimine conjugates for gene transfer. *Bioconjug. Chem.*, 13(5):937–944, 2002.

[39] D.A. Wang et al. Novel branched poly(ethylenimine)-cholesterol water-soluble lipopolymers for gene delivery. *Biomacromolecules*, 3(6):1197–1207, 2002.

[40] M. Lee, J. Rentz, S.O. Han, D.A. Bull, and S.W. Kim. Water-soluble lipopolymer as an efficient carrier for gene delivery to myocardium. *Gene Ther.*, 10(7):585–593, 2003.

[41] G.P. Tang et al. Polyethylene glycol modified polyethylenimine for improved CNS gene transfer: effects of PEGylation extent. *Biomaterials*, 24(13):2351–2362, 2003.

[42] L. Shi et al. Repeated intrathecal administration of plasmid DNA complexed with polyethylene glycol-grafted polyethylenimine led to prolonged transgene expression in the spinal cord. *Gene Ther.*, 10(14):1179–1188, 2003.

[43] T. Merdan et al. Pegylated polyethylenimine-Fab' antibody fragment conjugates for targeted gene delivery to human ovarian carcinoma cells. *Bioconjug. Chem.*, 14(5):989–996, 2003.

[44] W. Guo and R.L. Lee. Receptor-targeted gene delivery via folate-conjugated polyethylenimine. *AAPS PharmSci.*, 1(4):E19, 1999.

[45] J.M. Benns, A. Maheshwari, D.Y. Furgeson, R.I. Mahato, and S.W. Kim, Folate-PEG-folate-graft-polyethylenimine-based gene delivery. *J. Drug Target*, 9(2):123–139, 2001.

[46] C.H. Ahn, S.Y. Chae, Y.H. Bae, and S.W. Kim. Synthesis of biodegradable multi-block copolymers of poly(L-lysine) and poly(ethylene glycol) as a non-viral gene carrier. *J. Control Rel.*, 97(3):567–574, 2004.

[47] Oupicky, D., R.C. Carlisle, and L.W. Seymour. *Triggered intracellular activation of disulfide crosslinked polyelectrolyte gene delivery complexes with extended systemic circulation in vivo*. Gene Ther. 2001. 8(9): p. 713–24.

[48] M.D. Brown et al. Preliminary characterization of novel amino acid based polymeric vesicles as gene and drug delivery agents. *Bioconjug. Chem.*, 11(6):880–891, 2000.

[49] D. Putnam, C.A. Gentry, D.W. Pack, and R. Langer. Polymer-based gene delivery with low cytotoxicity by a unique balance of side-chain termini. *Proc. Natl. Acad. Sci. U.S.A.*, 98(3):1200–1205, 2001.

[50] T.G. Kim et al. Gene transfer into human hepatoma cells by receptor-associated protein/polylysine conjugates. *Bioconjug. Chem.*, 15(2):326–332, 2004.

[51] J.H. Jeong, S.W. Kim, and T.G. Park. Novel intracellular delivery system of antisense oligonucleotide by self-assembled hybrid micelles composed of DNA/PEG conjugate and cationic fusogenic peptide. *Bioconjug. Chem.*, 14(2):473–479, 2003.

[52] Y.B. Lim et al. Biodegradable polyester, poly[alpha-(4-aminobutyl)-L-glycolic acid], as a non-toxic gene carrier. *Pharm. Res.*, 17(7):811–816, 2000.

[53] A. Maheshwari et al. Soluble biodegradable polymer-based cytokine gene delivery for cancer treatment. *Mol. Ther.*, 2(2):121–130, 2000.

[54] M. Lee, K.S. Ko, S. Oh, and S.W. Kim. Prevention of autoimmune insulitis by delivery of a chimeric plasmid encoding interleukin-4 and interleukin-10. *J. Control Rel.*, 88(2):333–342, 2003.

[55] A. Maheshwari, S. Han, R.I. Mahato, and S.W. Kim. Biodegradable polymer-based interleukin-12 gene delivery: role of induced cytokines, tumor infiltrating cells and nitric oxide in anti-tumor activity. *Gene Ther.*, 9(16):1075–1084, 2002.

[56] M. Kramer et al. Dendritic polyamines: simple access to new materials with defined treelike structures for application in nonviral gene delivery. *Chembiochem*, 5(8):1081–1087, 2004.

[57] N.A. Jones, I.R. Hill, S. Stolnik, F. Bignotti, S.S. Davis, and M.C. Garnett. Polymer chemical structure is a key determinant of physicochemical and colloidal properties of polymer-DNA complexes for gene delivery. *Biochim. Biophys. Acta*, 1517(1):1–18, 2000.

[58] Z.Y. Zhang and B.D. Smith, High-generation polycationic dendrimers are unusually effective at disrupting anionic vesicles: membrane bending model. *Bioconjug. Chem.*, 11(6):805–814, 2000.

[59] J.H. Lee et al. Polyplexes assembled with internally quaternized PAMAM-OH dendrimer and plasmid DNA have a neutral surface and gene delivery potency. *Bioconjug. Chem.*, 14(6):1214–1221, 2003.

[60] J.Y. Cherng, P. van de Wetering, H. Talsma, D.J. Crommelin, and W.E. Hennink. Effect of size and serum proteins on transfection efficiency of poly ((2-dimethylamino)ethyl methacrylate)-plasmid nanoparticles. *Pharm. Res.*, 13(7):1038–1042, 1996.

[61] D.W. Lim, Y.I. Yeom, and T.G. Park. Poly(DMAEMA-NVP)-b-PEG-galactose as gene delivery vector for hepatocytes. *Bioconjug. Chem.*, 11(5):688–695, 2000.

[62] A. Akinc, D.M. Lynn, D.G. Anderson, and R. Langer. Parallel synthesis and biophysical characterization of a degradable polymer library for gene delivery. *J. Am. Chem. Soc.*, 125(18):5316–23, 2003.

[63] A. Akinc, D.G. Anderson, D.M. Lynn, and R. Langer. Synthesis of poly(beta-amino ester)s optimized for highly effective gene delivery. *Bioconjug. Chem.*, 14(5):979–988, 2003.

[64] D.G. Anderson, D.M. Lynn, and R. Langer. Semi-automated synthesis and screening of a large library of degradable cationic polymers for gene delivery. *Angew. Chem. Int. Ed. Engl.*, 42(27):3153–3158, 2003.

[65] S. Jon, D.G. Anderson, and R. Langer. Degradable poly(amino alcohol esters) as potential DNA vectors with low cytotoxicity. *Biomacromolecules*, 4(6):1759–1762, 2003.

[66] Y.B. Lim, S.M. Kim, H. Suh, and J.S. Park. Biodegradable, endosome disruptive, and cationic network-type polymer as a highly efficient and nontoxic gene delivery carrier. *Bioconjug. Chem.*, 13(5):952–957, 2002.

[67] J. Luten et al. Water-soluble biodegradable cationic polyphosphazenes for gene delivery. *J. Control Rel.*, 89(3):483–497, 2003.

[68] N.C. Bellocq, S.H. Pun, G.S. Jensen, and M.E. Davis. Transferrin-containing, cyclodextrin polymer-based particles for tumor-targeted gene delivery. *Bioconjug. Chem.*, 14(6):1122–1132, 2003.

[69] S.J. Hwang and M.E. Davis. Cationic polymers for gene delivery: designs for overcoming barriers to systemic administration. *Curr. Opin. Mol. Ther.*, 3(2):183–191, 2001.

[70] S.R. Popielarski, S. Mishra, and M.E. Davis. Structural effects of carbohydrate-containing polycations on gene delivery. 3. Cyclodextrin type and functionalization. *Bioconjug. Chem.*, 14(3):672–678, 2003.

[71] S.H. Pun et al. Cyclodextrin-modified polyethylenimine polymers for gene delivery. *Bioconjug. Chem.*, 15(4):831–840, 2004.

[72] T.M. Reineke and M.E. Davis. Structural effects of carbohydrate-containing polycations on gene delivery. 2. Charge center type. *Bioconjug. Chem.*, 14(1):255–261, 2003.

[73] T.M. Reineke and M.E. Davis. Structural effects of carbohydrate-containing polycations on gene delivery. 1. Carbohydrate size and its distance from charge centers. *Bioconjug. Chem.*, 14(1):247–254, 2003.

[74] M.E. Davis et al. Self-assembling nucleic acid delivery vehicles via linear, water-soluble, cyclodextrin-containing polymers. *Curr. Med. Chem.*, 11(2):179–197, 2004.

[75] Z. Zhao, J. Wang, H.Q. Mao, and K.W. Leong. Polyphosphoesters in drug and gene delivery. *Adv. Drug. Deliv. Rev.*, 55(4):483–99, 2003.

[76] J. Wen, H.Q. Mao, W. Li, K.Y. Lin, and K.W. Leong. Biodegradable polyphosphoester micelles for gene delivery. *J. Pharm. Sci.*, 93(8):2142–57, 2004.

[77] Wang, J., S.J. Gao, P.C. Zhang, S. Wang, H.Q. Mao, and K.W. Leong, Polyphosphoramidate gene carriers: effect of charge group on gene transfer efficiency. *Gene Ther.*, 11(12):1001–1010, 2004.

[78] J. Wang, S.W. Huang, P.C. Zhang, H.Q. Mao, and K.W. Leong. Effect of side-chain structures on gene transfer efficiency of biodegradable cationic polyphosphoesters. *Int. J Pharm.*, 265(1–2):75–84, 2003.

[79] J. Wang, P.C. Zhang, H.Q. Mao, and K.W. Leong. Enhanced gene expression in mouse muscle by sustained release of plasmid DNA using PPE-EA as a carrier. *Gene Ther.*, 9(18):1254–1261, 2002.

[80] J. Wang et al. New polyphosphoramidate with a spermidine side chain as a gene carrier. *J. Control Rel.*, 83(1):157–168, 2002.

[81] S.W. Huang, J. Wang, P.C. Zhang, H.Q. Mao, R.X. Zhuo, and K.W. Leong. Water-soluble and nonionic polyphosphoester: synthesis, degradation, biocompatibility and enhancement of gene expression in mouse muscle. *Biomacromolecules*, 5(2):306–311, 2004.

[82] J. Wang, H.Q. Mao, and K.W. Leong. A novel biodegradable gene carrier based on polyphosphoester. *J. Am. Chem. Soc.*, 123(38):9480–9481, 2001.

[83] S. Wang, N. Ma, S.J. Gao, H. Yu, and K.W. Leong. Transgene expression in the brain stem effected by intramuscular injection of polyethylenimine/DNA complexes. *Mol. Ther.*, 3(5 Pt 1):658–664, 2001.

[84] Y. Li et al. CNS gene transfer mediated by a novel controlled release system based on DNA complexes of degradable polycation PPE-EA: a comparison with polyethylenimine/DNA complexes. *Gene Ther.*, 11(1):109–114, 2004.

[85] F.C. MacLaughlin et al. Chitosan and depolymerized chitosan oligomers as condensing carriers for in vivo plasmid delivery. *J. Controll. Rel.*, 56(1–3):259–272, 1998.

[86] K.Y. Lee, I.C. Kwon, Y.H. Kim, W.H. Jo, and S.Y. Jeong. Preparation of chitosan self-aggregates as a gene delivery system. *J. Controll. Rel.*, 51(2–3):213–220, 1998.

[87] S.C. Richardson, H.V. Kolbe, and R. Duncan. Potential of low molecular mass chitosan as a DNA delivery system: biocompatibility, body distribution and ability to complex and protect DNA. *Int. J. Pharm.*, 178(2):231–43, 1999.

[88] M. Thanou, J.C. Verhoef, and H.E. Junginger. Chitosan and its derivatives as intestinal absorption enhancers. *Adv. Drug Deliv. Rev.*, 50 Suppl 1:S91–S101, 2001.

[89] H.Q. Mao et al. Chitosan-DNA nanoparticles as gene carriers: synthesis, characterization and transfection efficiency. *J. Control Rel.*, 70(3):399–421, 2001.

[90] K. Roy, H.Q. Mao, S.K. Huang, and K.W. Leong. Oral gene delivery with chitosan–DNA nanoparticles generates immunologic protection in a murine model of peanut allergy. *Nat. Med.*, 5(4):387–91, 1999.

[91] K.W. Leong, H.Q. Mao, V.L. Truong-Le, K. Roy, S.M. Walsh, and J.T. August. DNA-polycation nanospheres as non-viral gene delivery vehicles. *J. Control Rel.*, 53(1–3):183–193, 1998.

[92] S. Gao et al. Galactosylated low molecular weight chitosan as DNA carrier for hepatocyte-targeting. *Int. J. Pharm.*, 255(1–2):57–68, 2003.

[93] T. Kiang, J. Wen, H.W. Lim, and K.W. Leong. The effect of the degree of chitosan deacetylation on the efficiency of gene transfection. *Biomaterials*, 25(22):5293–301, 2004.

[94] M. Huang, E. Khor, and L.Y. Lim. Uptake and cytotoxicity of chitosan molecules and nanoparticles: effects of molecular weight and degree of deacetylation. *Pharm. Res.*, 21(2):344–353, 2004.

[95] M. Koping-Hoggard et al. Improved chitosan-mediated gene delivery based on easily dissociated chitosan polyplexes of highly defined chitosan oligomers. *Gene Ther.*, 11(19):1441–1452, 2004.

[96] M. Koping-Hoggard, Y.S. Mel'nikova, K.M. Varum, B. Lindman, and P. Artursson. Relationship between the physical shape and the efficiency of oligomeric chitosan as a gene delivery system in vitro and in vivo. *J. Gene Med.*, 5(2):130–141, 2003.

[97] M. Thanou, B.I. Florea, M. Geldof, H.E. Junginger, and G. Borchard. Quaternized chitosan oligomers as novel gene delivery vectors in epithelial cell lines. *Biomaterials*, 23(1):153–159, 2002.
[98] J.L. Chew, C.B. Wolfowicz, H.Q. Mao, K.W. Leong, and K.Y. Chua. Chitosan nanoparticles containing plasmid DNA encoding house dust mite allergen, Der p 1 for oral vaccination in mice. *Vaccine*, 21(21–22):2720–2729, 2003.
[99] M. Kumar et al. Intranasal gene transfer by chitosan-DNA nanospheres protects BALB/c mice against acute respiratory syncytial virus infection. *Hum. Gene. Ther.*, 13(12):1415–1425, 2002.
[100] M. Kumar, X. Kong, A.K. Behera, G.R. Hellermann, R.F. Lockey, and S.S. Mohapatra. Chitosan IFN-gamma-pDNA Nanoparticle (CIN) Therapy for Allergic Asthma. *Genet. Vaccines Ther.*, 1(1):3, 2003.
[101] V.L. Truong-Le, J.T. August, and K.W. Leong. Controlled gene delivery by DNA-gelatin nanospheres. *Hum. Gene Ther.*, 9(12):1709–1717, 1998.
[102] V.L. Truong-Le et al. Gene transfer by DNA-gelatin nanospheres. *Arch. Biochem. Biophys.*, 361(1):47–56, 1999.
[103] J. Wang et al. Evaluation of collagen and methylated collagen as gene carriers. *Int. J. Pharm.*, 279(1–2):115–126, 2004.
[104] J. Bonadio, E. Smiley, P. Patil, and S. Goldstein. Localized, direct plasmid gene delivery in vivo: prolonged therapy results in reproducible tissue regeneration. *Nat. Med.*, 5(7):753–759, 1999.
[105] C. Wang et al. Molecularly engineered poly(ortho ester) microspheres for enhanced delivery of DNA vaccines. *Nat. Mater.*, 3(3):190–196, 2004.
[106] C.G. Oster, M. Wittmar, F. Unger, L. Barbu-Tudoran, A.K. Schaper, and T. Kissel. Design of amine-modified graft polyesters for effective gene delivery using DNA-loaded nanoparticles. *Pharm. Res.*, 21(6):927–931, 2004.
[107] J.W. Hong, J.H. Park, K.M. Huh, H. Chung, I.C. Kwon, and S.Y. Jeong. PEGylated polyethylenimine for in vivo local gene delivery based on lipiodolized emulsion system. *J. Control Rel.*, 99(1):167–176, 2004.
[108] Rhaese, S., H. von Briesen, H. Rubsamen-Waigmann, J. Kreuter, and K. Langer. Human serum albumin-polyethylenimine nanoparticles for gene delivery. *J. Control Rel.*, 92(1–2):199–208, 2003.
[109] W. Guo and R.J. Lee. Efficient gene delivery via non-covalent complexes of folic acid and polyethylenimine. *J. Control Rel.*, 77(1–2):131–138, 2001.
[110] S.V. Vinogradov, T.K. Bronich and A.V. Kabanov. Nanosized cationic hydrogels for drug delivery: preparation, properties and interactions with cells. *Adv. Drug Deliv. Rev.*, 54(1):135–147, 2002.
[111] P. Lemieux et al. Block and graft copolymers and NanoGel copolymer networks for DNA delivery into cell. *J. Drug Target*, 8(2):91–105, 2000.
[112] K. McAllister et al. Polymeric nanogels produced via inverse microemulsion polymerization as potential gene and antisense delivery agents. *J. Am. Chem. Soc.*, 124(51):15198–15207, 2002.
[113] A.V. Kabanov, E.V. Batrakova, and V.Y. Alakhov. Pluronic block copolymers as novel polymer therapeutics for drug and gene delivery. *J. Control Rel.*, 82(2–3):189–212, 2002.
[114] A.V. Kabanov and V.Y. Alakhov. Pluronic block copolymers in drug delivery: from micellar nanocontainers to biological response modifiers. *Crit. Rev. Ther. Drug Carrier Syst.*, 19(1):1–72, 2002.

10

Dip-Pen Technologies for Biomolecular Devices

Debjyoti Banerjee, Ph.D.
Group Leader and Staff Mechanical Engineer, Applied Biosystems Inc.
(formerly Microfluidics Engineer, NanoInk Inc.)

10.1. INTRODUCTION

Since the 1950s, Scanning Electron Microscopy (SEM) has been commercially available and used to measure feature sizes below 1 micron. Modified SEMs have been employed since the 1960s to perform sub-micron lithography, which then made rapid advances in the 1990s to a process, known as electron beam lithography (EBL). Since the 1980s, Surface Tunneling Microscopy (STM) and Atomic Force Microscopy (AFM) have ushered the era of nanotechnology where it is possible to measure and control the manipulation of matter on the 100nm scale and below. These techniques are broadly classified as "Scanning Probe Microscopy (SPM)". The earliest forms of nanofabrication using STM based approaches were used to pattern "hard" materials (such as silicon-dioxide; as opposed to "soft" materials such as polymers or biological materials) and restricted to single layer processing. These methods were initially motivated by applications in the semi-conductor industry.

The nanofabrication processes pioneered by EBL and STM were followed by development of diverse nanofabrication approaches in the 1990s, which included micro contact printing (which is derived from soft lithography), step and flash imprint lithography (SFIL), nano imprint lithography (NIL), etc. However, these methods are only capable of single layer registration and the feasibility of registering multi-layers in alignment with the previously fabricated features are yet to be demonstrated. Also, these methods are not capable of spatially addressing biological materials in chemically distinct arrays. Of these methods, micro contact printing is the most flexible in its ability to pattern different materials while

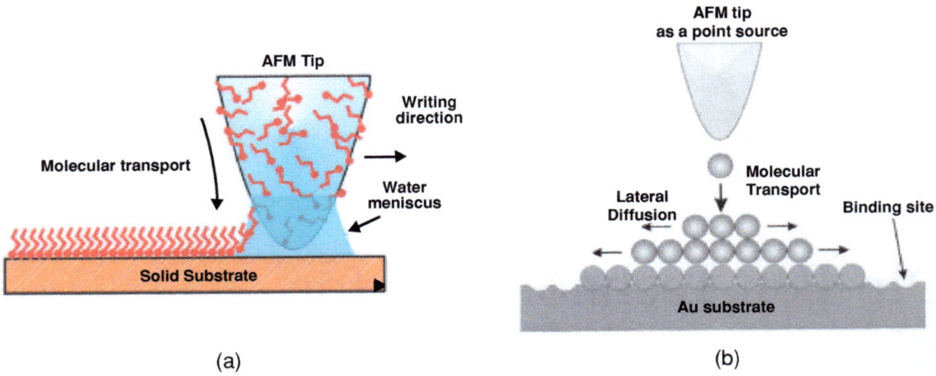

FIGURE 10.1. (a) Schematic representation of the dip-pen nanolithography process. (Reprinted with the permission from Piner *et al., Science* **283**:661–63, © 1999, AAAS). (b) Proposed transport mechanism of ink molecules from the tip to the substrate. The incoming molecular flux from the tip creates a concentration gradient around the tip, and ink molecules subsequently diffuse over the region already occupied by other ink molecules (drawn as filled circles) to be finally trapped by the bare surface of the substrate. (Reprinted with the permission from Jang *et al., Journal of Chemical Physics* **115** (6):661–63, © 2001, American Institute of Physics).

others are fairly restrictive in patterning a narrow range of materials. So far, NIL has been demonstrated to have the highest precision of close to 20 nm. The speed of fabrication for NIL, SFIL and micro contact printing is quite high compared to EBL at the expense of accuracy. Currently, the equipment cost and operating cost of these methods are estimated to range from approximately $500,000 (NIL) to about $20 million for EBL.

Dip Pen Nanolithography™ (DPN™) was invented by Chad Mirkin, Richard Piner and Seunghun Hong at North Western University in 1999 (US Patent 6635311). DPN uses chemically coated scanning probe microscopy tips to deposit nano-scale chemical patterns on a substrate in a direct write process. Feature sizes as small as 15 nm have been demonstrated with 5 nm spatial resolution for small molecules [30]. The genesis of the invention is the investigation of small capillary meniscus ("capillary bridge") that was found to nucleate on scanning probe tips when brought in close proximity to a substrate [32]. Experimental pictures showing the existence of these capillary bridges have been reported by Schenk *et al.*, 1998. Molecular dynamic simulations have also indicated the existence of such a nucleating process [17]. Also, a thin aqueous film (of the order of 1–2 molecule thickness) is believed to adhere on all surfaces under ambient conditions, which can also aid in the formation of a capillary bridge.

Initial experiments using DPN were demonstrated by forming self assembled monolayer (SAM) of different small molecules primarily on gold and oxidized silicon surfaces. Typically ODT (octa-decane-thiol) and MHA (mercapto-hexadecanoic-acid) were used in these initial experiments. These deposited chemistries were used as etch masks to pattern oxidized silicon and metals to demonstrate their efficacy for semi-conductor applications [43].

In the DPN terminology—the chemically coated scanning probe tip is called the "pen", the chemical used to coat the tip is known as the "ink" and the substrate for writing is known as the "paper". The DPN process is similar to writing using quill pens. Incidentally, quill pens were used as late as the 18th century until the invention of fountain pen, which were followed by invention of the ball point pen in 1940. Prior to quill pens, pens made out of reeds were used as early as 2000 B.C. for writing on parchments (or papyrus) in the ancient Egyptian civilization, the Roman civilization and the civilizations in South-East Asia. In a close resemblance to this development, the development of DPN has been followed by the development of Fountain Pen Lithography (FPN), which was proposed by Kim *et al.*, 2003 (Espinosa group, Northwestern University). The desired mode of FPN operation would require a continuous supply of ink to the pen tip from a reservoir that would obviate the need for loading the pen with ink by dipping. Espinosa's group has fabricated a FPN device that consists of an ink supply reservoir connected by micro-channels to a hollow scanning probe tip. Successful demonstration of an FPN device that does not need dipping for loading ink and is able to write patterns with feature size less than 100 nm is yet to be demonstrated. Similar approaches to using fountain pen nano-patterning are also described by Lewis *et al.* [22] and Shalom *et al.* [37].

A number of experiments with small molecules (e.g., MHA, ODT) show that the deposition process is diffusion limited [39, 42]. By dragging the pen tip on a surface, larger feature sizes were obtained when dragged at a lower velocity or at elevated temperatures. Similarly, by touching the tip on a substrate larger spot sizes were obtained when the dwell times were increased for a particular temperature or when the experiments were performed for a fixed dwell time and at an elevated temperature. These are evidences that prove that the deposition process is diffusion controlled. Also, the relative humidity of the ambient conditions around the pen tip has a significant effect on the writing process. At higher relative humidity larger feature sizes are obtained, keeping all other factors constant. Within the bounds of experimental error it was observed that the deposited features were independent of the force applied on the tips [30]. This behavior enables an array of scanning probes employed for DPN to have a lower variability in spot sizes due to different contact forces being applied to the tips. This is of great significance for a multi-pen DPN architecture targeted for a reliable industrial scale instrument since it enables a more repeatable process with lower variability.

Thus, the DPN process is independent of the contact forces and is affected by the dwell time (or speed of writing), ambient humidity, the surface diffusivity of the deposited chemical species (which is a function of the molecular weight of the chemical species, activation energy for detachment from scanning probe tip and temperature), surface roughness of the substrate, material properties of the substrate, material properties of the scanning probe tip and the tip radius of the scanning probe tip.

DPN has significant advantages over other nano-lithographic processes mentioned earlier. DPN can be used to deposit multi-layer patterns in close registration with the different features fabricated in the previous steps. The features deposited in the prior fabrication steps can be registered by scanning the chemically coated probe tips on the substrate at faster scan speeds that prevents diffusional deposition of the ink. After registering the location of the previous features new features can then be aligned and deposited using different sets of inks. Thus it has been demonstrated as a direct write tool with high precision, accuracy and resolution. DPN can be used to deposit biological and soft polymeric materials at room temperature and under ambient conditions without exposing them to harsh physical environments (e.g., vacuum) and harsh chemical environments (e.g., etching solutions, photoresist, etc.). The fabrication steps are performed at the desired locations with minimal risk for cross contamination. It is also possible to perform these fabrication steps under a water droplet to prevent denaturing of sensitive proteins. DPN process has been demonstrated for a variety of inks and substrates. The inks demonstrated range from small molecules (ODT, MHA) to polymers to biological materials (proteins, peptides, oligo-nucleotides). The substrates used for performing DPN range from silicon, glass, germanium, gallium-arsenide to metals (such as gold). Using DPN it is possible to integrate small molecule chemistry and large bio-molecules on a single substrate. The lithography process does not need a costly clean room infrastructure for operation. The whole operation can be performed on a tabletop instrument using a humidity control chamber. The DPN platform is flexible, digitally programmable and reconfigurable. The DPN process has enabled a rapid proto-typing platform with a fast turn around. However, a single pen platform can be quite slow and currently has limited small-scale application only in research laboratories. The process can be scaled to a high throughput industrial tool. This can be achieved by a multi-pen (or "plotter") configuration where an array of probe tips is used with each probe tip coated with unique ink chemistry. NanoInk Inc. has licensed the core DPN technology for commercializing it as an industrial scale process. Zhang *et al.* (2002) and research groups at NanoInk [36] have developed various multi-pen DPN plotters. The development of high throughput DPN platform is a separate subject area and can be found in a separate reference. Software called DPNWrite™ is also available from NanoInk for the automated operation of the DPN instrument (commercially available as NScriptor™). The DPN technology commercially available from NanoInk requires a relatively low investment for infrastructure and operation, which is currently at a fraction of the cost required for other nano-lithographic platforms of comparable resolution, accuracy and precision.

10.2. GENERAL APPLICATIONS

DPN applications have been envisioned in micro/nano-fluidics, nanoelectronics, molecular electronics (organic and bio-organic circuits), ultra high density oligonucleotide arrays (gene biochips, and protein/peptide biochips), solid state nano-resists, cryptography, security and authentication, photo-mask repair, semi-conductors, materials discovery, combinatorial synthesis, nano-printed catalysts, nano-crystals (colloidal crystals, bio-structures), nano-sensors (ultra small sensors with high sensitivity and selectivity) and bio-synthesis platforms (for nucleic acids, peptides, etc.).

DPN technologies have been used in various applications: patterning of small molecules (such as MHA, ODT, silanes, etc. as mentioned before), polymers (including conducting polymers, Lim and Mirkin, 2002), sol-gel chemistry, metal inks [35], nano-wires for optical applications [31], dendrimers [29], gold nano-particles [25], and beads using electro-static inter-actions or redox potentials [16].

This report is limited to the discussion of the bio-molecular applications of DPN and therefore the inter-related subject areas covered include: DPN applications in biomolecular patterning (e.g., proteins/peptides, nucleic acids, lipids, etc.), DPN enabled molecular structures for viral and cellular capture, and bio-molecular patterning using microfluidics enabled DPN. The related subjects involving molecular electronics (for future applications in bio-organic circuits), nano-bio sensors and biosynthesis enabled by dip pen technologies will be inherently covered as a part of this discussion.

Dip pen technologies have been adapted and modified in various forms which include: E-DPN or Electrochemical DPN [1, 2, 23], electroless DPN [35], cathodic alkyne electrografting [14], DPN using diels alder reaction [28], SP-CP or scanning probe contact printing [41], static plowing [34] and nano-grafting [3] which has also been referred to as "NPRW" (nano pen reader and writer). With the exception of E-DPN and nano-grafting, which have been demonstrated for patterning of bio-molecular devices the other derivative methods have been demonstrated for only small organic, inorganic and polymer molecules and therefore are left out of the scope of the current discussion. Interested readers are advised to consult the relevant references quoted above.

10.3. BIO-MOLECULAR PATTERNING USING DPN

Chips fabricated using plastics; silicon and glass with patterned arrays of bio-molecules on them are known as microarrays. Currently, the feature sizes and feature spacings in commercial microarrays are order of tens of microns. Microarrays have grown to be valuable tools for life-sciences research. Micro-arrays made of functionalized nucleic acids, such as DNA and RNA, are used to quantify both gene expression and genomic structure. Typical techniques include single-nucleotide-polymorphism (SNP) detection and are commonly used for diagnostics and drug development research. Protein microarrays are also under development for similar purposes.

Micro and nano arrays fabricated using DPN have a number of advantages. By reducing the spot sizes from tens of microns to hundreds of nano-metres, DPN would enable 10,000 to 100,000 times more features to be fabricated in the same foot-print on a microarray [9]. Such bio-molecular devices can be classified as nano-arrays where the sub-micron feature sizes (and feature spacing) are of the order of tens of nanometers or less. This would allow larger amount of information to be stored and probed on a given microarray chip. With suitable detection schemes this would enable larger number of disease samples or drug candidates to be sampled in a given assay.

Similar to the advantages gained in the micro-electronics industry from increased miniaturization, the micro-array platforms would also benefit from faster, sophisticated, high throughput, multiplexed systems with reduced reagent consumption and lower production costs. Other benefits include: reduced sample requirements, novel label free detection schemes (which is also currently a disadvantage—since such detection schemes are not yet well developed), and promise of higher detection sensitivities. Using DPN it is possible to fit the whole human genome using a 2 cm $\times$ 2 cm chip with feature sizes of 150 nm, while current micro-array technology with 20 micron spot size would require a foot print of a few square meters [13]. The detection schemes for these nano-arrays are not mature yet. However, almost every property (e.g., size, shape, mechanical or chemical or electrical properties, hydrophillicity) of these bio-molecules change upon reaction with an analyte. These changes can be monitored in situ by using near field or proximal probe technologies or on-chip electronic detection circuit.

Ginger, *et al.* [13] proposed that such devices can be used to control the "attachment of proteins and virus particles in specific orientation and thus to study reactivity as a function of structural configuration". The authors also identified non-specific binding (NSB) of

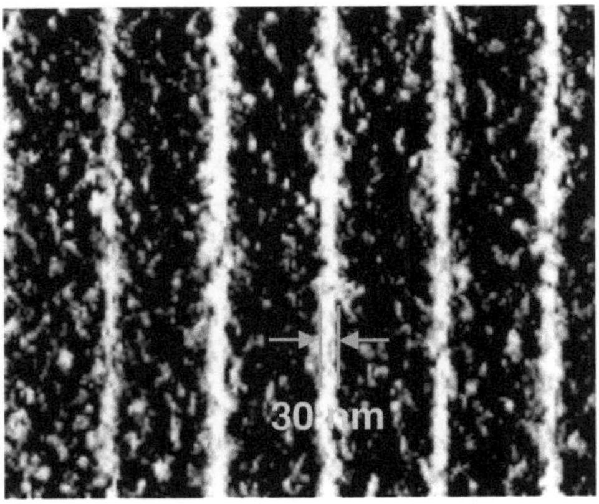

FIGURE 10.2. 30 nm wide lines of alkanethiol functionalized oligo-nucleotide drawn on gold substrate using DPN. (Reprinted with the permission from Mirkin, 2000, *Inorganic Chemistry*, 39 (11): 2258–2272, © 2000, American Chemical Society).

analytes to be a potential impediment to the success of nano-arrays. The authors propose the incorporation of redundancy in the design of nano-arrays for error checking and improving the reliability of the experiments.

Various patterning techniques have been developed to fabricate bio-molecules using DPN. Typically these techniques can be broadly classified related to fabrication of nano-arrays for nucleic acids, proteins/peptides and composite bio-molecules (e.g., peptide functionalized nucleic acids).

10.3.1. Nano-Pattering of Oligonucleotides Using DPN

Mirkin [30] initially demonstrated the power of DPN for bio-molecular patterning by writing an array of lines using thiol-functionalized oligonucleotide (12-nucleotide sequence) by chemisorption on a gold substrate and achieved 30 nm line width in the lithographic process (Figure 10.2).

This study was followed by development of a DNA nano-array with 100–200 nm diameter spots. The fabrication scheme is shown in Figure 10.4. Initially, MHA patterned areas are chemisorbed on a gold substrate by DPN and the un-patterned areas are blocked by immersing the substrate in a solution of ODT. A single strand of alkylamine modified DNA is then coupled to the MHA pattern by forming an amide bond (marked as "a" in the figure). Subsequently, another set of MHA patterns are formed using DPN on the areas blocked by ODT without chemically modifying the first pattern of MHA. It was found that MHA replaces ODT in the patterned areas. Another alkyamine modified DNA strand was immobilized on the second MHA pattern (marked as "b" in the figure). This enabled fabrication of two sets of nano-patterns comprising of two distinct oligo-nucleotide sequences.

The substrate was then washed with a linker sequence. This linker strand was designed to bind in an orthogonal manner to the two different oligo-nucleotide patterns (3'-5- and 5'-3', respectively from the surface) generating floppy ends that are complementary to the

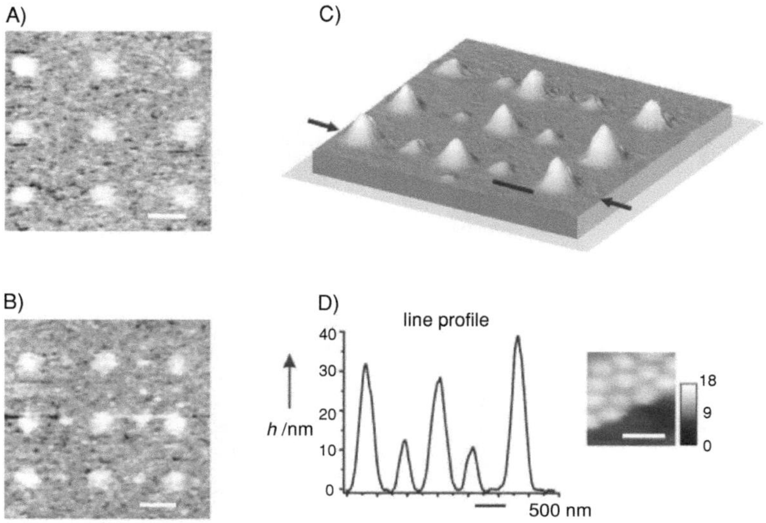

FIGURE 10.3. AFM images of patterned gold substrates. A) Topographical image (contact mode) of the substrate after DNA has been coupled to the DPN-generated MHA pattern. B) Topographical image of the two-component DNA pattern. C) Noncontact AFM topography image of particles after the orthogonal assembly process. Scale bars in A, B, and C are 1 micron. D) Line scan of the first row of particles in the image in Figure 1C. The inset shows a high resolution, tapping-mode image of one of the nanostructures comprising 13 nm particles (scale bar is 30 nm). (Reprinted with the permission from Demers *et al.*, 2001, *Agnew Chem. Intl. Ed.*, 40(16):3071–3073, © 2001, WILEY-VCH).

pre-designed oligo-nucleotide-modified particles. The intermediate assay steps involved rinsing with PBS buffer. Finally, to demonstrate orthogonal assembly of the nanostructures, a droplet of solution containing DNA-derivatized 13 and 30 nm particles was placed on the substrate. AFM images of the resulting structures show that the 13 and 30 nm particles are only immobilized on the complementary oligo-nucleotide patterns. Control experiments performed without the linker show that no particle assembly takes place, and that with the linker, each particle can be introduced separately to analogous substrates to generate one component nanostructures (either 13 or 30 nm particles) on the complementary regions of the substrate.

A similar result was reported by Ivanisevic *et al.* [16]. Redox active ferrocenylalkylthiol ink was patterned on a gold surface using DPN and appropriately applied changes in electrode potential resulted in ink oxidation, to trigger and guide the assembly of polyanionic oligonucleotide-modified particles in an orthogonal manner. Alkyl-thiol terminated DNA strands were functionalized on gold nanoparticles (5 nm and 13 nm diameter) and immobilized on redox-active ferricyanide inks using linear sweep voltametry to control the assembly of charged nanostructures with sub-100 nm dimensions.

Zhang *et al.* [46] used a slightly different approach to functionalize oligonucleotides on a surface using DPN and demonstrated orthogonal assembly of DNA functionalized gold nanoparticles using hybridization reactions (Figure 10.5). A gold substrate was patterned using MHA deposited by DPN followed by wet etching in ferri/ferrocyanode etchant

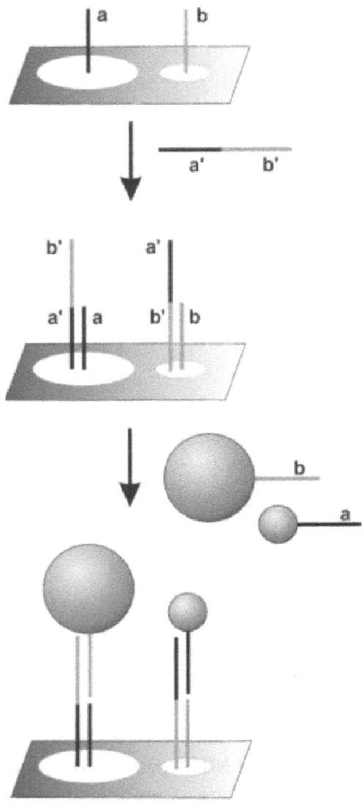

FIGURE 10.4. A schematic representation of orthogonal assembly with DNA nano-structures and nano-particles functionalized with comple-mentary DNA. From the 5' end the sequences are: a: TCTCAACTCGTAA$_{10}$, b: A$_{10}$CGCATTCAGGAT, a'b':TACGAGTTGAGAATCCTGAATGCG; white elipsoid=MHA pattern. (Reprinted with the permission from Demers *et al.*, 2001, *Agnew Chem. Intl. Ed.*, 40(16):3071–3073, © 2001, WILEY-VCH).

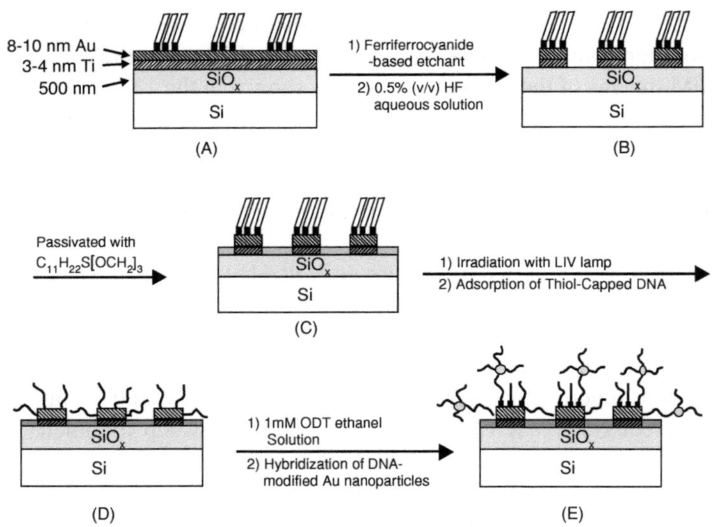

FIGURE 10.5. DPN combined with wet etching to generate arrays of gold nanostructures functionalized with oligonucleotides. (Reprinted with the permission from Zhang *et al.*, 2002, *Advanced Materials*, 14(20): 1472–1474, © 2002, WILEY-VCH)

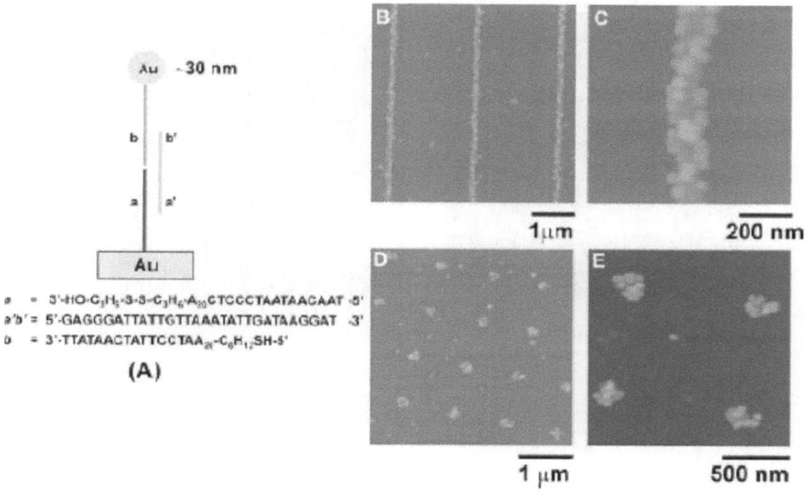

FIGURE 10.6. (**A**) Illustration of three strand DNA hybridization scheme. (**B**) TMAFM image of line features after hybridization with DNA (b)-modified nanoparticles. (**C**) High resolution image of one of the lines in (B). (**D**) TMAFM image of dot features after hybridization with DNA (b)-modified nanoparticles. (**E**) High resolution image of four dots of the features in (D). (Reprinted with the permission from Zhang *et al.*, 2002, *Advanced Materials*, 14(20): 1472–1474, © 2002, WILEY-VCH).

and passivation using ODT. The MHA was stripped from the gold surfaces by irradiating with UV light to oxidize MHA and subsequently rinsing with DI water. The exposed gold surfaces were then functionalized with oligo-nucleotide by immersing in a solution of disulfide functionalized oligo-nucleotide. The substrate was treated with ODT to remove non-specifically bound oligo-nucleotides and to increase the efficiency of the subsequent hybridization reaction. To demonstrate the binding of oligonucleotides to the exposed gold region of the substrate and not in the etched region, the hybridization properties of the surface were studied using complementary linker strands and using complementary DNA strands functionalized on gold nano-particles (30 nm diameter). The arrays modified with DNA (***a***) were first treated with linker DNA (***a'b'***) which had part of its sequence complementary to DNA (***a***) and part of its sequence complementary to DNA (***b***) which was functionalized on the gold nano-particles. After rinsing with PBS buffer the substrate was washed with DNA (***b***) modified gold nano-particles and again rinsed with PBS buffer. Subsequent TMAFM images showed that gold nano-particles were immobilized on the gold part of the substrate and not in the surrounding areas. A separate control experiment was performed to demonstrate that gold nano-particles could also be hybridized to the nano-features using a two-stranded scheme, however the specificity of binding was much lower with the two strand scheme.

The methods mentioned before are indirect methods that either involve resists or prefabricated chemical affinity templates that direct the assembly of a single oligo-nucleotide structure from solution onto a set of nanoscopic features on a surface of interest. To be able to generate nanoarrays of multicomponent systems the device should be capable of directly printing a set of different oligo-nucleotide structures on a surface of interest with

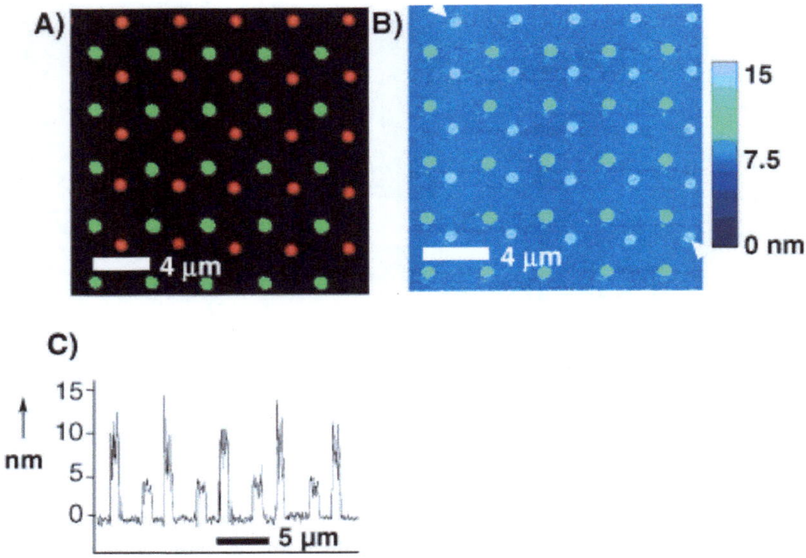

FIGURE 10.7. Direct patterning of multiple-DNA inks by DPN. (A) Combined red-green epifluorescence image of two different fluorophore-labeled sequences (Oregon Green 488-X and Texas Red-X) simultaneously hybridized to a two-sequence array deposited on an SiOx substrate by DPN. (B) Tapping-mode AFM image of 5 (dark)- and 13 (light)-nm–diameter gold nanoparticles assembled on the same pattern after dehybridization of the fluorophore-labeled DNA. The scale bars represent 4 microns. (C) The line plot was taken diagonally through both nanoparticle patterns, and the start and finish are indicated by the arrows in (B). (Reprinted with the permission from Demers *et al.*, 2002, *Science*, 296: 1836–38, © 2002, AAAS).

nano-scale resolution, high registration alignment and control over the biological activity of the resulting structures. Subsequently, Demers *et al.* [10] used direct write DPN to pattern oligo-nucleotides on both metal and insulating substrates (Figure 10.7). DNA patterns were written on gold using DPN by using hexanethiol-modified oligonucleotides as ink. Silicon wafers modified with 3'-mercaptopropyltrimethoxysilane layers were used as substrate for writing with acrylamide modified (Acrydite) oligo-nucleotide inks using DPN. The resulting patterns could be used as hybridization arrays using both fluorophore-labeled complementary oligonucleotides as well as complementary-DNA functionalized gold nano-particles. This demonstrated that size of the nano-particles could be used as tags for AFM based screening procedure—much in a similar manner to the multi-colored fluorophore labels used for optical screening.

Belabure *et al.* [6] fabricated oligonucleotide spots using a technique similar to DPN. They used an array of four silicon micro-cantilevers with passivated aluminum electrodes. By immersing the cantilevers into a droplet and applying an electrostatic field, a combination of electrowetting and dielectrophoretic forces were used to load bio-molecules on to the cantilever array. Cyanine-3 functionalized oligonucleotide solution was loaded on the cantilevers and deposited by reacting with the amino groups of a dendrimer coated glass slide. At the expense of larger size (~70 microns spot size) this platform has several advantages over DPN. The cantilevers have a good loading efficiency, can be used in parallel to maximize throughput (compared to conventional pin spotters used in micro-array

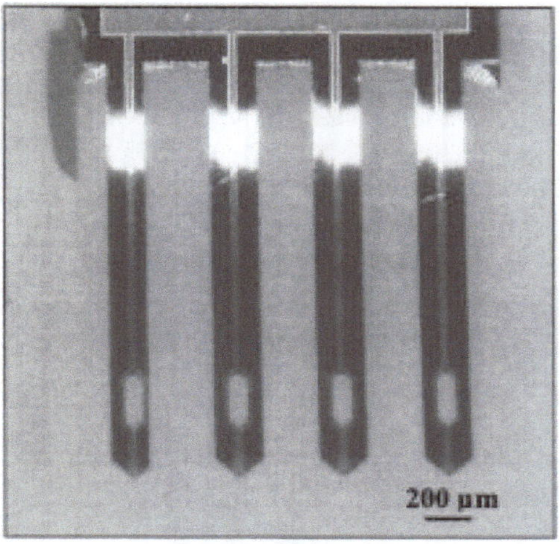

FIGURE 10.8. Arrays of four cantilevers with passivated aluminum electrodes properly designed for electro wetting. The cantilevers are 2 mm long, 210 microns wide, 5 microns thick with 450 micron spacing. (ref. Belaubre *et al.*, 2003). (Figure 3.7 and 3.8 reprinted with the permission from Belaubre *et al.*, 2003, *Applied Physics Letters*, 82(18): 3122–3124 © 2003, American Institute of Physics)

fabrication), can be easily cleaned and therefore reused without significant cross contamination. Two sets of oligo-nucleotides (labeled with Cyanine-3 and Cyanine-5) were repeatedly deposited in a microarray format. A cleaning protocol was optimized and implemented in this study.

Bruckbauer *et al.* [7] used voltage-controlled nano-pipet to deposit antibodies and DNA on a surface to create multi-component and sub-micron features, which has the flexibility to deposit single molecules. Zhou *et al.* [48] demonstrated the patterning of DNA combined with biotin-streptavidin interaction could be used to fabricate three-dimensional bio-nano structures. These two studies will be discussed in more detail section 10.3.3.

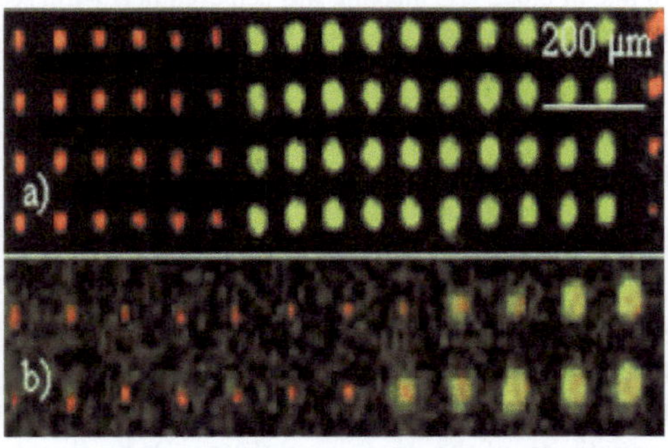

FIGURE 10.9. Fluorescence image of spots obtained with a solution containing Cyanine3-labeled oligonucleotides (15 mers) deposited on a glass slide. Deposition of two different solutions with the same cantilevers: (**a**) Oligonucleotides (15 mers) spots labeled with Cyanine3 (lighter spots) and Cyanine5 (darker, smaller spots) using the cleaning procedure: no cross contamination can be observed. (**b**) Same experiment without the cleaning procedure: cross contamination is clearly visible.

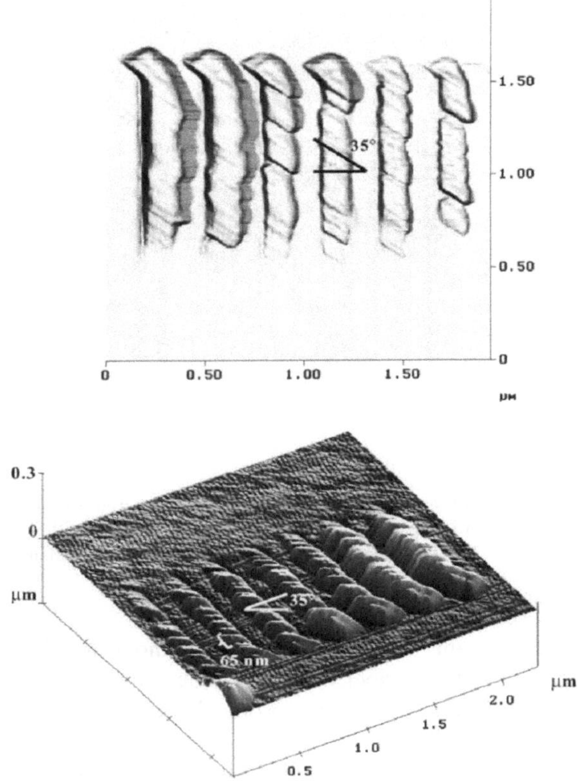

FIGURE 10.10. Top and surface plot views of a topography image of modified collagen molecules deposited on Au substrates at varying forces by using the same tip inked once in collagen solution. Note the 35°, 65-nm periodicity observed along the long axis of the lines, as well as the 300-nm height of the molecules. Rotated TESP AFM tips have a cantilever length of 225 μm and a spring constant of 1–5 N/m rotated at a 15° angle so that high-aspect ratio features can be easily visualized. Parallel lines were made at forces varying from 0.50 nN and a step of 0.05 nN with a scan rate of 0.002 Hz. Imaging was achieved in a 3-μm scan width at a rate of 2.0 Hz. Scan angle did not change the directional orientation of the patterns.

10.3.2. Nano-Patterning of Protein and Petides Using DPN

Proteins are more challenging to implement compared to oligo-nucleotides in nano-arrays. The challenges arise since proteins get easily denatured (due to loss of hydration) and may lose their biological active state due to structural modification on surface adsorption. Other challenges include non-specific adsorption of proteins to the cantilevers which may compromise the sensitivity of AFM operation, non-specific adsorption of proteins to the substrate and the cross reactivity of the protein analytes. Both direct and indirect patterning of protein nano-arrays has been demonstrated in the literature where the retention of their bio-recognition properties has been successfully demonstrated. However, with peptide nano-arrays it is difficult to reliably demonstrate that they are biologically active due to their surface adsorption (or chemi-sorption) after printing them on a substrate.

A
B
C

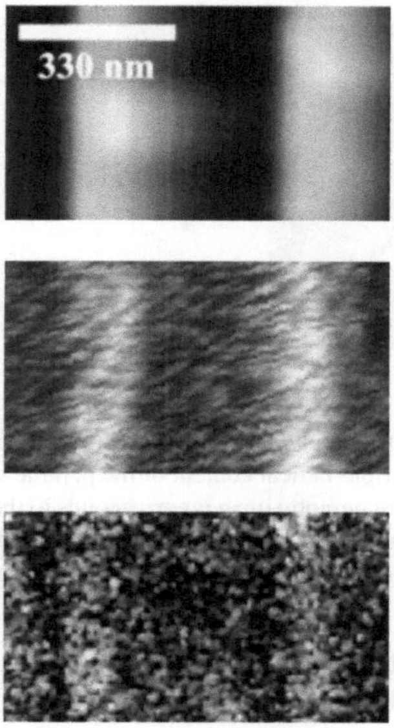

FIGURE 10.11. AFM topography image of collagen patterns without Ab binding (*A*); an NSOM topography image of same collagen pattern with Ab binding (*B*); and an NSOM image of same collagen pattern with Ab binding (*C*). NSOM fluorescence image was obtained at an excitation wavelength of 488 nm, and lines observed in *C* correspond to Ab fluorescence. (Figure 10.10 and 10.11 reprinted with the permission from Wilson *et al.*, 2001, *Proceedings of the National Academy of Sciences*, 98(24): 13660–64 © 2001, PNAS).

Wilson *et al.* [44] demonstrated protein and peptide nano-patterning using DPN by separately printing 30–50 nm wide lines of collagen and collagen-like peptide molecules on a gold surface. This demonstrated the "direct-write" capability of biologically relevant molecules using DPN while simultaneously preserving their structure and biological functionality. This provides tremendous flexibility in biological device applications and proteomics assays as well as a new strategy to study the important hierarchical assembly processes of biological systems. The patterned proteins and peptides were analyzed using Circular Dichroism (CD) spectroscopy and Near-Field Scanning Optical Microscopy (NSOM). Thiolated collagen molecules were patterned by using tapping mode AFM on a thin film of gold (formed by vapor deposition on a mica substrate). The collagen-patterned substrate was then blocked with BSA in PBS. Immuno-affinity studies were performed on the patterned collagen by incubating the substrate in a Rabbit anti-mouse collagen primary antibody solution. The AFM image of the patterned collagen showed a helical architecture of the assembled proteins with a characteristic periodicity of 653.4 nm and angle of $35 \pm 0.7^{\circ}$ along the axis of the patterned lines. This shows a self-assembly mechanism that preserves the characteristics of the native collagen and therefore retains the reactivity, specificity and epitope display after thiolation. This was verified by NSOM images of the fluorescence from the specifically bound antibodies to the collagen structures. In a separate experiment an N-terminal cysteine was incorporated (for chemisorption to the gold substrate) with a collagen-like peptide and patterned using DPN on the gold substrate. CD analyses verified that the collagen-like peptides spontaneously assembled into their

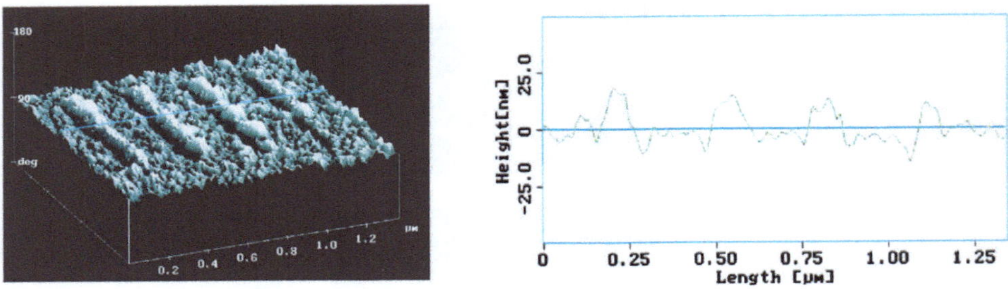

FIGURE 10.12. Topography image of a collagen-like peptide for lines 1 μm in length and 30-nm in width. The collagen-like peptide [Cys(Glu)$_5$(GlyAlaHypGlyProHyp)$_6$(Glu)$_5$, where Hyp is hydroxyproline] was dissolved in 40 mg/ml of aqueous solution. The triple-helical content of the peptide was determined by CD before patterning. The uniform height observed in the topography images corresponds to the 15-nm height of the peptide triple helix. DPN was performed in the same manner as described for the collagen molecule. (Reprinted with the permission from Wilson *et al.*, 2001, *Proceedings of the National Academy of Sciences*, 98(24): 13660–64 © 2001, PNAS).

characteristic homotrimeric triple helices in aqueous solution. The authors argue that the patterned peptides therefore behave in a similar manner to the collagen protein and can be modeled as rigid rods.

Lee *et al.* [20] used DPN to fabricate protein nano-arrays with 100–350 nm feature sizes. Initially MHA patterns were printed using DPN on gold thin-film substrate in the form of dots or grids. The exposed gold substrate was blocked using an ethanolic solution of 11-mercaptoundecyl-tri(ethylene glycol) surfactant. The substrate was then rinsed in ethanol and NANO-pure water. Proteins were immobilized by immersing the substrate in a solution containing the desired protein. The substrate was then rinsed with Tris buffer, Tween-20 surfactant solution and finally with NANO-pure water. The proteins had strong affinity for carboxylic acid-terminated monolayers and weak affinity for surfaces coated with 11-mercaptoundecyl-tri(ethylene glycol). The proteins were then characterized by AFM. For immunoglobin G (IgG) patterns, the reaction of the array with rabbit anti-body

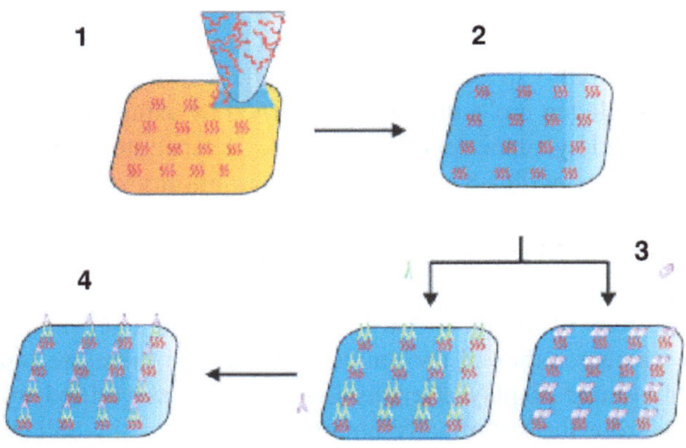

FIGURE 10.13. Diagram of proof-of-concept experiments, in which proteins were absorbed on preformed MHA patterns. The resulting protein arrays were then characterized by AFM. (Reprinted with the permission from Lee *et al.*, 2002, *Science*, 295: 1702–1705, © 2002 AAAS).

to mixtures of proteins was studied by AFM. Tapping mode AFM images showed that Lysozyme cleanly assembled on the MHA nanopatterns with almost no evidence of non-specific protein adsorption and the height profiles suggest that between one and two layers of protein adsorb to each MHA site. The authors also attempted to study cell adhesion using protein functionalized nano-arrays. This will be discussed in Section 10.4.

Noy *et al.* [31] demonstrated a direct write method using DPN by coating the pen with human chorionic gonadotropin (HCG) antibody tagged with tetramethylrhodamine (TMR) dye and printing the protein on glass surface. The glass surface was pre-treated with 3-glycidoxypropyltrimethoxysilane to introduce expoxy groups that facilitate protein adhesion to the surface. The authors report using a high concentration of the protein "ink". Also, the pattern was over written 10 times to enhance the surface concentration of the protein. Confocal fluorescence microscopy was used to detect the printed proteins. This resulted in feature sizes larger than 200 nm. According to Ginger *et al.* [13], the authors "did not report the characterization of the biological activity of the protein patterns after deposition, so it is unknown whether the antibody retained its structure and function after deposition and surface coupling". Cheung *et al.* [8] fabricated chemoselective protein-to-surface linkers to create nanoscale chemical templates for capturing virus and for forming virus nano-arrays. This work will be discussed in Section 10.4.

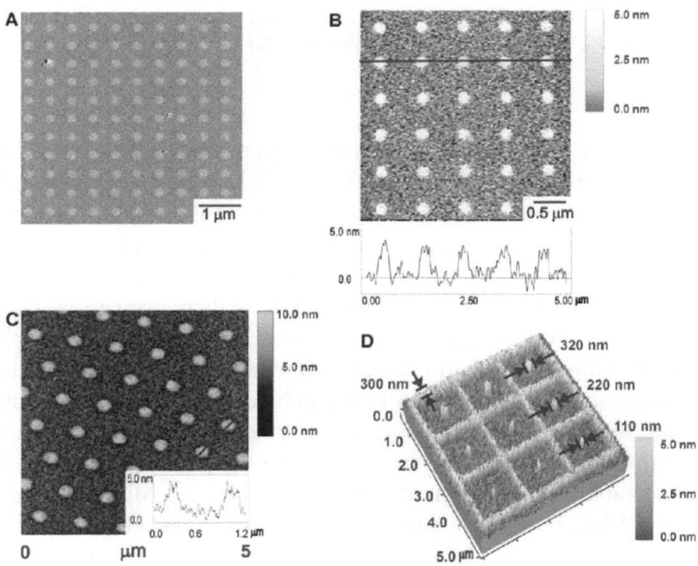

FIGURE 10.14. AFM images and height profiles of lysozyme nanoarrays. (A) Lateral force image of an 8 μm by 8 μm square lattice of MHA dots deposited onto an Au substrate. The array was imaged with an uncoated tip at 42% relative humidity (scan rate = 4 Hz). (B) Topography image (contact mode) and height profile of the nanoarray after lysozyme adsorption. A tip-substrate contact force of 0.2 nN was used to avoid damaging the protein patterns with the tip. (C) A tapping mode image (silicon cantilever, spring contant = ~40 N/m) and height profile of a hexagonal lysozyme nanoarray. The image was taken at a scan rate of 0.5 Hz to obtain high resolution. (D) Three-dimensional topographic image of a lysozyme nanoarray, consisting of a line grid and dots with intentionally varied feature dimensions. Imaging was done in contact mode as described in (B). (Reprinted with the permission from Noy *et al.*, 2002, *Nano Letters*, 2(2): 109–112 © 2002, American Chemical Society).

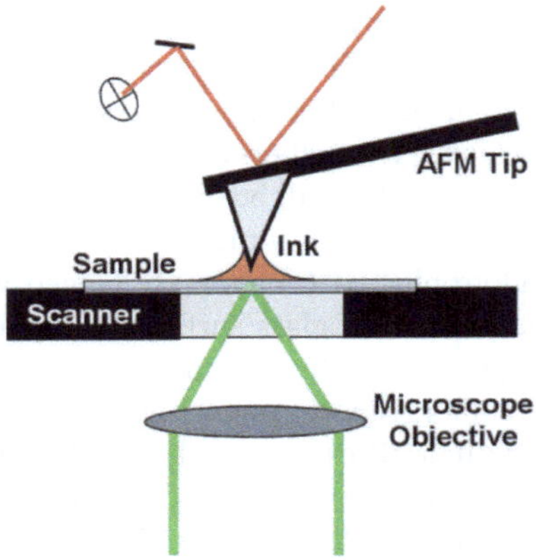

FIGURE 10.15. Schematic of the experimental setup used for dip-pen nanolithography. The instrument combines atomic force microscopy and scanning confocal microscopy functionalities. A Bioscope AFM head (Digital Instruments, Santa Barbara, CA) is mounted on a Nikon Eclipse 300 inverted microscope equipped with a custom-built stage that incorporates a closed-loop piezo scanner (PolytecPI, Germany). The scanning stage is used in both the lithography procedure and confocal imaging. The fluorescence excited by a 514 nm line of Ar-Ion laser (Coherent Innova-70) is collected with 100 × 1.4 NA lens, passed through a 50 micron pinhole to reject stray light and then detected by an avalanche photodiode (Perkin-Elmer Optoelectronics, Santa Clara, CA). Excitation light was rejected by a long-pass filter (Omega Optical, VT). (Reprinted with the permission from Noy *et al.*, 2002, *Nano Letters*, 2(2): 109–112, © 2002, American Chemical Society).

Zhang *et al.* [47] reported the generation of arrays of nanoscale features functionalized with inorganic nanoparticles and proteins (rabbit IgG). In the case of rabbit IgG, the bioactivity of the array was demonstrated by monitoring its reaction with fluorophore-labelled anti-rabbit IgG. The fabrication strategy was similar to that used by Zhang *et al.* [46] for fabricating DNA nano-arrays. MHA nano-resist is used to etch nanofeatures on a gold substrate using ferricyanide wet etchant. The MHA is then stripped by oxidation using prolonged exposure to UV light and rinsing in water. The exposed silicon oxide substrate is then blocked by immersing it in n-octadecyltrimethoxysilane (OTS) solution in toluene which prevents non-specific adsorption of protein by chemisorption to silicon oxide. This forms a template for protein absorption to form protein nano-arrays. The topography of the template surface is characterized by AFM. The template is immersed in buffer solution containing rabbit IgG and rinsed with PBS buffer and Milli-Q water. Rabbit IgG and a broadclass of IgG proteins in general are known to have a high affinity for COOH-terminated SAMs at pH 7 and a relatively weak affinity for hydrophobic surfaces coated with alkanesilane as compared to the carboxylic acid-terminated surfaces. AFM characterization of the protein nano-array shows an increase of height of the patterned areas, which is consistent with the formation of an IgG monolayer on the template. To test the bioactivity of the IgG array, the sample was immediately immersed into a PBS buffer solution of fluorophore (TRITC)-labeled anti-rabbit IgG. After rinsing the substrate with PBS buffer, Tween-20 surfactant solution and Milli-Q water, confocal fluorescent microscopic images are obtained. The images showed that the TRITC-labeled anti-rabbit IgG specifically binds to the IgG-immobilized nano-features. TMAFM image of the substrate also showed an

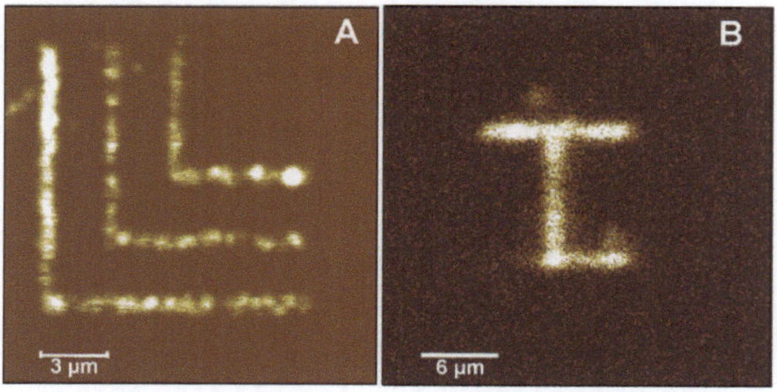

FIGURE 10.16. (A) Scanning confocal microscopy image of a Rhodamine 6G patterned in a series of lines on glass surface. The pattern was fabricated using Ultralever contact AFM probes (Park Scientific Instruments). Probe translation speed was 0.2 μm/s. (B) Scanning confocal microscopy image of a HCG antibody pattern fabricated on glass. Probe translation speed was 0.2 μm/s. Each line in the pattern was overwritten 10 times to maximize the protein surface density. (Reprinted with the permission from Noy *et al.*, 2002, *Nano Letters*, 2(2): 109–112, © 2002, American Chemical Society).

increase in height of the nanopatterns which is consistent with the absorption of a monolayer of the anti-rabbit IgG.

Hyun *et al.* [15] demonstrated an indirect stepwise method for fabrication of patterned protein nanostructures with feature sizes of the order of 200 nm. First a SAM of MHA is patterned onto gold by DPN and the un-patterned regions are passivated with a protein-resistant oligoethylebe glycol-terminated alkanethiol SAM. This is followed by covalently conjugating an amine-terminated biotin derivative on the chemically activated MHA SAM nanopattern. The surface is then incubated with streptavidin to form strepatavidin nanostructures, mediated by molecular recognition between biotin and streptavidin. Finally, protein nanopatterns are fabricated by molecular recognition-mediated immobilization of biotinylated protein (BSA) from solution. The authors argue that the fabrication methodology is generically applicable because of the prevalence of the biotin-tagged molecules.

Jung *et al.* [18] also demonstrated an indirect method for fabricating protein nanoarrays. In this study MTPMS (3'-mercapto-propyl-trimethoxy-silane) was patterned on a clean glass substrate using DPN. The un-patterned areas were passivated with trialkoxysilane containing poly(ethylene glycol). The availability of the pendant thiol groups of DPN-patterned MPTMS for further coupling reactions was tested by introducing biotin and Cy3-streptavidin. Fluorescence was observed using an inverted epi-fluorescence microscope and images of three 2.5 microns by 10 microns regions with Cy3-streptavidin patterns were obtained in this study.

The methods mentioned before are indirect procedures that either involves resists or prefabricated chemical affinity templates that direct the assembly of a single protein structure from solution onto a set of nano-scopic features on a surface of interest. Such affinity arrays do not allow immobilization of more than one set of proteins reliably on a substrate. To be able to generate nanoarrays of multicomponent systems the device should be capable of directly printing a set of different protein/peptide structures on a surface of interest with nano-scale resolution, high registration alignment and control over the biological activity of the resulting structures.

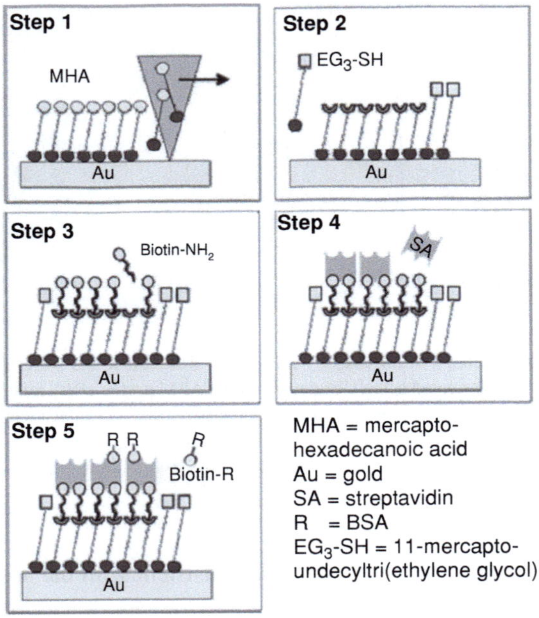

FIGURE 10.17. Stepwise Fabrication Process to Create Molecular Recognition-Mediated Protein Nanostructures. (Reprinted with the permission from Hyun *et al.*, 2002, *Nano Letters*, 2(11): 1203–1207, © 2002, American Chemical Society).

To demonstrate proof of concept experiments, Lysozyme (Lyz) and rabbit immunoglobulin-gamma (IgG) nanodot arrays were constructed using DPN in direct-write fashion by Lee *et al.* [21]. This study demonstrated successful fabrication and testing of a two-component protein nano-array that was biologically active and capable of detecting a biological complement in solution. The pen tips were chemically modified to achieve

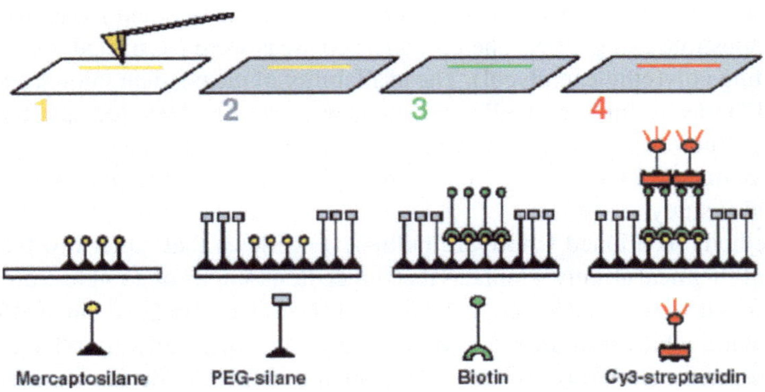

FIGURE 10.18. Schematic of biotin-streptavidin conjugation to patterned MTPMS. (Reprinted with the permission from Jung *et al.*, 2003, *Journal of the American Chemical Society*, 2003, 125(40): 12096–12097, © 2003, American Chemical Society).

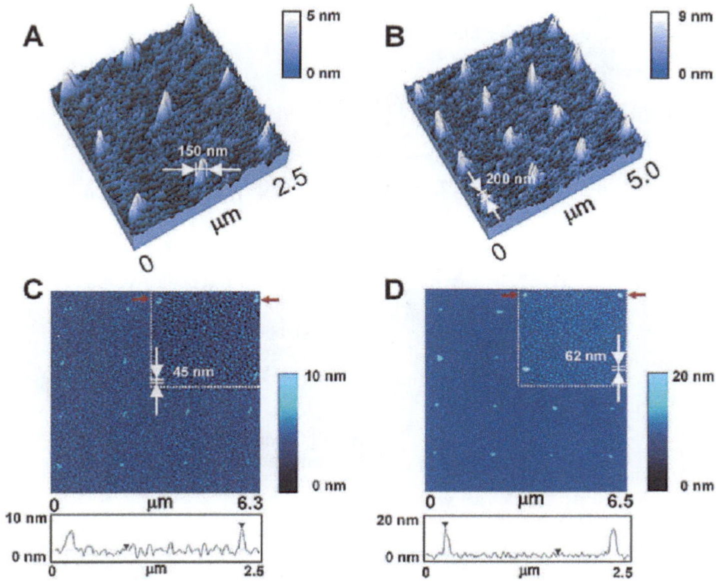

FIGURE 10.19. (A) Tip Modification Scheme; (B) Protein Patterning Schemes. (Reprinted with the permission from Lee *et al.*, 2003, *J. Am. Chem. Soc.*, 2003, 125(19): 5588–5589, © 2003, American Chemical Society).

reliable DPN operation with inks containing protein solutions in PBS. The DPN steps were performed in a humidity control chamber where the relative humidity was maintained at 80–90%. Relative humidity values below 70% resulted in inconsistent transport properties probably due to denaturing of the proteins.

The proteins in solution are likely to bind non-specifically on the gold coated reflective backside of the cantilevers if they are dipped into the protein solution. This would reduce the reflectivity of the cantilevers compromising the quality of AFM operation and thus adversely affect the AFM scanning and DPN writing operation. Therefore, to prevent protein adsorption on the reflective gold layer on the back side of a commercial AFM cantilever, the pens were dipped in an ethanolic solution of a symmetric 11-mercapto-undecylpenta(ethylene glycol)disulfide (PEG). PEG is known to repel proteins and significantly reduce protein adsorption on surfaces treated with PEG. The pens were then rinsed in ethanol, dried and the tip side was coated with a very thin layer of gold (with Ti adhesion layer) using thermal evaporation methods. The pens were then treated with a solution of thiotic acid in ethanol (a strong oxidizer) to make them hydrophilic and aid in protein adsorption on the tips.

The chemically treated pens were dipped in the desired protein solution and were used immediately after dipping to print protein nano-arrays on a gold substrate. The gold substrate was surface treated according to the method outlined in Weinberger *et al.* [43]. Gold substrates were chosen due to the strong affinity of cysteine residues (in the chosen proteins) for adsorption on gold and also because PEG can be used as a passivating layer on gold for the areas not occupied by the proteins. Gold passivated with PEG is known to resist nonspecific adsorption of proteins. The substrate was subsequently blocked with PEG and rinsed with Nano-pure water. The PEG solution was also prepared in NANO-pure

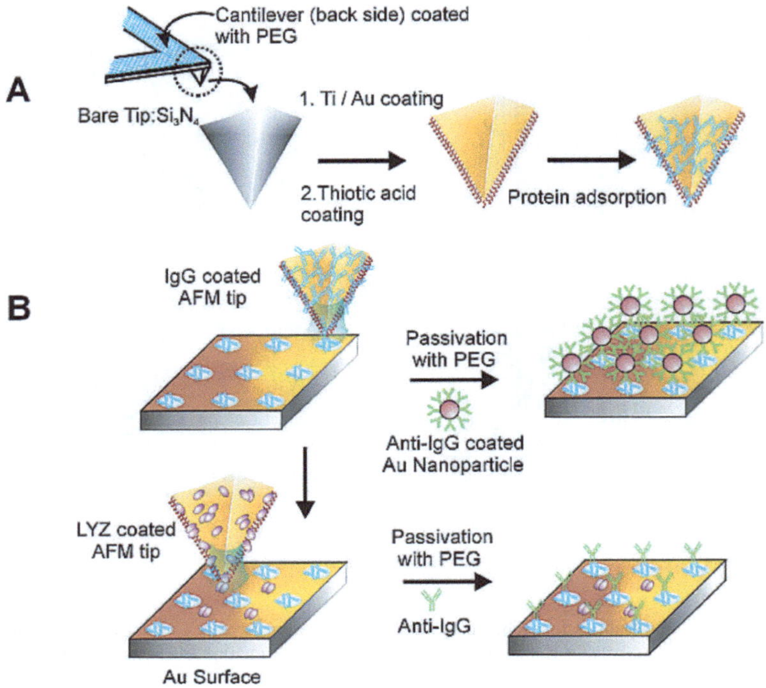

FIGURE 10.20. Protein nanoarrays prepared via direct-write DPN. (A) Contact mode image (contact force 0.1 nN) of lysozyme nanodot arrays. Each dot took 20 s to form. (B) Contact mode image (contact force 0.1 nN) of IgG nanodot arrays. Each dot took 30 s to form. An IgG nanodot array before (C) and after (D) treatment with a solution anti-IgG coated Au nanoparticles: Images were taken at 0.5-Hz scan rate in tapping mode. Each dot took 5 s to form. (Reprinted with the permission from Lee *et al.*, 2003, *J. Am. Chem. Soc.*, 2003, 125(19): 5588–5589, © 2003, American Chemical Society).

water (instead of conventional organic solvents like ethanol) to prevent the denaturing of proteins patterned on the substrate.

To test the bio-recognition properties of the IgG nanoarray, it was incubated in a solution of gold nanoparticles coated with antirabbit IgG (in PBS). A comparison of the AFM height profiles of the array before and after treatment with this solution showed a height increase in the active area of the array with little nonspecific binding to the passivated inactive areas. This is consistent with the formation of monolayer on the active areas of the substrate.The advantage of DPN direct write patterning (over indirect affinity arrays) is the ability to fabricate complex multi-component protein nano-arrays without the risk of cross contamination.

To demonstrate the concept, the authors fabricated a two-component protein nano-array. Initially, IgG functionalized nano-arrays were fabricated using the approach mentioned above. In the second step, Lysozymes features were printed between the IgG features. The array was incubated in a solution containing anti-rabbit IgG. Significant height increase (almost double) due to anti-rabbit IgG binding was observed only on the rabbit IgG features and not the area patterned with Lysozyme. This near doubling of feature height is attributed

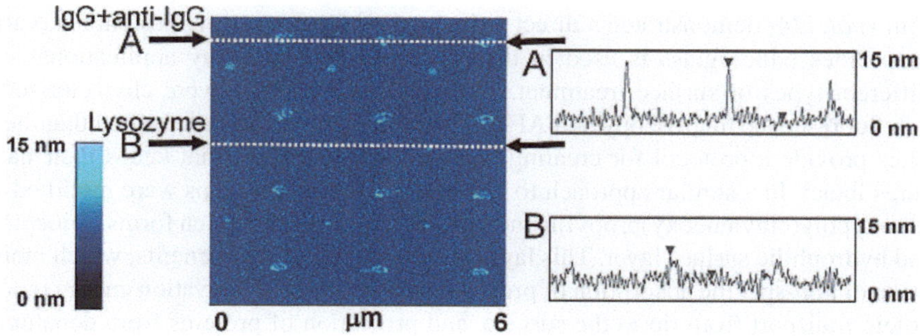

FIGURE 10.21. Two-component protein pattern after reaction with anti-IgG. (A) A height increase of 5.5 ± 0.9 nm ($n = 10$) in the IgG features is observed by AFM. (B) No height increase is observed. The image was taken under the same conditions in Figure 1 (C) and (D). (Reprinted with the permission from Lee *et al.*, 2003, *J. Am. Chem. Soc.*, 2003, 125(19): 5588–5589, © 2003, American Chemical Society).

to a 1:1 reaction between the two protein structures and further demonstrates the feasibility of probeless detection with nanoscale systems. The authors noted the density of the anti-IgG bound to the IgG features to be non-uniform and attributed this to the random orientation of the IgG epitopes and their partial denaturation after adsorption on the Au surface, which was observed in other studies. This work demonstrates a convenient method for generating protein nano-arrays on a surface in a direct-write fashion that is amenable for massive parallelization and multiplexing with multi-component functional proteomic arrays. The slow diffusion rates of the large protein molecules, the need for chemically modified tips and surface treated gold substrates are the inherent deficiencies of this method.

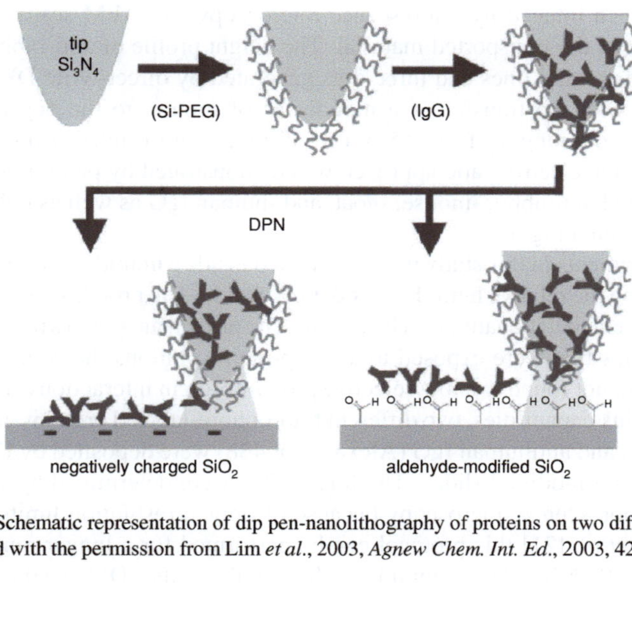

FIGURE 10.22. Schematic representation of dip pen-nanolithography of proteins on two different modified substrates. (Reprinted with the permission from Lim *et al.*, 2003, *Agnew Chem. Int. Ed.*, 2003, 42: 2309–12, © 2003, Wiley-VCH).

Lim *et al.* [24] demonstrated a direct write method for fabricating protein nano-arrays on glass slides (since glass is used in the conventional microarray applications) using two different types of surface treatment. The fabricated features were characterized using both fluorescence microscopy and AFM detection methods. It was verified that the two strategies provide a protocol for creating protein nano-structures that keeps their natural structures intact. In a similar approach to Lee *et al.* [21] the AFM tips were modified with 2-[methoxypoly(ethyleneoxy)propyl]trimethoxysilane (Si-PEG) which forms a biocompatible and hydrophilic surface layer. This layer provides a number of benefits, which include: inhibition of non-specific adsorption of proteins, reduction of the activation energy required for protein transport from tip to the surface, and protection of proteins from denaturation on the tip surface. The authors observed that in the absence of this layer the protein inking solutions in PBS do not wet the silicon nitride cantilevers used as AFM tips and produce inconsistent low density protein patterns. In addition to cantilever modification, the authors also developed two strategies for efficient ink transport on silicon oxide substrate. In the first approach a negatively charged surface was created by treatment with a base (which would enable the positively charged parts of the protein to interact electro-statically). In the second approach an aldehyde-modified substrate was made (to enable covalent coupling of the proteins).

The glass slides were then rinsed with ethanol and cured under flowing nitrogen. After surface treatment of the substrate and the chemical modification of the tip, the AFM tip was immersed into a protein solution. The protein solution was dissolved in PBS and stabilized with glycerol. The tip was then mounted on an AFM and used to write patterns of protein by direct-write DPN. The un-patterned regions of the aldehyde-modified surface was passivated by immersing the glass slides in an aqueous solution of amine terminated oligo(ethylene glycol) to prevent non-specific adsorption of proteins from solution.

Rabbit IgG and fluorophore-labeled anti-rabbit IgG nano-arrays were fabricated by direct write DPN strategy mentioned above. The fluorophore-labeled anti-rabbit IgG nano-structures were then imaged by fluorescence microscopy and AFM scans to confirm the chemical identity of the transported material. The height profile of anti-rabbit IgG pattern consisting of two parallel lines and three dots generated by direct-write DPN were found to be consistent with the transfer of a monolayer of protein to the negatively charged surface. Protein spots ranging from 55 nm to 550 nm were printed using this protocol. The generality of the electro-static approach was demonstrated by patterning and imaging fluorophore-labeled antirabbit, -mouse, -goat, and -human IgG as well as unlabeled rabbit, mouse, goat and human IgG.

All the experiments in this study were conducted inside a humidity-controlled chamber and by maintaining the relative humidity at 60–90% inside. The proteins were kept hydrated at all times to prevent their denaturation by drying. The protein ano-arrays made from protein solution coated tips that were exposed to air or protein solutions that were not stabilized with glycerol were not found to show selective protein-protein interactions. To demonstrate multiple protein-ink capabilities, two different fluorophore-labeled proteins, anti-rabbit IgG (Alexa Fluor 594) and antihuman IgG (Alexa Fluor 488) were deposited by DPN in a serial fashion on aldehyde-modified slides. The line widths were determined by writing speeds rather than by fluorescence microscopy because of optical resolution limits. Also, unlike the work of Lee *et al.* [21] this protocol can be performed for a longer range of ambient humidity (over a 60–90% relative humidity). Also, unlike other DPN studies the spots did

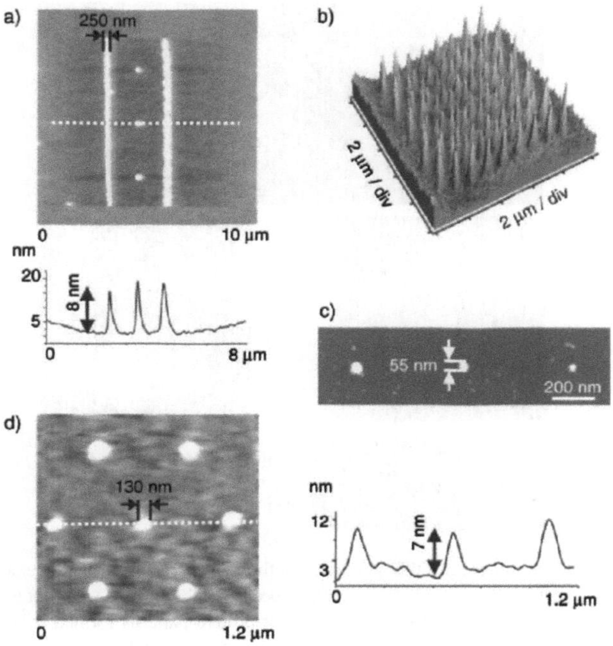

FIGURE 10.23. AFM topography images of protein nanostructures. (a) Antirabbit IgG protein nanostructures on a negatively charged SiO2 substrate written at 0.08 mms 1 and corresponding line profile; (b) 3D topography image of antirabbit IgG protein dot arrays (contact time = 3 s; div = division); (c) tapping-mode image of antirabbit IgG protein dots (contact time = 0.5 s, set point = 0.5 nN); (d) rabbit IgG nanostructures on an aldehyde-modified SiO2 substrate (contact time = 2 s). AFM images were collected withan uncoated tip at a scan rate of 2 Hz. (Reprinted with the permission from Lim *et al.*, 2003, *Agnew Chem. Int. Ed.*, 2003, 42: 2309–12, © 2003, Wiley-VCH).

not have uniform edges and ink transport seemed to be dominated by physical stamping rather than by diffusion from the tip to the substrate.

To test the bio-recognition properties of the IgG structures generated by DPN, a rabbit IgG and human IgG pattern were incubated in PBS solution containing a mixture of fluorophore labeled anti-rabbit IgG (Alexa 594, red) and anit-human IgG (Alexa 488, green). The substrate was then rinsed successively in PBS buffer, Tween-20 surfactant solution and Milli-Q water. Fluorescence microscopy images showed that anti-rabbit IgG had bound selectively to rabbit IgG patterns and the anti-human IgG had bound selectively to human IgG patterns. Cross-reactivity between the rabbit IgG and the human IgG was estimated by the authors to be approximately 20% based on two-color reading in each of the features.

Agarwal *et al.* [1] demonstrated that the DPN process can be employed to deposit a synthetic peptide driving the AFM cantilever in tapping mode at high amplitudes. The advantage of this method is that it does not exert harsh disruptive forces on surfaces (unlike in contact mode), which makes it amenable for printing bio-molecules on soft substrates and in situations where the molecules are not strongly immobilized. The synthetic peptide contained an N-terminal c-Myc tag (an antigenic epitope) and a C-terminal histidine (His) tag. The c-Myc tag allows detection of the peptide using indirect immunofluorescence

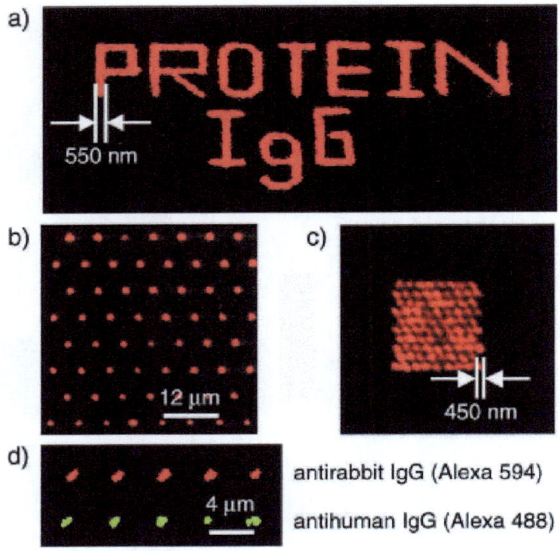

FIGURE 10.24. Fluorescence images of DPN-generated antirabbit IgG (labeled withAlexa 594) structures on a negatively charged SiO2 surface (a–c): (a) words (writing speed = 0.06 mms−1); (b) dot arrays (contact time = 5 s); (c) dot arrays (contact time = 3 s); (d) two-component protein patterns formed by DPN on an aldehyde-derivatized SiO2 surface (contact time = 5 s). (Reprinted with the permission from Lim *et al.*, 2003, *Agnew Chem. Int. Ed.*, 2003, 42: 2309–12, © 2003, Wiley-VCH).

while the C-terminal His-tag can be used in binding to Ni-chelating substrates. The motivation for using a synthetic peptide was to subsequently detect and/or build upon initially pattern. However, the authors report that initial experiments to detect the synthetic peptide via immuno-labeling were unsuccessful, probably due to gold substrate quenching the fluorescently-labeled anti-body and/or a weak signal. Height scans also indicated the peptide was laying flat on the substrate—thus partially contributing to the gold mediated quenching and/or unsuitable epitope presentation. The authors proposed a future study using nickel-histadine (Ni-His) binding. In the initial experiments nickel nitriloacetic acid (Ni-NTA) slides were used as substrates for the DPN experiments but were found to be unsuitable because of scratching during contact mode AFM scans and also during deposition. Peptide deposition was not observed even when the experiments were conducted in tapping mode, probably due to the inherent high surface roughness of the slides. Hence gold substrates were prepared by template stripping. This is achieved by sputter coating a freshly cleaved mica with gold followed by attachment of gold coated surface of the mica substrate to double sided tape. The mica was then peeled off just before use to expose an atomically flat gold surface (surface roughness <0.1nm). Some interesting observations were made in this study. The drive amplitude for the scanning probe was found to significantly reduce from the natural value to a tuned preset target amplitude when coated with peptide—which provided a way to determine if the probe had been coated with peptide. The deposition process required 2–5 times higher drive amplitude than the imaging mode and was found to increase to the natural value upon depletion of peptide. The authors also studied the diffusion characteristics in tapping mode DPN and found that diffusion spots similar to

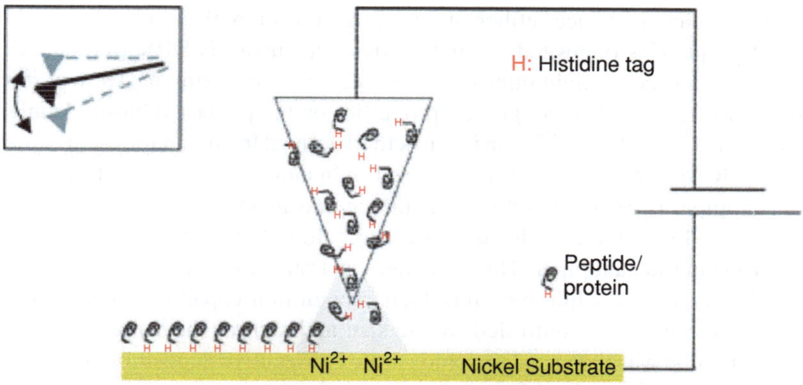

FIGURE 10.25. Schematic of E-DPN process used in tapping mode. (Reprinted with the permission from Lee *et al.*, 2003, *J. Am. Chem. Soc.*, 2003, 125: 7408–12, © 2003, American Chemical Society).

contact mode DPN are obtained. The authors were able to write 70–200 nm wide lines using tapping mode DPN. The results showed indirect evidence of water meniscus nucleating and growing on the scanning probe tips that increased the drive amplitude for imaging with each successive writing operation. The authors also speculated that high driving amplitudes during DPN can blunt the AFM probe.

The protein surface adsorption techniques used in the previously mentioned studies are inherently unreliable because the substrates need to undergo multiple solution/buffer treatments that can disrupt or weaken the electrostatic binding. So, a stronger bond is needed in the assays used for protein nano-arrays. Ni-His bonds are commonly employed in immobilized metal-affinity chromatography (IMAC) for protein purification. Proteins with exposed histidine and cysteine side chains have a high affinity for certain metal ions like Ni^{2+}, Co^{2+}, Zn^{2+}, Cu^{2+}, Fe^{3+} and Mn^{2+}. The Ni-His bond can be used to provide strong linker chemistry by using the hex-His repeat incorporated in recombinant bio-molecules.

Agarwal *et al.* [2] subsequently used combination of electrochemical DPN (or E-DPN) and tapping mode AFM to demonstrate precise, patterned and stable immobilization of peptides, proteins and small synthetic molecules (porphyrins) on a metallic substrate. The results indicate that electrochemistry at the nanoscale can be used to immobilize molecules via metal chelation (e.g., Ni-His) and/or metallochemistry. In this technique, the AFM tip is coated with peptide/proteins and scanned at slow speeds with a negative bias applied to the tip and a nickel-coated substrate held at ground potential. A small meniscus forms on the AFM tip due to capillary condensation. The authors argue that the applied potential casues the ionization of the nickel surface. The histidine tagged proteins or peptides or free base porphyrins (used as inks in this study) coated on the ink tips dissolves in the water meniscus and attaches to the ionized nickel surface. In this study different ink materials were considered, including synthesized, processed and purified solutions of the histidine tagged proteins TlpA, its truncated mutant TlpA-8, the green fluorescent protein fused with R5 peptide (GFP-R5), and terta(N-methylpyridinium)porphyrin chloride salt. The AFM probes were dipped into these inks and used to write on a substrate made from freshly cleaved mica coated with metallic nickel. It was observed that no protein deposition occurred on the nickel surface, etiher at zero potential or with native uncoated probes. GFP-R5 and porphyrins deposited on nickel substrates using E-DPN were subsequently detected using fluorescence light microscopy. However, no experiments were performed in this study to ascertain the bio-recognition properties on the patterned bio-molecules.

As mentioned in Section 10.3.1, using a method derived from scanning-ion conductance microscopy (SICM) Bruckbauer *et al.* [7] used a "nanopipet" to fabricate protein nano-arrays. A nanopipet is manipulated over a surface and maintained at a constant distance from the substrate by monitoring the ion current that flows between an electrode in the bath and an electrode in the nanopipet. The study demonstrates that antibodies can be delivered and attached to a surface while retaining their recognition capability. The deposition of molecules on the surface is controlled by the size and sign of the voltage applied to the pipet. This technique was then used to demonstrate consecutive deposition of biomolecules at the same point on the surface for creating spatially varying fluorescent and biological properties. Nanopipets were fabricated from glass capillaries using a laser based pipet puller from Sutter Instrument, CA. The nanopipets were typically 3 cm long, 5mm long tapered tip, inner diameter of 90–130 nm and outer diameter of 240–280 nm. The ink materials were Rabbit IgG labeled with Alexa Fluor 488 dye and Biotin labeled with Alexa fluor 647 dyes. The substrate for writing is prepared by coating glass bottomed surfaces with different proteins. The different coatings used were: streptavidin, aminosilanes, biotin, streptavidin and rabbit IgG. The writing experiments were performed using an inverted microscope with an epi-fluorescence filter and two xy stages (one for moving the pipet and one for scanning the substrate). A voltage range of ± 1V was applied across Ag/AgCl electrodes in the bath and the ion current was amplified to modulate the nanopipet tip to ± 50 nm by using feedback control applied to the sample stage. The authors report that the distance between the nanopipet and the substrate was 120 nm during deposition. Different pipets were used to deposit two different molecules with an alignment accuracy of 250 nm. Alignment was

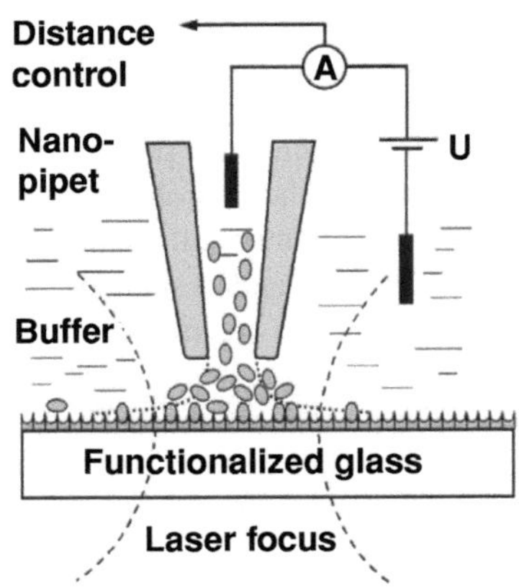

FIGURE 10.26. Schematic of the experiment. A voltage (*U*) is applied between two Ag/AgCl electrodes, one inserted into the nanopipet and the other inserted into the bath of buffer solution. The pipet is filled with biomolecules, and the ion current (*A*) is used as a fine control of the molecule delivery as well as for tip-surface distance control. Laser illumination and confocal detection are used to image the deposited features. (Reprinted with the permission from Bruckbauer *et al.*, 2003, *J. Am. Chem. Soc.*, 2003, 125(32): 9834–39, © 2003, American Chemical Society).

achieved by scanning the tip over a previously written feature. An optical apparatus was used to obtain fluorescence images of the deposited features. This study demonstrated the ability to deliver goat anti-rabbit IgG antibody by controlling the applied voltage from −0.5V to −1.5V.

Experiments were performed to check if the deposited biomolecules (proteins and DNA) could retain their functionality. A 2 × 2 array of Alexa Fluor 488-labeled IgG (green fluorescence) was deposited on a positively charged glass surface which was then treated with human IgG followed by goat anti-rabbit IgG labeled with Alexa Fluor 647 (red fluorescence) and with intermediate rinsing steps using PBS (to remove excess IgG). The fluorescent images showed successful binding between rabbit IgG and anti-rabbit IgG which is consistent with the behavior of rabbit IgG since it has less affinity for the surrounding human IgG. This confirmed that the deposited rabbit IgG maintains its biological function. The authors theorized that background signal from non-specifically bound proteins could be reduced by using BSA (bovine serum albumin) to block the surface and to prevent non-specific adsorption. Experiments were also performed to successfully co-localize molecules in different layers on top of each other. Indirect evidence of delivering individual molecules on a substrate by controlling the voltage was also presented in this study.

10.3.3. Nano-Patterning of Composite Bio-Molecular Structures

Composite biomolecular devices (e.g., combination of proteins/peptides and nucleic acids, etc.) were also fabricated using different dip pen technologies. In addition to the results of Bruckbauer *et al.* [7] mentioned in the previous section, they also reported the deposition of biotinylated ssDNA molecules on a streptavidin-coated substrate (and subsequently blocked with non-complementary biotinylated ssDNA with a different sequence) using nanopipets and demonstrated their successful hybridization by treatment with cDNA molecules.

Zhou *et al.* [48] demonstrated an interesting example of fabricating three-dimensional DNA-protein complexes using micro-contact printing and nanoshaving. It was suggested that similar structures with much smaller features could be built using DPN. DNA was chosen as a building block because it can be easily functionalized and the structural layers can be functionalized with different sequences of DNA increasing the diversity of the fabricated structures. A templated sequential assembly of streptavidin and bis-biotinylated DNA was performed with fine control over pattern size and height. The fabrication steps involved: a gold surface patterned using self assembled monolayer capped with biotin (for specific binding of biotin) and blocking the remainder of the substrate with SAM terminated hexa-ethylene glycol followed by treatment with streptavidin to produce a structured streptavidin surface. Since each streptavidin molecule has four biotin binding sites, (two on each side of the protein) only the bottom biotin sites are used leaving the upper ones available. This surface is then treated with a short rigid duplex (ds) DNA, with both ends modified with biotin groups. The surface is then sequentially treated with streptavidin followed by bis-biotinylated dsDNA to build up multilayered streptavidin-DNA structures selectively on the unpatterned surface.

Apart from DNA and proteins, lipids and liposomes would be interesting biomolecular devices to fabricate using dip pen technologies. No examples exist in the literature as yet to demonstrate the feasibility of fabricating these devices using DPN. Moraille and Badia

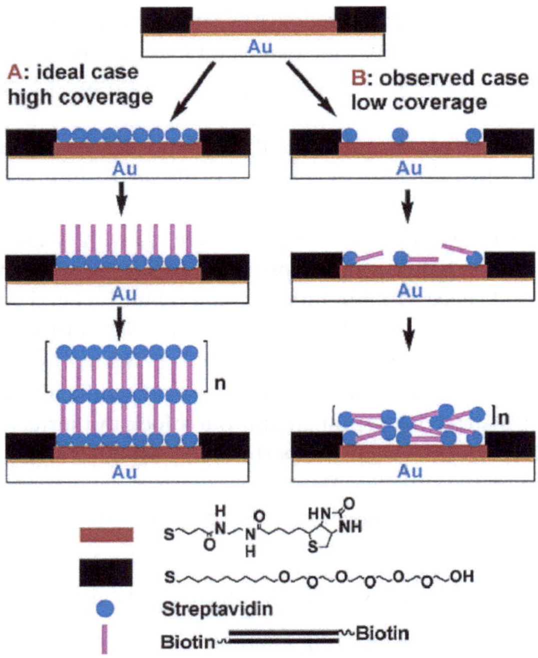

FIGURE 10.27. Strategy for the fabrication of 3D DNA-Streptavidin Surface Assemblies. (Reprinted with the permission from Zhou *et al.*, 2003, *Nano Letters*, 2003, 3:1517–20, © 2003, American Chemical Society).

(2003) demonstrated an interesting experiment for forming phosphor-lipid bi-layers stripes (~150–250 nm) on mica using dynamic wetting instabilities that can arise in Langmuir-Blodgett technique. Possibly, a better control of the self-assembly of the phospho-lipid bi-layer stripes can be achieved by surface nano-patterning using DPN. Fabrication of lipids using direct write DPN is an intriguing challenge and the next frontier for DPN research.

10.4. DPN BIO-MOLECULAR DEVICES FOR CELL AND VIRUS CAPTURE

Dip pen technologies have been successfully employed to fabricate bio-molecular nano-arrays to demonstrate specific adhesion of cells and virus particles. These studies relied on fabricating nano-arrays with proteins or inks using DPN that have affinity for the cell membranes or virus capsids. However, the studies did not demonstrate if the biological activity of the captured cells or virus particles were retained after surface immobilization.

Lee *et al.* [20] were the first to attempt cell adhesion studies using a protein nano-array composed of a cell binding protein called Retronectin. Cell adhesion on Retronectin and subsequent spreading of cell occurs by interaction with the extra-cellular matrix protein on cell membranes by focal adhesion. Earlier studies of immobilizing proteins were limited to to 1 micron patches due to lack of higher resolution patterning methods and thus limited the investigation of size and distribution of the focal adhesion during cell adhesion since

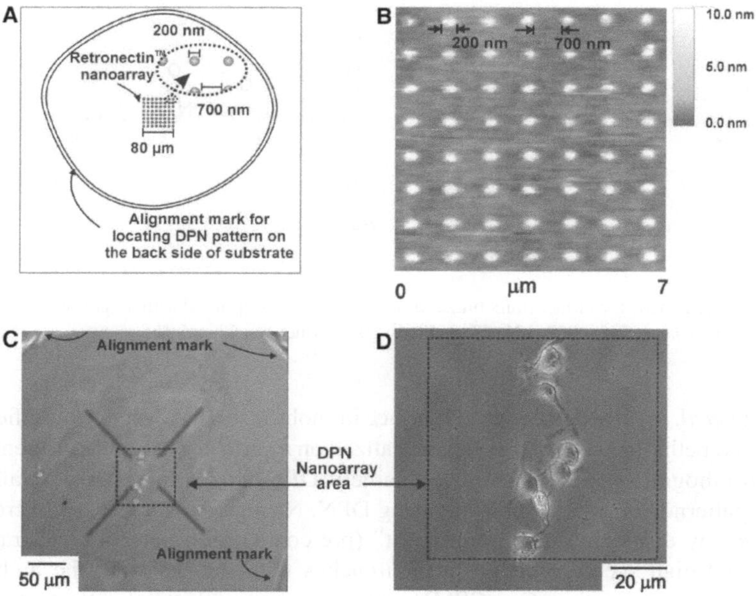

FIGURE 10.28. Diagram describing the cell adhesion experiment on the DPN-generated pattern. The total patterned area is 6400 mm2. The alignment marks were generated by scratching a circle into the backside of the Au-coated glass substrate. (B) Topography image (contact mode) of the Retronectin protein array. Imaging conditions were the same as in Fig. 1B. (C) Large-scale optical microscope image showing the localization of cells in the nanopatterned area. (D) Higher resolution optical image of the nanopatterned area, showing intact cells. (Reprinted with the permission from Lee *et al.*, 2002, *Science*, 295: 1702–1705, © 2002 AAAS).

such interactions occur on the sub-micron scale. MHA was patterned into a square array of dots 200 nm in diameter and separated by 700 nm using DPN. AFM images showed that on subsequent treatment Retronectin was adsorbed almost exclusively to the MHA nanopatterns. The protein nano-array was then incubated with bovine calf serum containing fibroblast cells followed by rinsing with PBS and studied under optical microscope. The cells were found to attach only to the patterned regions of the substrate and spread, but not completely, into a more flattened morphology than in their unbound state.

Smith *et al.* [40] used a binary ink pattern for selective chemo-specific immobilization of cysteine-mutant cowpea mosaic virus capsid particles (cys-VCPs). A gold-coated substrate was patterned with 150 nm diameter MHA SAM using DPN. By treating the substrate with 11-mercaptoundecyl-tri(ethylene glycol) the unoccupied areas were blocked by forming a monolayer. The substrate was then incubated in the virus solution in PBS and subsequently rinsed with PBS and NANO-pure water. TM-AFM images of the substrates revealed that the virus particles were immobilized on the MHA patterns—showing that they were inert to the ethylene-glycol derivative. Subsequently mutant CPMV particles engineered with cysteine inserts on the capsid were used to test an affinity assay on a substrate. The substrate was fabricated using two sets of ethylene glycol derivative inks. Using surface plasmon resonance spectroscopy the binding properties of Cys-CPMV and wt-VCP were studied to show specific adsorption of Cys-CPMV.

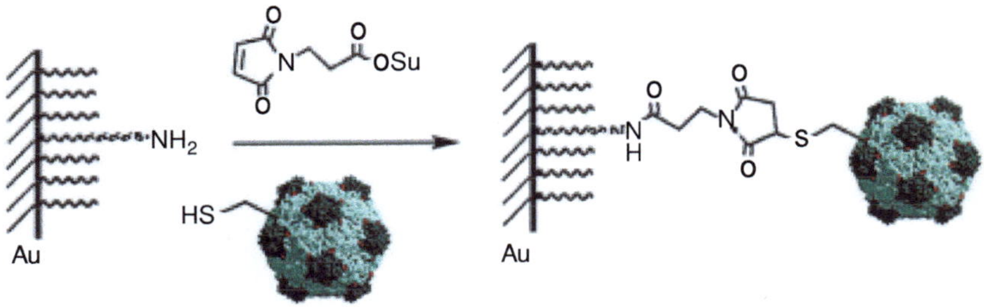

FIGURE 10.29. Schematic for surface immobilization of Cys-CPMV. (Reprinted with the permission from Cheung *et al.*, 2003, *J. Am. Chem. Soc.*, 2003, 125: 6848–49, © 2003, American Chemical Society).

Cheung *et al.* [8] also demonstrated surface immobilization of cysteine modified CPMV capsid on alkanethiol nano-patterns functionalized on a surface using two different Scaning Probe Nanolithography (SPN) techniques, namely, DPN and nano-grafting. An alkanethiol SAM was patterned on a gold substrate using DPN. Nanografting was used to create large area patterns by displacing a "protein resist" (pre-coated on a flat gold substrate) under high load and high speed to create linear trenches with thiols. Using the highly selective thiol-maleimide reaction, Cys-CPMV virus was chemoselectively attached to on thiol templates containing the maleimido functionality. The nano-grafted substrate was tested for virus assembly by treating it with PBS solution containing the CPMV virus. It was observed that bigger micron scale patches were composed of randomly assembled virus

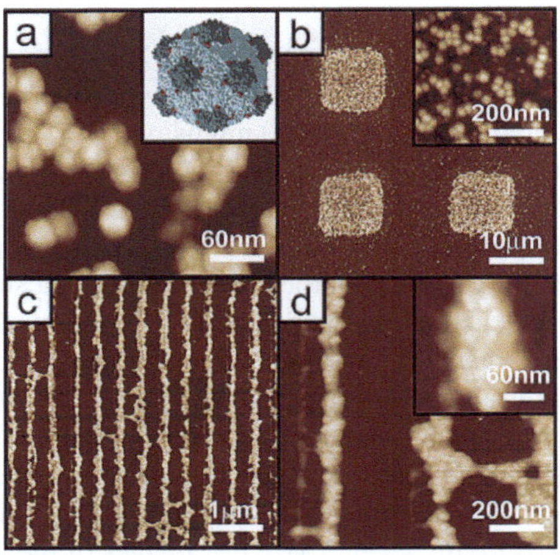

FIGURE 10.30. AFM height image of cow pea mosaic virus. (Inset) Model of genetically modified CPMV virus with unique cysteine residues (Cys-CPMV). Red dots indicate the locations of mutated cysteine residues. (b) AFM height image of Cys-CPMV virus assembled on micrometer-sized template. (inset) Zoom-in section of the functionalized square shown in (b). (c) AFM height image of a monolayer-thick virions assembled on a parallel line pattern created by nanografting with the chemoselective linkers. (d) Zoom-in section of (c). (inset) Zoom-in image of another section of the same sample for (c). (Reprinted with the permission from Cheung *et al.*, 2003, *J. Am. Chem. Soc.*, 2003, 125: 6848–49, © 2003, American Chemical Society)

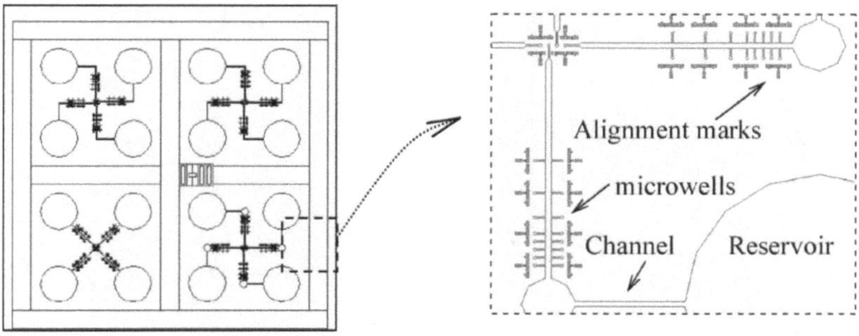

FIGURE 10.31. Mask layout showing 4 different layouts for delivery of 4 different inks. The inset shows the location of reservoirs (diameter: 1mm), satellite reservoirs, micro-channel, alignment marks, microwells and tributaries connecting the microchannels.

particles. However, when the dimension of the chemical template was reduced sub-micron scale the morphology of virus assembly changed dramatically. The authors theorize that probably due to size exclusion and a significant intervirion attractive interaction, the virions started to pack together into a close-packed morphology. The nano-grafted lines acted as templates for attachment of the virions which subsequently then also acted as templates for lateral growth of the viral assembly. "As other virions "stuck" to these lines, the lines "grew" laterally at the resulting step-edge and, in some cases, contacted adjacent lines of virions to form a pseudo-2D assembly" [8]. However, the lateral expansion did not proceed in a smooth and well-ordered manner. The authors propose that "By taking a lesson from the science of bulk crystallization of macromolecules, modulation of this strong intervirion interaction by altering solution conditions such as pH and ionic strength could assist the creation of 1D nucleation templates for growth of ordered 2D crystalline films".

10.5. USING MICROFLUIDICS FOR DPN APPLICATIONS IN BIOMOLECULAR PATTERNING[1]

To enable parallel writing of multiple patterns using the DPN process, the NScriptor^{TM} system (manufactured by NanoInk Inc., Chicago, IL) integrated with a multi-pen array of inked pens was developed to simultaneously write multiple patterns. This configuration requires selective deposition of ink chemistries onto the individual pens, which is accomplished by "dipping" the multi-pen array into an array of "inkwells." The device design was implemented to deliver between 4 aqueous inks (for genomic applications) and 10 aqueous inks (for proteomics applications) into appropriately spaced inkwell array. Fluid actuation occurs by open channel meniscus driven flow (wicking) in micro-channels, which distribute liquid from reservoirs into an array of terminal inkwells connected by tributaries (Figures 10.31 and 10.32).

[1] Excerpt courtesy of the SPIE © 2004, Banerjee *et al.*, 2004, Proceedings of the SPIE Vol 5345, Paper No. 5345–28.

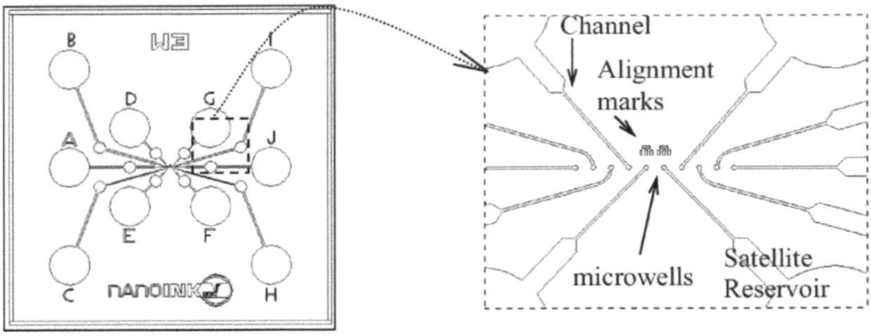

FIGURE 10.32. Mask layout showing a different layout for delivery of 10 different inks. The inset shows the location of reservoirs (diameter: 1 mm), satellite reservoirs, micro-channel, alignment marks, microwells and tributaries connecting the microchannels.

10.5.1. Analysis

To understand the ink-delivery it was necessary to calculate the filling times of microchannels of different sizes and the meniscus division from microchannels into tributaries. The flow rates in the micro-channels were calculated from a balance of the capillary and viscous forces (and neglecting inertial forces):

$$\Delta p = \frac{4\sigma\, Cos\theta}{D_h} = \frac{1}{2}\rho\, u^2 f \frac{L}{D_h} \tag{10.1}$$

Where, p: pressure drop in channel flow, μ: coefficient of surface tension, θ: contact angle, D_h: hydraulic diameter of the channel (where $D_h = 4A/P_w$, A: cross sectional area for flow direction; P_w is the wetted perimeter of the cross section), ρ: density of liquid flowing in the microchannels, u: average velocity in the microchannels, f: Fannings friction factor, L: length of the channel. For laminar flow in circular channels, $f = 64/Re_{Dh}$, where $Re_{Dh} = \rho\, u\, D_h/\mu$, and μ is the kinematic viscosity of the working liquid. Using the above formulation for f (valid for $Re_{Dh} < 2000$) Equation (10.1) can be simplified as:

$$uL = D_h \sigma\ Cos\theta/8\mu \tag{10.2}$$

For properties of water, the initial velocity of fluid flow in the microchannels can be calculated to be of the order of 1m/s for $L/D_h > 10$. This estimate matched very well with experimental data described later. Equation (10.2) can be used to estimate the filling time (τ) as:

$$\tau = \int_{L_1}^{L_2} \partial L/u = \frac{1}{2}\left(L_2^2 - L_1^2\right)\frac{8\mu}{D_h\sigma\ Cos\theta} \tag{10.3}$$

Equation (10.3) can be written in non-dimensional form as:

$$\left[\frac{\sigma\ Cos\theta/D_h}{4\mu}\right]\tau = \frac{L_2^2 - L_1^2}{D_h^2} \tag{10.4}$$

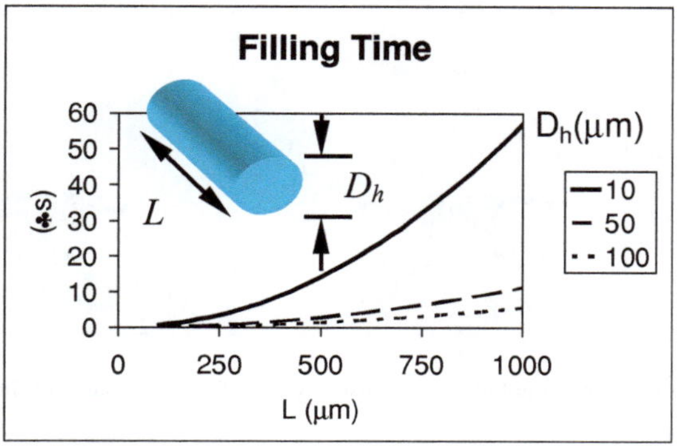

FIGURE 10.33. Filling time (τ) in micro-seconds for capillary pumped flow of water in microchannels of different length (L) and hydraulic diameter (D_h) using Equation 10.3.

Equation (10.3) is used to calculate filling time of microchannels of different hydraulic diameters using the properties of water and is plotted in Figure 10.33. The results show that for the same length the wider channels (larger D_h) fill up faster than the narrow channels. This is due to capillary forces ($\sim 1/D_h$) being retarded by even larger pressure drop ($\sim {D_h}^2$) in narrow microchannels. The time calculated by the above analyses would over-predict the filling time of the microchannels by less than 1% if inertia effects are included (for $L/D_h > 10$). The error in the estimated fill time is thus less than 1%. The analysis also shows that filling time is lower for open channel flow (D_h is lower) than for pipe flow.

10.5.2. Computational Fluid Dynamic (CFD) Simulation

Volume of Fluids (VOF) method was used to verify the analysis mentioned above. Also, a design choice to use closed channel flow (or "pipe flow") or open channel flow was made based on the CFD simulation results. The models were created and tested using the VOF method in the CFD-ACE+ design and simulation multi-physics software (CFD Research Corporation, Hunstville, AL). In open channel flow there is less susceptibility to bubble formation for liquid flow in the microchannels and during the filling of the microwells. However, the design trade off for these types of flow is the enhanced evaporative loss of the inks, especially for liquids with high vapour pressure under ambient conditions. The evaporative losses in closed channel (or pipe flow) is lower however, the microwells, microchannels and reservoirs have a greater propensity to form bubbles during filling. Bubble formation in these components can impede capillary flow and can lead to catastrophic failure for fluid actuation in these devices. It was verified from the simulations that the filling action is faster in open channel than pipe flow and is consistent with the analytical results. Figure 6 and 7 shows simulation of meniscus break-up from a 5 µm wide channel into a 2 µm wide tributary (both 10 mm deep) using the Volume of Fluids (VOF) method. Figure 10.34 shows that the meniscus breakup processs is initiated at the top and bottom wall in pipe flow. Figure 10.35 shows that the meniscus break-up process is initiated at the bottom wall for open-channel flow and therefore has less susceptibility to trap bubbles during the filling

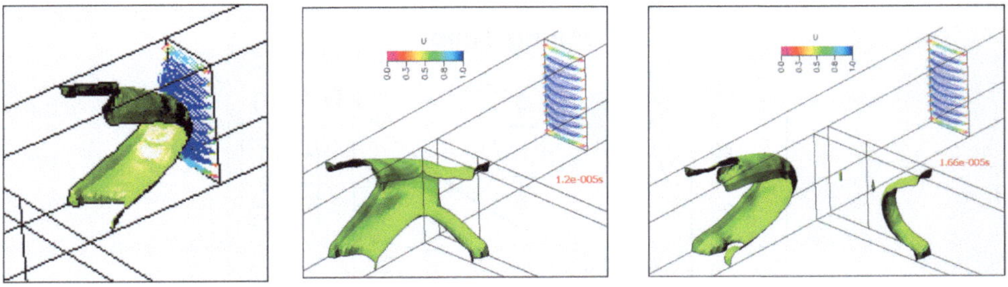

FIGURE 10.34. Temporal simulation of meniscus shapes for microchannel filling in pipe flow.

process. The results also show that the filling process is faster in open channel flow. The simulation data also validate the analytical prediction that the filling is process is faster in the wider channel compared to the narrower tributary channel.

10.5.3. Fabrication

Different wet and dry etching methods were investigated. Figure 10.36(a) demonstrates the initial process flow for fabricating the Inkwells using reactive ion etching (RIE). In this process, a <100> silicon wafer was oxidized in a Thermco Minibrute Oxidation furnace for 4hours to obtain a 1 micron oxide layer. Photoresist was patterned on the oxidized silicon wafer and the underlying Silicon substrate was exposed by a using timed wet etch in HF solution. A Technic's RIE was then used to etch the microfluidic components in the silicon substrate. Inkwells fabricated using this process flow were limited to a depth of 10 microns.

Due to rapid evaporative loss of aqueous inks these inkwells could be used only for less than a minute from the point of loading to the point of complete evaporative loss of the inks. Also, there was no contamination control for different inks used in these inkwells. It was decided that high aspect ratio (10:1) microfluidic structures were needed for minimizing evaporative loss of liquid. Also, hydrophobic barriers were designed to contain the inks in their respective microwells and prevent cross contamination. Deep Reactive Ion Etching (DRIE) using the Bosch process (STS Limited, Imperial Park, Newport, UK) was found to be

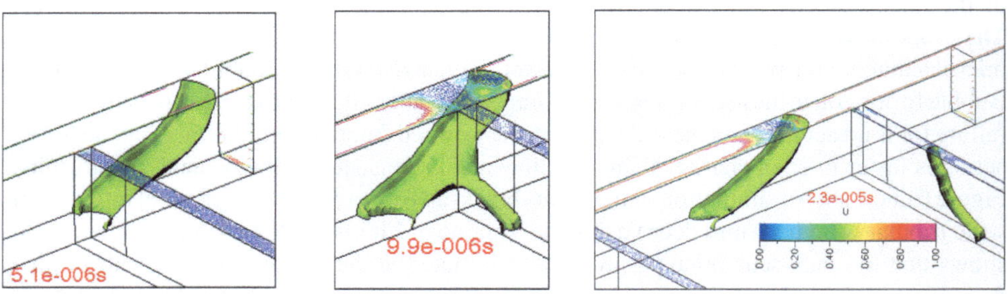

FIGURE 10.35. Temporal simulation of meniscus shapes for filling in open-channel flow.

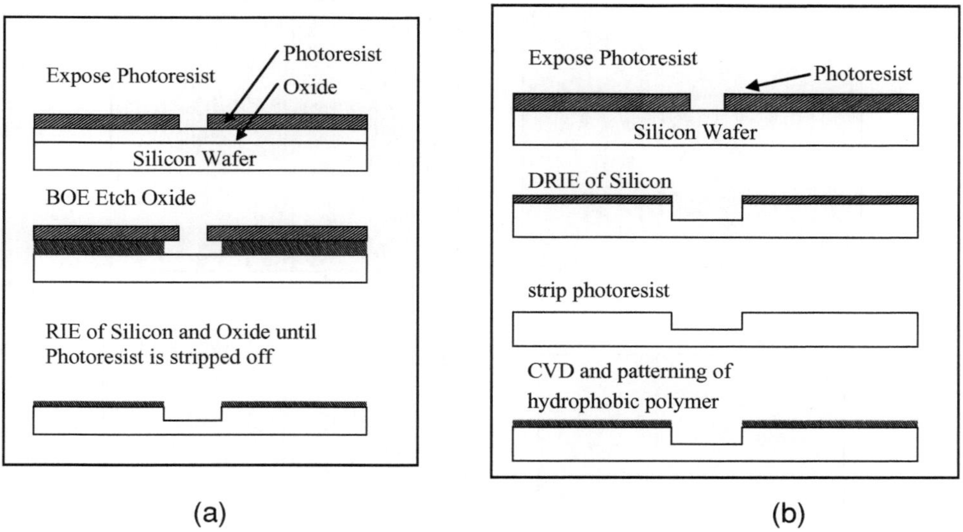

FIGURE 10.36. Process development flow diagram.

the most suitable method (Figure 10.36(b)). In this optimized process flow, a photo-resist was patterned on a <100> silicon wafer and used to form the microfluidic components using DRIE. The remaining photo-resist was stripped off. A hydrophobic polymer was selectively deposited and patterned using the Bosch process on the surface of the wafer to form barriers for controlling cross-contamination. Figure 10.37 shows the SEM images of different inkwell layouts.

10.5.4. Experimental Apparatus

Nano-liter volume droplets were deposited into the reservoirs using a syringe with a micro-needle (Hamilton Company, Reno, NV) and Model P-10 3-axis micropositioner with an inker bracket (Miller Design, San Jose, CA) for filling the reservoirs. The movements of the menisci in the micro-channels were recorded using an imaging apparatus, comprising of a high resolution Pullnix-TM1300 camera (1280 $\times$ 1028 pixels), a Road Runner R12 frame grabber, Nikon-SMZ10A trinocular microscope (coaxial illumination) and PC for data acquisition. Image Pro Express software (Media Cybernetics) was used for image analysis. Figure 10.38 shows the apparatus for loading the inklwells and recording the images for microwell filling.

10.5.5. Results and Discussion

The Inkwells were loaded with various types of inks and tested for their efficacy in filling, dipping of pens and writing. Figure 10.39 shows temporal sequence images of meniscus movement from a reservoir into the array of inkwells after loading the reservoir with ink. The reservoir in these images was loaded with a solution of ethylene glycol

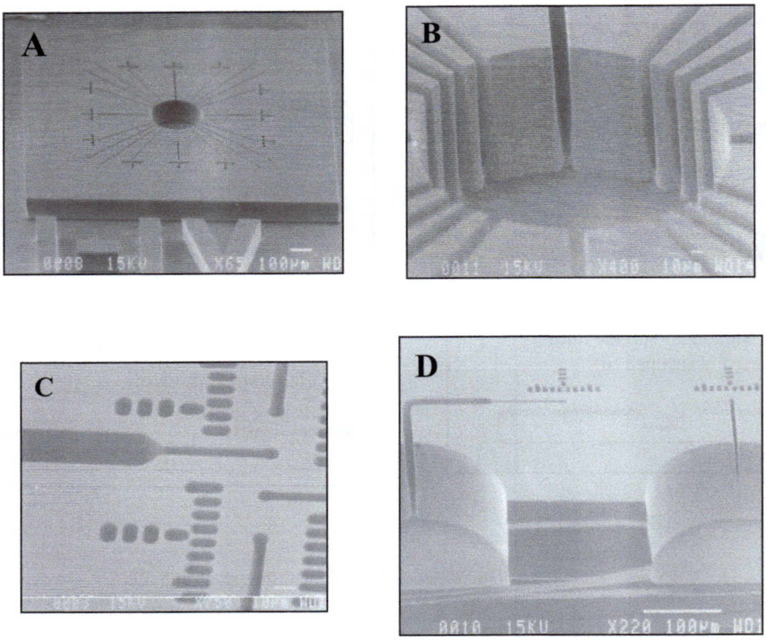

FIGURE 10.37. SEM images of inkwells. (A) Overview of reservoir and microwells. (B) Close-up view of reservoir. (C) Close up of microwells with alignment marks. (D) Cross-section of reservoir and microchannel.

and water. From the images it was estimated that the speed of meniscus movement in the microchannels reached 1 m/s which is consistent with the analytical calculations and the CFD simulation data.

Figure 10.40 shows SEM images of two different DPN pen array aligned with microwells in two different inkwells. Figure 10.41 shows an image showing the alignment of the

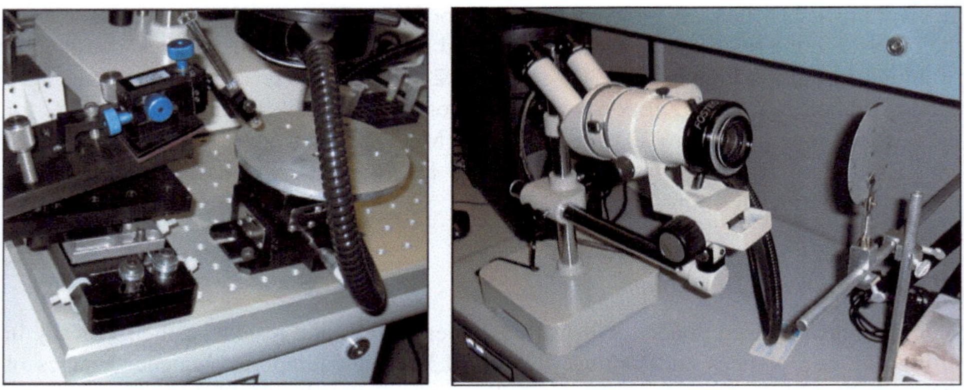

FIGURE 10.38. Apparatus for loading Inkwells and viewing the filling process.

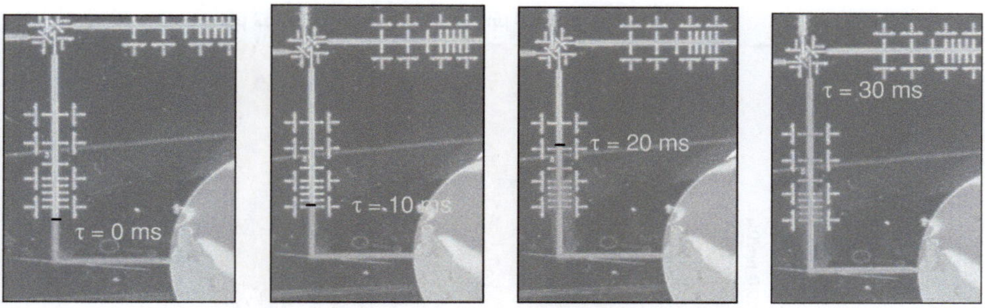

FIGURE 10.39. Time sequence images of meniscus motion in microchannel from reservoir to microwells.

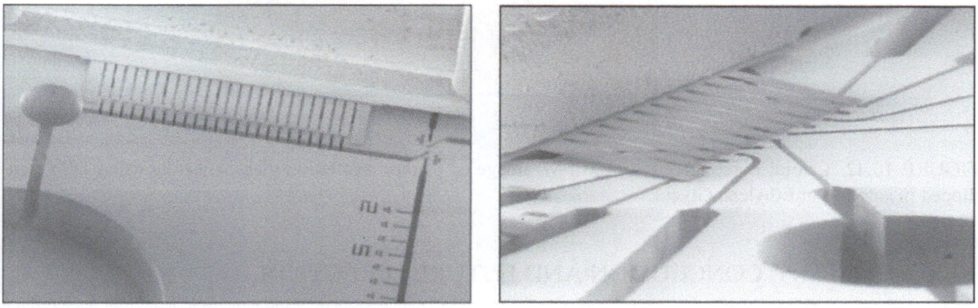

FIGURE 10.40. SEM images showing DPN pens aligned with microwells in Inkwells.

multi-pen array and the dipping of the pens into the inkwells filled with EasyInk™. The dipped pens are recognized by the different intensity of light reflected from the cantilevers. The dipped pens were used to write nanoscale patterns on a glass surface. The written patterns were imaged using an M5 Atomic Force Microscope (Park Scientific, currently part of Veeco Instruments, Santa Barbara, CA). Figure 10.42 shows a Lateral Force Microscopy (LFM) image of a written pattern. From the experiments it was found that the designs incorporating a satellite reservoir (which also serves as a bubble trap) had the most desirable characteristics for inkwell filling, dipping and writing (Figure 10.43).

FIGURE 10.41. (A) Multi-pen array is aligned with the inkwells (light on). (B) Tips are dipped into the wells (light off). (C) Illuminated image showing dipped pens reflecting light differently than the un-dipped pens.

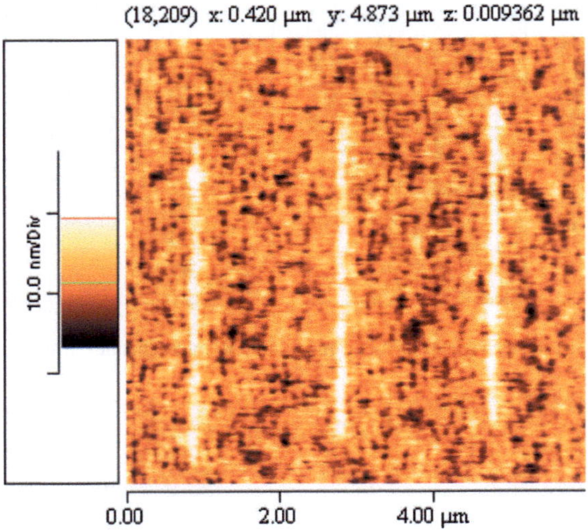

FIGURE 10.42. Lateral Force Microscopy (LFM) image of pattern written on glass substrate using DPN pen dipped in solution of Ethylene Glycol.

10.6. SUMMARY, CONCLUSION AND FUTURE DIRECTION

Dip pen technologies and their derivative methods are versatile and have powerful applications. The application of dip pen technologies to fabrication of bio-molecular devices has raised interesting possibilities in life sciences research. The biological applications have been in fabricating nano-arrays of oligo-nucleotides, proteins/peptides and composite bio-molecules. The techniques rely on indirect affinity patterning as well direct write approaches. The application of these methods to oligo-nucleotide synthesis is going to be challenging and exciting. The above studies also show a lot of promise for development of protein nano-arrays and for their ability to make fundamental contribution to clinical, process and research

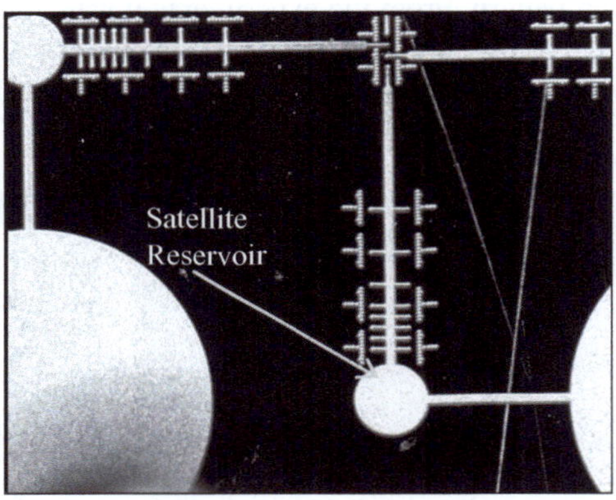

FIGURE 10.43. Optimal design selected from this study incorporating a satellite reservoir as a bubble trap.

proteomics. The nano-arrays and the AFM based scanning methods have a lot of promise for revealing structural information of proteins and the nature of protein-protein interactions. These devices have also been used for surface immobilization and specific adhesion of cells and virus particles. Such studies are expected to yield valuable information about focal interactions in cell adhesion that occur in the sub-micron scale. This nascent technology will take some time to achieve its full potential and many interesting aspects (such as lipid deposition) still remain unexplored.

REFERENCES

[1] G. Agarwal, L.A. Sowards, R.R. Naik, and M.O. Stone. Dip-pen nanolithography in tapping mode. *J. Am. Chem. Soc.*, 125:580–583, 2003a.

[2] G. Agarwal, R.R. Naik, and M.O. Stone. Immobilization of histidine-tagged proteins on nickel by electrochemical dip pen nanolithography. *J. Am. Chem. Soc.*, 125:7408–7412, 2003b.

[3] N.A. Amro, S. Xu, and G.Y. Liu. Patterning surfaces using tip-directed displacement and self-assembly. *Langmuir*, 16:3006–3009, 2000.

[4] G. Agarwal, L.A. Sowards, R.R. Naik, and M.O. Stone. Dip-pen nanolithography in tapping mode. *J. Am. Chem. Soc.*, 125:580–583, 2003.

[5] D. Banerjee, N. Amro, and J. Fragala. Optimizing Microfluidic Ink Delivery Apparatus for Dip Pen Nanolithography. *Proceedings of the SPIE Vol. 5345, Photonic West 2004 Symposium on Microfluidics, BioMEMS and Medical Microsystems II*, Paper No. 5345-28, Jan 24–29, San Jose, CA, 2004.

[6] P. Belaubre, M. Guirardel, G. Garcia, J.B. Pourciel, V. Leberre, A. Dagkessamanskaia, E. Trévisiol, J.M. François, and C. Berguad. Fabrication of biological microarrays using microcantilevers. *Appl. Phys. Lett.*, 82 (18):3122–3124, 2003.

[7] A. Bruckbauer, D.J. Zhou, and L.M. Ying *et al.* Multicomponent submicron features of biomolecules created by voltage controlled deposition from a nanopipet. *J. Am. Chem. Soc.*, 125(32):9834–9839, 2003.

[8] C.L. Cheung, J.A. Camarero, B. Woods, T. Lin, J.E. Johnson, and J.J. DeYoreo. Fabrication of assembled virus nanostructures on templates of chemoselective linkers by scanning probe nanolithography. *J. Am. Chem. Soc.*, 125:6848–6849, 2003.

[9] L.M. Demers and G. della Cioppa. Nanotechnology to advance discovery r&d—tutorial: Dip pen nanolithography as a next-generation, massively parallel nanoarray platform. *Genet. Eng. News*, 23(15), 2003.

[10] L.M. Demers, D.S. Ginger, S.J. Park, Z. Li, S.W. Chung, and C.A. Mirkin. Direct patterning of modified oligonucleotides on metals and insulators by dip-pen nanolithography. *Science*, 296:1836–1838, 2002.

[11] L.M. Demers, S.J. Park, T.A. Taton, Z. Li, and C.A. Mirkin. Orthogonal assembly of nanoparticle building blocks on dip-pen nanolithographically generated templates of DNA. *Angew. Chem. Int. Ed.*, 40(16):3071–3073, 2001.

[12] L.M. Demers and C.A. Mirkin. Combinatorial templates generated by dip-pen nanolithography for the formation of two-dimensional particle arrays. *Angew. Chem. Int. Ed.*, 40(16):3069–3071, 2001.

[13] D.S. Ginger, H. Zhang, and C.A. Mirkin. The evolution of dip-pen nanolithography. *Angew. Chem. Int. Ed.*, 43(1):30–45, 2004.

[14] P.T. Hurley, A.E. Ribbe, and J.M. Buriak. Nanopatterning of alkynes on hydrogen-terminated silicon surfaces by scanning probe-induced cathodic electrografting. *J. Am. Chem. Soc.*, 125(37):11334–11339, 2003.

[15] J. Hyun, S.J. Ahn, W.K. Lee, A. Chilkoti, and S. Zauscher. Molecular recognition-mediated fabrication of protein nanostructures by dip-pen lithography. *Nano Letters*, 2(11):1203–1207, 2002.

[16] A. Ivanisevic, J.M. Im, K.B. Lee, S.J. Park, L.M. Demers, K.J. Watson, and C.A. Mirkin. Redox-controlled orthogonal assembly of charged nanostructures. *J. Am. Chem. Soc.*, 123:12424–12425, 2001.

[17] J.Y. Jang, G.C. Schatz, and M.A. Ratner. Liquid meniscus condensation in dip-pen nanolithography. *J. Chem. Phys.*, 116(9):3875–3886, 2002.

[18] H. Jung, R. Kulkarni, and C.P. Collier. Dip-pen nanolithography of reactive alkoxysilanes on glass. *J. Am. Chem. Soc.*, 125(40):12096–12097, 2003.

[19] K.-H. Kim, N.M. Ke, and H.D. Espinosa. "Massively Parallel Multi-tip Nanoscle Writer with Fluidic Capabilities—Fountain Pen Nanolithography (FPN)", *Proceedings of the 4th International Symposium on MEMS and Nanotechnology*, the 2003 SEM Annual Conference and Exposition on Experimental and Applied Mechanics, June 2–4, Charlotte, North Carolina, Session 52, Paper 191, pp. 235–238, 2003.

[20] K.B. Lee, S.J. Park, C.A. Mirkin, J.C. Smith, and M. Mrksich. Protein nanoarrays generated by dip-pen nanolithography. *Science*, 295:1702–1705, 2002.

[21] K.B. Lee, J.H. Lim, and C.A. Mirkin. Protein nanostructures formed via direct-write dip-pen nanolithography. *J. Am. Chem. Soc.*, 125(19):5588–5589, 2003.

[22] A. Lewis, Y. Kheifetz, E. Shambrodt, A. Radko, E. Khatchatryan, and C. Sukenik. Fountain Pen nanochemistry: Atomic force control of chrome etching. *App. Phys. Lett.*, 75(17):2689–2691, 1999.

[23] Y. Li, B.W. Maynor, and J. Liu. Electrochemical AFM "dip-pen" nanolithography. *J. Am. Chem. Soc.*, 123:2105–2106, 2001.

[24] J.H. Lim, D.S. Ginger, K.B. Lee, J.M. Nam, and C.A. Mirkin. Direct-write dip-pen nanolithography of proteins on modified silicon oxide surfaces. *Angew. Chem. Int. Ed.*, 42:2309–2312, 2003.

[25] X. Liu, L. Fu, S. Hong, V.P. Dravid, and C.A. Mirkin. Arrays of magnetic nanoparticles patterned via "dip-pen" nanolithography. *Ad. Mat.*, 14(3):231–234, 2002.

[26] J.H. Lim and C.A. Mirkin. Electrostatically driven dip-pen nanolithography of conducting polymers. *Ad. Mat.*, 14 (20):1474–1477, 2002.

[27] P. Manandhar, J. Jang, G.C. Schatz, M.A. Ratner, and S. Hong. Anomalous surface diffusion in nanoscale direct deposition processes. *Phys. Rev. Lett.*, 90(11):115505-1–115505-4, 2003.

[28] S. Matsubara, H. Yamamoto, K. Oshima, E. Mouri, and H. Matsuoka. Fabrication of nano-structure by Diels-Alder reaction. *Chem. Lett.*, 9:886–887, 2002.

[29] R. McKendry, W.T.S. Huck, B. Weeks, M. Florini, C. Abell, and T. Rayment. Creating nanoscale patterns of dendrimers on silicon surfaces with dip-pen nanolithography. *Nano Lett.*, 2(7):713–716, 2002.

[30] C.A. Mirkin. Programming the assembly of two- and three-dimensional architectures with DNA and nanoscale inorganic building blocks. *Inorg. Chem.*, 39(11):2258–2272, 2000.

[31] A. Noy, A.E. Miller, J.E. Klare, B.L. Weeks, B.W. Woods, and J.J. DeYoreo. Fabrication of luminescent nanostructures and polymer nanowires using dip-pen nanolithography. *Nano Lett.*, 2(2):109–112, 2002.

[32] R.D. Piner and C.A. Mirkin. Effect of Water on Lateral Force Microscopy in Air. *Langmuir*, 13:6864–6868, 1997.

[33] R.D. Piner, J. Zhu, F. Xu, S. Hong, and C.A. Mirkin. "Dip-pen" nanolithography. *Science*, 283:661–663, 1999.

[34] L.A. Porter, A.E. Ribbe, and J.M. Buriak. Metallic nanostructures via static plowing lithography. *Nano Lett.*, 3(8):1043–1047, 2003.

[35] L.A. Porter, H.C. Choi, J.M. Schmeltzer, A.E. Ribbe, L.C.C. Elliott, J.M. Buriak. Electroless nanoparticle film deposition compatible with photolithography, microcontact printing, and dip-pen nanolithography patterning technologies. *Nano Lett.*, 2(12):1369–1372, 2002.

[36] B. Rosner, T. Duenas, D. Banerjee, R. Shile, N. Amro, and J. Rendlen. Active Probes and Microfluidic Ink Delivery for Dip Pen Nanolithography, *Proceedings of SPIE Symposium on Microelectronics, MEMS and Nanotechnology, Paper #* Perth, Australia, 5275-33,10–12 December, 2003.

[37] S. Shalom, K. Lieberman, and A. Lewis. A micropipette force probe suitable for near-field scanning optical microscopy. *Rev. Sci. Inst.*, 63(9):4061–4065, 1992.

[38] M. Schenk, M. Futing, and R. Reichelt. Direct visualization of the dynamic behavior of a water meniscus by scanning electron microscopy. *J. App. Phy.*, 84(9):4880–4884, 1998.

[39] P.V. Schwartz. Molecular transport from an atomic force microscope tip: A comparative study of dip-pen nanolithography. *Langmuir*, 18:4041–4046, 2002.

[40] J.C. Smith, K.B. Lee, Q. Wang, M.G. Finn, J.E. Johnson, M. Mrksich, and C.A. Mirkin. Nanopatterning the chemospecific immobilization of cowpea mosaic virus capsid. *Nano Lett.*, 3(7):883–886, 2003.

[41] X.F. Wang, K.S. Ryu, D.A. Bullen, J. Zou, H. Zhang, C.A. Mirkin, and C. Liu. Scanning probe contact printing. *Langmuir*, 19(21):8951–8955, 2003.

[42] B.L. Weeks, A. Noy, A.E. Miller, and J.J. De Yoreo. Effect of dissolution kinetics on feature size in dip-pen nanolithography. *Phys. Rev. Lett.*, 88(25):255505-1–255505-4, 2002.

[43] D.A. Weinberger, S. Hong, C.A. Mirkin, B.W. Wessels, and T.B. Higgins. Combinatorial generation and analysis of nanometer- and micrometer-scale silicon features via "dip-pen" nanolithography and wet chemical etching. *Adv. Mat.*, 12(21):1600–1603, 2000.

[44] D.L. Wilson, R. Martin, S. Hong, M. Cronin-Golomb, C.A. Mirkin, and D.L. Kaplan. Surface organization and nanopatterning of collagen by dip-pen nanolithography. *Proc. Nat. Acad. Sci. U.S.A.*, 98(24):13660–13664, 2001.

[45] M. Zhang, D. Bullen, S.W. Chung, S. Hong, K.S. Ryu, Z.F. Fan, C.A. Mirkin, and C. Liu. A MEMS nanoplotter with high-density parallel dip-pen nanolithography probe arrays. *Nanotechnology*, 13:212–217, 2002.

[46] H. Zhang, Z. Li, and C.A. Mirkin. Dip-pen nanolithography-based methodology for preparing arrays of nanostructures functionalized with oligonucleotides. *Adv. Mat.*, 14(20):1472–1474, 2002.

[47] H. Zhang, K.B. Lee, Z. Li, and C.A. Mirkin. Biofunctionalized nanoarrays of inorganic structures prepared by dip-pen nanolithography. *Nanotechnology*, 14(10):1113–1117, 2003.

[48] D. Zhou, A. Bruckbauer, and L.M. Ying. Building three-dimensional surface biological assemblies on the nanometer scale. *Nano Lett.*, 3(11):1517–1520, 2003.

11

Engineered Inorganic-Binding Polypeptides for Bionanotechnology[1]

Candan Tamerler†* and Mehmet Sarikaya*†

*Materials Science & Engineering, University of Washington, Seattle, WA 98195, USA

†Molecular Biology & Genetics, Istanbul Technical Univ., Maslak, Istanbul, 80626, Turkey

11.1. INTRODUCTION

Future biomimetic systems, developed either for nanobiotechnology or nanotechnology, could include protein(s) in its assembly, formation, and, perhaps, in its final structure as an integral component leading to specific and controllable functions. In the new field of molecular biomimetics, a true marriage of traditional physical and biological fields, hybrid materials could potentially be assembled from the molecular level using the recognition and assembly properties of proteins that specifically bind to inorganics [1]. Molecular biomimetics offers three simultaneous solutions to the problem of the control and fabrication of large-scale nanostructures and ordered assemblies of materials in two- and three-dimensions. The *first* is that inorganic-binding peptides and proteins are selected and designed at the molecular level and through genetics. This allows control at the lowest dimensional scale possible. The *second* is that such proteins can be used as *linkers* or *molecular erector sets* to join synthetic entities, including nanoparticles, functional polymers, or other nanostructures on molecular templates Finally, the *third* solution is that the biological molecules self- and co-assemble into ordered nano-structures. This ensures a robust assembly process for the construction of complex nano-, and possibly hierarchical-structures, similar to those found in nature.

[1] Some parts of this chapter are adapted from M. Sarikaya *et al.*, *Ann. Rev. Mater. Res.*, **34**, (2004).

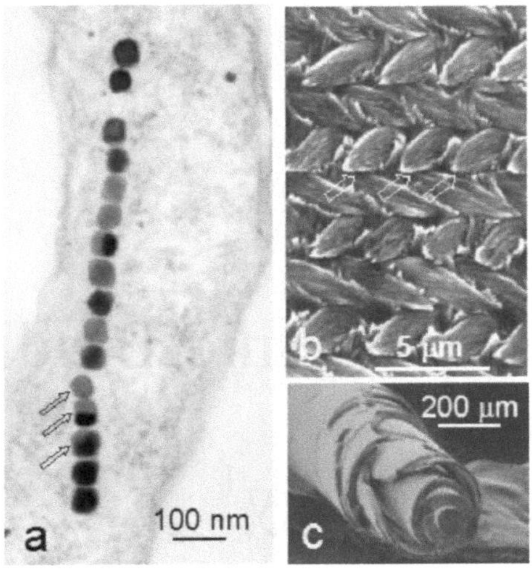

FIGURE 11.1. Examples of biological hard tissues in which proteins are in the key for fabrication. (a) A string of magnetite nanoparticles in a magnetotactic bacterium (TEM image) providing a bio-compass. (b) Highly organized, woven fully ceramicized enamel of mouse tooth showing enamel rods that consist of thousands of hydroxyl-apatite nanocrystallites leading to hardest/toughest bio hard tissue. (c) Laminated, glassy siliceous fibers of sponge spicule from *Rosella* that has excellent wave guide properties of an optical fiber.

Biological materials are highly organized from the molecular to the nano-, micro-, and macro- scales, with intricate architectures, often manifesting themselves in a hierarchical manner that ultimately make up a myriad of different functional units [2–5]. They are "smart" or self-directed in their organization and formation, dynamic in their interaction with the surroundings, complex in their structures and functions, self-healing in damage control, multifunctional in their physical and chemical properties, and have characteristics difficult to achieve in purely synthetic systems, even with recently developed molecular processes. Using closely controlled molecular, nano- and micro-structures through molecular recognition, templating, and self-assembly, biological materials have properties of technological and medical interest that surpass synthetic systems with similar phase compositions. Examples of biomaterials include soft tissues, and all macromolecular materials, or hard tissues, i.e., macromolecule/inorganic hybrid systems [3–5]. Muscle, membranes, skin, tendon, spiders' silks, and cuticles, to cite a few, belong to the former group and are of great interest in soft tissue engineering [6,7]. Hard tissues include bones, dental tissues (i.e., dentin and enamel), spicules, spines, shells, skeletal units of single-celled organisms (e.g., radiolarian) or plants, bacterial thin films and nanoparticles [3]. A common denominator, in these hard tissues is that, in addition to the presence of an inorganic component, there is at least one (and often several) proteinaceous phase (Figure 11.1). The inorganic phase could include magnetite (Fe_3O_4,) particles in magnetotactic bacteria or teeth of chiton [8]; silica (SiO_2) as skeletons of radiolarian [3] or tiny light-gathering lenses and optical wave guides in sponges [9], hydroxyapatite ($Ca_2C(OH)_3$ in bones (10) and dental tissues of mammalians [11], calcium carbonate ($CaCO_3$) in the shells of mollusks as armor [12] or as thin protective films in some species of S-layer bacteria [13], and spines and tests of sea-urchins [3].

These hard tissues could have a wide range of physical properties and are used as effective functional devices from single celled organisms to higher organisms. All of these biological tissues are synthesized in aqueous environments in mild physiological conditions using biomacromolecules, primarily proteins but also carbohydrates and lipids, under the genetic control of the organisms [2–5, 14]. In addition to enzymatic reactions in the synthesis of inorganic phases, proteins both collect and transport raw materials, and consistently and uniformly self- and co-assemble subunits into short- and long-range ordered nuclei and substrates [3–5]. Whether in controlling or participating in tissue formation and regeneration, or being an integral part of the tissue in its biological functions and physical performance, proteins are indispensable part of the biological structures and systems. Therefore, proteins are the "work-horses" that control the fabrication of biological structures by orchestrating the assembly of materials in two- and three-dimensions.

There are several ways to obtain inorganic surface specific proteins. These include extraction from hard tissues, designing them via theoretical approaches or utilizing the already existing ones. Each of these approaches has its own major limitations and may not be practical enough to serve in all nano scale engineering applications. With the recent developments in recombinant DNA technology, the preferred approach to obtain inorganic surface specific proteins is to use combinatorial biology-based molecular libraries. Here, a large number of random peptides are screened for their inorganic binding property. In addition to their identification, the inorganic surface specific proteins can now be also engineered, modified or redesigned for the synthesis and construction of nanostructured materials.

In this chapter, we discuss the unique advantages of genetically engineered polypeptides for inorganics (GEPI) for controlled assembly and formation of functional and hybrid materials and systems in nanobiotechnology applications. First we will briefly summarize combinatorial biology based methods for selection of inorganic polypeptides and present some of the binding sequences that we have identified for materials with selective functionality. Then we will discuss assessment of the binding characteristics using experimental and modeling approaches. And, finally, we will demonstrate utility of engineered polypeptides by giving case studies to represent a variety of potential applications. In the first example, we will discuss the use of inorganic-binding polypeptides attached to cell or phage in sorting cells and phages, respectively. In the second, we will demonstrate the heterofunctionality offered by inorganic binding polypeptides with potential application in sensing systems by showing immobilization of an enzyme, e.g., alkaline phosphatase, onto gold surface via gold binding polypeptide with retained catalytic activity of the enzyme. In a third example, the gold binding polypeptide will be used as a specific immobilization agent for target proteins, an essential property for achieving the nanoscale effects. In the final example, we will present a pathway for the biofabrication offered by inorganic binding polypeptides. Here, we will discuss silica synthesis in the presence of one of the strong silica specific polypeptide containing M13 phage.

11.2. SELECTION OF INORGANIC BINDING POLYPEPTIDES

Since the invention of phage display nearly two decades ago [15], display technologies have been proven to be an extraordinarily powerful tool for a myriad of biotechnological and biological applications. These include the characterization of receptor and antibody binding sites, the study of protein-ligand interactions, and the isolation and evolution of proteins or enzymes exhibiting improved or altered binding characteristics for their ligands. The three most common approaches, phage display (PD), cell surface display (CSD) and ribosome display (RD) have recently been reviewed [16–20]. All technologies are based on the common theme of linking phenotype and genotype. In our laboratory as well as in number of other laboratories, both CSD and PD approaches have been adapted for the selection of polypeptide sequences. These approaches have both advantages and disadvantages, and one is preferred over the other depending on a number of factors. The combinatorial biology based peptide libraries are generated by inserting randomized oligonucleotides into phage genomes or into plasmids in such a way that the foreign gene products (7–15 amino acid long polypeptides) will be displayed on the surface of the phage or bacterium, respectively (Figure 11.2). The randomized phage genomes or plasmids are then packed into phage or transformed into cells, respectively to display a different but random peptide unit. The next step is the biopanning, which involves the interaction of the library with the substrate, i.e., an inorganic surface. Several washing steps eliminate the peptides that nonspecifically bound to the substrate, and the specific ones are eluted and pooled together for further interactions with the substrate to identify the real binders. Following the amino acid composition characterization of the individual clones by DNA sequencing, binding characterization including several microscopic and spectroscopic techniques is carried out on the selected ones. These display techniques have been established in the last decades to identify antibodies for desired antigens; in materials science applications, however, they need special care in which inorganic materials are used as the substrates. During biopanning experiments, the inorganic substrate should be treated in such a way to retain surface properties, and prevent any physical or chemical modification or deterioration.

Phage display and cell surface techniques have been useful in the selection of polypeptides that are capable of binding inorganic materials. To date, CSD has been used to identify peptides recognizing iron oxide, cuprous oxide, gold, zinc oxide, and zeolites whereas phage display has been employed to isolate sequences binding to gallium arsenide, silica, silver, zinc sulfide, calcite, cadmium sulfide and among noble metals to platinum and palladium [1]. In our laboratory, we have adapted both techniques to identify peptides, which are specifically bind to metals, semiconductors and dielectrics. For CSD, we used the FliTrx CSD system [21], which positions random sequences of 12 amino acids as disulfide-constrained loops within Thioredoxin 1 (Trx) which is itself inserted into FliC, the major *E. coli* flagellar protein [22]. The resulting fusion proteins are exported to the cell surface where they assemble into partially functional flagella (an extended surface feature used for cell motility that is 20 nm in diameter and several micrometers in length in its authentic form). The FliTrx library has an estimated diversity of $\approx 1.8 \times 10^8$ and does not contain any predefined structural motif. A unique feature of the FliTrx system is that elution of cells binding to the target ligand is accomplished by imposition of shear stress (typically by vortexing), which leads to physical breakage of the flagella (Figure 11.2).

In phage display approach, we used the both the constrained and linear forms of hepta- or dodecapeptide libraries exposed on the surface of filamentous phage M13 by virtue of its fusion to the minor coat protein pIII (g3p) (Figure 11.2). The constrained form was constructed by flanking the random sequence by two cysteine residues that form a disulfide bond under oxidizing conditions and lead to the display of the heptapeptide as a loop. We

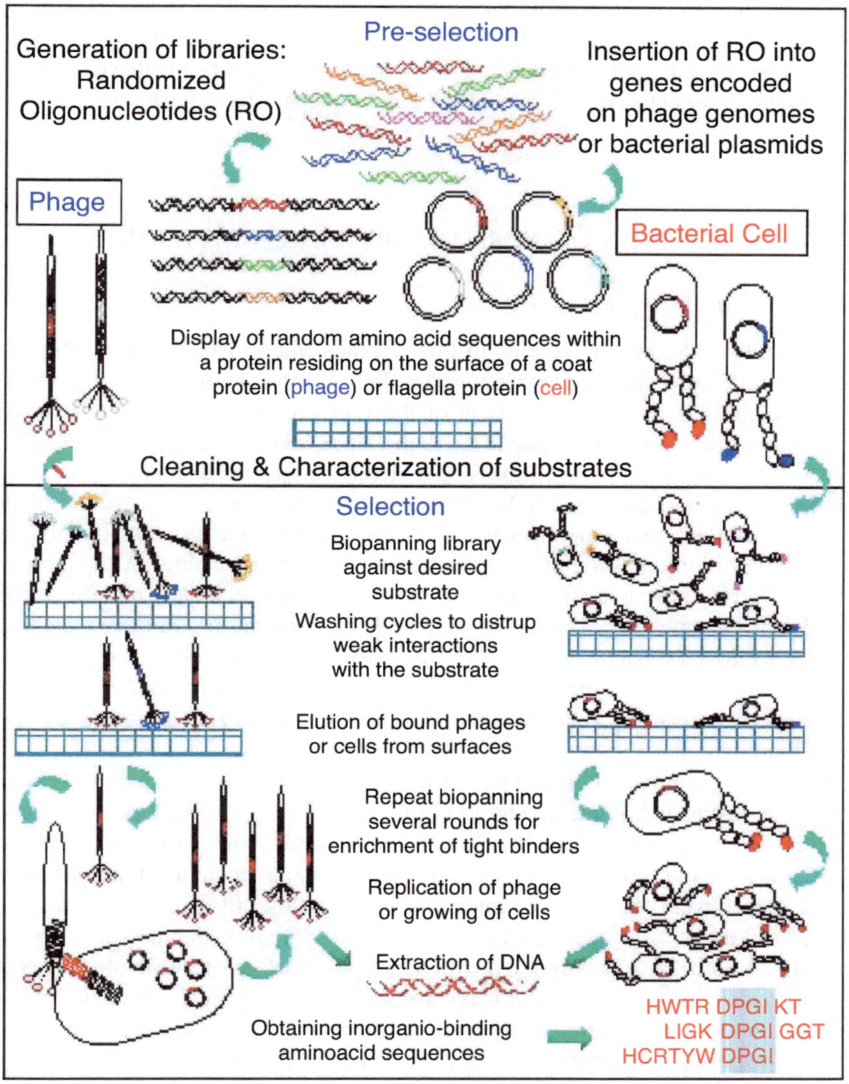

FIGURE 11.2. Schematics of phage display (PD) (left) and cell-surface display (CSD) (right) protocols adapted from molecular biology for the selection of inorganic-binding polypeptides for use as linkers and templating agents in nano- and naon-biotechnology applications.

have selected and identified the binding sequences for platinum, palladium surfaces among metals, quartz and sapphire among metal oxides, and also gallium nitrate and mica. When working with inorganic surfaces as substrates, surface could be well defined, such as a single crystal or a nanostructure or it might be rough, or totally non-descriptive such as a powder where the sequence space is the largest. In our studies, we often work with powders with various sizes (ranging from a few nms to sub-μm) and morphologies (sharp corners, rods, spherical particles, etc.) as well as the crystal face specific substrates to understand the recognition mechanism of the polypeptides. Usually, we have carried out the selection always by using materials (e.g., noble metals and metal oxides) that can be synthesized and stable in aqueous environment under physiological conditions and have fairly inert and smooth surfaces.

11.3. BINDING AFFINITY OF INORGANIC-BINDING POLYPEPTIDES

11.3.1. Molecular Adsorption of GEPI

The selective binding of proteins to their ligands often invokes complementarity in molecular architectures and surface chemistries. For example, in one of strongest receptor-ligand interactions found in nature, each of the four biotin molecules binds to each of pockets of tetrameric protein streptavidin via the "lock and key" mechanism [23, 24]. By analogy, inorganic-binding polypeptides could ideally be selected using inorganic materials exhibiting specific morphology, crystallography, or surface stereochemistry. In the design and assembly of functional inorganics, it is essential to understand the nature of polypeptide-inorganic recognition and binding to optimize and tailor these events. Although protein-surface interactions have been the object of considerable research, how proteins recognize inorganics remains unclear [24–26]. The specificity of a protein for a surface may originate from both chemical [26–28] (e.g. H-bonding, polarity and charge effects) and structural [29] (size and morphology) recognition mechanisms. An additional problem is that inorganic surface properties must be well characterized (often difficult to do so) to understand the nature of binding.

Fluorescence microscopy (FM) has proven to be very useful to qualitatively rank a population of polypeptides selected by PD or CSD in terms of affinity and specificity for a particular inorganic surface [30–35]. Similar to thiol and silane SAMs [36, 37] scanning probe microscopy (atomic force microscopy and scanning tunneling microscopy, AFM and STM, respectively) may also be integrated to the study of polypeptide assembly and binding onto inorganic surfaces (M. H. Zareie and M. Sarikaya, unpublished, 2004). The SPM results could be closely correlated with data from quartz crystal microbalance (QCM) and surface plasmon resonance (SPR) spectroscopy to quantitatively characterize protein adsorption kinetics under various solution conditions such as protein concentration, pH, salinity and variety of inorganic surfaces [38]. Traditional spectroscopy techniques, such as X-ray photoelectron spectroscopy (XPS) and time-of-flight-secondary ion mass spectroscopy (TOF-SIMS) techniques have recently proven useful in providing the fingerprint of specific peptide adsorption onto solid surfaces [39,40]. For the evaluation of the molecular structure and conformation on solid surfaces, liquid state and solid-state nuclear magnetic resonance (NMR) spectroscopy, provide essential data towards understanding molecular recognition and binding mechanisms [29]. As part of the theoretical tools, predictive molecular dynamics (MD) and statistical physics protocols could provide valuable insights into the three dimensional structure of inorganic binding peptides on atomically flat surfaces [41, 42]. It is clear that parallel studies on the same polypeptide/inorganic system with modeling and experimental techniques (e.g., MD and NMR in addition to SPR) would be the most powerful approach to understanding the binding. Some of these techniques will be

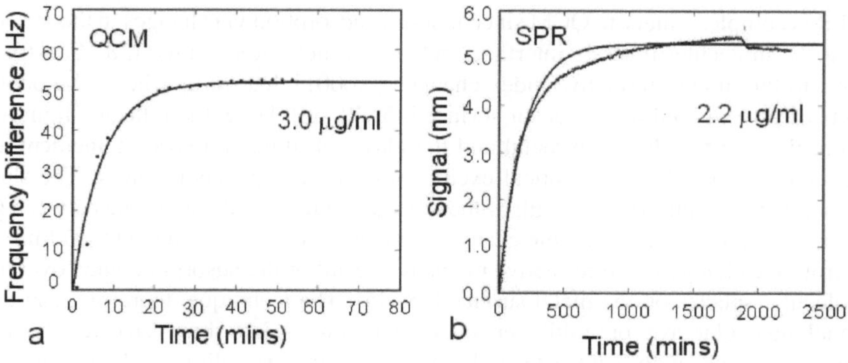

FIGURE 11.3. Experimental (dotted line) and model (continuous line) adsorption kinetics determined by (a) QCM and (b) SPR using 3-repeat gold-binding protein on flat gold substrates. The model is based on Langmuir monoplayer adsorption (M. Duman *et al.*, unpublished, 2004).

briefly discussed throughout this section by presenting part of our ongoing studies (1,42, 43, E. E. Oren *et al.*, unpublished, 2004).

For robust applications of the inorganic-binding polypeptides, determinations of their more fundamental binding parameters such as adsorption and desorption rate constants, or equilibrium adsorption constants are necessary. There is a scarcity of literature reporting quantitative adsorption properties of the inorganic binding proteins on inorganic substrates although these could be achieved using a number of well-established tools including QCM and SPR measurements [43–46]. For example, QCM is based on a mechanical measurement to study the adsorption of proteins and synthetic molecules (such as self-assembling monolayers) on surfaces [43, 44]. Here, a quartz crystal disk mounted with an electrode (typically gold), senses the change in resonance frequency due to the change of deposited molecular mass on the substrate surface. When operating in air, QCM can accurately measure submonolayer films but viscosity and thickness become important parameters in liquids. We used QCM to evaluate the binding kinetics of GEPI on gold surface that was coated on the quartz crystal (M. Duman *et al.*, unpublished, 2004). The experiments were carried out in water under neutral pH conditions in the buffers used in the selection process and various concentrations of the polypeptide at constant temperature conditions (Figure 11.3(a)). In these experiments, it was found that a gold binding polypeptide, GBP-1 (MHGKTQATSGTIQS (1,31)) binding to Au followed Langmuir kinetics which provided adsorption rate constant (k_a) as well as free energy of binding, $2.4 \times 10^3 (M)^{-1}(sec)^{-1}$ and, -9.4 $(kcal)(mol)^{-1}$ respectively [47]. Compared to those of alkanethiols ($2 \times 10^3 (M)^{-1}(sec)^{-1}$ and -5.6 $(kcal)(mol)^{-1}$), these could be considered as moderate values [48]. The structural and assembly characteristics of the GBP-1 and the potential of its genetic manipulation, however, make it a highly useful molecular linker as discussed in the application section of this manuscript. We have carried out binding experiments with Pt-coated quartz substrates to assess the cross-specificity of GBP-1 for Pt under the same experimental conditions used for Au [47]. Overall, the equilibrium constant (K_d) of GBP-1 for Pt was found to be 20-fold lower compared to Au, suggesting that GBP-1 adsorbs on Au more efficiently and stably than on Pt metal substrate [47, 49].

SPR is complementary to QCM since it detects adsorption via changes in the refractive index at the molecule/substrate interface, rather than net mass change [38, 45, 46]. It is sensitive to the minute refractive index changes (0.0001) that occur when submonolayer amounts of protein bind to the sensor surface [45, 48, 50]. Here, the reflected light intensity from the interface between metal and the glass substrate at a specific incident angle is measured as a result of the optical excitation of surface plasmon waves. The shift in position of the adsorption wavelength of the reflected light from the interface between glass prism and substrate is caused by the change in the evanescent wave in the metal film. This arises from the change in the refractive index as a result of the adsorption and coverage of the molecular species on the metal surface [38, 45]. The technique, therefore, can detect fractional molecular layer or multilayer coverage. Figure 11.3(b) shows an example of SPR experiments carried out to evaluate GBP-1 binding on a polycrystalline gold film. The experimental conditions here were similar to those used in the QCM experiments except that the polypeptide solution was released in the SPR chamber via a fluid cell with a predetermined, optimum flow rate [47]. Rapid assembly of GBP-1 onto the sensor surface was observed as a sharp increase in SPR shift. This initial adsorption could be fitted with a single exponential term leading to Langmuir kinetics similar to those found in the QCM experiments (Figure 11.3(a)). In fact the binding parameters found with SPR and QCM corroborate with one another, with the SPR values for K_d and ΔG being $-6.5 \times 10^3 (M)^{-1}(sec)^{-1}$ and -8.9 (kcal)(mol)$^{-1}$, respectively [49].

11.3.2. Physical Specificity and Molecular Modeling

The specificity of a polypeptide for an inorganic surface is likely to be rooted in its molecular conformation. Structural information is, therefore, essential not only to elucidate the fundamentals of the recognition process, but also for practical applications. Such knowledge would allow genetic or chemical modifications to create additional functionalities via molecular linking (e.g., attachment of conducting or light-emitting polymers to create hybrid, heterofunctional molecules, ability to bind DNA or proteins, etc.), thereby yielding a molecular “tinker-toy”. Ideally, using molecular dynamics/simulated annealing protocols and solution and/or solid-state NMR constraints, one could obtain an averaged lowest energy structure for as many GEPI as is feasible, and utilize these structures, along with simulation program in modeling the orientation and binding energetics at specific interfaces. These data could then be used to rank peptides by interfacial interaction energies, allowing the identification of important side chains and preferred chain alignments for each polypeptide with a given inorganic substrate surface. The experimental findings through specific measurements combined with the structural information from simulations would provide a coherent understanding of GEPI-inorganic surface interactions.

To date, only limited NMR studies have been performed to understand the molecular recognition principles of inorganic binding polypeptides. These include, for example, the adsorption of salivary proteins on hydroxyapatite [51], experiments of $CaCO_3$-bound polypeptides from mollusk shells [52], and recently conducted NMR analysis of CSD-selected single-repeat polypeptide binding to gold [53]. Molecular modeling of GEPI-binding to inorganics has also been limited. We have recently performed computer structure modeling studies to predict the shape of several GEPIs (gold binding and platinum binding polypeptides) in solution, and present a summary here as an illustration (Figure 11.4)

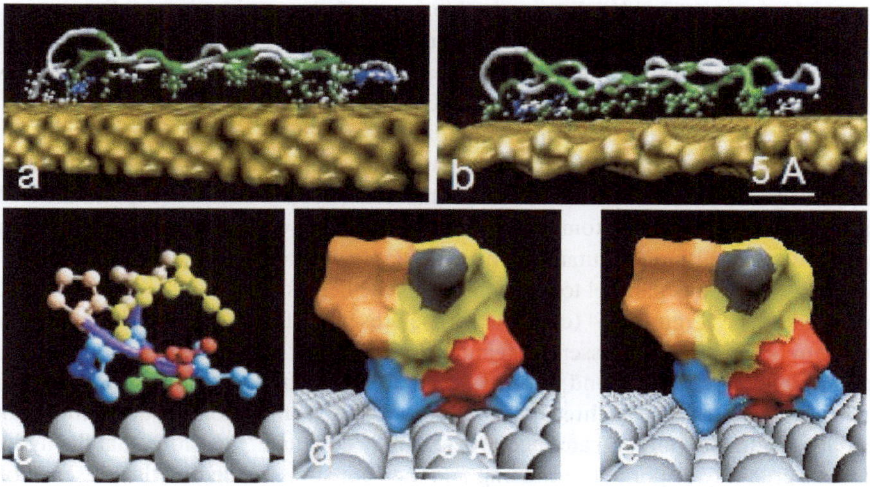

FIGURE 11.4. (a) and (b) are molecular dynamic rendering of gold binding protein (3-repeat GBP-1-1) on Au{111} and {112} surfaces viewed edge-on, respectively. Coloring corresponds to residue type: polar residue are highlighted in green, charged in blue, and hydrophobic in white [42]. Molecular illustrations are produced using VMD [42]. (c) The molecular conformation of a Pt-binding polypeptides on Pt(110). (d) and (e) are a setereo pair of same binder, of which the side chains are colored according to **CPTSTGQAC**. The Pt atoms are rendered by overlapping spheres whereas the protein atoms are rendered by balls and cylinders model with the coil representation of the secondary structure of the protein [49].

[42, 49]. The preliminary research was carried out in the hope that any coincidence between the amino acid residues of the peptides and the spacing of the inorganic lattices would shed light on how a GEPI binds to an metal surface. First, raw amino acid sequence data for 3 repeats of the 14 amino acid GBP-1 was compared to all known protein structures using FASTA searches and various first order secondary structure prediction algorithms (Chou-Fasman and Holley/Karplus). Figure 11.4(a) shows the three-repeat GBP-1 above a {111} Au atomic lattice that highlights the correspondence of OH^- groups to gold atom positions. These initial results suggest that binding repeats of GBP1 form an anti-parallel β-pleated sheet conformation, which places OH^- groups from serine and threonine residues into a regular lattice based on energy minimized in vacuo using X-PLOR (Figure 11.4 (a) [42]. We also showed in these preliminary studies that GBP-1 does not bind to Au{112} surface as tightly (Figure 11.4(b)), because of the migration of water molecules through the atomic grooves on this crystallographic surface which then decouples the protein from the surface.

Following the competitive binding experiments performed by one of the strong Pt-binders, i.e., SD152 (PTSTGQA), the conformational analysis of the same sequence was performed to obtain more insight on the binding mechanism. The energy minimized constrained peptide structures were placed on top of different symmetry centers of the Pt crystallographic surfaces, starting at 4 Å separation. Consequently, various initial configurations were considered by rotating and tilting the peptide on the rigid substrate which was chosen to be the three of the highest density atomic planes of platinum. The molecular conformations produced this way were then energy minimized by only allowing the peptide to translate on the surface. The comparison of each of the configurations in terms of the overall energy of the system provided the most stable protein-substrate binding configuration (see, e.g., Figure 11.4(c)). The geometrical configurations of the Pt-binding polypeptide on all three major crystallographic Pt surfaces reveals that there are certain amino acids having close contacts with the metal surface, i.e., G, Q and TST triplet in SD152 [49]. The simulations seem to agree with the experimental findings that the presence of threonine and serine, with hydroxyl groups on their side chains, may be the amino acids responsible for binding. A detailed further conformational analysis of the hydroxyl and amine groups on the Pt(111) surface reveals that oxygen atoms of hydroxyl groups of two threonines and the nitrogen atom of the amine group of glutamine form a triangle with sub-nm sized edge lengths. The plane of this triangle is parallel to and constitutes the nearest polypeptide atoms (excluding H's) to the surface (Figure 11.4 (c) and (d)). The presence of this type of *polypod molecular architecture* seems to be an essential feature for molecular surface recognition among the strong Pt binders [49]. We found that all of the remaining strong Pt-binders had amino acid sequences containing at least three hydroxyl (preferably from threonine) and amine groups, potentially leading to polypod architecture. One may conclude, therefore, that the hydroxyl groups are crucially important in binding and they are strengthened by the presence of the amine or a derivative of amine containing amino acids (e.g., glutamine, asparagines, and histidine) [49].

11.4. POTENTIAL APPLICATIONS OF MOLECULAR BIOMIMETICS IN BIO- AND NANOBIOTECHNOLOGY

Controlled binding and assembly of proteins onto inorganic substrates is at the core of biological materials science and engineering with wide ranging applications [54, 55] (Figure 11.5). Protein adsorption and macromolecular interactions at solid surfaces play key roles in the performance of implants and hard-tissue engineering. DNA and proteins adsorbed specifically onto probe substrates are used to build microarrays suitable for modern genomics [56], pharmogenetics [57], and proteomics [58, 59]. All potential applications of GEPI form the core of biological materials science and engineering applications by

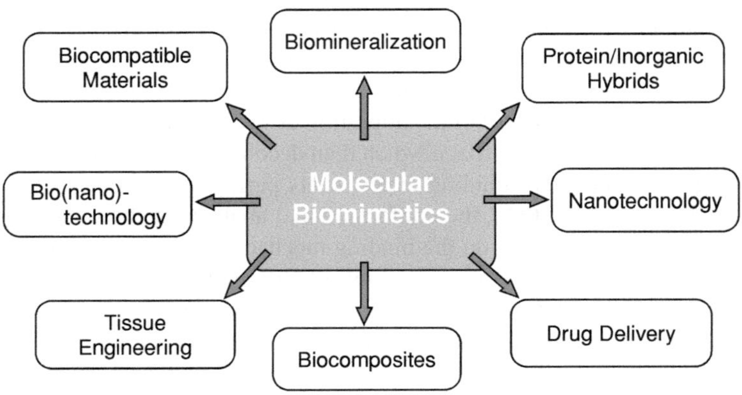

FIGURE 11.5. Potential applications of engineered polypeptides through molecular biomimetics.

providing controlled binding and assembly on inorganics. Once the inorganic binding domain was identified, then different binding motifs could be used via genetic fusion of these segments such that they would link two different materials which will permit assembly of complex nanocomposite and hybrid materials. Based on its recognition and self-assembly characteristics, the role of the engineered protein in these hybrid structures would be to provide the essential molecular linkage between the inorganic components, and at the same time, be an integral component of the overall structure providing to it functional (e.g., mechanical) durability. Engineered polypeptides hybridized with functional synthetic molecules could be used as hybrid building blocks in molecular electronics and photonics [1, 2, 60]. One could genetically fuse GEPI onto another protein or a macromolecule as functional ligands to create heterofunctional molecular constructs. The unique advantages of engineered polypeptides, created through molecular biomimetics discussed here, include highly specific molecular surface recognition of inorganics, self-assembly into ordered structures, and tailoring of their molecular structures and functions through molecular biology and genetics protocols. The examples given below represent a typical variety of potential applications.

11.4.1. GEPI-Assisted Cell and Phage Sorting and Differentiation

Cell sorting and separation are critical issues in the study of cell differentiation, growth-factor interaction, apoptosis, and cell proliferation, to name a few [61–63]. Existing techniques, such as fluorescent activated cell sorting, sedimentation, and centrifugal elutriation, require batch processing, mostly rely on diffusion, and require large volumes [61]. As a first step in cell separation and spatial organization, we have demonstrated a novel approach in which cells with genetically fused inorganic-binding polypeptides on their flagella are directed attached on micropatterned substrates. We isolated *E. coli* variant CN48 (ADRTRGRIRGNC) in our laboratory due its binding activity to Cu_2O using the flagellar display protocol described above [35]. As substrates, stripped surfaces of Cu_2O and $Cu(OH)_2$ were prepared using an electrochemical process. The variant bacteria labeled with viable cell fluorescent dye was contacted with this micropatterned substrate as illustrated in Figure 11.6(a) and (b). The $Cu(OH)_2$ region in the micrograph is distinguished as it adsorbs free dye, and as a result, exhibits a mild fluorescent background. On the other hand, a large number of CN48 cells adhere to the Cu_2O stripes, evidenced as bright rods of approximately 2 μm length (size of the cells) and only a few adsorbs to $Cu(OH)_2$ (possible due to non-specific adsorption). The results indicate that flagella displayed GEPI (CN48) exhibits a strong affinity to Cu_2O and not to $Cu(OH)_2$ regions. In these experiments, negative control *E. coli* cells do not adhere to either material. Similar protocol can be carried out for cell sorting or differentiation in heterocellular, viral, and other macromolecular systems, e.g., in separation of DNA and RNA in genomics, nuclear antigens in cell cycle, and proteins in proteomics [61–64].

For phage sorting experiment, we used Pt-specific phages which were also tested for their competitive specificity versus a metal oxide using patterned substrates. By competitive we mean binding of selected polypeptides to a given substrate (the one used for the selection) in the presence of another one under the same environmental conditions. To prepare the substrate, we used an electrochemical approach in which platinum macrodots were printed on a silica (glass) substrate [49]. Relatively flat and circular regions of nominally 50-μm-diameter area containing metallic platinum regions are shown in the light optical image in

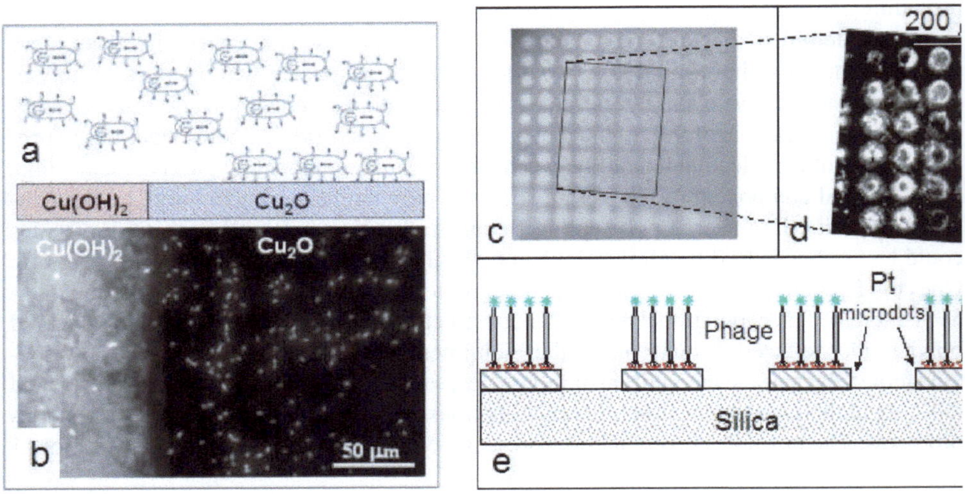

FIGURE 11.6. (a) Schematic and (b) experimental image showing selective adsorption of E. coli on cuprous oxide, rather than on cupric hydroxide, section of a striped substrate. The cell flagella were encoded with a dodecapeptide cuprous oxide (Cu_2O) binder that was selected using the flagella display [35]. (c) An array of platinum discs electrochemically printed on a glass (silica) substrate. (d) The tilted box indicates the region of the fluorescence microscopy image, showing SD152 phage selectively binding to the platinum regions and not the surrounding glass substrate. (e) A schematic representation of the phage sorting experiment [49].

Figure 11.6(c). Both the size of the macrodots and their Pt contents could be easily controlled using the electrochemical conditions during deposition. The phage with a strong Pt-binding property was prepared and immunolabelled similar to those used in the binding experiments that were carried out using FM. Specifically, we used SD152 and fluorescently tagged it via the protocol discussed elsewhere (see 49). Exposing the patterned substrate to buffer solution containing the SD152 phage results in the phage binding onto the Pt dots preferably over the silica regions. Figure 11.6(d) is a FM image of the patterned substrate showing a reversal of contrast compared to the bright field image (Figure 11.6(b)), i.e., light Pt regions (Figure 11.6(e) schematically depicts the phage sorting process). These results, and based on the modeling studies discussed next, have significant implications showing that only a few amino acids are actually enough to recognize a metal surface. These results demonstrate that very short polypeptides are capable of selecting, with programmed specificity, between two different materials, a metal and an oxide, in our case. As swhon here, this result provides a potential for the development of micropatterned multimaterial platforms for phage sorting in biotechnology.

11.4.2. Target Immobilization via Engineered Polypeptides as Molecular Erector Films

Protein adsorption and immobilization on inorganic materials have been critical issues in biosensing and biochips used in genomics and proteomics [56–59, 64]. A common practical way of immobilizing a probe protein on an inorganic material has been to use self-assembled monolayers (SAM) as linker molecules with, e.g., thiol or silane linkages, that preferentially attach to a noble metal (gold) or an oxide (silica), of rigid, rod-like molecules. Typically the free ends of these molecules are chemically modified to link to a probe

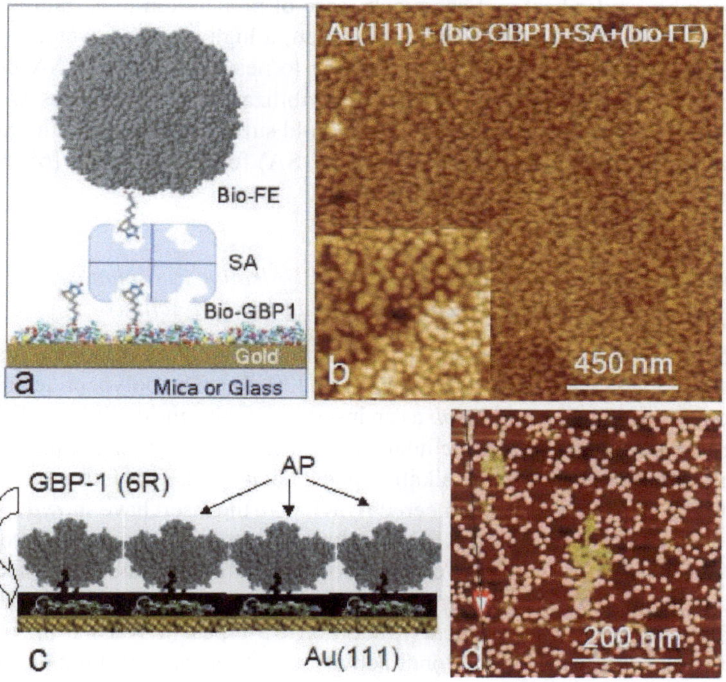

FIGURE 11.7. Experimental demonstrations of GEPI as possible molecular linkers in nanobiotechnology. (a) A schematic illustration of targeted binding of biotinylated ferritin (bio-FE) on strepavidin (SA, probe), immunobilized on bioreactive, biotinylated 3R-GBP-1 monomolecular film assembled on Au(111). (b) AFM image of the assembled molecular constructs (M. H. Zareie & M. Sarikaya, 2004, unpublished). (c) Schematic representation of alkaline phosphotase genetically linked with 9-repeat GBP-1 and assembled on a gold substrate) (d) AFM image of the self-assembled construct on Au(111) (T. Kacar *et al.*, unpublished, 2004).

protein to create functional surfaces to which target is selected [65, 66]. This approach has significant shortcomings [36, 37, 43, 46]. These include that thiolates are narrow, rigid, and 1-D restricting their versatility; they are non-specific to a given material, restricting their use in desired substrates; probe biomolecules can only be chemically linked (not genetically), and, more significantly, SAMs are prepared in non-biological environments preventing their full functionality in *in vivo* applications.

We demonstrate the utility of the self-assembled GBP-1 as a functional molecular erector set by using GBP-1 and immobilize a probe protein onto it (M. H. Zareie and M. Sarikaya, unpublished, 2004). Because of its high binding affinity to biotin, streptavidin is often used in biotechnology as a molecular linker, e.g., in biological assays, sensors, synthesis, and purification [65]. For immobilization (Figure 11.7(a)), first, biotinylated GBP1 (bio-GBP1) is assembled onto gold surface, with the same ordered assembly as GBP1 alone. Streptavidin (SA, the probe molecule) was then immobilized onto this ensemble with a high surface coverage. We next immobilized biotinylated ferritin (FE, a model target protein) onto this functional molecular substrate to assess that the immobilized SA has retained its specificity and biological activity. In addition to its significance in medicine, ferritin was chosen as a target protein in our experiments because of its well-characterized structure, stability under heat and chemicals, ease of preparation (e.g., biotinylation) and imaging with AFM [67]. Upon FE immobilization, a high surface coverage (determined from the SPR shift) was obtained corresponding to nearly 95% of the SA binding sites (Figure 11.7(b)). This is a highly efficient immobilization and compares well with the self-assembled biotin-terminated alkylthiols on gold surface that are traditionally used for biosensor applications immobilizing probes (e.g., SA) for target protein [65–66] or DNA selection [64].

11.4.3. Genetically Engineered Bifunctional GEPI-Alkaline Phosphatase Molecular Construct: Expressing both Catalytic and Inorganic-Binding Activity

Gold binding polypeptides selected by cell surface display are one of the first examples of engineered polypeptides for inorganic surfaces [30, 31]. The polypeptide sequence was displayed on the outer surface of *E.coli* as an insertion in the gene coding for the λ-receptor, lamB. They were isolated as extracellular loops of maltoporin which were subsequently fused to the amino terminus of the alkaline phosphotase (ALKP) with retention of gold-binding activity. The genetically engineered ALKP with the insert have shown specific binding for gold compared to native ALKP. More than 50 sequences were identified and one of them studied in detail, GBP-1 (31). We have incorporated further genetic engineering strategies for increased binding affinity to gold surface via insertion of multiple repeats of GBP-1 into ALKP including 5, 6, 7 and 9 repeats in addition to 3-repeat (T. Kacar *et al.*, unpublished, 2004). Here we present the bifunctional activity the 9 repeat gold binding polypeptide-ALKP constructs (Fig. 11.7(c and d)) [68]. We have tested the possible use of gold binding polypeptide as a molecular linker when it is part of ALKP and examined the enzyme to see whether it still retains its catalytic activity in the presence of p-nitrophenylphosphate in an assay buffer (Figure 11.8). Here, the protein solutions having wild type and ALKP- 9R GBP-1 were left for overnight for binding in the presence of a gold substrate. After a period of incubation time, the substrate was taken out of solution and the remaining supernatant

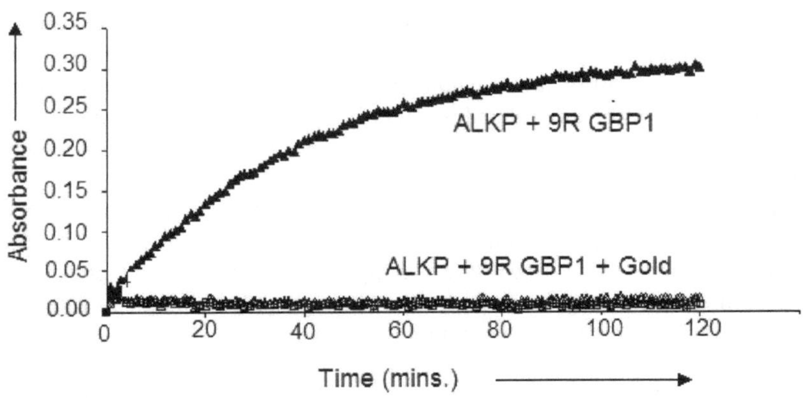

FIGURE 11.8. Catalytic activity profile of ALKP+9R GBP-1 construct versus that of supernatant examined after the adsorption of the molecular construct onto a gold substrate (see Figure 11.6 c–d) showing that the catalytic activity is reduced to nil in the supernatant which now lacks ALKP in solution.

was checked for the alkaline phosphatase activity. While the catalytic activity of the wild type protein solution could be observed, no activity was observed in the supernatant of the recombinant construct (Figure 11.8). The significance of this result is that the construct has a bifunctional property; i.e., while the gold binding property of the recombinant biomolecule resulted in the protein construct to be linked to the gold surface, the ALKP activity was still retained (data not shown). On the other hand, no activity could be observed in the supernatant solution which was stripped of the molecular construct because of its binding to the gold substrate.

Among many other activities, ALKP is an essential enzyme for biomineralization in the tooth pulp in cementum fabrication as well as in dentin formation. The enzyme exerts major effects on the extracellular phosphate concentration by catalyzing the pyrophosphate degradation. Alkaline phosphatase-gold binding bifunctional construct could be a promising hybrid molecular construct especially, e.g., in the area of periodontal regeneration. ALKP constructs provide excellent probing properties to follow the mechanism via gold binding property, and therefore provide the controlled delivery of ALKP to the desired location and, consequently, promote the mineralization, a potential utility in tooth regeneration.

11.4.4. Bionanofabrication: Silica Synthesis Using Inorganic Binding Polypeptides

In nature, silica exhibit diverse and extraordinarily designed shapes and structures in simple single-celled organisms (e.g., diatom skeletons [25, 69]) as well as in multicellular ones (such as sponge spicules [70]). The proteins directing silica synthesis in biological systems have been studied extensively [25, 70]. For example, silaffins extracted from the cell wall of a diatoms *Cylindrotheca* [25] or silicatein extracted from the spounge spicules of *Terhya aurantia* [69] are good examples of these proteins. Silica binding peptides have also been identified using silica as the substrate using PD protocol [71]. Using the catalytic activities of three sources of silica-binding polypeptides, it is now possible to fabricate silica at room temperature in aqueous solutions. In general, peptide clones were examined for silica precipitation assay and it was observed that the peptides rich in basic and hydroxy amino acids exhibited more silica precipitating activity. We also selected silica binding dodecapeptides via an M13 phage display library on single-crystal silica (quartz) wafer substrates [72]. In our studies, we carried out 5 rounds of biopanning and 50 sequences were obtained presenting different affinities for quartz substrate. Binding characterization of the selected phage samples were examined by AFM and immunofluorescence microscopy experiments, and categorized into three different classes as strong, medium, and weak binders. We noticed, firstly that there are significant differences between the amorphous silica binding polypeptide sequences [71] and those that bind to single crystal quartz substrate [72]. The sequences (or domains) were also different between quartz binding polypeptides and silicateins and sillafins [25, 69]. The high silica affinity presenting selected phage samples were tested for their effect on silica synthesis. One of the strong binders, namely DS202 (RLNPPSQMDPPF), was used in the silica synthesis. The solutions containing the proper buffer was incubated with the peptide-bound phage and in freshly prepared tetra methyl ortho silicate (TMOS) solution for 3–4 min at room temperature. The samples for transmission electron microscopy were collected before gellation by centrifuging the phage containing TMOS solution [72]. As shown in Figure 11.9, the TEM image shows highly branched, web-like open structure of silica which actually forms large bulk material (SEM image in Figure 11.7(a)).

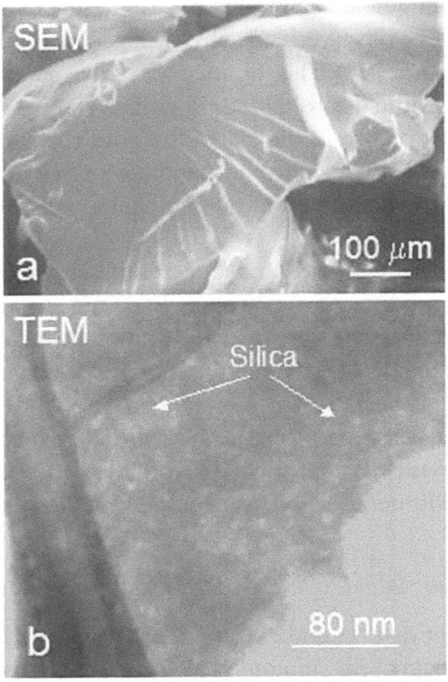

FIGURE 11.9. Biofabricated silica formed using silica-binding polypeptide selected using phage display. (a) Bulk materials revealed by scanning electron microscope secondary electronimaging and tramnsmisson electron microscope, bright fild image (D, Sahin *et al.*, unpublished, 2004).

As discussed in the introduction [3–5], proteins, in biological systems, can be nucleating agents, growth modifiers (or otherwise, templates), and catalyzers in the synthesis and structural build up of inorganic materials. As we demonstrate here, the use of inorganic binding polypeptides, selected via biological combinatorial methodologies, is a significant step in the controlled fabrication and formation of biomimetic materials of practical interest. Not only selection, but tailoring is possible via recombinant DNA technologies to further engineer binding and multifunctionality properties of these short polypeptides, such as by developing multiple repeats, shuffling binding/non-binding domains, or creating hybrids. These new technologies (GEPI-technologies) are likely to lead to potential applications of inorganic-binding polypeptides as coatings in creating biocompatible substrates for therapeutic devices, as agents of biomineralization in tissue engineering for hard tissue regeneration, as linkers of nanoparticles and carriers for drug delivery and release, and in creating multicomponent molecular substrates for proteomics and genomics.

11.5. FUTURE PROSPECTS AND POTENTIAL APPLICATIONS IN NANOTECHNOLOGY

Engineered proteins hold great promise for the creation of architectures at the molecular or nanoscale levels [1, 73]. Genetically engineered proteins for inorganics (GEPI) represents a new class of biological molecules that are combinatorially selected to bind to specific inorganic surfaces [1, 21, 30–35, 47, 73]. The ordered assembly of GEPI on inorganic surfaces could have a significant impact in molecular biotechnology applications. Our results described above are the first demonstrations that combinatorially selected polypeptides can self-assemble specifically on a selected inorganic single crystal surface, and that a GEPI may be molecularly recognizing an inorganic surface. Realizing the fact that thiol and silane linkages are the other two major molecular linkers for noble metal and oxide (silica) surfaces that have constituted the field of self-assembled molecules up until now [36, 37], it is naturally expected that self assembled GEPI monolayers used as "molecular erector sets" will open up new avenues for designing and engineering new and novel functional surfaces [1, 73]. We have already demonstrated that inorganic materials can be assembled at the nanoscale by proteins that have been genetically-engineered to bind to selected materials surfaces. It is also the first time that engineered proteins were shown to affect crystal morphology [31]. The combinatorial genetic approach is a general one, which should be applicable to numerous surfaces [21, 30–35, 47, 49]. The modularity of binding motifs should allow genetic fusion of peptide segments recognizing two different materials. The resulting heterobifunctional molecules could be used to attach different materials to each other and may permit assembly of complex nanocomposite and hybrid materials. These results could lead to new avenues in nanotechnology, biomimetics, biotechnology, crystal and tissue engineering such as in the formation, shape-modification, and assembly of materials, and development of surface-specific protein coatings. One particular potential utility of GEPI as a molecular linker is in nanobiotechnology. The examples given here are some of the few applications demonstrating the potential of these genetically engineered polypeptides.

Although significant advances have been made in developing protocols for surface-binding polypeptides, many questions remain to be answered before their robust design and widespread practical applications are affectively realized. These questions include: What are the physical and chemical bases for recognition of inorganic surfaces by the genetically engineered polypeptides? What are the long-range assembly characteristics, kinetics, and stability of the binding? What are the molecular mechanisms of engineered polypeptide binding onto (noble) metals compared to those on non-metals? How do surface characteristics affect binding? Based on the insights achieved, can we develop a "road map" in using GEPIs as molecular linkers and open new avenues in self-assembled molecular systems in nanobiotechnology based on biology? Considering the rapid developments in the inorganic-binding polypeptide selection protocols and the increased variety of materials utilized as substrates, many of these questions are expected to find answer in the near future with significant impact in broad multidisciplinary fields with potential wide ranging applications [73].

ACKNOWLEDGEMENTS

This research was supported by the US Army Research Office through the DURINT (Defense University Initiative on NanoTechnology) Program (Grant No: DAAD19-01-1-0499). We would like to thank our colleagues; Drs. F. Baneyx, A. Jen, D. Schwartz, J. Evans, and post-doctoral and graduate students; H. Fong, E. E. Oren, M. H. Zareie, S. Dincer, M. Duman, D. Sahin, T. Kacar, D. Heidel, and E. Venkatasubramanian.

REFERENCES

[1] M. Sarikaya, C. Tamerler, A.Y. Jen, K. Schulten, and F. Baneyx. Molecular biomimetics: nanotechnology through biology. *Nature Mater.*, 2:577–585, 2003.

[2] P. Ball. Life's lessons in design. *Nature*, 409:413–416, 2001.

[3] H.A. Lowenstam and S. Weiner. *On Biomineralization*. Oxford: Oxford Univ. Press, 1989.

[4] M. Sarikaya and I.A. Aksay. (ed.) *Biomimetics: Design & Processing of Materials*. New York, American Institute of Physics, 1995.

[5] S. Mann. (ed.) *Biomimetic Materials Chemistry*. New York, VCH Publishers, 1996.

[6] D. Kaplan, W.W. Adams, B. Farmer, and C. Viney. Silk-Biology, Structure, Properties, and Genetics. *Silk Polymers ACS Symposium Series*, 544:2–16, 1994.

[7] S.E. Sakiyama-Elbert and J.A. Hubbell. Design of novel biomaterials. *Ann. Rev. Mater. Res.*, 31:183–201, 2001.

[8] R.B. Frankel and R.P. Blakemore. (ed.) *Iron Biominerals*. New York: Plenum Press, 1991.

[9] H. Fong, M. Sarikaya, S. White, and M.L. Snead. Nanomechanical properties profiles across DEJ in human incisor teeth. *J. Mater. Sci. & Eng.-C.*, 7:119–28, 2000.

[10] M. Glimcher and M. Nimni. Collagen cross-linking and biomineralization. *Connect Tissue Res.*, 27:73–83, 1992.

[11] M.L. Paine, D.H. Zhu, W. Luo, P. Bringas Jr., M. Goldberg, S.N. White, Y.P. Lei, M. Sarikaya, H.K. Fong, and M.L. Snead. Enamel biomineralization defects result from alterations to amelogenin self-assembly. *J. Struct. Biol.*, 132:191–200, 2000.

[12] G. Mayer and M. Sarikaya. Rigid biological composite materials: Structural examples for biomimetic design. *Exp. Mechanics.*, 42:395–403, 2002.

[13] S. Schultze, G. Harauz, and T.I. Beveridge. Participation of a cyanobacterial S layer in fine-grain mineral formation. *J. Bacteriol.*, 174:7971–7981, 1992.

[14] M. Sarikaya. Biomimetics: Materials fabrication through biology. *Proc. Natl. Acad. Sci.* USA., 96:14183–14185, 1999.

[15] G.P. Smith. Filamentous fusion phage: Novel expression vectors that display cloned antigens on the virion surface. *Science*, 228:1315–1317, 1985.

[16] M. Dani. Peptide display libraries: design and construction. *J. Recept. Signal Transduct. Res.*, 21:469–488, 2001.

[17] K.D. Wittrup. Protein engineering by cell surface display. *Curr. Opin. Biotechnol.*, 12:395–399, 2001.

[18] H.M.E. Azzazy and W.E. Highsmith Jr. Phage display technology: clinical applications and recent innovations. *Clin. Biochem.*, 35:425–445, 2002.

[19] W.J. Dower and L.C. Mattheakis. In vitro selection as a powerful tool for the applied evolution of proteins and peptides. *Curr. Opin. Chem. Biol.*, 6:390–398, 2002.

[20] P. Samuelson, E. Gunneriusson, P.Å. Nygren, and S. Ståhl. Display of proteins on bacteria. *J. Biotechnol.*, 96:129–154, 2002.

[21] T.K. Thai, M.S.R. Sastry, M. Sarikaya, D.T. Schwartz, and Baneyx. Identification and characterization of Cu_2O and ZnO-binding polypeptides by Escherichia coli cell surface display: Towards an understanding of metal-oxide binding, *Biotechn. and Bioeng.*, 87:129–137, 2004.

[22] Z. Lu, K.S. Murray, V. Van Cleave, E.R. LaVallie, M.L. Stahl, and J.M. McCoy. Expression of thioredoxin random peptide libraries on the *Escherichia coli* cell surface as functional fusions to flagellin: a system designed for exploring protein-protein interactions. *Biotechnology*, 13:366–372, 1995.

[23] S. Izrailev, S. Stepaniants, M. Balsera, Y. Oono, and K. Schulten. Molecular dynamics study of unbinding of avidin-biotin complex. *Biophys. J.*, 72:1568–1581, 1997.

[24] H.V. Perez-Luna, M.J. O'Brien, K. Opperman, P.D. Hampton, P.S. Stayton, L. Klumb, and G.P. Lopez. Molecular recognition between genetically engineered streptavidin and surface-bound biotin. *J. Amer. Chem. Soc.*, 121:6469–6478, 1999.

[25] N. Kruger, R. Deutzmann, C. Bergsdork, and M. Sumper. Species-specific polyamines from diatoms control silica morphology. *Proc. Natl. Acad. Sci. USA.*, 97:14133–14138, 2000.

[26] C.A. Knight, C.C. Cheng, and A.L. DeVries. Adsorption of a-helical antifreeze peptides on specific ice crystal surface planes. *Biophys. J.*, 59:409–418, 1991.

[27] S. Mann. Molecular recognition in biomineralization. *Nature*, 332:119–123, 1988.

[28] I. Weissbuch, L. Addadi, M. Lahav, and L. Leiserowitz. Molecular recognition at crystal interfaces. *Science*, 253:637–645, 1991.

[29] J.S. Evans. 'Apples' and 'oranges': Comparing the structural aspects of biomineral- and ice-interaction proteins. *Curr. Opin. Coll Interf. Sci.*, 8:48–54, 2003.

[30] S. Brown. Engineering iron oxide-adhesion mutants of *Escherichia coli* phage λ-receptor. *Proc. Natl. Acad. Sci. USA*, 89:8651–8655, 1992.

[31] S. Brown, M. Sarikaya, and E. Johnson. Genetic analysis of crystal growth. *J. Mol. Biol.*, 299:725–732, 2000.

[32] S.R. Whaley, D.S. English, E.L. Hu, P.F. Barbara, and M.A. Belcher. Selection of peptides with semiconducting binding specificity for directed nanocrystal assembly. *Nature*, 405:665–658, 2000.

[33] R.R. Naik, L. Brott, S.J. Carlson, and M.O. Stone. Silica precipitating peptides isolated from a combinatorial phage display libraries. *J. Nanosci. Nanotechnol.*, 2:95–100, 2002.

[34] C.M. Li, G.D. Botsaris, and D.L. Kaplan. Selective in vitro effects of peptides on calcium carbonate crystallization. *Cry. Growth & Design*, 2:387–393, 2002.

[35] H. Dai, C. Nguyen, M. Sarikaya, F. Baneyx, and D.T. Schwartz. Through-mask anodic patterning of copper surfaces and film stability in biological media. *Langmuir*, 20:3483–3486, 2004.

[36] G.M. Whitesides, J.P. Mathias, and C.T. Seto. Molecular Self-assembly and nanochemistry-A chemical strategy for the synthesis of nanostructures. *Science*, 254:1312–1319, 1991.

[37] R. Schreiber. Structure and growth of self-assembled monolayers. *Prog. Surf. Sci.*, 65:151–256, 2000.

[38] H. Raetner. *Surface plasmon on smooth and rough surfaces and on gratings*. Springer and Verlag, Berlin, 1982.

[39] M.C. Coen, R. Lehman, P. Groning, M. Bielmann, C. Galli, and L. Schlapbach. Adsorption and bioactivity of protein A on silicon surfaces studied by AFM and XPS. *J. Colloid Interface Sci.*, 233:180–189, 2001.

[40] N. Xia and D.G. Castner. Preserving the structure of adsorbed protein films for time-of-flight secondary ion mass spectrometry analysis. *J. Biomed. Mater. Res. A.*, 67:179–190, 2003.

[41] Y. Dai and J.C. Evans. An energy-based mapping method for identifying the in-plane orientations of the polypeptides and other macromolecules at crystalline interfaces. *J Chem. Phys.*, 112:144–157, 2000.

[42] R. Braun, M. Sarikaya, and K.S. Schulten. Genetically engineered gold-binding polypeptides: Structure prediction and molecular dynamics. *J. Biomater. Sci.*, 13:747–758, 2002.

[43] L.E. Bailey, D. Kambhampati, K.K. Kanazawa, W. Knoll, and C.W. Frank. Using surface plasmon resonance and the quartz crystal microbalance t monitor *in situ* the interfacial behaviour of thin organic films. *Langmuir.*, 18:479–489, 2002.

[44] B.S. Murray and C. Deshaires. Monitoring protein fouling of metal surfaces via a quartz crystal microbalance. *J. Colloid Interface Sci.*, 227:32–41, 2000.

[45] J. Homola, S.Y. Sinclair, and G. Gauglitz. Surface plasmon resonance sensors: a review. *Sens. Actua. B.*, 54:3–15, 1999.

[46] L.S. Jung, K.E. Nelson, P.S. Stayton, and C.T. Campbell. Binding and dissociation kinetics of wild-type and mutant streptavidins on mixed biotin-containing alkylthiolate monolayers. *Langmuir.*, 16:9421–9432, 2000.

[47] S. Dincer S, C. Tamerler, D. Heidel, E.E. Oren, and M. Sarikaya. Pt-binding septapeptides from Phage Display Libraries, *J. Amer. Chem. Soc.*, submitted, 2004.

[48] D.S. Karpovich and G.J. Blanchard. Direct measurement of the adsorption kinetics of alkanethiolate self assembled monolayers on a microcrystalline gold surface, *Langmuir.*, 10:3315–3322, 1994

[49] C. Tamerler, M. Duman, E.E. Oren, and M. Sarikaya. Specificity of an engineered inorganic binding polypeptide: A case study on gold and platinum. *Angew. Chem.*, submitted, 2004.

[50] A.J. Haes and R.P. Van Duyne. A nanoscale optical biosensor: Sensitivity and selectivity of an approach based on the localized surface plasmon resonance spectroscopy of triangular silver nanoparticles. *J. Am. Chem. Soc.*, 124:10596–10604, 2002.

[51] J.R. Long, W.J. Shaw, P.S. Stayton, and G.P. Drobny. Structure and dynamics of hydrated statherin on hydroxyapatite as determined by solid-state NMR. *Biochemistry-US.*, 40:15451–15455, 2001.

[52] B. Zhang, B. Wustman, D. Morse, and J.E. Evans. Model peptide studies of sequence regions in he elastromeric biomineralization protein Lustrin A. I. The C-domain concensus -PG-, -NVNCT-motif. *Biopolymers*, 64:352–362, 2002.

[53] J.L. Kulp III, M. Sarikaya, and J.T. Evans. Characterization of the integral sequence repeat from the *E. coli* gold binding protein, GBP-1. *J. Mater. Chem.*, 14:2325–2332, 2004.

[54] Sakiyama-Elbert SE and Hubbell JA. 2001 Design of novel biomaterials. *Ann. Rev. Mater. Res.*, 31:183–201, 2001.

[55] B. Ratner, F. Schoen, A. Hoffman, and J. Lemons. (eds.) *Biomaterials Science: Introduction to Materials in Medicine*., Academic Press, San Diego/USA, 1996.
[56] J.R. Epstein, I. Biran, and D.R. Walt. Fluorescence based nucleic acid detection and microarrays. *Anal. Chim. Acta*., 469:3–36, 2002.
[57] M.L. Yarmush and A. Yarayam. Advances in proteomics technologies. *Ann. Rev Biomed Eng*., 4:349–373, 2002.
[58] P. Cutler. Protein Arrays: The current state-of-the-art. *Proteomics*, 3:3–18, 2003.
[59] M.E. Chicurel and D.D. Dalma-Weiszhausz. Microarrays in pharmagenomics—advances and future promise. *Pharmacog*. 3:589–601, 2002.
[60] C.M. Niemeyer. Nanoparticles, proteins, and nucleic acids: Biotechnology meets materials science. *Angew. Chem. Int. Ed*., 40:4128–4158, 2001.
[61] K.F. Sullivan and S.A. Kay. (eds.) *Methods in Cell Biology*., Academic, New York, 58: pp. 315–41, 1999
[62] A. Athanasopoulou, A. Kolliadima, and G. Karaiskadis. New methodologies of field-flow fractionation for the separation and characterization of dilute colloidal samples. *Instrum. Sci. Technol*., 24:79–94, 1996.
[63] M.P. Macdonald, G.C. Spalding, and K. Dholokia. Microfluidic sorting in an optical media. *Nature*, 426:421–424, 2003.
[64] S. Panda, T.K. Sato, G.M. Hampton, and J.B. Hogenesh. An array of insights: Application of DNA-Chip technology in the study of cell biology. *Trends in Cell Biol*., 13:151–156, 2003.
[65] L.E. Bailey, D. Kambhampati, K.K. Kanazawa, W. Knoll, and C.W. Frank. Using surface plasmon resonance and the quartz crystal microbalance t monitor *in situ* the interfacial behaviour of thin organic films. *Langmuir*., 18:479–489, 2002.
[66] C.D. Hodneland, Y.S. Lee, D.S. Min, and M. Mrksich. Selective Immobilization of to Self-assembled Monolayers Presenting Active Site-directed Capture Ligands. *Proc. Natl. Sci. USA*., 99:5048–5052, 2002.
[67] M. Harrison and P. Arosio. Ferritins: Molecular properties, iron storage function and cellular regulation. *Biochim. Biophys. Acta-Bioenerget*., 1275:161–203, 1996.
[68] T. Kacar, C. Tamerler, and M. Sarikaya,
[69] J.N. Cha, K. Shimizu, Y. Zhou, S.C. Christiansen, B.F. Chmelka *et al*. Silicatein filaments and subunits from a marine sponge direct the polymerization of silica and silicones *in vitro*. *Proc. Natl. Acad. Sci. U.S.A*., 96:361–365, 1999.
[70] D. Sahin, C. Tamerler, E.E. Oren, and M. Sarikaya. Silica-binding Polypeptides, *J. Adv. Mater*., unpublished, 2004.
[71] E.V. Amburst *et al*. Genome of Thalassiosira pseudonana: Ecology, evolution, and metabolism. *Science*, 306:79–85, 2004.
[72] R.R. Naik, L.L. Brott, S.J. Clarson, and M.O. Stone. Silica-precipitating peptides from combinatorial phage display peptide library, *J. Nanosci. and Nanotechn*., 2, 95–100, 2002.
[73] M. Sarikaya, C. Tamerler, D.T. Schwartz, and F. Baneyx. Materials Assembly and Formation using Engineered Polypeptides. *Ann. Rev. Mater. Res*., 34:373–407, 2004.

12

Dynamic Nanodevices Based on Protein Molecular Motors

Dan V. Nicolau

Department of Electrical Engineering and Electronics, University of Liverpool, Liverpool, UK

12.1. INTRODUCTION

Most of the present micro/nano biodevices are designed for a single use, as opposed to 'classical' non-biodevices (e.g., from the steam engine to the microchip). Once their function, be that simple molecular recognition like in microarrays or even biomolecular computation as in DNA computation arrays, is fulfilled and the information is passed further to signal and information processing systems, the product becomes functionally obsolete. There are indeed a few notable exceptions, e.g., biosensors and charge-controlled DNA hybridization arrays, but even these function for a limited period of time. This one-use character of micro/nano-biodevices is more an expression of the lack of robustness of their components (e.g., proteins, cells) rather than one of economic sense. Moreover, in advanced biodevices the biomolecular recognition will help to achieve their function, rather than being their function, which would allow these devices to have a continuous instead of one-off mode of operation.

Another characteristic feature of most micro/nano biodevices is their static nature, again in opposition to the dynamic modus operandi of most abiotic devices (with notable exceptions, e.g., microchip). With the possible and mild exceptions of microfluidics, where bio-objects move (by external means!) to perform the designed functions, and laser and magnetic tweezers, 'classical' micro/nano-biodevices are not dynamic per se. Moreover, the static nature of these devices reinforces the 'one-use' character. Again this static modus operandi is not an expression of much economic sense, but one of the lack of robust operation of the components once taken out of their natural environment and integrated in hybrid devices.

These differences between abiotic devices, which are in general designed to operate in dynamic conditions and frequently for longer periods of time, and biodevices, which are in general designed for one-off and static operation modes, is even more striking if we take into consideration that *natural* biodevices, including those with micron and nanoscale dimensions, are "designed" to operate in dynamic conditions and certainly for "multiple" if not continuous use. Moreover, these natural biodevices have self-repairing functions and, once worn-out or obsolete, "disintegrate" to allow others to take their place. Without anticipating the further discussion on opportunities (and challenges) offered by protein molecular motors, this short analysis suggests that it would be possible and it would certainly make economic sense, for future biodevices to have dynamic and continuous use features. These devices, which would most likely operate in highly parallel, possibly hierarchical architectures, would comprise moving elements that are self-propelled. Fortunately, Nature offers several working models of molecular motors, many tested in primitive hybrid dynamic nano-devices.

This contribution reviews the state of the art in the design, fabrication and operation of primitive bio-nano-devices based on protein molecular motors, as well as the challenges and the opportunities offered by alternative architectures and possible applications. The focus of the review will be on self-propelled machines, which have the highest chances to develop in autonomous dynamic nanodevices and systems of devices based on protein molecular motors.

12.2. PROTEIN MOLECULAR MOTORS—BIOPHYSICAL ASPECTS

Protein molecular motors, which work either in tandem as a pair of proteins (i.e., linear motors), or single (i.e., rotary motors), transform chemical energy, through the hydrolysis of adenosintriphosphate (ATP) in mechanical energy. Molecular motors are ubiquitous proteins responsible for biological functions as diverse as cell movement and division, transport of vesicles and muscle function (linear motors) and cell motility and enzymatic activity (rotary motors).

12.2.1. Rotary Motors

Of the rotary motors, the **ATPase enzyme** appears to be the smallest (approximately 10–12nm); [25, 67], the most efficient (generating up to 100pN nm with almost 100% efficiency [48, 118]) and the quickest (a rotational velocity of approximately 17r.p.s. at no loading [5]). All of these advantages made this system quite attractive for its use in hybrid nanodevices. ATP synthase is a large enzyme, which synthesizes ATP in the mitochondria. Similar enzymes can be found in e.g., plant cholorplasts and bacterial cell membranes, the latter being specifically appropriate for robust hybrid nanodevices. The structure of the protein comprises the actual engine (F1 unit) mounted on a 'pedestal' (F0 unit) as in Figure 12.1 [25].

Dynein belongs to a class of cytoskeletonal motor proteins, together with myosin and kinesin, but unlike these and despite the similarity in *local* operation, the global arrangement of dynein and associated proteins makes it a rotary motor. Despite certain advantages, the

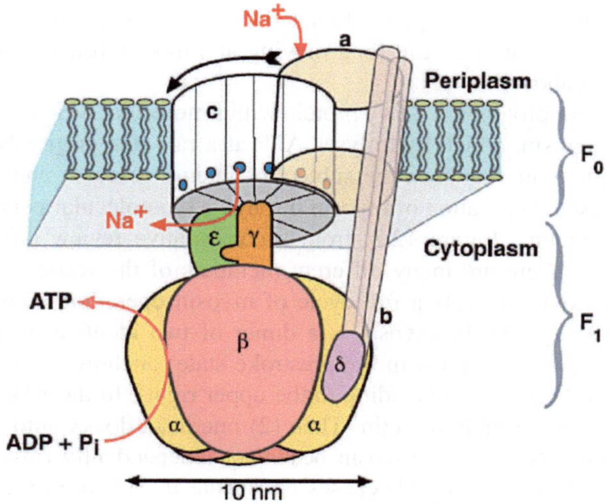

FIGURE 12.1. Architecture and modus operandi of the Na^+ FoF1 ATPase. The arrow indicates the direction of rotation during ATP synthesis. [From reference [25], reprinted with permission]

synthesis and handling of dynein is not trivial and so far dynein-based nanodevices have not been reported.

12.2.2. Linear Motors

Although it has been demonstrated that actin filaments perform a rotational motion during their motion on myosin-functionalized surfaces [48, 79], the general direction of motion is linear along their axis. Arguably, linear motors have been studied more intensively than rotary ones as candidates for hybrid dynamic nanodevices, because of the directed motion which can be used for transport of cargo. Linear motors operate as a tandem of two proteins, one being the ‘rail’ (F-actin filament or microtubule) and the other being the actual motor (myosin, or kinesin, respectively).

12.2.2.1. Myosin-Actin System The **actin-myosin** system comprises *actin filaments*, which have a cable-like structure with a diameter of about 6 nm. The building block of the actin filament is the actin monomer (G-actin), which is a globular protein of molecular weight 45 kDa. An actin filament is a one-start, left-handed helix of actin monomers [58] with a full period of 72 nm. It can be more appropriately viewed as a two-stranded, right-handed helix with the strands called protofilaments wrapping round each other with a period of 72 nm. There are extensive contacts between the protofilaments. Because the actin monomers are asymmetrical, the actin filament is polar with structurally different ends. The direct consequence of this polarity is that the polymerization is faster at one end and slower at the other. The faster-growing end is called the plus (+) end, whereas the slower-growing end is the called minus (−) end. Actin filaments range in length from 35 nm to 100 μm in cells and muscles, whereas in *in vitro* experiments the filaments are 5 μm to 20 μm. Therefore, the filaments are modeled as slender rods and mechanical properties of the filaments are computed and expressed related to this physical model. For instance, the viscous drag on a filament of 20 μm long moving at 1 μm/s through a solution with the viscosity of water is about 0.14 pN.

The *myosin* motor protein is a heavy protein with a molecular weight of about 500 kDa. The operation of myosin, which hydrolyses ATP at a rate that is greatly accelerated by the interaction with actin, has been the subject of intense research and debate. One approach is to describe the operation of myosin through a biomolecular perspective, through successive improvements. Figure 12.2, from a comparative review [109] of 'classical' myosin and kinesin (there are many different members of the respective families, with very distinct behaviour) presents a full cycle of myosin operation from a biomolecular perspective. Essentially, (1) the myosin is a dimer of two identical motor heads (blue: catalytic cores; yellow: lever arms in the prestroke state) anchored to the thick filament (top) by a coiled coil (gray rod extending to the upper right). In the ADP-Pi-bound state, the catalytic core binds weakly to actin. Then (2) one head docks onto an actin binding site (green). Importantly, the two myosin heads act independently (myosin II is a non-processive motor). Actin docking (3) causes phosphate release from the active site. The lever arm then swings to the poststroke, ADP-bound state (red), which moves the actin filament by ~10nm. After completing the stroke, (4) ADP dissociates and ATP binds to the active site, which rapidly reverts the catalytic core to its weak-binding actin state. The lever arm will then recock back to its prestroke state (i.e., back to 1). The second model of actomyosin motors, which relies on thermodynamic reasoning, has been denominated as a "thermal ratchet model" ([40]; and followed by many contributions recently reviewed by Cooke [22]). Essentially, a thermodynamically-favourable position of myosin, which is in general a result of thermal-generated fluctuations, is captured by formation of the bond with actin. Subsequent relaxation of the spring provides the enthalpy necessary to perform mechanical work in the power stroke. The tight myosin-actin bond at the end of the power stroke is broken by ATP-binding. In this way the binding cycle, controlled by ATP, rectifies (hence thermal ratchet character) random thermal fluctuations into unidirectional mechanical work.

Whatever the model, from the device design point of view the operational parameters of the motor system are more relevant. Elaborate experiments derived from motility assays (e.g., Toyoshima et al. [101]), involving optical traps (e.g. [27]), microneedles [42] or atomic force microscopy [3] provided general engineering-relevant parameters. Apart from distance-relevant parameters mentioned above [109], the velocity, the force generated and ATP consumption are of prime interest. Among the myosin family, the fastest motor is myosin XI that can achieve a velocity of 60 μm/s [59], but myosin II, which is used in many experiments, can move at velocities up to 6 μm/s and consumes 20 ATP molecules per second [39]. A lower limit for the force generated by the myosin crossbridge has been determined to be 1pN, but also forces up to 10 pN have been measured, with a value of 1.5 pN per crossbridge most likely for myosin II [39].

12.2.2.2. Kinesin-Microtubule System The kinesin-tubulin system utilizes microtubules as tracks. Microtubules are formed by the polymerisation of the globular protein tubulin, instead of actin. The kinesin-tubulin motor is responsible for, among other things, the transport of material inside cells, with vesicles being attached to the two-headed kinesin molecule as it moves along the microtubule.

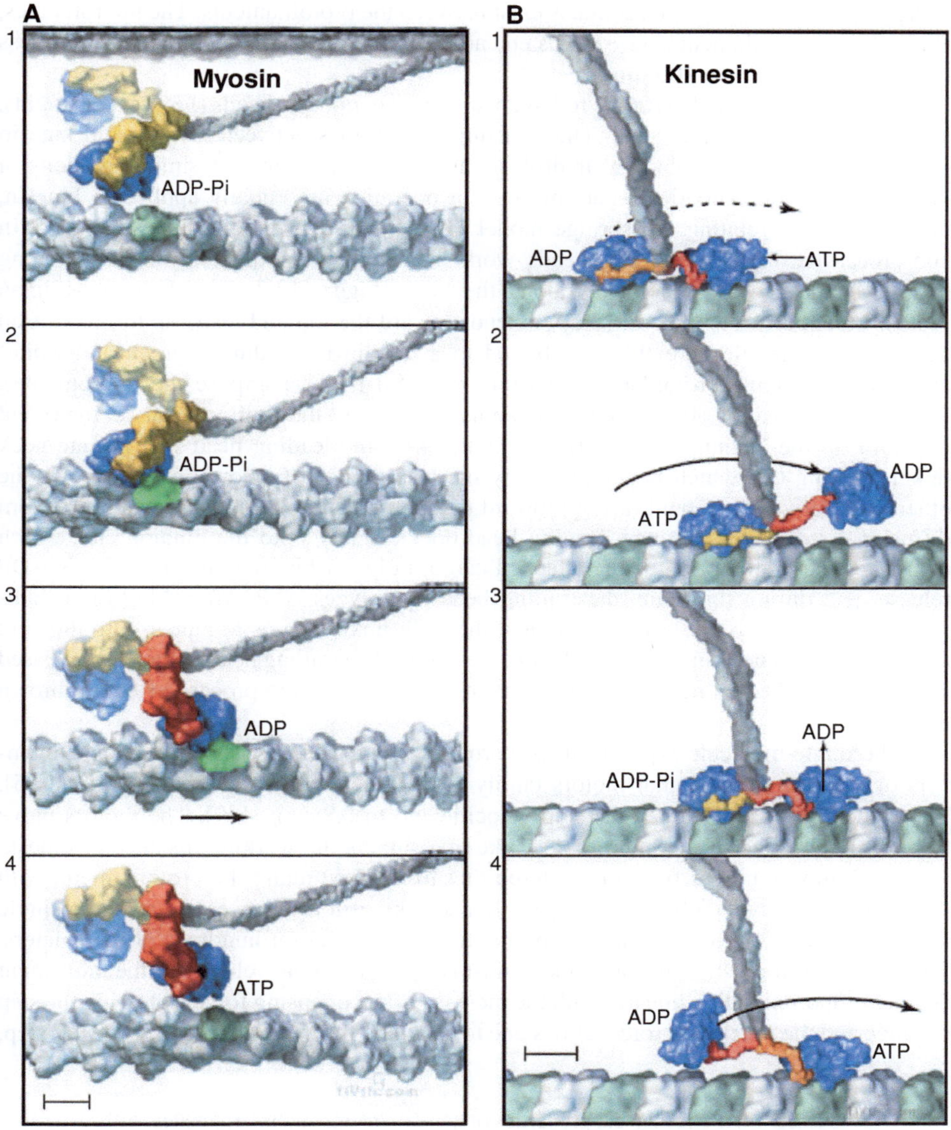

FIGURE 12.2. Detailed cycles of the operation of non-processive myosin (A) and kinesin (B) (from Reference [109], reprinted with permission; see explanation in the text).

Microtubules have a structure resembling a pipe with an inner diameter of about 18 nm and an outer diameter of about 25 nm [16]. The building block of the microtubule is the $\alpha\beta$ tubulin heterodimer with a molecular weight of 50 kDa and a length of 8nm. The dimers associate head to tail to form a protofilament. Most microtubules have 13 protofilaments which associate laterally to form a sheet that closes to form a cylindrical tube: the microtubule. As is the case for actin filaments, contacts exist between the protofilaments. The microtubules, similarly to actin filaments, have a plus end and a minus end and range in length from less than 1 μm to more than 100 μm.

The *kinesin* molecule [106] is rod-shaped with two globular heads (Figure 12.2, right), similar to the myosin molecule. The structural similarities between myosin and kinesin suggest that nucleotide binding, hydrolysis and release may trigger similar motions in the motor domain. Theories regarding myosin operation are equally applied to kinesin, in particular the rotating crossbridge model [37, 38]. In conventional kinesin, unlike in non-processive myosin, the two heads work in a highly coordinated manner, moving hand-in-hand, [39] or processively along the track (Figure 12.2, from Vale & Milligan [109]). Essentially, (1) the coil (gray) extends toward the top and leads up to the kinesin cargo. Each catalytic core (blue) is bound to a tubulin heterodimer (green and white) along a microtubule protofilament (13 protofilament tracks compose the microtubule). To adopt this position, the neck linker points forward on the trailing head (orange) and rearward on the leading head (red). ATP binding to the leading head will initiate neck linker docking (2), which is completed by the leading head (yellow), which throws the other head forward by 16nm (arrow) toward the next tubulin binding site. After a random diffusional search (3), the new leading head docks tightly onto the binding site, which completes the 8nm motion of the attached cargo. Polymer binding also accelerates ADP release, and during this time, the trailing head hydrolyzes ATP. After ADP dissociates (4), an ATP binds to the leading head and the neck linker begins to zipper onto the core (partially docked neck indicated by the orange color). The trailing head, which has released its Pi and detached its neck linker (red) from the core, is in the process of being thrown forward.

The single-molecule experiments performed with kinesin have shown that some members of the kinesin molecular motors family can develop velocities up to 1.8 μm/s [95]. The conventional kinesin can develop velocities up to 0.8 μm/s [39]. The motors move toward the plus end of the microtubule, like myosin. As the working distance of kinesin is 8 nm it moves moves from one heterodimer to the next along the protofilament, with the binding sites being about 8 nm apart. Because kinesin is a processive motor, a single kinesin molecule is able to pull a 1 μm diameter glass bead for hundreds of nanometers. The maximum force kinesin can work against is ~6 pN and the velocity of the movement decreases approximately linearly with the increase of the opposing force. Knowing the step distance and the maximum force allows one to calculate the maximum work done per step,

TABLE 12.1. Mechanical properties of actin filaments and microtubules [69].

	Young's modulus ($\times 10^9$ Nm^{-2})	Flexural rigidity[1] ($\times 10^{-27}$ Nm^2)	Persistence length[2] (μm)
Actin filaments	2.3	~60	15
Microtubules	1.9	~30000	6000

[1] Flexural rigidity is a material constant which is an analogue for the spring constant but for bending moment, not for the elastic force: the bigger the flexural rigidity it is, the more difficult to bend the slender rod, (which in this case is a physical model for the filament) it is

[2] Persistence length has the intuitive meaning of the length over which the thermal bending becomes appreciable or the length over which the bending of one point influences the bending of another point

TABLE 12.2. Experimental data on myosin II and conventional kinesin [39].

Parameter	Myosin II	Kinesin
Force/head at V = 0 μm/s	1.5 pN	3 pN
Speed (high ATP, no force)	6 μm/s	0.8 μm/s
Mechanical Work	50 pN nm	50 pN nm
Cycle time (at V_{max})	40 ms	20 ms
Distance per ATP	240 nm	16 nm
Work distance	5 nm	8 nm
Next binding site	36 nm	8 nm

i.e., 48 pN · nm. The free energy of ATP hydrolysis under cellular conditions is 100 pN · nm, so the maximum efficiency of kinesin is ~50% [39].

The main parameters of actin filaments and microtubules, and myosin and kinesin motors, are compared in Table 12.1 and Table 12.2, respectively.

12.2.3. Actin/Microtubule Polymerisation

The polymerization of actin and tubulin do not constitute a molecular motor per se, but can yield mechanical work [22, 72, 100] that can be useful in future nanodevices.

The assembly and disassembly of filaments and microtubules occur by the addition and subtraction of subunits at one or both ends. This has been demonstrated by *in vitro* experiments visualized by electron microscopy of actin filaments and by video-enhanced microscopy of microtubules ([71, 111] respectively). The spontaneous breakage and annealing of actin filaments and microtubules *in vitro* is very slow. Because of the contacts between protofilaments, it is energetically less favorable to break the filament in the middle than to remove a subunit from the end. The theoretical model of two-stranded filaments predicts that the lengths of the filaments are exponentially distributed with the average depending on the monomer concentration. For example, if the total concentration of actin monomer is 10 μM, than the average actin filament is 2.75 μm long [39]. There exists a critical concentration of subunits, below which no apparent filaments are formed. Figure 12.3 (from Cooke [22]) describes schematically the mechanism of force generation occasioned by actin polymerisation.

An interesting 'natural demonstration' of the actin polymerisation motor is offered by the bacterium Lysteria (and few other bacteria, e.g., Shigella), which manages to propel itself by surfing on top of a wave of polymerised actin ('actin rocket'), with myosin being irrelevant in the process.

12.3. NANODEVICES BASED ON PROTEIN MOLECULAR MOTORS—OPERATIONAL ASPECTS

12.3.1. Motility Assays and Single Molecule Techniques

Two experimental techniques, i.e., (i) *in vitro* motility assays; and (ii) single molecule visualization, manipulation and measurement; resulted in important advances in the understanding and quantification of the functions of molecular motors.

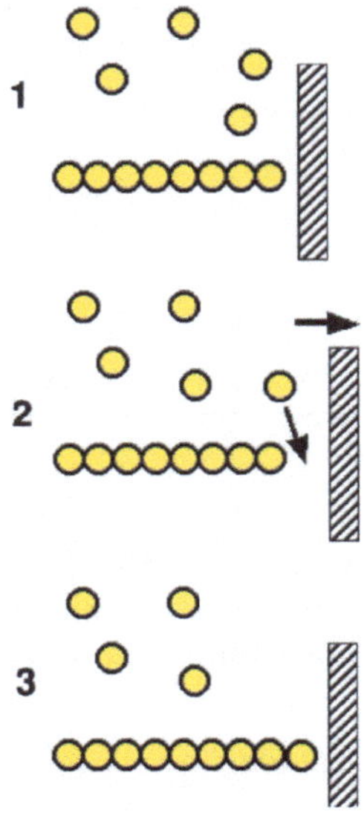

FIGURE 12.3. Force production by actin polymerization ([22], reprinted with permission). A single actin filament is shown near and pushing towards a wall. In the transition from State 1 to 2 the gap between the end of the filament and the load is increased by either a fluctuation in the position of the load, or in the position of the actin filament. This allows the insertion of an additional actin monomer, thus extending the filament, and leading to a new equilibrium position that has moved 2.7 nm to the right in State 3.

The motility assay, with various architectures presented in Figure 12.4, is based on observation of the motile element of the motor system in the presence of ATP on top of a surface, which is functionalized (for dual protein systems) with its pair protein. There are two possible geometries used for *in vitro* motility assays of linear motors: (i) the *gliding assay* consisting of actin filaments or microtubules sliding on myosin or kinesin functionalized surfaces; and (ii) the *bead assay* consisting of objects (e.g., a microbead, typical diameters of 1 μm) functionalized with the motor protein (i.e., myosin/kinesin) sliding along actin filaments or microtubule that immobilized on surfaces. Because of its simplicity, the gliding assay is far more used for both actin-myosin [49, 117] and kinesin-tubulin system [36] than the bead assay, which has also been proposed for actin-myosin [82, 87] and kinesin-tubulin system [37]. Motility occurs in a flow cell which comprises two parallel surfaces, at least one being transparent, and usually both, which are separated by a thin (approximately 100μm) spacer (e.g., grease or double sided tape). The motility flow cell is mounted on a microscope, which is used to observe the movement of the motile element, either by fluorescence if the motile element is fluorescently tagged, or directly via differential interference contrast (DIC) for microtubules which are larger than actin filaments. An intermediate case is the gliding motility assay with bead-tailed actin filaments [94]. In order to achieve the specific positioning of the bead at one (barbed) end of actin filament, a 1 μm polystyrene bead was coated with gelsolin, which specifically binds at the barbed end of the filament. In principle the same procedure can be applied to microtubules. More recently, the motility has been observed using the Atomic Force Microscope (AFM) and a special fast procedure [3].

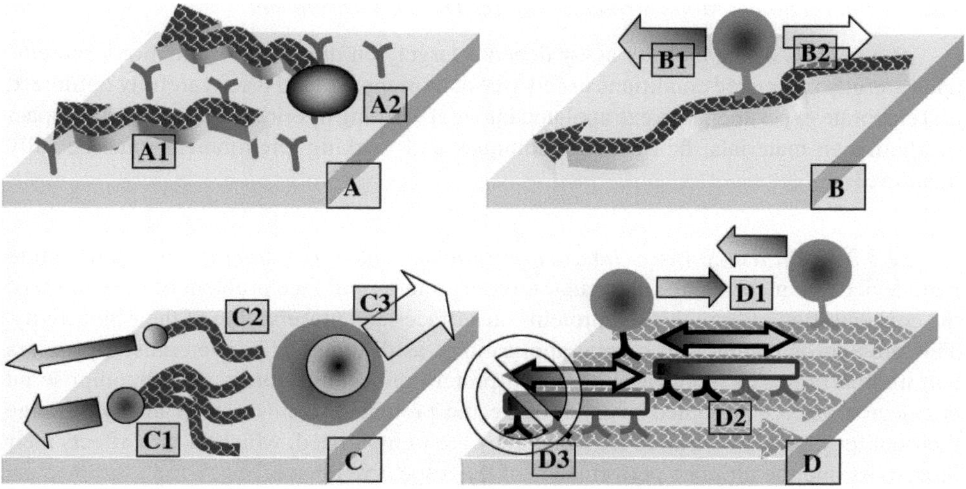

FIGURE 12.4. Types of motility assays. A. Gliding motility assays. A1. Classical gliding motility assay in which actin filaments or microtubules slide on a surface covered with molecular motors (a protein in the myosin or kinesin family). A2. Bead tailed gliding motility assays in which a bead functionalised with gelsolin is attached at the trailing, barbed end of the filament. Small beads are more likely to be bound to a single filament, but bigger beads will statistically have more than one filament attached, effectively arresting the motility. B. Bead motility assays. B1. Classical bead motility assay in which a bead is functionalised with motor protein (proteins in myosin or kinesin family, or motor fragments) and runs unidirectionally, according to motor type, on single filaments or microtubule immobilised on the surface. B2. Single molecule studies using optical traps or microneedles apply a force in the opposite direction of the movement, thus offering the possibility to measure the force generated by the motor. C. Actin polymerisation motility assay. C1. Actin polymerisation on anisotropically coated beads, i.e., one half coated with proteins (Arp) that trigger actin polymerisation. Beads move in a manner similar to Shigella, Listeria etc. C2. Actin polymerisation on small beads inherently progresses anisotropically. C3. Anisotropic polymerisation around a bead creates a gel with uneven distribution which in turn moves the bead. D. Possible bidirectionality assays. D1. Beads with functionalised with anti-parallel directionality, e.g., kinesin and ncd, move in opposite directions on parallel tracks (microtubules). D2. Payloads functionalised with antiparallel motors move in one direction according to the result of the "tug of war" between motors. D3. The "tug of war" mechanism is replaced by the coordination between motors, in the sense that one motor works (right) the other is blocked (left).

In vitro motility assays, although not always denominated as such, have been also developed for other systems that yield mechanical work at the nanoscale, i.e., rotary motors (e.g. [67]) and actin and microtubule polymerization (e.g. [14] and [26], respectively).

Single molecule techniques, e.g. [28, 96], which are modifications of the *in vitro* motility assays, have been used to measure the distances moved by single molecules, as well as the forces they generate—parameters briefly reviewed above. In the filament assay, the filament or the microtubule is held in a force transducer and presented to a motor that is fixed to a surface. In the bead assay, the motor is attached to a bead held in a force transducer and the bead is presented to a filament that is fixed to a surface. The force transducers, e.g., cantilever rods, AFM tips or optical tweezeers, must be able to produce and monitor forces in the pN range. The motion produced when the motor interacts with the filament is measured using photodiodes capable of sensing subnanometer displacements with millisecond resolution [39].

12.3.2. Interaction of Motor Proteins with the Device Environment

The success of the motility assay depends largely on the purity of the motor proteins and carefully optimized conditions in cell-free conditions. Should these carefully optimized and elaborate experiments be extrapolated into real working hybrid nanodevices, the impact of alternative materials, fabrication techniques and working environment must be fully quantified.

12.3.2.1. Protein-Surface Interaction—Biomolecular Considerations Any future nanodevice that incorporates molecular motors proteins will face problem of the control of the concentration of immobilised proteins on surfaces and preservation of their bioactivity. This problem is common to several applications, as different as biomaterials, biosensors and microarrays. All biodevices based on protein molecular motors would comprise an engineered surface designed to immobilize the protein. Proteins almost always change their conformation on surfaces on which they are immobilized, which in turn affects their bioactivity and the ultimate performance of the respective device.

In the case of devices based on protein molecular motors, the above problem is exacerbated by the well-known sensitivity of motor proteins when taken out of their natural, cell environment. However, the problem is even further exacerbated by a similar complication faced by microarray technology, namely how to find an optimum single surface that preserves protein activity, or at least reasonable bioactivity levels, when many different proteins are immobilised on the *same* surface. Although molecular motor proteins are nominally a single molecule, one fact that is often overlooked in the context of real and robust artificial devices is that the protein molecular motors are in a sense 'multiple' or at the very least 'dual' proteins, i.e., proteins that present different molecular structures for each stage in the power stroke cycle. These molecular structures are separated in time by intervals on the order of ns. Figure 12.5 presents the structure of the molecular surface of the S1 unit of myosin only in two states (structures collected from the Protein Data Base, at www.pdb.org), i.e., rigor state (2MY, left) and working state (1DFK, right) as probed by a probe ball with a large radius (1nm) using Connolly's method [19–21]. It was found [63] that at this large value the properties of the protein molecular surface do not vary with the ball radius, and hence this radius the probing with a flat surface. The two structures of the same biomolecule presented in Figure 12.5A are starkly different. First, the S1 region of myosin-rigor state presents a much larger molecular area, particularly in the neck region. Second, the molecular surface of the S1 region of myosin in working state presents 'hot spots' of hydrophobicity and hydrophilicity (red and blue, respectively), particularly in the ATP pocket region. Possibly, a similarly large difference can be found for other motor proteins (e.g., kinesin).

But how different are these two structures? Can we have some measure of their dissimilarity? The method used to visualize the distribution of hydrophobicity on the molecular surface can also be used to quantify other properties (e.g., charges) as mean values and densities. Figure 12.5 B [68] compares the similarity of the molecular surfaces of 42 proteins according to a cluster analysis (nearest neighbour, simple Euclidian distance) using the average properties, their density (e.g., total hydrophobicity per total area) and their specific density (e.g., hydrophobicity per hydrophobic area) for hydrophobicity, hydrophilicity, positive and negative charges. The dissimilarity between S1 myosin-rigor (2MYS) and S1

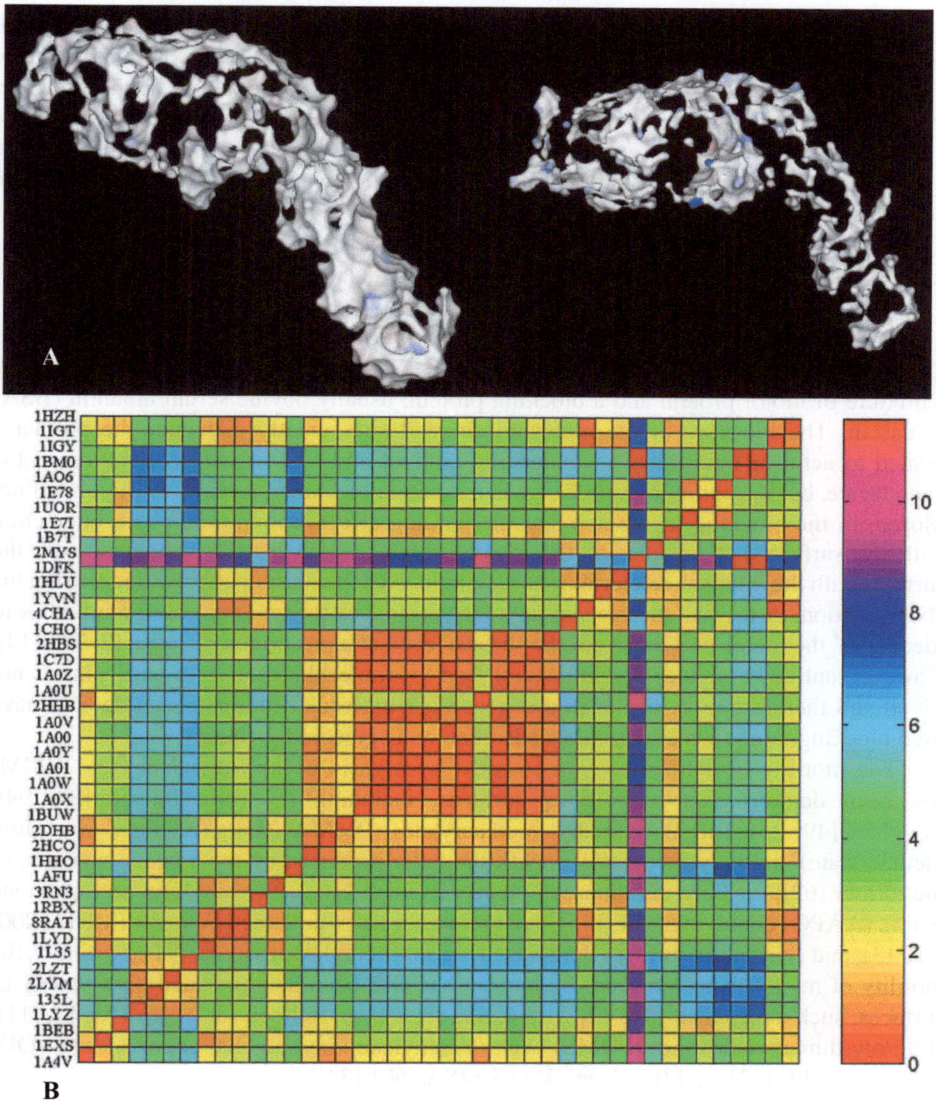

FIGURE 12.5. Molecular surface of the S1 unit of myosin head in the rigor and working state (top panel, left and right, respectively) and the similarity comparison between these two S1 states (1DFK and 1MYS, respectively, from the Protein Data Base, www.pdb.org) and other 40 proteins. Perfectly similar proteins (self-similar, identical proteins -on the diagonal-) coloured in red; most dissimilar proteins in purple.

myosin-working (1DFK) is clearly very large (red rectangles on the first diagonal represent self-identical structures; white circle focuses on S1 structures). More importantly, this difference is as large as, or larger than the difference between the former structure and all other 40 proteins, which are indeed extremely different (i.e., lysozyme, haemoglobin in the middle of the cluster, IgG, albumin and so on)!

To complicate things further, a study of motor proteins in electric fields (kinesin-tubulin; [83]) demonstrated that the actual net charges of the proteins, and therefore the charges on the molecular surface, as measured experimentally are very different than those derived from molecular structures, perhaps close to an order of magnitude difference! Consequently, the electrostatic interactions, which are long range but reversible, will be damped and the hydrophobic interactions, which are short range and irreversible hence causing protein denaturation, will be amplified.

It appears that the search for a 'magic surface' that is non-denaturating for both structures of the motor protein is very much warranted, but also extremely difficult.

12.3.2.2. Impact of Device Materials and Surfaces—Experimental Observations Traditionally, the gliding motility assays have been run on surfaces that are covered by a mixture of motor protein and a blocking protein, usually bovine serum albumin (BSA) or casein. The blocking of the motor-unoccupied sites greatly alleviates the problems related to actin or microtubule non-specific binding and subsequent deactivation of the nanodevice, but the optimum ratio between motor and blocking proteins is difficult to find. Moreover, the blocking of the non-specific binding affects actin/microtubule interaction with the surface, but not that of the motor protein, which is still in contact with the surface, with the inherent deleterious surface-induced denaturation effects suggested in the above section. A logical and elegant approach would be to use engineered antibodies to 'decouple' the motor proteins from the surface, even at a prescribed distance [8, 74, 114]. However, antibody engineering in general, and for molecular motors in particular, is not trivial and therefore most of the studies regarding molecular motors-based devices have used blocking/motor protein mixtures adsorbed on engineered surfaces.

The motility of actin filaments powered by myosin or its fragments (e.g., HMM) has been demonstrated on various surfaces, including (i) nitrocellulose [84, 94], (ii) glass [49, 74]; (iii) Poly(methyl methacrylate) (PMMA [93]); (iv) Poly (*tert* butyl methacrylate) and Poly(methacrylic acid), the latter produced by the e-beam exposure of the former [62]; (v) Poly(tetrafluoroethylene) (PTFE [92]); (vi) O-acryloiyl acetophenone oxime (AAPO) copolymer [54, 115]; (vii) printable cross-linkable UV-resist (MRL-6000) [12, 13]; and (viii) glass surfaces derivatized with trimethylchlorosilane [91]. Similarly, the motility of microtubules powered by kinesin has also been demonstrated on a variety of surfaces, such as (i) glass [11, 17, 24, 36, 43, 88, 89]; (ii) silicon [88]; (iii) PMMA [11]; (iv) Poly(dimethylsiloxane) (PDMS) [11]; (v) ethylene-vinyl alcohol copolymer (EVOH) [11]; (vi) PTFE [23]; and (vi) deepUV resist (SAL601 [34]).

Several comments are in place here. First, different surfaces will induce differential adsorption of proteins (i.e., BSA and motors) and therefore tests have to be run to find an optimum concentration of proteins in the buffers. Second, several materials mentioned above are commonly used, or are designed to be used as micro- and nano-lithography resists, which eases the fabrication of test devices. Moreover, these materials with radiation-controlled surface properties, can be tuned by the exposure energy to optimise motility parameters. Third, minute variations in experimental conditions can lead to totally different results. For instance, several authors (e.g. [62, 93]) demonstrated the motility of actin filaments on patterned PMMA, whereas others (e.g. [12, 13, 74]) used PMMA for patterns that do not support motility. This difference in experimental results, which is less evident for hard surfaces (e.g., glass and silicon), derives from the complex character of polymers and their surfaces. Indeed the same polymer, but with different polymer physico-chemistry (e.g., molecular weights and distribution of molecular weights), processed in different conditions (e.g., different cast solvents; baking temperatures and durations), and exposed to different buffer conditions (e.g., duration, temperature, ionic strength) will present different surfaces to the motor proteins and therefore induce different motility behaviour. Fourth, and related to the above point, the polymer surfaces, especially for non-crosslinked polymers and polymers with low glass transition temperature (i.e., in rubber state) will swell in contact with the buffer, leading to the interaction of top polymeric chains with the motor proteins. For this reason, PDMS and PU do not support motility for actin-myosin (our work, unpublished results) and microtubules-kinesin [11], respectively. Finally, the critical failure mechanism for motility assays is the denaturation of the motor protein [54, 62] and degradation of the microtubules [11] for actin-myosin and for kinesin-microtubule systems, respectively.

Because the bead motility assay is far less used, the data regarding the surfaces that support motility is less extensive. Moreover, the optimum bead motility assay would require a compact carpet of aligned actin filaments or microtubules, if possible immobilized on the surface, which will have therefore a far lesser impact on motility. Indeed the first motility assays used naturally self-assembled actin filaments in the algae Nitella (e.g. [82, 83]) and later artificially self-assembled F-actin paracrystals on lipid surfaces [90]. Also early [95, 102] and recent [43] work demonstrated that microtubules adhere strongly to amine-terminated surfaces while retaining the ability to act as substrates for kinesin-coated beads. The above mentioned criticality of microtubule deactivation has also been addressed through stabilization with taxol [107] or glutaraldehyde [103].

The assays using rotary motors have been also used as primitive nanodevices, but as this approach is seldom used the relevant data is less extensive. Metallic surfaces including nickel [6, 67, 81, 85, 86], copper and gold [5, 57] have been used for the anchoring of histidine-tagged F1 ATPase without apparent deleterious effects on its function. It should be noted that this approach is far more appropriate for protein immobilization due to its inherent capacity to orient the biomolecule with the working region upwards, but equally more achievable in the case of F1 ATPase because of its small dimensions. Returning to linear motors, the nature of the surface, as well as the concentration of motor protein in the solution used for immobilization, will modulate the active (i.e., non-denatured) and the total number of motor molecules on the device surface. The impact of the surface density of active motors is markedly different for processive (e.g., kinesin, myosin V) than for non-processive motors (e.g., myosin II). For instance, myosin V velocities are found not to vary over several decades of myosin density [75], a behaviour similar to kinesin. This density independence is in contrast to the behavior of myosin II, which exhibits a drop in velocity as the density is decreased [105]. Another study [50] demonstrated that the sliding velocity of actin filaments decreased nonlinearly with reduced density of HMM from myosin II molecules, which is consistent with unloaded filament sliding velocity being limited by the number of cycling cross-bridges so that maximal velocity is attained at a critical, low level of actomyosin interactions. Such data are consistent with myosin II having a low duty ratio (ratio of the strongly bound state time to the total actin activated ATPase cycle time) and myosin V having a high duty ratio. A high duty ratio for myosin V is consistent with the property of processivity, in which case the duty ratio for a given myosin head need be at least 0.5.

12.3.2.3. Impact of Fluid Environment Conditions As mentioned before, protein molecular motors are quite robust in cell environment but they are quite sensitive in more realistic, device-relevant conditions. However, all possible applications envisaged for molecular motors-based nanodevices ask for a working environment that is far from cell conditions. The search for a solution regarding robustness of molecular motors powered devices can strategically progress in two directions: (i) search or engineer molecularly robust proteins; or (ii) find environmental conditions that stabilise the functions of the proteins *in vitro*. Triggered by the increased interest in molecular motors based nanodevices, the more strategic search for robust protein molecular motors has started in earnest only recently. On the other hand, the large interest in understanding the operation of 'classical' protein molecular motors in motility assay conditions provides a large amount of information regarding the later option, which has been further augmented by experiments regarding primitive prototypes of nanodevices.

Firstly, the extreme sensitivity of protein molecular motors proteins asks for a careful minimization of chemical species toxic to motility. In particular, radical species (inherently produced in situ by the photobleaching of fluorophores) have to be totally removed from the fluid system. This is achieved by adding oxygen scavengers (e.g., amines) in controlled conditions. Other toxic chemical species, e.g., small organic molecules, can ex-diffuse from the polymeric walls of the flow cell, especially if these are made of polymers in elastomeric state. Indeed, a recent comparative study [11] regarding the motility of microtubules in flow cells made of different polymers (i.e., PU, PMMA, PDMS and EVOH), demonstrated the complex relationship between the fluid environment, building materials and operation conditions. Essentially, without illumination, only PU had a substantial negative impact on microtubule motility, while PMMA, PDMS and EVOH showed stabilities comparable to glass. Under the influence of light, however, the microtubules degraded rapidly on PDMS or PMMA, similarly with motility experiments in glass flow cells if oxygen scavengers were not added to the medium. Strong photobleaching of the fluorophores, which occurs mainly on the polymer surface, coincided with accelerated microtubule depolymerization. While the sensitivity of the microtubule-kinesin system (which is more related to microtubules) can be addressed by correct operating conditions and appropriate materials or fixation with glutaraldehyde for gliding and bead motility assays, respectively, the sensitivity of the actin-myosin system, which is more related to the motor protein, is more difficult to address. Traditionally, the use of extremely controlled motility assay conditions and addition of phalloidin (a pentapeptide present in a toxic mushroom, which blocks the depolymerisation of actin [110]) allows for several hours of gliding motility assays.

Of course one can consider applications where the problem of environmental sensitivity is a non issue, for instance devices that operate in fluid environments that have very few, not deactivating, chemical species (e.g., oligonucleotides [24]) and which operate under very controlled parameters (e.g., temperature, pH). Conversely, one can consider applications where it is precisely the deactivation of the motility by the fluid environment that constitutes function of the molecular motors-based nanodevice (e.g., detection by heavy metal ions, [55]). Finally, one might consider re-creating the conditions in the cell to allow molecular motors devices operate in a robust manner, rather than searching for robust motor proteins that operate in motility-adverse conditions—but this endeavour is some time away.

Temperature is another environmental parameter that heavily impacts on the operation of molecular motors based nanodevices. Intuitively, as motility is a chemical reaction, the rate of operation (e.g., motility speed) of protein motors will increase with the operating temperature. However, as molecular motors are essentially mechanical nanodevices the rate of operation will plateau at some temperature. Moreover, as protein molecular motors are actually very sensitive nanodevices, excessive temperature will result in a decrease of the operation rate due to protein denaturation. Other operational parameters (e.g., force) could, in principle, also be ffected by temperature, but the processivity of the motors is an important parameter. Indeed, a study [44] using bead motility assays of kinesin (a processive motor) demonstrated that the kinesin-coated beads gliding velocity increases with temperature following an Arrhenius law (with an activation energy of 50kJ/mol, consistent with the temperature dependence of the microtubule-dependent ATPase activity) between 15 and 35°C, whereas the force generated remained essentially the same (7.3pN). The authors concluded that the force generation can be attributed to the temperature insensitive nucleotide-binding state(s) and/or conformational change(s) of the kinesin-microtubule complex, whereas the gliding velocity is determined by the ATPase rate. A similar study [45] regarding the impact of temperature on the motility of the non-processive actin-myosin system demonstrated that the sliding force increase moderately with temperature (4 to 6pN), whereas the velocity increased significantly (from 1 to 10μm/s for 15°C and 35°C, respectively). The authors concluded that the temperature-driven moderate increase in force is caused by the increased number of available cross-bridges for actin interaction, because the cross-bridge number similarly increased with temperature. Another study [9] which used gliding motility assays for kinesin also demonstrated a 10-fold increase in velocity (maximum of 1.8μm/s) with temperature from 5 to 40°C.

12.4. DESIGN, FABRICATION AND OPERATION OF PROTEIN MOLECULAR MOTORS-BASED NANODEVICES

The aspects discussed above set the limits for possible materials for device fabrication and the limits for the correct operation conditions, but they say very little about the how these devices operate, let alone how they will be fabricated or designed. As mentioned before, motility assays are effectively primitive hybrid dynamic nanodevices, but more advanced devices will need to incorporate 'smarter' features. Moreover, and apart from realistic endurance of the devices (an issue addressed in the above section), the success of future dynamic nanodevices based on linear molecular motors will depend on successful resolution of all or some of the following technological problems: (i) precise positioning of the motors on designed areas—essential for the gliding assay geometry devices and devices based on rotary motors; (ii) confinement of the movement of motile elements exclusively on fabricated paths for gliding assays; (iii) achievement of unidirectional polarity of the movement of the motile elements, when linear motors are used; and (iv) the on-off control of their operation.

12.4.1. Lateral Confinement of Movement for Motile Elements

For rotary motors, the lateral confinement of movement is inherently achieved by the precise positioning of rotary motors [5, 6, 57, 67, 81, 85, 86] via, e.g., the nanofabrication of

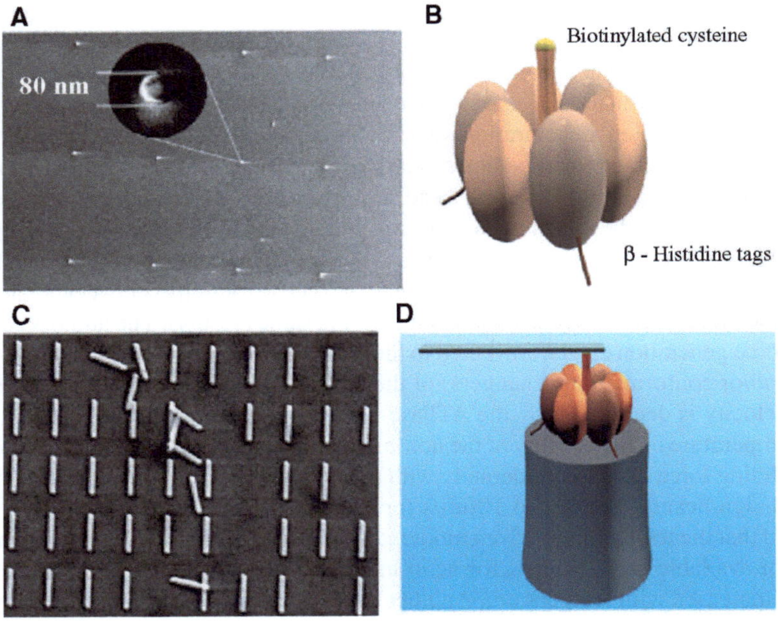

FIGURE 12.6. Schematic diagram of the F1-ATPase biomolecular motor-powered nanomechanical device (from Soong et al., [85], reprinted with permission). The device consisted of (A) a Ni post (height 200 nm, diameter 80 nm), (B) the F1-ATPase biomolecular motor, and (C) a nanopropeller (length 750 to 1400 nm, diameter 150 nm). The device (D) was assembled using sequential additions of individual components and differential attachment chemistries.

nickel nano-posts, followed by specific immobilization of F1 ATPase molecules that have histidine tags at the opposite end of the motor domain (Figure 12.6).

For devices using linear molecular motors and the gliding architecture, the lateral confinement of the filaments/microtubules motility can be achieved via (i) mechanical confinement in micro- or nano-fabricated conduits (e.g., channels, etc.), (ii) selective patterning of protein motors, assuming that the filaments or microtubule will preferentially stay on motor tracks, or (iii) using both methods at the same time.

Initial studies focused on mechanical confinement only. For instance, it has been demonstrated that parallel nano-scratches fabricated in PTFE by dragging a metal rod in controlled conditions (weight on the rod, temperature) can effectively confine the movement of actin filaments [92] and microtubules [23]. More recent studies have demonstrated confinement of microtubules in micro-channels fabricated in kinesin-adhesive PU and AZ5214 photoresist [18] and [32] respectively). Mechanical-only confinement, which has been found to be more effective for deeper channels and for walls with retrograde slopes [32], is suitable for microtubules (large and rigid compared with actin filaments), but is unlikely to work for small and very flexible actin filaments, although the retrograde slope should in principle be effective.

The patterning of linear motors on micro- and nano-fabricated tracks, which is expected to significantly improve the confinement of the motility of motile filaments or microtubules, has been achieved by preferential adsorption of motor proteins (i) on flat, adhesive patterns in a non-adhesive background; (ii) on adhesive microfeatures elevated from a non-adhesive base; and (iii) in microchannels with non-adhesive walls. The motor patterning on flat tracks has been achieved for HMM on hydrophobic Poly(*tert* butyl methacrylate) tracks on a less-adhesive Poly(methacrylic acid) background, i.e., unexposed areas left after deep-UV or e-beam negative tone lithography [62], and for kinesin on adhesive glass areas on a non-adhesive Poly(ethylene oxide) (PEO) coated background [18]. The second technological choice was implemented in an early study [93] via preferential adsorption of HMM on PMMA microfabricated features on a non-adhesive glass background. Most studies, however, couple the patterning of motor proteins with the mechanical confinement of motility in profiled microfabricated micro-channels, which has been achieved for HMM/myosin for the following architectures: (i) floor: UV-resist (MRL-6000); walls: PMMA [12, 13]; (ii) floor: glass; walls: PMMA [74]; (iii) floor: ablated O-acryloyl acetophenone oxime (AAPO) copolymer; walls: BSA coated AAPO [54, 115]; and (iv) floor: glass; walls: metal electrodes [4]. Similarly, kinesin immobilization in micro-channels has been demonstrated for (i) floor: SiO_2; walls: PEO-coated SU8 (a photoresist) [18]; (ii) floor: glass; walls: SAL601 (an e-beam resist) [34]; and (iii) floor: glass; walls: SU8 [43]. Figure 12.7 presents the alternative strategies to ensure the lateral confinement of motile elements in gliding motility assays.

The difficulty of confining the motility of motor-coated beads derives from the difficulty of patterning actin filaments and microtubules. Still, an early study [102] demonstrated confinement of the movement of kinesin-coated beads on microtubules which had been pre-aligned in a flow field and immobilized on aminosilane patterns. More recently [4] used microfabricated electrodes and flow fields to transversally align actin filaments. Also, studies regarding tubulin polymerisation demonstrated the transversal alignment [26], and star-shaped features in microfabricated chambers [35].

12.4.2. Control of Unidirectional Movement by External Means

The second desideratum of molecular motors-powered nano-devices is the control of the directionality of movement. While directionality is inherent to rotary motors (which rotate in one sense only; [25, 67]), to polymerization of actin (and tubulin) and in bead architecture assay, it has to be achieved by external means for the more popular nanodevice using linear motors in a gliding architecture.

The easiest method for the control of directionality of motile elements is the alignment of actin filaments and microtubules in the fluid flow that occurs in microfabricated structures either intentionally (e.g., microfluidics) or unintentionally (e.g., higher evaporation rate at one end of the flow cell). For instance, microtubules [88] and actin filaments (Nicolau et al., unpublished results) have been aligned by a fluid flow when exiting a motility chamber connected with the flow current area. Of course, the control of unidirectional movement by flow can also be used for the *operation* of bead geometry assays, but then the more specific unidirectionality control exerted by the polarity of the actin filament or microtubule is hindered. It is therefore more likely that the alignment of filaments/microtubules will be useful for the *fabrication* of devices based on bead geometry.

The most used method for achieving directionality control in gliding geometry devices is based on the mechanical guidance by microfabricated 'molecular selectors' [63]. The

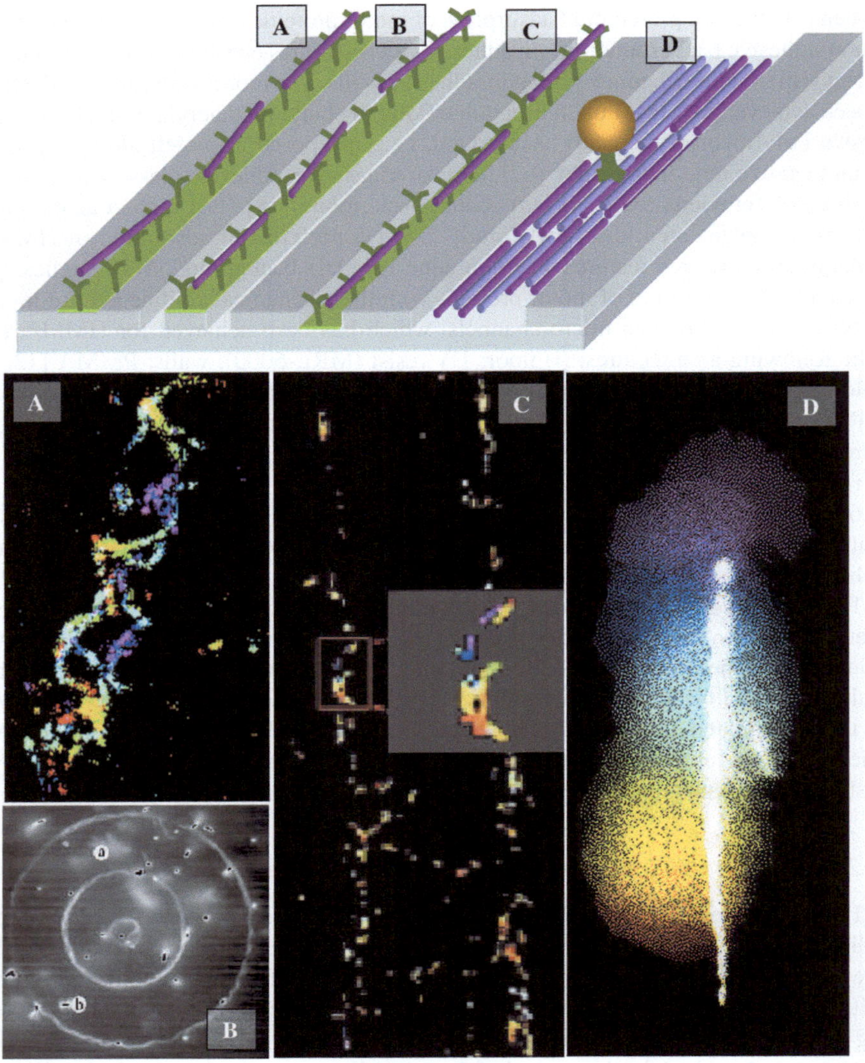

FIGURE 12.7. Mechanisms for lateral confinement of motile elements using micro/nano-fabrication. A. Actin filaments and microtubules slide on motor-functionalised stripes (green) on flat surfaces. Color coded trajectory of actin filaments on HMM-functionalised tracks (adapted from [62]; red—beginning of the sequence; purple end of the 4s sequence). B. The filaments (potentially, microtubule too, although the lateral confinement has not been demonstrated yet) slide on top of motor-functionalised microstructures. The inset presents an overlap of images of actin filaments (from [93], reprinted with permission). C. Confinement of filaments or microtubules in microchannels with motor-functionalised floor. The inset presents the color coded trajectories, with an enlargement, for actin filaments sliding in 3μm channels in a 5s sequence (adapted from [54]). D. Confinement of the movement of motor-functionalised beads in micro-channels that have filaments or microtubules aligned on the floor of the channel. The filaments/microtubules are parallel aligned, but have random directionalities. The inset presents the color coded trajectory of a HMM-coated bead on actin bundle self-assembled in a microchannel (adapted from [2]).

first demonstration of this principle [34] used e-beam fabricated arrow-shaped channels connecting two reservoirs of microtubules to statistically select and redirect the motility in a preferential direction, depleting one 'pool' of microtubule and enriching the other. This elegant work has been followed by several fabrication improvements, e.g., use of SU8 and UV lithography instead of SAL600 and e-beam lithography [43]. The implementation of mechanically-controlled unidirectionality is also supported by extensive statistical work regarding the mechanics of microtubules colliding with microfabricated obstacles [17, 18]. It is very likely that mechanically-controlled unidirectionality for actin filaments will be less successful, or at least it will have to use different, more rounded shapes of the molecular selector. Indeed, early work regarding motility confinement of actin filaments [93] demonstrated the smooth movement of filaments on ∞-shaped features. Figure 12.8 presents several molecular selector geometries.

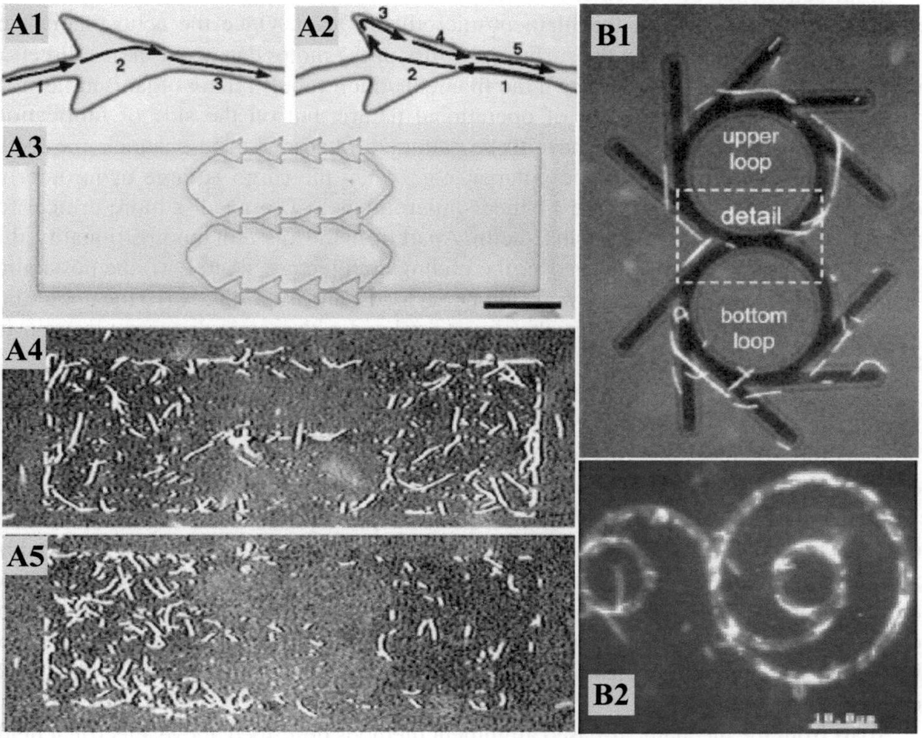

FIGURE 12.8. Control of directionality with microfabricated structures (molecular selectors) functionalised with motor proteins. A. Microfabricated arrowheads rectify microtubule movements (from Reference [34]; reproduced with permission). A microtubule entering the arrowhead feature from the left (A1) passes through the rectifier, but when entering from the right (A2) it is statistically turned back. Two microfabricated 'pools' separated by three rectifiers (A3) are equally filled with microtubules A4) but after a while the left pool has the majority of microtubules. B. Microfabricated rounded routers. B1. The "figure 12.8"-shaped track is comprised of two circular loops that form a tangential crossing junction and is decorated with unidirectional reflectors for microtubules (Clemens [17]; reproduced with permission). B2. Spiral shaped routers conduits for actin filaments [64].

The second method for unidirectionality control proposes the use electric fields to orient the movement of electrically charged actin filaments or microtubules. The method uses electrodes to create a current in the flow cell through the motility buffer. Care should be taken to set the current/voltage at low level to avoid electrochemical reactions in the buffer. This method used to enforce the unidirectionality has been demonstrated for the motility of actin filaments, especially when coupled with the lateral mechanical confinement of motility in microfabricated channels [74]. The method has also been used to enforce the unidirectionality for microtubules [89]. Recently, it has been demonstrated that the motility of actin filaments can be directed along electric field lines between micropatterned electrodes [4].

12.4.3. Control of Unidirectional Movement by Self-Assembled Tracks

Unidirectionality of movement of the motile elements is an inherent feature of the bead architecture assays—at the filament/microtubule level. Once the actin-filament or microtubule is immobilised on the surface, motility of the motor can occur in one direction only, but the challenge is laying down the filaments/microtubules in an organised fashion. Therefore, the challenges are not of operational nature, but on the side of fabrication, as filaments and/or microtubules have to be patterned in stripes or in channels, perfectly aligned, parallel and with the same polarity, e.g., when the cargo is large or in order to counter the possible defects in the filaments/microtubules. The need is more critical for non-processive systems, e.g., myosin II-actin. Apart from the inherent unidirectionality, this approach has important advantages over the gliding architecture, namely (i) the possibility to engineer the motor protein to carry different payloads; and importantly (ii) the possibility to run objects in opposite directions when functionalized with motors that move in opposite polarities (e.g., kinesin and Ncd).

Conceptually, the simplest method to fabricate parallel, unidirectional tracks with microtubules is to use the flow-induced alignment in a combination with surface-immobilization of filaments or microtubules. Importantly, early work demonstrated that the flow-oriented actin filaments has nearly the same high degree of alignment as the actin filaments in muscle fibers [70]. This general approach has been used to orient microtubules by end-specific immobilization with minus-end specific antibodies followed by flow-induced alignment [51]. Also an array of aligned microtubules has been fabricated via the immobilization of short microtubule polymerization seeds at the beginning of microchannels followed by microtubule polymerization in flow [10]. Recently, we also used this method to fabricate bundles of actin filaments on which HMM-functionalised beads moved unidirectionally [2].

Apart from the mechanical-based alignment methods described above, electrical fields can be used to align filaments or microtubules in a similar fashion as for motility gliding assays. This approach has been demonstrated recently using micropatterned electrodes [4]. A recent review [31] discussed the (im)possibility of using magnetic fields to orient microtubules, due to the excessive magnetic fields required.

However, all of the above methods lose sight of an important element: the self-assembling properties of actin filaments and microtubules. Given these properties, a much better option would be to self-assemble actin filaments or microtubules on molecularly or nano-organized surfaces. This has been achieved by deposition of actin filaments on lipid monolayers through a simplified Langmuir-Blodgett technique [84, 90] and more recently on mica and HOPG [113]. It has been found that, apart from the unidirectionality of movement, the velocity of myosin-coated beads increased linearly (up to 100μm/s) with bead diameter (up to 60μm diameter), a counterintuitive finding with important potential for future nanodevices.

Certainly the most exiting technological avenue would be to design and 'write' complex tracks comprising individual actin filaments and/or microtubules and let motor-functionalised objects run on tracks and perform designed functions (e.g., transport, upload-download) in particular locations. Several works demonstrated that actin filaments self-assemble in 2-D paracrystal circular shapes when deposited on phospholipid layers [97–99]. More recently almost perfect circular shapes have been demonstrated when depositing F-actin from a solution on mica [113]. Some of these self assembled features made of actin are presented in Figure 12.10. The self-assembly of individual microtubules in tracks with small curvature radius would be more difficult given their rigidity and dimensions. Figure 12.9 presents some methods for patterning and self-assembly of actin filaments and microtubules.

12.4.4. On-Off Control of the Operation of Protein Molecular Motors Devices

The discussion regarding the sensitivity of protein molecular motors suggests that there are many 'off buttons', but most of these will shut down the devices permanently and therefore are not useful, unless a one-off function is envisaged, e.g., some biosensors [24, 55]). The inhibition of microtubule-based kinesin motility using local anaesthetics has been demonstrated [56]. 'On buttons', in particular those based on caged-ATP, have also been proposed [31, 33].

On-off control could be achieved, conceptually, through the variation of ATP concentration. ATP, as well as ionic strength (in particular Mg^{2+} and Ca^{2+}) and pH of the buffer solution, control the operation of molecular motors, but while ATP is depletable, ionic strength and pH are largely constant during the lifetime of the nanodevice. A study mentioned above [9] found that raising the molar Mg^{2+}/ATP ratio can substantially elevate the gliding velocity of microtubules in gliding motility assays, with optimal conditions at 2.5 mM ATP, 12.5 mM Mg^{2+}, 37°C, and 450 kinesin molecules/μm^2 (the latter parameter being linked to ATP consumption). A similar study mapped the response of the actomyosin system versus ATP concentration, ionic strength and pH [112].

The use of ATP-concentration as an on-off switch has been demonstrated for rotary and linear motors. The circular motion of the rotary motors can be controlled by the balance between ATP and sodium azide, the former being the 'gas pedal' and the latter being the 'brake pedal' [85]. Also the molecular engineering of a mutant F1-ATPase motor containing a metal-binding site provided a Zn^{2+}-dependent, reversible on/off switch [52]. The actomyosin system has also been on-off controlled by the interplay between ionic strength and ATP level. At high ionic strength (60mM KCl) the arrest of motility of the actomyosin system is reversed by the *decrease* of ATP concentration to submicromolar levels [47]. Finally, microtubule motility can be turned on by the UV-induced release of caged ATP and off through the enzymatic ATP degradation by hexokinase [31].

However, despite their operational importance, ATP, ionic strength and pH are unlikely to be the most effective 'knobs' for on-off control; the ATP concentration is either practically constant (when the motors operate in relatively large volumes compared to the mass of protein) or it is regulated in nanoconfined spaces *by* the functioning of the motors, rather

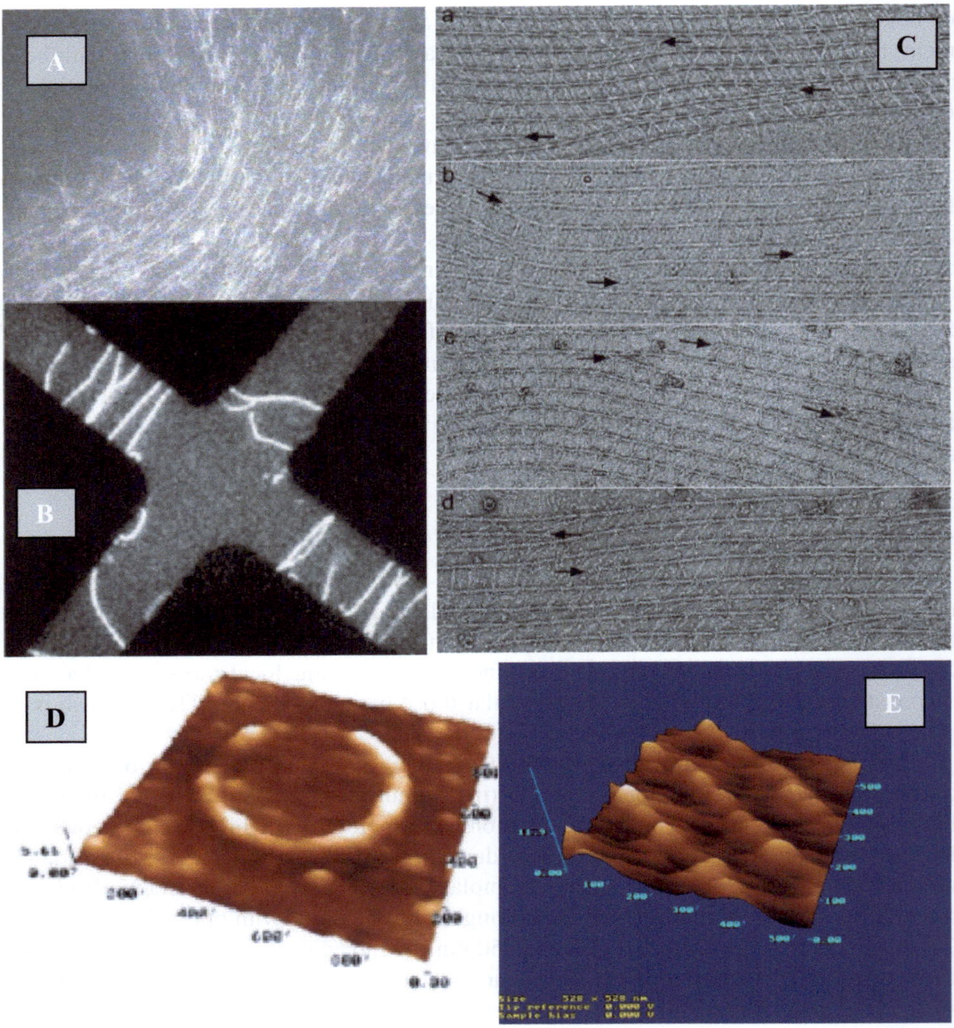

FIGURE 12.9. Methods for patterning and self-assembly of actin filaments and microtubules. A. Alignment of actin filaments in flow conditions. B. Alignment of actin filaments between microfabricated electrodes (from [4]; reproduced with permission). C. Arrays of actin filaments ([99]; reproduced with permission) cross-linked with a-actinin from (a) skeletal muscle, (b) smooth muscle, (c) cardiac muscle, and (d) Dictyostelium discoidium, a nonmuscle isoform. In all three cases, the morphology of the bundles is similar. D. Circular shape of actin filaments (imaged with AFM) deposited on hygrophilic mica [113]. E. Parallel actin features on hydrophobic HOPG, aligned (~ 200nm) longitudinally [113].

than vice versa. Also the ionic strength and pH are supposed to remain at an optimum constant level. Moreover, the variation of the concentration of these parameters, if desirable, will be slow controllers of the functions of the motors. The reliable, quick, robust, 'smart' and nano-localised on-off control of the molecular motors-powered devices is perhaps one of the most difficult problems to be solved.

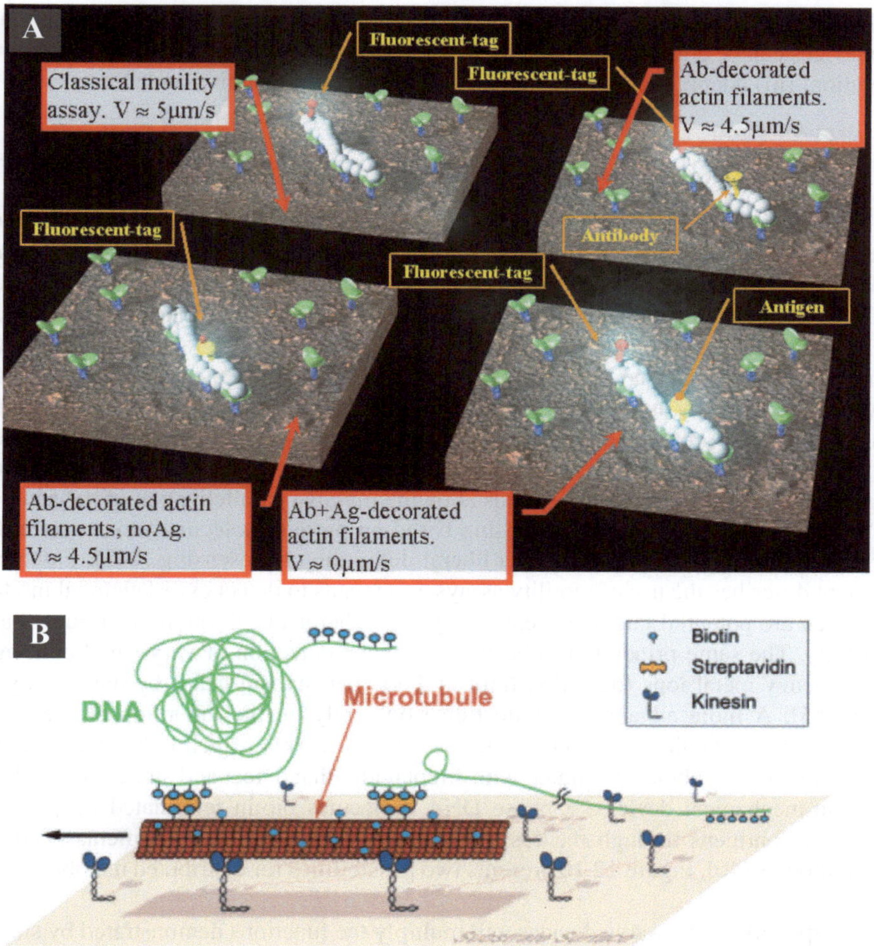

FIGURE 12.10. Distributed bionanosensing devices powered by molecular motors. A. The motility of an antibody* decorated, actin* filaments (* = fluorescently tagged) is stopped by the binding of antibodies with their anti-antibodies (from [66]). B. Schematic diagram of the microtubule-DNA transport and manipulation system (from [24]; reprinted with permission). Microtubules decorated with biotin-streptavidin stretch DNA molecules immobilised on surfaces when they pick up the biotinylated end of a DNA molecule.

12.5. PROTOTYPES OF NANODEVICES BASED ON PROTEIN MOLECULAR MOTORS

Apparently, the 'epic' stage of developing nanodevices based on protein molecular motors, which consisted in a large number of fundamental studies based on motility assays as well as numerous studies towards design, fabrication and operation of these devices, is poised to be followed by an engineering, more application oriented stage. Still many of the primitive devices demonstrated in the last years do have a targeted application—be that involuntarily or by design. The possible applications for future molecular motors-based nanodevices can be clustered in few targeted areas: (i) biosensing devices; (ii) force/power generation; (iii) nanomanufacturing; and (iv) information processing and storage.

12.5.1. Sensing Devices

The working principle of several sensing devices based on protein molecular motors, which are arguably the most advanced in terms of development, relies on the sensitive, large and often quick change of the motility upon molecular recognition of biochemical species either directly by protein molecular motors or molecular receptor decorated motile elements. Unlike all other devices, biosensing devices can operate in a global, randomly 'distributed' manner, or in a spatially-addressable manner. The distributed sensing devices are essentially classical motility assays with motility modulated by some biomolecular recognition event, e.g., protein-protein, antibody-antigen, or DNA-DNA interaction. Spatially addressable sensing devices rely on the same principle, but the motility is confined on purposefully designed microfabricated structures, thus allowing massive parallelization of the analysis.

The distributed devices for biosensing based on protein molecular motors have their first patent as early as 1998, albeit in a liberal definition of biosensing [108]. Essentially the patent describes the use of motility assays as a means to detect cytoskeletonal modulators, which are potential targets for e.g., drugs, agrochemicals, through the abrupt change of motility. The same principle has been used for the detection of very small concentrations of heavy metal ions (detection limit of 1 ion per myosin head) by the actomyosin system [55]. A more advanced architecture would rely on the mounting of 'detectors' (e.g., ssDNA, antibodies, etc.) on the motile elements. This principle has been demonstrated for (i) microtubules decorated with streptavidin that detect and stretch, hence being also nano-mechanical devices, λ-phase DNA molecules biotin terminated [24] and (ii) detection of antigens through the cessation of the motility of actin filaments decorated with antibodies [46]. Figure 12.10 presents two possibilities for distributed nanobiosensing devices.

The spatially addressable devices will multiply the functions demonstrated by simpler devices, and therefore have not been yet developed to the same extent as distributed devices. One device that relies on the modulation of motility by antibody-antigen recognition and the transduction of motility in electromagnetic signals has also been proposed [29, 63].

12.5.2. Nanomechanical Devices

Because of their 'punctual' nature the rotary motors are perfect candidates for arrayed devices performing mechanical work. Montemagno and co-workers' crucial work provided the proof of principle for the building of a hybrid nanodevice based on a rotary motor [85]. Building on some previous work [67], their hybrid nanodevice powered by a rotary molecular motor consisted of three major elements: (i) a microarray of nano-sized nickel posts fabricated by e-beam lithography; (ii) a thermostable form of Ni-selective F1-ATPase which selectively attaches on the Ni nano-posts; and (iii) Ni nanopropellers (Ni rods) with functionalized surface that allow specific attachment of the lever of the motor. The design, the fabrication concept and the microarray organization of the hybrid nanodevice are presented in Figure 12.6. Despite the low fabrication yield (only 5 out of 400 propellers rotated) no backward steps have been observed (possibly due to the high ATP concentration). Also the device showed a 2.5 hours long endurance cycle. Subsequent work [5, 80] discussed the many engineering issues raised by the difficult interfacing between inorganic nano-engineered objects and very delicate proteins.

The inherent directionality built in microtubules and actin filaments, as well as their polymerization, recommend linear motors as prime candidates for nanodevices performing *directional* mechanical work.

In principle, most experimental devices for single molecules studies are molecular motors-powered devices, but these devices use an external restraint to control the movement or measure the force exerted *by* the motors rather than use the motors *for* a useful function. However, a forcemeter *using* molecular motors (rather than being designed and operated *for* their probing) has been recently proposed [32]. This forcemeter consists of two microtubules, one motile sliding on kinesin-coated surface and another one mounted at one end on top of a bead immobilized on the surface (i.e., cantilevered microtubule). Both microtubules are decorated with complementary biomolecules, e.g., the motile one with streptavidin-coated beads and the cantilevered one with biotin. Upon contact between complementary biomolecules on different microtubules, the movement of the motile microtubule in association with the binding force will bend the cantilevered microtubule. The mechanical stiffness of the microtubule and its deflection before rupture/detachment will allow the calculation of the binding force. This device aims to measure the force associated with molecular recognition rather than perform molecular work (as proposed by [24]), so in a sense it is also a biosensing device too.

Most bead geometry motility assays are primitive self-propelled nanodevices with inherent directionality control. An early study demonstrated the high speed of large beads functionalized with HMM running on paracrystal actin filaments self-assembled on a lipid monolayer mounted on glass [90]. This study, which unfortunately was not followed up, is important because it demonstrates the benefits of using self-aligned nano-tracks in terms of both directionality and amplification of force. Also, the transport of micro-objects has been demonstrated for: (i) myosin-coated (e.g., magnetic) beads walking on actin bundles of Nitella [116]; (ii) gelsolin-functionalised 40nm gold nanoparticles attached at one end of actin filaments has also been reported [61]; (iii) HMM-functionalised beads travelling on actin bundles self-assembled in microfabricated channels [2]. The microtubule-kinesin system was also used for (i) the transport, rotation and flip-over of kinesin-powered microchips made of silicon along flow-aligned microtubules immobilized on the surface of a flow cell [51], (ii) the capture and transport of streptavidin-coated beads by biotin-decorated microtubules [30], (iii) transport and assembly of quantum dots mounted on microtubules via biotin-streptavidin binding; and (iv) formation of membrane tubes and tubular networks by lipid giant unilamellar vesicles, decorated with kinesin-functionalised polystyrene beads (no tube formation was observed when kinesins were directly bound to the membrane, [76]). Finally, actin polymerization was used to move beads in proto-devices using (i) 0.5 μm diameter beads uniformly coated with ActA, which is a bacterial protein which induces local polymerization of actin [14]; (ii) 2 μm diameter beads half-coated with ActA to ensure directionality of movement [14]; and (iii) beads coated with VCA, which is a subdomain of the Wiskott-Aldrich syndrome protein (WASP) [7]. The last study demonstrated the size modulated behaviour of bead movement, with the average velocity increasing linearly with

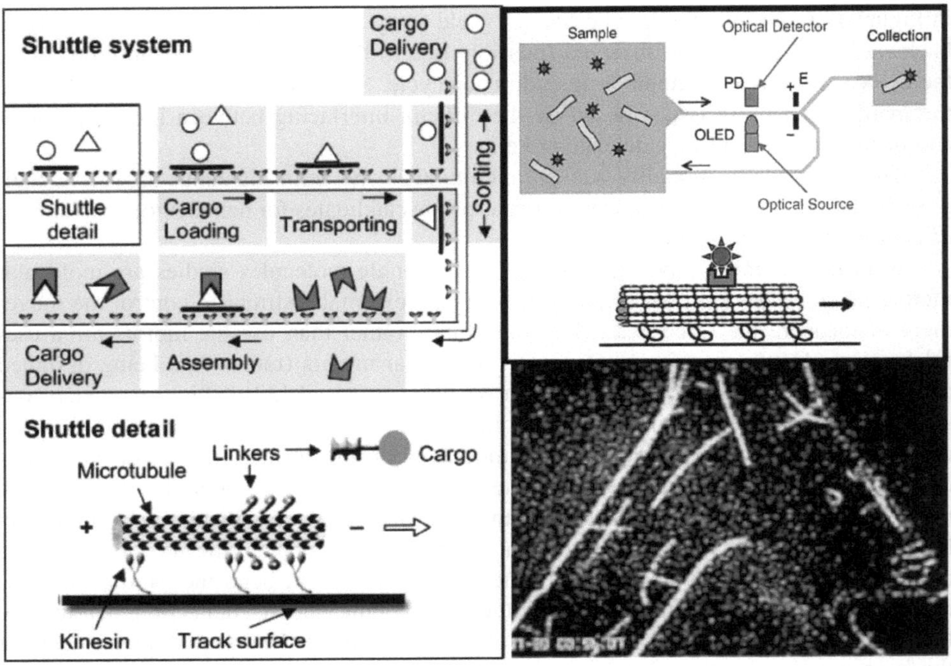

FIGURE 12.11. Planar nanomechanical devices. A. Projected system for loading, transport, sorting and assemble (from Reference [32]; reproduced with permission). A1. Molecular shuttle system with subsystems connected by tracks. A2. Shuttle design 'hooks' for cargo. B. Projected system for detection and purification (from Reference [43]; reproduced with permission). B1. Integrated system, comprising sample chamber with microtubules and analyte molecules, motor-functionalized channels, organic LED exciter and photodiode (PD) detector, and electrodes (E) to direct cargo-laden microtubules to the collection chamber. B2. Image of the electrode controlled gate.

the inverse of bead diameter (approx 1 μm/min for ~1 μm beads to approx 0.3 μm/min for approx. 3 μm beads). More importantly, small beads (<2.5 μm) move smoothly, but large beads (>4.5 μm) move in a periodic, start-and-stop fashion, and medium sized beads, e.g., around 3 μm, move in an intermittent or even chaotic manner.

Bi-directional devices have not been demonstrated *in vitro* yet, but they have been recently examined and described in vivo. There are two possible mechanisms to explain the bidirectional movement of cargo in natural processes, e.g., axonal vesicles, mitochondria, melanosomes, etc. (a comprehensive list with references provided by Gross et al. [30]): the "tug of war" and coordination [30]. In the tug of war model opposite-polarity motors are active simultaneously (Figure 12.4, D1). Net motion results when one set of motors successfully competes against the opposing motors. On the other hand, in the motor coordination model, competition is avoided because when plus-end motors are active, minus-end motors are turned off and vice versa (Figure 12.4, D2). For clarity, in Figure 12.4 D2 only the cargo and the motors are depicted; hypothetical molecules that allow the motors to assemble into complexes and that mediate interactions between motors are not shown. Although it has been found [30] that, at least for opposite polarity motors on lipid droplets in Drosophila

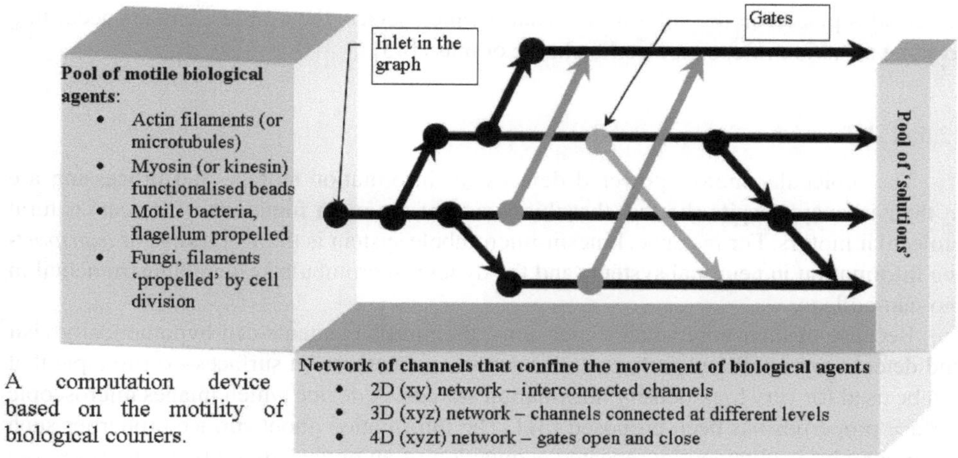

FIGURE 12.12. A possible scheme of a biocomputation device using biological agents, e.g., actin filaments or microtubules; or molecular motor-propelled whole organisms (bacteria or fungi).

embryos, the likely mechanism is the coordination performed by dynactin which enables dynein to stay "on" during minus-end motion and keeping it "off" during plus-end motion, both mechanisms constitute models for bidirectional movement in hybrid nanodevices based on protein molecular motors.

Finally, the planar devices, i.e., devices that operate on a micro/nano-fabricated area rather than on a linear dimension, are perhaps the most advanced in conceptualization, if not also in device implementation. Essentially, planar devices address the problem of continuous operation of nanodevices based on protein molecular motors. Indeed, most of 'single dimension' devices described above (exception being the natural or hypothetical bi-directional devices and those based on rotary motors) cannot continuously operate as the unidirectional movement has to stop when the motile element (e.g., microtubule, or bead) reaches the end of the micro/nano-fabricated pathway. Some possibilities to 'recharge' these devices do exist, e.g., the use of magnetic beads as loads (as suggested by Yamasaki & Nakayama [116]; Uchida et al. [104]; Fulga et al. [29]) and the application of a directional electromagnetic field that will collect the beads at one end of the tracks. Such a system, however, will address the motile elements globally and therefore much of the functionality of individual nano-devices will be lost. An alternative is to run the motile elements in a circular fashion on micro- or nano-tracks on larger areas, perhaps with functions being sequentially performed along the way. Most 'molecular selectors' proposed and demonstrated [6, 32, 63, 93]; as well as the one proposed by Hiratsuka et al., [34] (Figure 12.8 A3) if it can be arranged in a circular concatenation of individual selectors) are essentially planar nanodevices. More elaborate schemes have been also proposed, for instance a device which would comprise a 'bay area' for microtubules decorated with molecular 'hooks'; a 'fishing' path, where the motile elements capture different payloads; a mechanical selector which sorts and redirects the motile elements according to their cargo and cargo delivery areas [32]. Another device that has also been proposed comprises an optical detector connected to an electrical gate (electrodes placed on the sides of a channel) which selects the motile elements according to some selection rule, e.g., electric charge or mass [43].

12.5.3. Information Storage and Processing

The molecular motors-powered devices for information storage and processing are in their infancy, despite the fact that this is one of the main functions of (linear) natural molecular motors. For instance, kinesin-microtubule system is the workhorse of *transporting* information in neuronal systems and the dynein-microtubule system is instrumental in bio-camouflage.

Because of their very small dimensions, the motile elements can hypothetically visit and detect very small and perhaps concealed nanostructures on surfaces—a principle that can be used for very high density information storage. A device which images microscopic surface properties has been proposed [31]. The information about surface properties such as topography is obtained by repeated acquisition of an optical signal (e.g., fluorescence) from a large number of motile elements (e.g., microtubules) moving on random paths across a micro/nano-structured surface. These self-propelled probes sample the surface in a statistical process in contrast to the deterministic, linear sampling performed by a scanning probe microscope. The spatial distribution of hydrophobicity on a nano-patterned surface, as opposed to topography, can also be mapped with very high resolution using the observation that the fluorescence of rhodamine-phaloidin-tagged actin filaments varies with the nature of the surface [62].

More opportunities, but also more challenges, will arise in the area of biocomputation devices. One prospect would be to use the -possible- unusual ferroelectric properties of microtubules and/or the formation of travelling kink solitons (e.g. [41, 77, 78]) for new quantum electronic devices. A more immediate application would arise from DNA-based computation [1] and from maze-solving with microfluidics devices (e.g. [15]) or by amoeboid cells [60]. Essentially, it is entirely possible that the motility of actin filaments (or microtubules) can be used to explore mazes or other more elaborate micro- or nano-fabricated networks that code a mathematical problem [64]. Many advancements in the area of nanomechanical devices, e.g., electrode-controlled gates [43]; molecular selectors [17, 32, 34] will be useful modules for a molecular motors-based calculator.

12.6. PERSPECTIVES

The author hopes that the above sections convinced the reader that the protein molecular motors offer many promises for the design, fabrication and operation of dynamic nanodevices, opportunities that many research groups are trying to capitalize on. Nevertheless, many challenges for the future molecular motors-based nanodevices still remain open, in particular a realistic answer to the following questions:

- *Competitive edge*. What is the competitive edge, exactly, of protein molecular motors compared to other alternative devices? For instance, biosensing devices based on molecular motors do have the obvious advantage of 'beating diffusion', i.e., the molecular sensors travel and fish for complementary biomolecules rather than waiting for these to diffuse to them. This advantage is indeed critical to bioterrorism applications where speed of detection is essential, and it could be that all other issues (e.g., cost) would be less relevant. Similar analysis has to be made for other groups of applications, i.e., nanomechanical or biocomputation devices. As a general rule of thumb, high-added value uses, where the costs are less important, are more likely to offer 'killer applications'.

- *Endurance*. How can the endurance of protein molecular motors be (greatly!) improved to a level where they would become competitive in fabrication and operation costs with other, non-protein-based dynamic devices, e.g., Nano Electro-Mechanical Systems (NEMS)? Would a 'quantum leap' search for inherently robust proteins, e.g., those present in living beings living in extreme conditions (thermophiles or arctic temperatures) or motor proteins working in extreme conditions, e.g., from sharks, yield the answer to this problem? Or would a more gradual improvement based on a better understanding of the impact of operating conditions, i.e., surfaces, temperature, etc. be more productive? Or would the use of molecular motors-based devices be better restricted to those applications which are inherently environmental-insensitive, e.g., lab-only applications, such as genomics or biocomputation?
- *Rotary motors*. Going into the specifics of the choices for dynamic nanodevices, how can rotary motors be best used for practical (and competitive) applications? The inherent architecture of rotary motors makes then destined to work in an array format on planar devices, but how can myriads of motors be independently addressed and controlled—an information storage and processing nightmare similar to the problem of addressing highly parallel-ised AFM tips? Can the rotary motors be used for local power generation at the nano-level, which will solve an important problem for autonomous nanodevices?
- *Linear motors: To bead or not to bead*? Is the simplicity of implementation of the gliding geometry a good trade-off against the loss of information regarding directionality, which is built in actin filaments/microtubules, for bead geometry? In certain applications, e.g., distributed sensing devices, the gliding geometry has important advantages, but for more advanced devices, where control of directionality is essential, the proposed molecular selectors means that these devices would work with a sizeable error rate. On the other hand, the bead geometry, despite long term advantages, raises important and immediate technological problems. If the inherently-unidirectional tracks are to be laid in a designed manner, what would be the methods to do that? A solution would be to locally control the actin and tubulin polymerization on a nano-structured surface (but then how to orient the structures on that surface in the desired manner?) using multi-photon stereo-lithography and caged compounds that initiate polymerisation. Alternatively manipulate the flow and (if possible) local electric fields to capture filaments or microtubules in desired positions and orientations. The bead geometry also raises the exciting opportunity, but also the challenge, of *bi*-directional control with antagonist motors that operate in a tug-of-war or coordination mechanism.
- *Linear motors: cargo blues*. In most if not all devices proposed so far, a cargo, i.e., a biomolecule (antibody, streptavidin or biotin, DNA), a bead or a rod, is transported (or rotated) by the motor or filament/microtubule. With the exception of devices based

on rotary motors and possibly others (e.g., those for power generation or detection of movement) the cargo has to be 'unloaded' sometime and in precise locations, in particular if a continuous mode of operation is envisaged. Although many schemes have been devised regarding the attachment of payloads, the much more difficult issue of controlled loading and unloading deserves much more work. A less important, but also less challenging issue is the use of rods instead of beads as carried objects. Rods have the advantage of being more easy to functionalise differently at their ends, and hence test the tug of war mechanism, as well as being able to improve the directionality of movement.

- *Fuel*. Where is the ATP coming from? If a continuous operation is envisaged, the fuel for the dynamic nanodevices has to be continuously provided. At the limit, these nanodevices can operate in large (relative to protein mass) volumes and therefore the ATP concentration would be quasi-constant. A more sustainable solution would be to synthesise ATP in a separate chamber ('ATP factory') and recirculate ADP through a microfluidics network. ATP can be provided by bacteria, in which case careful filtering has to be arranged between the 'ATP factory' and the operational area comprising the molecular motors-powered devices. Perhaps the most sustainable, but also the most challenging, solution is to integrate ATP-ases not only from the mechanical point of view (see below) but also from the fuel production and recirculation point of view.
- *Integration*. Most of the primitive devices being demonstrated so far use the kinesin-microtubule system, with few important exceptions that use rotary motors, acto-myosin and actin polymerization systems. The (relative!) robustness and processive nature of kinesin and the rigidity and larger dimensions of microtubules help an easier demonstration of prototypes, but the myosin-actin filament system has advantages too, in particular higher velocities. Also in some applications (e.g., biocomputation) actin filament flexibility is a big bonus. And what about dynein? Presently the purification and handling of dynein are not trivial, but this protein has many advantages too. Moreover, if continuous operation is envisaged, which will require bi-directionality, we need at least two types of motor proteins working in the same device, possibly on the same cargo. Finally, there are applications where actin and possibly microtubule polymerization, e.g., fabrication of tracks, is essential. It appears that the future efforts have to be directed towards the integration of several types of motors on the same device. For instance, it is possible that 'micro' rotary motors are likely to be used *in conjunction*, not in competition, with the 'micro' linear motors, just as 'macro' mechanical rotary and linear devices are synergistically used in explosion engines.

12.7. CONCLUSION

Although several aspects regarding the fundamentals of molecular motors and associated proteins are still the subject of intense study, the research has reached a critical mass where many fundamental parameters necessary for the engineering design of molecular motors-powered nanodevices are already available. A separate line of research, which developed in the last half a decade developed, also provided some answers regarding the possible engineering options for the design, fabrication and operation of future dynamic devices based on protein molecular motors. Many important challenges still remain to be resolved, but the primitive prototypes already demonstrated offer useful benchmarks in the development of robust and continuous-operation devices and systems of devices for biosensing, mechanical work and possibly biocomputation.

ACKNOWLEDGEMENTS

This paper is the result of many discussions with colleagues working in molecular motors area. In particular, many of the ideas presented here have been sparked by hearty discussions with Hitoshi Suzuki and Kazu Oiwa at Kansai Advanced Research Center; and Taka Taguchi and Noburu Yumoto at AIST in Japan; Rob Cross at Marie Curie Research Institute in UK; Cris dos Remedios and Murat Kekic at University of Sydney in Australia; Jacques Prost and Cecille Sykes at Curie Institute in France; Carlo Montemagno at UCLA; Abe Lee at UCI; Gerry Pollack and Henry Hess at Washington University; Bruce Gaber, Joel Schnur and Fran Ligler at NRL; and at last but not least, with present and former members of my research team—Chitty Mahanivong, Jon Wright, Kristi Hanson and Dan Nicolau Jr.

REFERENCES

[1] L.M. Adleman. Molecular computation of solutions to combinatorial problems. *Science*, 266:1021–1023, 1994.

[2] Y.V. Alexeeva, E.P. Ivanova, D.K. Pham, V. Buljan, and D.V. Nicolau. Controlled self-assembly of actin filaments for dynamic biodevices. *Nanobiotechnology*, 1:1551–1294, 2006.

[3] T. Ando, N. Kodera, E. Takai, D. Maruyama, K. Saito, and A. Toda. A High-speed Atomic Force microscope for Studying Biological Macromolecules. *Proc. Natl. Acad. Sci. U.S.A.*, 98:12468–12472, 2001.

[4] S.B. Asokan, L. Jawerth, R.L. Carroll, R.E. Cheney, S. Washburn, and R. Superfine. Two-dimensional manipulation and orientation of actin-myosin systems with dielectrophoresis. *NanoLetters*, 3:431–437, 2003.

[5] G.D. Bachand and C.D. Montemagno. Constructing Organic/Inorganic NEMS Devices Powered by Biomolecular Motors. *Biomed. Microdevices*, 2:179–184, 2000.

[6] G.D. Bachand, R.K. Soong, H.P. Neves, A. Olkhovets, H.G. Craighead, and C.D. Montemagno. Precision attachment of individual F1-ATPase biomolecular motors on nanofabricated substrates. *NanoLetters*, 1:42–44, 2001.

[7] A. Bernheim-Grosswasser, S. Wiesner, R.M. Golsteyn, M.-F. Carleir, and C. Sykes. The dynamics of actin-based motility depend on surface parameters. *Nature*, 417:308–311, 2002.

[8] L. Bourdieu, M.O. Magnasco, D.A. Winklemann, and A. Libchaber. Actin filaments on myosin beds: The velocity distribution. *Phy. Rev. E*, 52:6573–6579, 1995.

[9] K.J. Böhm, R. Stracke, and E. Unger. Speeding up kinesin-driven microtubule gliding *in vitro* by variation of cofactor composition and physicochemical parameters. *Cell Biol. Intl.*, 24:335–341, 2000.

[10] T.B. Brown and W.O. Hancock. A polarized microtuble array for kinesin-powered-nanoscale assembly and force generation. *Nanoletters*, 2:1131–1135, 2002.

[11] C. Brunner, K.-H. Ernst, H. Hess, and V. Vogel. Lifetime of biomolecules in polymer-based hybrid nanodevices. *Nanotechnology*, 15:540–548, 2004.

[12] R. Bunk, J. Klinth, J. Rosengren, I. Nicholls, S. Tagerud, P. Omling, A. Mansson, and L. Montelius. Towards a 'nano-traffic' system powered by molecular motors. *Microelect. Eng.*, 67–68:899–904, 2003

[13] R. Bunk, J. Klinth, L. Montelius, I.A. Nicholls, P. Omling, S. Tågerud, and A. Månsson. Actomyosin motility on nanostructured surfaces. *Biochem. Biophys. Res. Comm.*, 301:783–788, 2003b.

[14] L.A. Cameron, M.J. Footer, A. Van Oudenaarden, and J.A. Theriot. Motility of ActA protein-coated microspheres driven by actin polymerization. *Proc. Natl. Acad. Sci. U.S.A.*, 96:4908–4913, 1999.

[15] D.T. Chiu, E. Pezzoli, H. Wu, A.D. Stroock, and G.M. Whitesides. Using three-dimensional microfluidic networks for solving computationally hard problems. *Proc. Natl. Acad. Sci. U.S.A.*, 2001, 98:2961–2966.

[16] D. Chretien and R.H. Wade. New data on the microtubule surface lattice. *Biol Cell*, 71:161–174, 1991.

[17] J. Clemmens, H. Hess, H. Howard, and V. Vogel. Analysis of microtubule guidance in open microfabricated channels coated with the motor protein kinesin. *Langmuir*, 19:1738–1744, 2003.

[18] J. Clemmens, H. Hess, R. Lipscomb, Y. Hanein, K.F. Bhringer, C.M. Matzke, G.D. Bachand, B.C. Bunker, and V. Vogel. Mechanisms of microtubule guiding on microfabricated kinesin-coated surfaces: chemical and topographic surface patterns. *Langmuir*, 19:10967–10974, 2003b.

[19] M.L. Connolly. Solvent-accessible surfaces of proteins and nucleic acids. *Science*, 221:709–713, 1983.

[20] M.L. Connolly. Analytical molecular surface calculation. *J. Appl. Crystall.*, 16:548–558, 1983.

[21] M.L. Connolly. Measurement of protein surface shape by solid angles. *J. Mol. Graph.*, 4:3–6, 1986.

[22] R. Cooke. The sliding filament model: 1972–2004. *J. Gen. Physiol.* 123:643–656, 2004

[23] J.R. Dennis, J. Howard, and V. Vogel. Molecular shuttles: directing the motion of microtubules on nanoscale kinesin tracks. *Nanotechnology*, 10:232–236, 1999.

[24] S. Diez, C. Reuther, C. Dinu, R. Seidel, M. Mertig, W. Pompe, and J. Howard. Stretching and transporting DNA molecules using motor proteins. *Nanoletters*, 3:1251–1254, 2003.

[25] P. Dimroth, H. Wang, M. Grabe, and G. Oster. Energy transduction in the sodium F-ATPase of *Propionigenium modestum. Proc. Natl. Acad. Sci. U.S.A.*, 96:4924–4929, 1999.

[26] M. Dogterom and B. Yurke. Measurement of the force-velocity relation for growing microtubules. *Science*, 278:856–860, 1997.

[27] D.E. Dupuis, W.H. Guilford, J. Wu, and D.M. Warshaw. Actin filament mechanics in the laser trap. *J. Muscle Res. Cell. Motil.*, 18:17–30, 1997.

[28] J.T. Finner, R.M. Simmons, and J.A. Spudich. Characterization of single actin-mysoin interactions. *Biophys. J.*, 68:291–297, 1995.

[29] F. Fulga, S. Myhra, Jr. D.V. Nicolau, and D.V. Nicolau. Interrogation of the dynamics of magnetic microbeads on the meso-scale via electromagnetic detection. *Smart Mat. Struct.*, 11:722–727, 2002.

[30] S.P. Gross, M.A. Welte, S.M. Block, and E.F. Wieschaus. Coordination of opposite-polarity microtubule motors. *J. Cell Biol.*, 156:715–724, 2002.

[31] H. Hess and V. Vogel. Molecular shuttles based on motor proteins: active transport in synthetic environments. *J. Biotechnol.*, 82:67–85, 2001.

[32] H. Hess, C.M. Matzke, R.K. Doot, J. Clemmens, G.M. Bachand, B.C. Bunker, and V. Vogel. Molecular shuttles operating undercover: A new photolithographic approach for the fabrication of structured surfaces supporting directed motility. *NanoLetters*, 3:1651–1655, 2003.

[33] H. Higuchi, E. Muto, Y. Inoue, and T. Yanagida. Kinetics of force generation by single kinesin molecules activated by laser photolysis of caged ATP. *Proc. Natl. Acad. Sci. U.S.A.*, 94:4395–4400, 1997.

[34] Y. Hiratsuka, T. Tada, K. Oiwa, T. Kanayama, and T.Q.P. Uyeda. Controlling the direction of kinesin-driven microtubule movements along microlithographic tracks. *Biophys. J.*, 81:1555–1561, 2001.

[35] T.E. Holy, M. Dogterom, B. Yurke, and S. Leibler. Assembly and positioning of microtubule asters in microfabricated chambers. *Proc. Natl. Acad. Sci. U.S.A.*, 94:6228–6231, 1997.

[36] J. Howard, A.J. Hunt, and S. Baek. Assay of microtubule movement driven by single kinesin molecules. *Meth. Cell Biol.* 39:137–147, 1993.

[37] J. Howard. The movement of kinesin along microtubules. *Annu. Rev. Physiol.*, 58:703–729, 1996

[38] J. Howard. Molecular motors: Structural adaptations to cellular functions. *Nature*, 389:561–567, 1997.

[39] J. Howard. *Mechanics of Motor Proteins and the Cytoskeleton*, Sinauer Associates, Inc., 2001.

[40] A.F. Huxley. Muscle structure and theories of contraction. *Progr. Biophys. BioPhys. Chem.*, 7:255–318, 1957.

[41] E. Insinna. Biocomputation and Nonlinear Dynamics in the Primitive Sensorimotor Mechanism of Euglena Gracilis. *Biocomputing and emergent computation: Proceedings of BCEC97*, Dan Lundh, Björn Olsson, Ajit Narayanan (Eds.). World Scientific, pp. 218–227, 1997.

[42] A. Ishijima, Y. Harada, H. Kojima, T. Funatsu, H. Higuchi, and T. Yanagida. Single molecule analysis of the actomyosin motor by nanometer-piconewton manipulation with microneedle: unitary steps and forces. *Biophysi. J.*, 70:383–400, 1996.

[43] L. Jia, S.G. Moorjani, T.M. Jackson, W.O. Hancock. Microscale transport and sorting by kinesin molecular motors. *Biomed. Microdev.*, 6:67–74, 2004.

[44] K. Kawaguchi and S. Ishiwata. Temperature dependence of force, velocity, and processivity of single kinesin molecules. *Biochem. Biophys. Res. Commun.*, 272:895–899, 2000.
[45] M. Kawai, K. Kawaguchi, M. Saito, and S. Ishiwata. Temperature change does not affect force between single actin filaments and HMM from rabbit muscles. *Biophys. J.*, 78:3112–3119, 2000.
[46] M. Kekic, G. Solana, D.V. Jr. Nicolau C.G. dos Remedios, and D.V. Nicolau. Nanosensing device based on actomyosin motility. (in preparation)
[47] M.S.Z. Kellermeyer and G.H. Pollack. Rescue of *in vitro* motility halted at high ionic strength by reduction of ATP to submicromolar levels. *Biochim. Biophys. Acta*, 1277:107–114, 1996.
[48] K. Jr. Kinosita, R. Yasuda, H. Noji, S. Ishiwata, and M. Yoshida. F1-ATPase; a rotary motor made of a single molecule. *Cell*, 93:21–24, 1998.
[49] S.J. Kron and J.A. Spudich. Fluorescent actin filaments move on myosin fixed to a glass surface. *Proc. Natl. Acad. Sci. U.S.A.*, 83:6272–6276, 1986.
[50] B. Liang, Y. Chen, C.-K. Wang, Z. Luo, M. Regnier, A.M. Gordon, and P.B. Chase. Ca^{2+} regulation of rabbit skeletal muscle thin filament sliding: Role of cross-bridge number. *Biophys. J.*, 85:1775–1786, 2003.
[51] L. Limberis and R.J. Stewart. Toward kinesin-powered microdevices. *Nanotechnology*, 11:47–51, 2000.
[52] H.Q. Liu, J.J. Schmidt, G.D. Bachand, S.S. Rizk, L.L. Looger, H.W. Hellinga, and C.D. Montemagno. Control of a biomolecular motor-powered nanodevice with an engineered chemical switch. *Nature Mat.*, 3:173–177, 2002.
[53] R.W. Lymn and E.W. Taylor. Mechanism of adenosine triphosphate by actomyosin. *Biochemistry*, 10:4617–4624, 1971.
[54] C. Mahanivong, J.P. Wright, M. Kekic, D.K. Pham, C. dos Remedios, and D.V. Nicolau. Manipulation of the Motility of Protein Molecular Motors on Microfabricated Substrates. *Biomed. Microdev.*, 4:111–116, 2002.
[55] R. Martinez-Neira, M. Kekic, V. Buljan, D.V. Nicolau, and C.G. dos Remedios. A Novel Biosensor for Mercuric Ions Based on Motor Proteins. *Biosens. Bioelect.*, 19:(in print), 2004.
[56] Y. Miyamoto, E. Muto, T. Mashimo, A.H. Iwane, I. Yoshiya, and T. Yanagida. Direct inhibition of microtubule-based kinesin motility by local anesthetics. *Biophys. J.*, 78:940–949, 2000.
[57] C. Montemagno and G. Bachand. Constructing nanomechanical devices powered by biomolecular motors. *Nanotechnology*, 10:225–231, 1999.
[58] P.B. Moore, H.E. Huxley, and D.J. DeRosier. Three-dimensional reconstruction of F-actin, thin filaments and decorated thin filaments. *J. Mol. Biol.*, 50:279–295, 1970.
[59] M. Morimatsu, A. Nakamura, H. Sumiyoshi, N. Sakaba, H. Taniguchi, K. Kohama, and S. Higashi-Fujime. The molecular structure of fastest myosin from green algae, Chara. *Biochem. Biophys. Res. Commun.*, 270:147–152, 2000.
[60] T. Nakagaki, H. Yamada, and A. Tóth. Maze-solving by an amoeboid organism. *Nature*, 407:470, 2000.
[61] H. Nakayama, T. Yamaga, and Y. Kunioka. Fine profile of actomyosin motility fluctuation revealed by using 40-nm probe beads. *Biochem. Biophys. Res. Comm.*, 246:261–266, 1998.
[62] D.V. Nicolau, H. Suzuki, S. Mashiko, T. Taguchi, and S. Yoshikawa. Movement of actin filaments on microlithographically-functionalized myosin tracks. *Biophys. J.*, 77:1126–1134, 1999.
[63] D.V. Nicolau and R. Cross. Protein profiled features patterned via bilayer microlithography and confocal microscopy. *Biosen. Bioelect.*, 15:85–91, 2000.
[64] D.V. Jr. Nicolau and D.V. Nicolau. Computing with the actin-myosin molecular motor system. In Biomedical Applications of Micro- and Nanoengineering. SPIE Proc. 4937, 219–225, 2002.
[65] D.V. Jr. Nicolau, F. Fulga, and D.V. Nicolau. Impact of protein adsorption on the geometry design of microfluidics devices. *Biomed. Microdev.*, 5:227–233, 2003.
[66] D.V. Nicolau. *Nanodevices based on linear protein molecular motors: Challenges and opportunities*. In: J. Reif (Ed.), Foundations of Nanoscience: Self-Assembled Architectures and Devices, 2004.
[67] H. Noji, R. Yasuda, M. Yoshida, and Jr, K. Kinosita. Direct observation of the rotation of F1ATPase. *Nature*, 386:299–302, 1997.
[68] K. Oiwa. Protein motors: Their mechanical properties and applications to nanometer-scale devices. *Mat. Sci. Forum*, 426–432:2339–2344, 2003.
[69] F. Oosawa and S. Asakura. *Thermodynamics of polymerization of protein*. New York Academic Press, 1975.
[70] E.M. Ostap, T. Yanagida, and D.D. Thomas. Orientational distribution of spin labeled actin oriented by flow. *Biophys. J.*, 63:966–975, 1992.

[71] T.D. Pollard. Rate constants for the reactions of ATP- and ADP-actin with the ends of actin filaments. *J. Cell Biol.*, 103:2747–2754, 1986

[72] T.D. Pollard, L. Blanchoin, and R.D. Mullins. Molecular mechanisms controlling actin filament dynamics in nonmuscle cells. *Annu. Rev. Biophys. Biomol. Struct.*, 29:545–576, 2000.

[73] I. Rayment, H.M. Holden, M. Whittaker, C.B. Yohn, M. Lorenz, K.C. Holmes, and R.A. Milligan. Structure of the actin-myosin complex and its implications in muscle contraction. *Science*, 261:58–65, 1993.

[74] D. Riveline, A. Ott, F. Julicher, D.A. Winkelmann, O. Cardoso, J.J. Lacapere, S. Magnusdottir, J.L. Viovy, L. Gorre-Talini, and J. Prost. Acting on actin: the electric motility assay. *Eur. Biophys. J.*, 27:403–408, 1998.

[75] R.S. Rock, M. Rief, A.D. Mehta, and J.A. Spudich. *In vitro* Assays of Processive Myosin Motors. *Methods*, 22:373–381, 2000.

[76] A. Roux, G. Cappello, J. Cartaud, J. Prost, B. Goud, and P. Bassereau. A minimal system allowing tubulation with molecular motors pulling on giant liposomes. *Proc. Natl. Acad. Sci. USA*, 99:5394–5399, 2002.

[77] M.V. Sataric and J.A. Tuszynski. Relationship between the nonlinear ferroelectric and liquid crystal models for microtubules. *Physical Rev. E.*, 67:011901–11, 2003.

[78] M. Sataric, J. Tuszynski, and R. Zakula. Kink-like excitations as an energy-transfer mechanism in microtubules. *Phys. Rev. E.*, 48:589–597, 1993.

[79] I. Sase, H. Miyata, S. Ishiwata, and K. Kinosita. Axial rotation of sliding actin filaments revealed by single-fluorophore imaging. *Proc. Natl. Acad. Sci. U.S.A.*, 94:5646–5650, 1997.

[80] M. Schliwa and G. Woehlke. Molecular motors. *Nature*, 422:759–765, 2003.

[81] J.J. Schmidt, X.Q. Jiang, and C.D. Montemagno. Force tolerances of hybrid nanodevices. *Nanoletters*, 11:1229–1233, 2002.

[82] M.P. Sheetz and J.A. Spudich. Movement of myosin-coated fluorescent beads on actin cables *in vitro*. *Nature*, 303:31–35, 1983.

[83] M.P. Sheetz, R. Chasan, and J.A. Spudich. ATP-dependent movement of myosin *in vitro*: Characterization of a quantitative assay. *J. Cell Biol.*, 99:1867–1874, 1984.

[84] D. Shi, A.V. Somlyo, A.P. Somlyo, and Z. Shao. Visualizing filamentous actin on lipid bilayers by atomic force microscopy in solution. *J. Microscopy*, 201:377–382, 2001.

[85] R.K. Soong, G.D. Bachand, H.P. Neves, A.G. Olkhovets, H.G. Craighead, and C.D. Montemagno. Powering a nanodevice with a biomolecular motor. *Science*, 290:1555–1558, 2000.

[86] R.K. Soong, H.P. Neves, J.J. Schmidt, G.D. Bachand, and C.D. Montemagno. Engineering Issues in the fabrication of a hybrid nano-propeller system powered by F1-ATPase, *Biomed. Microdev.*, 3:71–73, 2001.

[87] J.A. Spudich, S.J. Kron, and M.P. Sheetz. Movement of myosin-coated beads on oriented filaments reconstituted from purified actin. *Nature*, 315:584–586, 1985.

[88] P. Stracke, K.J. Böhm, J. Burgold, H.J. Schacht, and E. Unger. Physical and technical parameters determining the functioning of a kinesin. *Nanotechnology*, 11:52–56, 2000.

[89] R. Stracke, K.J. Böhm, L. Wollweber, J.A. Tuszynski, and E. Unger. Analysis of the migration behaviour of single microtubules in electric fields. *Biochem. Biophys. Res. Comm.*, 293:602–609, 2002.

[90] H. Suda and A. Ishikawa. Accelerative sliding of myosin-coated glass beads under suspended condition from actin paracrystal. *Biochem. Biophys. Res. Comm.*, 237:427–431, 1997.

[91] M. Sundberg, J.P. Rosengren, R. Bunk, J. Lindahl, I.A. Nicholls, S. Tågerud, P. Omling, L. Montelius, and A. Månsson. Silanized surfaces for *in vitro* studies of actomyosin function and nanotechnology applications. *Analyt. Biochem.*, 323:127–138, 2003.

[92] H. Suzuki, K. Oiwa, A. Yamada, H. Sakakibara, H. Nakayama, and S. Mashiko. Linear arrangement of motor protein on a mechanically deposited fluoropolymer thin film. *Jap. J. Appl. Phys.*, Part 1, 34:3937–3941, 1995.

[93] H. Suzuki, A. Yamada, K. Oiwa, H. Nakayama, and S. Mashiko. Control of actin moving trajectory by patterned poly(methylmethacrylate) tracks. *Biophys. J.*, 72:1997–2001, 1997.

[94] N. Suzuki, H. Miyata, S. Ishiwata, and K. Jr. Kinosita. Preparation of bead-tailed actin filaments: estimation of the torque produced by the sliding force in an *in vitro* motility assay. *Biophys. J.*, 70:401–408, 1996.

[95] K. Svoboda, C.F. Schmidt, B.J. Schnapp, and S.M. Block. Direct observation of kinesin stepping by optical trapping inteferometry. *Nature*, 365:721–727, 1993.

[96] H. Tanaka, A. Ishijima, M. Honda, K. Saito, and T. Yanagida. Orientation dependence of displacements by a single one headed myosin relative to the actin filament. *Biophys. J.*, 75:1866–1894, 1998.

[97] K.A. Taylor and D.W. Taylor. Formation of 2-D paracrystals of F-actin on phospholipid layers mixed with quaternary ammonium surfactants. *J. Struct. Biol.*, 108:140–147, 1992.

[98] K.A. Taylor and D.W. Taylor. Formation of two-dimensional complexes of F-actin and crosslinking proteins on lipid monolayers: demonstration of unipolar alpha-actinin-F- actin crosslinking. *Biophys. J.*, 67:1976–1983, 1994.

[99] K.A. Taylor and D.W. Taylor. Structural studies of cytoskeletal arrays formed on lipid monolayers. *J. Struct. Biol.*, 128:75–81, 1999.

[100] J.A. Theriot. The polymerization motor. *Traffic*, 1:19–28, 2000.

[101] Y.Y. Toyoshima, S.J. Kron, and J.A. Spudich. The myosin step size: Measurement of the unit displacement per ATP hydrolyzed in an *in vitro* motility assay. *Proc. Natl. Acad. Sci. U.S.A.*, 87:7130–7134, 1990.

[102] D.C. Turner, C. Chang, K. Fang, S.L. Brandow, and D.B. Murphy. Selective adhesion of functional microtubules to patterned silane surfaces. *Biophys. J.*, 69:2782–2789, 1995.

[103] D. Turner, C. Chang, K. Fang, P. Cuomo, and D. Murphy. Kinesin movement on glutaraldehyde-fixed microtubules. *Anal. Biochem.*, 242:20–25, 1996.

[104] G. Uchida, Y. Mizukami, T. Nemoto, and Y. Tsuchiya. Sliding motion of magnetizable beads coated with Chara motor protein in a magnetic field. *J. Phys. Soc. Japan*, 67:345–350, 1998.

[105] T.Q.P. Uyeda, H.M. Warrick, S.J. Kron, and J.A. Spudich. Quantized velocities at low myosin densities in an *in vitro* motility assay. *Nature*, 352:307–311, 1991.

[106] R.D. Vale, T.S. Reese, and M.P. Sheetz. Identification of a novel force generating protein, kinesin, involved in microtubule based motility. *Cell*, 42:39–50, 1985.

[107] R.D. Vale and H. Hotani. Formation of membrane networks *in vitro* by kinesin-driven microtubule movement. *J. Cell Biol.*, 107:2233–2241, 1988.

[108] R. Vale, D. Pierce, J. Spudich, and L.S.B. Goldstein. Assays for detecting modulators of cytoskeleton function. WO 99/11814, 1998.

[109] R.D. Vale and R.A. Milligan. The way things move: Looking under the hood of molecular motor proteins. *Science*, 288:88:95, 2000.

[110] P. VanBuren, K. Begin, and D.M. Warshaw. Fluorescent phalloidin enables visualization of actin without effects on myosin's actin filament sliding velocity and hydrolytic properties *in vitro*. *J. Mol. Cell. Cardiol.*, 30:2777–2783, 1998.

[111] R.A. Walker, E.T. O'Brian, N.K. Pryer, M.F. Soboeiro, W.A. Voter, H.P. Erickson, and E.D. Salmon. Dynamic instability of individual microtubules analyzed by video light microscopy: rate constants and transition frequencies. *J. Cell Biol.*, 107:1437–1448, 1988.

[112] F. Wang, L. Chen, O. Arcucci, E.V. Harvey, B. Bowers, Y. Xu, J.A. Hammer, 3rd, and J.R. Sellers. Effect of ADP and ionic strength on the kinetic and motile properties of recombinant mouse myosin V. *J. Biol. Chem.*, 275:4329–4335, 2000.

[113] G.S. Watson, C. Cahill, J. Blach, S. Myhra, Y. Alekseeva, E.P. Ivanova, and D.V. Nicolau. Actin Nanotracks for Hybrid Nanodevices Based on Linear Protein Molecular Motors. In: J.T. Borenstein, P. Grodzinski, L.P. Lee, J. Liu, Z. Wang, D. McIlroy, L. Merhari, J.B. Pendry, D.P. Taylor (eds.), Nanoengineered Assemblies and Advanced Micro/Nanosystems. *Proceedings of MRS Spring Meeting*, San Francisco, April 2004, (in print).

[114] D.A. Winkelmann, L. Bourdieu, A. Ott, F. Kinose, and A. Libchaber. The flexibility of attachment of myosin on surfaces influences F-actin motion, *Biophys. J.*, 68:2444–2453, 1995.

[115] J.P. Wright, D.K. Pham, C. Mahanivong, M. Kekic, C.G. dos Remedios, and D.V. Nicolau. Micropatterning of Myosin on O-Acryloyl Acetophenone Oxime (AAPO), Layered with Bovine Serum Albumin (BSA). *Biomed. Microdev.*, 4:205–211, 2002.

[116] H. Yamasaki and H. Nakayama. Fluctuation analysis of myosin-coated bead movement along actin bundles of nitella. *Biochem. Biophys. Res. Comm.*, 221:831–835, 1996.

[117] T. Yanagida, M. Nakase, K. Nishiyama, and F. Oosawa. Direct observation of motion of single F-actin filaments in the presence of myosin. *Nature*, 307:58–60, 1984.

[118] M. Yoshida, E. Muneyuki, and T. Hisabori. ATP synthase–a marvelous rotary engine of the cell. *Nat. Rev. Mol. Cell. Biol.*, 2:669–677, 2001.

13

Nanodevices in Biomedical Applications

Bryan Ronain Smith[1], Mark Ruegsegger[2], Philip A. Barnes[3], Mauro Ferrari[4], and Stephen C. Lee[5]

[1]*Bryan Ronain Smith Biomedical Engineering Center, 1080 Carmack Road, Columbus, Ohio 43210*
[2]*Dorothy M. Davis Heart and Lung Research Institute, Cardiology Division, Biomedical Engineering Center, 1080 Carmack Road, Columbus, Ohio 43210*
[3]*Biomedical Engineering Center, 1080 Carmack Road, Columbus, Ohio 43210*
[4]*Professor, Brown Institute of Molecular Medicine Chairman, Department of Biomedical Engineering, University of Texas Health Science Center, Houston, TX; Professor of Experimental Therapeutics, University of Texas M.D. Anderson Cancer Center, Houston, TX; Professor of Bioengineering, Rice University, Houston, TX; Professor of Biochemistry and Molecular Biology, University of Texas Medical Branch, Galveston, TX; President, the Texas Alliance for NanoHealth, Houston, TX*
[5]*Dorothy M. Davis Heart and Lung Research Institute, Department of Cellular and Molecular Biology, Department of Chemical Engineering, Biomedical Engineering Center, 1080 Carmack Road, Columbus, Ohio 43210*

13.1. INTRODUCTION

In the early 21st century, nanotechnology is a field in rapid flux and development, and definition of its boundaries can be elusive. Aspects of multiple disciplines, ranging from physics to computer science to biotechnology, legitimately contribute to the endeavor. This breadth of field allows many interested parties to contribute to nanotechnology, but the same ambiguity can effectively render the field indistinct. The precise definition of nanotechnology remains debatable, so consideration of the present scope of the field may be useful.

13.1.1. Defining Nanotechnology and Nanodevices

We consider nanotechnology to be the discipline that works toward satisfying desired objectives using materials and devices whose valuable properties owe to some specific nanometer-scale element(s) of their structures. Drawing from the foundations and latest discoveries of the basic sciences of physics, chemistry, and biology, the field is highly application oriented, so its main thrust is fulfillment of tasks/needs: Technical information is primarily of interest to the extent that it bears on device design, function, or application.

Our definition of nanotechnology engenders some corollaries. For instance, it intentionally includes macroscale structures whose useful properties derive from nanoscale aspects of their structures. Second, the modifier "specific" (from "specific nanometer scale elements") intentionally excludes materials whose utility is solely derived from being finely divided (e.g., from high surface-to-volume ratios), or other bulk chemical and physical properties. We consider therapeutic nanodevices to be more intriguing than nanomaterials *per se* and in view of our definition, regard biotechnology as a subdiscipline of nanotechnology. Individual biological macromolecules (proteins, nucleic acid molecules, supramolecular complexes, etc.) often exhibit the coordinated, modular multifunctionality that is characteristic of purpose-built, engineered devices. Biological macromolecule functions depend on their precise nanoscale structures, so inclusion of many aspects of biotechnology under the nanotechnology rubric is justifiable. Frequently, no comparably "tunable" synthetic analogue of biological macromolecules yet exists, so therapeutic nanodevices often absolutely require inclusion of active biomolecules for their intended function.

Devices are integrated functional structures as opposed to mixtures of materials, and thus they exhibit desirable emergent properties that are inherent in their design. These properties emerge as a result of the spatio-temporal organization of components, as well as coordination and regulation of action of individual components and their surface properties. The organization of components in devices allows them to perform multistep, cohesive work processes in aggregate. Our device definition excludes nanomaterials used solely as drug formulation excipients (i.e., pharmacologically inert materials included in formulations that improve pharmacophore uptake, biodistribution, pharmacokinetic, handling, storage or other properties), although it adopts those same materials if they are integral components of therapeutic devices. Our discussion will be confined primarily to hybrid biological-synthetic nanotherapeutic devices used *in vivo*, because *in vivo* applications clearly benefit from the minimal invasiveness and other advantages that ultra-small, but multipotent, nanotherapeutics might offer.

Biotechnology historically focuses on production of individual soluble protein and nucleic acid molecules for pharmaceutical use, paying only limited attention to functional supramolecular structures [95–99, 104]. This bias toward free molecules may be surprising considering the importance of integrated supramolecular structures in biology and the potential for using object-oriented approaches to design increasingly complex nanodevices [92]. However, the bias toward single molecule therapeutics has been driven by commercial considerations: single protein drugs are among the easiest biotechnological therapeutics to realize from both technical and regulatory perspectives. However, this is changing, and actual supramolecular therapeutic devices are now being introduced.

Synthetic materials derived using micro/nanotechnology techniques afford the opportunity to complement the traditional limits of biotechnology. These materials can provide scaffolds that support higher-level organization of multiple biomolecules into devices that perform work activities that mixtures of their protein components as soluble molecules could not. Such supramolecular structures are generally called nanobiotechnological devices [77], nanobiological devices [95–99, 104] or semi-synthetic (bio-synthetic) nanodevices, and we will focus on these systems and their therapeutic applications.

13.1.1.1. Integrating Biomolecules Into Devices Nanodevice design is comparable to the design of other engineered structures, providing that the unique properties of the materials relating to their nanoscale aspects, (i.e., quantum, electrical, mechanical, chemical, and biological properties, etc), as well as the impact of these properties in therapy, are considered. Hybrid nanodevices may interact with patients on multiple levels, ranging from organismal to molecular, but it is reasonable to expect that most nanotherapeutics will interface with patients at the nanoscale at least to some extent [4, 5, 14, 35, 37, 50, 95–98, 101]. This generally means interaction between therapeutic devices and patient biological macromolecules, supramolecular structures, and organelles. Therefore, incorporation of biological macromolecules (and other biostructures) into nanodevices is often prescribed ([6, 50]; Hess et al., 2004; [95, 98, 101]). However, incorporating biological structures into devices imparts distinct challenges unseen in other aspects of engineering.

Semi-biological nanodevices obligately incorporate pre-fabricated biological components in contrast to synthetic devices. Knowledge of the properties of these biological device components is often incomplete at best, as they were not made by human design; thus, the range of activities inherent in any nanobiological device design is usually much less well-defined than for fully synthetic devices. Further complications ensue in consideration of the activities of biological molecules, which are often multifaceted (many genes and proteins display plieotropic behavior), and the entire scope of functionality and emergent properties of individual biomolecules in interactions with other biological systems (as in nanotherapeutics) is often not known. This makes design, prototyping, and testing of biological nanodevices an experimentally intensive, iterative process [96–98, 101] that may become more facile using tools borrowed from systems engineering and object-oriented design [94].

In general, proteins, nucleic acids, lipids, and other biomolecules are more labile with respect to physical damage than are synthetic materials. However, with the possible exceptions of topical agents or oral delivery and endosomal uptake of nanotherapeutics (both involving exposure to low pH), conditions encountered by nanobiological therapeutics *in vivo* are usually tolerable to the patient, and device lability in the face of physical insult is generally a major consideration only in *ex vivo* settings (relating to storage, sterilization, *ex vivo* cell culture, etc.). However, living organisms have various metabolic facilities responsible for normal tissue remodeling and responses to pathogens (circulating and tissue-bound proteases and other enzymes, various clearance organs, the immune system, etc.) that can potentially degrade biological components of hybrid nanotherapeutic devices, leading to partial or complete loss of nanobiological device structure and/or function. As will be discussed, numerous strategies are available to stabilize/isolate biomolecules and structures in *in vivo* environments ([29, 30, 63, 79; Ferrari et al., 1996; 83, 96, 111]). On the other hand, instability of active biological components can offer a simple way to define or restrict the activity of nanodevices that contain biomolecules.

13.1.1.2. Nanobiological Device Design Paradigms The canonical properties of ideal hybrid bio-synthetic nanodevices [6, 50, 98, 107] begin with the concept that biological components of devices that must retain their function in new device contexts. One must abstract a sufficient (and necessary) fraction of a functional biological unit from its native context to allow it to perform the function for which it was chosen within the device. Further, device function emerges from the summed activities of individual biological and synthetic device components *plus* the result of the organization of the components relative to each other as entailed in design. The organizational control of components not only allows disparate biomolecules incorporated into hybrid nanodevices to contribute to a coherent work process, but in principle allows design of devices consisting of native biological components whose overall device functions are nonetheless unprecedented in the biological world.

One vision of nanoscale therapeutic platforms arose from collaboration between National Aeronautics and Space Agency (NASA) and the National Cancer Institute (NCI). While NASA is chiefly concerned with minimal mass therapeutics, the NCI is concerned with early detection of disease for improved prognosis. Because this requires screening a population of predominantly healthy patients, the screening approach must be minimally invasive to non-invasive. The institutions' needs are complementary, and the partnership aims to demonstrate that the needs can be met with ultra-small (micro- or nanoscale), multifunctional hybrid bio-synthetic devices. Additionally, explicit and integral to the collaboration's objectives are that proposed platforms must not only ameliorate undesired physiological conditions, but also recognize (sense) and report them. Extensive capacity for molecular recognition and communication with external clinicians/operators is vital to the NASA/NCI vision, which we will show is eminently feasible using hybrid nanodevices. Such capabilities would allow drugs or other therapeutic interventions to be provided in a controlled manner, maximizing advantages and minimizing side effects, thereby giving way to "smart" therapeutics (see the discussion of targeting and triggering below).

Significant progress has been made in many facets of device design and realization, but no multifunctional nanoscale hybrid therapeutic platform has yet been commercialized. However, the idea of segregation of individual therapeutic functions into individual modules allowed in the therapeutic platform paradigm may be very powerful. Classes of broadly similar devices could be tailored by interchanging modules (targeting modules, sensing modules, drug dispensing modules, etc.) of the hybrid device as appropriate to particular disease states or therapeutic courses [92]. Interchangeability of functional modules will allow tuning of nanotherapeutics to needs of individual patients or conditions. Standardization of modules could potentially mitigate regulatory burdens for variants of individual therapeutic platforms.

13.2. OPPORTUNITIES FOR BIOMEDICAL NANOTECHNOLOGY: TECHNOLOGICAL AND BIOLOGICAL

Enabling technologies for therapeutic nanodevices are the "tools of the trade" for scientists in the realm of nanobiotechnology. Some selected enablers for hybrid nanodevices are technological in nature: nanomaterial self-assembly properties, surface and bulk electrical and thermal properties, bioconjugation methods, engineered materials for conditional release of therapeutics, external triggering strategies, etc. Others relate to clinically relevant aspects of nanobiotechnology, such as disease-state tissue or cell-specific biology that can be exploited by nanotherapeutics, e.g., the emerging vascular address system as well as intrinsic triggering approaches. Additional opportunities for hybrid nanoscale approaches specific to particular disease states will be examined in association with those applications.

13.2.1. Device Assembly

The assembly of components into functional devices is amenable to diverse approaches, from microfabrication techniques to self-assembly to chemical methods. In the case of devices comprising a single molecule or processed from a single crystal (e.g., single polymers, certain microfabricated structures, grafted polymeric structures, or precipitated particulates), assembly is often not an issue. However, integration of multiple, individually microfabricated components is periodically necessary and may at times drive the need for assembly, even for silicon and other semiconductor-based devices. Indeed, in cases of semiconductor and polymeric devices, standard top-down techniques can sometimes be employed in concert with bottom-up device assembly methods to achieve better and/or more efficient device construction approaches, and in some instances this combination of approaches may be the only feasible way to create a device. Furthermore, many hybrid nanodevices contain multiple, chemically diverse components that must be precisely assembled to sustain their functional, cooperative contributions to device performance.

13.2.1.1. Low Throughput Construction Methods and "One Off" Nanostructures Electron-beam (e-beam) lithography and force microscopy approaches (including dip-pen nanolithography or DPN) comprise the methods amenable to direct-write technology, which involve direct transcription of information or materials to a surface. These methods often require no mask and can yield resolutions on the order of tens of nanometers [26, 49, 53, 91, 113]. The DPN method, for instance, involves dipping a scanning probe microscope tip into "ink" (the molecule to be "written"), and dragging the tip across a substrate, akin to dragging a pen across paper. Direct-write technologies such as DPN can be combined with technologies such as computer aided design (CAD) to generate repeatable patterns (e.g., if designed appropriately, circuits) with bio- or other small molecules. However, despite efforts to the contrary, a key limitation of all these methods is throughput: the direct-writing devices must be controlled by or programmed by a human operator in order to produce specific nanostructures. A common expedient invoked to render these approaches suitable for manufacturing involves multiplexing the direct writing implements to perform highly parallel processes [26, 49, 53, 91, 113]. While significant advances have been made, it remains to be seen whether the parallel process tact will improve device construction throughput sufficiently to support consumer nanotherapeutic manufacture. Since therapy for a single patient may sometimes involve billions of individual nanodevice units, each individual nanotherapeutic structure must require only minimal input from a human synthesis/manufacturing technician. However, as standard of care increasingly evolves toward "personalized" courses of therapy [43, 114, 115] and individual therapeutics become more powerful, relatively low throughput synthesis/assembly methods may become more attractive. For the time being, ideal manufacturing approaches for nanotherapeutic devices resemble industrial polymer chemistry (occurring in bulk), in convenient buffer systems, or in massively parallel approaches.

13.2.1.2. Self-Assembly of Synthetic Polymers Self-assembly, driven by thermodynamics, has long been recognized as a potentially critical labor-saving approach to construction of nanostructures [33], and many organic and inorganic materials have self-assembly properties that can be exploited by properly devised design features to build structures with controlled configurations. Final configurations of these structures are limited by the ability to "tune" the properties of the individual subunits (e.g., by manipulation of the self-assembly intra/intermolecular forces) and control the assembly environment to generate the specific structures that were desired.

Micelles are self-assembled micro-nanostructures commonly used for drug delivery [18, 80, 114, 115, 188, 189]. They are formed from the association of block co-polymer subunits, each individual subunit containing hydrophobic and hydrophilic domains. Micelles spontaneously form when the concentration of their subunits exceeds the critical micelle concentration (cmc) in a solvent in which one of the polymeric domains is immiscible (Figure 13.1). The cmc is determined by the immiscible polymeric domain, and can be adjusted by control of the chemistry and length of the immiscible domain, as well as by control of solvent conditions. Micelles formed at low concentrations from low cmc polymers are stable at high dilution. If necessary, micelles can be further stabilized (also often enhancing solubility) by covalent cross-linking to generate shell-stabilized structures [7, 80, 184, 188, 189]. Micelles formed from monomers with high cmcs can dissociate upon dilution, a phenomenon that might be exploited to control release of therapeutic cargos.

Properties of micelles, such as size dispersity, can be manipulated by control of solvent conditions, incorporation of excipients (to modulate polymer packing properties), temperature, and agitation. Reasonably size-monodisperse preparations (polydispersity of 1–5%) of nanoscale micellar structures can be prepared [18, 80, 114, 115, 184, 188, 189]. The extensive versatility of industrial polymer chemistry allows micellar structures to be finely tuned to be applicable to the task at hand, such as nanodiscs that are designed to self-assemble to a controlled size and composition in solution [27]. Modifications to the surface or bulk can be made to generate structures that are amenable to targeting or to support higher order assembly properties. Moreover, they can be loaded with a therapeutic or other molecules for delivery and caused to dissociate or expel their payloads at preferred times or bodily sites, under the influence of local physicochemical or externally controlled conditions. The tunability of these and other properties at the base level of monomeric subunits, as well as the level of assembled higher order structures, impart potentially powerful properties to micelles as nanoscale drug delivery and imaging vehicles.

PAMAM (polyamidoamine) dendrimers and other fractal-like materials display packing properties that can be utilized to assemble higher order aggregate structures termed "tecto(dendrimers)" [190]. These self-assembly properties are often exploited in oncological nanotherapeutics, as illustrated by Figure 13.2 (Roy et al., 2003; [6, 147]). These self-assembling therapeutic complexes may be pre-formed prior to administration, or they may be designed to assemble once in an appropriate solution or at the site of intended therapeutic action. Individual functional modules of the therapeutic assembly could be administered sequentially in order to tailor therapies more precisely to individual patient reactions and side effects.

Self-assembly approaches frequently do not feature precise control of subunit identity and stoichiometry in the assembled complexes, although quantitative methods are

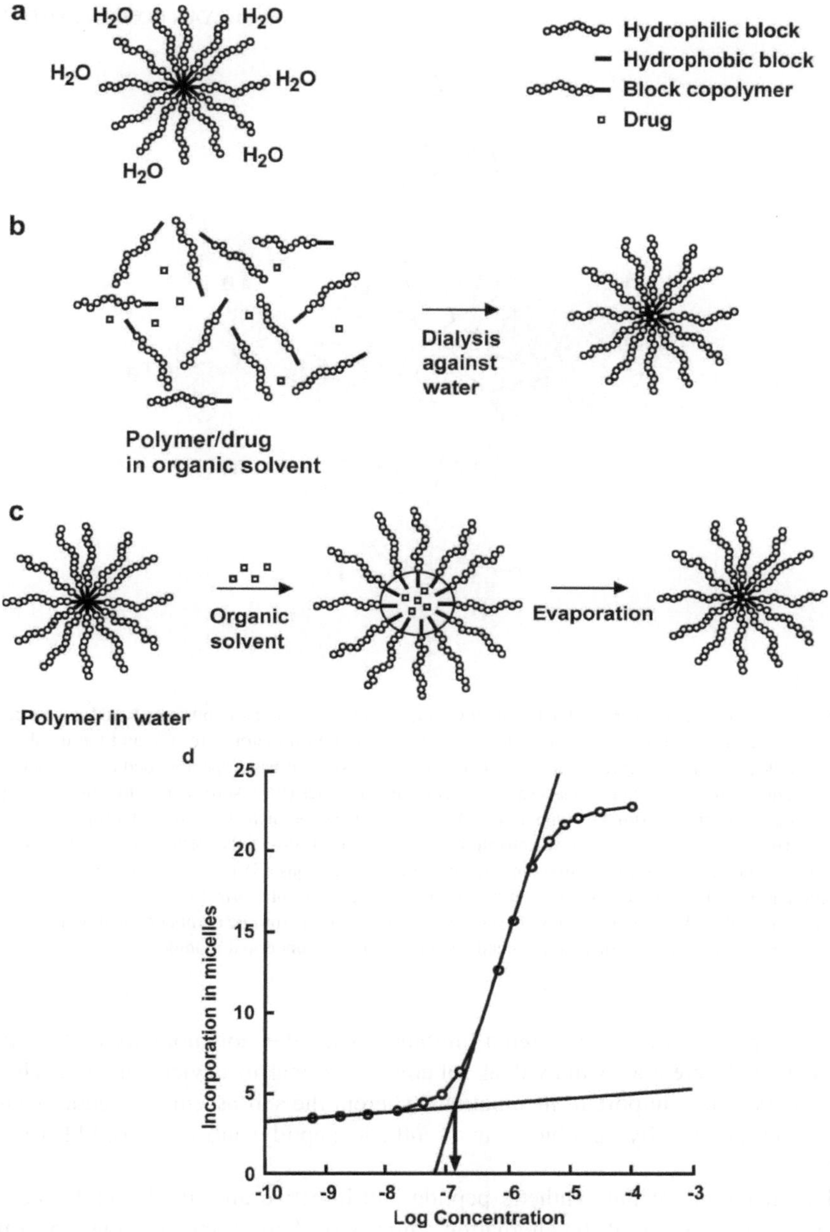

FIGURE 13.1. Micellar drug delivery vehicles and their self-assembly from block copolymers [80, 188, 189]. a) Morphology of a micelle in an aqueous buffer. Hydrophobic and hydrophilic polymer blocks, copolymers enclosing the blocks, micelles generated from the block copolymers, and (hydrophobic) drugs for encapsulation in the micelles are indicated. b) Micelle self-assembly and charging with drug occurring simultaneously when the drug-polymer formulation is transitioned from organic to aqueous solvent by dialysis. c) Pre-formed micelles can be passively imbibed with drugs in organic solvent. Organic solvent is then removed by evaporation, resulting in compression of the (now) drug-bearing hydrophobic core of the micelle. d) An illustration of concentration-driven micelle formation. At and above the critical micelle concentration (cmc) block copolymer monomers assemble into micelles, rather than exist as free block copolymer molecules. The arrow indicates the cmc for this system.

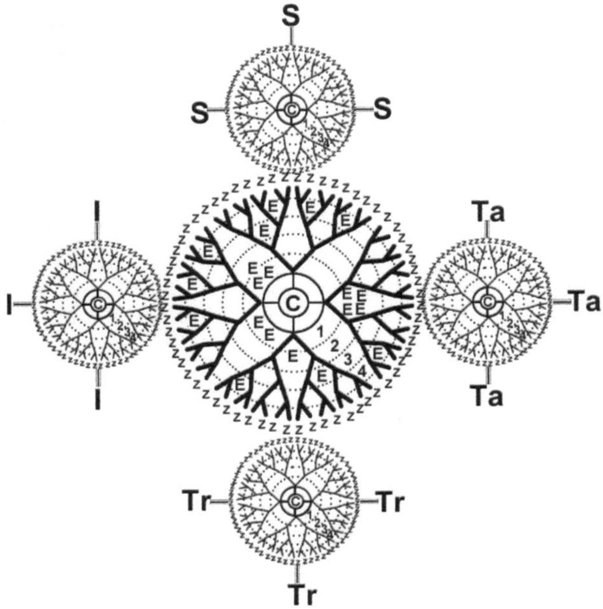

FIGURE 13.2. A hypothetical, modular nanotherapeutic patterned after the dendrimer-based cluster agent for oncology from Baker [6]. Each dendrimer subunit is grown from an initiator core (C), and the tunable surface groups of the dendrimers are represented by Z. Each dendrimer subunit has a specific, dedicated function in the device: the central dendrimer encapsulates small molecule therapeutics (E), whereas other functional components are segregated to other dendrimer components. These include biochemical targeting/tethering functions (Ta), therapeutic triggering functions to allow activation of prodrug portions of the device by an external operator (Tr), metal or other constituents for imaging (I) and sensing functions (S) to mediate intrinsically controlled activation/release of therapeutic. This design represents a therapeutic platform [6, 50] because of its modular design. The depicted device is only one possible configuration of a nearly infinite number of analogous therapeutics that can be tuned to fit particular therapeutic needs by interchanging functional modules.

being developed [176]. This is often a limitation when the stoichiometry and relative arrangement of differentiable individual subunits is critical to device function. However, stoichiometry is less important in situations wherein the self-assembling components are identical and functionally fungible, as in a synthetic, peptidyl anti-infective [42, 168], as in Figure 13.3.

Monomer units of the synthetic peptide anti-infective are circular molecules made planar as a consequence of the alternating chirality of their amino acid (aa) constituents. The R-groups (which are of varying hydrophobic or hydrophilic chemical specificities [21]) of the alternating D, L aas are arranged in the plane of the closed peptide rings, extending out from the center. Hydrogen bonds between individual rings drive self-assembly of the toroids into rod-like stacks, while the R-groups' interactions drive formation of multiple stacks. R-groups are generally hydrophobic and are selected to allow stacked toroids to intercalate preferentially in specific lipid bilayers (i.e., in those of pathogen membranes vs. those of host membranes). Membrane intercalation by stacked toroids selectively reduces the integrity of pathogen membranes, so that particular toroid species exhibit selective toxicity to specific

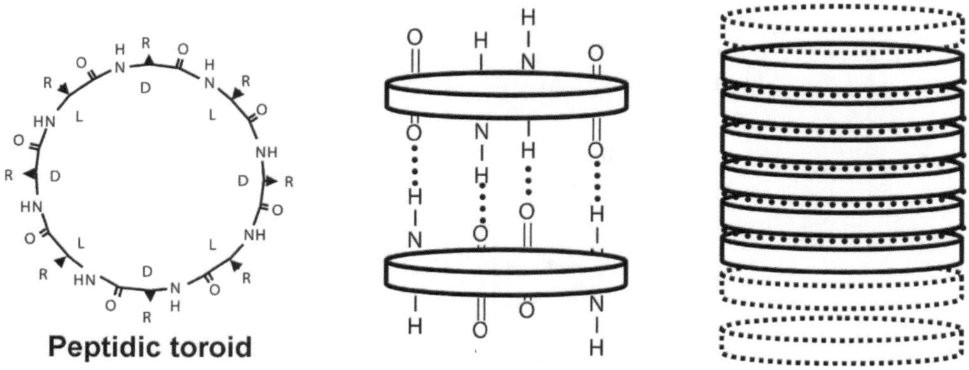

FIGURE 13.3. A self-assembling peptide antibiotic nanostructure [42, 168]. Peptide linkages and the □-carbons and their pendant R groups are indicated. The synthetic peptide rings are planar due to the alternating chirality (D or L) of their amino acid (aa) constituents. R groups of aas radiate out from the center of the toroid structure. Individual toroids self-assemble (stack) as the result of hydrogen bonding interaction between amine and carboxy groups of the peptide backbones of adjacent toroids. The surface chemistry of multi-toroid stacks is tuned at the level of the aa sequence, and therefore, R group content of the synthetic peptide rings. The chemical properties of the stacked toroid surfaces allow them to intercalate into the membranes of pathogenic organisms with lethal consequences. The specific membrane preferences for intercalation of the compound can be tuned by control of R group content of toroids.

pathogens. Nanoantibiotics like the peptide toroids [42, 168] and the N8N antimicrobial nanoemulsion [60] may become increasingly significant in human therapy as infection by bacteria resistant to traditional antibiotics continues to emerge as a major cause of morbidity and mortality in the developed world.

13.2.1.3. Biomolecules in Therapeutic Nanodevices: Self-Assembly and Orthogonal Conjugation Like all self-assembly processes, biological self-assembly is driven by thermodynamic forces. Biological macromolecules undergo self-assembly at multiple levels, beginning at the level of intramolecular self-assembly (e.g., protein folding from linear peptide sequences). Furthermore, higher order structures are typically constructed by the intermolecular self-assembly of smaller self-assembled subunits (for instance, structures assembled by hybridization of multiple oligonucleotides, enzyme complexes, *etc.*).

Proteins are copolymers of 20 chemically differentiable aa subunits [21]. Linear polypeptide chains form specific secondary structures via interactions between aa side chains. These secondary structures associate among themselves to form tertiary and quaternary structures that constitute final folded protein structures. This self-assembly process generates consistent nanostructures that derive their biological properties from their strict control of the deployment of chemical specificities (the aa side chains) in three-dimensional space. Indeed, peptides and proteins can be designed to self-assemble into functional three-dimensional nanoscale structures [154]. Proteins' useful biological functions, such as electron-transfer (e.g., metalloproteins such as azurin), can be harnessed and applied in fledgling subdisciplines such as nanobioelectronics, in which proteins and other biomolecules may be exploited under non-physiological conditions to create solid state hybrid electronic nanodevices using self-assembly mechanisms [149, 159].

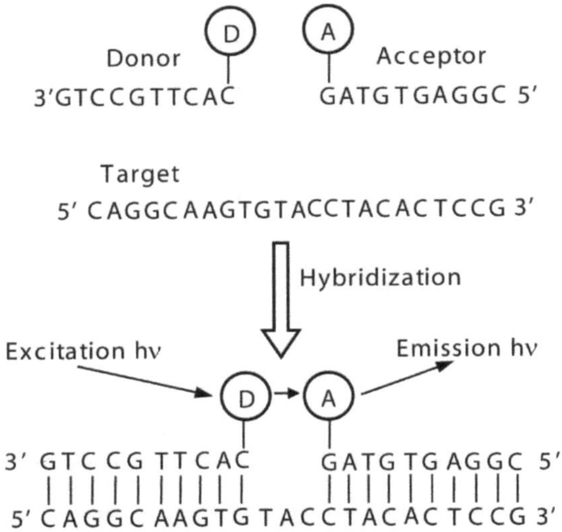

FIGURE 13.4. Self-assembly and biological macromolecules: a fluorescence transfer device that depends on self-assembly of biomolecules. The device is composed of donor and acceptor molecules brought into close proximity (within a few angstroms) by the base pair hybridization of complementary oligonucleotides. When the structure is assembled, acceptor and donor are energetically coupled, and fluorescence transfer may occur [65, 121, 127, 172].

In turn, biomolecules may also be used as affinity reagents to drive assembly of assorted nanostructures, both as free molecules and when conjugated to heterologous nanomaterials. For instance, three-dimensional nanostructures can be made by DNA hybridization [65, 121, 127, 145, 172]. DNA can also be used as a template for controlled construction of chains of nanocrystals and nanowires [32, 182]. However, DNA nanostructures exhibit tightly controlled topographies, but limited geometries [172], due to the flexibility of DNA strands. Oligonucleotides, antibodies, and other specific biological affinity reagents can also be used to assemble higher order configurations of nanomaterials appended to them (Figure 13.4, see also targeting and triggering discussions below). The domains of proteins responsible for assembly and recognition are often sufficiently small, continuous, and discrete to be abstracted from their native contexts as modules. Thus, binding domains of proteins (such as the variable regions of antibodies) can be appended to nanomaterials to drive specific nanostructure assembly.

Numerous orthogonal bioconjugation approaches have been devised for protein semisynthesis [46, 67, 83, 109, 137, 138]. These protein synthetic chemistries use the exclusive reactivities of electrophile-nucleophile pairs to permit site-specific conjugation of polypeptides to heterologous materials. They have been applied to the synthesis of multiple therapeutic nanodevices [95–98, 104, 105, 109, 137, 138]. They are valuable because protein function is intensely reliant on the three-dimensional conformations: chemical derivatization at critical aa sites can profoundly impact, and even neutralize, protein bioactivity. Because orthogonal approaches direct conjugation to pre-selected sites, and since the sites of conjugation in the protein can be chosen for their tolerance of adducts at those positions, proteins coupled to nanomaterials by orthogonal methodologies often retain their biological activity to a large extent. In contrast, protein bioactivity in conjugates generally is lost or severely impaired when proteins are coupled to nanomaterials using highly promiscuous conjugation chemistries. Proteins such as cytokines (and other protein hormones), for instance, elicit their effects by interacting with a receptor, and a large fraction of their surfaces (20% or more) is directly or indirectly involved in receptor binding. Such proteins are often exceptionally sensitive to the positions at which nanoscale heterologous materials are appended to them; appending nanomaterials to their receptor-binding domains can cause steric problems sufficient to inactivate them entirely. Proteins for which a minor fraction of their surfaces contribute to the interesting aspects of their bioactivities (from the standpoint of the nanodevice designer), such as certain enzymes or intact antibodies, are usually somewhat less sensitive to the promiscuity of the bioconjugate strategy used, but the benefits of orthogonal conjugation strategies can also apply to bioconjugates involving enzymes, antibodies, and analogous molecules in terms of obtaining greater yields of active molecules [46, 67, 83, 109, 137, 138]. The potentially substantial advantages of orthogonal conjugation for incorporation of active biological structures into hybrid semi-synthetic nanodevices is becoming more fully recognized, and can not be overrated [109, 137, 138].

13.2.2. Targeting: Delimiting Nanotherapeutic Action in Three-Dimensional Space

The delivery of therapeutics to desired sites of action is a key strategy to enhance clinical benefit, especially for drugs only useful within narrow windows of concentration because of their systemic toxicity. Diverse targeting approaches exist, ranging from more passive methods that exploit differential extravasation limits of vasculature of various tissues (see the discussion of oncology below), to size and surface chemistry preferences for cellular uptake (see discussion of vaccines below), preferential partition of nanomaterials by virtue of their extent of opsonization (see below), or the affinity of biological molecules adorning the nanodevice for counter-receptors on the cells or tissues of interest. Relatively minor modulation of only surface chemistry or material type, for instance, can give rise to widely disparate parenterally-administered particulate biodistributions [197].

13.2.2.1. The Reticuloendothelial System and Clearance of Foreign Materials. Surface chemistry, particle size, and other physical properties can drive targeting of nanomaterials (and presumably the nanodevices containing them) to clearance organs and tissues. For instance, clearance in urine is governed by the filtration preferences of the kidney. Most molecules passing into the urine are less than 25–50 kilodaltons (kDa; 25–50 kDa particles corresponding *loosely* to effective diameters of about 5 nm or less) in mass and are preferably positively charged. These parameters are routinely varied to control clearance rates of administered drugs. The clearance of low molecular weight nanomaterials in urine can be suppressed by tuning their molecular weights and effective diameters, generally by conjugating polymers such as poly(ethylene glycol) (PEG) using various chemistries [63].

The first step in the clearance of foreign materials is coating non-native particles with serum proteins (opsonization, [24]). Tissue dendritic cells (DCs) and the organs of the reticuloendothelial system, or RES (i.e., thymus, liver and spleen), recognize and take up opsonized particles. The RES extracts materials from circulation using both passive diffusion and active processes (e.g., receptor-mediated endocytosis), which often results in the partition of positively charged nanoparticles (e.g., PAMAM dendrimers) into the RES.

Legitimate applications of targeting to the kidney and the RES (for instance in glomerular disease [132]) notwithstanding, intrinsic targeting to clearance sites is principally a technical problem that hinders therapeutic delivery to other sites. Numerous targeting strategies exist that minimize uptake of nanotherapeutic devices by clearance systems and maximize delivery to desired sites. The most direct approach is targeting via biological affinity reagents decorating the surfaces of therapeutic nanodevices.

13.2.2.2. Applying Biological Affinity Properties for Nanotherapeutic Targeting Biological affinity-mediated, tissue-specific delivery requires the presence of tissue-specific surface features (usually proteins or glycoproteins called tissue-specific antigens). The hunt for tissue-specific surface antigens for drug targeting has been arduous, yet unsatisfying. Principal obstacles have been specificity (not many antigens are uniquely present in any single tissue at a practical concentration), and availability (some tissues do not have their own unique antigenic signature or markers, or they may not be accessible from the vasculature, which prohibits parenteral administration). The discovery of a vascular address system [2, 87, 166, 167], including antigens specific to organs and tumors [13], is potentially exciting.

Identification of the vascular address system was achieved by administration of a peptide phage display library to animals, resecting individual organs from the animals and extracting phage from the vasculature of the isolated organs (Figure 13.5 [2, 13, 87, 166, 167]; see [67] and [81] for a discussion of display technology). More recently, small molecules other than peptides have been used to interrogate selective affinities for particular organs and tissues [13], significantly increasing the overall diversity of potential targeting molecules and thereby amplifying the potential for device targeting. The vasculature of individual organs presents unique constellations of receptors that can be recognized by short (10 amino acids or less) peptide sequences from a phage display library or other small molecule/chemical collection. Furthermore, the affinity-selected peptides may be covalently or otherwise bound to nano- to microscale particles by conjugation, and thereby tether peptide/particle devices to their cognate receptor sites, as illustrated by the binding phage particles presenting the peptides to specific vascular locations. The vascular address system has been used to target a specific organ (the prostate) for destruction *in vivo* [1], establishing the utility of the address system in drug delivery [1, 40]. Additionally, peptides to antigens specific to tumor vasculature have been selected and applied: liposomes targeted to tumor vasculature were shown to slow tumor growth markedly by cutting off the tumor's blood supply [58]. Mapping of the vascular address system continues [2] and it remains to be seen whether each organ has a single molecular marker constituting its address that will be amenable to delivery of drugs tethered to the cognate peptide(s). Vascular addresses deliver materials to the organ vasculature: extravasation and access of organ tissue spaces by nanotherapeutics remains a separate issue.

13.2.3. Triggering: Spatially and Temporally Delimiting Nanotherapeutic Action

Safe and effective controlled triggering of therapeutic action can complement therapeutic targeting. If adequate control of the site and time of nanotherapeutic delivery by other means is unfeasible or impractical, the site of therapeutic action can be spatially or

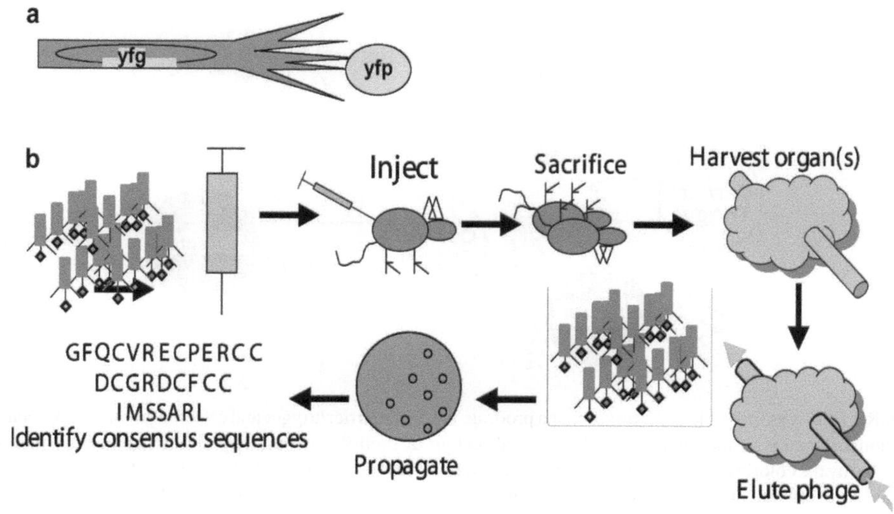

FIGURE 13.5. a) A schematic of a filamentous phage particle encapsulating DNA, which encodes a gene of interest (your favorite gene, or YFG) and presents the corresponding protein (your favorite protein, or YFP) on its surface. The linkage between YFP and YFG provided by the phage allows simultaneous affinity isolation and recovery of proteins of interest along with replicable genetic elements encoding them. When variant protein libraries are built by this method, they can be sorted for their ability to bind receptors by affinity. b) A library of filamentous phage presenting random peptides is shown schematically. The library is injected into the vasculature of an animal (in this case, a mouse), where individual phage bind specific receptors present in the vasculature of different tissues. The animals are sacrificed and organs of interest are harvested. Phage particles bound to receptors of vasculature of the organ at hand can be eluted by any of several methods (low pH elution is schematically shown here). *E. coli* is infected with the eluted phage particles and clonally propagated, allowing identification of the peptides encoded by the eluted phage which mediated binding to the vascular receptors. Ultimately, this permits the identification of consensus peptide sequences that recognize specific receptors on the vasculature of the organ or tissue involved. These receptors constitute the molecular addresses for these organs, and the peptides isolated from phage eluted from them are biological affinity reagents that can be used to direct nanotherpeutic devices to vasculature of particular organs [2, 87, 167].

temporally delimited by specific triggering. The triggering event could involve release of active therapeutic from a reservoir, or chemical/physical processing of drug materials from an inert to an active form (inert administrations that are converted to active form at a specific time or location are called prodrugs, Figure 13.6). Three main triggering strategies are commonly used: external stimuli, intrinsic triggering, and secondary signaling. Triggering strategies require predictable sensitivity of therapeutic nanodevices to a controlled triggering event, or a spatially/temporally intrinsic triggering event mediated by the host.

13.2.3.1. Intrinsic Physiological Conditions for Nanotherapeutic Triggering The prodrug strategy of intrinsic triggering depends on the conditions at the desired site of action to trigger the functionality of the therapeutic nanodevice. Proteins can be engineered to make their conformations sensitive to intrinsic conditions at their desired sites of activity, very similar to how the three-dimensional conformation of, and therefore the function of, proteins can be controlled using certain external triggering strategies [59]. Diphtheria

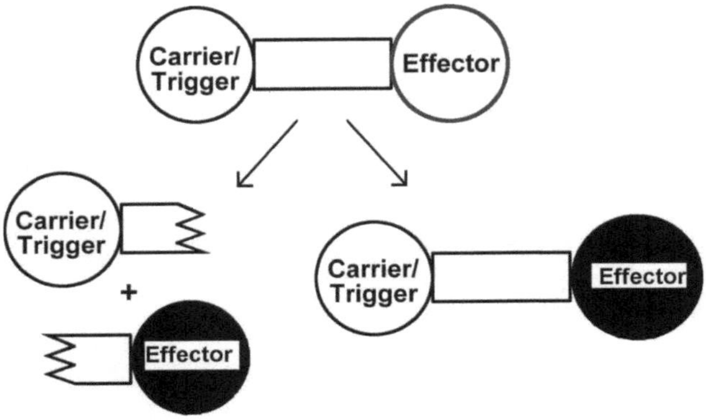

FIGURE 13.6. One possible configuration of a prodrug, in which carrier/trigger and effector functions are separate functional domains of the therapeutic. The crucial feature of prodrugs is that they are therapeutically inert (as indicated by the colorless effector domain) until an activation event occurs (mediated here through the carrier/trigger domain, with effector activation indicated by its change to a filled in circle). Activation events can involve cleavage of inhibitory carrier/trigger domains from effectors. Other activation strategies involve a chemical change or shift in conformation of the effector, mediated through the carrier/trigger domain and in response to an environmental condition.

toxin (DT), for example, has been engineered to require passage at low pH to achieve its toxic configuration. When the variant DT is associated with a receptor ligand, it can only be taken up and transported into low-pH lysozomal vesicles by cells with the appropriate counter-receptor on their surfaces [156]. Thus, toxicity of the variant DT is limited to cells expressing a specific receptor.

Another triggering approach in which the properties of a nanoscale drug delivery device are modulated at the site of action by an endogenous enzymatic activity is called enzyme-activated delivery (EAD [125]). Typically associated with this mechanism are liposomal, micellar, or other nanostructure from which synthetic pendent groups can be cleaved by metabolic enzymes (e.g., alkaline phosphatase, phospholipases, proteases, glycosidic enzymes, etc.) that are known to exhibit elevated levels of expression at the site of therapy. Such nanostructures can be designed so that cleavage of the pendent group causes a conformational or electrostatic change in the polymeric components of the delivery device, rendering the micellar structure fusogenic, leaky, or causing partial or complete dissociation of the structure to release its therapeutic payload at pre-selected sites [37]. Furthermore, a recent triggering approach entails the use of transfection into cells to realize intracellular reduction (based upon interior cellular conditions) of a reducible (disulphide) bond between a chemical-DNA nanostructure, which has implications for site-specific gene delivery [17].

13.2.3.2. External Stimuli for Nanotherapeutic Triggering Most frequently, an external operator/clinician imparts external stimuli as a mode of site-specific energy input. This energy is usually input as one or more of light, ultrasound, neutrons, or magnetic or electrical fields. Organic polymeric structures such as micelles, for instance, are amply suitable for interaction with these energy sources. Ultrasound can be use to dissociate them reversibly so that they expel their contents or expose their internal spaces to the environment by interrogation with ultrasound pulses (20 to 90 kilohertz range). This process has

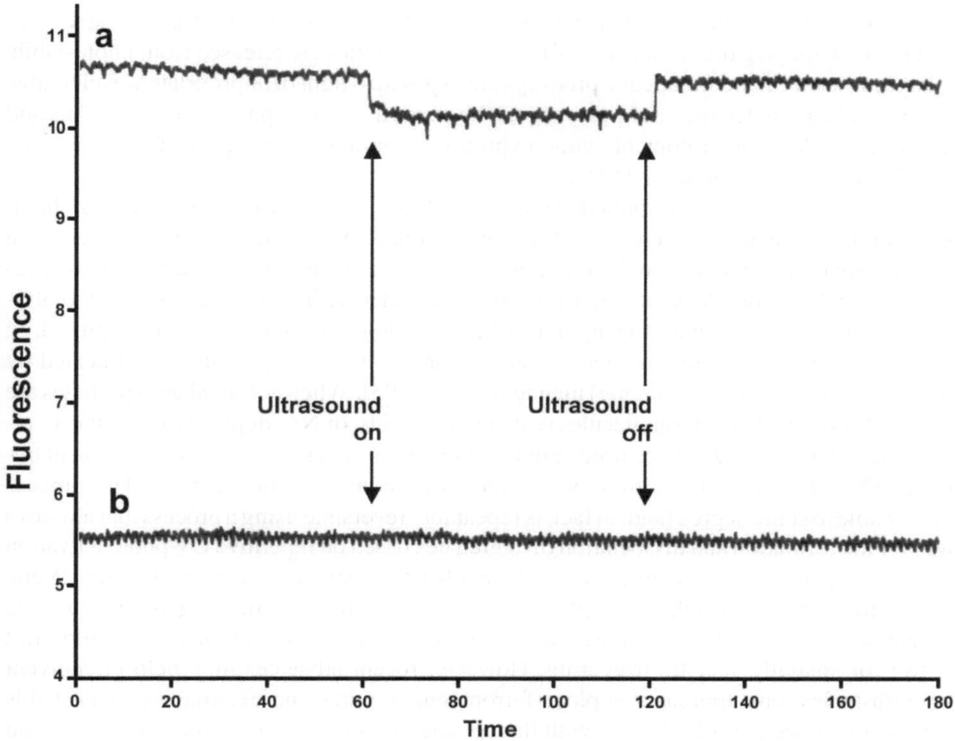

FIGURE 13.7. Externally applied stimuli can trigger drug release: hydrophobic drug molecules are reversibly deployed from micelles in response to acoustic stimuli [72]. Intact pluronic micelles maintain the hydrophobic cytotoxin doxorubicin in a hydrophobic environment. Doxorubicin fluorescence is quenched in aqueous envioronments, and hence changes in integrity of pluoronic micelles carrying doxorubicin can be monitored by doxorubicin fluorescence. Here, an ultrasound pulse is applied to suspensions of such micelles or to doxorubicin in solution. Application of ultrasound triggers exposure of micelle-encapsulated doxorubicin to the aqueous solvent (i.e., drug release), as demonstrated by the reduction in fluorescence (in arbitrary units, over time, in seconds). After cessation of the acoustic pulse, doxorubicin is repackaged into the micelles, as evidenced by the increase in fluorescence after the ultrasound administration (Curve a). Ultrasound has no impact on fluorescence of doxorubicin in solution, as expected (Curve b).

been used to control the release of cytotoxin (doxorubicin) from micelles (Figure 13.7), and short ultrasound transients could be used for pulsatile or intermittent delivery of therapeutics to patients [72]. Ultrasound-mediated gene transfection is another promising method for physical localization of gene delivery, except in this case ultrasound is employed [136]. In a different technique, after diseased cells have been targeted with particles loaded with an appropriate boron isotope, thermal neutron bombardment is employed to destroy targeted cells. In this approach, termed boron neutron-capture therapy (BNCT), neutron irradiation alone is insufficient for deleterious heating of cells, but when cells are loaded with the relevant boron isotope and concentrations (and targeting method to diseased tissue, described below) the combination is lethal to targeted cells [124].

Another external triggering modality is light. Bioactive materials can be covalently conjugated to nanoscale delivery vehicles by photo-labile linkages [6, 152, 177] and/or photosensitizers which, for instance, can generate reactive oxygen molecules capable of selectively damaging diseased cells [34]. Active therapeutics are released from photo-labile linkages when the nanotherapeutic prodrugs are exposed to light of appropriate wavelengths. Alternatively, particles such as liposomes can be constructed with photoactive materials and designed to release their contents upon exposure to certain wavelengths of photons, e.g., ultraviolet or gamma radiation [141].

Furthermore, externally applied magnetic fields can be used to control activity of therapeutic biological macromolecules. Eddy currents induced by alternating magnetic fields can heat nanometallic particles and their immediate vicinity to control the bioactivity of individual biological macromolecules [59]. Liposomal particulates can be configured as externally-triggered drug delivery agents using magnetic technology [134] and as dual imaging/drug delivery reagents by designing them to release their contents (e.g., drug) when heated by MRI (magnetic resonance imaging) magnetic fields [193]. When colloidal gold particles are covalently conjugated to biomolecules (e.g., nucleic acids, or NA, duplexes, or proteins), alternating magnetic fields can be used to induce heating sufficient to cause dissociation of hybridized NA structures or denaturation of three-dimensional protein structures. This process is reversible to some degree (and, in fact, is repeatably reversible using a process that employs light for the formation and dissociation of aggregates based on repetitive UV photoactivation [144]), and permits the construction of bio-synthetic assembly or release switches. While this approach has unmistakable implications for temporally-specific triggered release, the current difficulties in directing magnetic fields to pre-selected tissue locations could limit its use for spatially-specific triggering. However, recent advances may help circumvent these obstacles: one approach employs ferromagnetic wires and external magnetic fields to guide magnetic particles loaded with therapeutic to preferred sites, and then retain them there [160]. Another is preferentially to target/tether particulates to surface antigens (e.g., selectively targeting of tumors, vasculature, etc.), and then trigger release (Hafeli, 2004).

13.2.3.3. Nanotherapeutic Triggering Using Secondary Signaling Delivery strategies such as secondary signaling, or multicomponent delivery systems, are considerably more complex (Figure 13.8). Both site-specific delivery and systemic delivery of one or

FIGURE 13.8. Secondary biochemical signals can trigger release of active therapeutics. a) Antibody directed enzyme-prodrug therapy (ADEPT) exploits enxyme-monoclonal antibody (mAB) bioconjugates to activate prodrugs [174]. Here, a mAB is covalently linked to a catalyst (C in rhombus, usually an enzyme). The binding specificity of the monoclonal antibody causes the bioconjugate to bind to a desired target site (possibly by recognition of a vascular address, as shown in Figure 13.5). The catalyst (enzyme) specificity of the bioconjugate is capable of triggering activation of a prodrug, similarly to Figure 13.6. The corresponding prodrug is then administered (represented as in Figure 13.6). The enzyme localized at the desired target site of therapeutic action then activates the prodrug (represented here by cleavage of carrier/trigger moiety and change in symbol of the effector to a filled in circle, as in Figure 13.6). b) Another secondary signaling triggering strategy [121]. Here, single stranded nucleic acid (SSNA) drives assembly of an activating complex for a prodrug. A catalyst (indicated again by C) is conjugated to an oligonucleotide that hybridizes to a single stranded nucleic acid (SSNA). A prodrug, containing an effector domain (E) and a second oligonucleotide, in this case homologous to a region of the same SSNA, and proximal to the site of hybridization of the first oligonucleotide, is also administered. Catalyst specificity is tuned to correspond to linkage between the prodrug effector and oligonucleotide. The catalyst in the assembled structure is sufficiently near to the prodrug to cleave the linkage between effector and oligonucleotide, liberating active drug (filled in circle). Potentially, prodrug and catalyst oligonucleotides might dissociate from the SSNA and the cycle could be repeated.

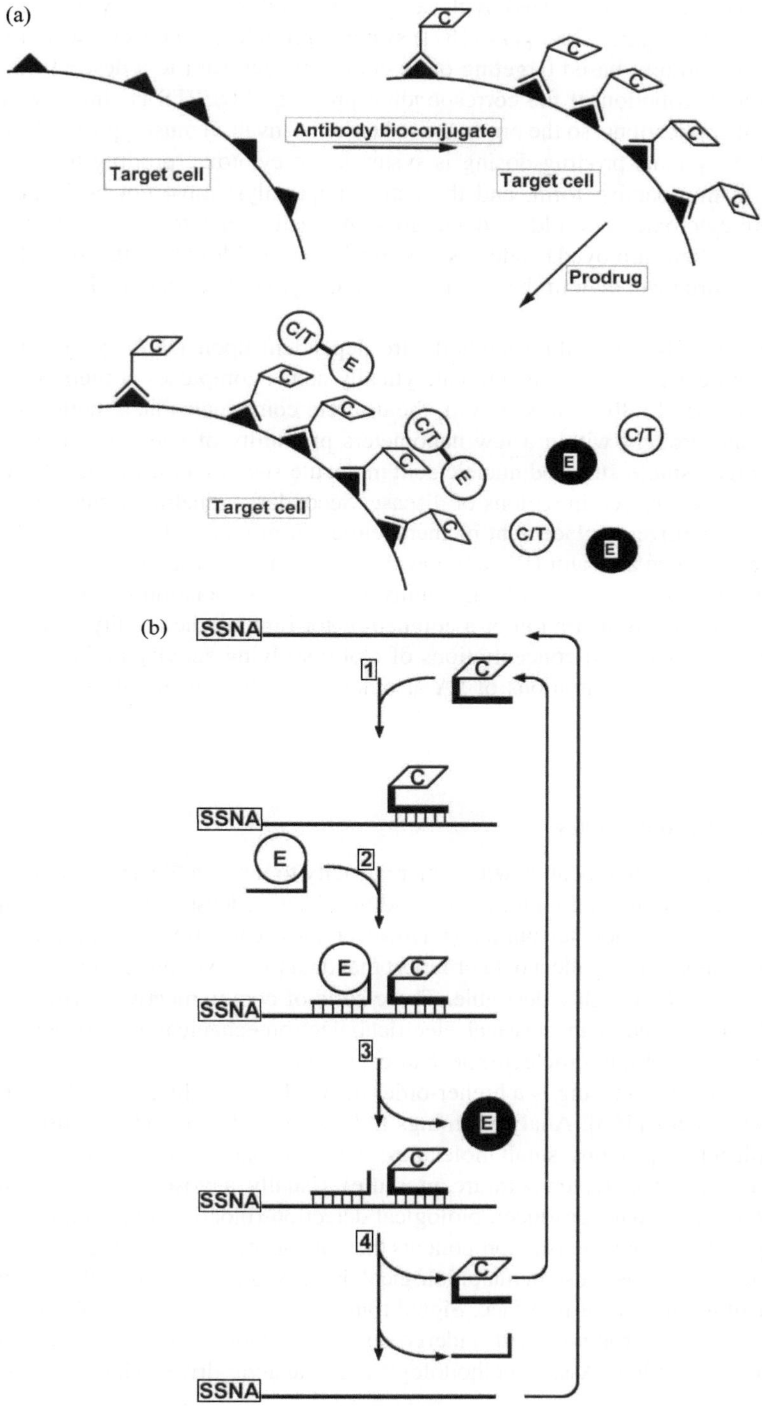

(a)
Antibody bioconjugate
Target cell
Target cell
Prodrug
C
C/T
E
Target cell
(b)
SSNA
1
2
3
4
E
C

more components can be realized by these systems. ADEPTS (antibody-directed enzyme-prodrug therapy, Figure 13.8a [174, 187]) systems provide examples of such approaches, which exhibit affinity-based targeting of an enzymatic activator to a desired site followed by systemic distribution of the corresponding prodrug. ADEPTS is chiefly employed in oncological applications, so the prodrug used (when activated) must typically be cytotoxic. In ADEPTS systems prodrug dosing is systemic, so cytotoxic prodrug toxicity must be very low in the inactive form, and the activating catalyst must not be in proximity at sites where cytotoxicity would be deleterious. Activating enzymes with no physiological analogue are often employed to address this consideration. Moreover, tuned, high lability of the activated drug can delimit the site and extent of cytotoxic effect mediated by ADEPTS systems.

Other secondary signaling methods are dependent upon DNA hybridization to assemble therapeutic components into catalytically active complexes at their sites of action [121], see Figure 13.8b. In this system, the discrete components lack therapeutic activity until they are brought within a few nanometers proximity of one another by hybridization to a target, single-stranded nucleic acid molecule such as a messenger RNA (mRNA) molecule over-expressed in regions of disease. Secondary signaling systems may also exploit the competitive displacement of therapeutics from a carrier structure [194]. For instance, plasminogen activator (PA, a blood clot dissolving enzyme) can be targeted to clots by non-covalent association with an affinity reagent. Dissociation of the targeted PA is effected by bolus administration of a competitor for binding the affinity reagent. This results in high (efficacious) concentrations of clot-dissolving activity in the vicinity of the clot, but minimal concentrations of PA at other sites where it might promote undesired bleeding.

13.2.4. Sensing Approaches

Arguably, nanotherapeutics with either capacity to target to desired sites or to be triggered by an environmental stimulus or condition have at least a rudimentary capacity to sense and respond to their surroundings. However, these capabilities do not necessarily imply a further capacity to collect data or report that data to an external operator, though these capabilities are often highly desirable. These sorts of communicative attributes are more commonly associated with classical electrical/electromechanical/electrochemical sensor systems (e.g., MEMS, microelectromechanical systems).

Conventionally, sensing is a higher-order device functionality, depending on multiple device components [163]. Analytes (things to be sensed) for nanotherapeutic application can be molecules, (proteins, small molecules, salts, hydrogen ions, etc.), or physical conditions (such as redox state, temperature, pressure). Usually, a biosensor system comprises a sensor interface, signal transducer, biological detection (bioaffinity) agent, and an associated assay methodology. Sensor components that interact directly with the analyte comprise the interfaces. For sensor use in nanobiological devices, immobilized biological molecules often constitute the sensor interface. Signal transducers are sensitive to physical/chemical changes in their environment and undergo some detectable change in state as the result of analyte recognition. Assay methodology reads analyte-driven changes in transducer state, and determines the sensitivity, robustness, complexity and rapidity of the sensing processes.

13.2.4.1. Cantilever and Field Effect Biosensors Microcantilevers comprise a mechanical mechanism that transduce sensed events [103, 153]; changes in both the resonant frequency and deflection of cantilevers resulting from analyte binding or dissociation can be conveniently and sensitively detected by optical, capacitive, interferometric, or piezoresistive/piezoelectric methods. The dimensions of cantilevers range from micron to submicron proportions [153]. Operationally, they are versatile and can be used in air, vacuum, or liquid, though they typically suffer some degradation in performance in liquid media. Under optimal conditions, cantilever systems have used changes in resonant frequency to resolve mass down to 0.39 ag (1 ag $= 10^{-18}$ g) assuming appropriate cantilever dimensions (here, 4 μm length, 500 nm width, and 160 nm thickness) [68]. The ability to resolve such masses enables detection and identification of chemical/biological species, e.g., single viruses, and further suggests that zeptogram levels of mass detection are attainable [55, 68]. Incorporation of biological receptors and/or affinity reagents on the cantilever surfaces can be used to detect specific biochemical binding and dissociation events. Microcantilevers are also highly temperature sensitive, capable of detecting changes in temperature as low as 10^{-5} K, as well as detecting minute variances in pH. Analogous to most micromachined structures, they can be batch fabricated and conveniently multiplexed.

A second form of microfabricated sensing architecture, Field Effect Transistors (FET) consist of a current source and a current drain with a conductive path (which comprises the sensing channel) between them and a sensing gate to which a bias can be applied. The binding of analyte to the modified gate induces charge transfer, which results in a dipole between the surface and the underlying conductive path. Parameters of current passing between source and drain of a semiconductor FET are very sensitive to the charge state and potential of the surface in the channel. As with cantilevers, the binding specificity of macromolecular analytes of interest can be provided by deployment of biological affinity reagents in the FET sensing channel. Submicron FETs are routinely constructed; design of FETS incorporating carbon nanotubes could offer still greater miniaturization [20, 25, 116, 123] as well as novel opportunities for nanoelectronics [20].

13.2.4.2. Conducting Polymers and Sensor Biocompatibility The biocompatibility of most metallic structures (as might be used in the sensors described) is inadequate at best. Most metals in contact with blood rapidly become opsonized (coated with serum proteins), undergo electrochemical degradation, and have other inferior qualities. However, polymer chemistries have the capability to tune compositions for improved biocompatibility properties and are commonly used to produce more biologically tolerable synthetic surfaces. Electrically conductive polymers are therefore attractive for electrical sensors. Moreover, these polymers are amenable to an assortment of deposition techniques with diverse benefits. For example, one deposition technique for these polymers allows them to be directly written onto substrates with uniform nanoscale widths very rapidly using atomic force microscopy (AFM) tips [76].

There exist polymeric materials with intrinsic conductivities comparable to that of metals (up to $1.5 \times 10^7 (\Omega\text{-m})^{-1}$, about twice that of copper on the basis of weight). Significant conductivity has been confirmed for approximately a dozen polymers, including polyacetylene, polyparaphenylene, polypyrrole, and polyaniline, all doped with various impurities [15, 175]. Precise control of doping parameters enables tuning of electrical conductivity properties over several orders of magnitude [47, 175]. The polymeric materials themselves are often compatible with biomolecules, such as proteins, in solution. Additionally, polymers are amenable to rather simple assembly of sensor transducer-interface components through deposition of the polymer and trapped protein (sensor interface) directly onto a metallic micro- or nanoelectrode surface [8, 175]. Combined with a basic nanoelectrode array microfabrication method [86], simultaneous conducting polymer-interface protein deposition tools may offer an exceedingly facile means to fabricate multiplexed sensor arrays. Further, novel biodegradable polymers which retain electrical conduction properties in physiological conditions should prove tremendously useful to fields such as bioelectronics and tissue engineering [175].

13.2.4.3. Communicating With the Outside Sensing modalities are of little use in and of themselves: they must either communicate with other device constituents or with external observers or operators. Sensor coupling in independently operating devices fabricated by microfabrication can be accomplished via direct means, with the coupling linkages incorporated into the assembly protocol. Linking with biological device components can also be direct, as when conductive materials are conjugated to biomolecules, which can finely modulate their activity [59].

Indirect sensor-device coupling, on the other hand, is often accomplished through electrochemically-produced mediator molecules [16]. Autonomous nanodevice sensors may report conditions at the site of therapeutic action to an external operator, who could choose any one of the external triggering strategies discussed above to initiate therapy via the nanoscale drug delivery moiety at appropriate site(s) and time(s). The communication/sensor interrogation can most crudely be accomplished by direct electrical wiring of sensors to an external observation station. Alternatively, if the event to be detected is optically transducible, as in colorimetric smart polymers whose optical properties can be modulated in the presence of analyte [179], sensors could be interrogated by fiber optics. These modalities are most applicable to sensor arrays conveyed to the site of interest by a catheter.

Multiple instances of prodrugs that are activated by cleavage of an inhibitory domain from the complex have been provided. Analogous non-invasive methods for communication with external operators could exploit this phenomenon by detecting the cleaved inhibitory domain in bodily fluids. If the cleaved moiety clears through urine, the extent of drug activation may be monitored non-invasively via urinalysis. However, there is no *a priori* need to associate "monitorable" cleavage events with drug activation. An operator might, for instance, administer a catalyst that cleaves a detectable substance from a hybrid nanodevice. If this secondary signaling moiety were targeted to the desired site of therapy autonomously, presence of the detectable cleaved product would provide data regarding the physical location of the therapeutic nanodevice in the patient. Cleaved product would be undetectable unless the therapeutic and secondary signaling moiety co-localized at a single site. Other sophisticated means of communication involve ultrasound or electromagnetic radiation to transmit information. These approaches are often exploited in *in vivo* imaging methods.

13.2.5. *Imaging Using Nanotherapeutic Contrast Agents*

Imaging (in particular MRI—our primary topic) enables non-invasive visualization of organs and tissues for multiple therapeutic applications. Administration of contrast agents greatly enhances the cross-sectional and volumetric images, allowing spatial resolution of considerably less than 1 mm. Nanoscale particulates (5–100 nm in diameter) have shown extraordinary utility as contrast-enhancing agents for medical imaging modalities such as MRI [12, 39, 78, 85].

13.2.5.1. Basics of Magnetic Resonance Imaging (MRI) In MRI, items to be imaged are exposed to a powerful magnetic field and well-defined radio-frequency pulses. The external magnetic field (B_o) serves to loosely align protons either with (lower energy level) or against (high energy level) the field; the difference between the two energy levels is proportional to B_o. Once the protons are separated into these two populations a short multi-wavelength burst (pulse) of radio frequency energy is applied. Any given proton will absorb only the frequency that corresponds to its distinct energy state. This resonance absorption is followed by the excitation of protons from the low to high energy level, and equivalent protons moving from high to low energy levels. After the radio frequency pulse, protons quickly return to their original equilibrium energy levels (termed "relaxation," involving detectable release of absorbed energy). Following re-establishment of equilibrium, another pulse can be applied. Data is collected by a receiver that detects and quantifies relaxation energy (i.e., amplitude, phase, and frequency). Imaging algorithms derive high-resolution images in two or three dimensions from this data.

Intrinsic factors such as the proton density of the tissues, local blood flow, as well as relaxation time constants (longitudinal relaxation time, T1, and transverse relaxation time, T2) influence image quality [143]. T1 relaxation measures energy transfer from an excited proton to its local environment while T2 relaxation gauges the duration of coherency between resonating protons after a pulse, prior to their return to equilibrium. Signal intensity is linearly affected by proton density, but changes in T1 or T2 result in exponential variances in signal intensity. Therefore, T1 and T2 can be manipulated to enhance imaging by administration of exogenous contrast-enhancing agents.

13.2.5.2. Nanoparticle Contrast Agents This section focuses on paramagnetic, ferromagnetic, and superparamagnetic contrast agents. Paramagnetics (gadolinium, Gd, iron, Fe, chromium, Cr and manganese, Mn) have permanent dipole moments, though the magnetic moments of individual domains are typically unaligned. When an external field aligns them, the material is magnetic. Contrast enhancement by paramagnetics owes to manipulation of T1 constants. A major drawback to these agents is the heavy metal toxicity associated with many of them.

While multiple substances can be ferromagnetic or superparamagnetic, the most common form of both materials in nanotherapeutics contain iron (Fe) clusters, which produce magnetic moments 10–1000 times greater than do individual iron ions. Clusters greater than approximately 30 nm in diameter are ferromagnetic, whereas smaller particles are superparamagnetic [143]. Ferromagnetic materials sustain their magnetic moment after the external field is removed, but superparamagnetic materials lose their magnetic field subsequent to field removal, much as do paramagnetics. Both ferromagnetic and superparamagnetic materials minimize the proton signal by shortening T2 [139], resulting in negative contrast effects (i.e., darkening of the image [82]). Both classes of contrast agents have the advantage that they are metabolizable, and therefore have relatively limited toxicity.

13.2.5.3. Contrast Agent Synthesis and Incorporation of Biomolecules Historically, first generation contrast agents frequently contained a single metal ion/complex, whereas emerging agents incorporate nanoscale metal clusters, crystals, or aggregates, sometimes encapsulated within a synthetic or biopolymer matrix or shell [78, 85, 135, 158, 178, 199]. Such metal cluster agents improve contrast effects and MR images relative to free ions (discussed above), and may be configured for longer blood half-life and/or greater degrees of biocompatibility.

Particles are typically produced from colloidal suspensions in which metallic cores are thoroughly mixed with the matrix material before being precipitated out of solution with a non-solvent. Typical of matrix materials used in commercial imaging reagents is Dextran (a polymer of 1,6-□-D-glucose); other polymeric materials such as PEG (polyethylene glycol) are also used. Nanoparticulate contrast agents must be purified under tightly controlled environments [185]. They are characterized for their size dispersity by methods such as electron microscopy (e.g., TEM, transmission electron microscopy) and AFM [9, 178, 199] for metallic cores ranging in size from about 4 to 20 nm in diameter [171, 199], while the complete coated particles can be up to 100 nm or more in diameter [118]. However, compared to larger cores, nanoparticles less than 20 nm in diameter generally display considerably longer blood half-life and improved T1 and T2 relaxivity effects [195].

Furthermore, surface chemical groups of the contrast agent polymer matrix can be derivatized to improve biocompatibility. As discussed above, rapid clearance of nanoparticles via the RES can be mitigated by surface modifications that reduce non-specific opsonization (e.g., by use of PEG modifications [199]). Neutral, hydrophilic surfaces tend to adsorb less serum protein than hydrophobic or charged surfaces, and bisphosphonate and phosphorylcholine-derived thin film coatings have been applied to iron oxide nanoparticles to stabilize them against pH, opsonization, aggregation, and other physiologically relevant properties [150]. These film coatings do not completely eradicate protein adsorption, as dense layers or thick brushes of polysaccharides or hydrophilic polymers are often more effective methods of evading opsonization [165].

The earliest stage contrast agents were also principally blood-pool agents that traveled freely through the whole vasculature with no specific affinity for any one bodily site or tissue. Targeting contrast-enhancing nanoparticles to sites of interest should mitigate heavy metal toxicity by diminishing the dose necessary to acquire a satisfactory image. Targeting of contrast agents may additionally furnish enhanced diagnostic information. For instance, contrast nanoparticles targeted to molecular fibrin at a clot site on a vessel wall have been created [44], potentially allowing differentiation between vulnerable and stable atherosclerotic plaques. Similarly prognostically valuable information regarding disruptions of the blood-brain barrier (BBB) may be visualized with MRI as a result of delivering contrast-enhancing nanoparticles for specific binding to the affected site [130]. In recent advances, MRI has proven to be useful in combination with optical approaches by using a dual fluorochrome technique to image protease activity *in vivo* for applications in precise imaging of, for instance, tumor size and depth [85].

13.3. SPECIFIC THERAPEUTIC APPLICATIONS OF HYBRID NANODEVICES

As with any novel therapy, nanobiotechnological therapeutic strategies will involve economic risks for their producers, and until they are well-understood, they will entail clinical risks. Consequently, their earliest applications will likely be in highly prevalent conditions for which current therapeutic modalities do not fully satisfy patient need. Here three areas will be discussed that appear to be promising early opportunities for hybrid nanotherapeutic devices: cancer, cardiovascular disease, and immunization.

13.3.1. Hybrid Nanotherapeutic Devices in Oncology

Cancer is an immensely expensive disease state for which current therapies often fail to provide desirable outcomes, and sometimes exhibit catastrophic side-effects. Moreover, the biology, chemistry, and physics of cancer provide opportunities that might be uniquely addressed with the novel properties of nanobiotechnological approaches.

13.3.1.1. Nanodevices to Exploit Novel Tumor Architectures Relative to healthy tissues, tumors are chaotic structures exhibiting vast structural heterogeneity spatially and temporally [3, 4, 74, 131, 191]. This heterogeneity of tumor tissues engenders conditions that can be exploited by nanoscale therapeutic devices. For instance, abnormally low vascular integrity and limited lymphatic drainage in tumors lead to the Enhanced Permeability and Retention effect (EPR [35–37, 122, 124]) that can be exploited to deliver nanoscale materials. EPR tumor targeting does not depend on biological affinity reagents but rather on the enhanced ability of high molecular weight materials to extravasate in tumors, and the diminished lymphatic clearance characteristic of tumors allows nanomaterials to be preferentially deposited and retained in tumor tissue. EPR has been applied to delivery of cytotoxins to tumoral sites, while minimizing systemic exposure, and provides the basis for a growing class of polymer therapeutics [35–37, 122, 124]. Systemic exposure is generally minimized by tuning the size of the nanodelivery device so that it is higher than the extravasation limits of healthy vasculature, but smaller than the size limit for excretion via the kidney. Effectively, these systems exploit competition between accumulation of cytotoxins in tumors and excretion in urine.

Figure 13.9 shows that conjugation of therapeutics to polymers/particles to produce delivery complexes that are subject to EPR can enhance tumoral accumulation by multiple orders of magnitude relative to free cytotoxins [35–37, 122, 124]. EPR allows more intensive therapy of tumors by widening the effective therapeutic window for cytotoxins, and can produce significantly enhanced clinical outcomes [35–37, 122, 124]. EPR strategies are widely applicable: almost any particulate material falling within the appropriate size range will partition into tumor tissue via EPR.

PK1 (Figure 13.10) is an early polymer therapeutic which exploits EPR [35–37, 122]. PK1 is composed of a polymeric backbone (HPMA, hydroxyl polymethacrylamide) that targets desired sites as a result of its specific nanoscale size (by EPR effect), and a cytotoxin (doxorubicin) covalently linked by a peptide. The device exploits both EPR targeting and intrinsic triggering. PK1 itself exhibits very low toxicity, and the full toxicity of doxorubicin is manifest when individual doxorubicin molecules are released from the PK1 complex. The peptide linker between doxorubicin and HPMA corresponds to a protease overexpressed in

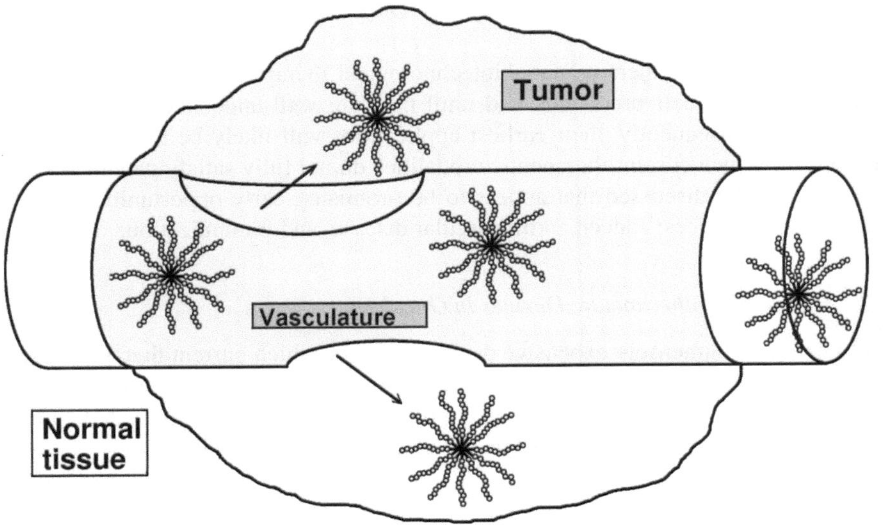

FIGURE 13.9. Therapeutic nanodevices can be specifically delivered to tumors by virtue of their size [35–37, 122]. An illustration of the relatively low integrity of tumor vasculature is depicted here in which therapeutic nanodevices (in this case, micelles carrying cytotoxins) cannot leave (i.e., extravasate from) the vasculature of normal tissue, but the vasculature of tumors is sufficiently leaky to allow specific delivery to tumoral interstitial spaces. This is the basis of size-dependent targeting to tumors by the enhanced permeability and retention, or EPR, effect.

the target tumor. Consequently, PK1 accomplishes both tumor-specific delivery of cytotoxin as well as preferential release of active toxin in the tumor.

It is unnecessary for therapeutics to be covalently linked complexes to be delivered via EPR. Drug delivery vehicles can be engineered to interact with blood constituents non-covalently so as to bring the apparent size of the complex into a range that allows preferential tumor deposition. Indeed, this is the delivery mechanism of the oldest EPR-exploiting therapeutic that is clinically applied, SMANCS (S-Methacryl-neocarzinostatin [122]). EPR-driven partitioning of non-covalent SMANCS-serum albumin complexes into tumors implies that other non-covalent complexes (e.g., the multi-dendrimer device of Figure 13.2, were it already assembled prior to administration) could also target to tumors via EPR. Indeed, generation of controlled-size delivery complexes by conjugation of therapeutics (see Figure 13.10) are able to improve tumoral accumulation considerably.

A more recent method for nanodevice-driven tumor targeting and destruction entails a combination of the use of an external triggering strategy, BNCT (boron neutron-capture therapy, described above), with EPR properties. EPR is used to target diseased tissue with transferrin-conjugated PEG liposomes filled with boron isotope, while BNCT is employed for preferential termination of disease sites using neutron irradiation [124].

13.3.2. Nanotherapeutics for Cardiovascular Applications

Theoretically, every cell in the body is accessible to the bloodstream via the vasculature. However, opsonization and immune clearance, as well as targeting and triggering of

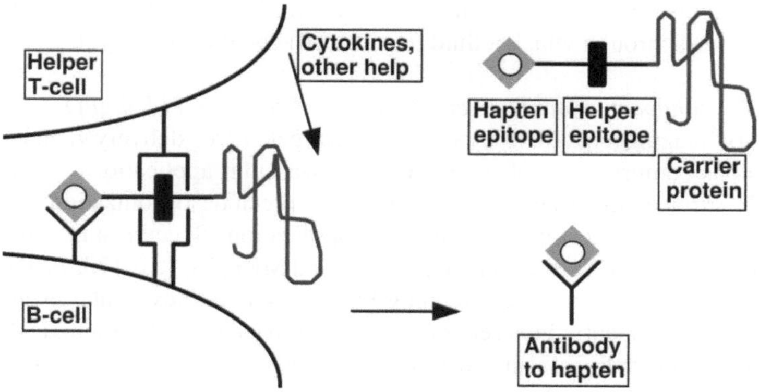

FIGURE 13.10. Antibody responses [75] are efficaciously triggered by antigens that contain both B-cell epitopes (diamond) and helper T-cell epitopes (rectangle). The B-cell epitope binds a receptor on the B-cell surface (the Y shape shown interacting with the B-cell epitope) that is essentially a membrane-bound form of the antibody species that particular B-cell can generate. The peptidic T-helper epitope engages another receptor complex on the B-cell surface (class II MHC, the two part receptor schematically depicted as interacting with the T-helper epitope, see text). When presented in the context of the class II MHC complex of the B-cell, the T-helper epitope can be recognized by a cognate T-cell receptor on the helper T-cell surface (shown as a receptor on the T-cell surface interacting with the T-helper epitope/class II MHC complex). In response to recognition of its appropriately presented, cognate T-helper antigen, the helper T-cell provides multiple stimulatory signals that cause the B-cell to proliferate, secrete antibody and differentiate. This is a classical hapten-carrier configuration of an immunogen [75], in which a synthetic B-cell epitope is the hapten (in our cases, a PAMAM dendrimer), and the protein covalently linked to the hapten is the carrier, providing the T-helper epitope. Note that the antibody produced recognizes *only* the B-cell epitope, and *not* the T-helper epitope. The T-helper epitope also need not be present for antibody recognition of carrier epitopes, nor is its presence necessary to maintain the ongoing antibody response once antibody production is triggered. Note that whether the T-helper epitope remains covalently linked to the B-cell epitope during antigen presentation (here, the intact carrier protein is shown linked to the B-cell epitope) is debatable. Proximity of the B-cell eptiope/receptor complex and T-helper epitope/class II MHC complex on the B-cell surface may be sufficient to trigger a robust antibody reaction.

therapeutics, persist as fundamental concerns that must be addressed to facilitate broad application of nanotherapeutic devices to cardiovascular disease. A brief overview is presented in this section of hybrid nanotherapeutic devices used in cardiovascular applications.

13.3.2.1. Tissue Engineering: Cardiovascular Applications Synthetic or semi-synthetic artery substitutes are highly valuable because functional, immunologically compatible arteries for coronary bypass are not always readily obtainable. Current tissue engineering methods (commonly microfabrication-based) entail synthesis of three-dimensional, porous scaffolds that facilitate adhesion, growth, and proliferation of seeded cells to create an operational vessel. Such approaches permit molecular-level control of cellular adhesion and propagation by control of surface topography, surface chemistry, and decoration with morphogenic and proliferative signaling ligands. One potentially powerful application of controlled cellular growth is to help develop artificial capillaries to remediate vascular disease ([10]; Moldovan, unpublished) through techniques such as electrospinning, in which polymers are spun as fibers ranging from micron-level diameters to less than 100nm [84]. Micromachined silicon or polymer channels of the same proportions as capillaries have shown promise as effective instruments for controlling endothelial cell growth for the formation of tubes through which a fluid (e.g., blood) could eventually flow.

13.3.2.2. Nanoparticulate Delivery Agents for Therapy and Imaging Targeted contrast agents for imaging (e.g., MRI, x-ray, etc.) and targeted drug delivery vehicles to manage vascular lesions comprise two of the major cardiovascular applications of nanoparticles. Nanoparticle surfaces must be suitably tailored to preclude opsonization and clearance and to drive binding at desired sites in either application. Specific imaging of unstable atherosclerotic plaques via targeted nanoparticles and MRI [73, 151, 170, 196] might eventually provide minimally invasive screening for incipient cardiovascular events. Coupled with appropriate therapeutic interventions, these approaches might dramatically diminish the morbidity and mortality associated with these events, and diminish the need for surgical intervention. Furthermore, hybrid nanotherapeutic devices have been designed to embrace the roles of both contrast and drug delivery (Figure 13.2). Multifunctional, single-platform particles for cardiovascular therapy signify the cutting edge in cardiovascular medicine, and are being constantly improved for maximal circulation time, targeting efficiency, degradability, and biocompatibility [73, 155, 167, 180].

13.3.3. Hybrid Nanotherapeutics and Specific Host Immune Responses

Immune responses that are aimed at immunogens/antigens are a pervasive feature of the responses of vertebrates to exposure to foreign macromolecules, proteins, particles, and organisms [11, 62, 75]. In particular, undesired antibody responses are the single most important obstacle to conquer for launching clinical use of recombinant proteins. Conversely, undesired cellular immune responses are key mediators of disorders that are often devastating for patients, whether the responses are directed to self (in autoimmune disease) or to foreign antigens (e.g., in transplant rejection). As with protein therapeutics, host immune responses to nanotherapeutic devices are a critical determinant of their effectiveness. There is thus substantial potential benefit to any strategy (including biotechnological strategies) that permits controlled manipulation of immune responses, either to augment desirable responses, as in vaccines, or to mitigate injurious responses.

13.3.3.1. Immunology Basics In general, specific immune responses are directed to individual biological molecules or structures and are classified as either cellular responses, mediated by cytotoxic T-lymphocytes (CTLs), or humeral responses, mediated by soluble proteins secreted by B-cells (antibodies). The effector notwithstanding (CTLs or antibodies), specific responses are directed to individual molecular characteristics of antigens termed epitopes. There exist three categories of epitopes: cytotoxic T-cell (CTL) epitopes, helper T-cell epitopes, and B-cell epitopes. The former two epitope categories are peptides which must be presented on cell surfaces in the context of the class I (in the case of CTL epitopes for CTL responses) or class II (in the case of T-helper epitopes, engaged in both antibody and CTL responses) major histocompatibility complex (MHC) molecules. B-cell epitopes include synthetic molecules as well as peptides and are a chemically and structurally heterogeneous group of epitopes. In the case of both class I and class II, the antigen-presenting complexes are relocated to the cell surface following loading (charging) with epitopes. The process of movement of molecules between different cellular locations (termed trafficking) is, by necessity, strictly regulated by and driven by specific cellular proteins, systems, and structures. These intricate antigen processing and presentation systems may be adjusted and exploited to offer novel opportunities in vaccine design.

13.3.3.2. Particulate Vaccines and Hybrid Nanodevice Design for Vaccination Vaccines are therapeutics that modulate immune responses (antibody, CTL responses, or both), most often safely to produce long-term desired immunity to disease or pathogens. Recent developments in immunobiology have exposed opportunities for nanobiological vaccine design [88, 133, 157, 162] and enhancement in synthetic vaccine half-lives [192]. For instance, it is now apparent that antigen-presenting cells (APCs) prefer to take up particulate antigens with diameters in the hundreds of nanometers to micrometer range. Particulate vaccine formulations can be orders of magnitude more immunogenic than concentration-matched doses of the same antigens in soluble forms and extensive effort is now applied to optimizing the formulation for maximal APC uptake and MHC presentation [88, 133, 157, 162].

Numerous proteins and peptides now known that mediate the trafficking of carried materials to specific intracellular sites are being exploited by nanobiological design strategies ([129]; Mrsny et al., 2001). These proteins and peptides are isolated from pathogens and biological toxins (which use them to invade and/or colonize host cells). Control of trafficking allows epitopes to be differentially shunted into either the MHC class I or class II presentation pathways, and would afford a degree of control of the specific response induced (e.g., whether CTLs are induced, whether T-cell help is available for antibody or CTL responses). Vaccines present an opportunity for multilevel targeting: both intercellularly, to the desired APCs and intracellularly, to desired APC cellular compartments. As we have seen, such targeting needs can be accommodated by nanobiotechnological approaches. Hybrid nanotherapeutic design strategies [6, 48, 50, 98, 110] can often combine expertise in polymer chemistry, physics, nanomaterials, mechanics, chemistry, and cell, molecular, and immunobiology to produce highly efficacious therapeutic vaccines nanodevices.

13.4. CONCLUSIONS

Expediency has driven the preliminary formulation of the nanobiological approach to device construction [95–99]. At the time, no technology was accessible to construct wholly synthetic nanocomponents with activities comparable to those existing in proteins. Use of biological macromolecules in nanodevices was a stopgap innovation intended to serve until robust nanodesign and synthetic capability became available. In the intervening years, there have been relatively modest improvements in nanocomponent assembly and creation, but incorporation of functional biomolecules frequently persists as the only practical approach to construct specific functional nanostructures. Therapeutic nanodevices could represent a special case that will always be best served by incorporation of biomolecules, regardless of advances in materials science, microfabrication, and polymer chemistry. Biomolecules can often provide the necessary interface between hybrid synthetic devices and biological systems (e.g., patients).

Biomedical nanotechnology is unlike many traditional biomedical engineering endeavors in the intensity of its dependence on basic (and emerging) biology. The relative intricacy and low definition of biological systems sometimes make traditional engineers uneasy, but these characteristics are presently an unavoidable fact of life for anyone hoping to engineer indwelling, nanoscale delivery and therapeutic devices. In fact, successful therapies as a rule interact with biological systems in a size appropriate manner. Disease states involving aberrant behavior of nanoscale components (like proteins or nucleic acids) often must be addressed with dimensionally appropriate therapeutics. The drive to miniaturization is also supported by the need for non-invasive therapeutic interventions. At some point, the distinction between materials science and pharmacology becomes blurred.

Another notable attribute of biomedical nanotechnology is its highly interdisciplinary nature. The field clearly requires expertise in materials science, diverse aspects from the engineering disciplines (e.g., electrical, mechanical, chemical, materials, and computer engineering, etc.), physics, polymer chemistry, biotechnology, specific disease states, etc. In part, this is because the orthodoxy of the field has yet to be formed and accepted. It is not known which approaches will prove to be most practical in any given set of applications. Still, the required breadth of experience is seldom resident in single investigators, so the field will likely remain a highly collaborative endeavor for the foreseeable future.

We believe that hybrid nanodevices have the potential to revolutionize health care in many arenas. Particular opportunities in oncology, cardiovascular disease, and vaccination have been highlighted that seem to be excellent candidates for nanobiological therapeutic intervention. However, we have not commented on the substantial broader implications of nanobiological device design in other disease states and beyond medicine. While that topic is a suitable subject for another article (or book), it seems likely that approaches being developed to support various platform therapeutics will be applicable to construction of other consumer nanodevices. The ultimate limits of this technological approach are certainly not immediately obvious, but it seems likely that humanity will ultimately use methods currently emerging for building hybrid therapeutic nanodevices in other economically and scientifically important endeavors.

ACKNOWLEDGEMENTS

The authors gratefully acknowledge many friends and colleagues whose input helped shape this article. We acknowledge the support services of Anita Bratcher in manuscript preparation and the artistic stylings of Vladimir Marukhlenko for the figures incorporated in the manuscript.

REFERENCES

[1] W. Arap, W. Haedicke, M. Bernasconi, R. Kain, D. Rajotte, S. Krajewski, M. Ellerby, R. Pasqualini, and E. Ruoslahti. Targeting the prostate for destruction through a vascular address. *Proc. Natl. Acad. Sci. U.S.A.*, 99, 2002a.

[2] W. Arap, M. Kolonin, M. Trpel, J. Lahdenranta, M. Cardo-Vila, R. Giordano, P.J. Mintz, P. Ardelt, V. Yao, C. Vidal, L. Chen, A. Flamm, H. Valtanen, L.M. Weavind, M.E. Hicks., R. Pollock, G.H. Botz, C.D. Bucana, E. Koivunen, D. Cahil, P. Troncosco, K.A. Baggerly, R.D. Pentz, K.-A. Do, C. Logothetis, and R. Pasqualini. Steps towards mapping the human vasculature by phage display. *Nature Med.*, 8:121–127, 2002b.

[3] D.F. Baban and L.W. Seymour. Control of tumor vascular permeability. *Adv. Drug Deliv. Rev.*, 34:109–119, 1998.

[4] J. Baish, Y. Gazit, D. Berk, M. Nozue, and L.T. Baxter. Role of tumor vascular architecture in nutrient and drug delivery: an invasion percolation-based network model. *Microvasc. Res.*, 51:327–346, 1996.

[5] J.R. Baker Jr. Therapeutic nanodevices. In S.C. Lee and L. Savage (eds.) *Biological molecules in nanotechnology: the convergence of biotechnology, polymer chemistry and materials science.* IBC Press, Southborough, MA, pp. 173–183, 1998.

[6] J.R. Baker Jr., A. Quintana, L. Piehler, M. Banazak-Holl, D. Tomalia, and E. Racka. The synthesis and testing of anti-cancer therapeutic nanodevices. *BMMD.*, 3:61–69, 2001.

[7] R. Barreiro-Iglesias, L. Bromberg, M. Temchenko, T.A. Hatton, A. Concheiro, and C. Alvarez-Lorenzo. Solubilization and stabilization of camptothecin in micellar solutions of pluronic-g-poly(acrylic acid) copolymers. *J. Contr. Rel.*, 97: 537–549, 2004.

[8] P.N. Bartlett and Y. Astier. Microelectrochemical enzyme transistors. *Chem. Comm.*, 2000.

[9] G. Biddlecombe, Y. Gun'ko, J. Kelly, S. Pillai, J. Coey, M. Ventatesan, and A. Douvalis. Preparation of magnetic nanoparticles and their assemblies using a new Fe(II) alkoxide precursor. *J. Mater. Chem.*, 11:2937–2939, 2001.

[10] J. Borenstein, H. Terai, K. King, E. Weinberg, M. Kaazempour-Mofrad, and J. Vacanti. Microfabrication technology for vascularized tissue engineering. *Biomed. Microdev.*, 4:167–175, 2002.

[11] F. Breitling and S. Dubel. *Recombinant Antibodies*. John Wiley & Sons, London, 1998.

[12] M. Brown and R. Semelka. MRI: *Basic Principles and Applications*. John Wiley and Sons, Inc., New York, 1999.

[13] D.M. Brown, M. Pellecchia, and E. Ruoslahti. Drug Identification through in vivo Screening of Chemical Libraries. *ChemBioChemical*, 5:871–875, 2004.

[14] C. Brunner, K.-H. Ernst, H. Hess, and V. Vogel. Lifetime of biomolecules in polymer-based hybrid nanodevices. *Nanotechnol.*, 15:S540–S548, 2004.

[15] W.D. Callister Jr. *Material Science and Engineering: An Introduction*. John Wiley & Sons, Inc., New York, 1997.

[16] C.N. Campbell. How far are we from detecting single bioconjugation events? In S.C. Lee and L. Savage, (eds.) Biological Molecules in Nanotechnology: the Convergence of Biotechnology, polymer chemistry and materials science. IBC Press, Southborough, MA. pp. 163–171, 1998.

[17] R.C. Carlisle, T. Etrych, S.S. Briggs, J.A. Preece, K. Ulbrich, and L.W. Seymour. Polymer-coated polyethylenimine/DNA complexes designed for triggered activations by intracellular reduction. *J. Gene Med.*, 6:337–344, 2004.

[18] G. Cevc. Lipid vesicles and other colloids as drug carriers on the skin. *Adv. Drug Del. Rev.*, 56:675–711, 2004..

[19] B.-X. Chen, S.R. Wilson, M. Das, D.J. Coughlin, and B.F. Erlanger. Antigenicity of fullerenes: antibodies specific for fullerenes and their characteristics. *Proc. Natl. Acad. Sci. U.S.A.*, 95:10809–10813, 1998.

[20] W.B. Choi, E. Bae, D. Kang, S. Chae, B.-h. Cheong, J.-h. Ko, E. Lee, and W. Park. Aligned carbon nanotubes for nanoelectronics. *Nanotechnology*, 15:S512–S516, 2004.

[21] D.J.A. Crommelin and R.D. Sindelar. *Pharmaceutical Biotechnology*. Harwood Academic Publishers, Amsterdam, 1997.

[22] J. Cumings and A. Zetti. Low-friction nanoscale linear bearing realized from multiwall carbon nanotubes. *Science*, 289:602–604, 2000.

[23] J. Davies. Aminoglycoside-aminocyclitol antibiotics and their modifying enzymes. In V. Lorian (ed.) *Antibiotics in Laboratory Medicine*. Williams and Wilkins, Baltimore. pp. 474–489, 1980.

[24] S.S. Davis. Biomedical applications of nanotechnology-implications fordrug targeting and gene therapy. *Trends Biotechnol.*, 15:217–224, 1997.

[25] C. Dekker. Carbon nanotubes as molecular quantum wires. *Phys. Today*:22–28, 1999.

[26] L.M. Demers, D.S. Ginger, S.-J. Park, Z. Li, S.-W. Chung, and C.A. Mirkin. Direct patterning of modified oligonucleotides on metals and insulatos by dip-pen nanolithography. *Science*, 296:1836–1838, 2002.

[27] I.G. Denisov, Y.V. Grinkova, A.A. Lazarides, and S.G. Sligar. Directed Self-Assembly of Monodisperse Phospholipid Bilayer Nanodiscs with Controlled Size. *J. Am. Chem. Soc.*, 126:3477–3487, 2004.

[28] W.A. Denny. The role of hypoxia activated prodrugs in cancer therapy. *The Lancet Oncol.*, 1:25–29, 2000.

[29] T. Desai, D. Hansford, L. Kulinsky, A. Nashat, G. Rasi, J. Tu, Y. Wang, M. Zhang, and M. Ferrari. Nanopore technology for biomedical applications. *BMMD*, 21:11–40, 1999.

[30] T.A. Desai, W.H. Chu, J.K. Tu, G.M. Beattie, A. Hayek, and M. Ferrari. Microfabricated immunoisolating biocapsules. *Biotechnol. Bioeng.*, 57:118–120, 1998.

[31] E. Di Fabrisio, A. Nucara, M. Gentili, and R. Cingolani. Design of a beamline for soft and deep lithography on third generation synchrotron radiation source. *Rev. Sci. Instrum.*, 70:1605–1613, 1999.

[32] W.U. Dittmer and F.C. Simmel. Chains of semiconductor nanoparticles templated on DNA. *Appl. Phys. Lett.*, 85:633–635, 2004.

[33] K.E. Drexler. Engines of Creation: the Coming era of Nanotechnology. Anchor Books, New York, 1986.

[34] H.Y. Du, M. Olivo, B.K.H. Tan, and B.H. Bay. Photoactivation of hypericin downregulates glutathione S-transferase activity in nasopharyngeal cancer cells. *Cancer Lett.*, 207:175–181, 2004.

[35] R. Duncan. Drug targeting: where are we now and where are we heading? *J. Drug Targ.*, 5:1–4, 1997a.

[36] R. Duncan. Polymer therapeutics for tumor specific delivery. *Chem. Ind.*, 7:262–264, 1997b.

[37] R. Duncan, S. Gac-Breton, R. Keane, Y.N. Sat, R. Satchi, and F. Searle. Polymer-drug conjugates, PDEPT and PELT: basic principles for design and transfer from the laboratory to clinic. *J. Cont. Rel.*, 74:135–146, 2001.

[38] G.M. Dykes. Dendrimers: A review of their appeal and applications. *J. Chem. Tech. Biotech.*, 76:903–918, 2001.

[39] A. Elster, S. Handel, and A. Goldman. *Magnetic Resonance Imaging: A Reference Guide and Atlas*. J.B. Lippincott Co., Philadelphia, 1997.

[40] M. Essier and E. Ruoslahti. Molecular specialization of breast vasculature: A breast homing phage displayed peptide binds to aminopeptidase P in breast vasculature. *Proc. Natl. Acad. Sci. U.S.A.*, 99:2252–2257, 2002.

[41] S. Fawell, J. Seery, Y. Daikh, C. Moore, L.L. Chen, B. Pepinsky, and J. Barsoum. Tat-mediated delivery of heterologous proteins to cells. *Proc. Natl. Acad. Sci. U.S.A.*, 91:664–668, 1994.

[42] S. Fernandez-Lopez, H.-S. Kim, E.C. Choi, M. Delgado, J.R. Granja, A. Khasanov, K. Kraehenbuehl, G. Long, D.A. Weinberger, K.M. Wilcoxen, and M. Ghardiri. Antibacterial agents based on the cyclic D,L-alpha-peptide architecture. *Nature*, 412:452–455, 2001.

[43] M. Ferrari and J. Liu. The engineered course of treatment. *Mech. Eng.*, 123:44–47, 2001.

[44] S. Flack, S. Fischer, M. Scott, R. Fuhrhop, J. Allen, M. McLean, P. Winter, G. Sicard, P. Gaffney, S. Wickline, and G. Lanza. Novel MRI contrast agent for molecular fibrin. *Circulation*, 104:1280–1285, 2001.

[45] T. Gardner, C.R. Cantor, and J.J. Collins. Construction of a genetic toggle switch in *Escherichia coli*. *Nature*, 403:339–342, 2000.

[46] I.E. Gentle, D.P.De Souza, and M. Baca. Direct Production of Proteins with N-Terminal Cysteine for Site-Specific Conjugation. *Bioconj. Chem.*, 15:658–663, 2004.

[47] M. Gerard, A. Chaubey, and B.D. Malhotra. Applications of conducting polymers to biosensors. *Biosens. Bioelectron.*, 17:345–349, 2002.

[48] M.E. Gillogly, N.L. Kallinteris, M. Xu, J.V. Gulfo, R.E. Humphreys, and J.L. Murray. li-Key/Her-2/neu MHC class-II antigenic epitope vaccine peptide for breast cancer. *Cancer Immunol. Immunother.*, 53:490–496, 2004.

[49] D.S. Ginger, H. Zhang, and C.A. Mirkin. The Evolution of Dip-Pen Nanolithography. *Ang. Chem. Int. Ed.*, 43:30–45, 2003.

[50] D.S. Goldin, C.A. Dahl, K.L. Olsen, L.H. Ostrach, and R.D. Klausner. Biomedicine. The NASA-NCI collaboration on biomolecular sensors. *Science*, 292:443–444,2001.

[51] T.J. Golentz, K. Klimpel, S. Leppla, J.M. Keith, and J.A. Berzofsky. Delivery of antigens to the MHC class I pathway using bacterial toxins. *Hum. Immunol.*, 54:129–136, 1997.

[52] R. Gordon. Computer controlled evolution of diatoms: design for a compustat. *Nova Hedwigia*, 112:215–219, 1996.

[53] T.R. Groves, D. Pickard, B. Rafferty, N. Crosland, D. Adam, and G. Schubert. Maskless electron beam lithography: propects, progress and challenges. *Microelect. Eng.*, 61:285–293, 2002.

[54] A. Guiseppi-Elie, C. Lei, and R.H. Baughman. Direct electron transfer of glucose oxidase on carbon nanotubes. *Nanotechnology*, 13:559–564, 2002.

[55] A. Gupta, D. Akin, and R. Bashir. Single virus particle mass detection using microresonators with nanoscale thickness. *Appl. Phys. Lett.*, 84:1976–1978, 2004.

[56] M. Guthold, R. Superfine, and R. Taylor. The rules are changing: force measurements on single molecules and how they relate to bulk reaction kinetics and energies. *BMMD*, 3:9–18, 2001.

[57] U.O. Hafeli. Magnetically modulated therapeutic systems. *Int. J. Pharm.*, 277:19–24, 2004.
[58] D. Hallahan, L. Geng, S. Qu, C. Scarfone, T. Giorgio, E. Donnelly, X. Gao, and J. Clanton. *Cancer Cell.*, 3:63, 2003.
[59] K. Hamad-Schifferli, J.J. Schwartz, A.T. Santos, S. Zhang, and J.M. Jacobson. Remote electronic control of DNA hybridization through inductive coupling to an attached metal nanocrystal antenna. *Nature*, 415:152–155, 2002.
[60] T. Hamouda, A. Myc, B. Donovan, A.Y. Shih, J.D. Reuter, and Jr. J.R. Baker. A novel surfactant nanoemulsion with a unique non-irritant topical antimicrobial activity against bacteria, enveloped viruses and fungi. *Microbiol. Res.*, 156:1–7, 2001.
[61] C.V. Harding and R. Song. Phagocytic processing of exogenous particulate antigens by macrophages for presentation by class I MHC molecules. *J. Immunol.*, 152:4925–4933, 1994.
[62] E. Harlow and D. Lane. *Antibodies: A Laboratory Manual*. Cold Spring Harbor Press, Cold Spring Harbor, 1989.
[63] J.M. Harris, N.E. Martin, and M. Modi. Pegylation: a novel process for modifying pharmacokinetics. *Clin. Pharmacokin.*, 40:539–551, 2001.
[64] J. Hasty, F. Isaacs, M. Dolnik, D. McMillen, and J.J. Collins. Designer gene networks: towards fundamental cellular control. *Chaos*, 11:107–220, 2001.
[65] M.J. Heller. Utilization of synthetic DNA for molecular electronic and photonic-based device applications. In S.C. Lee and L. Savage (eds.), *Biological Molecules in Nanotechnology: the Convergence of Biotechnology, Polymer Chemistry and Materials Science*. IBC Press, Southborough, MA. 59–66, 1998.
[66] G.T. Hermanson. Bioconjugate chemistry. Academic Press, San Diego, 1996.
[67] A.S. Hoffman and P.S. Stayton. Bioconjugates of smart polymers and proteins: synthesis and applications. *Macromol. Symp.*, 207:139–152, 2004.
[68] B. Ilic, H.G. Craighead, S. Krylov, W. Senaratne, C. Ober, and P. Neuzil. Attogram detection using nanoelectromechanical oscillators. *J. Appl. Phys.*, 95:3694–3703, 2004.
[69] J.A. Hoffman, P. Laakkonen, K. Porkka, M. Bernasconi, and E. Ruoslahti. In T. Clackson, H. Lowman (eds.), *Phage Display: A Practical Approach*. Oxford University Press, Oxford, 2004b.
[70] D. Hogemann, L. Josephson, R. Weissleder, and J. Basilion. Improvement of MRI probes to allow efficient detection of gene expression. *Bioconjugate Chem.*, 11:941–946, 2000.
[71] D.J. Hornbaker, S.-J. Kahng, S. Mirsa, B.W. Smith, A.T. Johnson, E.J. Mele, D.E. Luzzi, and A. Yazdoni. Mapping the one-dimensional electronic states of nanotube peapod structures. *Science*, 295:828–831, 2002.
[72] G.A. Husseini, G.D. Myrup, W.G. Pitt, D. Christensen, and N.Y. Rapoport. Factors affecting acoustically triggered release of drugs from polymeric micelles. *J. Cont. Rel.*, 69:43–52, 2000.
[73] F.A. Jaffer and R. Weissleder. Molecular Imaging of the Cardiovascular System. *Circ. Res.*, 94:433, 2004.
[74] R.K. Jain. Delivery of molecular and cellular medicine to tumors. *J. Cont. Rel.*. 53:49–67, 1998.
[75] C.A. Janeway, P. Travers, M. Walport, and J.D. Capra. *Immunobiology*. Elsevier Science, London, 1999.
[76] S.-Y. Jang, M. Marquez, and G.A. Sotzing. Rapid Direct Nanowriting of Conductive Polymer via Electrochemical Oxidative Nanolithography. *J. Amer. Chem. Soc.*, 126:9476–9477, 2004.
[77] L. Jelinski. Biologically related aspects of nanoparticles, nanostructured materials and nanodevices. In R.W. Siegel, E. Hu, and M.C. Roco (eds.), *Nanostructure Science and Technology*. Kluwer Academic Publishers, Dordrecht. 113–130, 1999.
[78] P. Jendelova, V. Herynek, J. DeCroos, K. Glogarova, B. Andersson, M. Hajek, and E. Sykova. Imaging the fate of implanted bone marrow stromal cells labeled with superparamagnetic nanoparticles. *Magn. Res. Med.*, 50:767–776, 2004.
[79] C. Jianrong, M. Yuqing, H. Nongyue, W. Xiaohua, and L. Sijiao. Nanotechnology and Biosensors. *Biotech. Adv.*, 22:505–518, 2004.
[80] M.-C. Jones and J.-C. Leroux. Polymeric micelles-a new generation of colloidal drug carriers. *Eur. J. Pharma. Biopharma.*, 48:101–111, 1999.
[81] B.K. Kay et al. *Phage Display of Peptides and Proteins*. Academic Press, San Diego, 1996.
[82] D. Kehagias, A. Gouliamos, V. Smyrniotis, and L. Vlahos. Diagnostic efficacy and safety of MRI of the liver with superparamagnetic iron oxide particles (SH U 555 A). *J. MRI.*, 14, 2001.
[83] S.B.H. Kent. Building proteins through chemistry: total chemical synthesis of protein molecules by chemical ligation of unprotected protein segments. In S.C. Lee and L. Savage (eds.), *Biological Molecules in Nanotechnology: the Convergence of Biotechnology, Polymer Chemistry and Materials Science*. IBC Press, Southborough, MA. pp. 75–92, 1998.

[84] S. Kidoaki, I.K. Kwon, and T. Matsuda. Mesoscopic spatial designs of nano- and microfiber meshes fortissue-engineering matrix and scaffold based on newly devised multilayering and mixing electrospinning techniques. *Biomaterials.*, 26:37–46, 2005.

[85] M.F. Kircher, R. Weissleder, and L. Josephson. A Dual Flourochrome Probe for Imaging Proteases. *Bioconj. Chem.*, 15:242–248, 2004.

[86] I. Kleps, A. Angelscu, R. Valisco, and D. Dascalu. New micro and nanoelectrode arrays for biomedical applications. *BMMD.*, 3:29–33, 2001.

[87] M. Kolonin, R. Pasqualini, and W. Arap. Molecular addresses in blood vessels as targets for therapy. *Curr. Opin. Chem. Biol.*, 5:308–313, 2001.

[88] M. Kovasovics-Bankowski, K. Clark, B. Benacerraf, and K.L. Rock. Efficient major histocompatibility complex class I presentation of exogenous antigen upon phagocytosis by macrophages. *Proc. Natl. Acad. Sci. U.S.A.*, 90:4942–4946, 1993.

[89] J. Kreuter, D. Shamenkov, V. Petrov, P. Ramge, K. Cychutek, C. Koch-Brandt, and R. Alyautdin. Apolipoprotein-mediated transport of nanoparticle-bound drugs across the blood-brain barrier. *J. Drug Target.*, 10:317–325, 2002.

[90] R.P. Lanza, S.J. Sullivan, and W.L. Chick. Perspectives in diabetes. Islet transplantation with immunoisolation. *Diabetes.*, 41:1503–1510, 1992.

[91] K.-B. Lee, S.-J. Park, C.A. Mirkin, J.C. Smith, and M. Mrksich. Protein nanoarrays generated by dip-pen nanolithography. *Science*, 295:1702–1705, 2002a.

[92] J.H. Lee, K.H. Yoon, K.S. Hwang, J. Park, S. Ahn, and T.S. Kim. Label free novel electrical detection using micromachined PZT monolithic thin film cantilever for the detection of C-reactive protein. *Biosens. Bioelect.*, 20:269–275, 2004.

[93] L.J. Lee. BioMEMS and micro-/nano-processing of polymers-an overview. *Chinese Journ. Chem. Eng.*, in press.

[94] S.C. Lee, K. Bhalerao, and M. Ferrari. Object-Oriented Design Tools for Supramolecular Devices and Biomedical Nanotechnology. *Ann. N.Y. Acad. Sci.*, 1013:1–14, 2004.

[95] S.C. Lee. Biotechnology for nanotechnology. *Trends Biotechnol.*, 16:239–240, 1998a.

[96] S.C. Lee. Engineering the protein components of nanobiological devices. In S.C. Lee and L. Savage, (eds.), *Biological Molecules in Nanotechnology: the Convergence of Biotechnology, Polymer Chemistry and Materials Science*. IBC Press, Southborough, MA. pp. 67–74, 1998b.

[97] S.C. Lee. How a molecular biologist can wind up organizing nanotechnology meetings. In S.C. Lee and L. Savage (eds.), *Biological Molecules in Nanotechnology: the Convergence of Biotechnology, Polymer Chemistry and Materials Science*. IBC Press, Southborough, MA. p. iv, 1998c.

[98] S.C. Lee. The nanobiological strategy for construction of nanodevices. In S.C. Lee and L. Savage (eds.), *Biological Molecules in Nanotechnology: the Convergence of Biotechnology, Polymer Chemistry and Materials Science*. IBC Press, Southborough, MA. pp. 3–14, 1998d.

[99] S.C. Lee. A biological nanodevice for drug delivery. In *National Science and Technology Council*. IWGN Workshop Report: Nanotechnology Research Directions. International Technology Research Institute, World Technology Division. Kluwer Academic Publishers, Baltimore. pp. 91–92, 2000.

[100] S.C. Lee. Antibody responses to nanomaterials. In Nanospace 2001: Exploring Interdisciplinary Frontiers. Institute for Advanced Interdisciplinary Research, Houston, TX, 2001a.

[101] S.C. Lee. Dendrimers in nanobiological devices. In D. Tomalia and J. Frechet (eds.), Dendrimers and Other Dendritic Polymers. John Wiley & Co., London. pp. 548–557, 2001b.

[102] S.C. Lee, R. Ibdah, C.D. van Valkenburgh, E. Rowold, A. Donelly, A. Abegg, J. Klover, S. Merlin, and J. McKearn. Phage display mutagenesis of the chimeric dual cytokine receptor agonist myelopoietin. *Leukemia*, 15:1277–1285, 2001a.

[103] S.C. Lee, M.S. Leusch, V.A. Luckow, and P. Olins. Method of producing recombinant viruses in bacteria. In US Patent and Trademark Office 5,348,886, USA, 1994.

[104] S.C. Lee, R. Parthasarathy, and K. Botwin. Protein-polymer conjugates: synthesis of simple nanobiotechnological devices. *Polymer Preprints*, 40:449–450, 1999.

[105] S.C. Lee, R. Parthasarathy, K. Botwin, D. Kunneman, E. Rowold, G. Lange, J. Zobel, T. Beck, T. Miller, and C.F. Voliva. Antibodies reponses to dendrimers: strategies and implications. *PMSE*, 84:824–825, 2001b.

[106] S.C. Lee, R. Parthasarathy, K. Botwin, D. Kunneman, E. Rowold, G. Lange, J. Zobel, T. Beck, T. Miller, and C.F. Voliva. Humeral immune responses to polymeric nanomaterials. In C. Carraher and G. Swift (eds.), Functional Condensation Polymers. Kluwer Publishers, New York. pp. 31–41, 2002b.

[107] S.C. Lee, R. Parthasarathy, T. Duffin, K. Botwin, T. Beck, G. Lange, J. Zobel, D. Jansson, D. Kunneman, E. Rowold, and C.F. Voliva. Antibodies to PAMAM dendrimers: reagents for immune detection, assembly and patterning of dendrimers. In D. Tomalia and J. Frechet (eds.), Dendrimers and Other Dendritic Polymers. John Wiley & Co., London. 559–566, 2001c.

[108] S.C. Lee, R. Parthasarathy, T. Duffin, K. Botwin, T. Beck, G. Lange, J. Zobel, D. Kunneman, E. Rowold, and C.F. Voliva. Recognition properties of antibodies to PAMAM dendrimers and their use in immune detection of dendrimers. *BMMD*, 3:51–57, 2001d.

[109] G. Lemieux and C. Bertozzi. Chemoselective ligation reactions with proteins, oligosaccharides and cells. *Trends Biotechnol.*, 16:506–512, 1998.

[110] A. Lenarczyk, T.T.T. Le, D. Drane, J. Malliaros, M. Pearse, R. Hamilton, J. Cox, T. Luft, J. Gardner, and A. Suhrbier. ISCOM® based vaccines for cancer immunotherapy. *Vaccine*, 22:963–974, 2004.

[111] L. Leoni and T. Desai. Micromachined biocapsules for cell-based sensing and delivery. *Adv. Drug Deliv. Rev.*, 56:211–229, 2004.

[112] M.S. Leusch, S.C. Lee, and P.O. Olins. A novel host-vector system for direct selection of recombinant baculoviruses (bacmids) in *Escherichia coli. Gene.*, 160:191–194, 1995.

[113] J.A. Lewis and G.M. Gratson. Direct writing in three dimensions. *Materials Today*, 7:32–39, 2004.

[114] K.C.P. Li, S.D. Pandit, S. Guccione, and M.D. Bednarski. Molecular Imaging Applications in Nanomedicine. *BMMD*, 6:113–116, 2004.

[115] Z. Li, Y. Zhang, P. Fullhart, and C.A. Mirkin. Reversible and Chemically Programmable Micelle Assembly with DNA Block-Copolymer Amphophiles. *Nano Lett.*, 4:1055–1058, 2004.

[116] Y.X. Liang and T.H. Wang. A double-walled carbon nanotube field-effect transistor using the inner shell as its gate. *Physica E*, 23:232–236, 2004.

[117] G. Lin, R. Palmer, K. Pister, and K. Roos. Miniature heart cell force transducer system implemented in MEMS technology. *IEEE Trans. Biomed. Engin.*, 48:996–1006, 2001.

[118] C. Liu and Z. Zhang. Size-dependent superparamagnetic properties of Mn spinel ferrite nanoparticles synthesized from reverse micelles. *Chem. Mater.*, 13:2092–2096, 2001.

[119] P. Lockman, R. Mumper, M. Khan, and D. Allen. Nanoparticle technology for drug delivery across the blood-brain barrier. *Drug Devel. Indust. Pharm.*, 28:1–13, 2002.

[120] V.A. Luckow, S.C. Lee, G.F. Barry, and P.O. Olins. Efficient generation of infectious recombinant baculoviruses by site-specific, transposon-mediated insertion of foreign DNA into a baculovirus genome propagated in *E. coli. J. Virol.*, 67:4566–4579, 1993.

[121] Z. Ma and S. Taylor. Nucleic acid triggered catalytic drug release. *Proc. Natl. Acad. Sci. U.S.A.*, 97:11159–11163, 2000.

[122] H. Maeda, T. Sawa, and T. Konno. Mechanism of tumor-targeted delivery of macromolecular drugs, including the EPR effect in solid tumor and clinical overview of the prototype polymeric drug SMANCS. *J. Cont. Rel.*, 74:47–61, 2001.

[123] R. Martel, T. Schmidt, H.R. Shea, T. Hertel, and P. Avouris. Single and multiwall carbon nanotube field-effect transistors. *Appl. Phys. Lett.*, 73:2447–2449, 1998.

[124] K. Maruyama, O. Ishida, S. Kasaoka, T. Takizawa, N. Utoguchi, A. Shinohara, M. Chiba, H. Kobayashi, M. Eriguchi, and H. Yanagie. Intracellular targeting of sodium mercaptoundecahydrododecaborate (BSH) to solid tumors by transferrin-PEG liposomes, for boron neutron-capture therapy (BNCT). *J. Cont. Rel.*, 98:195–207, 2004.

[125] P. Meers. Enzyme-activated targeting of liposomes. *Adv. Drug Deliv. Rev.*, 53:265–272, 2001.

[126] M. Mendoca-Dias, E. Gaggelli, and P. Lauterbur. Paramagnetic contrast agents in nuclear magnetic resonance medical imaging. *Sem. Nuclear Med.*, 13:364–376, 1983.

[127] R.C. Merkle. Biotechnology as a route to nanotechnology. *Trends Biotechnol.*, 17:271–274, 1999.

[128] G. Morón, G. Dadaglio, and C. Leclerc. New tools for antigen delivery to the MHC class I pathway. *Trends Immunol.*, 25:92–97, 2004.

[129] R. Mrsny, A.L. Daughtery, M. Mckee, and D. Fitzgerald. Bacterial toxins as tools for mucosal vaccination. *Drug Disc. Today*, 7:247–257, 2002.

[130] L.L. Muldoon, M.A. Pagel, R.A. Kroll, S. Roman-Goldstein, R.S. Jones, and E.A. Neuwelt. A physiological barrier distal to the anatomic blood-brain barrier in a model of transvascular delivery. *AJNR Am. Journ. Neuroradial.*, 20:217–222, 1999.

[131] J. Nagy. Competition and natural selection in a mathematical model of cancer. *Bull. Mathem. Biol.*, 66:663–687, 2004.

[132] N.S. Nahman, T. Drost, U. Bhatt, T. Sferra, A. Johnson, P. Gamboa, G. Hinkle, A. Haynam, V. Bergdall, C. Hickey, J.D. Bonagura, L. Brannon-Pappas, J. Ellison, A. Mansfield, S. Shiwe, and N. Shen. Biodegradable microparticles for in vivo glomerular targeting: implications for gene therapy of glomerular disease. *BMMD*, 4:189–196, 2002.

[133] M. Nielson, C. Lundegaard, P. Worning, C.S. Hvid, K. Lamberth, S. Buus, S. Brunak, and O. Lund. Improved prediction of MHC class I and class II epitopes using a novel Gibbs sampling approach. *Bioinformatics*, 20:1388–1397, 2004.

[134] H. Nobuto, T. Sugita, T. Kubo, S. Shimose, Y. Yasunaga, T. Murakami, and M. Ochi. Evaluation of systemic chemotherapy with magnetic liposomal doxorubicin and a dipole external electromagnet. *Inter. J. Cancer*, 109:627–635, 2004.

[135] C. Nolte-Ernsting, G. Adam, A. Bucker, S. Berges, A. Bjornerud, and R. Gunther. Abdominal MR angiography performed using blood pool contrast agents. *AJR*, 171:107–113, 1998.

[136] T. Nozaki, R. Ogawa, L.B. Feril, G. Kagiya, H. Fuse, and K. Kondo. Enhancement of ultrasound-mediated gene transfection by membrane modification. *J. Gene Med.*, 5:1046–1955, 2003.

[137] R. Offord and K. Rose. Multicomponent synthetic constructs. In S.C. Lee and L. Savage (eds.), *Biological Molecules in Nanotechnology: the Convergence of Biotechnology, Polymer Chemistry and Materials Science*. IBC Press, Southborough, MA. pp. 93–105, 1998.

[138] T. Olafsen, C.-W. Cheung, P.J. Yazaki, L. Li, G. Sundaresan, S.S. Gambhir, M.A. Sherman, L.E. Williams, J.E. Shively, A.A. Raubitschek, and A.M. Wu. Covalent disulfide-linked anti-CEA diabody allows site-specific conjugation and radiolabeling for tumor targeting applications. *Prot. Eng. Des. Selec.*, 17:21–27, 2004.

[139] M. Ollsen, B. Persson, and L. Salford. Ferromagnetic particles as contrast agent in T2 NMR imaging. *Magn. Reson. Imag.*, 4:437–440, 1986.

[140] C. Oseroff, A. Sette, P. Wentworth, E. Celis, A. Maewal, C. Dahlberg, J. Fikes, R.T. Kubo, R.W. Chestnut, H.M. Grey, and J. Alexander. Pools of lapidated HTL-CTL constructs prime for multiple HBV and HCV CTL epitope responses. *Vaccine*, 16:823–833, 1998.

[141] O.C. Ozlem and V. Hasirci. UV-induced drug release from photoactive REV sensitized by suprofen. *J. Contr. Rel.*, 96:85–96, 2004.

[142] R. Roy and J.M. Kim. Cu(II)-Self-assembling bipyridyl-glycoclusters and dendrimers bearing the Tn-antigen cancer marker: syntheses and lectin binding properties. *Tetrahedron*, 59:3881–3893, 2003.

[143] H. Paajanen and M. Kormano. Contrast agents in magnetic resonance imaging. In J. Skucas (ed.), *Radiographic Contrast Agents*. Aspen Publishers, Rockville, MD. pp. 377–406, 1989.

[144] A. Pal, S.K. Ghosh, K. Esumi, and T. Pal. Reversible Generation of Gold Nanoparticle Aggregates with Changeable Interparticle Interactions by UV Photoactivation. *Langmuir*, 20:575–578, 2004.

[145] S.-J. Park, A.A. Lazarides, J.J. Storhoff, L. Pesce, and C.A. Mirkin. Structural Characterization of Oligonucleotide-Modified Gold Nanoparticle Networks Formed by DNA Hybridization. *J. Phys. Chem. B* (in press), 2004.

[146] J. Parkinson and R. Gordon. Beyond micromachining: the potential of diatoms. *Trends Biotechnol.*, 17:190–196, 1999.

[147] A.K. Patri, I.J. Majoros, and Jr, J.R. Baker. Dendritic polymer macromolecular carriers for drug delivery. *Curr. Opin. Chem. Biol.*, 6:466–471, 2002.

[148] K. Pearce, B. Cunningham, G. Fuh, T. Teeri, and J.A. Wells. Growth hormone affinity for its receptor surpasses the requirements for cellular activity. *Biochemistry*, 38:81–89, 1999.

[149] P.P. Pompa, A. Biasco, R. Cingolani, R. Rinaldi, M. Ph. Verbeet, G.W. Canters. Structural stability of protein monolayers in air. *Phys. Rev. E*, 69:032901–032904, 2004.

[150] D. Portet, B. Denizot, E. Rump, J.-J. Lejeune, and P. Jallet. Nonpolymeric coatings of iron oxide colloids for biological use as magnetic resonance imaging contrast agents. *J. Coll. Inter. Sci.*, 238:37–42, 2001.

[151] H. Quick, J. Debatin, and M. Ladd. MR imaging of the vessel wall. *Eur. Radiol.*, 12:889–900, 2002.

[152] A. Quintana, E. Raczka, L. Piehler, I. Lee, A. Myc, I. Majoros, A.K. Patri, T. Thomas, J. Mule, and Jr., J.R. Baker. Design and function of a dendrimer-based therapeutic nanodevice targeted to tumor cells through the folate receptor. *Pharma. Res.*, 19:1310–1316, 2002.

[153] R. Raiteri, M. Grattarola, H.-J. Butt, and P. Skladl. Micromechanical cantilever-based biosensors. *Sensors. Actuat. B*, 79:115–126, 2001.

[154] K. Rajagopal and J.P. Schneider. Self-assembling peptides and proteins for nanotechnological applications. *Curr. Opin. Struc. Biol.*, 14:480–486, 2004.

[155] D. Ranney. Biomimetic transport and rational drug delivery. *Biochem. Pharm.*, 59:105–114, 2000.

[156] V. Raso, M. Brown, and J. McGrath. Intracellular triggering with low pH-triggered bispecfic antibodies. *J. Biol. Chem.*, 272:27623–27628, 1997.

[157] S. Raychanduri and K.L. Rock. Fully mobilizing host defense: building better vaccines. *Nature Biotech.*, 16:1025–1031, 1998.

[158] F. Rety, O. Clement, N. Siauve, C.-A. Cuenod, F. Carnot, M. Sich, A. Buisine, and G. Frija. MR lymphography using iron oxide nanoparticles in rats: pharmacokinetics in the lymphatic system after intravenous injection. *J. MRI*, 12:734–739, 2000.

[159] R. Rinaldi and R. Cingolani. Electronic nanodevices based on self-assembled metalloproteins. *Physica. E*, 21:45–60, 2004.

[160] J.A. Ritter, A.D. Ebner, K.D. Daniel, and K.L. Stewart. Application of high gradient magnetic separation principles to magnetic drug targeting. *J. Magnetism Magn. Mater.*, 280:184–201, 2004.

[161] J.C. Roberts, M.K. Bhalgat, and R.T. Zera. Preliminary biological evaluation of polyamidoamine (PAMAM) starburst dendrimers. *J. Biomed. Mater. Res.*, 30:53–65, 1996.

[162] K. Rock, S. Gamble, and L. Rothstein. Presentation of exogenous antigen with class I major histocompatibility complex molecules. *Science*, 249:918–921, 1990.

[163] K. Rogers. Principles of affinity-based biosensors. *Mol. Biotechnol.*, 14:109–129, 2000.

[164] H.W. Rohrs and R.S. Ruoff. The use of carbon nanotubes in hybrid nanometer-scale devices. In S.C. Lee and L. Savage (eds.), *Biological Molecules in Nanotechnology: the Convergence of Biotechnology, Polymer Chemistry and Materials Science*. IBC Press, Southborough, MA. pp. 33–38, 1998.

[165] M. Ruegsegger and R. Marchant. Reduced protein adsorption and platelet adhesion by controlled variation of oligomaltose surfactant polymer coatings. *J. Biomed. Mater. Res.*, 56:159–167, 2001.

[166] E. Ruoslahti. Drug targeting to specific vascular sites. *Drug Disc. Today*, 7:1138–1143, 2002.

[167] E. Ruoslahti. Special delivery of drugs by targeting to tissue-specific receptors in vasculature. *Pharmaceutical News*, 7:35–40, 2000.

[168] A. Saghatelian, Y. Yokobayashi, K. Soltani, and M.R. Ghadiri. A chiroselective peptide replicator. *Nature*, 409:777–778, 2001.

[169] D. Satake, H. Ebi, N. Oku, K. Matsuda, H. Takao, M. Ashiki, and M. Ishida. A sensor for blood cell counters using MEMS technology. *Sensors Actuat. B—Chem.*, 83:77–81, 2002.

[170] S. Schmitz, M. Taupitz, S. Wagner, K.-J. Wolf, D. Beyersdorff, and D. Hamm. Magnetic resonance imaging of atherosclerotic plaques using supermagnetic iron oxide particles. *J. MRI*, 14:355–361, 2001.

[171] J. Schnorr, M. Taupitz, S. Wagner, H. Pilgrimm, J. Hansel, and B. Hamm. Age-related blood half-life of particulate contrast materials: experimental results with a USPIO in rats. *J. MRI*, 12:740–744, 2000.

[172] N.C. Seeman, J. Chen, Z. Zhang, B. Lu, H. Qiu, T.-J. Fu, Y. Wang, X. Li, J. Qi, F. Liu, L.A. Wenzler, S. Du, J.E. Mueller, H. Wang, C. Mao, W. Sun, Z. Shen, M.H. Wong, and R. Sha. A bottom-up approach to nanotechnology using DNA. In S.C. Lee and L. Savage (eds.), *Biological Molecules in Nanotechnology: the Convergence of Biotechnology, Polymer Chemistry and Materials Science*. IBC Press, Southborough, MA. pp. 45–58, 1998.

[173] A.H. Sehon. Suppression of antibody responses by conjugates of antigens and monomethoxypoly(ethylene glycol). In J.M. Harris (ed.), *Poly(ethylene glycol) Chmeistry*. Plenum, New York. pp. 139–151, 1992.

[174] P.D. Senter and C.J. Springer. Selective activation of anticancer prodrugs by monoclonal antibody-enzyme conjugates. *Adv. Drug Deliv. Rev.*, 53:247–264, 2001.

[175] G. Shi, M. Rouabhia, Z. Wang, L.H. Dao, and Z. Zhang. A novel electrically conductive and biodegradable composite made of polypyrrole nanoparticles and polylactide. *Biomaterials*, 25:2477–2488, 2004.

[176] D. Shu, L. Huang, and P. Guo. A simple mathematical formula for stoichiometry quantification of viral and nanobiological assemblage using slopes of log/log plot curves. *J. Virol. Meth.*, 115:19–30, 2004.

[177] P. Shum, J.-M. Kim, and D.H. Thompson. Phototriggering of liposomal delivery systems. *Adv. Drug Deliv. Rev.*, 53:273–284, 2001.

[178] Skotland, T.P. Sontum, and I. Oulie. In vitro stability analyses as a model for metabolism of ferromagnetic particles (Clariscan^{TM}), a contrast agent for magnetic resonance imaging. *J. Pharm. Biomed. Anal.*, 28:323–329, 2002.

[179] J. Song, Q. Cheng, S. Zhu, and R.C. Stevens. "smart" materials for biosensing devices:cell-mimicking supramolecular assemblies and colorometric detection of pathogenic agents. *BMMD*. 4:213–222, 2002.

[180] K. Soppimath, T. Aminabhavi, A. Kulkarni, and W. Rudzinski. Biodegradable polymeric nanoparticles as drug delivery devices. *J. Cont. Release*, 70:1–20, 2001.

[181] R. Spindler. PAMAM starburst dendrimers: designed nanoscopic reagents for biological applications. In S.C. Lee and L. Savage (eds.), *Biological Molecules in Nanotechnology: the Convergence of Biotechnology, Polymer Chemistry and Materials Science*. IBC Press, Southborough, MA. pp. 15–32, 1998.

[182] R.M. Stoltenberg and A.T. Woolley. DNA-Templated Nanowire Fabrication. *BMMD*, 6:105–111, 2004.

[183] L. Thunus and R. Lejeune. Overview of transition metal and lanthanide complexes as diagnostic tools. *Coord. Chem. Rev.*, 184:125–155, 1999.

[184] K.B. Thurmond II, H. Huang, and K.L. Wooley. Stabilized micellar structures in nanodevices. In S.C. Lee and L. Savage, (eds.), *Biological Molecules in Nanotechnology: the Convergence of Biotechnology, Polymer Chemistry and Materials Science*. IBC Press, Southborough, MA. pp. 39–43, 1998.

[185] L. Tiefenauer, G. Kuhne, and R. Andres. Antibody-magnetite nanoparticles: in vitro characterization of a potential tumor-specific contrast agent for magnetic resonance imaging. *Bioconjugate Chem.*, 4:347–352, 1993.

[186] D. Tomalia and H.M. Brothers II. Regiospecific conjugation to dendritic polymers to produce nanodevices. In S.C. Lee and L. Savage (eds.), *Biological Molecules in Nanotechnology: the Convergence of Biotechnology, Polymer Chemistry and Materials Science*. IBC Press, Southborough, MA. pp. 107–120, 1998.

[187] R.A. Tromp, S.S.G.E. van Boom, T.C. Marco, S. van Zutphen, G.A. van der Marel, H.S. Overkleeft, J.H. van Boom, and J. Reedijk. The β-glucuronyl-based prodrug strategy allows for its application on β-glucuronyl-platinum conjugates. *Bioorg. Med. Chem. Lett.*, 14:4273–4276, 2004.

[188] I. Uchegbu. Parenteral drug delivery: 1. *Pharma. Journal*. 263:309–318, 1999a.

[189] I. Uchegbu. Parenteral drug delivery: 2. *Pharma. Journal*. 263:355–359, 1999b.

[190] S. Uppuluri, D.R. Swanson, L.T. Peihler, J. Li, G. Hagnauer, and D.A. Tomalia. Core shell tecto(dendrimers). I. Synthesis and characterization of saturated shell models. *Adv. Mater.*, 12:796–800, 2000.

[191] P. Vaupel, D.K. Kelleher, and O. Thews. Modulation of tumor oxygenation. *Int. J. Rad. Oncol. Biol. Phys.*, 42:843–848, 1998.

[192] S. Vichier-Guerre, R. Lo-Man, V. Huteau, E. Deriaud, C. Leclerc, and S. Bay. Synthesis and immunological evaluation of an antitumor neoglycopeptide bearing a novel homoserine Tn antigen. *Bioorg. Med. Chem. Lett.*, 14:3567–3570, 2004.

[193] B.L. Viglianti, S.A. Abraham, C.R. Michelich, P.S. Yarmolenko, J.R. MacFall, M.B. Bally, and M.W. Dewhirst. In Vivo Monitoring of Tissue Pharmacokinetics of Liposome/Drug Using MRI: Illustration of Targeted Delivery. *Magn. Reson. Med.*, 51:1153–1162, 2004.

[194] H. Wang, H. Song, and V.C. Yang. A recombinant prodrug type approach for triggered delivery of streptokinase. *J. Cont. Rel.*, 59:119–122, 1999.

[195] R. Weissleder, G. Elizondo, J. Wittenberg, C. Rabito, H. Bengele, and L. Josephson. Ultrasmall superparamagnetic iron oxide: characterization of a new class of contrast agents for MR imaging. *Radiology*, 175:489–493, 1990.

[196] S. Wickline. Plaque characterization: surrogate markers or the real thing? *J. Amer. Coll. Cardio.*, 43:1185–1187, 2004.

[197] G. Wunderlich, T. Gruning, B.-R. Paulke, A. Lieske, and J. Kotzerke. ^{99m}Tc labelled model drug carriers—labeling, stability and organ distribution in rats. *Nucl. Med. Biol.*, 31:87–92, 2004.

[198] B.I. Yacobson and R.E. Smalley Fullerene nanotubes: C1,000,000 and beyond. *Amer. Scient.*, 85:324–337, 1997.

[199] Y. Zhang, C. Sun, N. Kohler, and M. Zhang. Self-Assembled Coatings on Individual Monodisperse Magnetite Nanoparticles for Efficient Intracellular Uptake. *BMMD*, 6:33–40, 2004.

14

Modeling Biomolecular Transport at the Nanoscale

A. T. Conlisk

Department of Mechanical Engineering, The Ohio State University, Columbus, Ohio 43210-1107

ABSTRACT

Biomolecular transport devices are now being used for drug development and delivery, single molecule manipulation, detection and transport and rapid molecular analysis. Many of these processes are illustrated by natural ion channels which are ion-selective nanoscale conduits in the body which allow nutrients in and waste products out. In this chapter we review the state of the art of modeling and computation of biomolecular transport in what we term synthetic ion nanochannels consisting of rectangular silicon channels for which the walls are negatively charged; we also consider the case where the walls are not charged. We consider computational techniques ranging from continuum models utilizing the Poisson-Nernst-Planck system to molecular dynamic simulations that allow tracking of individual molecules. Biomolecular transport can be modeled by incorporating hindered diffusion concepts and the methods are employed to predict the transport of albumin and glucose in silicon nanochannels. Brownian Dynamics and Molecular Dynamics methods represent techniques that must be used when continuum methods break down and these methods are also reviewed. It should be noted that the exact boundary between continuum methods and molecular simulation methods is not always clear.

14.1. INTRODUCTION

Transport of ionic and biomolecular species has become an important problem with applications to rapid molecular analysis, drug development and delivery, separation and

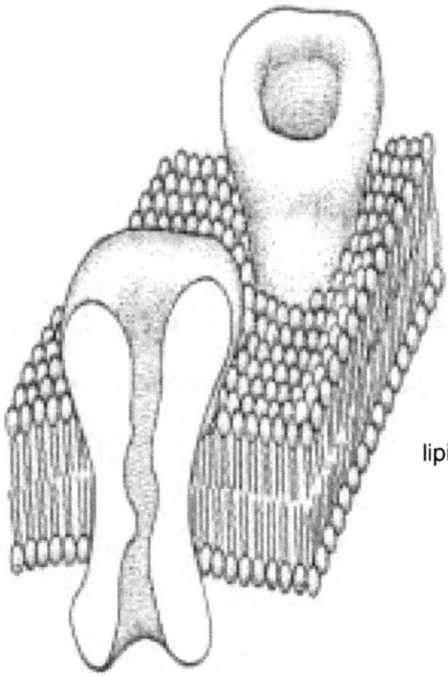

FIGURE 14.1. Image of ion channels within the lipid bilayer. From *(http://hebb.mit.edu/courses/8.515/lecture1/sld013.htm, 3/11/03)*

mixing and other problems. These synthetic devices operate in a manner similar to natural ion channels, although the parameter ranges of operation may be different. Natural ion channels play a crucial role in the transport of biofluids into and out of cells [1–3].

The basic units of all living organisms are cells. In order to keep the cells functioning properly a continuous flux of ions in and out of the cell and the cell components is required. The cell is surrounded by a plasma membrane which provides selective transfer of ions. The cell membrane is made up of a double layer of lipid molecules (lipid bilayer) in which proteins are embedded. The lipid molecules are made up of a charged polar end (spheres in Figure 14.1), oriented towards the surface of the membranes, which are attached to two nonpolar fatty acid chains (lines attached to the spheres in Figure 14.1). The polarity of the membrane makes it challenging for molecules to move in and out of cells and its components. Ions (Ca^{2+}, Cl^-, K^+, Na^+, H^+, Mg^{2+}, HCO_3^-, PO_4^{2-}) selectively move through the membrane via ion channels made of proteins. These protein channels are made up of a polypeptide chain with two polar regions connected by a nonpolar region which associates with the nonpolar region of the phospholipids membrane. The channel walls are charged and the ions are transported by a transmembrane voltage drop. Natural ion channels are roughly circular, although the cross-sectional area varies in the primary flow direction [2], synthetic channels are usually rectangular in cross-section and often much bigger than natural ion channels whose diameter is on the order of 1 *nm* or so.

Nanopumps which we term *synthetic ion channels* and used to transport fluids through nanochannels are essentially ion channels; these pumps may have walls that are charged or uncharged and may or may not have a voltage drop across them. A nanopump fabricated by iMEDD, Inc. of Columbus, Ohio(Sketch supplied by A. Boiarski) is depicted on

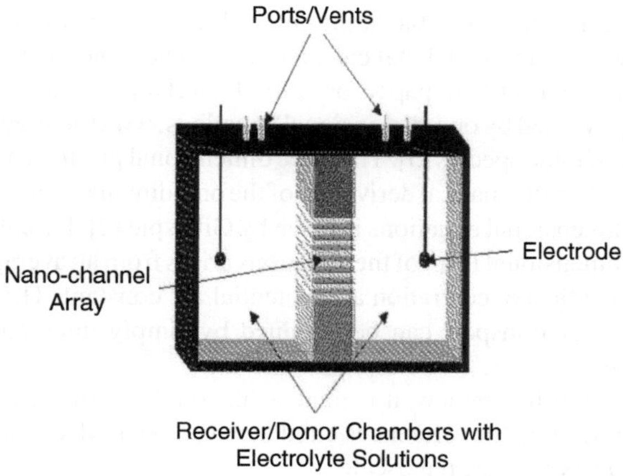

FIGURE 14.2. iMEDD, Inc of Columbus electroosmotic pump.

Figure 14.2. The channels are of dimension $W \sim 44\mu m$, $L \sim 3.5\mu m$ and channel heights $h \sim 4 - 50nm$. Thus the channels are nano-constrained in one dimension and are somewhat bigger than natural ion channels.

Synthetic ion channels are of interest because they can be used as pumps to deliver a drug, for example, to a target site. The iMEDD, Inc. nanochannel membrane is essentially an array of synthetic ion channels. In this chapter we discuss the basic principles of biomolecular transport in nanoscale channels with the particular application to drug delivery and biomolecular sensing. The simplest way of modeling biomolecular transport is to treat the biomolecule similar to an ionic species, its transport characterized by a diffusion coefficient and imbedded in a buffer solution such as Phosphate Buffered Saline (PBS). The strength of this approach is that continuum methods may be used; however a major weakness is that conformation and shape changes cannot be incorporated.

In very small channels, where the ratio of the biomolecular length scale to smallest channel dimension is of O(1), the diffusion coefficient is much smaller than what is measured in the bulk. Consequently, accurate rates of transport depend on the ability to determine these "hindered diffusion" coefficients and two models are discussed.

The characterization of the transport of a biomolecule as another species in a fluid mixture ignores the possibility of deformation of the molecule as it is transported into a channel or tube; this deformation is critical in some cases determining the selectivity properties of a given device. Thus continuum models which treat the biomoleule specifically have been developed and these methods are called Brownian Dynamics. Brownian Dynamics methods treat the solvent as a continuum and an explicit model of the biomolecule is developed. Finally, the use of molecular techniques such as Molecular Dynamics (MD) simulations in which both the solvent and solute are treated as individual molecules will also be presented.

Biomolecules are usually charged and the electrostatic interactions with the walls is complex. While many proteins are globular, they can deform significantly and the nature and precise determination of the shape of a protein is difficult to quantify. Thus modeling efforts often use only the overall charge and size of the molecule. Experimental studies have shown that once the ratio of the biomolecular length scale to smallest channel dimension is of a certain size, the molecule may even block the channel resulting in a sudden drop of current [4]. The author is aware of no model that can predict this phenomenon from first principles.

There are perhaps thousands of papers on natural ion channels and most of them assume that the transport is governed by one-dimensional equations, with the single dimension in the direction of the flow of ionic species [5]. The three dimensional problem has been considered by Cardenas *et al* [6]. A mathematical derivation of the one-dimensional governing equations from the full three-dimensional equations is given by Gillespie [2]. Essentially, Gillespie [2] shows that the one-dimensional form of the equations arises from an averaging process along given curves on which the concentration and potential are constant. These same equations describing biomolecular transport can be obtained by simply integrating over the cross section of the channel.

Given the breadth of this review, it is impossible to cite all of the relevant references; nevertheless, it is hoped that the references cited here can be used as a starting point for an in-depth study of transport at the nanoscale.

In this chapter we discuss the properties of natural and synthetic ion channels and review the computational methods used to evaluate transport through these channels. The focus of this review is on the three main classes of methods corresponding to the Poisson-Nernst-Planck system, on Brownian Dynamics methods and on Molecular Dynamics methods. After an initial discussion of the nature of natural ion channels, we discuss rectangular synthetic ion channels having one nanoconstrained dimension. These channels are appropriate for use in single molecule detection and analysis, and drug delivery.

Based on the continuum methods of analysis we then apply the results to the transport of albumin and glucose and compare the theoretical results with the experimental data from iMEDD. The comparisons are surprisingly good considering that the only physical parameter of the protein accounted for in the model is the globular size.

14.2. BACKGROUND

It is useful to begin the discussion with a review of the modeling procedures employed for natural ion channels since the theory of such models has been significantly improved in recent years. Natural ion channels are one class of membrane transport proteins and are narrow, water-filled tunnels, permeable to the few ions and molecules small enough to fit through the tunnels (approximately 10 Å in diameter) [7]. Ion channels are responsible for electrical signaling in nerves and muscles [8]. The ions responsible for the majority of nervous signaling are Ca^{2+}, Cl^{-}, K^{+}, and Na^{+}. The channels are responsive to different stimuli: a membrane potential change, a neurotransmitter or other chemical stimulus, a mechanical deformation, and more. The channel's response to these stimuli, called gating, corresponds to an opening or closing of the pore. The open pore has a selective permeability to ions, only allowing certain ions to flow through its electrochemical activity gradients. The ions flow at a very high rate, greater than 10^6 ions per second, with a single ion taking on the order of one microsecond to traverse the channel. Malfunctioning ion channels are responsible for the onset of cystic fibrosis, cholera, and may be the cause of schizoprenia and bipolar disorders.

According to Levitt [5], five major methods of modeling natural ion channels have appeared in the literature: molecular dynamics, three-dimensional Brownian dynamics, three-dimensional Poisson-Nernst-Planck (PNP), one-dimensional Brownian dyanamics, and the one-dimensional Poisson-Nernst-Planck approach. It has been suggested often in the literature that if the channel properties are known to an atomic level then the most direct method is molecular dynamic simulations, because they take into consideration the atomic structure of the water molecules and channel proteins. The weakness of this method is the computation time; a 1998 simulation of an acetylcholine channel took approximately one year for a single ion crossing [5]. This long computation time is due to the fact that molecular simulation can only describe nanoseconds of real time, which is not long enough for modeling ion movement through the entire channel which takes place on the microsecond scale. Molecular dynamic simulations can be simplified by using advanced computational techniques. These techniques include alchemical free energy perturbation and umbrella sampling.

In three-dimensional Brownian dynamics, the protein structure is held fixed and the water molecules within the natural ion channel channel are treated as a continuum. The solutes are treated discretely and the ion-ion interactions are included. According to Levitt [5], the best method of modeling natural ion channels is a combination of molecular dynamics and Brownian dynamics. The three-dimensional Poisson-Nernst-Planck approach simplifies the Brownian dynamic approach by averaging the electric field on an ion over all the possible positions of the other ions. This average electric field is calculated using Poisson's equation. The Poisson equation is coupled with the Nernst-Planck equation, which models ions undergoing Brownian motion in the electric field. This approach is much faster than Brownian dynamics because it replaces a sequence of time step calculations with a global numerical solution.

The use of Poisson-Nernst-Planck equations, a continuum theory, to model ion channels began in the early 1990s with Chen *et al.* [10] This model has been used to model Ca^{2+} channels [12, 13] as well as Gramicidin channels [15]. In addition to simply modeling the flow within the channel, these models usually incorporate the flow within the baths on each end of the channel [2, 16]. Non-ideal effects which occur at higher electrolyte concentration levels have recently been examined by Gillespie *et al.* [17] The disadvantage of this method is that the ion-ion interactions are reduced to a ion-mean field interaction. Recall that natural ion channels are around 1 nm in diameter and molecular effects can become important at $\sim 5nm$.

The three-dimensional Brownian Dynamics and Poisson-Nernst-Planck approaches can be further simplified by averaging the concentration over the cross section. By comparison with experimental data we will be able to quantify the accuracy of this approximation. It has been observed that in most circumstances this assumption does not generate much error, because it often cancels out the error introduced by uncertainties in atomic channel structure and this is the basis for the use of the Poisson-Nernst-Planck system to model flow in ion channels [2, 3, 9, 10, 14]. An extensive reading list on the use of MD simulations to model natural ion channels is given at the end of the paper by Roux [11].

14.3. GOVERNING EQUATIONS FOR SYNTHETIC ION CHANNELS IN THE CONTINUUM REGIME: THE POISSON-NERNST-PLANCK SYSTEM

The equations governing transport of a given species in an electrolyte buffer solution are given by the Navier-Stokes equations along with an electrical potential equation, and equations for the concentration of each of the species [18]. The flux of a given species in an electrolyte solution is due in general to Fickian diffusion, electrical migration and to bulk fluid motion. Thus most biomolecular transport studies begin with the expression for the dimensional flux of species A in the form

$$\vec{n}_A^* = -cD_A \nabla X_A + u_A z_A F X_A \vec{E}^* + cX_A \vec{u}^* \tag{14.1}$$

Here D_A is the diffusion coefficient, c the total concentration, X_A is the mole fraction of species A, which can be either an anion or a cation, $u_A = \frac{D_A}{RT}$ is the mobility, where R is the gas constant, T is the temperature, z_A is the valence, F is Faraday's constant, $\vec{E}^*$ is the total electric field and $\vec{u}^*$ is the mass average velocity of the fluid. In almost all cases of practical interest, the mixture is dilute in the electrolytes and so the diffusion coefficient can be approximated by its value for the solvent. The mole fraction distribution is governed by taking $\nabla\cdot$ of the flux equation (1) and the result is

$$\frac{\partial}{\partial y}\left(\frac{\partial D_A X_A}{\partial y}\right) + \epsilon_1^2 \frac{\partial}{\partial x}\left(\frac{\partial D_A X_A}{\partial x}\right) + \epsilon_2^2 \frac{\partial}{\partial z}\left(\frac{\partial D_A X_A}{\partial z}\right)$$
$$= z_A \left(\epsilon_1 \frac{\partial X_A E_x}{\partial x} + \frac{\partial X_A E_y}{\partial y} + \epsilon_2 \frac{\partial X_A E_z}{\partial z}\right) \tag{14.2}$$

Here we have assumed that the mole fractions are small and so the mixture is dilute; in this case the diffusion coefficient of an ionic species can be approximated by its value with respect to the solvent, water. Also we have not assumed that the diffusion coefficient is a constant; D_A is a scaled dimensionless diffusion coefficient, the scaling factor corresponding to say its value in the bulk. We will discuss the possible form of a variable diffusion coefficient later.

The dimensionless equations governing bulk fluid motion for a mixture containing two electrolytes in the absence of the nonlinear convective terms which at nanoscale will be small are given by

$$0 = -\epsilon_1 \epsilon^2 \frac{\partial p}{\partial x} + \beta \sum_{i=1}^{N} z_i X_i + \epsilon^2 \nabla^2 u \tag{14.3}$$

$$0 = -\epsilon^2 \frac{\partial p}{\partial y} + \Lambda \beta \epsilon^2 \frac{\partial \phi}{\partial y} \sum_{i=1}^{N} z_i X_i + \epsilon^2 \nabla^2 v \tag{14.4}$$

$$0 = -\epsilon_2 \epsilon^2 \frac{\partial p}{\partial z} + \Lambda \beta \epsilon_2 \epsilon^2 \frac{\partial \phi}{\partial z} \sum_{i=1}^{N} z_i X_i + \epsilon^2 \nabla^2 w \tag{14.5}$$

where the convective terms have been neglected since the Reynolds $Re = \frac{\rho U_0 h}{\mu}$ is small. Conservation of mass requires

$$\epsilon_1 \frac{\partial u}{\partial x} + \frac{\partial v}{\partial y} + \epsilon_2 \frac{\partial w}{\partial z} = 0 \tag{14.6}$$

Finally, the electrical potential is governed by

$$\epsilon^2 \left(\frac{\partial^2 \phi}{\partial y^2} + \epsilon_1^2 \frac{\partial^2 \phi}{\partial x^2} + \epsilon_2^2 \frac{\partial^2 \phi}{\partial z^2} \right) = -\beta \sum_{i=1}^{N} z_i X_i \tag{14.7}$$

where

$$\beta = 1 + \frac{c_3}{I}$$

where c_3 is the concentration of the solvent, $\epsilon = \frac{\lambda}{h}$, $\epsilon_1 = \frac{h}{L}$, $\epsilon_2 = \frac{h}{W}$ and $I = sum_i z_i^2 c_i)^{1/2}$ is the ionic strength. Here λ is the electric double layer thickness which will be present o near any charged surface and this quantity is defined by

$$\lambda = \frac{\sqrt{\epsilon_e RT}}{F I^{1/2}} \tag{14.8}$$

where F is Faraday's constant, ϵ_e is the electrical permittivity of the medium, c_i the concentrations of the electrolyte constituents, R is the gas constant, z_q is the valence of species i and T is the temperature. The electric double layer is the region near the charged surface where there is an excess of counterions which is balanced by the surface charge density [18].

The coordinates (x, y, z) are nondimensional; for example, $x = \frac{x^*}{L}$ and the scaling lengths in the three directions are (L, h, W) as depicted on Figure 14.3. Also (u, v, w) are the dimensionless velocities in each of the coordinate directions (x, y, z); for example $u = u^*/U_0$ where u^* is dimensional. The dimensionless pressure and potential

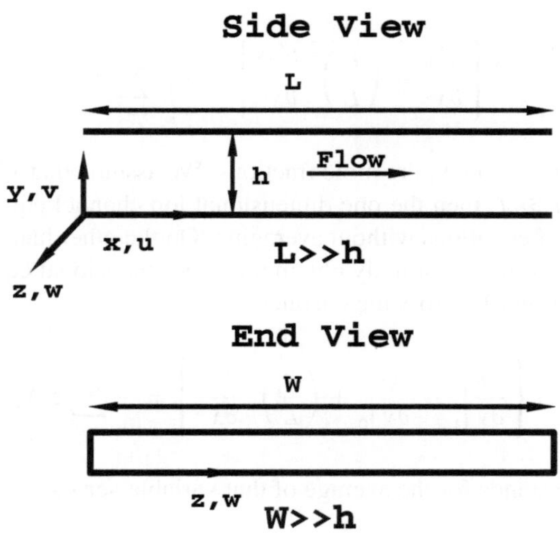

FIGURE 14.3. Geometry of the channel. Here it is only required that $h \ll W$ where W is the width of the channel; the length L in the primary flow direction can be of the order of the channel height or much smaller as depicted in the figure. u, v, w are the fluid velocities in the x, y, z directions.

are defined by [18]

$$p = \frac{p^* h}{\mu U_0}, \phi = \frac{\phi^*}{\phi_0}$$

and the velocity scale $U_0 = \frac{\epsilon_e E_0 \phi_0}{\mu}$ and $\phi_0 = \frac{RT}{F}$. The scaling of the pressure is typical for low Reynolds number problems.

In the general case this is a set of N+5 governing equations in the N+5 unknowns, velocity, potential and pressure, plus the N species mole fractions or concentrations. All of these equations are subject to boundary conditions. Generally Dirichlet or Neumann boundary conditions for the mole fractions and the potential and the velocity field is usually assumed to vanish at non-porous surfaces.

Often, the geometry of channels in applications allow approximate forms of the equations to be used in modeling. For example, in the case of electroosmotic pumps, rapid molecular analyzers and drug delivery systems, fabrication techniques do not permit nanoscale dimensions(under $50nm$) in more than one dimension. Thus the three dimensionality of the general problem can be reduced to one or two dimensions. How to do this is considered next.

14.4. THE ONE-DIMENSIONAL POISSON-NERNST-PLANCK EQUATIONS

Most synthetic ion channels considered for use in these applications are rectangular in cross-section as depicted in Figure 14.3. For very wide channels ($\epsilon_2 \ll 1$) the side walls have only a local effect. In this case, the potential through a very wide ion channel is controlled by the two-dimensional Poisson equation,

$$\epsilon^2 \left\{ \frac{\partial^2 \phi}{\partial y^2} + \left(\frac{h}{L}\right)^2 \frac{\partial^2 \phi}{\partial x^2} \right\} = -\beta \sum_{i=1}^{N} z_i X_i \tag{14.9}$$

where z_i are the valences and X_i are mole fractions. We assume that x is the flow direction and we note that if $h \gg L$ then the one dimensional ion channel equations emerge as an approximate system of equations without averaging. On the other hand if $h \sim L$ or $h \ll L$ then averaging is necessary to formally obtain the one-dimensional equations.

Integrating equation (9) across the channel,

$$\epsilon^2 \left\{ \left.\frac{\partial \phi}{\partial y}\right|_1 - \left.\frac{\partial \phi}{\partial y}\right|_0 + \left(\frac{h}{L}\right)^2 \frac{\partial^2 \bar{\phi}}{\partial x^2} \right\} = -\sum_{i=1}^{N} z_i \bar{X}_i \tag{14.10}$$

A bar over a variable stands for the average of that variable across the channel in y,

$$\bar{s} = \int_0^1 s dy \tag{14.11}$$

where s is any function. Equation (14.10) may be simplified to

$$\epsilon_1^2 \frac{\partial^2 \bar{\phi}}{\partial x^2} = -\sum_{i=1}^{N} z_i \bar{X}_i - 2\epsilon^2 \sigma_0 \tag{14.12}$$

where we redefine $\epsilon_1 = \frac{\lambda}{L}$ where λ is the electric double layer thickness. Also σ_0 is the surface charge density at $y = 0$

$$\sigma = \sigma_0 = -\frac{\partial \phi}{\partial y}(0) = \sigma_1$$

if the flow is symmetric about the centerline of the channel. Here σ is the dimensionless surface charge density and

$$\sigma = \frac{\sigma^*}{\frac{\epsilon_e RT}{F\lambda}}$$

σ^* is the dimensional surface charge density in C/m^2.

The ion concentrations within the channel are controlled by the Nernst-Planck equation, and for species A

$$\frac{\partial}{\partial y}\left(\frac{\partial X_A}{\partial y} + z_A X_A \frac{\partial \phi}{\partial y}\right) + \epsilon_1^2 \frac{\partial}{\partial x}\left(\frac{\partial X_A}{\partial x} + z_A X_A \frac{\partial \phi}{\partial x}\right) = 0 \tag{14.13}$$

where z_A is the valence of species A. Integrating equation (13) over y,

$$\left(\frac{\partial X_A}{\partial y} + z_A X_A \frac{\partial \phi}{\partial y}\right)\Bigg|_0^1 + \epsilon_1^2 \left(\frac{\partial^2 \bar{X}_A}{\partial x^2} + z_A \frac{\partial}{\partial x} \int_0^1 X_A \frac{\partial \phi}{\partial x} dy\right) = 0. \tag{14.14}$$

The first set of terms vanishes because there is no flux of species A at the walls $y = 0, 1$. The third term may be approximated by

$$\int_0^1 X_A \frac{\partial \phi}{\partial x} dy = \bar{X}_A \frac{\partial \bar{\phi}}{\partial x} \tag{14.15}$$

if the variables are not functions of y. This will be the case in most of the channel if $\frac{\lambda}{h} \ll 1$; that is the electric double layer is thin and it is readily shown that the core velocity is constant [18]. On the scale of the channel height, both X_A and ϕ are functions of y; however for small h/L, on the scale of the channel length, the channel height approaches zero, and from Taylor series considerations, both X_A and ϕ can be considered constant. Thus,

$$\frac{\partial}{\partial x}\left(\frac{\partial \bar{X}_A}{\partial x} + z_A \bar{X}_A \frac{\partial \bar{\phi}}{\partial x}\right) = 0. \tag{14.16}$$

For $\sigma_0 = 0$, these are the equations used by Barcilon *et al.* [9] Equation (14.16) is the Nernst-Planck equation for the species A and thus the system of equations corresponding to equations (14.12) and (14.16) is called the Poisson-Nernst-Planck system.

In the case when the walls of the channel are not charged, there is no electric double layer and the integrated equations will be exact since the potential and mole fraction distributions will be independent of y. In this case the ions and biomolecules move under the action of diffusion and electrical migration and if there is no voltage drop across the channel the process is called *electrophoresis*.

For these channels, the concentrations of the ions at the inlet and outlet are determined by the concentrations within the baths on either side of the membrane. We will consider a mixture of two ionic species plus a third species corresponding to the biomolecule. Therefore, the boundary conditions for the system are (dropping the bars for convenience)

$$X_A = X_{AL}, \quad X_B = X_{BL}, \quad X_C = X_{CL}, \quad \phi = V, x = 0 \tag{14.17}$$

$$X_A = X_{AR}, \quad X_B = X_{BR}, \quad X_C = X_{CR}, \quad \phi = 0, x = 1, \tag{14.18}$$

where the subscripts L and R refer to the left ($x = 0$) and right ($x = 1$) boundaries. It is useful to point out that averaging the continuity equation leads to the result that the average velocity is constant in x to maintain constant mixture flow rate.

14.5. HINDERED DIFFUSION CONCEPTS

The nature of the solution for the potential and the concentrations or mole fractions for the PNP system has been discussed in the literature [2, 3, 17]. As an ion or biomolecule approaches the entry to the small channel from the large bath, there is a resistance to motion from the walls. Thus the diffusion coefficient in the bath is likely to be much larger than that in the channel due to the proximity of the walls of the channel. Specifically, as a species enters the channel the diffusion coefficient should rapidly drop as shown on Figure 14.4.

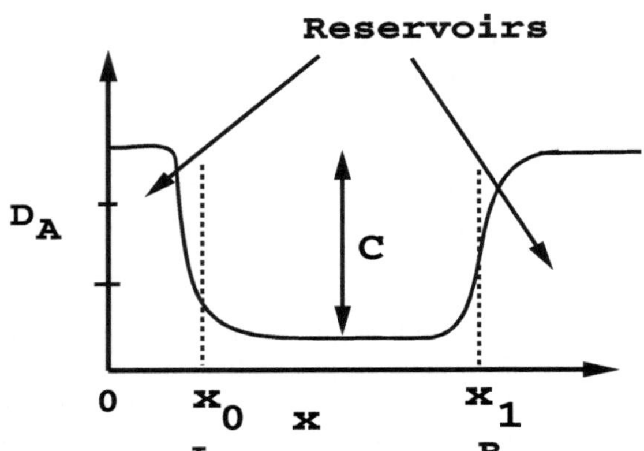

FIGURE 14.4. Sketch of the expected variation of the diffusion coefficient in as a function of distance from the upstream bath through the channel and into the downstream bath.

The diffusion coefficient should be constant in the bath and in the channel. The diffusion coefficient in the channel can be over an order of magnitude smaller in the channel than in the upstream and downstream reservoirs.

For dilute and moderately dense gases, excellent theories exist by which diffusion coefficients can be calculated from information about intermolecular forces [19]. However, no such theories exist for ordinary diffusion in dense gases and liquids, which results in numerous approximations and empirical relations.

An estimate of the diffusion coefficient is provided by the Nernst-Einstein equation. According to the Nernst-Einstein equation, the diffusion coefficient of a single particle or solute molecule A through a medium B may be described by [19]

$$D_{AB} = k_b T \left(\frac{m_A}{F_A} \right) \tag{14.19}$$

where m_A is the mobility of the component A, defined as the velocity scale divided by the drag force, and F_A is the drag between the solute and solvent molecules, $k_0 = R/N_A$ is Boltzmann's constant and N_A is Avogadro's number. The relation between force and velocity may be obtained for a rigid sphere moving in the creeping flow limit ($Re \ll 1$). If slip at the surface of the sphere is taken into account, then [20]

$$F_A = 6\pi \mu_B m_A a \left[\frac{2\mu_B + a\beta_{AB}}{3\mu_B + a\beta_{AB}} \right] \tag{14.20}$$

in which μ_B is the viscosity of the pure solvent, a is the radius of the diffusing particle, and β_{AB} is the coefficient of sliding friction. The coefficient of sliding friction β_{AB}, can be exploited in two limiting cases. In the case where there is no tendency for the fluid to stick at the surface of the diffusing particle (i.e. a hydrophobic particle in the case of a water solvent), then $\beta_{AB} = 0$, and equation (3.5) becomes $F_A = 4\pi \mu_B m_A a$. This slip limit is approached when a molecule diffuses through a medium consisting of molecules of comparable size [20] The other limiting case is for the no-slip condition of $\beta_{AB} = \infty$, which results in equation (14.20) reducing to $F_A = 6\pi \mu_B u_A a$; this is the case of a hydrophilic particle and in both these cases the result is known as Stokes' Drag Law. This would be the case for a sphere in a continuum or for a large spherical particle in a solvent of low relative molecular mass [20].

According to Peters [21], human serum albumin has a hydrophilic surface and is very soluble in water. Glucose molecules have polar hydroxyl groups (OH groups) that attract water molecules, which form a hydration shell[1]; hence, glucose is also hydrophilic. Therefore, both molecules should be described by the no-slip case of equation (14.20).

Substitution of equation (14.20) into equation (14.19) results in the following equation for the diffusion coefficient for the no-slip case:

$$D_A = \frac{k_b T}{6\pi \mu_B a} = \frac{RT}{6\pi N_A \mu_B a} \tag{14.21}$$

which is recognized as the Stokes-Einstein equation. The Stokes-Einstein equation has been shown to be fairly good for describing the diffusion of large spherical particles or

[1] http://staff.jccc.net/pdecell/chemistry/hydrophilic.html (04,2004)

TABLE 14.1. Ionic diameters and Diffusion coefficient in bulk. The symbol '*' denotes data from Cussler [29], '**' from Granicka, *et al.* [30], '***' from the web site. http://oto.wustl.edu/cochlea/model/diffcoef.htm.

Ion	$D_A(\text{x}10^{-5}cm^2/sec)$	Ionic Diameter(nm)
Li^+	1.03*	0.133
Na^+	1.33*	0.200
K^+	1.96*	0.276
Cs^+	2.06*	0.340
Mg^{2+}	0.71*	0.178
Ca^{2+}	0.79*	0.214
Cl^-	2.032*	0.362
H_2O	—	0.280
Albumin	0.061**	7
Glucose	0.94***	1
I_gG	0.04**	10

molecules in cases where the solvent appears to the diffusing species as a continuum [19]. The hydrodynamic theory further suggests that the shape of the diffusing species is very important. Tyrell and Harris [20] note that the validity of equation (14.21) is also restricted by the fact that the motion of only one diffusing particle is considered, and so the relationship can strictly only apply in the limit of infinite dilution.

The Stokes-Einstein equation assumes that the diffusing particles are isolated. That is, it does not take interactions with other particles or with boundaries such as walls into account. When a particle is diffusing near a wall, the Stokes drag will increase due to wall effects and the diffusion coefficient will be smaller than when the particle is far from the wall. In the small channels of interest in this work, the walls of the channel may affect transport. One value that can be used to quantify wall effects is the ratio of the diffusing particle radius a to its distance from the wall y_p. The expectation is that for finite a/y_p, wall effects will become important and the diffusion coefficient will decrease according to

$$D_H = CD_A \tag{14.22}$$

where D_H is the so-called hindered diffusion coefficient and $C < 1$ is some function of $\frac{a}{y_p}$. As the particle radius decreases compared with y_p, *i.e.* in the limit as $\frac{a}{y_p}$ becomes very small, the D_H should approach D_A as given in equation (14.21). Some representative ionic radii and diffusion coefficients are depicted on Table 14.1.

Several models have been presented in the literature that attempt to correct the diffusion coefficient for wall interactions. Lin, Yu, and Rice [22] examine hindered diffusion of an isolated uncharged sphere in a density-matched fluid confined between two parallel flat walls. The measured diffusion coefficients are used to test the predictions of the wall drag effect predicted by several approximate theories. The text by Happel and Brenner [23] examines the motion of a sphere moving relative to plane walls in the context of low Reynolds number hydrodynamics (see also Ref. 24). Levitt [25, 26] uses the pioneering work of Haberman and Sayre [28] in order to apply continuum fluid mechanics to the nonelectrolyte and volume flux across a membrane. The work of Haberman and Sayre [28] provides the important result of a series solution for the hindered diffusion coefficients for several special cases using the method of images; this is the solution on which the Levitt model [25, 26] is based. Unlike the previously discussed models [22, 23], the Levitt model is based on analysis of a biological system. Bevan and Prieve [27] have used experimental method called total internal reflection microscopy (TIRM) to deduce an average diffusion coefficient of a charged sphere in the presence of a wall experimentally.

The hindered diffusion models are of the form

$$D_H = C(a, y_p)D_A$$

where $C < 1$ and a is the molecular size, and y_p is the distance of the particle from the wall. In this work we assume that $y_p = h/2$. Here we discuss in detail the models of Levitt [25, 26] and Bevan and Prieve [27] because they seem to give the best results in comparison with experimental data as shown below. In the Levitt [25, 26] model the constant C for the hindered diffusion coefficient is given by

$$C = \frac{D_H}{D_{AB}} = \frac{1 - 2.1050\frac{a}{y_p} + 2.0865\left(\frac{a}{y_p}\right)^3 - 1.7068\left(\frac{a}{y_p}\right)^5 + 0.72603\left(\frac{a}{y_p}\right)^6}{S\left(\frac{a}{y_p}\right)} \tag{14.23}$$

where

$$S\left(\frac{a}{y_p}\right) = 1 - 0.75857\left(\frac{a}{y_p}\right)^5$$

The Bevan and Prieve model [27] based on their experimental measurements is depicted on Figure 14.5. A curve is given for the diffusion coefficient and a best-fit curve was developed after taking several points from the graph. The result is

$$C = \frac{D_H}{D_A} = \frac{4.2 + 3\left(\frac{a}{y_p}\right)}{6 + 9\left(\frac{a}{p_y}\right) + 2\left(\frac{a}{y_p}\right)^2} + 0.3 \tag{14.24}$$

Note that as $a/y_p \to 0$, the value of C approaches unity because the value of the hindered diffusion coefficient in this case will approach the Stokes-Einstein value of the diffusion coefficient.

The transport properties of the channels are determined by the diffusion coefficient and the potential within the channel. Thus as a first approximation the boundary layers at the entrance and exit of the channel can be neglected and the flux of solute may be determined from the values of the hindered diffusion coefficients in the interior of the channel along with the potential. We discuss the calculation of the potential next.

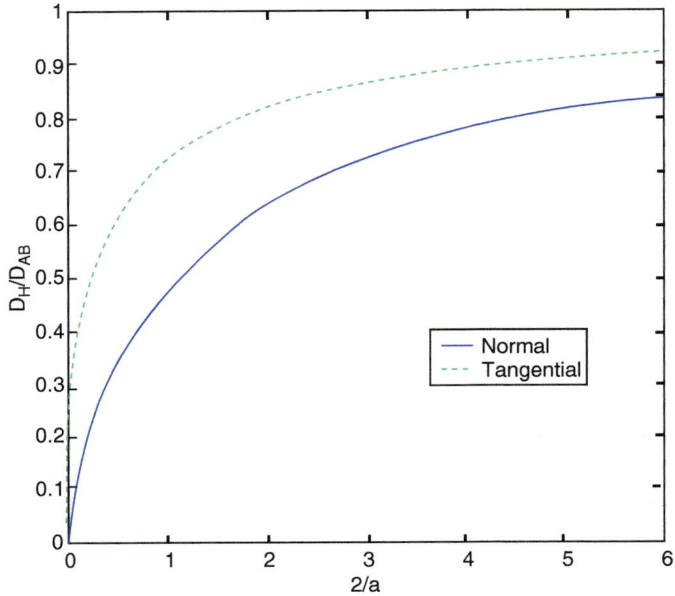

FIGURE 14.5. Hindered diffusion coefficient of Bevan and Prieve [27].

14.6. CALCULATING THE ELECTRICAL POTENTIAL

Gillespie [2] presents detailed results for a $(-1, 2, 1)$ electrolyte system. We may consider a Phosphate Buffered Saline (PBS) solution along with a protein, say albumin or glucose. In PBS, Na and Cl are the two most populous species and if we consider this case we have for the potential

$$\epsilon_1^2 \phi'' + X_A - X_B + z_C X_C = -2\epsilon^2 \sigma_0, \tag{14.25}$$

where $\phi'' = \frac{d^2\phi}{dx^2}$, X_A is the mole fraction of the sodium, X_B is the mole fraction of the chloride and X_C is the mole fraction of the biomolecule. For simplicity, the average bars are removed from equation (14.25) and all of the following equations, but the variables are all still averages over y. A single prime denotes a first derivative with respect to x, while a double prime denotes a second derivative with respect to x. Applying the same scaling factor and simplification to equation (14.16) and the other ions yields

$$X_{A'} + X_A \phi' = -\frac{n_A}{D_A} \tag{14.26}$$

$$X_{B'} - X_B \phi' = \frac{n_B}{D_B} \tag{14.27}$$

$$X_{C'} + z_C X_C \phi' = \frac{n_C}{D_C}. \tag{14.28}$$

where n_i is the flux of species i and is constant. Note that here the diffusion coefficients can be a function of x.

The diffusion coefficients are often used as fit parameters in numerical simulations [15]. In fact as we have seen, $D_A = D_A(x)$ with the diffusion coefficient in the bath being the normal bulk value and in the channel a "hindered" value which is much less than in the bulk as depicted on Figure 14.4.

We can obtain solutions for the mole fractions easily. After multiplying equation (14.26) by an integrating factor e^{ϕ} we find the solution for the mole fraction is given by

$$X_A = -n_A e^{-\phi} \int_0^x e^{\phi} \frac{dx}{D_A} + X_L e^{\phi(0)-\phi(x)} \tag{14.29}$$

where to satisfy the boundary conditions

$$n_A = \frac{X_L e^{\phi(0)} - X_R e^{\phi(1)}}{\int_0^1 e^{\phi} \frac{dx}{D_A}} \tag{14.30}$$

and we have written $n_A = n_{Ax}$. The problem now reduces to obtaining the electric potential. Note that in the case where there is no potential drop and the diffusion coefficient is constant then the flux n_A is determined solely by the values of the mole fractions in the two baths.

In the case where $\epsilon_1 \ll 1$ there will be boundary layers near the entrance and exits due to the rapid change diffusion coefficient. Away from these boundary layers in the channel where the diffusion coefficient is constant and for small ϵ_1 the core of the channel is electrically neutral leading to the result

$$X_A - X_B + z_c X_C = -2\epsilon^2 \sigma_0 \tag{14.31}$$

for a $(1, -1, z_C)$ system. If the perturbation from the ζ-potential at the wall is computed, then the Debye-Huckel approximation may be employed to compute an analytical solution for the surface charge density in terms of the wall mole fractions [31].

To calculate the surface charge density, we can assume that the Debye-Huckel approximation, valid for potential below $26mV$, holds as described earlier. If we compute the perturbation from the ζ-potential (the ζ potential is the electric potential at the wall), then we can use the Debye-Huckel approximation to compute an analytical solution for the surface charge density in terms of the wall mole fractions [31]. It is easily shown that the dimensionless surface charge density is [31]

$$\sigma_0 = -\frac{1}{\epsilon^2} \frac{\sum_{i=1}^{N} z_i X_i^0}{\sum_{i=1}^{N} z_i^2 X_i^0}\, tan\, h \left(\frac{1}{2\epsilon}\right) \tag{14.32}$$

where the X_i are the concentrations scaled by ionic strength and N is the number of species. If $\epsilon \sim 1$ the double layers are overlapping and if $\epsilon \ll 1$ in equation (14.32), the EDLS are thin. This equation holds in either case. Thus the correction term is O(1). As seen in equation (14.12) the presence of the fixed charge term leads to a parabolic correction. In practice, however as noted by Chen *et al.* [33] the surface charge distribution is not constant

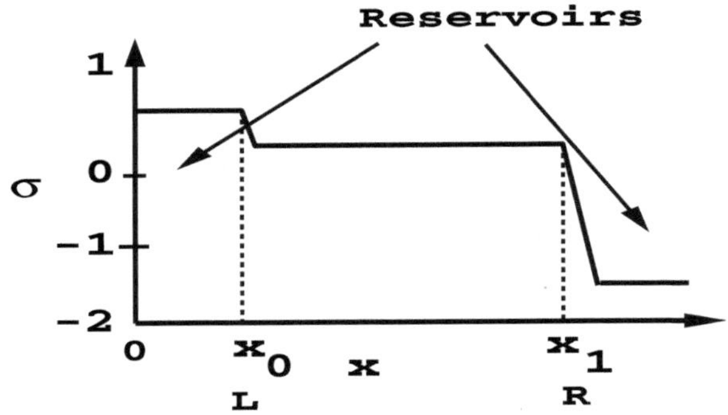

FIGURE 14.6. Sketch of the fixed charge on the side wall as depicted in Chen *etal*.[33] The units of the charge are in Molar.

but varies along the channel. It turns out that for many channels, the surface charge density is piecewise constant as shown on Figure 14.6 [33]. The scale shown is from Chen [33] for a synthetic protein channel called the LS channel. From this distribution it is seen that the correction term results in boundary layers near the entrance and exit of the channel. For the purposes of this chapter we consider only the case where the fixed charge distribution is constant, effectively for $x_0 \leq x \leq x_1$.

We first present results for uncharged side walls; to gain an appreciation of the basic behavior, in all of these results the potential and mole fractions are computed numerically and the diffusion coefficients are assumed equal. Figure 14.7(a) shows the result for an additional species of valence -1; in these results we assume the correction is zero. The dash dot line in the potential curve is the asymptotic result in the interior. The channel length is $L = 20nm$ and the EDL thickness is based on the ionic strength at $x = 0$. Note that the core of the channel is electrically neutral as noted by the net volume charge density curve.

Figure 14.8 shows results for the same parameters as on Figure 14.7 except now there is a potential drop of $V = 2$ across the channel. On Figure 14.9 are results for the parameters of Figure 14.7 except that the valence of the third species is $z_{n2} = -3$. On Figure 14.10 are results for the parameters of Figure 14.7 except the length of the channel is $L = 5nm$. Note now that the core of the channel is not electrically neutral.

It is useful to note that if the upstream and downstream concentrations are chosen to satisfy electroneutrality at $x = 0, 1$ there are no boundary layers and all of the quantities are linear with x.

We now look at the case where the surface charge density correction is not zero. On Figure 14.11 are results for the case of Figure 14.7 with the correction added in. Here we note that the sidewall mole fractions are not that different yet there is a large difference in the potential profile.

On the other hand, when there is significant difference in the mole fractions there is both a significant difference in the potential and the mole fractions as noted in Figure 14.12. These results show that the potential is extremely sensitive to the fixed charge distribution [12].

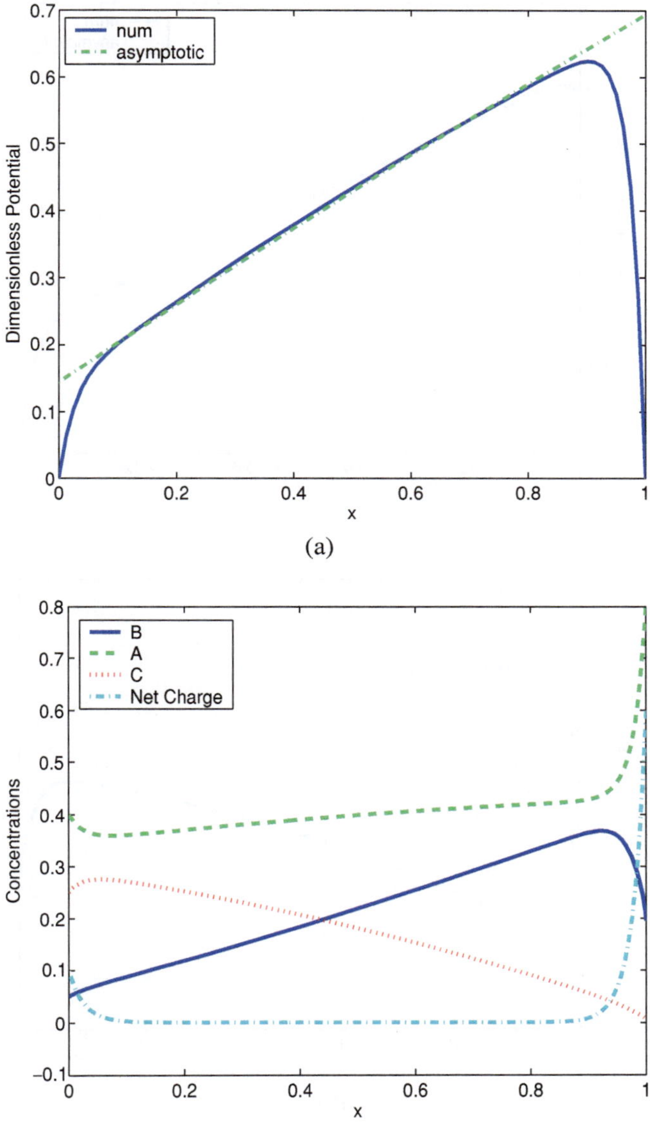

(b)

FIGURE 14.7. Concentration and potential results for the three component solution for constant surface charge density for the length of the channel $L = 20$ nm. Boundary conditions: $X_{AL} = 0.4(p)$, $X_{BL} = 0.05(n)$, $X_{CL} = 0.25(n2)$, $X_{AR} = 0.8$, $X_{BR} = 0.195$, $X_{CR} = 0.005$, $V = 0$. (a) Potential (b) Concentrations.

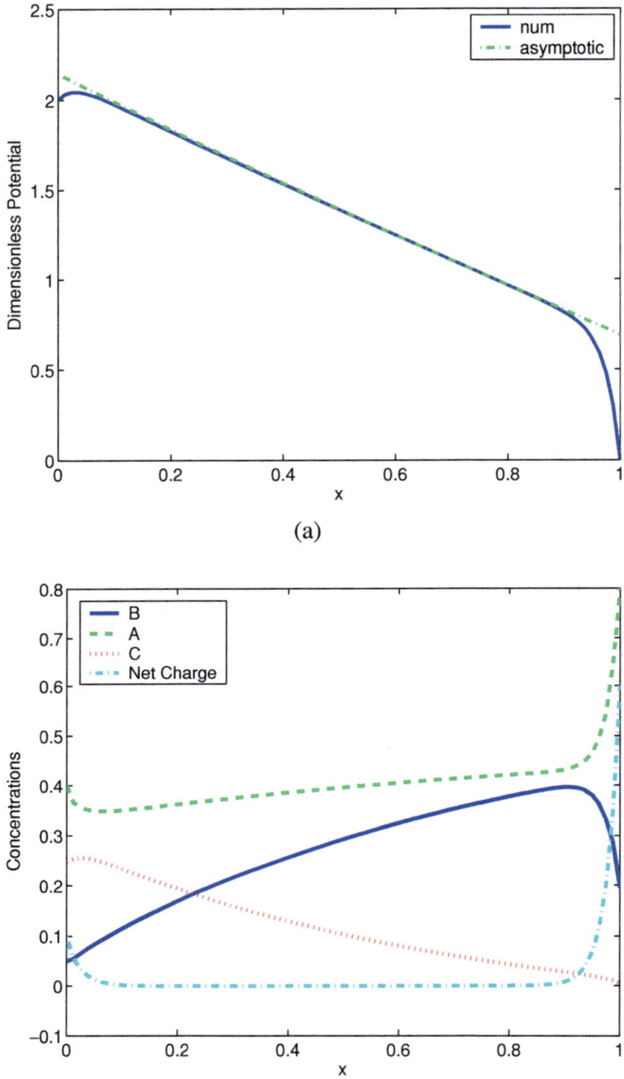

(a)

(b)

FIGURE 14.8. Concentration and potential results for the parameters of Figure 14.7, except $V = 2$. (a) Potential (b) Concentrations.

14.7. IONIC AND BIOMOLECULAR TRANSPORT: COMPARISON WITH EXPERIMENT

As mentioned earlier, current voltage relationships have been predicted for a number of natural ion channels including the sodium, potassium, calcium [12, 13], gramicidin [15],

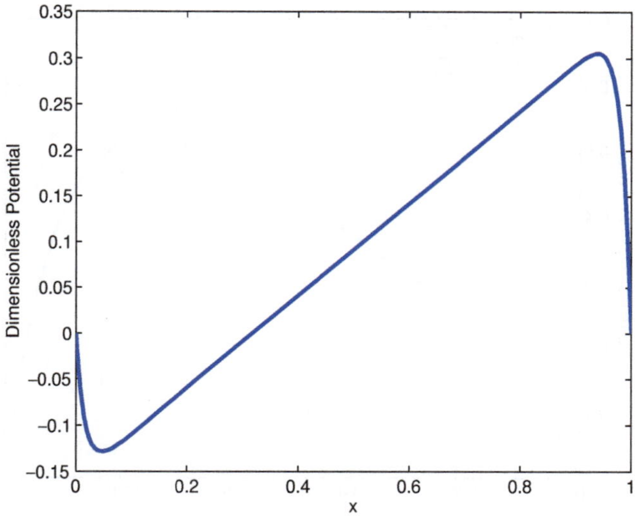

(a)

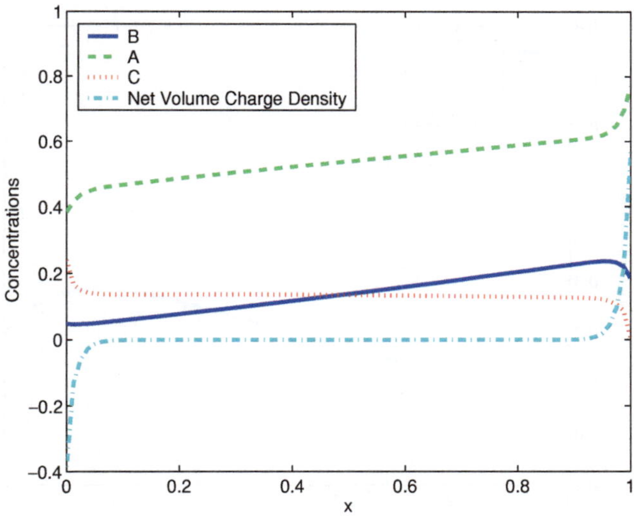

(b)

FIGURE 14.9. Concentration and potential results for the paremeters of Figure 14.7 except the valence $z_C = -3$. (a) Potential (b) Concentrations.

chloride, and porin and its mutant [14] using the one-dimensional PNP equations. Ionic transport is also well predicted by one-dimensional PNP theory [13]. Nonner [14] suggests that the reason the PNP equations seem to model biological ion channels is that the fixed charges on the walls of the channels is fairly large ($\sim 3 \times 10^{21} \frac{1}{cm^3}$ compared to the concentration outside the channel ($\sim 2 \times 10^{19} \frac{1}{cm^3} - 1 \times 10^{21} \frac{1}{cm^3}$. Moreover, the range of voltages

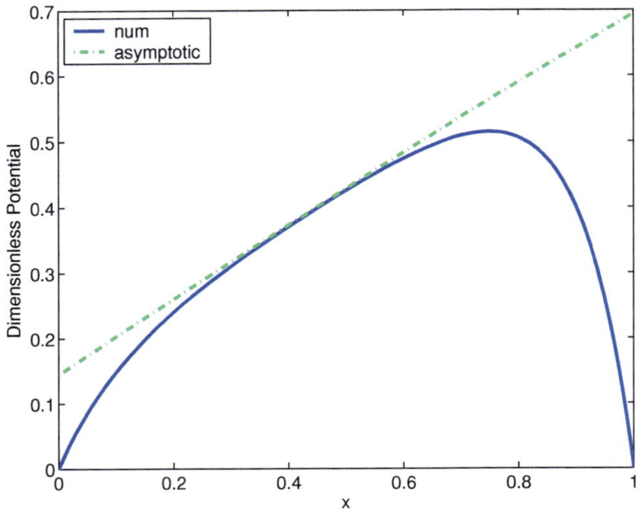

(a)

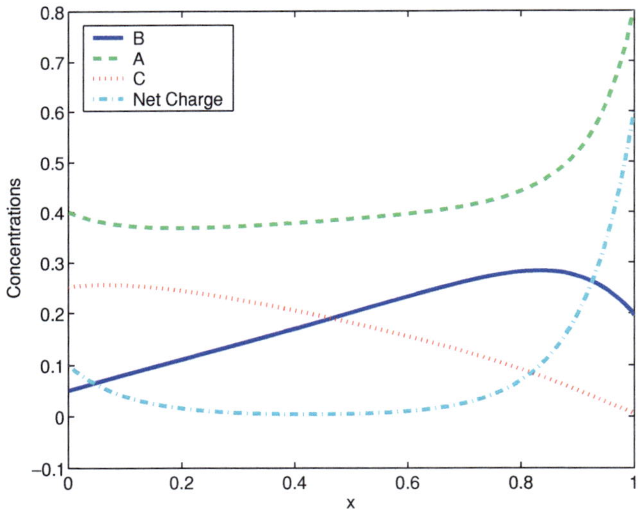

(b)

FIGURE 14.10. Concentration and potential results for the paremeters of Figure 14.7 except the length of the channel $L = 5nm$. (a) Potential (b) Concentrations.

in a biological ion channel is rather narrow being on the order of $\pm 200mV$. It should be note that diffusion coefficient and the dielectric constant of the medium are usually adjusted to fit the experimental data.

The three-dimensional PNP equations have been solved by Hollerbach *et al.* [15] and a qualitative sketch of their results and comparison with experimental data is shown on

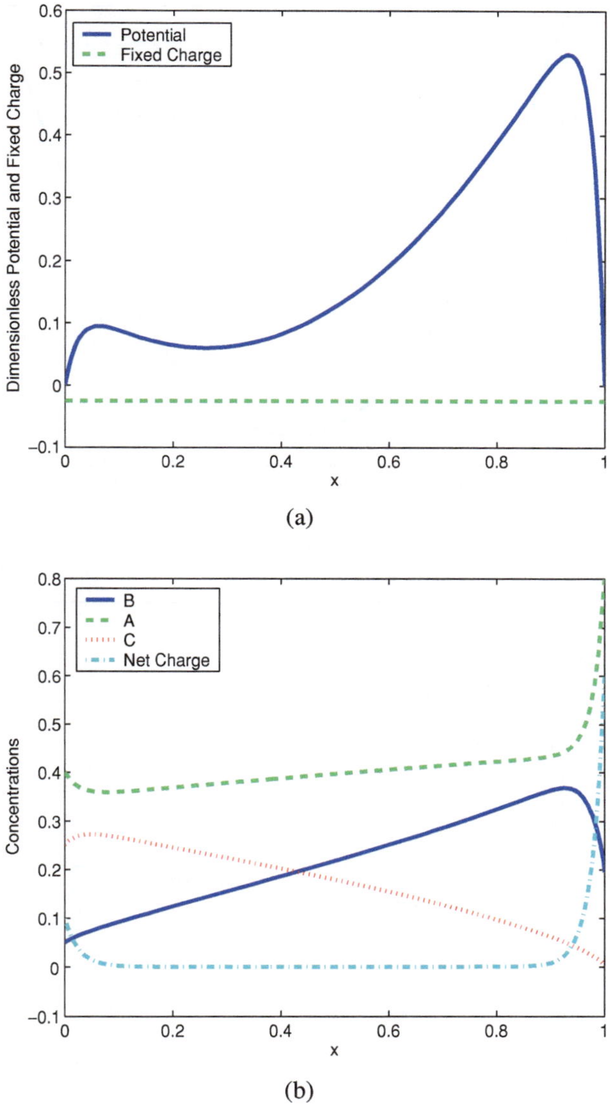

FIGURE 14.11. Concentration and potential results for the paremeters of Figure 14.7 including the effect of permanent or fixed charge. Here the wall concentrations scaled by ionic strength at $x = 0$, $p^0 = 2.0$, $n^0 = 1.9$. The length $L = 20nm$. (a) Potential (b) Concentrations.

Figure 14.13. Note that the current is linear with voltage and the comparison with the data is extremely good. Some fitting of the diffusion coefficients of sodium and chloride ions was peformed although the authors note that the results are not a strong function of these parameters.

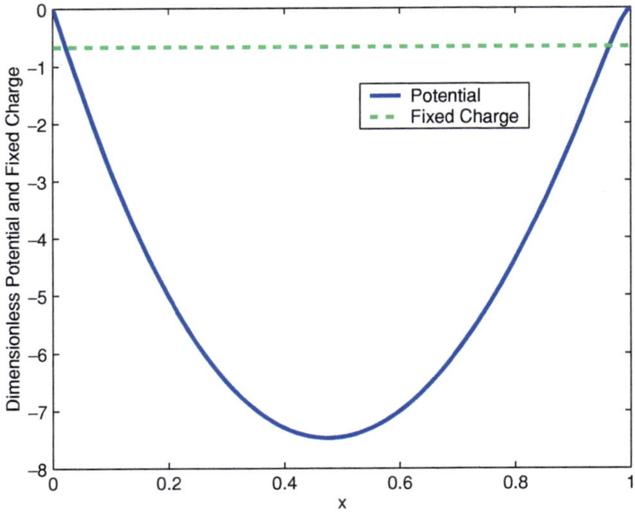

(a)

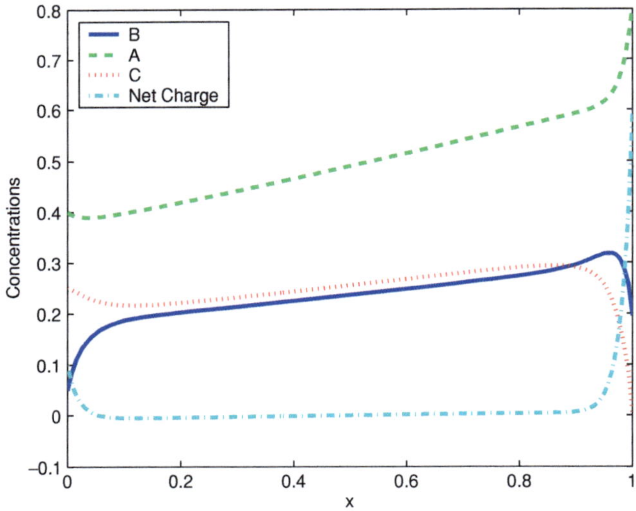

(b)

FIGURE 14.12. Concentration and potential results for the paremeters of Figure 14.7. including the effect of permanent or fixed charge. Here the wall concentrations scaled by ionic strength at $x = 0$, $p^0 = 2.0$, $n^0 = 0.4$. The length $L = 20nm$. (a) Potential (b) Concentrations.

We now consider a simpler case of the diffusional transport of glucose and albumin in a synthetic ion channel. The experimental data is from a series of experiments conducted by iMEDD, Inc; the device consists of a small donor region separated by a nanochannel membrane from a large receiver region as depicted on Figure 14.2. The volume of the donor

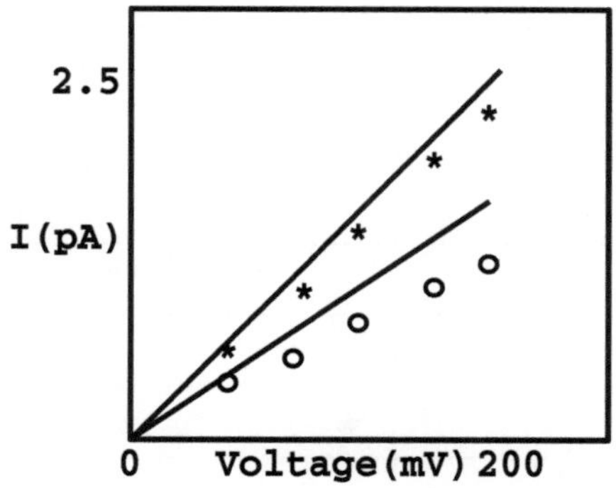

FIGURE 14.13. Sketch of typical current-Voltage curves and the comparison with experimental data as computed with three-dimensional PNP model for a Gamaicidin channel. Numbers denote approximate renge and differents lines denote different concentrations in the baths upstream.

region is 0.22 mL, while the volume of the receiver region is 11 mL. The nanoscale pores in the membranes are assumed to be of uniform length L and width W, but have different values for channel height h. Specifically, all of the nanopores have $L = 5\ \mu m$ and $W = 45\ \mu m$. The values of h are 7, 13, 20 and 27 nanometers. The nanopore membranes have about 10^6 pores/cm^2, so for the purpose of modeling, the entire membrane is treated as a single channel with a length of 5 μm and an effective cross-sectional area A_{eff} given by

$$A_{eff} = \frac{A_m p}{10^4}$$

where A_m is the physical area of the membrane (taken as 2 mm^2), and P is the membrane porosity. The porosity is given by

$$P = \frac{42.5\ (h)}{10^4}$$

The models for the hindered diffusion coefficients will be used to calculate the flux of albumin and glucose and the results will be compared with experimental data from iMEDD, Inc. The nanopore membranes are made from silicon and are therefore biologically, thermochemically, and mechanically stable[2]. Essentially the same device was used for these experiments for which there is no transmembrane potential as for the electroosmotic pumping problem [32].

Both the Levitt and the Bevan and Prieve hindered diffusion models were used to predict the flux. Since there is no potential drop across the membrane and the transport is by diffusion only, through uncharged channels the dimensional flux of species i is given by Fick's Law as

$$n_i = \frac{D_i}{L}(C_{iL} - C_{iR}) \tag{14.33}$$

[2] http://www.imeddinc.com/

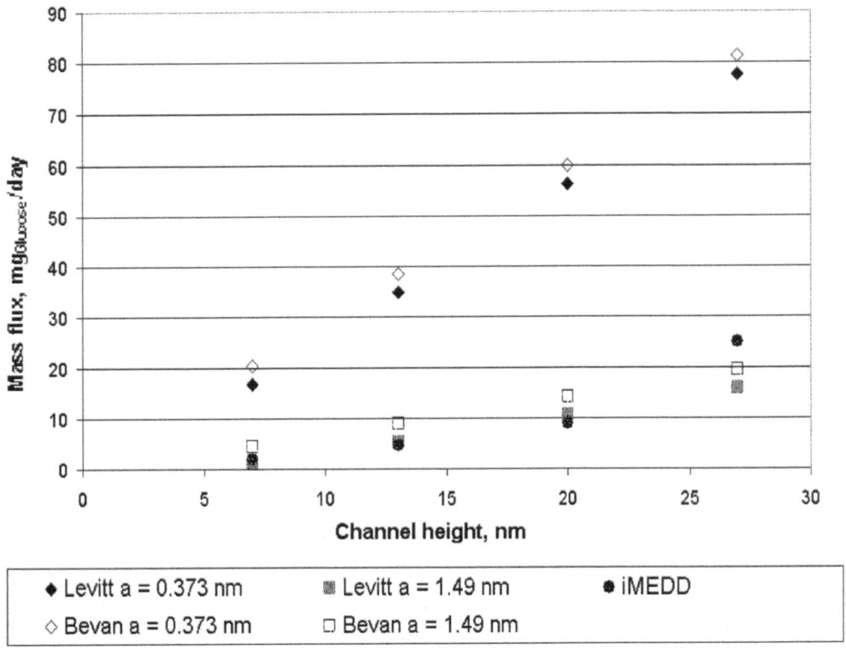

FIGURE 14.14. Results for the mass flow of glucose through the iMEDD membrane.

where D_i is the hindered diffusion coefficient. and L and R denote the values of the concentration in the left and right reservoirs.

The experimental data is calculated from a correlation developed by iMEDD. In principal the experimental results represent a time-averaged flux. Both the Levitt and the Bevan and Prieve models were tested. For all of the glucose results $C_L = 300\frac{gm}{liter}$ and $C_R = 0$. The results for Glucose are shown in Figure 14.14. Notice that when the molecular length scale a is given by the nominal radius of glucose ($a = 0.373$ nm), both the Levitt and the Bevan and Prieve models predict a much larger flux than the experimental results. If the length scale is increased by a factor of four ($a = 1.49$ nm), then both models predict a much closer flux to that of the experimental value. One could argue that increasing the molecular length scale corresponds to the hydration of the glucose molecules. It is surprising how close this simple model is to the experimental data.

The results for albumin are shown in Figure 14.15. As in the case of glucose, the experimental data represents a time-averaged flux. Here $C_L = 80\frac{gm}{liter}$ for $h = 7nm$ and $C_L = 5\frac{gm}{liter}$ for all other channel heights; $C_R = 0$ for all channel heights. Note that when the molecular length scale a is taken as the nominal molecular radius ($a = 3.55$ nm) both hindered diffusion models yield poor predictions of the flux at the $h = 7$ nm channel height. In the case of the Levitt model, a negative flux is predicted, which comes from the fact that the hindered diffusion coefficient is negative. A negative flux could be interpreted as the inability of of a molecule to enter the channel, but this interpretation is tenuous at best.

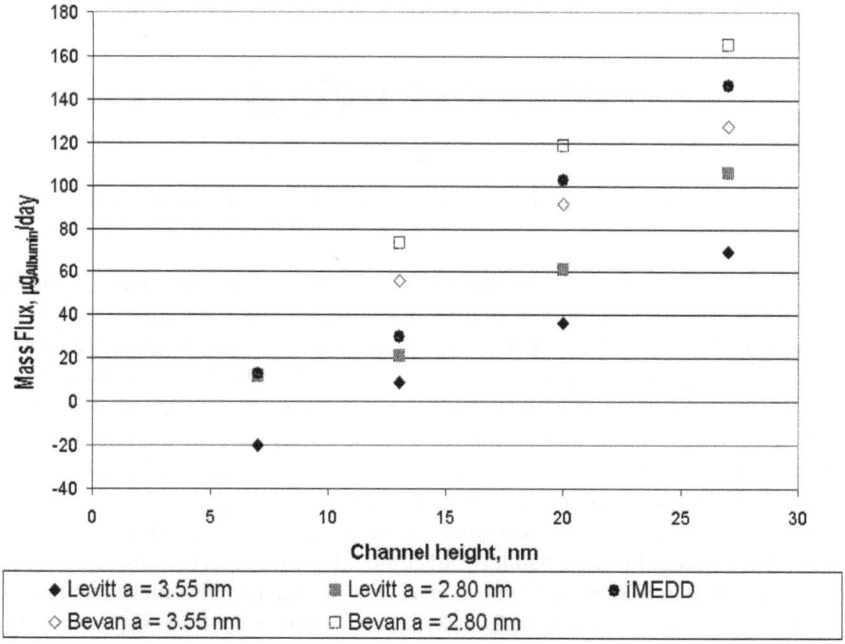

FIGURE 14.15. Results for the mass flow of albumin through the iMEDD membrane.

In the case of the channel heights of $h = 13, 20$, and 27 nm, the Bevan and Prieve model clearly yields a better prediction than the Levitt model when the nominal molecular radius ($a = 3.55$ nm) is used for the molecular length scale. In order for the Levitt model to better predict the experimental data, the molecular length scale must be decreased to $a = 2.80$ nm. Davidson and Dean [24] suggests that large proteins are subject to random changes in configuration and flow-induced deformation. Thus, the albumin molecules may be deformed, and their nominal molecular radii are decreased. This eliminates the problem of a negative hindered diffusion coefficient at $h = 7$ nm. For comparison the Bevan and Prieve model was also used for $a = 2.80$ nm, but the results do not improve from the $a = 3.55$ nm case.

14.8. BROWNIAN DYNAMICS

Brownian Dynamics is the next step toward a full molecular simulation; in this procedure the solvent is treated as a continuum. The solute whether it be an ionic species or long chain polymer is treated on an individual molecule basis.

In modeling natural ion channels the equation governing the motion of an ion in solution is of the Langevin type and is given by

$$m_i \frac{d^2\vec{r}_i}{dt^2} = -F_D \frac{d\vec{r}_i}{dt} + z_i \vec{E}_i + \vec{F}_R \tag{14.34}$$

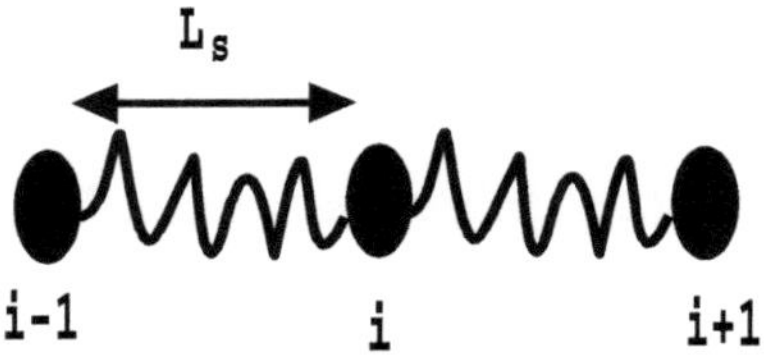

FIGURE 14.16. Sketch of a bead-spring chain model of a long chain polymer.

where F_D is the viscous drag coefficient (not dimensionless!) on the ion, $\vec{E}_i$ is the electric field on the ion and $\vec{F}_R$ is a random thermal force representing the interaction of the ion with the solvent and the walls [34]. The random force and the viscous force are related by the fluctuation-dissipation theorem as noted by Li *et al.* [35] This is a second order ordinary differential equation in time which may be integrated using any number of methods subject to specification of an initial distribution of ions. The key step in the calculation is the specification of the three-dimensional electric field and its variation in space. The one-dimensional equations are obtained by taking averages over space of each of the forces. As is the case for the use of PNP, Brownian Dynamics methods reproduce current-voltage curves for the Gramicidin ion channel [36] although some reservations remain for very small channels [36]. In particular, Corry *et al.* [37] show that the chloride current in a model channel essentially vanishes in the BD calculations whereas in the PNP computation the chloride current is approximately linear ranging from -20 pA to about 5pA when the voltage varies from -125 mV to 125 mV.

The modeling of a long chain biopolymer is considered in a similar way. Theoretical modeling of polymer chains began with the work of Kuhn [38] and several others [39, 40] and many other models of polymer chains have been developed over the years. The polymer, say a biomolecule is assumed to consist of a number of nodes connected by a structure such as a rigid rod (bead-rod model) or a flexible srping (bead-spring chain) or a combination of structures. Different methods of connecting the beads is described by Hur *et al.* [41] and Larson *et al.* [42]

We will describe in detail the bead-spring chain discussed in the work of Panwar and Kumar [43] and depicted on Figure 14.16. While acknowledging that the field of polymer modeling is vast, this analysis is typical of how such models are developed. Each bead simulating a node of the polymer is separated by a spring to simulate the flexible nature of the polymer. In many situations, the accelaeration of the biomolecule may be neglected and so the motion of the molecule is described by following the motion of the beads whose positions evolve in time according the force balance

$$\vec{F}_i^{drag} + \vec{F}_i^{brown} + \vec{F}_i^{spring,tot} + \vec{F}_i^{ext} = 0 \tag{14.35}$$

where the subscript i denotes a node on the polymer. The right side of the force balance, the acceleration term is zero since the inertia time scale is much smaller than the polymer relaxation time. The drag force is defined by the Stokes Drag Law defined above; we write the drag law as

$$\vec{F}_i^{drag} = F_D \left(\vec{u} - \frac{d\vec{r}_i}{dt} \right)$$

where F_D is here the drag force divided by velocity (units of $\frac{kg}{sec}$) and $\vec{u}$ is the bulk fluid velocity. In this formulation the disturbance to the bulk velocity due to the polymer has been neglected.

The Brownian motion force is given by the formula

$$\vec{F}_i^{brown} = \left(\frac{6F_D k_B T}{\Delta t}\right)^{1/2} \vec{n}_i$$

where k_B is Boltzmann's constant, T is the temperature, Δt is the time step in the simulation and $\vec{n}_i$ is a random vector distribution such that $|\vec{n}_i| \leq 1$. The quantity F_D is usually taken to be uniform for each bead.

The springs connecting the beads are assumed to be finitely extensible. The force that the spring exerts on the beads is proportional to the spring displacement so that

$$\vec{F}_i^{spring,tot} = k_s \vec{R}_i = k_s(\vec{r}_{i+1} - \vec{r}_i)$$

where k_s is the spring constant. Actually k_s can depend on the displacement and Panwar and Kumar [43] take

$$k_s = \frac{k_B T}{b_k} \lambda_i \frac{3 - \lambda_i^2}{1 - \lambda_i^2}$$

which is Cohen's approximation to the inverse Langevin function. The function

$$\lambda_i = \frac{R_i}{N_{K,s} b_K}$$

where b_K is the Kuhn length and $N_{K,s}$ is the number of Kuhn steps per spring. The total length of the spring is $L_s = N_{K,s} b_K$.

The relevant time scale for the motion of the biomolecule is given by

$$\tau = \frac{F_D b^2}{k_b T}$$

Using b as the relevant length scale then we can nondimensionalize the equation governing the motion of the polymer and the result is

$$\frac{d\vec{b}}{dt} = \frac{F_D b U_0}{k_b T} \vec{u} + \left(\frac{6}{\Delta t}\right)^{1/2} \vec{n}_i + \lambda_i \frac{3 - \lambda_i^2}{1 - \lambda_i^2}(\vec{b}_{i+1} - \vec{b}_i) - \lambda_{i-1} \frac{3 - \lambda_{i-1}^2}{1 - \lambda_{i-1}^2}(\vec{b}_i - \vec{b}_{i-1}) + \vec{F}^{ext} \tag{14.36}$$

where $\vec{F}^{ext}$ has been made dimensionless on $\frac{k_b T}{b}$ and Δt is the *dimensionless* time step. This is a first order nonlinear initial value problem for all of the beads which may be solved using a standard numerical technique such as Euler, modified Euler or the more accurate Adams-Moulton and Runge-Kutta methods subject to the specified initial position of the beads.

In Figure 14.17 the motion of a DNA strand in various channel geometries is compared [44]. On Figure 14.17(a) two DNA strands having the same inital conformation

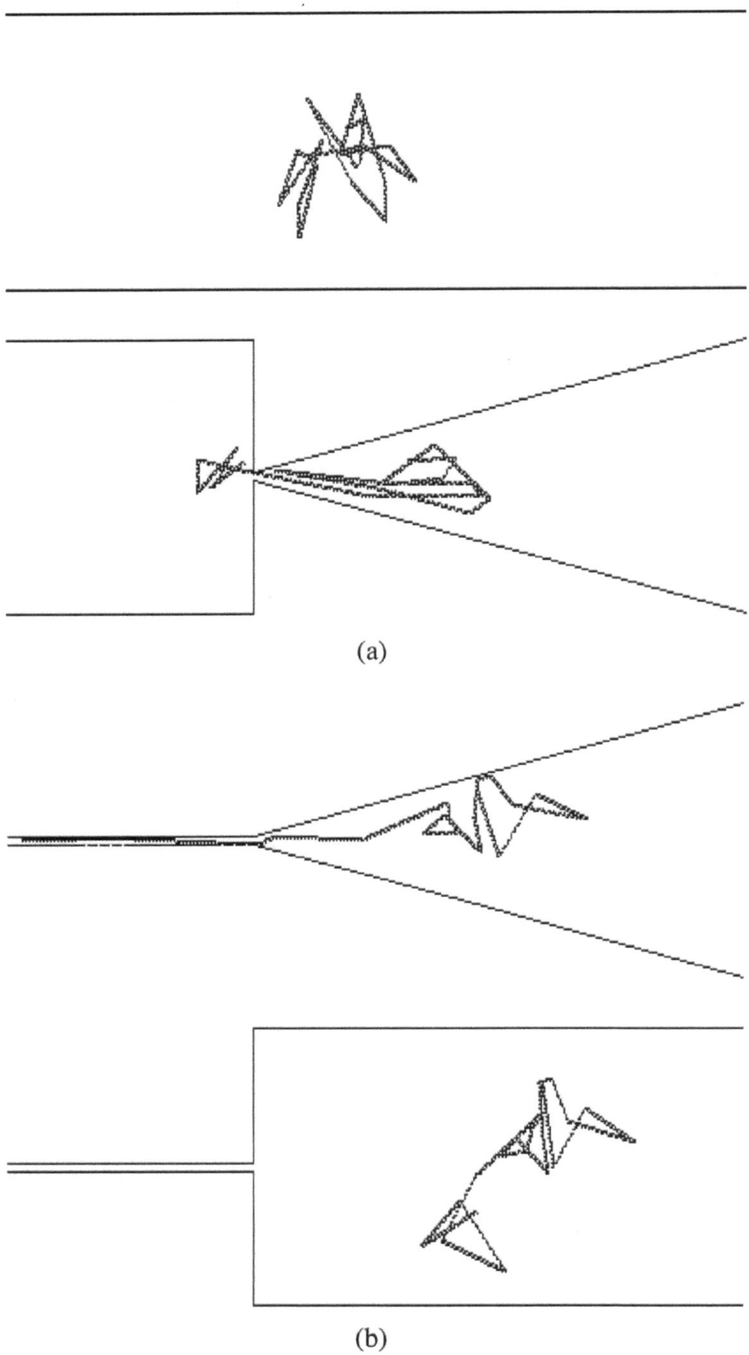

FIGURE 14.17. Motion of a DNA strand in two different geometries using a Brownian Dynamics simulation. (a) Motion in a straight and a converging channel. (b) Motion in a converging channel upstream of a tube and in a sudden contraction. The motion in all cases is right to left.

passes through a straight channel and a converging channel under the action of an electrical field; since DNA is negatively charged it moves from right to left. In the top plot, we see that the DNA does not unravel because the walls are sufficiently far away; on the other hand, the proximity of the walls in the converging channel causes the DNA to stretch itself out to get through the nano-pore.

A similar result is seen in Figure 14.17(b). Here the case of a sudden contraction is compared with the converging channel downstream of which is a nanotube. Here it is seen that the DNA strand will not go through the sudden contraction but again will stretch itself out to get through the converging channel and into the nanotube.

14.9. MOLECULAR DYNAMICS SIMULATIONS

Even though it seems to apply to any sort of molecular level theory, "molecular dynamics" often is used to designate the simulation of many-particle systems by solving classical equations of motion. Molecular dynamics (MD) can be used to obtain equilibrium thermal properties, which are also accessible through Monte Carlo (MC) methods. Unlike MC techniques, MD can also provide dynamical information such as rates of diffusion, relaxation or reaction. Typical MD simulations involve trajectories of $10^2 - 10^5$ particles evolving for times of $10^1 - 10^3 ps$ in simulations cells of $10^1 - 10^3 nm$.

The dynamical properties of systems under the action of external forces — electric fields, shear stress, etc. – are often of great interest. This is the province of non-equilibrium molecular dynamics (NEMD). Application of external forces of either realistic magnitude, or magnitude sufficient to observe sufficient signal-to-noise in simulations, is likely to cause drastic departure from equilibrium conditions for simulations sizes of the scale described above. Therefore, NEMD techniques involve regulating the simulated system to keep it close to target conditions. For example, NEMD simulations appropriate for predicting shear viscosity would lead to rapid frictional heating of a fluid unless precautions were taken. In such simulations, energy is withdrawn from the system in a manner to enforce a steady-state constant temperature.

The use of NEMD simulations to describe the trajectories of molecules in multicomponent mixtures characteristic of problems in biology and chemistry in a bulk fluid flow is in its relative infancy because of the relatively large computation time required. In this section we will describe the basic computational methodology for the Lennard-Jones potential, while recognizing that there are many other potentials describing intermolecular potentials in the literature [45–47].

The Lennard-Jones (LJ) potential is perhaps the most popular description of the intermolecular force field between two particles and is given by

$$\phi_{LJ} = 4\epsilon_W \left[\left(\frac{\sigma}{r}\right)^{12} - \left(\frac{\sigma}{r}\right)^{6} \right] r < r_c = 0 r \geq r_c \qquad (14.37)$$

where ϵ_W is the well depth at $\frac{d\phi_{LJ}}{dr} = 0$ and σ is the particle diameter. Here $r_c \sim \sigma$ is a cut-off radius. A sketch of this potential is given on Figure 14.18. The first term in the potential is the repulsive term and the second term represents the attractive part of the force field. An LJ fluid is a collection of smooth colliding balls. Note here that the "balls" are

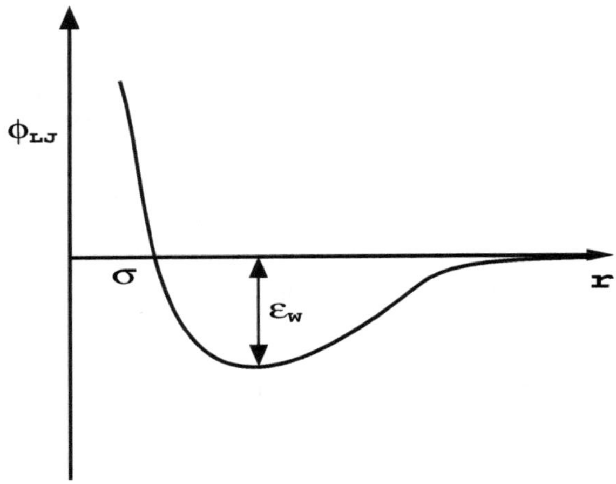

FIGURE 14.18. Sketch of the Lennard-Jones potential.

not polar and so strictly speaking the LJ potential cannot describe water; nevertheless an LJ calculation often gives useful results for aequous solutions.

Now the force corresponding to the potential is given by

$$\vec{F} = -\nabla\phi_{LJ}$$

and performing the integration we find that

$$\vec{F}_{ij} = \left(\frac{48\epsilon_W}{\sigma^2}\right)\left[\left(\frac{\sigma}{r_{ij}}\right)^{14} - \frac{1}{2}\left(\frac{\sigma}{r_{ij}}\right)^{6}\right]\vec{r}_{ij} \tag{14.38}$$

for $r < r_c$ and zero otherwise. Note that the potential is continuous at $r = r_c$ but higher order derivatives are discontinuous there; however experience has shown that the discontinuity does not adversly affect the computation. Then Newton's Law is given by

$$m_i\frac{d^2\vec{r}_i}{dt^2} = \vec{F}_i + \vec{F}_{exti} = \sum_{j=1}^{N}\vec{F}_{ij} + \vec{F}_{exti} \tag{14.39}$$

where $\vec{F}_{exti}$ is an external force field, perhaps an electric field(electroosmotic flow), N is the number of atoms or molecules and m is the mass of each molecule. The sum in equation (14.39) excludes the value $i = j$.

The problem can also be formulated in terms of the Hamiltonian which is defined as

$$H = \sum_{i=1}^{N}\frac{p_i^2}{2m} + \sum_{i<j}\phi_{LJij} \tag{14.40}$$

and the force associated with the Hamiltonian is given by

$$\vec{F} = -\nabla H + \vec{F}_{ext} = m\vec{a} \tag{14.41}$$

Note that the we have described the interaction potential between the atoms or molecules in the fluid; however the interaction potential between the the walls and fluid also need to be specified. If the walls are taken to be smooth round spheres then the interaction potential can also be of the LJ type but with possibly different different values of the LJ parameters although they are taken to be the same in some cases [48].

Equation (14.39) is usually put in dimensionless form to reduce the number of parameters required to vary to describe the system; chemists often refer to these dimensionless variables as reduced units. We define dimensionless variables by chosing σ, m and ϵ_W to be the units of length mass and energy respectively. Thus, for example the dimensionless length is

$$r^* = \frac{r}{\sigma}$$

The unit of time is $\sqrt{\frac{m\sigma^2}{\epsilon_W}}$ and substituting into equation (14.39) we have

$$\frac{d^2\vec{r}_i}{dt^2} = 48 \sum_{j=1}^{N} \left[r_{ij}^{-14} - \frac{1}{2} r_{ij}^{-6} \right] \vec{r}_{ij} + \vec{F}_{exti} \tag{14.42}$$

where again the sum excludes the value $i = j$.

Because of the computational intensity of the MD simulations, boundary conditions are important. In simulations in nanochannels, the nanoconstrained dimension can exhibit realistic boundary conditions; typically in a 6 nm channel, there will be about 20 water molecules and so this is a resonable number [48]. On the other hand, periodic boundary conditions must be used in the other two directions to reduce the computation time to a reasonable amount.

Ionic species are charged and so a potential must be specified between the individual ionic species. Consider a pair of ions, then the Coulombic interaction is of the form [48]

$$\phi_C = \zeta \frac{z_i z_j}{r_i - r_j} \tag{14.43}$$

where the variables are in dimensionless form and

$$\zeta = \frac{e^2}{\epsilon_r \epsilon_W \sigma}$$

is the dimensionless interaction energy, e is the electron charge and ϵ_r is the dielectric constant of the medium. Typical Lennard-Jones parameters for a sodium-chloride aqueous solution are shown on Table 14.2.

TABLE 14.2. Lennard-Jones parameters for a sodium-chloride aqueous solution [50].

Pair	σ(Å)	ϵ_W(KJ/mol)	Pair	σ(Å)	ϵ_W(KJ/mol)
$O-O$	3.151	0.6366	Na^+-Na^+	2.73	0.3576
Na^+-O	2.876	0.5216	$Cl^- - CL^-$	4.86	0.1679
$Cl^- - O$	3.785	0.5216	$Na^+ - CL^-$	3.87	0.1706

It is important to note that when an external force is applied to a system, a simulation cell will rapidly heat up if heat is not removed. The term "thermostat is given to the procedure used to remove the excess energy to keep the temperature of the cell constant. Various means have been used to do this and Berendson, Nose-Hoover and Isokinetic thermostats [46, 47] have been employed.

Once the problem is set up the equations, since they are nonlinear, need to be solved numerically. This can be done by any standard numerical method such as the lower order Euler and Modified Euler methods, or the more accurate single step Runge-Kutta methods and multistep Adams-Moulton method. A particularly useful method for MD simulations is to replace the second derivative by a discrete central difference approximation; this approximation may be derived simply from Taylor series approximations and the approximation at time level k for any position x is defined by

$$\frac{d^2x_i}{dt^2} = \frac{x_i^{k+1} - 2x_i^k + x_i^{k-1}}{\Delta t^2} + O(\Delta t) \tag{14.44}$$

Such an approximation in the context of MD simulations is called the Verlet algorithm [49]. This method is multi-step because it involves evaluation of the right side of equation (14.42) at two time levels. The Verlet algorithm is particularly attractive because it conserves energy much better than its accuracy would suggest.

Zhu *et al.* [48] have performed molecular dynamics study of electroosmotic flow in a nanochannel and compare their results with the classical Poisson-Boltzmann theory. This work is discussed elsewhere in this volume and we summarize the results briefly as follows. The MD solutions show that ions are excluded from the wall and so the bulk velocity is much higher than that from Poisson-Boltzmann theory which allows ions right up to the wall. The PB theory can be fixed up by redefining the boundary of the channel.

Cui [50] has studied the transport of ssDNA in 2.4 and 3.0 nm hydrophilic silica channels. He found that the water and ion density profiles are unaffected by the ssDNA and thus depend only on the pore diameter. Four types of ssDNA are modeled (G10,A10,C10,T10) where the "10" denotes that the strand consists of ten nucleotides. In each case the end-to-end distance or length of the ssDNA oscillates about a mean value just over $3nm$ as shown on Figure 14.19. Additional computations for a $2.4nm$ pore show increased fluctuation amplitude. The calculations are carried out over approximately $10-12ns$. Adorption of ions onto the ssDNA is also observed. These results show that the ssDNA remains extended in the nanopore as expected because it is negatively charged as is the wall of the channel.

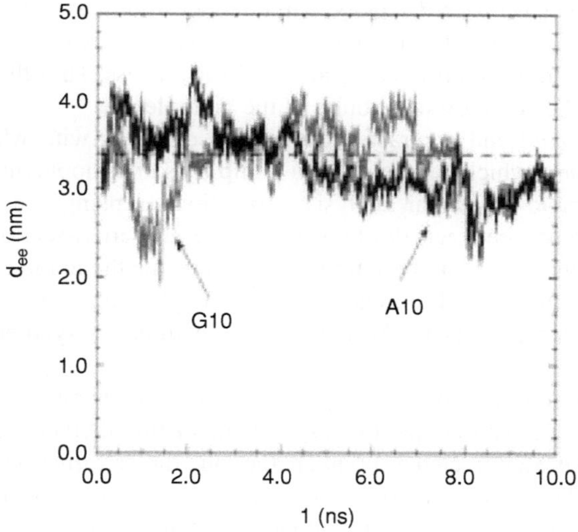

(a)

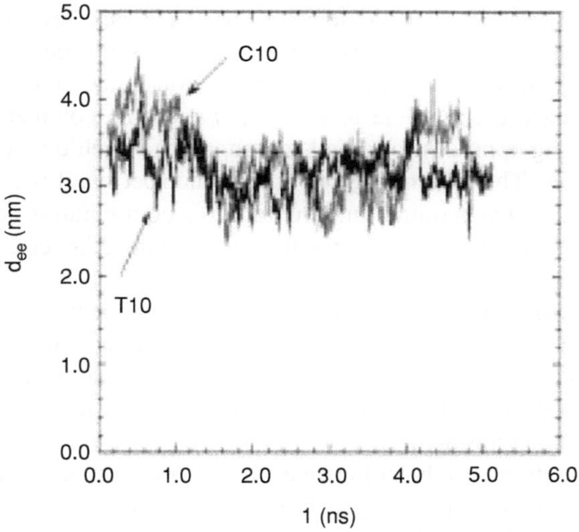

(b)

FIGURE 14.19. End to end distance of a ssDNA molecule. Dashed line indicates the length of a single pitch double helix B-form DNA. (a) poly (G_{10}) and poly (A_{10}). (b) poly (T_{10}) and poly (C_{10}).

14.10. SUMMARY

We have reviewed the state-of-the-art of modeling biomolecular transport in micro and nanochannels. We have focussed on those methods which can be used to predict both transport of a polymeric species and the structure of the background bulk flow. Thus only deterministic models have been discussed while recognizing that there is a large literature on stochastic techniques. The techniques discussed here are appropriate for the design of devices for which fluid transport is a crucial part of the process. Thus these methods do not, in general, resolve the complex substruture of the biomolecule.

We have compared and contrasted natural ion channels with what we have called synthetic ion channels which can be used to pump ions and biopolymers into and out of nanochannels for rapid molecular analysis, separations, sensing and other applications. The primary difference between the two is the size and cross-section of the channels; due to fabrication limitations, and often the application, synthetic ion channels are rectangular and bigger, while natural ion channels have roughly circular cross-section and are smaller. In many other respects, the basic physics of natural and synthetic ion channels are similar.

We have discussed three types of modeling techniques beginning with a continuum approach using the Poisson-Nernst-Planck system. In this method, ionic species and a simulated biomolecule are treated as point charges and the deformation of the biomolecule is therefore not modeled. This method seems to work well for cases where the biomolecule is not too large and this simple diffusion model compares well with a set of experimental data for the transport of glucose and albumin.

Of course, strictly continuum approaches do not incorporate the size of the biomolecule explicitly and thus do not account for shape changes. The next level of approximation is the Brownian Dynamics approach in which the solvent is treated as a continuum. This procedure leads to a first or second order order differential equation in time for the position of the solute molecules. If the solute is a large polymer, the polymer is then discretized using a ball and spring approach, or a ball-rod approach which defines how the polymer moves within the flow. The solvent and the other ionic species, if present are treated as a continuum. Treating the background fluid mixture as a continuum is necessary if the flow is in a microchannel since the domain is too large to treat the other constituents of the fluid on the molecular level.

The final level of approximation is to treat all of the constituents as individual molecules; this is the Molecular Dynamics approach and this method may be used as the domain of flow approaches nanoscale. However, it must be pointed out that the actual computation of the transport of a biopolymer in an imposed mean flow is in its infancy and much more work needs to be done in this area.

All devices no matter what their objective must involve system level modeling; the MD simulation technique is difficult to apply to the system design because of the computationally intensive nature of the solution. Thus it is desireable to perform system design studies with a continuum model of a given component; this can only be done if a lower order continuum model can be developed from the results of an MD simulation.

ACKNOWLEDGEMENTS

Portions of this work is supported by DARPA, Defense Sciences Office. The authors are grateful to the contract monitors Dr. Anantha Krishnan (DARPA), Mr Clare Thiem and Duane Gilmour of the Air Force Research Lab (IFTC) for their support. ATC is grateful to students Kelly Evers, Jennifer McFerran and Drew Kebbel some of whose work appears in this review. Thanks also to Sherwin Singer and Shengting Cui who read a preliminary draft of this paper. Thanks also to Prof. Derek Hansford and Dr. Mauro Ferarri both of whom got me started in this field.

REFERENCES

[1] B. Hille. *Ionic Channels of Excitable Membranes*, 2nd Edition, Sinauer Associates, Sunderland, MA, 1992.

[2] Dirk Gillespie. A Singular Perturbation Analysis of the Poisson-Nernst-Planck System: Applications to Ionic Channels, PhD Thesis, Rush Medical School, Chicago, 1999.

[3] Dirk Gillespie and S. Robert Eisenberg. Modified donnan potentials for ion transport through biological ion channels. *Phys. Rev. E*, 63:061902-1–06192–8, 2001.

[4] Braha, Orit, Gu, Li-Qun, Zhou, Li Lu, Xiaofeng, Cheley, Stephen and Bayley, Hagan. Simultaneous sensing of divalent metal ions. *Nat. Biotechnol.*, 18:1005–1007, 2000.

[5] D.G. Levitt. Modeling of Ion Channels. *J. Gen. Physiol.*, 113:789–794, 1999.

[6] Alfredo E. Cardenas, Rob D. Coalson, and Maria G. Kurnikova. Three-Dimensional Poisson-Nernst-Planck Theory Studies: Influence of Membrane Electrostatics on Gramicidin A Channel Conductance. *Biophys. J.*, 79(1):80–93, 2000.

[7] Alberts, Bruce et al. *Essential Cell Biology*, Garland Publishing, New York, 1998.

[8] Eisenberg, Bob. Ion Channels in biological membranes-electrostatic ananysis of a natural nanotube. *Contemp. Phys.*, 39(6):447–466, 1989.

[9] Barcilon, Victor, D.P. Chen, R.S. Eisenberg, and J.W. Jerome. Qualitative Properties of Steady-State Poisson-Nernst-Planck Systems: Perturbation and Simulation Study. *SIAM J. Appl. Math.*, 57(3):631–648, 1997.

[10] D.P. Chen, W. Nonner, and R.S. Eisenberg. PNP theory fits current-voltage (IV) relations of a neuronal anion channel in 13 solutions. *Biophys. J.*, 68:A370, 1995.

[11] Roux, Benoit. Theoretical and computational models of Ion channels. *Curr. Opin. Struct. Biol.*, 12:182–189, 2002.

[12] Nonner, Wolfgang and Eisenberg, Bob. Ion Permeation and Glutamate Residues Linked by Poisson-Nernst-Planck Theory in L-Type Calcium Channels. *Biophys. J.*, 75:1287–1305, 1998.

[13] Nonner, Wolfgang, Gillespie, Dirk, Henderson, Douglas, and Eisenberg, Bob. Ion Accumulation in a Biological Calcium Channel: Effects of Solvent and Confining Pressure. *J. Phys. Chem. B*, 105:6427–6436, 2001.

[14] Nonner, Wolfgang and Eisenberg, Bob. Electrodiffusion in Ionic Channels of Biological Membranes. *J. Mole. Liq.*, 87:149–162, 2000.

[15] Hollerbach, Uwe, Chen, Duan P., and Robert Eisenberg. Two- and three-dimensional Poisson-Nernst-Planck simulations of current flow through Gramicidin A. *J. Sci. Comp.*, 16(4):373–409, 2001.

[16] Z. Schuss, B. Nadler, and R.S. Eisenberg. Derivation of Poisson and Nernst-Planck equation in a bath and channel from a molecular model. *Phys. Rev. E*, 64:036116–1–036116–14, 2001.

[17] Gillespie, Dirk, Nonner, Wolfgang and Eisenberg, S. Robert. Coupling Poisson-Nernst-Planck and density functional theory to calculate ion flux. *J. Cond. Mat.*, 14:12129–12145, 2002.

[18] A.T. Conlisk, J. McFerran, Z. Zheng, and D. Hansford. Mass Transfer and Flow in Electrically Charged Micro- and Nanochannels. *Anal. Chem.*, 74(9):2139–2150, 2002.

[19] R.B. Bird, T.B. Drew, and J.W. Hoopes. (ed.) Theory of Diffusion. *Advances in Chemical Engineering*, Academic Press, New York, vol. 1, pp. 156–239, 1956.

[20] H.J.V. Tyrell and K.R. Harris. *Diffusion in Liquids: A Theoretical and Experimental Study*, Butterworth & Co. Ltd., 1984.

[21] T. Peters, Jr. *All About Albumin: Biochemistry, Genetics and medical Applications*, 3rd Ed., Academic Press, San Diego, 1996.

[22] B. Lin, J. Yu, and S.A. Rice. Direct measurements of constrained Brownian motion of an isolated sphere between two walls. *Phys. Rev. E.*, 62(3):3909, 2000.

[23] J. Happel and H. Brenner. *Low Reynolds Number Hydrodynamics: with Special Applications to Particulate Media*, Kluwer, Boston, 1983.

[24] M.G. Davidson and W.M. Dean. Hydrodynamic theory for the hindered transport of flexible macromolecules in porous membranes. *J. Memb. Sci.*, 35:167–192, 1988.

[25] D.G. Levitt. General continuum analysis of transport through pores, I. Proof of Onsager's Reciprocity postulate for uniform pore," *Biophys. J.*, 15:533–551, 1975.

[26] D.G. Levitt. General continuum analysis of transport through pores, II. nonuniform pores. *Biophys. J.*, 15:553–563, 1975.

[27] M.A. Bevan and D.C. Prieve. Hindered diffusion of colloidal particles very near to a wall: revisited. *J. Chem. Phys.*, 113(3):1228, 2000.

[28] W.L. Haberman and R.M. Sayre. *Motion of Rigid and Fluid Spheres in Stationary and Moving Liquids Inside Cylindrical Tubes*, David Taylor Model Basin (Dept. of the Navy), Report 1143, U.S. Department of Defense, Navy Department, Washington, D.C., 1958.

[29] E.L. Cussler. *Diffusion: Mass Transfer in Fluid Systems*, Second Ed., Cambridge University Press, 1997.

[30] Ludomira H. Granicka *et al.* Polypropylene hollow fiber for cells isolation: methods for evaluation of diffusive transport and quality of cells encapsulation. *Artifi. Cells, Blood Subs. Biotechnol.*, 31(3):249–262, 2003.

[31] A.T. Conlisk. The Debye-Huckel Approximation: Its Use In Describing Electroosmotic Flow in Micro and Nanochannels. *Electrophoresis*, 26:1896–1912, 2005.

[32] Zheng, Zhi, Hansford, J. Derek, and A.T. Conlisk. Effect of multivalent ions on electroosmotic flow in micro and nanochannels. *Electrophoresis*, 24(17):3006–3017, August 2003.

[33] Duan P. Chen, James Lear, and Robert Eisenberg. Permeation through an open channel: Poisson-Nernst-Planck theory of a synthetic ion channel. *Biophys. J.*, 72:97–116, 1997.

[34] S.H. Chung, M. Hoyles, T. Allen, and S. Kuyucak. Study of ionic currents across a model membrane channel using Brownian dynamics. *Biophys. J.*, 75:893–809, 1998.

[35] S.W.C. Li, M. Hoyles, S. Kuyucak, and S.H. Chung. Brownian dynamics study of ion transport in the vestibule of membrane channels. *Biophys. J.*, 74:37–47, 1998.

[36] Edwards, Scott, Corry, Ben, S. Kuyucak, and S.H. Chung. Continuum electrostatics fails to describe ion permeation in the gramicidin channel. *Biophys. J.*, 83:1348–1360, 2002.

[37] Corry, Ben, S. Kuyucak, and S.H. Chung. Tests of continuum theories as models of ion channels, II. Poisson-Nernst-Planck theory versus Brownian dynamics. *Biophys. J.*, 78:2364–2381, 2000.

[38] W. Kuhn and F. Grun. Relationships between elastic constants and stretching double refraction of highly elastic substances. *Kolloid-Z*, 101:248–253, 1942.

[39] Yamakawa, Hiromi. *Modern Theory of Polymer Solutions*, Harper and Row, New York, 1971.

[40] R.B. Bird, R.C. Curtiss, R.C. Armstrong, and O. Hassager. *Dynamics of Polymeric Liquids*, 2nd Ed., Wiley, New York, Vol. II, 1987.

[41] Joe S. Hur, Shaqfeh, S.G. Eric, and G. Larson Ronald. Brownian dynamics simulations of single DNA molecule in shear flow. *J. Rheology*, 44(4):713–742, 2000.

[42] R.G. Larson, Hua Hu, D.E. Smith, and S. Chu. Brownian dynamics simulations of a DNA molecule in an extensional flow field. *J. Rheology*, 43(2):267–304, 1999.

[43] Ajay S. Panwar and Kumar, Satish. Brownian dynamics simulations of polymer stretching and transport in a complex electroosmotic flow. *J. Chem. Phys.*, 118(2):925–936, 2003.

[44] Y-J. Juang, S. Wang, S. Wu, and L.J. Lee. Dynamics of single polymers in a stagnation flow induced by electrokinetics. *Phys. Rev. Lett.*, 93:26–105, 2004.

[45] Richard J. Sadus. *Molecular Simulation of Fluids: Theory, Algorithms and Object-Orientation*, Elsevier, Amsterdam, 1999.

[46] D.C. Rapaport. *The Art of Molecular Simulation*, 2nd Edition, Cambridge, 2004.

[47] David M. Heyes. *The Liquid State: Applications of Molecular Simulation*, Wiley, Chichester, 1998.

[48] Zhu, Wei, Singer, Sherwin, Zheng, Zhi, and A.T. Conlisk. Electroosmotic flow of a model electrolyte. to appear *Phys. Rev. E.*, 2005.

[49] L. Verlet. Computer experiments on classical fluids I. Thermodynamical properties of Lennard-Jones molecules. *Phys. Rev.*, 159:98, 1967.

[50] Cui, shengting. Molecular dynamics study of single-stranded DNA is aqueous solution confined in a nanopore. *Mol. Phys.*, 102(2):139–146, 2004.

15

Nanotechnology in Cancer Drug Therapy: A Biocomputational Approach

Hermann B. Frieboes[1], John P. Sinek[2], Orhan Nalcioglu[3],
John P. Fruehauf[4], and Vittorio Cristini[1,2,*]

Departments of Biomedical Engineering[1], Mathematics[2], Radiological Sciences and Tu & Yuen Center for Functional Onco-Imaging[3], Medicine—Hematology/Oncology[4], University of California, Irvine.

**Corresponding author: Department of Biomedical Engineering, REC 204, University of California, Irvine, CA 92697-2715.*

15.1. INTRODUCTION

15.1.1. Challenges with Chemotherapy

Although the clinical arsenal in treating cancer has been greatly extended in recent years with the application of new drugs and therapeutic modalities, the three basic approaches continue to be (in order of success) surgical resection, radiation, and chemotherapy. The latter treatment modality is primarily directed at metastatic cancer, which generally has a poor prognosis. A significant proportion of research investment is focused on improving the efficacy of chemotherapy, which is often the only hope in treating a cancer patient. Yet the challenges with chemotherapy are many. They include drug resistance by tumor cells, toxic effects on healthy tissue, inadequate targeting, and impaired transport to the tumor. Determination of proper drug dosage and scheduling, and optimal drug concentration can also be difficult. Finally, drug release kinetics at the tumor site is an important aspect of chemotherapy.

In this chapter we consider each of these hurdles and examine how nanotechnology can help to address them. The role of biocomputation will be explored as a means to specify cancer drug therapy, with the goal of applying the results in the clinical setting, especially the modeling of drug delivery via nanoparticles. Biocomputation could save lives and enhance the quality of cancer treatment by making it possible to tailor therapy to the individual patient and reduce the time and costs involved. With these goals in mind, we will look in more detail at the system-level biocomputation of tumor growth and cancer therapy, and raise considerations for future research. We begin by briefly reviewing the advantages of nanotechnology, its application to cancer chemotherapy, and its challenges in a biological setting.

15.1.2. Possibilities of Nanotechnology

Nanotechnology applied to cancer treatment may offer several promising advantages over conventional drugs. Nanoscale devices are two orders of magnitude smaller than tumor cells, making it possible for them to interact directly with intracellular organelles and proteins. Because of their molecule-like size, nanoscale "tools" may be capable of early disease detection using minimal amounts of tissue, even down to a single malignant cell [60]. These "tools" may not only prevent disease by monitoring genetic damage, but also treat cells *in vivo* while minimizing interference with healthy tissue. By combining different kinds of nanoscale "tools" on a single device, it may be possible to run multiple diagnostic tests simultaneously [56]. In particular, it is hoped that cancer drug therapy involving nanotechnology will be more effective in targeting malignant cells and sparing healthy tissue. In this regard, the role of nanoparticles loaded with chemotherapeutic drugs has been receiving much attention. Research and development in this area is expected to dramatically increase in importance in the coming years.

15.1.3. Chemotherapy via Nanoparticles

In general, nanoscale drug delivery systems for chemotherapy can be divided into two categories: polymer- and lipid-based [46]. Polymers, which are usually larger than lipid molecules, form a solid phase, such as polymeric nanoparticles, films, and pellets, while lipids form a liquid (or liquid crystalline phase), such as liposomes, cubersomes, micelles and other emulsions [22]. While polymer-based systems are considered biologically more stable than lipid-based systems, the latter are generally more biocompatible. Polymer-based systems might possess good drug targeting ability because their uptake may be different for cells in different tissues [53]. In fact, Feng and Chien [22] have suggested that a combination of polymer- and lipid-based systems could integrate their advantages while avoiding their respective disadvantages. An example of such a nanoparticle would be a liposomes-in-microspheres (LIM) system, where drugs are first loaded into liposomes, and then encapsulated into polymeric microspheres. This way both hydrophobic and hydrophilic drugs can be delivered in one nanoparticle. The bioactivity of peptides and proteins would be preserved in the liposomes, whose stability is protected by the polymeric matrix [22].

Chemotherapy using nanoparticles has been studied in clinical trials for several years and numerous studies have been published in this regard ([43], pp. 283–290). Two liposomally delivered drugs are currently on the market: daunorubicin and doxorubicin [51]. These encapsulated drugs can be formulated to maximize their half-life in the circulation.

For example, a "stealth" version of liposomal doxorubicin, coated with polyethylene glycol to reduce its uptake by the reticuloendothelial system, can extend its half-life in blood for up to 50–60 hours [10].

15.1.4. Challenges of Nanotechnology

The difficulties facing nanotechnology in the service of clinical medicine are numerous. These difficulties should be kept in mind when considering chemotherapeutic treatment involving nanotechnology and the potential role of biocomputation. First, there are basic physical issues with matter at such a small scale. Since matter behaves differently on the nano than it does at micro and macro levels, most of the science at the nanoscale has been devoted to basic research, designed to expand understanding of how matter behaves on this scale [56]. Because nanomaterials have large surface areas relative to their volumes, phenomena such as friction are more critical than they are in larger systems. The small size of nanoparticles may result in significant delay or speed-up in their intended actions. They may accumulate at unintended sites in the body. They may provoke unexpected immune system reactions. Cells may adapt to the nanoparticles, modifying the body's behavior in unforeseen ways [56]. The efficacy of nanoparticles may be adversely affected by their interaction with the cellular environment. For instance, the reticuloendothelial system (RES) may clear nanoscale devices, even "stealth" versions, too rapidly for them to be effective because of the tendency of the RES to phagocytose nanoparticles ([43], p. 259). Nanoparticles can be taken up by dendritic cells [18] and by macrophages [16]. RES accumulation of nanoparticles could potentially lead to a compromise of the immune system. On the other hand, larger nanoparticles may accumulate in larger organs, leading to toxicity [56]. Perhaps the biggest issue of all is that the physically compromised tumor vasculature may prevent most of the nanodevices from reaching the target cells by vascular transport or diffusion. Alterations in the tumor vasculature may adversely affect the convection of the nanodevices in the blood stream [9]. Local cell density and other stromal features may hamper drug or nanodevice diffusion through tumoral tissue. This topic will be examined in more detail when we consider the issue of chemotherapeutic drug transport and the system-level biocomputation of cancer therapy.

15.1.5. Biocomputation in Cancer Treatment

The challenges of nanotechnology may be better evaluated through the use of biocomputational methods that examine the fundamental physical principles that affect delivery and degradation of nanoparticles in cancer treatment. Biocomputation, in general, provides a means of mathematically modeling these physical principles so that basic truths about the interaction of nanotechnology and living tissue may be better understood. This knowledge could save time and resources by providing guidance to the experimentalist and the clinician, support a coherent framework for further research, and offer the potential to predict experimental outcomes. The main challenge of biocomputation is to be able to incorporate these physical principles into a biologically relevant model while retaining the capability to numerically solve for concrete results. It is difficult to model from the nanoparticle (10^{-9} m) to the tumor (10^{-3} m) scale, not only because matter behaves very differently in each, but because of the enormous computational cost associated with having to span six orders of magnitude of length scales over a significant period of biological time. In fact, simulation may require integration of multiple hierarchies of models, each differing in several orders of magnitude in terms of scale and qualitative properties [40].

Modeling of drug delivery encompasses the formulation of quantitative descriptions for drug transport in biological systems to evaluate feasibility of new drug delivery methods, to estimate dose response and toxicity, and to speed experimental and clinical evaluation [61]. Modeling principles apply to both procedures and technologies. For example, local drug administration, targeted drug delivery, and controlled drug release polymers should all be considered [61]. In the treatment of cancer, it is hoped that biocomputation will facilitate formulation of optimal treatment models that enable administration strategies for chemotherapy that maximize benefit while minimizing side effects [22]. Biocomputation-based generation of theoretical results could potentially be validated by correlation of numerical predictions with *in vitro* and *in vivo* data of a particular patient's cancer response to chemotherapy. In turn, these experimentally and clinically validated biocomputation results may be used to design personalized therapy protocols *in silico* using computer simulations.

Biocomputation of targeted and controlled drug delivery via nanoparticles is not only expected to offer insight into *in vivo* drug delivery, but also simulate the therapeutic effects of the delivery device and stipulate its preparation specifications in order to better address the challenges of nanotechnology. This approach may offer a means to optimize existing products and enhance new product development for cancer chemotherapy and disease treatment. The types of drug, excipient, and composition of the device could be essential components of a model [22]. Since there are no encompassing mathematical models that can apply to all conceivable physical and chemical processes in product development, it is important to develop an adequate theory grounded in physical considerations for specific systems. For instance, physical considerations that apply to polymer devices include drug delivery and diffusion, polymer swelling and degradation/erosion. It may also be necessary to consider osmotic, steric, magnetic, and charge effects [22].

15.2. ISSUES WITH CHEMOTHERAPY: HOW NANOTECHNOLOGY CAN HELP AND THE ROLE OF BIOCOMPUTATION

15.2.1. Drug Resistance

One of the major challenges that prevents most patients from benefiting from chemotherapy is the presence of tumor cell mechanisms that cause drug resistance. A tumor may evolve mechanisms to avoid damage by chemotherapeutic agents via the acquisition of mutations that confer a drug-resistant status. Nanoparticles with an appropriate surface coating could possibly overcome some mechanisms of cellular drug resistance, thereby improving the value of chemotherapy [22]. In fact, multidrug resistance (MDR) might be treatable with liposomes that enhance molecular MDR modulating strategies in addition to improving therapeutic activity through pharmacological optimization [54]. However, constant release of drug by nanoparticles at a tumor site could potentially exacerbate cellular resistance by exposing cells to a predictable (steady) level of stimulation. In fact, there is evidence that a single drug exposure can induce cellular resistance [80]. Biocomputation could help to quantify a nanoparticle drug release regimen that minimizes drug resistance. For example, Jackson and Byrne [33] proposed a mathematical model that described the chemotherapeutic response of a spherical vascular tumor containing two cell species. They contrasted the tumor response to continuous intravenous drug infusion versus intravenous bolus injection, and found that bolus injection decreased the time to cure when the drug resistant cell population was present. Biocomputation might also help to identify drug-resistant tumors via nanotechnology. Dordal *et al.* [17] analyzed fluorescent drug uptake by tumor cells using a three-compartment model in which rapid diffusion from extracellular fluid into a cell was followed by uptake into a non-exchangeable pool (where the drug bound with its intra-cellular target). By using a flow cytometric assay of drug uptake, the kinetic parameters of drug transport may be specified. The model could thus identify the presence of drug-resistant cells in a tumor by the reduced cellular uptake or increased cellular efflux of drug. This model could be a starting point to study the effects on drug resistance of drugs delivered with nanoparticles.

15.2.2. Drug Toxicity

Another challenge in chemotherapy is the use of potentially toxic side-groups that enhance the hydrophilicity of typically hydrophobic drugs. The addition of such side-groups may not be necessary with nanoparticles of biodegradable polymers that are small enough to allow intracapillary or transcapillary passage, and that possess a surface coating that evades macrophage uptake [22]. Thus, nanoparticles could be used to deliver traditional chemotherapy without toxic adjuvants to cancerous cells, and to treat conditions that may arise over time with anticancer therapy.

Toxicity could be considered in a biocomputational model as a constraint to preserve the white blood cell (leukocyte) number at a certain level while maximizing the reduction of the tumor cell population [4]. The goal would be to optimize the nanoparticle drug regimen under this constraint. On the other hand, Parker and Doyle [68] point out that through modeling of leukopenia, optimal delivery profiles could be constructed to minimize toxic effects. The method of delivery (e.g. bolus or continuous infusion) should also be considered, as it can lead to differences in toxicity.

15.2.3. Drug Targeting

Another issue with chemotherapy is that the drug may be delivered to tissues other than the tumor, affecting organs such as the heart and liver. Nanoparticles could provide a controlled and targeted means to deliver encapsulated drugs, resulting in lower side effects and higher efficacy [47]. A purely "chemical" strategy that relies on the molecular recognition of unique surface signatures of tumor tissue by chemical ligands (such as antibody-drug conjugates and immunoliposomes) may not work well with tumors because other tissue could also bear these signatures [71]. Controlled delivery may be achievable instead via a "physical" strategy because macromolecular transport across tumor microvessels can occur via fenestrations, vesicular vacuolar organelles, and transendothelial channels and inter-endothelial junctions [22]. The pore cutoff size of many tumor vessel models is between 380 and 780 nm, so nanoparticles in this size range should preferentially extravasate from tumor vessels [29, 89]. Since nanoparticles could also exit the circulation through the liver and bone marrow, the amount of particles needs to account for extravasation at these areas. A more unique signature of cancer cells is their abnormal DNA. Nanoparticles capable of screening DNA sequences of individual cells could recognize and kill cancerous ones.

Biocomputation could quantify the various means to target nanoparticles to specific sites within the body under various treatment scenarios. In fact, a combination of "chemical" and "physical" strategies may work best. One way of targeting is to conjugate cell-targeting agents on the nanoparticle surface [23, 25, 41]. Research is underway on ligand-targeted "stealth" liposomes that utilize moieties attached to the liposome surface to selectively bind the liposome to specific cancer cells [72]. A targeting ligand is chosen based on the ability of the target cell to internalize the liposome. Another way of targeting is through a magnetic field by using magnetite nanoparticles [32].

15.2.4. Drug Transport

A principal barrier to chemotherapy delivery can occur at the level of the compromised tumor vasculature. This barrier may prevent the delivery of adequate doses of drug to tumor cells [35]. Blood flow in tumor vessels is abnormal, since the flow is intermittent, periodically abating and reversing. These effects are caused by a chaotic arterial organization and impaired venous and lymphatic drainage [27]. The full consequences of an abnormal tumor vasculature on drug transport are not well understood. In the past some researchers (e.g. [37]) believed that it might take from days to months for a macromolecule to diffuse into the center of a tumor, mainly due to the high tumoral interstitial pressure and the collapsed tumor vessels. The high hydrostatic pressure in the tumor interstitium-see also Jain and Baxter [36] and the mathematical model by Sarntinoranont *et al.* [73]—would create an outward convective interstitial flow and cause drug resistance [38]. Recently researchers of this same group [67] have found evidence that proliferating cancer cells can cause intratumor vessels to compress and collapse, especially vessels without supportive stromal structures. Interstitial fluid pressure, on the other hand, is about the same as the microvascular pressure in the tumor, which makes it unlikely that the collapse of intratumor vessels is due to fluid pressure. It's important to note that this vessel collapse and transport limitation occurs on a timescale of hours and days, based on the rate of cell proliferation. However, the pharmacokinetics of a drug can be effected on a timescale of seconds, especially in highly perfused tissues such as the central nervous system, and as evidenced by radiographic scans showing drug delivery throughout brain tumors [7, 90]. Perhaps the drug is cleared from a tumor site on a timescale that precludes a full effect on all tumor cells, especially quiescent cells, and this effect, rather than a compromised tumor vasculature, is the main reason for the inadequate dosing. The tumor extracellular matrix assembly and composition could also be factors limiting drug transport [59].

The extent that a compromised tumor vasculature affects the bioavailability of larger molecular agents into the interstitium may depend on tumor type [81, 82]. By using contrast agents of different molecular weights in dynamic contrast enhanced MRI, it was shown that interstitial availability of macromolecular agents in different animal tumor models may be a function of tumor growth rate. (Fig. 15.1, top). In a fast growing tumor (top left), there was sufficient amount of macromolecular contrast agents in the interstitium within the measurement window of 16 minutes. In fact, the larger molecular weight agent started reaching equilibrium at the end of this period and resided in the interstitium for a

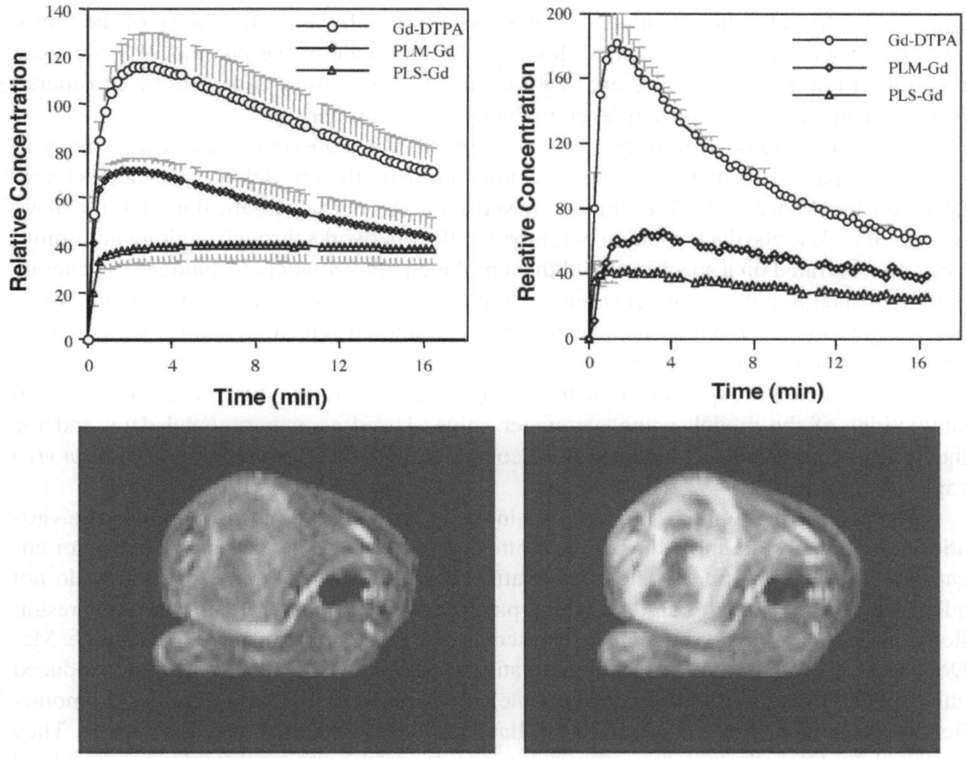

FIGURE 15.1. Top: contrast enhancement curves for different contrast agents (labeled) with hydrodynamic diameters 1-2.3 nm from a fast (left) and a slow (right) growing tumor. Bottom: Albumin-Gd-DTPA enhanced image at 10 minutes (left), and Gd-DTPA enhanced image at 30 sec (right). The former has higher molecular weight and 3 nm hydrodynamic diameter. Adapted from Su *et al.*, [81, 82], Magnetic Resonance in Medicine Vol. 34 and Vol. 39. Copyright © 1995 and 1998. Reprinted with permission of Wiley-Liss, Inc., & subsidiary of John Wiley & Sons, Inc.

considerable duration. Su *et al.* [81] showed that an even larger molecular weight contrast agent could have considerable interstitial uptake after 10 minutes (Fig. 15.1, bottom). The slower growing tumor demonstrated a very different behavior. In particular, the larger molecular weight agent did not leak into the extravascular space (Fig. 15.1, top right). In the faster growing tumor, vascular permeability was determined to be larger, resulting in higher accumulation of larger molecular weight agents in the extravascular space. It is conceivable that the behavior observed with macromolecular weight MR contrast agents also applies to similar size therapeutic drugs or nanoparticles.

Regardless of the potential mechanics that may affect drug delivery, very recently computer simulations in two spatial dimensions have demonstrated that nanoscale drug delivery systems could in principle be affected by similar limitations as traditional chemotherapy [79]. Nanoparticles first have to be transported in the blood stream to the vicinity of the tumor and extravasate from the blood vessels into the interstitial space; then the drug needs to be released and diffuse through or around the tumor cells [34]. In addition, nanoparticles must avoid protein binding in serum and in the extravascular space, metabolism in the blood, and phagocytosis by the reticuloendothelial system. Finally, the irregularity of the tumor vasculature with its abnormal blood flow and the timescale of the pharmacokinetics as a function of tumor type may present major obstacles. These limitations will be considered further when we address system-level modeling of cancer therapy.

Modeling of tumor vasculogenesis could be used as a possible angiogenesis assay to study the impact of an altered tumor vasculature on chemotherapy delivery via nanodevices. Chaplain and Anderson [11] recently reviewed a number of mathematical models that have been used to describe the formation of tumor capillary networks through angiogenic stimuli. They concentrated on a specific model that employed mathematical techniques to generate both two- and three-dimensional vascular structures. The model focused on the main events in angiogenesis, i.e., the migratory response of endothelial cells to cytokines secreted by a tumor, endothelial cell proliferation, endothelial cell interactions with extracellular matrix macromolecules, and capillary sprout branching and anastomosis. They presented numerical simulations of the model, using parameter values based on experimental data, and the theoretical structures thus generated were compared with the morphology of actual *in vivo* capillary networks.

The heterogeneity that may exist within the tumor (e.g. blood flow and pressure variation) complicates the modeling of chemotherapeutic agent delivery as a free drug or encapsulated in nanoparticles. The mathematical representation of vessel trees that do not adhere to normal diameter and branching patterns can become very complex. As a result, flows and pressures inside an abnormal vasculature become more difficult to calculate. McDougall *et al.* [52] used a discrete mathematical model to specifically study tumor-induced angiogenesis that described how the endothelial cell proliferative and migratory chemotactic responses led to the formation of a capillary sprout network of abnormal structure. They analyzed fluid flow through this network by considering the effects of fluid viscosity, blood vessel size and network structure on the rate of fluid flow, the amount of fluid present in the complete network at any given time, and the amount of fluid reaching the tumor. The incorporation of fluid flow through the generated vascular networks identified transport issues that may have implications for both nutrient supply and drug delivery to a tumor, echoing the earlier results of Jain [37]. In fact, under some conditions, the model showed that an injected chemotherapy drug could bypass the tumor altogether (Figure 15.2). Whether this effect would occur *in vivo* is unclear, since, as we have seen, there is evidence that drugs can be delivered throughout certain tumors. In general, though, as these and other researchers have noted, the normalization of the tumor vasculature could enhance the flow to a tumor mass, and thus aid the delivery of nanodevices as well.

15.2.5. Drug Dosage and Scheduling

Another determinant of drug efficacy is delivery of the optimal drug dosage. Considerations in this regard include tumor type and size, and the patient's physical parameters (e.g. body surface area, m^2). Chemotherapy drug dosages are selected in part based on the competing goals of maximizing death of malignant cells while minimizing damage to healthy cells. Because of more precise targeting by nanoparticles, drugs in nanocrystalline form may require smaller doses for equal effect. Since they could be delivered directly to the desired tissue while minimizing uptake by other tissues, the harm to healthy tissue

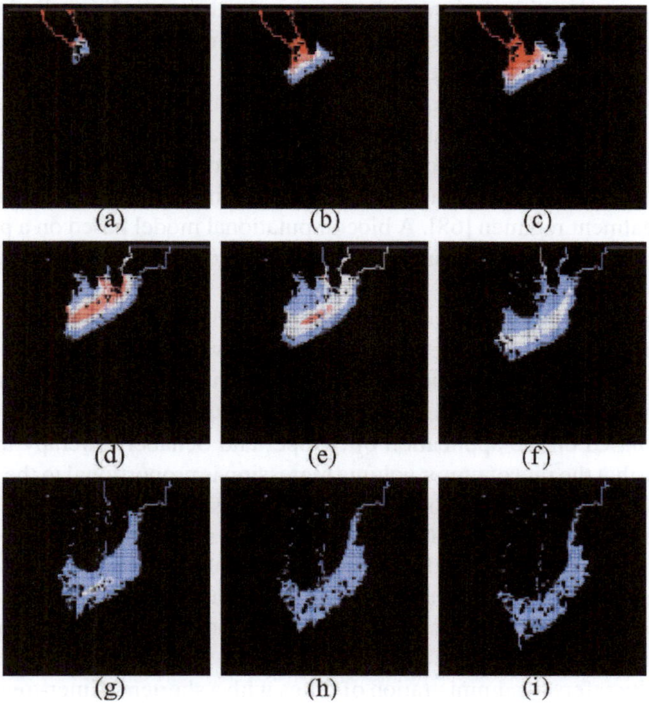

FIGURE 15.2. Effects of bolus injection through a computer simulated vasculature showing how most of the drug does not reach the tumor. Snapshots of drug concentration as it flows from parent vessel (situated at top edge of each picture) through the vascular network towards a tumor (situated at middle of bottom edge of each picture) over a physiological time duration (a)–(i). Colors: red = highest concentration; dark blue = lowest. Reprinted from Bulletin of Mathematical Biology, Vol 64, McDougall *et al.*, page 697, Copyright (2002), with permission from Elsevier.

would be reduced, although uptake by the liver and bone marrow might remain an issue. Better targeting also allows for more precise doses because the drug delivery will fluctuate less. For instance, future nanoparticles could achieve precise control over drug release via nanopores that act as particle membrane channels [3].

The fact is that the determination of drug doses and delivery schedules for a particular patient is a difficult process that relies on a series of trial-and-error procedures to determine the maximum tolerable dose and effective treatment regimen [68]. The variation of tumors in individual patients compounds the difficulty in determining an effective treatment based on the partial knowledge about the pharmacodynamics of the drug. The frequency of chemotherapeutic treatment has generally been based on the interval of time required for the myelopoietic cells to regenerate adequate numbers of lymphocytes, platelets, and erythrocytes, rather than being based on the effects on tumor cells, which may continue to proliferate faster than the recovery time of healthy tissue. A better understanding of the issues affecting chemotherapy dosage determination is needed to formulate dosages and schedules for drugs delivered by nanodevices.

From a biocomputational viewpoint, cancer growth has traditionally been defined as exponential (based on work by Skipper and Schabel in the 1970s). This tradition led to the log-kill hypothesis as the underpinning of current drug dosage and scheduling principles [66]. These principles include simultaneous combination chemotherapy, maximum tolerated dose within the combination, and equally spaced cycles of equal intensity. For example, chemotherapy has been routinely administered in 3-week intervals for metastatic breast cancer patients. Biocomputation could enable a more systematic approach to defining the drug treatment regimen [68]. A biocomputational model based on a patient's cancer characteristics could be defined, leading to treatment acceleration and less damage through ineffective dosages.

The Gompertzian growth curve [24] applied to a tumor shows that cell gain is greater than cell loss in the early part of the curve but slows down as the tumor gains mass [44, 88]. The curve is applicable to normal and malignant growth and has its origin in the molecular regulation of mitosis, tissue geometry, and apoptosis [66]. The Norton-Simon hypothesis [63], which is based on the application of Skipper and Schabel's therapy to Gompertzian computation, is that the rate of tumor volume regression is proportional to the rate of growth. The log-kill is greater when the tumor is treated at a smaller size, and its growth rate is higher if the cells are not destroyed. Since two drugs at single-agent dose could be toxic to a patient, Norton and Simon [63] determined that the alternatives were full dose, reduced dose, sequential dosing, and alternating dosing. The mathematical model thus allowed the development of dose density and sequential therapy, based on the theory of combination chemotherapy [65].

Dose density refers to administration of drugs with a shortened inter-treatment interval. It is based on the observation that in experimental models, a given dose always kills a certain fraction, rather than a certain number, of exponentially growing tumor cells [12]. Regrowth of cancer cells between cycles of chemotherapy is quicker in volume-reduced Gompertzian cancer models than in exponential models. The Norton-Simon model predicted that dose density would improve therapeutic results, and that sequential chemotherapy that maintains dose density would preserve efficacy while reducing toxicity [65]. The model explained how cancers that follow Gompertzian kinetics (e.g. breast cancer) respond to treatment, and how they differed from the exponentially growing models often used in the laboratory [64]. These considerations indicated that therapeutic results should be the same, even if the sequential pattern was less toxic [66].

Indeed, Citron *et al.* [12] recently reported that dose density can considerably improve clinical outcomes, and that sequential chemotherapy can be as effective as concurrent chemotherapy. As predicted by the model, sequential chemotherapy was better compared to a strictly alternating pattern [8, 62]. Various dose-dense drug regimens have been under investigation in recent years [56]. Clinical trials have further confirmed the model's prediction, leading to early breast cancer treatment that is shorter, less toxic, and more effective [66]. Future research into the biologic etiology of Gompertzian growth and the molecular mechanisms of its perturbation could generate new hypotheses for dose-schedule regimens that are empirically verifiably [12], and that could take into account drugs delivered by nanodevices.

Quantification of optimal chemotherapy profiles, usually assuming continuous drug delivery, can motivate the development of tumor growth models. An understanding of these models may be useful when considering cancer therapy via nanoparticles. Parker and Doyle [68] classified and described modeling approaches to cancer growth into two major groups: lumped parameter models and cell-cycle models. Lumped parameter models define tumor growth macroscopically in terms of cell count and selected tumor- or patient-dependent parameters, whereas cell-cycle models describe tumor behavior based on the number of cells in a given cell-cycle stage. Examples of lumped parameter modeling include logistic, Malthusian, Bertalanffy, and Gompertz equations. Each model is based on a growth function that is a continuous, monotonically rising function, describing increase per unit time in tumor cell count or tumor size. One major benefit of lumped parameter models is their advantage for controller design purposes due to their low order and monotonic cell growth behavior. However, the assumption of a homogeneously growing tumor cell population may not match real life, and the inability to account for cells in different growth stages prevents the use of lumped parameter models to study the effects of certain chemotherapeutic agents.

Although cell-cycle models may provide superior insight into the behavior of the tumor at the cellular level, they are very complex because each cell-cycle stage needs to have its own mathematical specification. In order to specify the exact number of cells in the various cell-cycle stages, direct measurements of the tumor cells would be necessary. Parker and Doyle [68] suggested that in the case when model parameters cannot be identified, approximate models of cell-cycle behavior can be constructed, and these models may be useful for analysis purposes. For instance, models of this type can handle the effects of cell-cycle specific drugs. Thus, if good estimates of model parameters are available, this model structure, with its additional detail, can provide a substantial advantage. Intermediate levels of model complexity and more detailed tumor growth models are also possible. In general, however, both lumped parameter and cell-cycle models share a number of shortcomings when it comes to their application to chemotherapy either as free drug or encapsulated in nanoparticles. For example, the assumption of continuous drug delivery may not be valid, since a metronomic regimen may not be appropriate. Moreover, the models are usually one-dimensional in space, disregarding the physical effects that the tumor three-dimensional heterogeneity can have on drug dosage and scheduling.

15.2.6. Drug Concentration

The therapeutic efficiency of a pharmaceutical product is determined by the proper concentration of drug at the lesion site, and biocomputation can describe the relevant pharmacokinetics, especially when considering devices on the nanoscale. Efforts in this area, for instance, have included the development of a mathematical model describing the microscopic profiles and biodistribution of drugs using enzyme-conjugated antibodies as part of a two-step method for cancer treatment [6]. The monoclonal antibodies by themselves may lead to heterogeneous uptake within the tumor, while the use of a low molecular weight agent may allow deeper penetration into the tumor. This mathematical model was used to describe concentration profiles surrounding individual blood vessels within a tumor, which allowed determination of the area under the curve and specificity ratios. Average tissue concentrations were determined by spatial integration and compared with experimental results. The model showed that the effective clearance of antibody inside the tumor is less efficient than outside the tumor, which may be due to the antibody accumulation at the tumor. The conclusion was that enzyme-conjugated antibodies could help to achieve a more uniform distribution and higher concentrations of the active agent, as well as greater specificity.

In another study of drug concentration at the tumor site, Quian *et al.* [71] developed a mathematical model that could specify the physical dimensions of polymer millirods, composed of PLGA (poly(lactic-co-glycolic acid)) microspheres, inserted directly at the ablation boundary of thermally ablated solid tumors. Based on the rod dimensions, the model showed how an initial loading dose of chemotherapeutic drug followed by a sustained release can provide optimal drug concentrations at the tumor site. Without a loading dose, it would take several days to attain a desired therapeutic concentration via a zero-order (constant) release device. This model may have relevance when establishing the optimal drug concentration via constant-release nanodevices.

Recently, Eliaz *et al.* [19] developed a cell kinetic model showing that the potency of a chemotherapeutic drug (doxorubicin) encapsulated in liposomes was 5 to 6 times higher than free drug. Targeted liposomes delivered more drug into the cell than the free form. In fact, drug delivery via targeted liposomes was more efficient in killing the cells per amount of intracellular drug. This efficiency may be due to the vascular trapping of liposomes in the peritumoral space generating a constant release of drug and creating a more uniform drug concentration. Clinical testing has confirmed that the plasma distribution and elimination half life of liposomal doxorubicin can be much longer than that of free drug, and response rates can be significantly higher [51].

In vivo, tumor response to therapy is governed by the pharmacodynamics and pharmacokinetics of the chemotherapy drug. The effective drug concentration in the tumor model is generally assumed to be equal to that in blood plasma because it is difficult to measure drug concentration within a tumor. However, given the possibility that transport limitations through the abnormally constituted tumor vasculature may cause the drug concentration to be lower within the tumor [37, 48, 67, 79], this assumption may not hold. Thus, the specification of optimal drug therapy is very complex, principally due to a poor understanding of the response of the tumor system to drug therapy [68], which can include variability in drug concentration. The concentration of drug delivered by nanoparticles would depend on the concentration of nanoparticles, which would also be subject to this complexity.

It is important to note that the clinical effect of a drug ultimately depends on the drug concentration inside individual tumor cells, not just the extracellular concentration in the tumor interstitium. For instance, to aid introduction of highly charged or macromolecular drugs into the cytoplasm, pH-sensitive liposomes have been developed that deliver their contents through penetration of the endosomal membrane [78]. Inner and outer cell space can be represented with a two-compartment model of drug concentration. Lankelma [48] described two bounds on the intracellular (inner compartment) concentration: low cellular drug influx with rapid efflux in a sparse cell cluster, and high cellular drug influx with low drug efflux in a tightly packed cluster. The former will lead to homogeneous drug distribution without a gradient, while drug gradients will last longer in the latter. Gradients thus depend on cellular influx and efflux, and on blood concentration as influenced by the tumor vasculature. Gradients of nanoparticles would depend on similar factors. Local drug or nanoparticle concentration could vary considerably due to intercapillary distances and heterogeneity in the tumor cell population.

15.2.7. Drug Release

In this section we review some fundamental concepts regarding drug release from nano- and micro-particles, and examine the role of biocomputational modeling in this area. Drugs can be designed for programmed release *in vivo* by encapsulating them in particles from a few nanometers to microns in size. Particles are usually ingested or implanted, and designed to deliver a controlled release of drug that may last for an extended period of time (weeks or months). In general, drug kinetics can be studied as material fluxes between conceptual units, called compartments [5]. Holz and Fahr [30] reviewed two main groups of biological compartment models, namely, physiological and mechanistic models. On the other hand, Veng-Pedersen [85] reviewed non-compartmentally based models. These utilize systems analysis, such as linear systems analysis (LSA). The wide array of available LSA-based kinetic analysis tools may offer an alternative to traditional kinetic modeling. Both compartmental and non-compartmental models, however, do not usually describe the complexity of multiple spatial dimensions that exist in the tumor environment.

Mathematical modeling can help to optimize the design of a therapeutic device to yield information on the efficacy of various release methods [22]. For example, Wang *et al.* [86] compared two types of drug formulations, namely, controlled release from polymers and systemic administration, to predict spatial and temporal variations of drug distribution at the tumor level in two dimensions. In contrast with bolus injection, polymer-based delivery imparted a longer exposure time, a higher mean concentration, and a reduced systemic toxicity. Drug release from a polymer nano- or microparticle has been traditionally classified based on the material erosion mechanism: surface or bulk erosion [45]. For either type of erosion, models developed to characterize the kinetics of drug release from spherical microparticles were described by Zhang *et al.* [91]. They pointed out three mechanisms that combine to control the overall drug release process: dissolution of drug from the solid phase, diffusion of dissolved drug, and erosion of the polymer matrix. These models can be solved under either a finite or infinite mass transfer condition. For bulk erosion of both hydrophobic and hydrophilic polymers, the models showed a reasonable match with experimental results reported in the literature. Results also indicated that the surrounding environment had a profound effect on the drug release pattern under a finite mass transfer condition. For various surface-eroding polymers, it was observed that the radius of the microsphere followed an approximately linear profile of reduction with respect to time. In some cases, erosion and dissolution appeared to be dominant factors for drug release patterns. For better application of these models, the proportion of amorphous and crystalline polymer, and free chain and rigid chain could be investigated to justify the corresponding parameter values. Furthermore, physical property data (such as diffusivity and porosity) for drugs and microspheres should be determined experimentally to improve simulation results. In particular, quantitative analysis on the experimental diffusivity coefficient, dissolution constant, and erosion constant might help in this regard.

Feng and Chien [22] provided a comprehensive list of mathematical models that have been developed to study drug release at the nanoparticle level. As Siepmann and Goepferich [74] pointed out, the modeling of bioerodible delivery systems is more complex than the modeling of diffusion or swelling-controlled devices. Chemical reactions (e.g. polymer chain cleavage) in bioerodible systems have to be taken into account in addition to physical mass transport phenomena. These reactions continuously change the conditions for mass transfer processes, complicating the modeling of erosion-controlled drug release. Siepmann and Goepferich [74] classified erosion-controlled drug release models into two categories: empirical models that commonly assume a single zero-order (constant) process to control rates of drug release, and models that consider physicochemical phenomena (such as chemical reaction processes or diffusional mass transfer). The latter category includes simulation of polymer degradation as a random event using direct Monte Carlo techniques (i.e., using computer-generated random numbers). The actual physics of the polymer dissolution process and the consequences for drug release have been modeled. For example, Narasimhan [55] described the main modeling contributions in this area using two broad approaches; phenomenological models and Fickian equations, and anomalous transport models and scaling law-based approaches.

Despite the phenomena complexity involved in drug release from nano- and microparticles, the two mathematical models commonly used to describe drug release kinetics from a large variety of devices are the Higuchi model [28] and the power model [69]. The Higuchi model is:

$$M_{\mathrm{t}}/\mathrm{A} = (D\,(2c_{\mathrm{o}} - c_{\mathrm{s}})c_{\mathrm{s}}\,t)^{1/2}$$

where M_{t} is cumulative amount of drug released at time t, A is surface area of the controlled release device exposed to the release medium, D is drug diffusivity, and c_{o} and c_{s} are initial drug concentration and drug solubility, respectively [76]. In general, the Higuchi model is valid for systems where drug concentration is much higher than drug solubility, whereas with the power model, the geometry of the system can be related to the drug release mechanism [42]. The power model is:

$$M_{\mathrm{t}}/M_{oo} = kt^{\mathrm{n}}$$

where M_{t} and M_{oo} are absolute cumulative amounts of drug released at times t and infinity, respetively, k is constant incorporating structural and geometrical device characteristics, and n is the release exponent, indicative of the mechanism of drug release [76]. For comparison to these two models, we note that drug release from a traditional matrix, as a result of a diffusion process that assumes excluded volume interactions between drug molecules, can be described by the Weibull function [87]:

$$M_{\mathrm{t}}/M_{oo} = 1 - \exp(-a\,t^{\mathrm{b}})$$

where a and b are empirical constants respectively defining the scale and shape of the response.

Various drug release/dissolution models were compared by Costa and Lobo [13]. They pointed out that models that in general describe drug release phenomena best are the Higuchi model, zero order model (as a special case of the power model), Weibull model and Korsmeyer-Peppas model. The Higuchi and zero order models represent two limit cases in the transport and drug release phenomena, while the Korsmeyer-Peppas model can be a decision parameter between these two models. Whereas the Higuchi model has a large application in polymeric matrix systems, the zero order model can be useful in describing membrane controlled dosage forms or coated dosage forms. Costa and Lobo [13] also suggested using the adjusted coefficient of determination (R^2_{adjusted}) to compare models with different numbers of parameters:

$$R^2_{\text{adjusted}} = 1 - \frac{(n-1)}{(n-p)}(1 - R^2)$$

where n is number of dissolution data points, p is number of parameters in the model, and R^2 is the (unadjusted) coefficient of determination.

As Siepmann and Peppas [76] pointed out, there are several important assumptions when applying the Higuchi model to controlled drug delivery systems. We will briefly mention them and consider how they apply in the case of drug release at a tumor. The first assumption is that initial drug concentration is much higher than drug solubility, which justifies a pseudo-steady state modeling approach. This assumption may not hold for a tumor undergoing intermittent exposure to chemotherapy, which means that a pseudo-steady state approach may not be sufficient. The second assumption is that the mathematical analysis is based on one-dimensional diffusion. Although the simplification afforded by one-dimensionality can lead to valid insights, it seems that the physical heterogeneity of the tumor environment would be better represented by a multi-dimensional analysis [79]. The third assumption is that the diameter of the suspended drug particles is much smaller than the thickness of the system, which would be true for chemotherapy via nanoparticles. The fourth assumption is that dissolution or swelling of the carrier system is negligible. This assumption may or may not hold in the tumor stroma. The fifth assumption is that drug diffusivity is constant, which may not necessarily hold in the case of chemotherapy. The final assumption is that perfect sink conditions are maintained. This assumption probably applies to the tumor environment because the flow of extracellular fluid carries the drug away.

Having discussed some of the fundamentals of nano- and micro-particle drug release and its associated modeling, we will now consider modeling efforts in system drug release kinetics. The pharmacokinetics and distribution of a drug can change substantially by encapsulation into nanoparticles. The drug will assume the pharmacokinetics of the carrier until its release [72]. Thus, from a system viewpoint, the use of nanoparticles changes the drug release so that it primarily consists of two phases. First phase involves delivery of nanoparticles to the tumor site, and the second phase involves drug release by the collection of nanoparticles. In fact, analysis of nanoparticle release profiles will usually display such a biphasic release pattern ([43], pp. 258–259). This behavior can be quantified in terms of drug release parameter values that become an input to the biocomputational modeling at the much larger millimeter-scale of the tumor.

Important work has been done on modeling system drug release kinetics of microparticles, whereas the modeling of nanoparticles has not been as extensive. A small selection of studies and review papers will be noted in this section to illustrate some of the main modeling aspects in this area. The hope is that most of this work could, with some further research, be extended to the nanoscale. In fact, this effort would aid in bridging the gap from the nano- to the macroscale, by providing a quantitative link that could serve as an input parameter for the modeling at the tumor scale.

In the area of drug release kinetics from microparticles, Siepmann *et al.* [77] described an applied mathematical model, considering drug diffusion with non-constant diffusivities (to account for polymer degradation), which was able to quantitatively describe

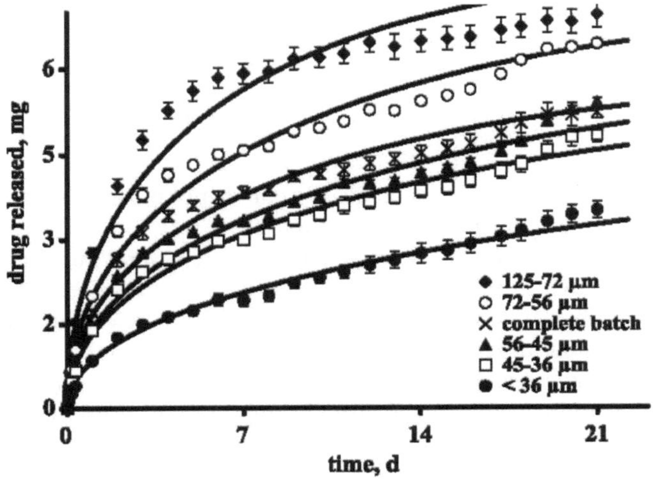

FIGURE 15.3. Fit of a mathematical model taking into account drug diffusion and polymer degradation (solid curves) to experimentally determined drug release from PLGA–based microparticles (symbols) in phosphate buffer pH 7.4. Particle size is given in the legend. Reprinted from Journal of Controlled Release, Vol 96, Siepmann *et al.*, page 32, Copyright (2004), with permission from Elsevier.

experimentally observed drug release patterns. An exponential relationship was established between the diffusion coefficient and the initial loading of drug, allowing resulting drug release kinetics for arbitrary microparticle sizes to be predicted in a quantitative way. Drug release was found to be independent of particle size (in the range of <36 μm to 125 μm), and drug transport was primarily controlled by diffusion (Figure 15.3). Hombreiro-Perez *et al.* [31] modeled drug release by non-degradable microparticles, proposing a means to predict the effect of different formulation parameters on resulting drug release patterns (such as the effect of microparticle size). Siepmann *et al.* [75] proposed a model quantifying drug release from bioerodible microparticles using Monte Carlo simulations. The model was able to describe observed drug release kinetics accurately over the entire period of time, including initial "burst" effects, subsequent zero-order drug release phases, and second rapid drug release phases (Figure 15.4). The evolution of drug concentration profiles within the microparticles could then be calculated. Finally, Faisant *et al.* [21] described a mathematical model that enabled a quantitative description of drug release patterns of PLGA microparticles. The release was biphasic (initial burst, followed by a zero-order phase) and mainly driven by drug diffusion. Coefficients for drug diffusion increased as the polymer absorbed water and the average molecular weight of molecules decreased. The polymeric network breakdown did not affect the release process because it occurred after the drug was depleted.

15.3. BIOCOMPUTATION AT THE SYSTEM LEVEL

15.3.1. Modeling at the Nanoscale

We will now examine the system-level biocomputation of cancer therapy and its relationship with nanotechnology by first evaluating system modeling at the nanoscale. In

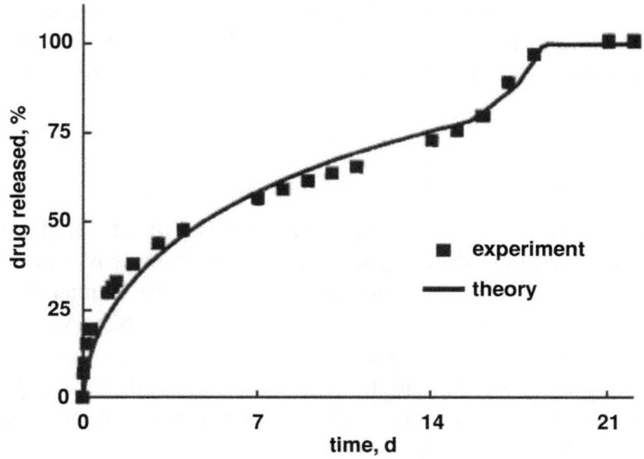

FIGURE 15.4. Triphasic drug release kinetics from PLGA-based microparticles in phosphate buffer pH 7.4: experimental data (symbols) and fitted theory (curve). Adapted from Pharmaceutical Research, Vol 19, 2002, p. 1887, Siepmann *et al.*, Figure 1, © 2002 Plenum Publishing Corporation. With kind permission of Springer Science and Business Media.

general, modeling requires a solid theoretical framework in order to produce results that can deliver insights into the phenomena under study (NINT, 2004). Materials modeling underscores most of nanoscience research so that the performance and characteristics of novel materials (such as polymers) may be predicted. General areas of research in fundamental nanoscience include the development of new theories (such as many-body quantum theories and mesoscopic theories) and modeling strategies such as multiscale modeling and multiphysics, data processing and analysis, and comparing theory with experimental results (EPSRC, 2004). These approaches may provide solid tools for modeling materials at the nanoscale. For example, Sumpter *et al.* [83] describe recent developments in the formation, characterization and simulation of nano- and micro-scale particles of amorphous polymer blends and semi-crystalline polymers, including the modeling of structural characteristics, thermal and mechanical properties, particle-surface interactions, and particle-particle interactions.

Nanoparticle technology is based on the physics of materials at the molecular level. In fact, mathematical and computational modeling of systems at the nanoscale requires a blend of quantum with classical mechanics. Since quantum mechanical models computed from first principles (i.e., without any empirical input) require a large amount of computational power, the size of a system that can be described by accurate quantum mechanical models is limited to about 50 atoms (using current computer technology) [84]. Classical models, on the other hand, can neglect important quantum effects that give nanoscale devices their unique properties [84]. These constraints make the modeling of systems at the nanoscale very challenging.

Although fundamental nanoscience biocomputation is not easy, it could be argued that modeling at the tumor scale should not be burdened by this complexity when considering nanotechnology in cancer treatment. The reason is that the bulk behavior of nanodevices is what matters the most at the tumor level—not the behavior of individual particles of sizes on the order of 10^{-9} m. It takes millions of nanoparticles and their combined effect to influence a tumor of size order 10^{-3} m. Thus, modeling of cancer drug treatment via nanotechnology could stay focused at the tumor scale by considering the collective behavior of nanodevices.

15.3.2. Modeling at the Tumor Scale

The foundation for models of nanoparticle delivery of cancer therapy depends on an accurate physico-chemical description of the tumor microenvironment, the parameters of which are derived from the study of cancer biology. Fundamental facts regarding tumor behavior have been well described ([1], pp. 1313–62). At a cellular level, tumor cells are transformed cells that have evaded natural cell senescence. All cells require a supply of energy to live and they produce metabolic waste. Cancer cells can survive where normal cells would die, such as under hypoxic, hypoglycemic and acidic conditions. They can adapt to changing micro-environmental conditions to develop resistance to therapy. They can sometimes thrive outside their natural environment, leading to metastases.

At a tumor (system) level, aggregates of cells will affect each other through mechanical forces in three-dimensional space in such a manner that some cells will have more access to nutrients than others. At first glance, the cell aggregate would be expected to expand as a perfect sphere if all cells on the periphery experienced the same mechanical forces. Experimental and clinical observation has shown that such perfect symmetry is usually not the case. The reason is that from a molecular perspective, the extracellular environment that each cell experiences (e.g. nutrient concentration) can vary quite dramatically, leading to favored cells entering the cell cycle more often than cells whose extracellular cues are more adverse ([1], pp. 985–6). Nutrient competition leads to the selection of cells favored for maximal proliferation in certain regions. Cells on the periphery of a tumor are favored and tend to proliferate faster than cells that are surrounded by other tumor cells. The morphology of a tumor can be seen as a function of response to various environmental fluctuations, including this nutrient diffusional instability [14, 15, 79, 92]. The goal of cancer modeling is to describe the actual detailed behavior of a cell aggregate as predicted by the proper physical formulation. The underlying hypothesis is that if the main components of this physical formulation are identified and abstracted to a mathematical level, then this formulation can be represented as an *in silico* system capable of predicting and shedding insight into the behavior of real tumors.

The history of the study of tumor biology via physical formulations has been long and insightful. For an excellent review, refer to Araujo and McElwain (2005). Cancer growth, angiogenesis, metastasis, etc. have all been abstracted to a mathematical level. A recent biocomputational implementation by Zheng *et al.* [92] encompassed some of the main physical characteristics of cancer growth and created an *in silico* system that exhibited combined two-dimensional tumor growth and angiogenesis. This system captured the complicated morphology and connectedness at the tumor/tissue interface, including invasive fingering, tumor fragmentation, and healthy tissue degradation. Implementation allowed for simulation of tumoral lesions through the stages of diffusion-limited dormancy, localized necrosis, vascularization and rapid growth, and tissue invasion in multiple spatial dimensions. Angiogenesis was included as a continuous feedback process involving tissue growth and nutrient demand. An application of this simulator to chemotherapy is described in the following section.

15.3.3. Modeling of Cancer Therapy

There are currently few biocomputational models that specifically consider nanotechnology as part of cancer treatment, especially taking into account the physical multidimensionality of the tumor mass. Recently, Sinek *et al.* [79] have studied nanoparticle mediated drug delivery and tumor response using the tumor simulator of Zheng *et al.* [92]. Their multi-scale and multi-dimensional simulations demonstrated the potential increased efficacy of nanoparticle-based therapy as well as its potential weaknesses, due principally to transport limitations. They assumed a best-case scenario involving an homogenous tumor with one cell type that was also assumed to be drug-sensitive, low host tissue toxicity due to targeted drug delivery, and a constant nanoparticle drug carrier delivery at levels calibrated to be lethal to *in vitro* cell culture on a time scale of hours. Two ends of a spectrum were considered: therapy involving very small (1–10 nm) nanoparticles that extravasate from tumor vasculature, diffuse within the interstitium, and target cells, and therapy involving large (100 nm) nondiffusing nanoparticles that are assumed to remain at their point of extravasation from the vasculature and to function as a constant source of drug. In both cases nanoparticles were assumed to be delivered only to the tumor due to large vasculature openings. Because of lower toxicity, larger and more uniform drug concentrations were delivered to tumor cells over longer time periods in comparison to traditional free-drug administration protocols.

However, their simulations also showed that nanoparticle-based chemotherapy could suffer from the same fundamental transport limitations as free-drug administration. Competition between vasculature density, which favors nutrient and nanoparticle extravasation, and intratumoral pressure, which may oppose it, could result in non-uniform delivery. Diffusion of nutrient molecules and drug carrier within the tumor interstitium may further contribute to this inhomogeneity. Figure 15.5 shows non-uniform intratumoral distributions in simulated chemotherapy involving a continuous blood-serum concentration of 1–10 nm nanoparticles.

Sinek *et al.* [79] also simulated the effects of antiangiogenic therapy on tumor vasculature. It had been previously proposed that this therapy could "normalize" tumor vasculature through more efficient and uniform delivery of molecules and particles [38]. In this simulation, larger 100 nm particles were assumed to extravasate uniformly along the normalized vasculature and to release drug at a constant rate. Although Figure 15.6 shows that tumor regression was significantly higher than in the previous simulation, Figure 15.7 indicates that there was drug concentration inhomogeneity. As a result of non-uniform delivery of chemotherapy, tumor regression was likewise non-uniform, being highest around areas of maximum drug extravasation. Average *in vivo* cell death rates as simulated *in silico* were several orders of magnitude lower than those calibrated *in vitro*. Perhaps more importantly, non-uniformity of tumor regression consistently led to fragmentation (Figure 15.7) and new stable tumor mass at significant levels (Figure 15.6).

In order to be more complete, modeling of chemotherapy using nanoparticles should include other factors that affect tumor growth. These include hypoxic cycling cells and a heterogeneous population of genotypes, some of which are resistant to the drug. Also, the

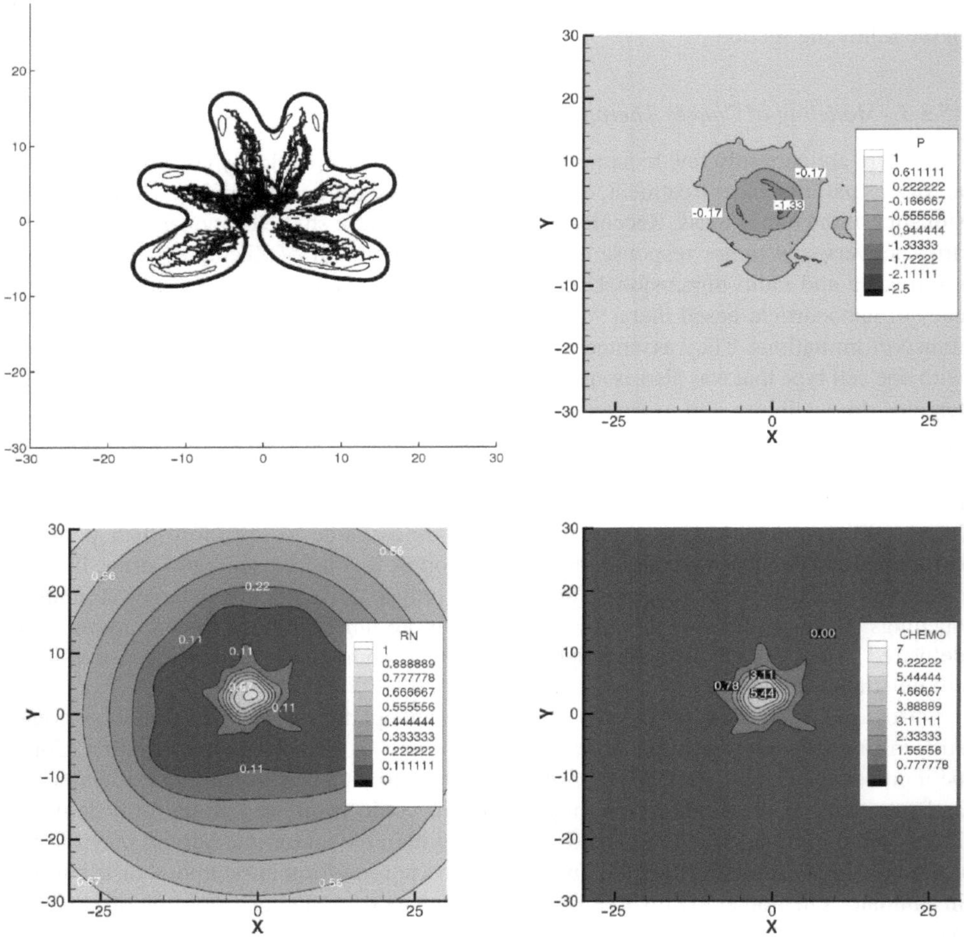

FIGURE 15.5. Top left: Stable highly perfused tumoral lesion proliferating around blood vessels. Solid thick perimeter: tumor boundary; solid thin perimeter: necrotic areas; solid: blood vessels. Top right: Tumor pressure distribution. Bottom left: Nutrient concentration. Bottom right: Nanoparticle distribution. All variables are dimensionless. Adapted from Biomedical Microdevices, Vol 6, 2004, p. 306, Sinek *et al.*, Figure 4b, © 2004 Kluwer Academic Publishers. With kind permission of Springer Science and Business Media.

contribution to pressure within the tumor by the mass of necrotic cells may not be negligible [79]. The nanoparticles themselves could be better modeled by augmenting the knowledge of how they work *in vivo*, considering issues of vessel extravasation, clustering, interstitial diffusion, interaction with tissue, and erosion at the tumor site.

For instance, the pH of the microenvironment is a significant aspect controlling the degradation kinetics of many pharmaceutically relevant polymers, since hydrolysis rates can vary by orders of magnitude at different pH values [49]. The tumor extracellular environment is more acidic than normal tissue because of lactic acidosis from glycolysis [26]. The poorly perfused tumor vasculature maintains the acidic environment as well. Polymer nanoparticle degradation will thus be affected. Furthermore, as these polymers degrade into acids, the pH

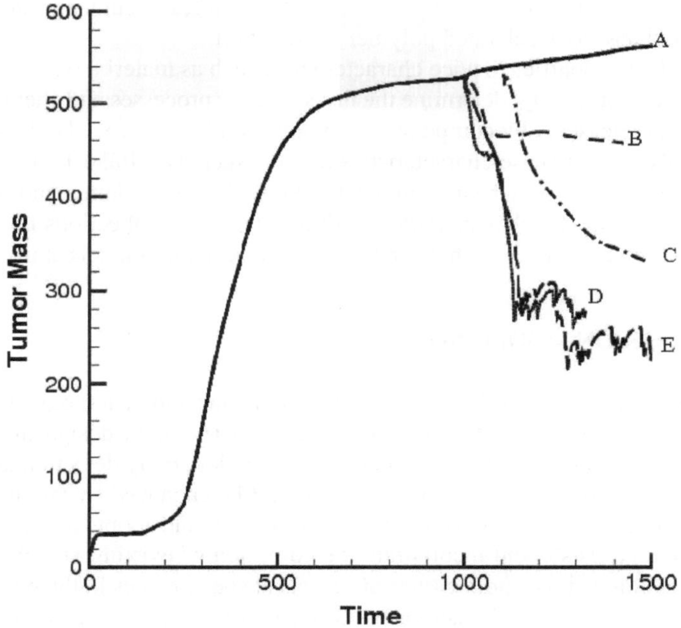

FIGURE 15.6. Simulation of tumor mass growth and regression as a function of time. The non-dimensionalized time unit is ≈3.3 days. (A) Without chemotherapy; (B) With chemotherapy via small, diffusing nanoparticles (Fig. 15.5); (C) With chemotherapy via large, non-diffusing nanoparticles and adjuvant anti-antiangiogenic therapy (Fig. 15.7); (D), (E) Simulations corresponding to cases (B) and (C) but assuming higher blood vessel mobility. Adapted from Biomedical Microdevices, Vol. 6, 2004, p. 303, Sinek *et al.*, Figure 3, © 2004 Kluwer Academic Publishers. With kind permission of Springer Science and Business Media.

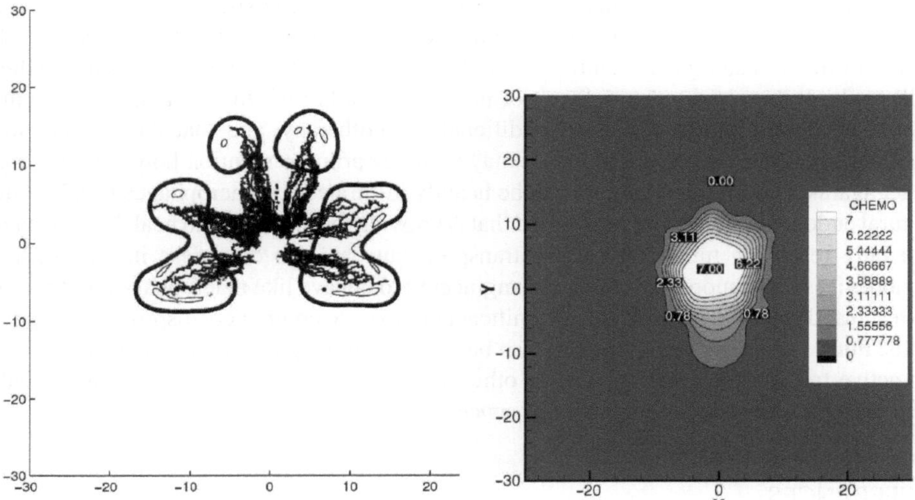

FIGURE 15.7. Left: Tumor regression with undesirable mass fragmentation after application of chemotherapy plus antiangiogenic therapy via nanoparticles. Right: Drug distribution. Adapted from Biomedical Microdevices, Vol 6, 2004, p. 307, Sinek *et al.*, Figure 5, © 2004 Kluwer Academic Publishers. With kind permission of Springer Science and Business Media.

of the device microenvironment will be lowered as drug is released, inducing an autocatalytic effect that could lead to accelerated polymer degradation.

The specific nanoparticle device characteristics, such as material type, shape and size, as well as encapsulated drug, determine the mass transfer processes and chemical reactions controlling drug release. A biocomputational model should not only take these factors into account, but also drug release characteristics *in vivo*, such as cellular tissue reactions and osmotic pressure, which may bear significantly on particle degradation and resulting drug release kinetics [74]. The characteristics of drug release for collections of nanoparticles may then yield input parameters that can be used in continuum models at the tumor scale.

15.4. OUTLOOK ON MODELING

Biocomputation at the service of cancer treatment via nanotechnology may be of value in dealing both with the issues of chemotherapy delivery and in the design and manufacture of nanodevices that could be effective in cancer prevention, early detection and diagnosis. At a system level, modeling at the tumor scale could be enhanced by taking into account cell genetic characteristics, such as mutations of oncogenes and apoptosis-suppressor genes. Models of angiogenesis could incorporate the co-option of existing vessels, and perhaps even the morphology, flow, and pressure of vessels at specific sites in the body affected by tumor growth. Drug release characteristics of nanoparticles could be quantified in terms of drug release parameter values that become part of the model at the tumor scale. The ultimate goal, and "holy grail" in this field, would be to enable the clinical application of biocomputation to design cancer treatment via nanotechnology based on a particular patient's physiological conditions. The results of the *in silico* model could then provide valuable diagnostic, prognostic, and therapeutic information.

We conclude by noting that biocomputation predicts that the transport of nanoparticles, aimed at therapy at the individual cell level, should be expected to be ruled by the same type of physical phenomena as similarly sized molecules in the human body. For example, the laws of diffusion along a concentration gradient and convection along a pressure gradient still apply. It thus seems that delivery of nanoparticles through the circulation to a tumor would encounter similar issues as traditional chemotherapy, and that these issues may challenge nanodevices of the future. It may be noteworthy to mention how the body may react against cancer: some tumors can be heavily infiltrated by macrophages [50, 70], in a natural endogenous response to a wound that is growing faster than it can heal. Macrophages not only reach the tumor by passive transport, but can also diapadese into it. Perhaps nanodevices and nanoparticles that are engineered to behave like macrophages will be able to reach and stay at the tumor site in significant numbers to completely dispose of malignant cells. Future work in biocomputation can be expected to help formulate the details of such an active transport homing, as well as other mechanisms, in order to achieve a clinically successful response in the treatment of cancer.

REFERENCES

[1] Alberts et al. *Molecular Biology of the Cell.* Taylor & Francis Group, New York, 2002.

[2] R.P. Araujo and D.L.S. McElwain. A history of the study of solid tumour growth: the contribution of mathematical modeling. *Bull. Math. Biol.*, June 2004. In press.

[3] The A to Z of Nanotechnology at website www.azonano.com.
[4] D. Barbolosi and A. Iliadis. Optimizing drug regimens in cancer chemotherapy: a simulation study using a PK-PD model. *Comp. Biol. Med.*, 31:157–172, 2001.
[5] L.A. Bauer. *Applied Clinical Pharmacokinetics*. Mc-Graw Hill, New York. pp. 26–45, 2001.
[6] L.T. Baxter and R.K. Jain. Pharmacokinetic analysis of the microscopic distribution of enzyme-conjugated antibodies and prodrugs: Comparison with experimental data. *Brit. J. Canc.*, 73(4):447–456, 1996.
[7] K.L. Black and N.S. Ningaraj. Modulation of brain tumor capillaries for enhanced drug delivery selectively to brain tumor. *Cancer Control.*, 11(3):165–173, 2004.
[8] G. Bonadonna, M. Zambetti, and P. Valagussa. Sequential or alternating doxorubicin and CMF regimens in breast cancer with more than three positive nodes: ten year results. *J. Am. Med. Assoc.*, 273:542–547, 1995.
[9] I. Brigger, C. Dubernet, and P. Couvreur. Nanoparticles in cancer therapy and diagnosis. *Adv. Drug Del. Reviews.*, 54(5):631–651, 2002.
[10] CCO Formulary, Liposomal Doxorubicin, October 2003.
[11] M. Chaplain and A. Anderson. Mathematical modelling of tumour-induced angiogenesis: network growth and structure. *Cancer Treat Res.*, 117:51–75, 2004.
[12] M.L. Citron, D.A. Berry, C. Cirrincione, C. Hudis, E.P. Winer, W.J. Gradishar, N.E. Davidson, S. Martino, R. Livingston, J.N. Ingle, E.A. Perez, J. Carpenter, D. Hurd, J.F. Holland, B.L. Smith, C.I. Sartor, E.H. Leung, J. Abrams, R.L. Schilsky, H.B. Muss, and L. Norton. Randomized trial of dose-dense versus conventionally scheduled and sequential versus concurrent combination chemotherapy as postoperative adjuvant treatment of node-positive primary breast cancer: first report of intergroup trial C9741/cancer and leukemia group B trial 9741. *J. Clin. Oncol.*, 21(8):1431–1439, 2003.
[13] P. Costa and J.M.S. Lobo. Modeling and comparison of dissolution profiles. *Eur. J. Pharm. Sci.*, 13:123–133, 2001.
[14] V. Cristini, J. Lowengrub, and Q. Nie. Nonlinear simulation of tumor growth. *J. Math. Biol.*, 46:191–224, 2003.
[15] V. Cristini, H.B. Frieboes, R. Getemby, S. Corenta, M. Ferrari, and S. Sinek. Morphological instability and cancer inussion. *Clin. Cancer Res.*, (in press) 2005.
[16] Z. Cui, C.H. Hsu, and R.J. Mumper. Physical characterization and macrophage cell uptake of mannan-coated nanoparticles. *Drug Dev. Ind. Pharm.*, 29(6):689–700, 2003.
[17] M.S. Dordal, A.C. Ho, M. Jackson-Stone, Y.F. Fu, C.L. Goolsby, and J.N. Winter. Flow cytometric assessment of the cellular pharmacokinetics of fluorescent drugs. *Cytometry*, 20:307–314, 1995.
[18] P. Elamanchili, M. Diwan, M. Cao, and J. Samuel. Characterization of poly(D,L-lactic-co-glycolic acid) based nanoparticulate system for enhanced delivery of antigens to dendritic cells. *Vaccine*, 22(19):2406–12, 2004.
[19] R.E. Eliaz, S. Nir, C. Marty, Jr. F.C. Szoka. Determination and modeling of kinetics of cancer cell killing by doxorubicin and doxorubicin encapsulated in targeted liposomes. *Canc. Res.*, 64:711–718, 2004.
[20] EPSRC Research Priorities and Opportunities at http://www.epsrc.ac.uk/ContentLiveArea/Downloads.
[21] N. Faisant, J. Siepmann, and J.P. Benoit. PLGA-based microparticles: elucidation of mechanisms and a new, simple mathematical model quantifying drug release. *Eur. J. Pharm. Sci.*, 15(4):355–366, 2002.
[22] S.S. Feng and S. Chien. Chemotherapeutic engineering: application and further development of chemical engineering principles for chemotherapy of cancer and other diseases. *Chem. Eng. Sci.*, 58:4087–4114, 2003.
[23] X. Gao, Y. Cui, R.M. Levenson, L.W. Chung, and S. Nie. In vivo cancer targeting and imaging with semiconductor quantum dots. *Nat. Biotechnol.*, 22(8):969–76, 2004.
[24] B. Gompertz. On the nature of the function expressive of the law of human mortality, and on a new mode of determining the value of life contingencies. *Philos. Trans. R Soc. Lond.*, 115:513–583, 1825.
[25] R. Gref, P. Couvreur, G. Barratt, and E. Mysiakine. Surface-engineered nanoparticles for multiple ligand coupling. *Biomaterials*, 24(24):4529–4537, 2003.
[26] C.J. Gulledge and M.W. Dewhirst. Tumor oxygenation: a matter of supply and demand. *Anticancer Res.*, 16:741–750, 1996.
[27] H. Hashizume, P. Baluk, S. Morikawa, J.W. McLean, G. Thurston, S. Roberge, R.K. Jain, and D.M. McDonald. Openings between defective endothelial cells explain tumor vessel leakiness. *Am. J. Pathol.*, 156(4):1363–1380, 2000.
[28] T. Higuchi. Rate of release of medicaments from ointment bases containing drugs in suspension. *J. Pharm. Sci.*, 50:874–875, 1961.

[29] S.K. Hobbs, W.L. Monsky, F. Yuan, W.G. Roberts, L. Griffith, V.P. Torchilin, and R.K. Jain. Regulation of transport pathways in tumor vessels: Role of tumor type and microenvironment. *Proceedings of the National Academy of Sciences of the United States of America*. vol. 95, pp. 4607–4612, 1998.
[30] M. Holz and A. Fahr. Compartment modeling. *Adv. Drug Del. Rev.*, 48(2–3):249–264, 2001.
[31] M. Hombreiro-Perez, J. Siepmann, C. Zinutti, A. Lamprecht, N. Ubrich, M. Hoffman, R. Bodmeier, and P. Maincent. Non-degradable microparticles containing a hydrophilic and/or a lipophilic drug: preparation, characterization and drug release modeling. *J. Control. Rel.*, 88(3):413–428, 2003.
[32] A. Ito, Y. Kuga, H. Honda, H. Kikkawa, A. Horiuchi, Y. Watanabe, and T. Kobayashi. Magnetite nanoparticle-loaded anti-HER2 immunoliposomes for combination of antibody therapy with hyperthermia. *Cancer Lett.*, 212(2):167–175, 2004.
[33] T.L. Jackson and H.M. Byrne. A mathematical model to study the effects of drug resistance and vasculature on the response of solid tumors to chemotherapy. *Math. Biosci.*, 164:17–38, 2000.
[34] R.K. Jain and L.E. Gerlowski. Extravascular transport in normal and tumor tissues. *Crit. Rev. Oncol. Hematol.*, 5(2):115–170, 1986.
[35] R.K. Jain. Determinants of tumor blood flow: a review. *Canc. Res.*, 48:2641–2658, 1988.
[36] R.K. Jain and L.T. Baxter. Mechanisms of heterogeneous distribution of monoclonal antibodies and other macromolecules in tumors: Significance of elevated interstitial pressure. *Cancer Res.*, 48:7022–7032, 1988.
[37] R.K. Jain. Physiological barriers to delivery of monoclonal antibodies and other macromolecules in tumors. *Cancer Res.* (Suppl.), 50:814s–819s, 1990.
[38] R.K. Jain. Delivery of molecular medicine to solid tumors: Lessons from in vivo imaging of gene expression and function. *J. Control. Rel.*, 74:7–25, 2001.
[39] R.K. Jain. Normalizing tumor vasculature with anti-angiogenic therapy: A new paradigm for combination therapy. *Nature Med.*, 7(9):987–989, 2001.
[40] H. Kitano. Computational systems biology. *Nature*, 420:206–210, 2002.
[41] N. Kohler, G.E. Fryxell, and M. Zhang. A bifunctional poly(ethylene glycol) silane immobilized on metallic oxide-based nanoparticles for conjugation with cell targeting agents. *J. Am. Chem. Soc.*, 126(23):7206–7211, 2004.
[42] K. Kosmidis, P. Argyrakis, and P. Macheras. A reappraisal of drug release laws using Monte Carlo simulations: the prevalence of the Weibull function. *Pharm. Res.*, 20(7):988–995, 2003.
[43] J. Kreuter. Nanoparticles. In J. Kreuter, (ed.), *Colloidal Drug Delivery Systems*, Marcel Dekker, Inc. New York, Basel, Hong Kong, 1994.
[44] A.K. Laird. Dynamics of tumor growth. *Br. J. Canc.*, 18:490–502, 1964.
[45] R. Langer and N.A. Peppas. Chemical and physical structure of polymers as carriers for controlled release of bioactive agents: A review. *Rev. Macromol. Chem. Phys.*, C23:61–126, 1983.
[46] R. Langer. Drug delivery and targeting. *Nature*, 392:6679, 5–10, 1998.
[47] R. Langer. Biomaterials in drug delivery and tissue engineering: One laboratory's experience. *Acc. Chem. Res.*, 33:94–101, 2000.
[48] J. Lankelma. Tissue transport of anti-cancer drugs. *Curr. Pharm. Des.*, 8:1987–1993, 2002.
[49] K.W. Leong, B.C. Brott and R. Langer. Bioerodible polyanhydrides as drug-carrier matrices. I: Characterization, degradation and release characteristics. *J. Biomed. Mat. Res.*, 19:941–955, 1985.
[50] A. Mantovani. Biology of disease. Tumor-associated macrophages in neoplastic progression: a paradigm for the in vitro function of chemokines. *Lab. Invest.*, 71:5–16, 1994.
[51] U. Massing and S. Fuxius. Liposomal formulations of anticancer drugs: selectivity and effectiveness. *Drug Resis. Updat.*, 3:171–177, 2000.
[52] S.R. McDougall, A.R.A. Anderson, and M.A.J. Chaplain et al. Mathematical modelling of flow through vascular networks: Implications for tumour-induced angiogenesis and chemotherapy strategies. *Bull. Math. Biol.*, 64(4):673–702, 2002.
[53] K. Maruyama. In vivo targeting by liposomes. *Biol. Pharm. Bull.*, 23(7):791–799, 2000.
[54] L.D. Mayer and J.A. Shabbits. The role of liposomal drug delivery in molecular and pharmacological strategies to overcome multidrug resistance. *Can. Metas. Rev.*, 20:87–93, 2001.
[55] B. Narasimhan. Mathematical models describing polymer dissolution: Consequences for drug delivery. *Adv. Drug Del. Rev.*, 48(2–3):195–210, 2001.
[56] National Cancer Institute (NCI) website at http://press2.nci.nih.gov/sciencebehind/nanotech.
[57] National Cancer Institute website at http://www.cancer.gov/clinicaltrials/results/dose-dense0604.

[58] National Institute of Nanotechnology (NINT) of Canada at www.industrymailout.com/industry/do/news/view?id=9003&print=1.

[59] P. Netti and R.K. Jain. Interstitial transport in solid tumors. In L. Preziosi (ed.), *Cancer Modeling and Simulation*, Chapman & Hall/CRC, Boca Raton, London, New York, Washington, pp. 62–65, 2003.

[60] NIH Publication No 04-5489 by National Cancer Institute. cancer *NANOTECHNOLOGY—Going Small for Big Advances. Jan. 2004*.

[61] National Institutes of Health (NIH) website at http://press2.nci.nih.gov/sciencebehind/nanotech.

[62] L. Norton. Implications of kinetic heterogeneity in clinical oncology. *Semin. Oncol.*, 12:231–249, 1985.

[63] L. Norton and R. Simon. The Norton-Simon hypothesis revisited. *Cancer. Treat. Res.*, 70:163–169, 1986.

[64] L. Norton. A Gompertzian model of human breast cancer growth. *Cancer. Res.*, 48:7067–7071, 1988.

[65] L. Norton. Theoretical concepts and the emerging role of taxanes in adjuvant therapy. *Oncologist*, 3 (suppl):30–35, 2001.

[66] L. Norton. Karnofsky Memorial Lecture. ASCO, 2004.

[67] T.P. Padera, B.R. Stoll, J.B. Tooredman, D. Capen, E. di Tomaso, and R.K. Jain. Cancer cells compress intratumour vessels. *Nature*, 427:695, 2004.

[68] R.S. Parker and F.J. Doyle III. Control-relevant modeling in drug delivery. *Adv. Drug Del. Rev.*, 48(2–3):211–228, 2001.

[69] N.A. Peppas. Analysis of Fickian and non-Fickian drug release from polymers. *Pharm. Acta. Helv.*, 60:110–111, 1985.

[70] P.J. Polverini. How the extracellular matrix and macrophages contribute to angiogenesis-dependent diseases. *Eur. J. Cancer*, 32A(14):2430–2438, 1996.

[71] F. Quian, G.M. Saidel, D.M. Sutton, A. Exner, and J. Gao. Combined modeling and experimental approach for the development of dual-release polymer millirods. *J. Control. Rel.*, 83:427–435, 2002.

[72] P. Sapra and T.M. Allen. Ligand-targeted liposomal anticancer drugs. *Prog. Lipid Res.*, 42:439–462, 2003.

[73] M. Sarntinoranont, F. Rooney, and M. Ferrari. Interstitial stress and fluid pressure within a growing tumor. *Ann. Biomed. Eng.*, 31:327–335, 2003.

[74] J. Siepmann and A. Goepferich. Mathematical modeling of bioerodible, polymeric drug delivery systems. *Adv. Drug Del. Rev.*, 48(2–3):229–247, 2001.

[75] J. Siepmann, N. Faisant, and J.P. Benoit. A new mathematical model quantifying drug release from bioerodible microparticles using Monte Carlo simulations. *Pharm. Res.* 19(12):1885–1893, 2002.

[76] J. Siepmann and N.A. Peppas. Modeling of drug release from delivery systems based on hydroxypropyl methylcellulose (HPMC). *Adv. Drug Del. Rev.*, 48(2–3):139–157, 2003.

[77] J. Siepmann, N. Faisant, J. Akiki, J. Richard, and J.P. Benoit. Effect of the size of biodegradable micro particles on drug release: experiment and theory. *J. Control. Rel.*, 96(1):123–134, 2004.

[78] S. Simoes, J.N. Moreira, C. Fonseca, N. Duezguenes, and M. Pedroso de Lima. On the formulation of pH-sensitive liposomes with long circulation times. *Adv. Drug Del. Rev.*, 56:947–965, 2004.

[79] J. Sinek, H.B. Frieboes, X. Zheng, and V. Cristini. Chemotherapy simulations demonstrate fundamental transport limitations involving nanoparticles. *Biomed. Microdev.*, 6(4):297–309, 2004.

[80] T. Sonoda, H. Kobayashi, T. Kaku, T. Hirakawa, and H. Nakano. Expression of angiogenesis factors in monolayer culture, multicellular spheroid and in vivo transplanted tumor by human ovarian cancer cell lines. *Cancer Lett.*, 196:229–237, 2003.

[81] M.Y. Su, A.A. Najafi, and O. Nalcioglu. Regional comparison of tumor vascularity and permeability parameters measured by albumin–Gd-DTPA and Gd-DTPA. *Magn. Reson. Med.*, 34(3):402–411, 1995.

[82] M.Y. Su, A. Muhler, X. Lao, and O. Nalcioglu. Tumor characterization with dynamic contrast-enhanced MRI using MR contrast agents of various molecular weights. *Magn. Reson. Med.*, 39(2):259–69, 1998.

[83] B.G. Sumpter, D.W. Noid, and M.D. Barnes. Recent developments in the formation, characterization, and simulation of micron and nano-scale droplets of amorphous polymer blends and semi-crystalline polymers. *Polymer*, 44(16):4389–4403, 2003.

[84] Trends in NanoTechnology (TNT) 2001 at nanoword.net/library/weekly/ aa091201a.htm.

[85] P. Veng–Pedersen. Noncompartmentally-based pharmacokinetic modeling. *Adv. Drug Del. Rev.*, 48(2–3):265–300, 2001.

[86] C.H. Wang, J. Li, and C.S. Teo et al. The delivery of BCNU to brain tumors. *J. Control. Rel.*, 61(1–2):21–41, 1999.

[87] W. Weibull. A statistical distribution of wide applicability. *J. Appl. Mechan.*, 18:293–297, 1951.

[88] C.P. Winsor. The Gompertz curve as a growth curve. *Proc. Natl. Acad. Sci.*, 18:1–8, 1932.

[89] F. Yuan, M. Dellian, D. Fukumura, M. Leunig, D.A. Berk, V.P. Torchilin, and R.K. Jain. Vascular permeability in a human tumor xenograft: molecular size dependence and cutoff size. *Cancer Res.*, 55(17):3752–3756, 1995.

[90] H.S. Zaki, M.D. Jenkinson, D.G. Du Plessis, T. Smith, and N.G. Rainov. Vanishing contrast enhancement in malignant glioma after corticosteroid treatment. *Acta Neurochir.*, 146:841–845, 2004.

[91] M. Zhang, Z. Yan, L.L. Chow, and C.H. Wan. Simulation of drug release from biodegradable polymeric microspheres with bulk and surface erosions. *J. Pharm. Sci.*, 92(10):2040–2056, 2003.

[92] X. Zheng, S.M. Wise, and V. Cristini. Nonlinear simulation of tumor necrosis, neo-vascularization and tissue invasion via an adaptive finite-element/level-set method. *Bull. Math. Bio.*, 67:211–259, 2005.

16

Nanomechanics and Tissue Pathology

Jason Sakamoto[1], Paolo Decuzzi[2], Francesco Gentile[3], Stanislav I. Rokhlin[4], Lugen Wang[5], Bin Xie[6], and *Senior Author:* Mauro Ferrari[7]

[1]*Biomedical Engineering, The Ohio State University, Columbus, Ohio*
[2]*Assistant Professor of Mechanical Engineering, CEMeC—Center of Excellence in Computational Mechanics, Politecnico di Bari, Department of Experimental Medicine, University Magna Graecia at Catanzaro, ITALY*
[3]*Department of Experimental Medicine, University Magna Graecia at Catanzaro, ITALY*
[4]*Professor, Nondestructive Evaluation Program, The Ohio State University, Columbus, Ohio*
[5]*Research Scientist, Nondestructive Evaluation Program, The Ohio State University, Columbus, Ohio*
[6]*Graduate Student, Nondestructive Evaluation Program, The Ohio State University, Columbus, Ohio*
[7]*Professor, Brown Institute of Molecular Medicine Chairman, Department of Biomedical Engineering, University of Texas Health Science Center, Houston, TX; Professor of Experimental Therapeutics, University of Texas M.D. Anderson Cancer Center, Houston, TX; Professor of Bioengineering, Rice University, Houston, TX; Professor of Biochemistry and Molecular Biology, University of Texas Medical Branch, Galveston, TX; President, the Texas Alliance for NanoHealth, Houston, TX*

16.1. INTRODUCTION

16.1.1. Background

Nanotechnology is an emerging field that has been embraced by those in clinical medicine. The most novel aspect of nanotechnology is the ability to precisely fabricate devices on a physical scale heretofore only realized in science fiction. Most notable medical applications have involved micro-sized devices with integrated micro- and/or nano-scale features used for controlled drug delivery or biomolecular analysis. BioMEMS (Biological Micro-Electro-Mechanical Systems) devices have served as conduits for nanotechnology to enter clinical medicine. However, new theoretical applications will further assert nanotechnology as a multifaceted biomedical discipline.

This chapter will introduce a nanomechanical field theory called doublet mechanics that delivers the ability to effectively reconstruct the material properties of tissue for diagnostic applications. This theoretical premise represents a medical breakthrough since current mechanical theories do not address modeling of material as heterogeneous as biological matter. Mathematical modeling of tissue involves physical domains too irregular to be addressed by continuum theories and too large and irregular for a discrete approach [1]. The details of the theories of doublet mechanics will be discussed later in this chapter.

The ability to quantitatively differentiate tissue types based upon biological and physical properties of materials should enable the precise detection and diagnosis of diseases and monitoring of their progression with unprecedented accuracy. Characterization-mode ultrasound is a new, non-imaging, tissue screening modality that utilizes the theories of doublet mechanics to reconstruct tissue material properties by analyzing tissue/ultrasound interactions. A characterization-mode ultrasound system is currently being developed for use as a pathologic evaluation tool. Histopathologic evaluation of tissue specimens is currently the preferred method for physicians to determine pathological basis of disease. This qualitative diagnostic methodology remains the "gold standard" for clinical medicine despite suffering from its inherent subjectivity. Characterization-mode ultrasound provides a rapid, quantitative screening alternative that can be integrated into clinical pathology laboratories to help quantify the stage of disease per biopsy specimen. This information can be used to augment a pathologist's diagnosis as well as provide an effective triage tool to highlight specimens of pathologic relevance.

The field of use for characterization-mode ultrasound is not limited to laboratory applications. Numerous clinical indications may be significantly impacted by this new ultrasound technology. A melanoma screening system is currently being developed to provide primary care physicians with a quantitative alternative for rating cutaneous malignancies. Similar to histopathologic evaluation, skin cancer detection suffers from subjective inspection criteria. Accuracy rates for the visual examination of suspicious cutaneous lesions are comparable to a coin toss. This low efficacy procedure promotes a clinical strategy where excessive numbers of skin biopsies are performed to rule out disease. Physicians are clinically motivated to continue this practice because studies have established the significant benefit of early detection for the case of malignant melanoma. The efficacy of current skin cancer screening procedures places the burden of diagnoses upon the pathologist and creates a significant financial responsibility for the patient. The design requirements for a characterization-mode ultrasound skin cancer detection system should provide primary care physicians with a rapid, noninvasive, screening tool for malignant melanoma that would assign a quantitative malignancy potential for specific cutaneous lesions.

The aim of this chapter is to expound on contributions that nanotechnology can possibly offer clinical medicine, specifically in the discipline of tissue diagnostics. A detailed introduction of doublet mechanics will be discussed comparing the nanomechanical field theory to other classical approaches for modeling tissue. The chapter will then continue to develop the theories of doublet mechanics by transitioning from theory to application, (i.e. characterization-mode ultrasound). The evolution of this new screening modality will be thoroughly presented to demonstrate its current and potential capabilities. A description of the ultrasonic theoretical foundation that enables tissue differentiation will also be included.

The chapter will conclude with potential applications of characterization-mode ultrasound that have promise of clinical significance.

16.1.2. The Diagnostic Conundrum

Surgical biopsy remains the "gold standard" for the clinical assessment of the pathologic basis of disease. The discipline of surgical pathology emerged in the 1800's once physicians realized that histologic changes presented in living tissue had the potential of providing reliable diagnosis. While morphologic presentation of tissue specimens continues to be the backbone of diagnostic surgical pathology, microbiologic, immunologic, and molecular analysis of tissue and cells are becoming indispensable clinical platforms for rendering pathologic evaluation. Collectively, these diagnostic modalities are critical for determining the appropriate medical treatment. However, the practice inherently suffers from being predominately a qualitative art. The quality of diagnoses is still entrusted with the pathologist's experience and knowledge.

Numerous studies have evaluated inter-observer variability between pathologists over the spectrum of malignant disease. Although the difficulty level to differentiate neoplastic growth within living tissue varies pending on tumor type, inter-observation studies have demonstrated that pathologists come to complete agreement as low as 9% of the time [2–6]. In defense of pathologists, this percentage is based upon the ability of a group to unanimously score a specimen according to a qualitative universal grading system. Pathologists have demonstrated over 98% accuracy when distinguishing between benign or malignant tissue. However, one cannot overlook the severe ramifications of inaccurately categorizing tissue biopsies within the inappropriate intermediate classification of disease [2, 7]. Histologic grading systems are often used to determine which patients will be administered survival enhancing adjuvant therapy.

16.1.3. Oncologic Opportunity: Breast Cancer

Accurate, non-subjective, pathologic evaluation is the paramount objective for applying characterization-mode ultrasound to definitively characterize neoplastic change in tissue. This new ultrasound modality is being developed to address one of the leading causes of cancer death for American women: breast cancer. Over 40,000 women are anticipated to succumb to breast cancer in the United States in 2004 [8]. The estimated 200,000 new cases of invasive disease brings the lifetime risk assessment to 1 in 8 women [8]. This staggering figure has prompted patient awareness and consequently increased the number of breast cancer screening procedures performed in the United States. Improved sensitivity of these screening modalities is responsible for the increasing number of breast biopsies being performed.

When a breast biopsy is taken, it is commonly pathologically evaluated with a histologic grading system called the Scarff-Bloom-Richardson criteria. This grading system scores specimens upon three distinct cellular features: tubule formation, frequency of cell mitosis, and nuclear pleomorphism [9, 10]. Each grading category is assigned a score ranging from 1–3 (Table 16.1). Once each feature is evaluated, patient prognosis can be inferred

TABLE 16.1. Scarff-Bloom-Richardson system is used to help determine breast cancer patient prognosis. The three cellular features (Tubule formation, Nuclear pleomorphism, and Mitosis count) are rated by the listed criteria. ***Courtesy of Imaginis.com***

Tubule Formation (% of Carcinoma Composed of Tubular Structures)	**Score**
>75%	1
10–75%	2
Less than 10%	3
Nuclear Pleomorphism (Change in Cells)	**Score**
Small, round, uniform nuclei, with the largest nuclei being no more then 3× the diameter of adjacent lymphocyte nuclei.	1
Moderate pleomorphism but lacking the nucleic uniformity of low nuclear grade lesions. The largest nuclei are up to 4× the diameter of adjacent lymphocyte nuclei.	2
Plemorophic, irregularly spaced nuclei, with very irregular nuclear outline and multiple nucleoli. The largest nuclei more than 5× the diameter of adjacent lymphocyte nuclei with frequent mitoses.	3
Mitosis Count (Cell Division)	**Score**
Up to 7	1
8 to 14	2
15 or more	3

from the composite score (Table 16.2). A sum of 3–5 is categorized as a Grade 1 tumor (well-differentiated), a sum of 6–7 a Grade 2 tumor (moderately-differentiated), and sum of 8–9 a Grade 3 tumor (poorly- differentiated) [9, 10].

Nuclear pleomorphism was found to be the most subjective of the three graded cellular features, [5, 11–15]. A generalized kappa (κ) test was applied to evaluate the level of inter-observer agreement for nuclear pleomorphism in several of the inter-observation studies. The κ values ranged from 0.24–0.50, which indicates fair to moderate degree of agreement between inter-observers (Table 16.3) [5, 11, 13–15]. Tubule formation and mitotic

TABLE 16.2. Summary of the Scarff-Bloom-Richardson System used to rate tumors. The composite score accumulated using the judging criteria in Table 16.1 determines the tumor grades. ***Courtesy of Imaginis.com***

Grade	Description	Score	5 yr survival.	7 yr survival.
Grade 1 (lowest)	Well-differentiated breast cells; cells generally appear normal and are not growing rapidly; cancer arranged in small tubules.	3,4,5	95%	90%
Grade 2	Moderately differentiated breast cells; have characteristics between Grade 1 and Grade 3 tumors.	6,7	75%	63%
Grade 3 (highest)	Poorly differentiated breast cells; Cells do not appear normal and tend to grow and spread more aggressively.	8.9	50%	45%

TABLE 16.3. Generalized kappa test.

Level of Agreement	Kappa (κ)
Poor	<0.21
Fair	0.21–0.40
Moderate	0.41–0.60
Good	0.61–0.80
Very Good	0.81–1.00

counts consistently ranged from 0.52–.0.64 [11, 14, 15]. Despite suffering from less defined evaluation criteria, the degree of nuclear pleomorphism is weighted equally when grading a specimen's malignancy status. The inter-observation studies have demonstrated that the main source of disagreement between pathologist evaluations arises from categorizing marginal specimens that lie on the borderline between nuclear pleomorphic grades 1 & 2 and 2 & 3. For these instances, cellular polarity, or tumor cell orientation towards luminal spaces, is used as the deciding classification factor. Specimens presenting with higher polarity correspond to a lower nuclear pleomorphic grade [11]. The final nuclear pleomorphic grade can significantly impact the clinical prognosis for a patient. A grading discrepancy between pathologists of 1 point on the Scarff-Bloom-Richardson grading system can alter a patient's 5 and 7-year survival percentages by 20–30% (Table 16.2). This situation establishes the clinical need for a quantitative method to augment pathological assessment.

Automated tissue screening is another benefit of this characterization-mode ultrasound system. Several mounted specimens are capable of being loaded into the device for evaluation. This provides a significant feature to enhance the efficiency of a pathology laboratory. The increased incidence of breast cancer has generated an annual estimate of 500,000–1,000,000 breast biopsies in the United States, of which, 60–90% of the lesions are benign at the time of excision [16, 17]. The magnitude of the number of breast biopsies is a direct consequence of the misconception that breast cancer screening modalities have a significant effect on mortality rates. Although the National Cancer Institute (NCI) has demonstrated this finding in extensive randomized controlled trials*, physicians are unlikely to abdicate biopsy in favor of erring on the side of caution [8]. Biopsies are also routinely performed even in cases where a low probability of malignancy is found from mammographic evaluations or other screening procedures for the purpose of alleviating patient anxiety. Characterization-mode ultrasound could fulfill the need to augment pathologists' efficiency by providing rapid quantitative diagnostic information.

16.1.4. Screening for Malignant Melanoma

Characterization-mode ultrasound technology can be configured to address numerous medical needs. Minor modifications of the above mentioned pathology specimen evaluation system can be made to transform the system for cutaneous applications. Incident rates for malignant melanoma suggest a need for a quantitative methodology for skin cancer

* The randomized controlled trials involved the participation of nearly half a million women from 4 countries.

screening. Currently, primary care physicians share the responsibility for diagnosing skin lesions. Diagnoses are based upon the qualitative morphological presentation of cutaneous lesions. The visual inspection criteria of suspicious lesions are commonly referred to as the ABCDE's of skin cancer (Asymmetry, Border irregularity, darkened Color, Diameter greater than 0.5cm, and Enlargement) [18]. Physicians apply these inspection guidelines to determine the necessity for pathologic examination. This subjective cutaneous evaluation suffers from its reliance upon visual inspection to differentiate between melanoma and benign pigmented lesions. In fact, primary care physicians have a relatively low accuracy rate when diagnosing melanoma based on visual characteristics. Internists have a 50% accuracy rate when diagnosing melanomas, 50–70% for family practice physicians, and 90% for dermatologists [19]. Unfortunately studies have found that a significant amount of early stage melanomas (50%) are misdiagnosed during clinical examination [20]. Furthermore, of all melanomas diagnosed in the clinic, only 35% are confirmed by histopathologic examinations [21]. This yields ample opportunity for any device capable of increasing rates of accurate diagnosis in the primary care setting.

In 2003, there were over 1,000,000 new cases of skin cancer nationwide, including 54,000 cases of melanoma [22]. Currently, melanoma is the fifth leading cause of cancer with 9,800 annual deaths [22]. Studies have shown that 1 in 63 men and 1 in 85 women in the United States will develop this disease at some point in their lives [19]. The rising incidence of all types of skin malignancies has outpaced that of all other cancers. Since 1973, skin cancer incidence has doubled, and mortality rates have jumped by 44% [23]. Approximately 40% of all cancer patients succumb to their illness, which accounts for 19% of all deaths on an annual basis [24].

Prognosis of melanoma is directly correlated to the thickness of the primary tumor. This thickness is clinically referred to as the Breslow thickness. The intrinsic growth pattern of melanoma is unique in that as long as the tumor is confined to radial growth, resection of the tumor yields greater than 95% cure rates (Breslow thickness <1mm.). However, if the tumor begins to grow vertically towards the underlying dermal layers towards the underlying dermal capillaries, the survivability drops dramatically. A Breslow thickness of 1.5mm–4mm has a 5 year survival rate of 70% and primary tumors with a Breslow thickness >4mm has a 5 years survival rate of 45% [23]. Once tumor cells have entered the dermal capillaries, ensuing metastasis to the liver, lungs, and brain may cause cure rates to fall to <10% [25]. This unique propensity to metastasize leads to a vast disparity in prognosis between early detection and late detection.

Primary care physicians are targeted to receive the maximum clinical benefit in the case of malignant melanoma identification utilizing nanomechanical characterization. The intent of creating a characterization-mode ultrasound skin cancer diagnostic device is to provide an objective screening modality to supplement visual inspection examinations. Design requirements would accommodate a noninvasive procedure that would feature rapid use and quantitative assessment of malignancy potential.

The case has been presented for the potential impact that characterization-mode ultrasound could have on tissue diagnostics. To this point, only claims have been made concerning the capabilities of characterization-mode ultrasound. In order to substantiate these claims, a detailed description of the theories of doublet mechanics must be explained. The following two Sections of this chapter are dedicated to the mathematical modeling of tissue response to impinging ultrasound waves. After a brief review of the classical approach based on

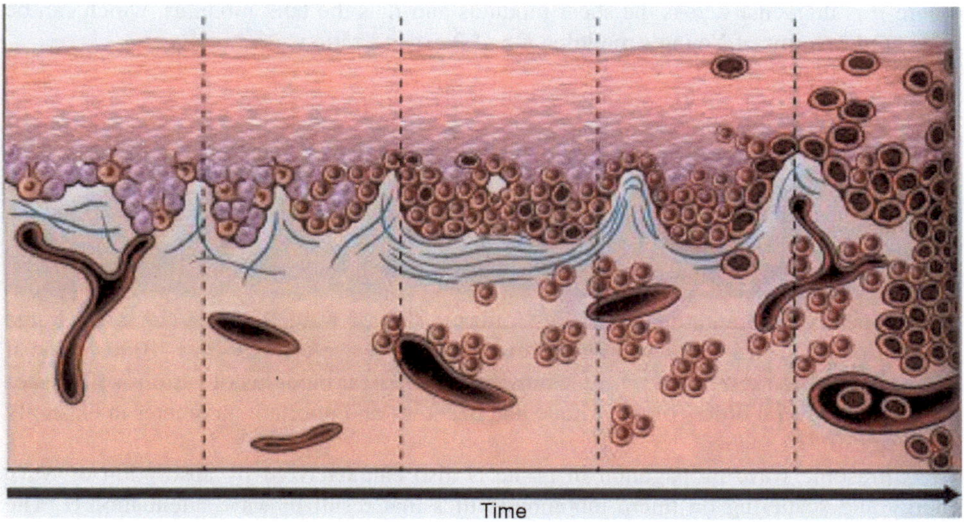

FIGURE 16.1. Disease progression for malignant melanoma. The dark spheres represent transformed melanocytes. Over time, melanocyte proliferation transitions from radial growth to vertical penetration into the dermis and underlying vessels. ***Courtesy of Robbins Pathology***

continuum mechanics, the foundations of doublet mechanics and its application to wave propagation in heterogeneous solids is presented.

16.2. THE *CLASSIC* APPROACH: CHARACTERIZATION-MODE ULTRASOUND AND CONTINUUM MECHANICS MODEL

16.2.1. Continuum Mechanics Description of Ultrasonic Wave Propagation

16.2.1.1. Bulk Wave: Velocity, Dispersion and Viscoelasticity Ultrasonic waves have been used extensively for tissue characterization [26, 27]. Ultrasonic waves produce small-amplitude mechanical vibrations and, depending on the mode being utilized, may induce both longitudinal and shear stresses in the tissue. Information on the elastic properties of a tissue can be obtained by measuring both the velocity and the attenuation of the ultrasonic wave. The phase velocity of the wave depends on the elastic constants and density of the body, while the attenuation depends on the microstructure. In an isotropic solid medium, which has only two independent elastic moduli, there exist two elastic waves: the longitudinal with velocity V_l and shear or transverse wave, with velocity V_t. They depend on the elastic moduli and density of the solid:

$$V_l = \sqrt{\frac{K + 4\mu/3}{\rho}}, \quad V_t = \sqrt{\frac{\mu}{\rho}}, \tag{16.1}$$

where ρ is the density, μ is the shear modulus and K is the bulk modulus, which can be expressed in terms of Young's modulus E and Poisson's ratio ν as:

$$K = \frac{E}{3(1-2\nu)}, \quad \mu = \frac{E}{2(1+\nu)}. \tag{16.2}$$

In soft tissue and polymers Poisson's ratio $\nu \approx 0.5$ and $E \approx 3\mu$. For liquids $\mu = 0$ and the velocity of the (longitudinal) wave is controlled only by the bulk modulus, (i.e., $V_l = \sqrt{\frac{K}{\rho}}$). Soft tissues have longitudinal velocity similar to that of water [27–29] (V_l in the range 1,400–1,600m/s) and very low highly attenuated shear wave velocity (in 1–10m/s range at frequency below 1kHz) [30, 31]. It is known that the shear modulus of various soft tissues ranges over several orders of magnitude and becomes an important parameter in elasticity imaging [30].

Ultrasonic wave propagation in tissue is also characterized by absorption of wave energy and scattering on micro-inhomogeneities that result in wave attenuation α. The effect of wave attenuation can be described phenomenologically by introducing a complex wave number and complex elastic moduli. The propagation of a plane harmonic wave in the x-direction can be represented as:

$$u = u_0 e^{i(kx-\omega t)} = u_0 e^{-\alpha x} e^{i(k'x-\omega t)}, \tag{16.3}$$

where $k = k' + i\alpha$ is the complex wave number, α the attenuation coefficient, $k' = \frac{\omega}{V} = \frac{2\pi}{\lambda}$ the wave number, $\omega = 2\pi f$ is the angular frequency, f frequency and λ wavelength. The attenuation of ultrasound in a soft tissue is highly dependent on frequency f and follows a power law [32, 33]:

$$\alpha(f) = \alpha_0 f^{\gamma}, \tag{16.4}$$

where α_0 is a constant and γ is usually between 1 and 2. The ultrasonic velocities (and elastic moduli) in polymers and soft tissues also depend on frequency; this phenomenon is called dispersion.

The classic Maxwell and Voigt relaxation models have been widely used in describing the viscoelastic behavior of tissues. However simple relaxation models cannot well describe the power law attenuation characteristics eqn. (16.4) and usually discrete or continuous spectra of different relaxation processes are needed to fit the experimental data in a given frequency range. More recently, in order to model the general power-law attenuation a more comprehensive form of constitutive behavior has been proposed [33] to relate the stress T and strain S tensors in biological materials and tissues:

$$T = c : S + \eta : r(S, t), \tag{16.5}$$

where c and η are elastic constants and viscosity tensors; $r(S, t)$ is a convolution operator which replaces the strain rate in the Newtonian viscosity relation. The operator form depends on the attenuation characteristic eqn. (16.4). Further generalization of this constitutive model that includes stress relaxation behavior is given in [32].

Due to causality and linearity, the frequency behaviors of attenuation and velocity (dispersion) are related and satisfy the Kramers-Krönig dispersion relations [34, 35]. In particular if the frequency dependence of attenuation is measured, the ultrasonic phase velocity can be calculated from:

$$\frac{1}{V(\omega)} = \frac{2}{\pi} \int_0^{\infty} [\alpha(\omega') - \alpha(\omega)] \frac{d\omega'}{\omega'^2 - \omega^2}, \tag{16.6}$$

where $V(\omega)$ is the ultrasonic velocity. This relation must be satisfied regardless of the mechanism of attenuation. The integral (16.6) is improper for $\gamma \geq 1$ in eqn. (16.4) and special methods of integration must be used [34], or time domain causal relationships [35] must be employed.

16.2.1.2. Modeling of Ultrasonic Interaction with Layer of Biological Tissue Ultrasonic spectroscopy has been considered promising for characterization of thin layers immersed in water or embedded between two known materials [36, 37]. The resonance spectrum of the adhesive layer at normal incidence has been applied to determine ultrasonic velocity and attenuation. However, because the reflection/transmission spectrum at normal incidence depends on two non-dimensional parameters (non-dimensional impedance and thickness), the three independent layer properties (density, thickness and longitudinal modulus) cannot be decoupled with sufficient accuracy. Additional relations necessary to obtain layer properties can be obtained from oblique measurement. Lavrentyev and Rokhlin [36] developed an ultrasonic method for the simultaneous determination of all properties of the layer: density, thickness, longitudinal and shear moduli, and attenuations, using ultrasonic measurements at two angles, normal and oblique. This method is very suitable for quantitative property measurement of soft tissue layers in the structure shown in Fig. 16.2.

Referring Fig. 16.2, we will discuss ultrasonic wave reflection from a sandwich structure. At normal incidence only longitudinal waves are present in this system; the reflection

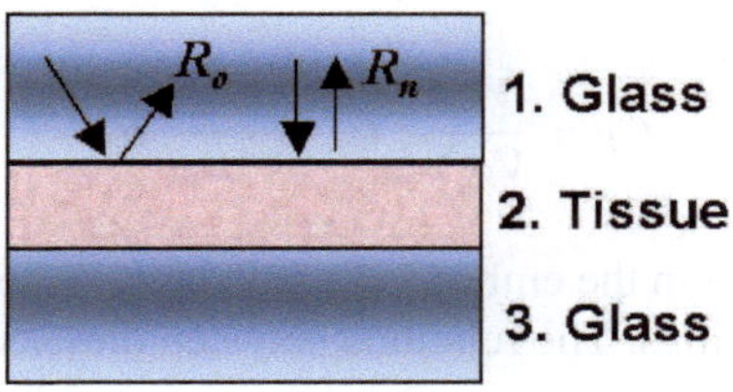

FIGURE 16.2. Normal and oblique incident ultrasound interaction with a tissue layer embedded between glass substrates.

coefficient for the layer is given by:

$$R = R_{12} + \frac{T_{12}T_{21}T_{23}e^{2ikh}}{1 - R_{21}R_{23}e^{2ikh}}, \tag{16.7}$$

where $k = k' + ik''$ is the wave number in the embedded layer, h is the layer thickness. R_{ij} is the reflection coefficient between media i and j: $R_{ij} = (Z_j - Z_i)/(Z_j + Z_i)$. R_{ij} is the reflection coefficient between media i and j: $R_{ij} = (Z_j - Z_i)/(Z_j + Z_i)$. T_{ij} is the transmission coefficient between media i and j: $T_{ij} = 2Z_j/(Z_j + Z_i)$. Z is the complex impedance. The longitudinal wave at normal incidence is independent of the shear properties of the embedded layer. For this reason, one also needs an obliquely incident ultrasonic wave for shear property measurement. In this case, the model employs the recursive compliance matrix method that is a general algorithm applicable for ultrasonic wave interaction with an arbitrary layered structure [38].

16.2.1.3. Definition of The Unique Set of Material Parameters The acoustic response from the embedded tissue layer depends on six parameters: elastic Lamé moduli $\lambda = K - 2/3\mu$, and μ; thickness h, density ρ, and longitudinal and shear attenuations α_l, α_t ($\alpha_{l,t}$ is defined in [36] as the ratio of the imaginary to the real part of the wave number):

$$\lambda, \mu, h, \rho, \alpha_l, \alpha_t. \tag{16.8}$$

To simplify the layer property determination, one must formulate a minimal set of independent parameters of the layer. Lavrentyev and Rokhlin [36] have proposed six non-dimensional parameters that can be separated into two groups of non-dimensional parameters and are uniquely determined by normal and oblique spectra. The normal reflection spectrum depends on three non-dimensional parameters:

$$Z_n = \frac{Z_l}{Z_0}, \quad H_n = \frac{h\omega_0}{V_l}, \quad \alpha_l, \tag{16.9}$$

where $\omega_0 = 1$ MHz for convenience as a normalization factor and V_l is the longitudinal velocity in the layer. $Z_l = \rho V_l$ and $Z_0 = \rho_0 V_{l0}$ correspond to the acoustic impedance of the layer and the substrate respectively, V_{l0} and ρ_0 are the substrate velocity and density. The oblique incident reflection coefficients, in addition to the three parameters defined by eqn. 16.12, depend on three more non-dimensional parameters defined in [39] as:

$$C_l = \frac{V_l}{V_{l0}}, \quad C_t = \frac{V_t}{V t_0}. \alpha_t \tag{16.10}$$

where V_t is the shear velocity in the embedded tissue layer; where V_{l0} and V_{t0} are substrate longitudinal and shear velocities. The relations between dimensional and non-dimensional parameters are determined as:

$$\rho = \frac{Z_n\rho_0}{C_l}; \quad h = \frac{H_n C_l V_{l0}}{\omega_0}; \quad \lambda + 2\mu = \rho C_l^2 V_{l0}^2; \quad \mu = \rho C_t^2 V_{t0}^2; \tag{16.11}$$

The non-dimensional parameters are determined from the normal and oblique reflection spectra of the imbedded layer. Since in some cases, the embedded layer reflection cannot be separated from the substrate reflections, the reflected spectrum from the layer depends on the gate size and position. In order to correctly account for the gate effect in the inversion model, one has to calculate the time domain signal in the same time-domain gate as that used for the data collection in the experiment [39].

16.2.1.4. Inversion Algorithm The inversion algorithm for the determination of the six non-dimensional parameters from normal and oblique time domain reflected signals includes two steps: first Z_n, H_n, α_l are determined from a reflection signal at normal incidence; then, considering Z_n, H_n, α_l as known, three more non-dimensional parameters are C_l, C_t, α_t determined from an oblique incident signal. The corresponding dimensional parameters are obtained using eqn. 16.11.

Both experimental and theoretical signals are bounded in time using the same time gate. To deconvolve the transducer frequency response and ultrasonic beam parameters, a reflected reference signal is used to calculate the transducer frequency response for the simulated reflection signal in the model. From the gated experimental and simulated signals, the spectrum is calculated and matched by least-squares minimization of the sum of squared deviations between the spectra, considering the non-dimensional parameters as independent variables in a multidimensional space:

$$\varepsilon = \min \sum_{f=f_1}^{f_2} (|R_t(\mathbf{X})| - |R_e(\mathbf{X}_0)|)^2. \tag{16.12}$$

Here $R_t(\mathbf{X})$ and $R_e(\mathbf{X}_0)$ are the spectra of the gated experimental and theoretical signals. $\mathbf{X}$ is an iterated set of non-dimensional parameters; $\mathbf{X}_0$ is the actual set of material properties. f_1 and f_2 are the bounds of the frequency range in which minimization is performed. The frequency bounds are selected based on the transducer bandwidth and the resonance frequency of the adhesive layer. After the parameters Z_n, H_n, α_l are obtained from the normal incident experiment measurement, a similar procedure is used to calculate the additional three parameters at oblique incidence.

There are inherent limitations of continuum mechanics that restrict its ability to be applied to non-homogenous materials such as tissue. The desire to model the intricate micro- and nano- structures of biological materials necessitates the employment of a new mechanical method that provides a critical scaling factor. The theories of doublet mechanics feature such scalability that enables the discrete modeling of granular or cellular structures endemic to tissue. The following Section will provide a detailed account of this new nanomechanical field theory.

16.3. AN INTRODUCTION TO "*DOUBLET MECHANICS*"

16.3.1. Connotations and Interpretation of Doublet Mechanics

This section is dedicated to the description of a new nanomechanical field theory called doublet mechanics (DM) [1]. This new modeling tool is capable of bridging the gap

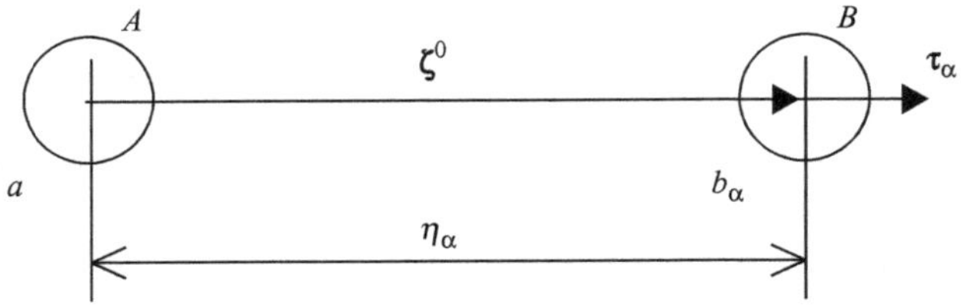

FIGURE 16.3. a particle doublet (A, B).

between continuum mechanics and discrete models without contradiction. In DM, solids are embodied by arrays of points, or nodes, at finite distance η_α and any pair of nodes constitutes a doublet (Fig. 16.3). It is not necessary for the nodes to be periodically located, as to form a lattice, although it may be convenient to assume so for certain applications. If the choice is made that the nodes be located as to reproduce one of the fourteen Bravais lattices, then the collection of nodes will consist of n basic vectors, where $n = m/2$ is the valence of the lattice, and m is the coordinate number of the array, (i.e. the number of closest lattice points to any given lattice point).

Two particular interpretations of DM exist. One is the granular, or sphere packing interpretation of doublet mechanics: within this perspective the material model underlying DM is a regular array of equal-sized elastic spheres of diameter *d*, the centers of which form a space Bravais lattice. In particular, the particle A is in contact with B_α if the doublet length η_α is equal to the true diameter of the sphere denoted by *d*. Such an α -doublet is called a contact doublet, the α-direction being the contact direction. There may be *s* contact directions in which all particles are in contact, and the underlying granule array may be called a regular s-contact n-valence array H_{sn}. If $s = n$, then the array H_{sn} becomes a regular completely contact array $H_{nn} \equiv H_n$, or a regular n-valence packing where all the adjacent particles are in contact with each other. This interpretation is certainly convenient, for it offers ease of visualization, but by no means necessary. DM is clearly valid for materials that exhibit a macroscopically evident lattice structure, but its usefulness still stands for macroscopically continuous media. This suggests a second, maybe more attractive, interpretation of DM that depicts the nodes of doublet mechanics as averages or representations of the complexity of actual particles or molecular interactions. Any arbitrary spatial distribution of points can be associated with a three dimensional covering, consisting of the Voronoi cells that include said points. The Voronoi cells for the Bravais lattices are regular polyhedra and are known as Fedorov polyhedra. The total amount of different Fedorov polyhedra for all the Bravais lattices is only five. The second interpretation for DM is then as follows: doublet mechanics actually models media as sets of Fedorov polyhedra. This corresponds to a representation of solid matter as a particulate assembly and gives rise to the so-called particulate interpretation of DM. Such a representation is extremely attractive in modeling biological tissues which considered as a spatial set of a large number of particulate (cells) dispersed in an extracellular matrix.

16.3.2. Microstrains and Microstresses: A Deeper Insight into Doublet Mechanics

In DM each lattice node is endowed with a rotation $\boldsymbol{\phi}$ and translation $\mathbf{u}$ vector with increment vectors $(\Delta\boldsymbol{\phi}, \Delta\mathbf{u})$ that may be expanded in a convergent Taylor series about the lattice nodal point. The order M at which the series is truncated defines the degree of approximation employed. The approximation theory with $M = 1$ does not contain any information on particle size or internodal distances and is called non-scale. Approximations with $M > 1$ include dependences on internodal distances. In DM nodes in any doublet may alter their axial distance, may rotate with respect to their common axis, and may relatively displace normally to their axis. So, doublet- or micro-strains of the longitudinal, torsional, and shear type can be reproduced. Simple geometry permits to express the microstrains via the displacements and rotation vectors and their derivates. Most importantly, the expressions relating microstrains to the displacement and rotation vectors also contain the lattice geometry and internodal distances, or particle dimensions: so that DM develops into a fully multi-scale theory.

Microstresses of the axial, torsional, and shear type are introduced in DM (in relation to the corresponding microstrains) as their energy conjugates. Energy methods can be employed to derive the microstress equations of conservation of linear and angular momenta. Furthermore, the natural boundary conditions, originating from equilibrium in the integral energy formulation, dictate the relationships between the microstresses and the macrostresses. Eventually, it is possible to express macrostresses in terms of the microstresses, the lattice geometry, and the typical internodal distance with the desired degree of approximation. While it is always possible to obtain the continuum-level stresses from the microstresses and microstructural information, it is not in general possible to derive microstresses and micro-geometry from the macrostresses alone. Herein lies the strength of the DM approach: additional, micro-level information is made available that cannot be reached through conventional mechanics.

The doublet microstrains and microstresses constitute the language of DM. Doublet microstresses have a clear physical meaning: they represent the elastic micro-forces and micro-couples of interaction between any two particles of a doublet. The doublet microstresses as mathematical quantities are much simpler than the conventional macrostresses that are second rank tensors. This quality considerably facilitates the formulation of invariant constitutive equations. As a whole, the novel concept of doublet microstrains and microstresses plays a crucial role in making DM different and competitive with other theories such as the classical theory of elasticity.

16.3.3. Comparison with Other Theories

16.3.3.1. Lattice Dynamics Lattice Dynamics (LD) studies vibrations of the atomic nuclei of solid crystals, the nuclei being considered as material points mutually bonded by elastic interatomic forces. The particle vibration can be small or large compared with interatomic distances. The corresponding approximations of the theory are called harmonic and anharmonic, respectively. The harmonic approximation leads to linear governing equations and is usually referred to as the classical LD. The hypothesis at the base of LD affects significantly its applicability. In particular it follows that classical LD is incapable of solving those vibration problems where forces are applied over the boundary of the body. Since LD was developed to study only elastic vibration, it is unable to address both inelastic and static problems of solid mechanics.

16.3.3.2. Classical and Cosserat's Theory of Elasticity Although all the governing equations and boundary conditions of DM are derived and written in terms of the doublet microstrains and microstresses, generalized macrostresses $\sigma_{ij}^{(M)}$, $M_{ij}^{(M)}$ are also introduced. In the first approximation of the theory $M = 1$, they coincide with the conventional Cauchy σ_{ij} and Cosserat M_{ij} stress tensors. At $M > 1$, they have no counterparts. The generalized macrostresses can be computed via the doublet microstresses by the macro-micro relations that have an essential feature: their inversion is, in general, impossible since the number of doublet microstresses is usually greater than that of the macrostresses. Nonetheless, if the doublet microstresses are known, it is always possible to determine the unknown macrotensors σ_{ij} and M_{ij}. This means that the doublet microstresses are more informative than the Cauchy σ_{ij} and Cosserat M_{ij} macrostresses and they thus provide deeper incite into the mechanical behavior of solids. Also, the above macro-micro relations enable one to express the equations of motion of DM in terms of both the doublet microstresses and the generalized macrostresses $\sigma_{ij}^{(M)}$ and $M_{ij}^{(M)}$. At $M = 1$, the latter relations reduce to the differential equations of motion of either the classical or the Cosserat elastic continua depending on whether the couple macrostresses M_{ij} are respectively equal to, or different from zero. Notice that whilst the classical theory of elasticity is explicitly scaleless, Cosserat's theory is scale dependent [1].

16.4. DOUBLET MECHANICS WITHIN THE LINEAR ELASTIC FRAMEWORK (MATHEMATICAL FORMULATION OF DOUBLET MECHANICS)

16.4.1. Microstructure

The fundamental equations of DM are now reviewed. For ease of visualization, the granular interpretation of DM is employed, whereby a solid is viewed as an assembly of equal spheres. A granular body is then described by a spatial set H of a large number N of identical particles of diameter d. To simplify the problem, the set H_0 is assumed to be regular: the centers of the nodes therefore form a Bravais lattice Γ.

A couple of adjacent particles A and B in H_0 represent a doublet with the vector axis ζ^0 emerging from the node $a \in$ A towards the node $b \in$ B. The corresponding unit vector is $\tau^0 = \zeta^0/\eta$; where $\eta \equiv |\zeta^0|$ is the distance between a and b, the internodal distance (Fig. 16.3).

16.4.2. Microstrains

When a granular body undergoes deformation, microstrains are developed within its doublets. The different typs of deformation a body can undertake are: i) relative separation of the doublet nodes leading to an elongation ε; ii) rotation of the particles about the doublet axis leading to a torsion μ; iii) slipping of the particles past their point of contact leading to a shear γ. The microstrains are induced by the translation of the nodes and by the rotation of the particles in the granular body. Therefore, it is possible to write the microstrains in terms of the displacement and rotational fields of the nodes. The two smooth mutually independent vector fields of the translations and of the rotations are, respectively, $\mathbf{u}(\mathbf{X}, t)$ and $\boldsymbol{\phi}(\mathbf{X}, t)$, where $\mathbf{X}$ is a position vector of an arbitrary point in Γ, and t is the time.

The analysis is now focused on the first of these vector fields: $\mathbf{u}(\mathbf{X}, t)$, on the basis of which a series of derivations will lead to the microstrains ε. The mathematical operations relative to $\boldsymbol{\phi}$ ($\mathbf{X}$, t), in order to obtain μ and γ, is, indeed, quite similar. The incremental vector $\Delta\mathbf{u}_\alpha$, which represents an increment of $\mathbf{u}$ in a transition from an arbitrary node $a \in$ A to the adjacent node $b \in$ B, is defined as follows:

$$\Delta\vec{u}_\alpha \equiv \vec{u}(\vec{x} + \vec{\zeta}^0_\alpha, t) - \vec{u}(\vec{x}, t). \tag{16.13}$$

The fundamental assumption, which lies at the basis of DM, is that the above increment vector may be expanded in a convergent Taylor series in a neighborhood of an arbitrary node $a \in \Gamma$:

$$\Delta\vec{u}_\alpha = \sum_{\chi=1}^{M} \frac{(\eta_\alpha)^\chi}{\chi!}(\vec{\tau}^0_\alpha \cdot \nabla)^\chi \vec{u}(\vec{X}, t); \tag{16.14}$$

the order at which the series is truncated being M, the degree of approximation. In the deformed doublet, the initial vector axis ζ^0 is mapped in the final vector ζ, where:

$$\vec{\zeta}_\alpha = \vec{\zeta}^0_\alpha + \Delta\vec{u}_\alpha, \tag{16.15}$$

and the corresponding director τ_α is:

$$\vec{\tau}_\alpha \equiv \frac{\vec{\zeta}_\alpha}{\zeta_\alpha} = \frac{1}{1+\varepsilon_\alpha}\left(\vec{\tau}^0_\alpha + \frac{\Delta\vec{u}_\alpha}{\eta_\alpha}\right); \tag{16.16}$$

where:

$$\varepsilon_\alpha = \frac{\zeta_\alpha - \eta_\alpha}{\eta_\alpha}. \tag{16.17}$$

If it is assumed that the relative displacements of the doublet nodes and the elongation microstrains are small, (i.e.$|\Delta\mathbf{u}_\alpha| \ll \eta_\alpha$ and $\varepsilon_\alpha \ll 1$), then the approximate relation follows:

$$\varepsilon_\alpha = \frac{\vec{\tau}^0_\alpha \cdot \Delta\vec{u}_\alpha}{\eta_\alpha}. \tag{16.18}$$

Combining expression (16.18) with eqns. from (16.14) to (16.17), the relation between the axial microstrain ε_α, and the vector fields of displacements of the nodes and the parameters of the geometry of the doublets is finally obtained:

$$\varepsilon_\alpha = \tau^0_{\alpha i} \sum_{\chi=1}^{M} \frac{(\eta_\alpha)^{\chi-1}}{\chi!} \tau^0_{\alpha k_1} \cdots \tau^0_{\alpha k_\chi} \left.\frac{\partial^\chi u_i}{\partial x_{k_1} \cdots \partial x_{k_\chi}}\right|_{x=x_0}. \tag{16.19}$$

The first approximation M $= 1$ for the elongation microstrain has the form:

$$\varepsilon_\alpha = \tau^0_{\alpha i}\, \tau^0_{\alpha j}\, \varepsilon_{ij}, \tag{16.20}$$

where:

$$\varepsilon_{ij} = \frac{1}{2}\left(\frac{\partial u_i}{\partial x_j} + \frac{\partial u_j}{\partial x_i}\right) \tag{16.21}$$

are the components of the classical linear strain tensor. Similarly to eqn. (16.19), it is possible to relate the microstrains of torsion μ and shear γ to the geometry parameters of the doublets, and both to the rotational and the translational vector fields of the nodes.

16.4.3. Microstresses and Transition to Macrostresses

The internal generalized microstrains engender within the granular medium generalized microstresses. The relation between the above quantities is written by means of the strain energy per unit volume of the medium (W), that is:

$$\begin{aligned} &\bullet \text{ Elongation microstress} \rightarrow p_\alpha = \frac{\partial W}{\partial \varepsilon_\alpha},\\ &\bullet \text{ Torsional microstress} \rightarrow m_\alpha = \frac{\partial W}{\partial \mu_\alpha},\\ &\bullet \text{ Shear microstress} \rightarrow t_{\alpha i} = \frac{\partial W}{\partial \gamma_{\alpha i}}. \end{aligned} \tag{16.22}$$

Employing energy methods, it is also possible to derive the fundamental equations of conservation of both linear momentum and momentum of momentum, and the natural boundary conditions at the external surface S of the medium (relations written in terms of the microstresses):

– *Conservation of linear momentum:*

$$\sum_{\alpha=1}^{n}\sum_{\chi=1}^{M}(-1)^{\chi-1}\frac{(\eta_\alpha)^{\chi-1}}{\chi!}\tau^0_{\alpha k1}\dots\tau^0_{\alpha k\chi}\frac{\partial^\chi(t_{\alpha i}+p_{\alpha i})}{\partial x k_1\dots\partial x k_\chi} + F_i = \rho\frac{\partial^2 u_i}{\partial t^2}, \tag{16.23}$$

– *Conservation of momentum of momentum:*

$$\sum_{\alpha=1}^{n}\left(\varepsilon_{ijq}\tau^0_{\alpha j}\tau_{\alpha q} + \sum_{\chi=1}^{M}(-1)^{\chi-1}\frac{(\eta_\alpha)^{\chi-1}}{\chi!}\tau^0_{\alpha k1}\dots\tau^0_{\alpha k\chi}\frac{\partial^\chi(m_{\alpha i}-\frac{1}{2}\eta_\alpha\varepsilon_{ijq}\tau^0_{\alpha j}\tau_{\alpha q})}{\partial x k_1\dots\partial x k_\chi}\right) = 0 \tag{16.24}$$

– *Natural boundary conditions (force boundary conditions):*

$$n_{kr}\sum_{\alpha=1}^{n}\tau^0_{\alpha kr}\sum_{\chi=r}^{M}(-1)^{\chi-1}\frac{(\eta_\alpha)^{\chi-1}}{\chi!}\tau^0_{\alpha kr+1}\dots\tau^0_{\alpha k\chi}\frac{\partial^{\chi-r}(t_{\alpha i}+p_{\alpha i})}{\partial x k_{r+1}\dots\partial x k_\chi} = T_i\delta_{r1}, \tag{16.25}$$

– *Natural boundary conditions (couple boundary conditions):*

$$n_{kr}\sum_{\alpha=1}^{n}\tau_{\alpha kr}^{0}\sum_{\chi=r}^{M}(-1)^{\chi}\frac{(\eta_\alpha)^{\chi-1}}{\chi!}\tau_{\alpha kr+1}^{0}\dots\tau_{\alpha k\chi}^{0}\frac{\partial^{\chi-r}(m_{\alpha i}-\frac{1}{2}\eta_\alpha\varepsilon_{ijq}\tau_{\alpha j}^{0}\tau_{\alpha q})}{\partial xk_{r+1}\dots\partial xk_{\chi}}=-M_i\delta_{r1}. \tag{16.26}$$

Where **T** and **M** are the force and couple vectors per unit area of the external surface S; **F** is the force per unit volume; ρ the density; and ε_{ijq} is the permutation tensor.

16.4.3.1. Transition to MacroStresses The surface force and couple vectors **T** and **M** can be rewritten in terms of the second order tensors of force and couple macrostresses, σ_{ki}, M_{ki}:

$$T_i=\sigma_{ki}n_k,\quad M_i=M_{ki}n_k. \tag{16.27}$$

By combining eqns. (4.15) with eqns. (4.13) and (4.14), the relations between the micro- and macro- stresses are finally found:

$$\sigma_{k1i}^{(M)}=\sum_{\alpha=1}^{n}\tau_{\alpha k1}^{0}\sum_{\chi=1}^{M}(-1)^{\chi+1}\frac{(\eta_\alpha)^{\chi-1}}{\chi!}\tau_{\alpha k2}^{0}\dots\tau_{\alpha k\chi}^{0}\frac{\partial^{\chi-r}(t_{\alpha i}+p_{\alpha i})}{\partial xk_2\dots\partial xk_\chi}, \tag{16.28}$$

$$M_{k1i}^{(M)}=\sum_{\alpha=1}^{n}\tau_{\alpha k1}^{0}\sum_{\chi=r}^{M}(-1)^{\chi+1}\left[\frac{(\eta_\alpha)^{\chi-1}}{\chi!}\tau_{\alpha k2}^{0}\dots\tau_{\alpha k\chi}^{0}\times\frac{\partial^{\chi-1}(m_{\alpha i}-\frac{1}{2}\eta_\alpha\varepsilon_{ijq}\tau_{\alpha j}^{0}\tau_{\alpha q})}{\partial xk_2\dots\partial xk_\chi}\right]. \tag{16.29}$$

M indicates the level of approximation at which the macrostresses are represented by the microstresses. From eqn. (16.28) one can see that, if scale effects are included, the macroscopic stress tensor need not to be symmetric, even for non-polar media.

Furthermore, if eqns. (16.28) and (16.29) are compared with the relations (16.23) and (16.24), the macrostress equations of motion of the granular medium can be derived:

$$\frac{\partial\sigma_{ij}^{(M)}}{\partial x_i}+F_j=\rho\frac{\partial^2u_j}{\partial t^2}, \tag{16.30}$$

$$\frac{\partial M_{ij}^{(M)}}{\partial x_i}+\varepsilon_{ijk}\sigma_{ik}^{(1)}=0. \tag{16.31}$$

In the first approximation of DM (M = 1), eqns. (16.30) and (16.31) describe the motion of Cosserat's continuum with asymmetric force microstresses. Under some conditions (see Section 16.3) the couple macrostresses vanish, and the relations reduce to the equations of motion of the symmetric classic continuum. In the subsequent approximations ($M > 1$) of DM, the macroscopic equations become quite different from the equations of motion of both classical and Cosserat continua.

16.4.4. Linear Elastic Doublet Mechanics

Up to now the fundamental equations of DM have been derived, but, with the exception of eqns. (16.22), nothing has been said on the relations between microstresses and microstrains (microconstitutive relations). Therefore in this section, the constitutive restrictions imposed on a class of microstructured media by the energy and entropy balance laws, and by Clausius inequality, are briefly exposed.

16.4.4.1. Balance Laws The differential formulation of energy balance is:

$$\rho r - \nabla \circ \vec{q} + P - \rho \dot{E} = 0, \tag{16.32}$$

where r is the volume rate of heat supplied per unit mass, **q** is the heat flux vector, P is the mechanical power per unit volume, E is the internal energy per unit mass, and ρ is the mass density. In addition, the differential law of entropy balance is:

$$\rho\left(\frac{r}{\theta} + \xi\right) - \nabla \circ \left(\frac{\vec{q}}{\theta}\right) = \rho \dot{s}, \tag{16.33}$$

where ζ is the internal rate of entropy production per unit mass, s is the entropy per unit mass, and θ is a function of the T, and other constitutive variables, such that $\theta \geq 0$, and $\partial\theta/\partial T \geq 0$. Combining eqns. (16.32) and (16.34), the following relation is obtained:

$$-\rho(\dot{\psi} + \dot{\theta} s) - \frac{\vec{q} \circ \vec{g}}{\theta} - \rho\theta\xi + P = 0, \tag{16.34}$$

where $\mathbf{g} = \nabla q$ and $\Psi = E - \theta\, s$, is the Helmhotz free energy. The expression of the unit mechanical power per unit volume P, in DM, is as follows:

$$P = \sum_{\alpha=1}^{n} (p_\alpha \dot{\varepsilon}_\alpha + m_\alpha \dot{\mu}_\alpha + \vec{t}_\alpha \circ \dot{\vec{\gamma}}_\alpha). \tag{16.35}$$

Furthermore, the microstresses are required to satisfy the balance of both linear momentum and momentum of momentum, eqns. (16.23) and (16.24). In the above balance laws, two sets of variables are contained:

$$\{\mathbf{u}, \boldsymbol{\phi}, \mathrm{T}\}, \tag{16.36}$$

and

$$\{\Psi, \theta, s, \zeta, \mathrm{p}_\alpha, \mathrm{m}_\alpha, \mathbf{t}_\alpha, \mathbf{q}\}; \tag{16.37}$$

Herein it is assumed that the variables (16.37) depend constitutively on the fields (16.36).

16.4.4.2. Linear Elastic Constitutive Assumptions If the material is assumed both linear and homogeneous, the constitutive relations take the form [1]:

$$\begin{aligned} p_\alpha &= \sum_{\beta=1}^{n}(A_{\alpha\beta}\varepsilon_\beta + B_{\alpha\beta}\mu_\beta) + J_\alpha\Theta, \\ m_\alpha &= \sum_{\beta=1}^{n}(D_{\alpha\beta}\varepsilon_\beta + E_{\alpha\beta}\mu_\beta) + K_\alpha\Theta, \\ t_{\alpha i} &= \sum_{\beta=1}^{n} I_{\alpha\beta ij}\gamma_{\beta j}. \end{aligned} \tag{16.38}$$

Where $\Theta = \theta - \theta_0$ is the increment in temperature, being θ_0 the temperature of the granular media in the initial state, J and K are the Fourier coefficients of heat conduction and $A_{\alpha\beta}$, $B_{\alpha\beta}$, $D_{\alpha\beta}$, $E_{\alpha\beta}$, $I_{\alpha\beta ij}$ at last, are the micro constants of elasticity of DM.

16.5. PLANE WAVES PROPAGATION WITHIN THE LINEAR ELASTODYNAMICS OF DOUBLET MECHANICS

16.5.1. Significance of the Analysis

In this section, the plane propagation of elastic waves in granular media is analyzed, with reference to a spatial arrangement of the particles that results in macroscopic-level isotropy in the plane of propagation and particle displacement. Doublet mechanics is shown to be fully compatible with crystal dynamics and continuum elastodynamics, yet intrinsically richer than the latter in that it affords the modeling of dispersion and retardation phenomena. Both of these effects are scale-related and disappear in the continuum or in the infinite wavelength limit. In view of its multi-scale nature and its capability to bridge the discrete and continuum viewpoints, DM represents a natural tool for the discussion of scaling effects in the dynamics of particular and granular media.

16.5.2. Dynamic Scaling Equations

A simplified version of the doublet mechanical governing equation is now derived, with the purpose of studying bi-dimensional propagation phenomena. It is assumed that the dynamic process is isothermal and the volume forces vanish. In addition, the particle interactions are assumed to be longitudinal (central), so that the shear and torsion microstresses vanish everywhere in the body:

$$m_{\alpha i} = t_{\alpha i} \equiv 0. \tag{16.39}$$

As already described in previous sections, the granular medium with such properties bears only conventional macrostresses σ_{ij} and does not sustain couple macrostresses M_{ij}, which are identically equal to zero in the volume V. With these assumptions, the only microstress engendered in the medium is the elongation microstress p_α, which can be expressed, according to eqns. (16.38) and (16.39), as:

$$p_\alpha = \sum_{\beta=1}^{n} A_{\alpha\beta}\varepsilon_\beta. \tag{16.40}$$

Assumption (16.39) identically satisfies the conservation of momentum of momentum (16.24); using relations (16.39) and (16.40) with the balance of linear momentum (16.23), the scale dynamic equations in terms of the microstrain (ε_β) is obtained:

$$\sum_{\alpha}^{n}\sum_{\beta}^{n}\sum_{\chi}^{m}(-1)^{\chi-1}A_{\alpha\beta}\tau_{\alpha i}^{0}\frac{(\eta_\alpha)^{\chi-1}}{\chi!}\tau_{k1}^{0}\ldots\tau_{k\chi}^{0}\frac{\partial^{\chi}\varepsilon_\beta}{\partial x_{k1}\ldots\partial x_{k\chi}} = \rho\frac{\partial^2 u_i}{\partial t^2}. \tag{16.41}$$

Combining eqn. (16.41) with the fundamental expression relating the microstrains to the displacements (16.20), the wave equations can be written in terms of the displacements and of the micro-parameters of the underlying medium only. In particular, in the first approximation of DM (M = 1), eqn. (16.41) reduces to:

$$C_{ijkl}\frac{\partial^2 u_j}{\partial x_k \partial x_l} = \rho\frac{\partial^2 u_i}{\partial t^2}, \tag{16.42}$$

these scaling dynamic equations are form-identical with the non-scaling equations of the conventional theory of anisotropic elasticity, within the DM context, however, the stiffness tensor is:

$$C_{ijkl} = \sum_{\alpha=1}^{n}\sum_{\beta=1}^{n} A_{\alpha\beta}\tau_{\alpha i}^{0}\tau_{\alpha j}^{0}\tau_{\beta k}^{0}\tau_{\beta l}^{0}. \tag{16.43}$$

For M = 2, instead, the equations of motion have the form:

$$C_{ijkl}\frac{\partial^2 u_j}{\partial x_k \partial x_l} - C_{ijklmn}\frac{\partial^4 u_j}{\partial x_k \partial x_l \partial x_m \partial x_n} = \rho\frac{\partial^2 u_i}{\partial t_2}, \tag{16.44}$$

where:

$$C_{ijklmn} = \sum_{\alpha=1}^{n}\sum_{\beta=1}^{n} A_{\alpha\beta}\frac{\eta_\alpha^2}{4}\tau_{\alpha i}^{0}\tau_{\alpha j}^{0}\tau_{\beta k}^{0}\tau_{\beta l}^{0}\tau_{\beta m}^{0}\tau_{\beta n}^{0}. \tag{16.45}$$

16.5.3. Plane Elastic Waves in Granular Media

A non-general solution of eqn. (16.41) is now sought focusing attention on a specific microstructure. To start with a set of lattice cosine directors τ_{jx} must be chosen, and a guess on the micro-constants of elasticity, $A_{\alpha\beta}$, has to be made. The three dimensional arrangement of nodes given in (Fig. 16.4) is considered. The valence of the lattice is therefore

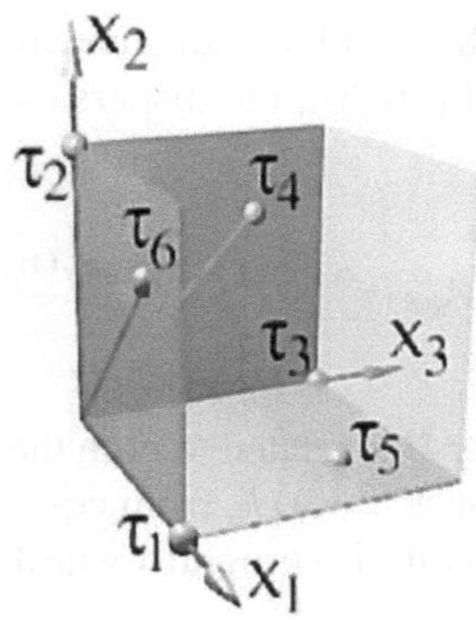

FIGURE 16.4. A particular array of doublets.

n = 6, while the direction vectors are as follows:

$$\begin{aligned} \tau_1 &= (1,0,0), \sqrt{2}\,\tau_4 = (0,1,1),\\ \tau_2 &= (0,1,0), \sqrt{2}\,\tau_5 = (1,0,1),\\ \tau_3 &= (0,0,1), \sqrt{2}\,\tau_6 = (1,1,0); \end{aligned} \tag{16.46}$$

the micro parameters matrix, furthermore, is embodied by:

$$\begin{bmatrix} A_{11} & A_{12} & A_{12} & 0 & A_{15} & A_{15} \\ & A_{11} & A_{12} & A_{15} & 0 & A_{15} \\ & & A_{11} & A_{15} & A_{15} & 0 \\ & & & A_{44} & 0 & 0 \\ & & & & A_{44} & 0 \\ Sym & & & & & A_{44} \end{bmatrix} \tag{16.47}$$

this particular choice of microstructure and physical parameters yields an isotropic constitutive relation at the continue scale, as long as the conditions:

$$A_{15} = -\frac{1}{2}A_{44},\ A_{12} = A_{11} - \frac{3}{4}A_{44}, \tag{16.48}$$

According to eqn. (16.43), the relation between the nano-parameters and the macroscopic Lamé's constants, for M = 1, is also found:

$$\lambda = A_{11} - A_{44},\ \mu = \frac{1}{4}A_{44}. \tag{16.49}$$

16.5.3.1. Longitudinal Waves A longitudinal plane wave is assumed to propagate in the x_1 direction of the medium:

$$\mathbf{u}_1 = \mathbf{u}_{10}\exp[i(\omega t - kx_1)], \tag{16.50}$$

where ω and k are, respectively, the angular frequency and wave number. Combining eqn. (16.50) with relations (16.41) and (16.20), the dispersion equation for plane longitudinal waves is obtained:

$$\sum_{\alpha=1}^{n}\sum_{\beta=1}^{n}\sum_{S=1}^{M}\sum_{\chi=1}^{M}(-1)^{\chi}(i)^{\chi+S}A_{\alpha\beta}\left(\tau_{\alpha 1}^{0}\right)^{\chi+1}\left(\tau_{\beta 1}^{0}\right)^{S+1}\frac{(\eta_{\alpha})^{\chi-1}(\eta_{\beta})^{S-1}}{\chi!S!}k^{\chi+S}=\rho\omega^{2}. \tag{16.51}$$

Since the expression of the velocity V is related to both the angular frequency and the wave number through the simple relation $V=\omega/k$, from eqn. (5.13) the expression of the Mth order longitudinal wave phase velocity is straightforwardly derived:

$$V_l^{(M)}=\left(\sum_{\alpha=1}^{n}\sum_{\beta=1}^{n}\sum_{S=1}^{M}\sum_{\chi=1}^{M}(-1)^{\chi}(i)^{\chi+S}\right.$$
$$\left.\times A_{\alpha\beta}\left(\tau_{\alpha 1}^{0}\right)^{\chi+1}\left(\tau_{\beta 1}^{0}\right)^{S+1}\frac{(\eta_{\alpha}/\lambda_w)^{\chi-1}(\eta_{\beta}/\lambda_w)^{S-1}}{\rho\chi!S!}(2\pi)^{\chi+S-2}\right)^{1/2} \tag{16.52}$$

Choosing M = 1, reduces eqn. (16.52) to:

$$V_l^{(1)}=\left(\sum_{\alpha=1}^{n}\sum_{\beta=1}^{n}A_{\alpha\beta}\left(\tau_{\alpha 1}^{0}\right)^{2}\left(\tau_{\beta 1}^{0}\right)^{2}\right)^{1/2}=\sqrt{\frac{A_{11}-A_{44}/2}{\rho}}\equiv\sqrt{\frac{\lambda+2\mu}{\rho}}, \tag{16.53}$$

that is in perfect agreement with the phase velocity obtained in the framework of classical continuum elastodynamics. The second approximation of DM (M = 2), instead, yields to:

$$V_l^{(2)}=V_l^{(1)}\sqrt{1+\frac{\delta}{\lambda_w^2}}, \tag{16.54}$$

where:

$$\delta=\frac{\pi^2(\eta_\alpha)^2\left(A_{11}-\frac{A_{44}}{\sqrt{2}}+\frac{A_{44}}{4}\right)}{A_{11}-A_{44}/4}. \tag{16.55}$$

It is evident that $V_1^{(2)}$ is a function of the wavelength. Moreover, assuming $\eta_\alpha=\eta_\beta=\eta$, and imposing $M\to\infty$, from eqn. (16.52) the following expression is obtained:

$$V_l^{(\infty)}=\sqrt{\frac{A_{11}}{\rho}\frac{\sin(\pi\eta/\lambda_w)}{\pi\eta/\lambda_w}}, \tag{16.56}$$

that is identical to the dispersion relation with periodicity $k=2\pi/\lambda_\omega$ for a one dimensional longitudinal wave propagating in a monoatomic lattice of spacing η. This fact confirms that, at the approximation level $M\to\infty$, DM is equivalent to lattice mechanics for an infinite, periodic and discrete crystal.

16.5.3.2. Shear Waves For the sake of brevity, herein only the shear wave phase velocity of the Mth order is given:

$$V_l^{(M)} = \left(\sum_{\alpha=1}^{n}\sum_{\beta=1}^{n}\sum_{S=1}^{M}\sum_{\chi=1}^{M}(-1)^{\chi}(i)^{\chi+S}\right.$$
$$\left.\times\, A_{\alpha\beta}\tau_{\alpha 2}^{0}\left(\tau_{\alpha 1}^{0}\right)^{\chi}\tau_{\beta 2}^{0}\left(\tau_{\beta 1}^{0}\right)^{S}\frac{(\eta_\alpha/\lambda_w)^{\chi-1}(\eta_\beta/\lambda_w)^{S-1}}{\rho\chi!S!}(2\pi)^{\chi+S-2}\right)^{1/2} \tag{16.57}$$

of a wave propagating in the (x_1) direction, and whose displacement is along the (x_2) axis:

$$\mathbf{u}_2 = \mathbf{u}_{20}\exp[i(\omega\, t - kx_1)]. \tag{16.58}$$

16.5.4. Discussion

The formulas (16.52) and (16.58) express the wave velocities to an arbitrary M^{th} approximation for M from 1 to ∞. The expression of the longitudinal wave velocity in the first approximation of DM is embodied by eqn. (16.53), which is independent by the non-dimensional scale parameter η/λ_ω. It is evident that eqn. (16.54), as well as any further expression of velocity obtained at $M \geq 2$, reduces to the non scale relation of velocity (16.53), via two different assumptions: i) the constituent granules of the solid are material points whose sizes are infinitesimal $\eta = 0$ and all the waves may have arbitrary but finite length, ii) the constituent particles of the solid may have an arbitrary but finite size ($\eta \neq 0$) and all the waves have an infinite length ($\lambda_\omega \rightarrow \infty$). Assumption i) is adopted in classical theory of elasticity, while the restriction ii) is employed in the special discrete-continuum theories of wave propagation in granular media and crystals. It is evident that neither of these approaches permits the analysis of scale effects, whereas DM is capable of modeling solids in view of their dual discrete-continuous nature. The developed micromechanical elastodynamics theory is applicable not only to the unbounded isotropic granular solids but also to any similar particulate media, (i.e. any media with microstructure characterized geometrically by finite particle sizes), such as a biological tissue.

Finally, it is of great importance to remark that the wave dispersion and wave retardation depend on the wavelength λ_ω and the characteristic distance η not separately, but in a non-dimensional combination $l = \lambda_\omega/\eta$. Thus the internal size η is significant only in comparison with the wavelength λ_ω. The latter may be arbitrary large and, consequently, the size η may be also arbitrary large provided the ratio l is greater than 2, ($\lambda_\omega/\eta > 2$). On these bases, the absolute dimension of the typical internal structure is not a determining factor concerning the applicability of the present theory.

16.6. REFLECTION AND TRANSMISSION OF PLANE WAVES (NUMERICAL APPLICATIONS OF DOUBLET MECHANICS TO MALIGNANT TISSUE)

In the upcoming paragraphs, a broad use will be made of the formulas and of the results already achieved in the previous sections, specifically of Sections 16.4. Mathematical Formulation of DM and 16.5. Plane Waves Propagation within the Linear Elastodynamics

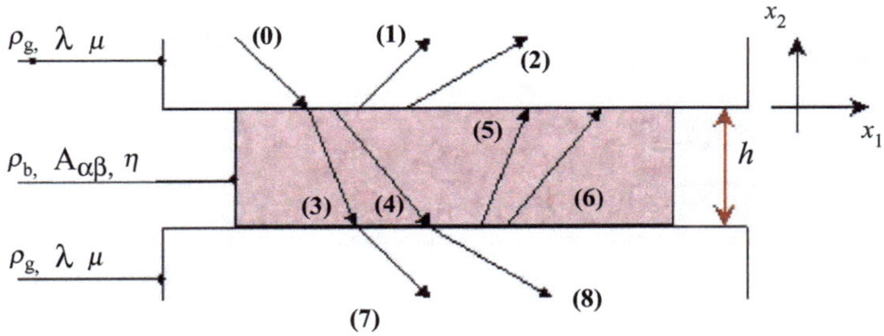

FIGURE 16.5. the DM model and the reflection-transmission phenomena at the interfaces.

of DM. In the ensuing analysis, the following common assumptions are made: i) no body forces are acting in the discrete domain, ii) the doublets are capable of axial microstrains only (the torsion and shear microstrains vanish everywhere), iii) the doublet constitutive response is linear elastic, and the dynamic process is isothermal, and iv) the nodes discrete packing is macroscopically isotropic in the non-scale limit.

16.6.1. The Reflection Equations

In order to develop the equations that embody the relations in the frequency domain between the reflection coefficients and the microparamenters of the tissue, the analysis is focused on the framework of Fig. 16.5, which gives a representation of the experimental specimen. Essentially, it is constituted of two glass substrates, between which a thin single layer of biological and granular tissue is embedded. For more details of the glass specimen fixture see Section 16.7.4. The heterogeneous nature of the tissue suggests it be modelled within the field theory of doublet mechanics. Both the glass substrates are completely identified by the density ρ_g, and the classical elasticity Lamé's constants λ and μ; conversely, the tissue layer is characterized, apart from its value of density ρ_b, by the DM microelastic moduli $A_{\alpha\beta}$ and the internodal distances η_α. The thickness of the layer h is also a parameter of the model.

The system is tested by an ultrasonic time harmonic plane wave (shear or longitudinal), which travels in the upper glass substrate with an angle θ_0 with respect to the x_2 axis. When the wave hits the first glass-layer interface, it divides up in four further waves: two waves (shear and longitudinal) transmit into the layer, while two are reflected in the upper glass substrate with the corresponding angles being θ_1, θ_2, θ_3 and θ_4 respectively. Other analogous reflection-transmission phenomena take place in the second layer-glass interface; consequently, the plane waves that propagate in the system are nine (Fig. 16.5). The propagation directions are also fully defined by nine non-independent angles that reduce to four independent angles when considering the second and the third laws of optics. In particular, the following relations hold:

$$\theta_1 = \theta_0, \theta_5 = \theta_3, \theta_6 = \theta_4, \theta_7 = \theta_0, \theta_8 = \theta_0. \tag{16.59}$$

When the material properties are known for both the layer and the glass and the initial incident angle θ_0 is set, the residual angles θ_i are completely determined by Snell's law.

Additionally, the time harmonic waves, $\mathbf{u}^{(n)}$, result analytically and are defined through the equations:

$$\mathbf{u}^{(n)} = A_n \mathbf{d}^{(n)} \exp[i\eta^{(n)}], \quad \eta^{(n)} = k_n(\mathbf{x} \cdot \mathbf{p}^{(n)} - c_n t), \tag{16.60}$$

where:

A_n : wave amplitude;
$\mathbf{d}^{(n)}$: particle motion unit vectors
$\mathbf{x}$: position vector;
$\mathbf{p}^{(n)}$: propagation unit vector;
k_n : wave numbers
c_n : wave propagation phase velocities.

For the intent here is to find out the reflection coefficients at the first interface, (i.e. the ratios $|A_1/A_O|$ and $|A_2/A_O|$ corresponding to the shear and longitudinal wave respectively), a set of equations that explicitly or implicitly contain the wave amplitudes A_i are necessary. These equations are embodied by the natural stress-displacement boundary conditions, imposed at every interface:

$$\sum_{i=1}^{n} u_1^{(i)} = 0, \sum_{i=1}^{n} u_2^{(i)} = 0, \sum_{i=1}^{n} \sigma_{22}^{(i)} = 0, \sum_{i=1}^{n} \sigma_{12}^{(i)} = 0; \tag{16.61}$$

where n is the number of waves which are relative to the interface, while $u_1^{(i)}, u_2^{(i)}, \sigma_{12}^{(i)}$ and $\sigma_{22}^{(i)}$ are respectively the shear and normal displacements, and the shear and normal macrostresses associated to the ith wave. The total amount of equations that can be written is therefore 8 (4 × 2), whereas the unknown variables are the 8 reflection coefficients $|A_i/A_O|$. Hence, once the material properties of both the glass substrates and the tissue layer are predetermined, the boundary conditions (BC) can be solved and the reflection coefficients consequently derived.

While the expression of the wave displacements $\mathbf{u}^{(n)}$ is explicitly formulated by eqn. (16.60), in order to fully write the BCs (16.61) a relation between the displacements and the macrostresses must be drawn. The classical theory of elasticity is employed when addressing waves propagating in the continuum glass substrates (i.e., the waves 0, 1, 2, 7, and 8), and DM is applied for waves transmitting within the discrete tissue (i.e., waves 3, 4, 5, and 6).

Given the expression of the displacements, eqns. (16.60), the elongation microstrains can be straightforwardly derived in accordance with the formula (16.19). Nonetheless, the latter is a general expression that relates the DM strains to the displacements at a generic order of approximation M. Now a hypothesis on the degree of approximation to be adopted must be made. In the following development of the analysis, the first and fundamental scale-accounting version of the doublet mechanical approach is employed. This version corresponds to the truncation of the displacement expansion at $M = 2$. This choice retains the advantage of analytical treatment, while sufficing to establish the already-mentioned qualitative features of the scale accounting treatment.

In accordance to this assumption, eqn. (16.19) reduces to the expression:

$$\begin{aligned}\varepsilon_\alpha = {} & \tau_{\alpha 1}^2 u_{1,1} + \tau_{\alpha 1}\tau_{\alpha 2}(u_{1,2} + u_{2,1}) + \tau_{\alpha 2}^2 u_{2,2} \\ & + \frac{1}{2}\eta_\alpha \big(\tau_{\alpha 1}^3 u_{1,11} + 2\tau_{\alpha 1}^2 \tau_{\alpha 2} u_{1,12} + \tau_{\alpha 2}^2 \tau_{\alpha 1} u_{1,22} \\ & \qquad + \tau_{\alpha 1}^2 \tau_{\alpha 2} u_{2,11} + 2\tau_{\alpha 2}^2 \tau_{\alpha 1} u_{1,12} + \tau_{\alpha 2}^3 u_{2,22}\big)\end{aligned} \tag{16.62}$$

which explicitly contains both the internodal distance η_α and the direction cosines of the unit vectors connecting two nodes (τ's). Once again, further conjecture is needed concerning the underlying lattice microstructure configuration. In particular, the three dimensional arrangement of nodes given in Fig. 16.4 is considered, as already postulated in Section (16.5.3). Consequently, the unit vector's and the micromoduli's expressions are also embodied by eqns. from (16.5.8) to (16.4.11). Notice that this particular choice of microstructure and physical parameters yields an isotropic constitutive relation at the continuum scale).

At this point, the relation (16.6.4) can be utilized and an explicit expression of the microstrains, as a function of the displacements, can be found. From the DM strains, the longitudinal overall microstresses (P_α) can be achieved for every (α) doublet according to the micro-constitutive relation (16.38). This then reduces to formula (16.40) since the particle interactions are longitudinal and the dynamic process is assumed isothermal.

Finally, by means of eqns. (16.28), the transition from macro- to micro- stresses is obtained. In particular, at the scale approximation $M = 2$, the latter micro-macro relations reduce to:

$$\sigma_{ij} = \sum_{\alpha=1}^{n} \left(\tau_{\alpha i}\tau_{\alpha j} P_\alpha - \frac{1}{2}\eta_\alpha \tau_{\alpha i}\tau_{\alpha j}\tau_{\alpha k} P_{\alpha,k} \right), \tag{16.63}$$

Substituting (16.60), (16.62) and (16.63) into (16.61), the macrostresses engendered in the biological and granular medium are related to the wave amplitudes A_i. This has been achieved through the fundamental kinematics and constitutive equations of DM. Eventually, a set of 8 linear equations is obtained, from which the 8 unknown A_i/A_0 can be derived.

16.6.2. Solution of the Equations: the Forward Problem

Arranging the sets of equations (16.60–16.63) in matrix form, the following relation is obtained:

$$[M]\{R\} = \{B\}, \tag{16.64}$$

where [M] is the [8×8] matrix of the coefficients, {B} the [6×1] vector of the known terms and {R}, the [6×1] vector of the unknown reflection coefficients: A_i/A_0. By investigating [M], one can establish the dependence of the reflection coefficients on the following quantities: i) the biological material properties (density, internodal distance, and the DM micromoduli ρ_b, η, A_{11}, A_{44}) ii) the glass substrates material properties (density, Lamé's constants ρ_g, λ, μ); iii) the layer's thicknesses $\{h\}$; iv) the propagation angle of the initial incident wave θ_0; v) the phase velocities and the angles of propagation of the waves in the media c_i, θ_i; vi) the time harmonic waves' angular frequency $\omega = 2\pi f$.

Solving the equations (16.64), require making an initial guess for the parameters i); setting the properties ii), iii) and iv); calculating the variables v); and inverting the system (16.64). It is important to explain how the above steps have to be carried out for every particular value of the angular frequency ω. To obtain a broad simulated reflection spectrum, therefore, the processes iii) and iv) must be iterated over a wide range of the frequencies. It is obvious that such a task cannot be accomplished without the assistance of a computer and a numerical code at this purpose developed.

16.6.3. The Inverse Problem and the Doublet Mechanics Parameters Identification

An example of simulated reflection spectrum is reported in Fig. 16.5. The numerical functions described in the previous paragraphs have been employed to fit a set of experimental points derived from the examination of a malignant biological breast tissue. Such fitting has led to the estimation of the DM parameters. In particular it has been derived that, $\rho = 0.95$ g cm^{-3}, $A_{11} = 2$ GPa, $A_{44} = 0.088$ GPa, and $\eta = 0.0067$ mm. The algorithm adopted in this study was built upon a routine least square method, (i.e., Levenberg-Marquadt minimization algorithm). The technique is fairly stable for initial guesses that are within 20% deviation from the true values, therefore it is critical that the first estimates of the DM parameters are reasonably assumptions. It is important to ensure that these estimates are consistent with previous results and/or based upon material properties of biological tissues presented in literature.

Since these properties represent classical mechanical properties, it is necessary to transfer the data into the framework of doublet mechanics. This is achieved by employing the transformation eqns. (16.49), that in the first approximation of DM, ($M = 1$), and for a particular configuration of the underlying lattice microstructure allows the nano- classical-parameter identification (see also Section 16.5.3).

The feasibility of this tissue diagnostic approach upon human breast cancer tissue has already been verified [40, 41]. Although the details of this study will be covered in detail in

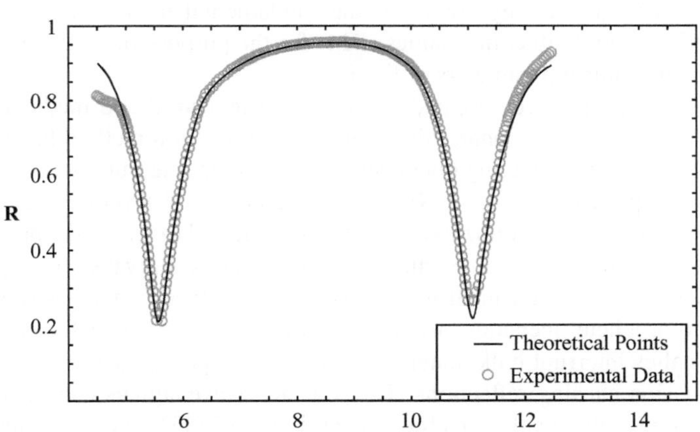

FIGURE 16.6. The reflection spectrum analytical-numerical model utilized to fit a set of experimental data from a specimen malignant tissue.

Section 16.7. Experimental Practice, it should be stated that the ultrasonic measurements demonstrated the ability to statistically differentiate normal and malignant tissue.

16.6.4. The Doublet Mechanics Approach: Final Marks

The innovation of the DM approach is the application of a microstructure-accounting mechanical field theory in combination with Non-Destructive Evaluate (NDE) techniques for the deconvolution of otherwise inaccessible information from human tissue samples. In addition, it has been demonstrated that [40–42]:

- There is a direct relationship between the existence of micro-architecture and the degree of dispersion. The general pattern of the dispersion is the high frequency retardation; the wave speeds decrease with increasing wave frequency. In particular, the direct relationship between the magnitude of the internodal distance and the severity of high-frequency retardation.
- The mechanical responses of tissue, as in the format of reflection spectra, are fully determined by the micro-level properties of tissue. Theoretical simulations showed that changes in any single parameter affected the overall response to a certain degree. The microelastic constant A_{11} appeared to be the most influential parameter. The effects of A_{44} and internodal distance η were less prominent.

16.7. EXPERIMENTAL PRACTICE

16.7.1. Characterization-Mode Ultrasound

The subjectivity of pathologic diagnoses has been presented in detail at the beginning of this chapter. On the basis of this "diagnostic conundrum," a new *ex vivo* procedure has evolved from conventional engineering techniques in conjunction with a new nanomechanical field theory to provide a quantitative tissue evaluation system. Characterization-mode ultrasound (CMUS) technology combines the fundamental theories of Non-Destructive Evaluation (NDE) and doublet mechanics (DM) for the purpose of mathematically reconstructing nanomechanical parameters of tissue.

NDE technology has been applied towards both industrial and medical applications since the 1920's. Inspired by sonar, this noninvasive inspection method has been used for flaw detection/evaluation, imaging diagnostics, dimensional measurements, material characterization, therapy, etc. A typical NDE system consists of several fundamental units: pulser/receiver, transducers, digitizers, and display devices. Driven by the pulser, the transducer generates high frequency ultrasonic energy. The sound energy then propagates through materials in the form of mechanical waves. Any material inconsistencies (i.e. discontinuity, crack, void or a heterogeneous interface) are detected as acoustical aberrations. These material anomalies interrupt bulk material attenuation properties thereby disrupting ultrasound transmission causing reflections. The reflected wave signals can then be captured and transformed into electrical signal for analysis Fig. 16.7 [43]. The amplitude-time signal can be directly related to the distance that the signal travelled. From this information, the discontinuity location, size, orientation and other features can be deduced. Today, quantitative theories have been developed to describe the interaction of the interrogating fields and material inhomogeneities. Objectives range from the determination of fundamental

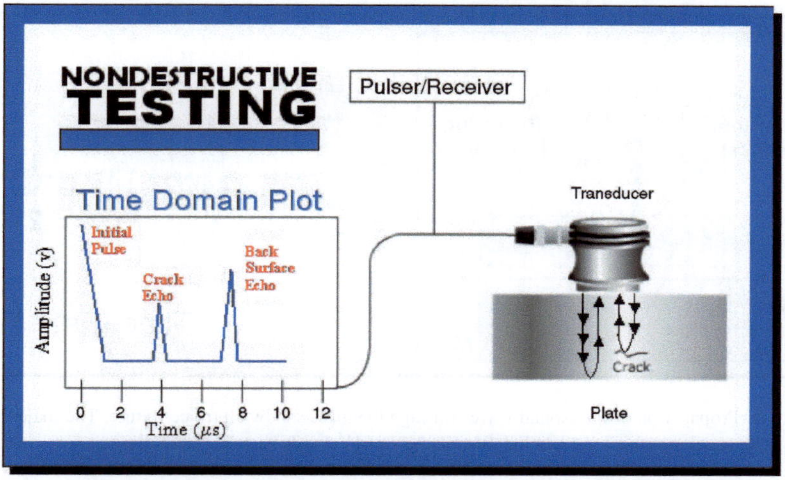

FIGURE 16.7. A demonstration of non-destructive evaluation testing upon a metallic block. A time domain plot is shown to illustrate the presence of a material inconsistency. ***Courtesy of NDT Resource Center***

microstructural characteristics (i.e. grain size, porosity, and texture) to material properties associated with failure mechanisms (i.e. fatigue, creep, fracture toughness, etc.).

16.7.2. Characterization-Mode Ultrasound System

The elements of the CMUS system are consistent with most NDE devices. A pulser-receiver unit provides the energy to generate a signal. The pulser emits very short (0.1 μsec), repetitive electrical pulses in intervals of approximately 1 msec. The high voltage (100 + V) electrical pulses then power unfocused broadband transducers (5–10 MHz) to produce mechanical waves that transmit as a beam of ultrasound. Current CMUS tissue diagnostic systems employ water as an ultrasound-coupling agent. Target specimens are submerged during ultrasound interrogation to ensure proper coupling between the specimen and the impinging ultrasound waves. The transducer transfers the electrical pulses into mechanical vibrations that are propagated through the water to the glass tissue-mounting fixture. The ultrasound/fixture interaction causes multiple reflections from the glass/tissue/glass sandwich structure that are distinguishable in the time domain (Fig. 16.8).

The portion of the time domain plot corresponding to the first glass/tissue interface is electronically gated. This section is highlighted in grey in Fig. 16.8. A Fast Fourier Transform (FFT) is then used to transform the time domain to the frequency domain. The data points from this frequency domain plot are then inputted into a mathematical model governed by the principles of doublet mechanics.

16.7.3. The Model

The theories of doublet mechanics and their application for tissue modeling have been previously discussed in previous sections. Once the time domain signal is recorded and properly gated, a subsequent FFT derives the reflection coefficient spectra of the tissue

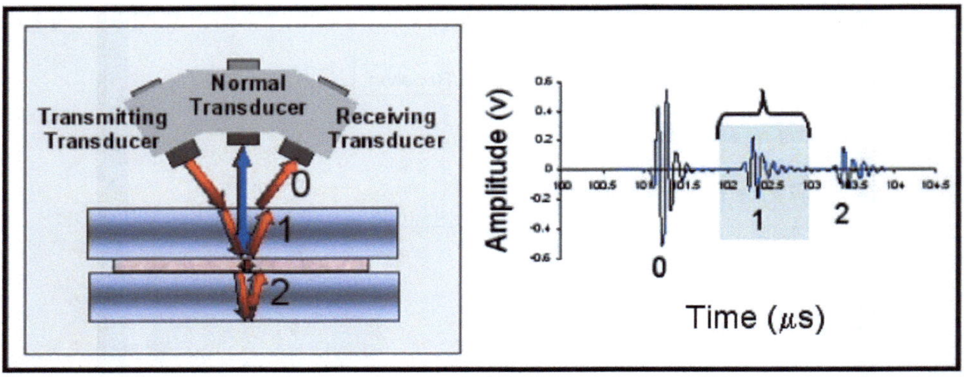

FIGURE 16.8. Propagation of ultrasound waves through the glass sandwich tissue fixture. The major sources of reflections at material interfaces are labeled: 0) water/glass; 1) glass/tissue; 2) tissue/glass.

specimen. These reflection spectra have been analytically-numerically modeled within the framework of DM and their dependence on the DM microparameters of the biological tissue has been fully established. Therefore, by inverting the aforementioned model on the experimentally gathered data, it is possible to quantitatively determine the tissue material properties. This data can then be finally utilized to carry out statistical and objective analysis.

16.7.4. Tissue Preparation

This testing system requires specific tissue preparation procedures. Tissue was obtained from a tissue procurement bank for the research and development of the CMUS system. Each specimen originated from fresh frozen block sections ($\sim$1 cm^3) of human breast tissue. For experimental purposes, frozen tissue sections were supplied in pairs: 1 normal and 1 malignant tissue section from a single patient. Previous studies demonstrated that the snap-freeze/thaw process did not significantly affect the acoustical properties of tissue [44]. For this reason, specimens preserved by snap-freezing were chosen for this study.

The current configuration of the CMUS system requires the testing fixture to be submerged in water. Water is utilized as an ultrasound coupling agent between the transducers and the targeted material. Each specimen is mounted and sealed between two planar pieces of glass to accommodate the aqueous environment. Tissue mounting procedures begin by preparing a 150 μm thick section from one fresh frozen tissue block. It is assumed that the cellular integrity of the 150 μm thick tissue section is maintained since the order of cell dimensions are approximately 10 μm [41]. The tissue is precisely cut into a 150 μm thick section using a cryostat microtome maintained at (-21 °C). The section is then carefully mounted onto a planar piece of glass ($-5.08 \times 7.62 \times 0.635$ cm). Mounting is performed inside the refrigerated working area of the cryostat microtome to prevent tissue curling as the result of thawing. A second piece of glass is then placed over the tissue section to effectively sandwich the specimen as seen in Fig. 16.9. The sandwich fixture is then clamped together and sealed with glass epoxy. Once the epoxy has cured ($\sim$20 mins @ room temperature), the clamps can be removed. The sandwich tissue mounting fixture is then ready for testing.

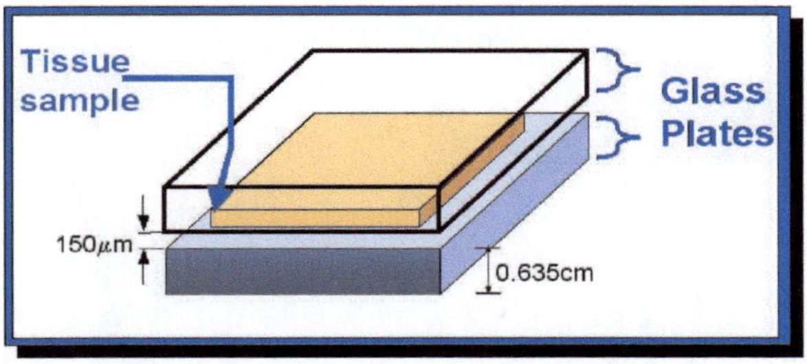

FIGURE 16.9. Glass sandwich tissue mounting fixture.

16.7.5. *Experimental Findings: Breast Cancer detection*

16.7.5.1. Breast Tissue Specimen Information The study by Liu & Ferrari (2002), utilized a CMUS tissue diagnostic system to interrogate human breast tissue. Each of five patients provided a pair of biopsy specimens (malignant & normal) to be evaluated. The tissue procurement bank provided the investigators with the following patient history (Table 16.4) [41].

16.7.5.2. Reconstruction Comparison: Experimental vs. Theoretical A comparison between experimental and theoretically derived reflection spectra demonstrates the integrity of the tissue reconstruction model. The spectral fit between the experimental and theoretical reconstructions infer that the initial estimates for the DM parameters were reasonably consistent with the true values.

16.7.5.3. Reconstruction Comparison: Normal vs. Malignant Tissue The ability of the CMUS tissue diagnostic system to differentiate between normal and malignant tissue can be shown visually in Fig. 16.12 [40]. Three distinct tissue samples were taken from a single specimen block of a certain tissue type. Both tissue specimen blocks originated from a single patient (Patient #1). A set of three samples from both normal and malignant tissue

TABLE 16.4. Tissue specimen information. ***Courtesy of Liu & Ferrari (2002)***

Patient#	Specimen ID #	Speclmen Size#	Cancer Type	Age	Race
1	0110C27lf	1 cm^2	IDC*	80	White
	0110C272f	1 cm^2			
2	020IC175j	1cm^2	IDC*	58	Caucasian
	020IC173v	0.4g			
3	0202C027a	1cm^2	IDC*	43	Unknown
	0202C026a	1cm^2			
4	0206C076d	1cm^2	IDC*	49	Black
	0206C075g	1cm^2			
5	0206C088a	0.54g	IDC*	50	Caucasian
	0206C087c	0.3g			

*IDC: Invasive Ductal Carcinoma

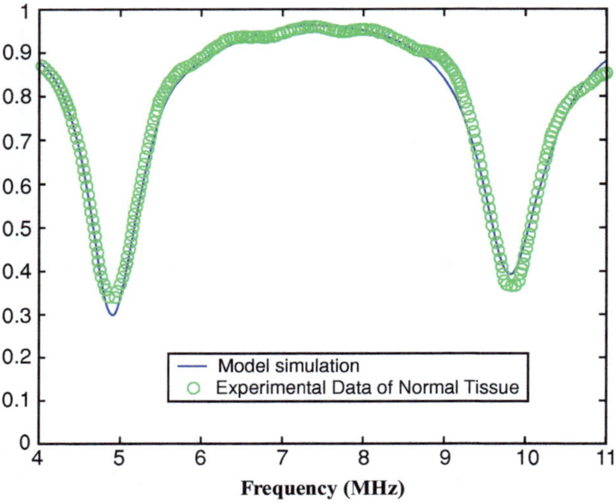

FIGURE 16.10. Experimental and theoretical reflection spectra for a normal tissue specimen from Patient #1. ***Courtesy of Liu, Rokhlin, & Ferrari (2002)***

specimens was evaluated for comparison. Fig. 16.12 demonstrates a difference in magnitude of relative minima and a minima shift between normal (N1, N2, N3) and malignant tissue (T1, T2, T3).

The ultrasonic measurement has shown that the spectral responses from normal and malignant tissue appear to be appreciably and consistently different. This suggests the practicability and reliability of the adopted NDE method for biological tissue testing. Specifically,

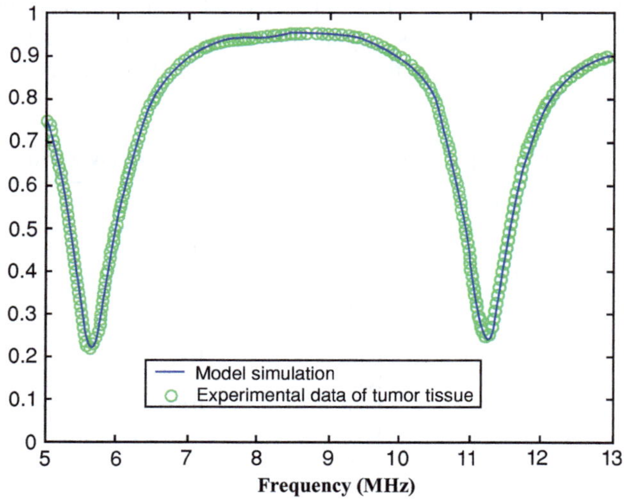

FIGURE 16.11. Experimental and theoretical reflection spectra for a malignant tissue specimen from Patient #1. ***Courtesy of Liu, Rokhlin & Ferrari (2002)***

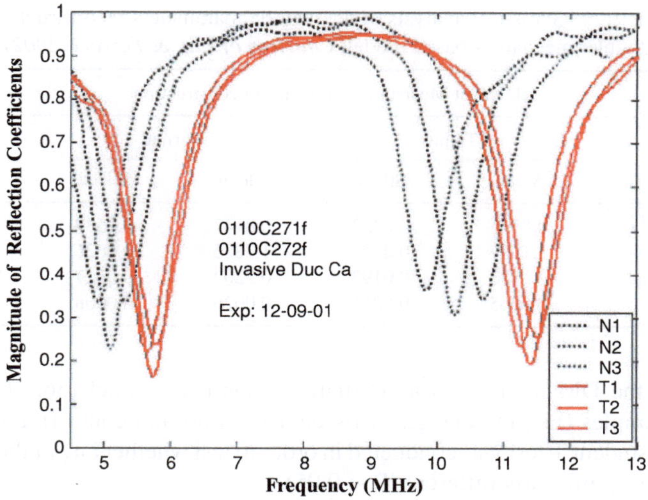

FIGURE 16.12. A comparison of reflection spectra of normal (N1, N2, N3) and malignant tissue (T1, T2, T3). Note the difference in magnitude of the local minima as well as the minima shift between the spectra from the two tissue types. ***Courtesy of Liu, Rokhlin & Ferrari (2002)***

the spectral differences included the following aspects:

- The locations of the spectral minima. Local minima indicate the existence of resonance within the tissue layer at a certain frequency that is determined by the layer geometry (thickness) and the physical properties of the tissue. The results showed that the minima in the spectra generated by the malignant specimens were located at higher frequencies as compared to those by the normal specimens.
- The distance between the two minima. The reflection spectra showed that the distance for malignant tissue was slightly greater than that for the normal.
- The depth of the minima, as defined by the magnitude of the minima. The measurement results showed that the reflection spectra of the malignant tissue generally had greater depth (smaller minima) than those of the normal tissue.

16.7.5.4. Reconstruction Comparison: Doublet Mechanics vs. Continuum Mechanics
For completeness, tissue properties were reconstructed by means of the inversion by

TABLE 16.5. Statistical analysis of the material parameters reconstructed by a continuum based model. ***Courtesy of Liu, Rokhlin & Ferrari (2002)***

Continuum Mechanic: Parameter Reconstruction					
	Tumor		Normal		T Test
Parameters	Mean	StDev	Mean	StDev	P value
$\rho\ (g/cm^3)$	0.9796	0.0788	0.8728	0.0469	0.1298
$\lambda\ (Gpa)$	2.2989	0.2201	1.6848	0.0530	0.0342
$\mu\ (Gpa)$	0.0438	0.0072	0.0381	0.0152	0.6046
$E\ (Gpa)$	0.1304	0.0214	0.1135	0.0449	0.5975

TABLE 16.6. Statistical analysis of the material parameters reconstructed by a doublet mechanics based model. ***Courtesy of Liu, & Ferrari (2002)***

Doublet Mechanic: Parameter Reconstruction					
	Tumor		Normal		T Test
Parameters	Mean	StDev	Mean	StDev	P value
$\rho\ (g/cm^3)$	0.8315	0.0233	0.8147	0.0589	0.6813
$A_{11}\ (Gpa)$	2.1637	0.0571	1.7836	0.0626	0.0015
$A_{44}\ (Gpa)$	0.0523	0.0192	0.2202	0.0327	0.0035
$\eta\ (mm)$	0.0065	0.0011	0.0026	0.0008	0.0091

applying both the DM model and a continuum (classical) model [36, 40, 42, 45]. For each pair of samples (i.e., the malignant tissue and its normal control), a two-tailed, unequal variance student T-test was conducted in order to test whether any of the reconstructed properties were significantly different ($P < 0.05$).

The differential analysis of continuum and doublet mechanics demonstrates that the discrete-based theoretical framework is advantageous in terms of its ability to differentiate normal and malignant tissue. The statistic analysis on the reconstructed values of tissue parameters for both continuum and doublet mechanics predictions shows that p-values for parameters in doublet mechanics reconstructions were lower than those in continuum mechanics [40]. More importantly, the study demonstrated that a greater number of parameters were significantly different ($P < 0.05$) when reconstruction was governed by doublet mechanics. These results suggest that doublet mechanics is more powerful and beneficial for the differentiation of normal and malignant tissue specimens. This can be attributed to material parameters that are more micro-structurally relevant for tissue evaluation.

16.8. NANOMECHANICAL METHOD FOR THE MOLECULAR ANALYSIS OF BREAST CANCER

16.8.1. Introduction

The development of the CMUS tissue diagnostic system represents a novel, quantitative, and high-throughput histopathological method to efficiently and objectively evaluate breast cancer biopsy specimens. A new study is being performed to advance the CMUS technology by improving the sensitivity of the technique through the employment of molecular targeted mechanical contrast agents. This allows the CMUS system to enhance its reconstruction abilities by analyzing molecular as well as architectural and mechanical changes associated with neoplastic disease. Biochemical information is captured through the use of conjugated nano-particulates targeted by their bioaffinity for the prognostic and predictive HER-2/neu breast cancer biomarker.

*16.8.2. The HER-2/*neu *Oncogene*

The HER-2/*neu* molecule is a transmembrane receptor tyrosine kinase of the epidermal growth factor receptor (EGF-R) family known as c-erbB-2/*neu* [46, 47]. Studies have shown

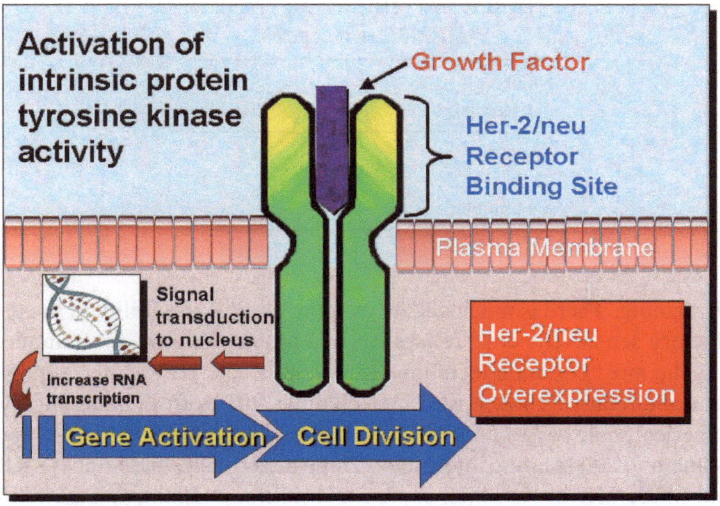

FIGURE 16.13. Ligand binding to the Her-2/neu receptor complex activates the intrinsic protein tyrosine kinase activity. This activation causes the Her-2/neu receptor to be overexpressed.

that the HER-2/*neu* protein is overexpressed in 15–30% of breast carcinomas [7, 18, 46–54]. It has been recognized that high levels of HER-2/*neu*-receptor expression is an indicator of poor prognosis in breast cancer patients, therefore, the HER-2/*neu* protein is a logical target for immunologically-based detection and anti-tumor therapy due to its significant amplification in malignant breast tissue.

HER-2/*neu* is a proto-oncogene found on chromosome 17 [7, 48]. It encodes a protein and functions as a cell membrane receptor. Alteration of proto-oncogenes can cause malignant transformation. Mutations that may result in uncontrolled cell proliferation include any changes in the genome (i.e. deletions, point mutations, insertions, translocations, and amplifications). The HER-2/*neu* family plays an important role in regulating cell growth, survival, and differentiation in a complex manner, and therefore, its malfunction has the potential for unregulated neoplastic growth.

Ligand binding to the HER-2/*neu* receptor complex on the cell surface leads to activation of intrinsic protein tyrosine kinase activity, triggering a cascade of events that leads to gene activation and overexpression. Fig. 16.13 depicts the steps in growth factor signal transduction. During oncogenic transformation, the number of gene copies per cell is increased, leading to an increase in mRNA transcription and a 10- to 100-fold increase in the number of HER-2/*neu* receptors on the cell's surface [53]. It is suspected that the overexpression of HER-2/*neu* leads to the constitutive activation of HER-2/*neu* receptors [52–54]. This abnormal amplification is attributed to unregulated cell growth and oncogenic transformation.

*16.8.3. HER-2/*neu *Exploitation*

Numerous studies have attempted to quantify the levels of expression of HER-2/*neu* using immunohistochemical (IHC) assays, fluorescence *in situ* hybridization (FISH), and

TABLE 16.7. IHC scoring criteria of HER-2/*neu* receptors.

Number of HER-2/ *neu* Receptors	IHC Score
22,000	0
92,000	+1
2.4 million	+3

trastuzumab staining. There is much debate over the clinical feasibility of these tests due to the unsatisfactory sensitivity, cost, reliability, and required expertise. Immunohistochemistry (IHC) is the most common technique used to evaluate HER-2/*neu* expression due to its ease of use and relatively low cost. The technique suffers from being highly subjective since the numerical scale (0, +1, +2, +3) used for tissue evaluation is based upon the qualitative assessment of cell staining intensities. Three breast cancer cell lines (SKBR3, MDA-MB-175, and MDA-MB-231), which express 2.4 million, 92,000, and 22,000 HER-2/*neu* receptor molecules respectively, were used to establish the IHC scores for the HER-2/*neu* expression [46]. See Table 16.7. HER-2/*neu* IHC scores of +2 and +3 are often associated with breast cancer patients having poor clinical outcomes.

The aim of this study is to develop a quantitative methodology that will allow the objective evaluation of tissue by analyzing the molecular as well as architectural and mechanical changes associated with neoplastic disease. Nano-particulates conjugated with HER-2/*neu* antibodies will be used to target and bind to cognate antigen receptors upon the tumor surface (Fig. 16.14). This will effectively tether nano-particulates to malignant tissue thus

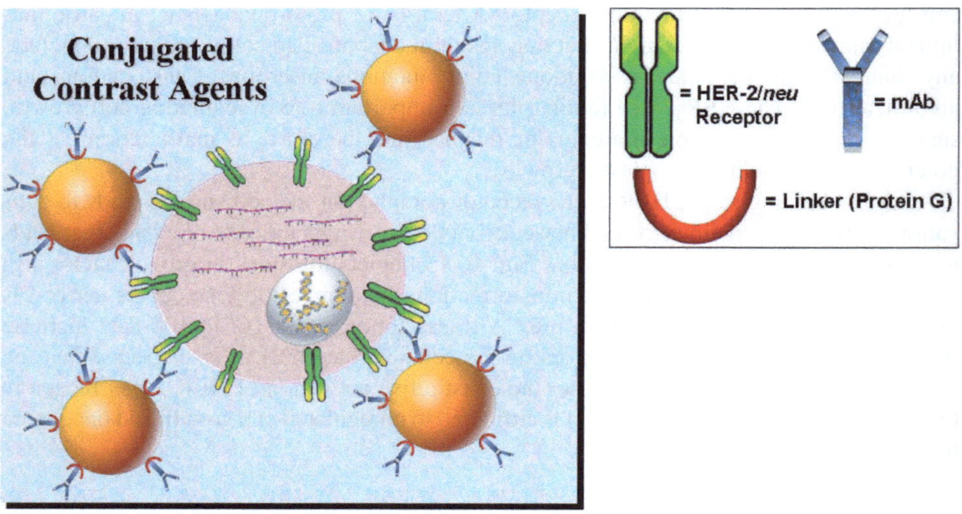

FIGURE 16.14. This drawing illustrates the overexpression of HER-2/*neu* receptors. Nanoparticles conjugated with monoclonal antibodies specifically bind to the cognate antigen receptors. The monoclonal antibodies are conjugated to the nanoparticles via a protein G linker.

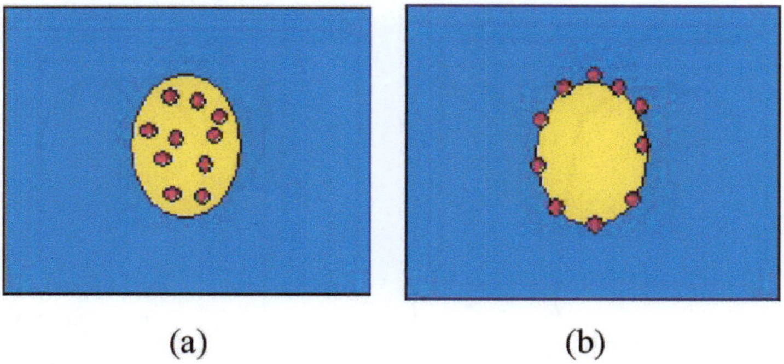

FIGURE 16.15. (a) Nanoparticles are randomly distributed in the targeted tissue. (b) Nanoparticles are attached to the interface between tissues.

locally altering the mechanical properties of the region. The presence of nano-particulates bound to the tumor will serve as mechanical ultrasonic contrast agents that will increase the sensitivity of the CMUS system, thereby enhancing the ability to detect even earlier onset of disease.

16.8.4. Ultrasound Interaction with Tissues with Targeted Nanoparticles

The object of an ultrasonic contrast agent is to change the tissue impedance $Z = \rho V$ by enhancing ultrasonic echo signatures (this is analogous to changing the density ρ by a radiographic contrast agent). Improved ultrasonic imaging due to such contrast agents has been demonstrated both in vivo and in vitro [55–60].

When nanoparticles migrate into the pathological tissue, two limiting ultrasonic/mechanical models can be considered: the particles are randomly distributed in the bulk of the targeted tissue (Fig. 16.15. (a)) or they are concentrated on the interface between the normal tissue and the pathological tissue (for example malignant tumor) (Fig. 16.15 (b)).

16.8.5. Preliminary Results: Randomly Distributed Particles in the Bulk

The CMUS technology was used to evaluate a 150 μm thick 1% agarose layer with dispersed gold nanoparticles of diameter between 15 nm to 20 nm. The reflection coefficients measured at 17° (oblique incidence) are shown as open circles in Fig. 16.16 for the layer without particles and Fig. 16.17 with particles. One can see two strong resonances in the reflection spectra at frequencies slightly above 5 and 10 MHz. The non-dimensional parameters were inversely determined by minimization using oblique incidence experimental spectra. The solid lines in Figs. 16.16 and 16.17 represent spectra calculated from the determined parameters indicating good match to the data. The reconstructed thickness of the slices is 147 μm, which agrees well with the thickness measured by a micrometer.

The CMUS mechanical model governed by doublet mechanics analyzed the reflection coefficient data for the case of randomly distributed particles in the bulk. The data for Table 16.8 was collected from 4 different tissue phantom specimens (2 × specimen with particles, 2 × specimen without particles). Each specimen was completely scanned by the

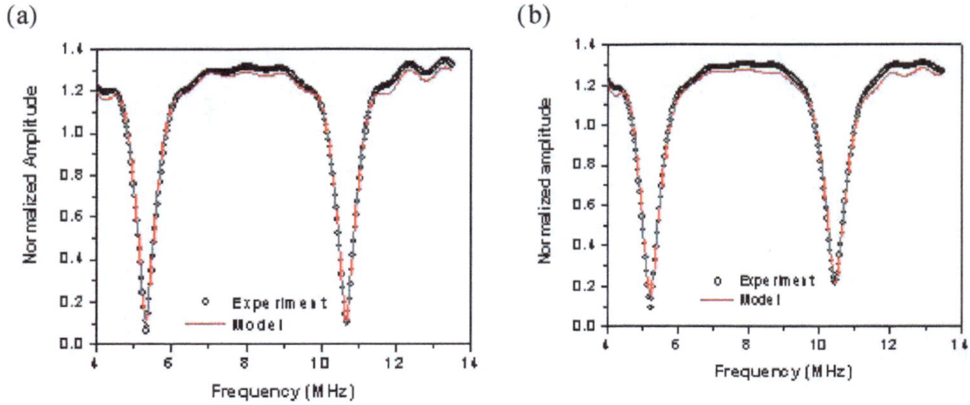

FIGURE 16.16. Reflection from the 1% agarose layer imbedded within the glass sandwich tissue mounting fixture. (a) Without particles (b) with particles. ***Courtesy of Sakamoto, Rokhlin, & Ferrari***

CMUS system. Nine readings from each scan were taken such that 18 total interrogation points were examined per specimen configuration. Table 16.8 displays the means and standard deviations for each DM parameter. A t-test was not applied to the values since the number of specimens interrogated is insufficient to relinquish any statistically significant findings. These preliminary results are only useful for displaying trends. All of the DM parameters agree with intuitive analysis, however, it is too early to make any significant conclusions.

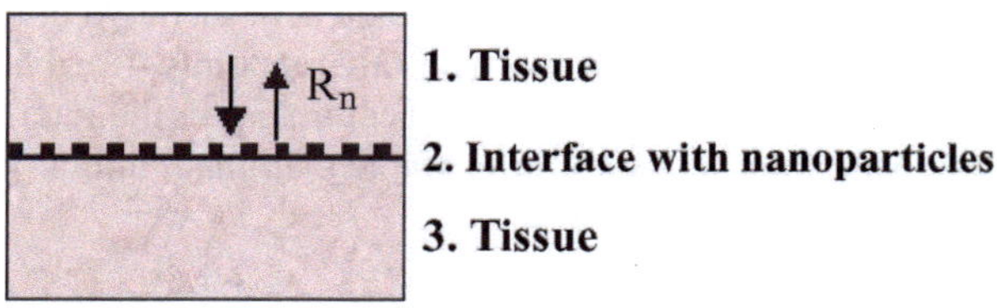

FIGURE 16.17. Normal and oblique incident ultrasound interaction with a tissue slice embedded in glass.

TABLE 16.8. The analysis of the material parameters reconstructed by a doublet mechanics based model for samples with and without particles randomly distributed throughout the bulk.

Double Mechanics: Parameter Reconstruction				
	With Particles		Without Particles	
Parameters	Mean	StDev	Mean	StDev
ρ (g/cm^3)	1.0701	0.0237	0.9769	0.0076
A_{11} (Gpa)	2.5555	0.0742	2.4908	0.0242
A_{44} (Gpa)	0.8737	0.0422	0.8021	0.0229
η (mm)	0.0023	0.0009	0.0037	0.0003

16.8.6. Preliminary Results: Randomly Distributed Particles Upon an Interface

As was discussed above, the tissue-targeted nanoparticles may be accumulated within the interface between normal and pathologic tissues. In this case, we can model the wave propagation using modified interfacial conditions (16.65). The fraction of ultrasonic wave energy transmitted through such an interface depends on the properties and distribution of the nanoparticles. If we neglect the mass M terms in Eq. (16.65), the reflection at the interface separating two tissues as shown in Fig. 16.17 is given by:

$$R_n = \frac{Z_2 - Z_1 + i\omega Z_1 Z_2/K_n}{Z_2 + Z_1 - i\omega Z_1 Z_2/K_n}. \tag{16.65}$$

When impedances (Z_1) and (Z_2) of the top and bottom tissues are close to each other ($Z_1 \approx Z_2 = Z = \rho V$), the energy reflection coefficient can be written as $|R_n|^2 = \dfrac{1}{1 + (2K_n/\omega Z)^2}$. When no nanoparticles are applied at the interface, (K_n) is infinity (the interface layer thickness $h_c = 0$), there is no ultrasonic reflection and regular boundary conditions are satisfied. When nanoparticles are accumulated at the interface, (K_n) decreases due to finite thickness (h_c) of the interface composite layer.

Fig. 16.18(a) shows experimental results obtained with a 50 MHz focus transducer for gold nanoparticles at the interface. Fig. 16.18(b) compares ultrasonic signatures from the interfacial region with and without nanoparticles. The region with nanoparticles is indicated by the dotted circle. As can be seen the reflection is larger at the region with the particles.

16.9. FUTURE OF CHARACTERIZATION-MODE ULTRASOUND

Characterization-mode ultrasound technology is a versatile methodology that has the ability to address many clinical indications. Such a technology can be used to standardize *ex vivo* histological exams of different tumoral tissues and for *in vivo* diagnosis of skin, mucosal, and vascular diseases. The next generation of CMUS systems in development involves a reconfiguration of the current hardware and software for the application of cutaneous lesion screening.

The new challenge for DM is to establish an ultrasound/tissue interaction model as such that the mechanical and architectural information concerning the microstructures of skin tissue can be made available. There is an apparent need to develop a mathematical model capable of inspecting multi-layered heterogeneous materials, as the skin. No longer will a simple single-layer interaction model suffice. Fig. 16.13 illustrates the additional complexity required to model a double-layered material. Also, the morphology of layer interfaces cannot be considered flat, but actually be modeled to appropriately reflect the crests and valleys of the dermal-epidermal junction.

In the case of malignant melanoma, the uncontrollable proliferation of melanocytes at the dermal/epidermal junction leads to malignant cell invasion of dermal blood vessels and subsequent metastasis (Figure 16.19). Once the melanocytes advance into the dermal layer, the propensity for metastatic spreading becomes critical. Although the initiation of the vertical growth phase may commence without obvious visual warnings, it is assumed that the

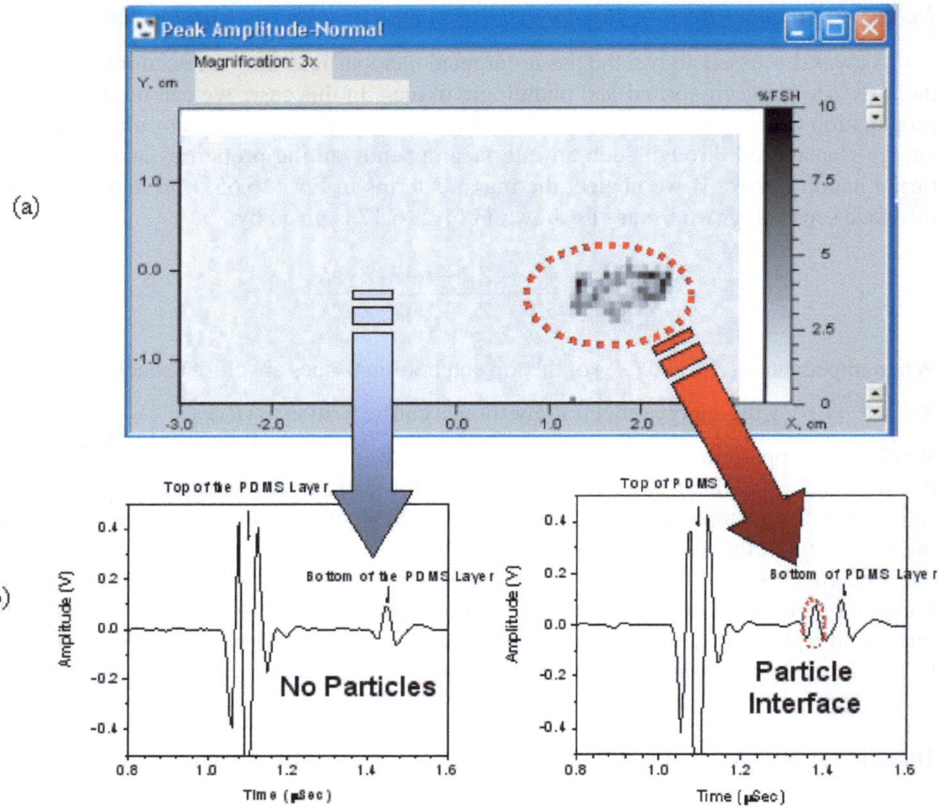

FIGURE 16.18. a) 50 MHz focused ultrasonic C-scan image of tissue phantoms (PDMS layer) with gold nanoparticles at the interface. The dotted circle indicates the area with nanoparticles. b) Comparison of ultrasonic echo signatures from interface with and without nanoparticles. ***Courtesy of Sakamoto, Rokhlin, & Ferrari***

changes in microstructure can be ascertained through DM to provide early detection. Future experiments are necessary to determine if a double-layer model is capable of reconstructing the significant details of the disease progression of melanoma. Although Figure 16.13b) illustrates a glass sandwich structure, the actual CMUS skin cancer screening device does not require such an experimental setup. Since only a few subcutaneous millimeters are relevant for melanoma and the layered structure of skin is relatively consistent, the bottom reference glass plate is not necessary as well as being impractical for noninvasive evaluation.

The CMUS system can be effortlessly adapted to numerous clinical indications once the challenges for addressing multi-layered material structures are overcome. The only limiting factor to its applicability will be the physical limits of current transducer technology. Today's transducers are small enough to be fitted upon catheters for oral, esophageal, colon, and vaginal indications. In the near future, transducer minimization may accommodate the detection and quantification of vulnerable plaque resident in coronary vessels. The clinical impact of CMUS technology has yet to be fully realized, however, the anticipation of its significance provides incentive for its continued development.

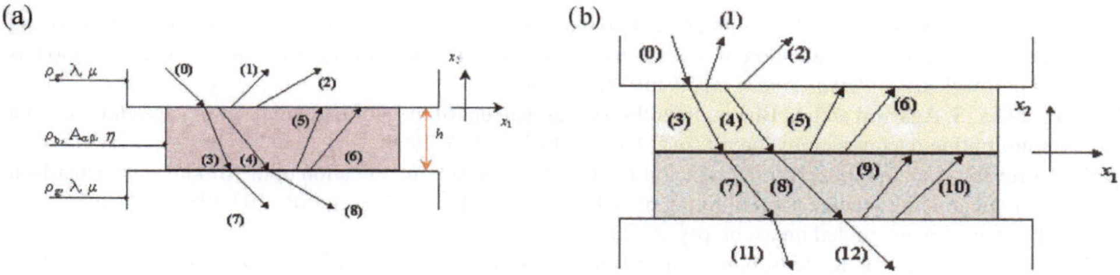

FIGURE 16.19. a) The original DM tissue reconstruction model was premised upon analyzing a single layered piece of tissue. b) For the proper evaluation of skin cancer, a new multi-layered model must be optimized.

ACKNOWLEDGMENTS

The authors wish to acknowledge financial support from the National Cancer Institute (R21-CA099089). The authors gratefully thank the support of others that have participated and have directly influenced the research and content presented herein: Stephen Lee, John Shapiro, Yoaav Krauthammer, Johanna Craig, Jim Coontz, Shawn Coontz, James Hamilton and Haruo Sakamoto.

REFERENCES

[1] M. Ferrari, V.T. Granik, A. Imam, and J.C. Nadeau. *Advances in Doublet Mechanics*, Springer-Verlag, New York, Inc., New York, 1997.

[2] H. Verkooijen et al. Interobserver variability between general and expert pathologists during the histopathological assessment of large-core needle and open biopsies of non-palpable breast lesions. *Eur. J. Can.*, 39:2187–2191, 2003.

[3] M. Piver et al. Comparative study of ovarian cancer histopathology by registry pathologists and referral pathologists: A study by the Gilda Radner Familial Ovarian Cancer Registry. *Gynecol. Oncol.* 78:166–170, 2000.

[4] R. Schlemper et al. Differences in diagnostic criteria for esophageal squamous cell carcinoma between Japanese and Western pathologists. *Cancer*, 88:996–1006, 2000.

[5] B. Dunne and J.J. Going. Scoring nuclear pleomorphism in breast cancer. *Histopathology*, 39:259–265, 2001.

[6] H. Tsuda et al. Evaluation of interobserver agreement in scoring immunohistochemical results of HER-2/neu (c-erbB-2) expression detected by HercepTest, Nichirei polyclonal antibody, CB11 and TAB250 in breast carcinoma. *Pathol. Internat.* 52:126–134, 2002.

[7] A. Paradiso et al. Interobserver reproducibility of immunohistochemical HER-2/neu evaluation in human breast cancer: the real-world experience. *Internat. J. Biol. Mark.* 19:147–154, 2004.

[8] National Cancer Institute; "*Breast Cancer (PDQ): Screening*", National Institute of Health; http://www.nci.nih.gov/cancertopics/pdq/screening/breast/HealthProfessional/page4:http://www.nci.nih.gov/cancertopics/pdq/screening/breast/HealthProfessional/page4 (2004).

[9] Czyz, A.H. "*Breast Cancer Diagnosis: Histologic Grades of Breast Cancer: Helping Determine a Patient's Outcome*", Imaginis.com; http://imaginis.com/breasthealth/histologic_grades.asp?mode=1:http://imaginis.com/breasthealth/histologic_grades.asp?mode=1; (2001).

[10] S. Amat et al. Scarff-Bloom-Richardson (SBR) grading: a pleiotropic marker of chemosensitivity in invasive ductal breast carcinomas treated by neoadjuvant chemotherapy. *Internat. J. Oncol.* 20:791–796, 2002.

[11] A. Douglas-Jones et al. Consistency in the observation of features used to classify duct carcinoma in situ (DCIS) of the breast. *J. Clin. Pathol.*, 53:596–602, 2000.

[12] A. Volpi et al. Prognostic significance of biologic markers in node-negative breast cancer patients: a prospective study. *Mod. Pathol.*, 17:1038–1044, 2004.

[13] H. Tsuda, E. Akiyama, M. Kurosumi, G. Sakamoto, and T. Watanabe. A quantitative model using mean and standard deviation for evaluation of interobserver agreement in nuclear atypia scoring of breast carcinomas in a protocol study. *Pathol. Internat.*, 50:119–125, 2000.
[14] M. Sikka, S. Agarwal and A. Bhatia. Interobserver agreement of the Nottingham histologic grading scheme for infiltrating duct carcinoma breast. *Ind. J. Can.*, 36:149–153, 1999.
[15] H. Frierson et al. Interobserver reproducibility of the Nottingham modification of the Bloom and Richardson histologic grading scheme for infiltrating ductal carcinoma. *Am. J. Clin. Pathol.*, 103:195–198, 1995.
[16] R.P. Burns. Image-guided breast biopsy. *Am. J. Surg.*, 173:9–11, 1997.
[17] F. Burbank. Stereotactic breast biopsy: Its history, its present, and its future. *Am. Surg.*, 62:128–150, 1996.
[18] R. Cotran, V. Kumar, and T. Collins. *Pathologic Basis of Disease*, 6th. Ed. 6th. W.B. Saunders, Philadelphia; 1999.
[19] M.S. Brady et al. Patterns of detection in patients with cutaneous melanoma. *Cancer*, 89:342–347, 2000.
[20] C.M.K. Grin, A, B. Welkovich, and R. Bart. Accuracy of clinical diagnosis of malignant melanoma. *Arch. Dermatol.*, 126:763–766, 1990.
[21] J.L. Bolognia, M. Berwick, and J.A. Fine. Complete follow-up and evaluation of a skin cancer screening in Connecticut. *J. Am. Dermatol.*, 23:1098–1106, 1990.
[22] American Cancer Society, Vol. 2004 1–50, American Cancer Society, Atlanta; 2003.
[23] M. Helfand, S.M. Mahon, K.B. Eden, P.S. Frame, and C.T. Orleans. Screening for skin cancer. *Am. J. Preven. Med.*, 20:47–58, 2001.
[24] T.E. Andreoli, J. Loscalzo, C.C.J. Carpenter, and R.C. Griggs. *Cecil Essentials of Medicine*, 5th. Ed. W.B. Saunders Co, Philadelphia; 2000.
[25] R. Cotran, V. Kumar, and T.C. *Pathologic Basis of Disease*, W.B. Saunders, Philadelphia; 1999.
[26] C.R. Hill, J.C. Bamber, and G.R.t. Haar (eds.) *Physical Principles of Medical Ultrasonics*, John Wiley & Sons, Hoboken, NJ, 2004.
[27] P.N.T. Wells. Ultrasonic imaging of the human body. *Rep. Prog. Phys.*, 62:671–722, 1999.
[28] F.A. Duck. Physical Properties of Tissue. Academic Press, London, 1990.
[29] S.A. Goss, R.L. Johnston, and F. Dunn. Comprehensive compilation of empirical ultrasonic properties of mammalian tissues. *J. Acoust. Soc. Am.*, 64:423–457, 1978.
[30] A.P. Sarvazyan, O.V. Rudenko, S.D. Swanson, J.B. Fowlkes, and S.Y. Emelianov. Shear wave elasticity imaging: a new ultrasonic technology of medical diagnostics. *Ultrasound Med. Biol.*, 24:1419–1435, 1998.
[31] L. Sandrin, M. Tanter, S. Catheline, and M. Fink. Shear modulus imaging with 2-D transient elastography. *IEEE Trans. Ultrason. Ferro. Freq. Contrl.*, 49:426–435, 2002.
[32] V.N. Alekseev and S.A. Rybak. Equations of state for viscoelastic biological media. *Acoust. Phys.*, 48:511–547, 2002.
[33] T.L. Szabo and J. Wu. A model for longitudinal and shear wave propagation in viscoelastic media. *J. Acoust. Soc. Am.*, 107:2437–2446, 2000.
[34] K.R. Waters, M.S. Hughes, J. Mobley, G.H. Brandenburger, and J.G. Miller. On the applicability of Kramers-Kronig relations for ultrasonic attenuation obeying a frequency power law. *J. Acoust. Soc. Am.*, 108:556–563, 2000.
[35] T.L. Szabo. Causal theories and data for acoustic attenuation obeying a frequency power law. *J. Acoust. Soc. Am.*, 97:14–24, 1995.
[36] A. Lavrentyev and S.I. Rokhlin. Determination of elastic moduli, density, attenuation, and thickness of a layer using ultrasonic spectroscopy at two angles. *J. Acoust. Soc. Am.*, 102:3467–3477, 1997.
[37] L. Adler, K.V. Cook, and W.A. Simpson. In Research Techniques in Nondestructive Testing, Vol. 3. R.S. Wang (ed.) Academic Press, New York, 1977, pp. 1–49.
[38] L. Wang and S.I. Rokhlin. Stable reformulation of transfer matrix method for wave propagation in layered anisotropic media. *Ultrasonics*, 39:413–424, 2001.
[39] L. Wang, B. Xie, and S.I. Rokhlin. Determination of embedded layer properties using adaptive time-frequency domain analysis. *J. Acoust. Soc. Am.*, 111:2644–2653, 2002.
[40] J. Liu. *Biomedical Engineering*, The Ohio State University, Columbus, 2002.
[41] J. Liu and M. Ferrari. A discrete model for the high frequency elastic wave examination on biological tissue. *CMES*, 4:421–430, 2003.
[42] J. Liu and M. Ferrari. Mechanical spectral signatures of malignant disease? A small-sample, comparative study of continuum vs. nano-biomechanical data analyses. *Dis. Mark.*, 18:175–183, 2002.

[43] NDT Resource Center; "*Basic Principles of Ultrasonic Testing*", Iowa State University:http://www.ndt-ed.org/EducationResources/communitycollege/ultrasonics/introduction/description.htm; (2001).

[44] A.F. van der Steen, M.H. Cuypers, J.M. Thijssen, and P.C. de Wilde. Influence of histochemical preparation on acoustic parameters of liver tissue: a 5-MHz study. *Ultrasou. Med. Biol.*, 17:879–891, 1991.

[45] S.I. Rokhlin and Y.J. Wang. Analysis of boundary conditions for elastic wave interaction with an interface between two solids. *J. Acoust. Soc. Am.*, 503, 1991.

[46] H.K.W. Koeppen et al. Overexpression of HER2/neu in solid tumours: an immunohistochemical survey. *Histopathology*, 38:96–104, 2001.

[47] N.E. Hynes and D.F. Stern. The biology of erbB-2/neu/HER-2 and its role in cancer. *Biochim. Biophys. Acta.*, 1198:165–184, 1994.

[48] I.F. Tannock and R.P. Hill. (eds.) *The Basic Science of Oncology*, 3rd Edn. McGraw-Hill, New York, 1998.

[49] J.M.S. Bartlett et al. Evaluating HER2 Amplification and Overexpression in Breast Cancer. *J. Pathol.*, 195:422–428, 2001.

[50] M. Pegram, G. Pauletti, and D.J. Slamon. HER-2/neu as a predictive marker of response to breast cancer therapy. *Breast Can. Res. Treat.*, 52:65–77, 1998.

[51] S. Masood, and M.M. Bui. Prognostic and Predictive Value of HER2/neu Oncogene in Breast Cancer. *Microsco. Res. Tech.*, 59:102–108, 2002.

[52] S. Menard, E. Tagliabue, M. Campiglio, and S.M. Pupa. Role of HER2 Gene Overexpression in Breast Carcinoma. *J. Cell. Physiol.*, 182:150–162, 2000.

[53] R.M. Neve, H.A. Lane, and N.E. Hynes. The role of overexpressed HER2 in transformation. *Ann. Oncol.*, 12:S9–S13, 2001.

[54] D.J. Slamon et al. Studies of the HER-2/neu Proto-Oncogene in Human Breast and Ovarian Cancer. *Sci., New Series*, 244:707–712, 1989.

[55] G.M. Lanza et al. In vitro characterization of a novel, tissue-targeted ultrasonic contrast system with acoustic microscopy. *J. Acoust. Soc. Am.*, 104:36–65, 1998.

[56] P.A. Dayton and K.W. Ferrara. Target imaging using ultrasound. *J. Mag. Reson. Imag.*, 16:362–377, 2002.

[57] C.S. Hall et al. Experimental determination of phase velocity of perfluorocarbons: applications to targeted contrast agents. *IEEE Trans. Ultraon. Ferroelect. Freq. Contr.*, 47:75–84, 2000.

[58] D.N. Patel, S.H. Bloch, P.A. Dayton, and K.W. Ferrara. Acoustic signatures of submicron contrast agents. *IEEE Trans. Ultraon. Ferroelect. Freq. Contr.*, 51:293–301, 2004.

[59] E. Unger, T.O. Matsunaga, P.A. Schumann, and R. Zutsh. Microbubbles in molecular imaging and theraphy. *Medicamundi*, 47:58–65, 2003.

[60] N. Marsh et al. Improvements in the ultrasonic contrast of targeted perfluorocarbon nanoparticles using an acoustic transmission line model. *IEEE Trans. Ultraon. Ferroelect. Freq. Contr.*, 49:29–38, 2002.

About the Editors

Professor Mauro Ferrari is a pioneer in the fields of bioMEMS and biomedical nanotechnology. As a leading academic, a dedicated entrepreneur, and a vision setter for the Nation's premier Federal programs in nanomedicine, he brings a three-fold vantage perspective to his roles as Editor-in-Chief for this work. Dr. Ferrari has authored or co-authored over 150 scientific publications, 6 books, and over 20 US and International patents. Dr. Ferrari is also Editor-in-Chief of Biomedical Microdevices and series editor of the new Springer series on Emerging Biomedical Technologies.

Several private sector companies originated from his laboratories at the Ohio State University and the University of California at Berkeley over the years. On a Federal assignment as Special Expert in Nanotechnology and Eminent Scholar, he has provided the scientific leadership for the development of the Alliance for Cancer Nanotechnology of the National Cancer Institute, the world-largest medical nanotechnology operation to date. Dr. Ferrari trained in mathematical physics in Italy, obtained his Master's and Ph.D. in Mechanical Engineering at Berkeley, attended medical school at The Ohio State University, and served in faculty positions in Materials Science and Engineering, and Civil and Environmental Engineering in Berkeley, where he was first tenured. At Ohio State he currently serves as Professor of Internal Medicine, Division of Hematology and Oncology, as Edgar Hendrickson Professor of Biomedical Engineering, and as Professor of Mechanical Engineering. He is Associate Director of the Dorothy M. Davis Heart and Lung Research Institute, and the University's Associate Vice President for Health Science, Technology and Commercialization.

Abe Lee has been working on micro/ and nanotechnology for biomedical and biotech applications since 1992. His recent research focuses on the development of integrated micro and nano fluidic chip processors for the following applications: point-of-care diagnostics, "smart" nanomedicine for early detection and treatment, stem cell biology and therapeutics, the synthesis of novel and pure materials, and biosensors to detect environmental and terrorism threats. He has over 30 peer reviewed publications and over 30 issued US patents. I am also subject editor for the Journal of Microelectromechanical Systems and member of the International Advisory Editorial Board for the Lab on a Chip journal.

Jim Lee's research interest includes BioMEMS/NEMS, and polymer micro/nanotechnology. In the last 4 years, he has over 20 refereed journal publications, 2 book chapters, and 5 patents in these areas. He is now leading an NSF Nanoscale Science and Engineering Center for Affordable Nanoengineering of Polymer Biomedical Devices at Ohio State University.

Index

Abbreviated Table of Contents

MIX
Papier aus verantwortungsvollen Quellen
Paper from responsible sources
FSC® C105338

If you have any concerns about our products,
you can contact us on
ProductSafety@springernature.com

In case Publisher is established outside the EU,
the EU authorized representative is:
Springer Nature Customer Service Center GmbH
Europaplatz 3, 69115 Heidelberg, Germany

Printed by Libri Plureos GmbH
in Hamburg, Germany